Mark S. Dept. of Integr. Biol. UC Berkeley

MUIRHEAD • Aspects of Multivariate Statistical Theory
PARZEN • Modern Probability Theory and Its Applications
PURI and SEN • Nonparametric Methods in General Linear Models
PURI and SEN • Nonparametric Methods in Multivariate Analysis
RANDLES and WOLFE • Introduction to the Theory of Nonparametric Statistics
RAO • Linear Statistical Inference and Its Applications, *Second Edition*
RAO • Real and Stochastic Analysis
RAO and SEDRANSK • W.G. Cochran's Impact on Statistics
ROHATGI • An Introduction to Probability Theory and Mathematical Statistics
ROHATGI • Statistical Inference
ROSS • Stochastic Processes
RUBINSTEIN • Simulation and The Monte Carlo Method
SCHEFFE • The Analysis of Variance
SEBER • Linear Regression Analysis
SEBER • Multivariate Observations
SEN • Sequential Nonparametrics: Invariance Principles and Statistical Inference
SERFLING • Approximation Theorems of Mathematical Statistics
SHORACK and WELLNER • Empirical Processes with Applications to Statistics
TJUR • Probability Based on Radon Measures
WILLIAMS • Diffusions, Markov Processes, and Martingales, Volume I: Foundations
ZACKS • Theory of Statistical Inference

Applied Probability and Statistics

ABRAHAM and LEDOLTER • Statistical Methods for Forecasting
AGRESTI • Analysis of Ordinal Categorical Data
AICKIN • Linear Statistical Analysis of Discrete Data
ANDERSON, AUQUIER, HAUCK, OAKES, VANDAELE, and WEISBERG • Statistical Methods for Comparative Studies
ARTHANARI and DODGE • Mathematical Programming in Statistics
BAILEY • The Elements of Stochastic Processes with Applications to the Natural Sciences
BAILEY • Mathematics, Statistics and Systems for Health
BARNETT • Interpreting Multivariate Data
BARNETT and LEWIS • Outliers in Statistical Data, *Second Edition*
BARTHOLOMEW • Stochastic Models for Social Processes, *Third Edition*
BARTHOLOMEW and FORBES • Statistical Techniques for Manpower Planning
BECK and ARNOLD • Parameter Estimation in Engineering and Science
BELSLEY, KUH, and WELSCH • Regression Diagnostics: Identifying Influential Data and Sources of Collinearity
BHAT • Elements of Applied Stochastic Processes, *Second Edition*
BLOOMFIELD • Fourier Analysis of Time Series: An Introduction
BOX • R. A. Fisher, The Life of a Scientist
BOX and DRAPER • Empirical Model-Building and Response Surfaces
BOX and DRAPER • Evolutionary Operation: A Statistical Method for Process Improvement
BOX, HUNTER, and HUNTER • Statistics for Experimenters: An Introduction to Design, Data Analysis, and Model Building
BROWN and HOLLANDER • Statistics: A Biomedical Introduction
BUNKE and BUNKE • Statistical Inference in Linear Models, Volume I
CHAMBERS • Computational Methods for Data Analysis
CHATTERJEE and PRICE • Regression Analysis by Example
CHOW • Econometric Analysis by Control Methods
CLARKE and DISNEY • Probability and Random Processes: A First Course with Applications, *Second Edition*

COCHRAN • Sampling Techniques, *Third Edition*
COCHRAN and COX • Experimental Designs, *Second Edition*
CONOVER • Practical Nonparametric Statistics, *Second Edition*
CONOVER and IMAN • Introduction to Modern Business Statistics
CORNELL • Experiments with Mixtures: Designs, Models and The Analysis of Mixture Data
COX • Planning of Experiments
DANIEL • Biostatistics: A Foundation for Analysis in the Health Sciences, *Third Edition*
DANIEL • Applications of Statistics to Industrial Experimentation
DANIEL and WOOD • Fitting Equations to Data: Computer Analysis of Multifactor Data, *Second Edition*
DAVID • Order Statistics, *Second Edition*
DAVISON • Multidimensional Scaling
DEGROOT, FIENBERG and KADANE • Statistics and the Law
DEMING • Sample Design in Business Research
DILLON and GOLDSTEIN • Multivariate Analysis: Methods and Applications
DODGE • Analysis of Experiments with Missing Data
DODGE and ROMIG • Sampling Inspection Tables, *Second Edition*
DOWDY and WEARDEN • Statistics for Research
DRAPER and SMITH • Applied Regression Analysis, *Second Edition*
DUNN • Basic Statistics: A Primer for the Biomedical Sciences, *Second Edition*
DUNN and CLARK • Applied Statistics: Analysis of Variance and Regression
ELANDT-JOHNSON and JOHNSON • Survival Models and Data Analysis
FLEISS • Statistical Methods for Rates and Proportions, *Second Edition*
FLEISS • The Design and Analysis of Clinical Experiments
FOX • Linear Statistical Models and Related Methods
FRANKEN, KÖNIG, ARNDT, and SCHMIDT • Queues and Point Processes
GALAMBOS • The Asymptotic Theory of Extreme Order Statistics
GIBBONS, OLKIN, and SOBEL • Selecting and Ordering Populations: A New Statistical Methodology
GNANADESIKAN • Methods for Statistical Data Analysis of Multivariate Observations
GOLDSTEIN and DILLON • Discrete Discriminant Analysis
GREENBERG and WEBSTER • Advanced Econometrics: A Bridge to the Literature
GROSS and CLARK • Survival Distributions: Reliability Applications in the Biomedical Sciences
GROSS and HARRIS • Fundamentals of Queueing Theory, *Second Edition*
GUPTA and PANCHAPAKESAN • Multiple Decision Procedures: Theory and Methodology of Selecting and Ranking Populations
GUTTMAN, WILKS, and HUNTER • Introductory Engineering Statistics, *Third Edition*
HAHN and SHAPIRO • Statistical Models in Engineering
HALD • Statistical Tables and Formulas
HALD • Statistical Theory with Engineering Applications
HAND • Discrimination and Classification
HILDEBRAND, LAING, and ROSENTHAL • Prediction Analysis of Cross Classifications
HOAGLIN, MOSTELLER and TUKEY • Exploring Data Tables, Trends and Shapes
HOAGLIN, MOSTELLER, and TUKEY • Understanding Robust and Exploratory Data Analysis
HOEL • Elementary Statistics, *Fourth Edition*

(continued on back)

MARKOV PROCESSES

MARKOV PROCESSES

CHARACTERIZATION AND CONVERGENCE

STEWART N. ETHIER

and

THOMAS G. KURTZ

JOHN WILEY & SONS

New York Chichester Brisbane Toronto Singapore

Library of Congress Cataloging-in-Publication Data:

Ethier, Stewart N., 1948–
Markov processes.

(Wiley series in probability and mathematical statistics)
Bibliography: p.
Includes index.
1. Markov processes. I. Kurtz, Thomas G. II. Title. III. Series.

QA274.7.E84 1985 519.2′33 85-12078
ISBN 0–471–08186–8

Printed in the United States of America

10 9 8 7 6 5 4 3 2 1

PREFACE

The original aim of this book was a discussion of weak approximation results for Markov processes. The scope has widened with the recognition that each technique for verifying weak convergence is closely tied to a method of characterizing the limiting process. The result is a book with perhaps more pages devoted to characterization than to convergence.

The Introduction illustrates the three main techniques for proving convergence theorems applied to a single problem. The first technique is based on operator semigroup convergence theorems. Convergence of generators (in an appropriate sense) implies convergence of the corresponding semigroups, which in turn implies convergence of the Markov processes. Trotter's original work in this area was motivated in part by diffusion approximations. The second technique, which is more probabilistic in nature, is based on the martingale characterization of Markov processes as developed by Stroock and Varadhan. Here again one must verify convergence of generators, but weak compactness arguments and the martingale characterization of the limit are used to complete the proof. The third technique depends on the representation of the processes as solutions of stochastic equations, and is more in the spirit of classical analysis. If the equations "converge," then (one hopes) the solutions converge.

Although the book is intended primarily as a reference, problems are included in the hope that it will also be useful as a text in a graduate course on stochastic processes. Such a course might include basic material on stochastic processes and martingales (Chapter 2, Sections 1–6), an introduction to weak convergence (Chapter 3, Sections 1–9, omitting some of the more technical results and proofs), a development of Markov processes and martingale problems (Chapter 4, Sections 1–4 and 8), and the martingale central limit theorem (Chapter 7, Section 1). A selection of applications to particular processes could complete the course.

As an aid to the instructor of such a course, we include a flowchart for all proofs in the book. Thus, if one's goal is to cover a particular section, the chart indicates which of the earlier results can be skipped with impunity. (It also reveals that the course outline suggested above is not entirely self-contained.)

Results contained in standard probability texts such as Billingsley (1979) or Breiman (1968) are assumed and used without reference, as are results from measure theory and elementary functional analysis. Our standard reference here is Rudin (1974). Beyond this, our intent has been to make the book self-contained (an exception being Chapter 8). At points where this has not seemed feasible, we have included complete references, frequently discussing the needed material in appendixes.

Many people contributed toward the completion of this project. Cristina Costantini, Eimear Goggin, S. J. Sheu, and Richard Stockbridge read large portions of the manuscript and helped to eliminate a number of errors. Carolyn Birr, Dee Frana, Diane Reppert, and Marci Kurtz typed the manuscript. The National Science Foundation and the University of Wisconsin, through a Romnes Fellowship, provided support for much of the research in the book.

We are particularly grateful to our editor, Beatrice Shube, for her patience and constant encouragement. Finally, we must acknowledge our teachers, colleagues, and friends at Wisconsin and Michigan State, who have provided the stimulating environment in which ideas germinate and flourish. They contributed to this work in many uncredited ways. We hope they approve of the result.

STEWART N. ETHIER
THOMAS G. KURTZ

Salt Lake City, Utah
Madison, Wisconsin
August 1985

CONTENTS

1 INTRODUCTION

The development of any stochastic model involves the identification of properties and parameters that, one hopes, uniquely characterize a stochastic process. Questions concerning continuous dependence on parameters and robustness under perturbation arise naturally out of any such characterization. In fact the model may well be derived by some sort of limiting or approximation argument. The interplay between characterization and approximation or convergence problems for Markov processes is the central theme of this book. Operator semigroups, martingale problems, and stochastic equations provide approaches to the characterization of Markov processes, and to each of these approaches correspond methods for proving convergence results.

The processes of interest to us here always have values in a complete, separable metric space E, and almost always have sample paths in $D_E[0, \infty)$, the space of right continuous E-valued functions on $[0, \infty)$ having left limits. We give $D_E[0, \infty)$ the Skorohod topology (Chapter 3), under which it also becomes a complete, separable metric space. The type of convergence we are usually concerned with is convergence in distribution; that is, for a sequence of processes $\{X_n\}$ we are interested in conditions under which $\lim_{n\to\infty} E[f(X_n)] = E[f(X)]$ for every $f \in \bar{C}(D_E[0, \infty))$. (For a metric space S, $\bar{C}(S)$ denotes the space of bounded continuous functions on S. Convergence in distribution is denoted by $X_n \Rightarrow X$.) As an introduction to the methods presented in this book we consider a simple but (we hope) illuminating example.

For each $n \geq 1$, define

(1) $$\lambda_n(x) = 1 + 3x\left(x - \frac{1}{n}\right), \qquad \mu_n(x) = 3x + x\left(x - \frac{1}{n}\right)\left(x - \frac{2}{n}\right),$$

and let Y_n be a birth-and-death process in $\mathbb{Z}_+$ with transition probabilities satisfying

(2) $$P\{Y_n(t+h) = j+1 \mid Y_n(t) = j\} = n\lambda_n\left(\frac{j}{n}\right)h + o(h)$$

and

(3) $$P\{Y_n(t+h) = j-1 \mid Y_n(t) = j\} = n\mu_n\left(\frac{j}{n}\right)h + o(h)$$

as $h \to 0+$. In this process, known as the Schlögl model, $Y_n(t)$ represents the number of molecules at time t of a substance R in a volume n undergoing the chemical reactions

(4) $$R_0 \underset{3}{\overset{1}{\rightleftharpoons}} R, \qquad R_2 + 2R \underset{1}{\overset{3}{\rightleftharpoons}} 3R,$$

with the indicated rates. (See Chapter 11, Section 1.)

We rescale and renormalize letting

(5) $$X_n(t) = n^{1/4}(n^{-1}Y_n(n^{1/2}t) - 1), \qquad t \geq 0.$$

The problem is to show that X_n converges in distribution to a Markov process X to be characterized below.

The first method we consider is based on a semigroup characterization of X. Let $E_n = \{n^{1/4}(n^{-1}y - 1) : y \in \mathbb{Z}_+\}$, and note that

(6) $$T_n(t)f(x) \equiv E[f(X_n(t)) \mid X_n(0) = x]$$

defines a semigroup $\{T_n(t)\}$ on $B(E_n)$ with generator of the form

(7) $$\begin{aligned} G_n f(x) = {} & n^{3/2}\lambda_n(1 + n^{-1/4}x)\{f(x + n^{-3/4}) - f(x)\} \\ & + n^{3/2}\mu_n(1 + n^{-1/4}x)\{f(x - n^{-3/4}) - f(x)\}. \end{aligned}$$

(See Chapter 1.) Letting $\lambda(x) \equiv 1 + 3x^2$, $\mu(x) \equiv 3x + x^3$, and

(8) $$Gf(x) = 4f''(x) - x^3f'(x),$$

a Taylor expansion shows that

(9) $$\begin{aligned} G_n f(x) = {} & Gf(x) + n^{3/2}\{\lambda_n(1 + n^{-1/4}x) - \lambda(1 + n^{-1/4}x)\}\{f(x + n^{-3/4}) - f(x)\} \\ & + n^{3/2}\{\mu_n(1 + n^{-1/4}x) - \mu(1 + n^{-1/4}x)\}\{f(x - n^{-3/4}) - f(x)\} \\ & + \lambda(1 + n^{-1/4}x)\int_0^1 (1 - u)\{f''(x + un^{-3/4}) - f''(x)\}\,du \\ & + \mu(1 + n^{-1/4}x)\int_0^1 (1 - u)\{f''(x - un^{-3/4}) - f''(x)\}\,du \\ & + \{(\lambda + \mu)(1 + n^{-1/4}x) - (\lambda + \mu)(1)\}\tfrac{1}{2}f''(x), \end{aligned}$$

for all $f \in C^2(\mathbb{R})$ with $f' \in C_c(\mathbb{R})$ and all $x \in E_n$. Consequently, for such f,

(10) $$\lim_{n\to\infty} \sup_{x\in E_n} | G_n f(x) - Gf(x) | = 0.$$

Now by Theorem 1.1 of Chapter 8,

(11) $$A \equiv \{(f, Gf) : f \in C[-\infty, \infty] \cap C^2(\mathbb{R}), Gf \in C[-\infty, \infty]\}$$

is the generator of a Feller semigroup $\{T(t)\}$ on $C[-\infty, \infty]$. By Theorem 2.7 of Chapter 4 and Theorem 1.1 of Chapter 8, there exists a diffusion process X corresponding to $\{T(t)\}$, that is, a strong Markov process X with continuous sample paths such that

(12) $$E[f(X(t)) \mid \mathscr{F}_s^X] = T(t-s)f(X(s))$$

for all $f \in C[-\infty, \infty]$ and $t \geq s \geq 0$. ($\mathscr{F}_s^X = \sigma(X(u) : u \leq s)$.)

To prove that $X_n \Rightarrow X$ (assuming convergence of initial distributions), it suffices by Corollary 8.7 of Chapter 4 to show that (10) holds for all f in a core D for the generator A, that is, for all f in a subspace D of $\mathscr{D}(A)$ such that A is the closure of the restriction of A to D. We claim that

(13) $$D \equiv \{f + g : f, g \in C^2(\mathbb{R}), f' \in C_c(\mathbb{R}), (x^2 g)' \in C_c(\mathbb{R})\}$$

is a core, and that (10) holds for all $f \in D$. To see that D is a core, first check that

(14) $$\mathscr{D}(A) = \{f \in C[-\infty, \infty] \cap C^2(\mathbb{R}) : f'' \in \hat{C}(\mathbb{R}), x^3 f' \in C[-\infty, \infty]\}.$$

Then let $h \in C_c^2(\mathbb{R})$ satisfy $\chi_{[-1,1]} \leq h \leq \chi_{[-2,2]}$ and put $h_m(x) = h(x/m)$. Given $f \in \mathscr{D}(A)$, choose $g \in D$ with $(x^2 g)' \in C_c(\mathbb{R})$ and $x^3(f - g)' \in \hat{C}(\mathbb{R})$ and define

(15) $$f_m(x) = f(0) - g(0) + \int_0^x (f - g)'(y) h_m(y)\, dy.$$

Then $f_m + g \in D$ for each m, $f_m + g \to f$, and $G(f_m + g) \to Gf$.

The second method is based on the characterization of X as the solution of a martingale problem. Observe that

(16) $$f(X_n(t)) - \int_0^t G_n f(X_n(s))\, ds$$

is an $\{\mathscr{F}_t^{X_n}\}$-martingale for each $f \in B(E_n)$ with compact support. Consequently, if some subsequence $\{X_{n'}\}$ converges in distribution to X, then, by the continuous mapping theorem (Corollary 1.9 of Chapter 3) and Problem 7 of Chapter 7,

(17) $$f(X(t)) - \int_0^t Gf(X(s))\, ds$$

is an $\{\mathscr{F}_t^X\}$-martingale for each $f \in C_c^2(\mathbb{R})$, or in other words, X is a solution of the martingale problem for $\{(f, Gf): f \in C_c^2(\mathbb{R})\}$. But by Theorem 2.3 of Chapter 8, this property characterizes the distribution on $D_{\mathbb{R}}[0, \infty)$ of X. Therefore, Corollary 8.16 of Chapter 4 gives $X_n \Rightarrow X$ (assuming convergence of initial distributions), provided we can show that

$$\lim_{\alpha \to \infty} \overline{\lim_{n \to \infty}} P\left\{ \sup_{0 \le t \le T} | X_n(t) | \ge \alpha \right\} = 0, \qquad T > 0. \tag{18}$$

Let $\varphi(x) \equiv e^x + e^{-x}$, and check that there exist constants $C_{n,\alpha} > 0$ such that $G_n \varphi \le C_{n,\alpha} \varphi$ on $[-\alpha, \alpha]$ for each $n \ge 1$ and $\alpha > 0$, and $\overline{\lim}_{\alpha \to \infty} \overline{\lim}_{n \to \infty} C_{n,\alpha} < \infty$. Letting $\tau_{n,\alpha} = \inf\{t \ge 0: | X_n(t) | \ge \alpha\}$, we have

$$\begin{aligned} e^{-C_{n,\alpha}T} \inf_{|y| \ge \alpha} \varphi(y) P\left\{ \sup_{0 \le t \le T} |X_n(t)| \ge \alpha \right\} \\ \le E[\exp\{-C_{n,\alpha}(\tau_{n,\alpha} \wedge T)\} \varphi(X_n(\tau_{n,\alpha} \wedge T))] \\ \le E[\varphi(X_n(0))] \end{aligned} \tag{19}$$

by Lemma 3.2 of Chapter 4 and the optional sampling theorem. An additional (mild) assumption on the initial distributions therefore guarantees (18).

Actually we can avoid having to verify (18) by observing that the uniform convergence of $G_n f$ to Gf for $f \in C_c^2(\mathbb{R})$ and the uniqueness for the limiting martingale problem imply (again by Corollary 8.16 of Chapter 4) that $X_n \Rightarrow X$ in $D_{\mathbb{R}^\Delta}[0, \infty)$ where $\mathbb{R}^\Delta$ denotes the one-point compactification of $\mathbb{R}$. Convergence in $D_{\mathbb{R}}[0, \infty)$ then follows from the fact that X_n and X have sample paths in $D_{\mathbb{R}}[0, \infty)$.

Both of the approaches considered so far have involved characterizations in terms of generators. We now consider methods based on stochastic equations. First, by Theorems 3.7 and 3.10 of Chapter 5, we can characterize X as the unique solution of the stochastic integral equation

$$X(t) = X(0) + 2\sqrt{2}W(t) - \int_0^t X(s)^3 \, ds, \tag{20}$$

where W is a standard, one-dimensional, Brownian motion. (In the present example, the term $2\sqrt{2}W(t)$ corresponds to the stochastic integral term.) A convergence theory can be developed using this characterization of X, but we do not do so here. The interested reader is referred to Kushner (1974).

The final approach we discuss is based on a characterization of X involving random time changes. We observe first that Y_n satisfies

$$Y_n(t) = Y_n(0) + N_+\left(n \int_0^t \lambda_n(n^{-1} Y_n(s)) \, ds\right) - N_-\left(n \int_0^t \mu_n(n^{-1} Y_n(s)) \, ds\right), \tag{21}$$

where N_+ and N_- are independent, standard (parameter 1), Poisson processes. Consequently, X_n satisfies

$$X_n(t) = X_n(0) + n^{-3/4}\tilde{N}_+\left(n^{3/2}\int_0^t \lambda_n(1 + n^{-1/4}X_n(s))\,ds\right) \tag{22}$$
$$- n^{-3/4}\tilde{N}_-\left(n^{3/2}\int_0^t \mu_n(1 + n^{-1/4}X_n(s))\,ds\right)$$
$$+ n^{3/4}\int_0^t (\lambda_n - \mu_n)(1 + n^{-1/4}X_n(s))\,ds,$$

where $\tilde{N}_+(u) = N_+(u) - u$ and $\tilde{N}_-(u) = N_-(u) - u$ are independent, centered, standard, Poisson processes. Now it is easy to see that

$$(n^{-3/4}\tilde{N}_+(n^{3/2}\,\cdot\,),\ n^{-3/4}\tilde{N}_-(n^{3/2}\,\cdot\,)) \Rightarrow (W_+, W_-), \tag{23}$$

where W_+ and W_- are independent, standard, one-dimensional Brownian motions. Consequently, if some subsequence $\{X_{n'}\}$ converges in distribution to X, one might expect that

$$X(t) = X(0) + W_+(4t) + W_-(4t) - \int_0^t X(s)^3\,ds. \tag{24}$$

(In this simple example, (20) and (24) are equivalent, but they will not be so in general.) Clearly, (24) characterizes X, and using the estimate (18) we conclude $X_n \Rightarrow X$ (assuming convergence of initial distributions) from Theorem 5.4 of Chapter 6.

For a further discussion of the Schlögl model and related models see Schlögl (1972) and Malek-Mansour et al. (1981). The martingale proof of convergence is from Costantini and Nappo (1982), and the time change proof is from Kurtz (1981c).

Chapters 4–7 contain the main characterization and convergence results (with the emphasis in Chapters 5 and 7 on diffusion processes). Chapters 1–3 contain preliminary material on operator semigroups, martingales, and weak convergence, and Chapters 8–12 are concerned with applications.

1 OPERATOR SEMIGROUPS

Operator semigroups provide a primary tool in the study of Markov processes. In this chapter we develop the basic background for their study and the existence and approximation results that are used later as the basis for existence and approximation theorems for Markov processes. Section 1 gives the basic definitions, and Section 2 the Hille–Yosida theorem, which characterizes the operators that are generators of semigroups. Section 3 concerns the problem of verifying the hypotheses of this theorem, and Sections 4 and 5 are devoted to generalizations of the concept of the generator. Sections 6 and 7 present the approximation and perturbation results.

Throughout the chapter, L denotes a real Banach space with norm $\|\cdot\|$.

1. DEFINITIONS AND BASIC PROPERTIES

A one-parameter family $\{T(t): t \geq 0\}$ of bounded linear operators on a Banach space L is called a *semigroup* if $T(0) = I$ and $T(s + t) = T(s)T(t)$ for all $s, t \geq 0$. A semigroup $\{T(t)\}$ on L is said to be *strongly continuous* if $\lim_{t \to 0} T(t)f = f$ for every $f \in L$; it is said to be a *contraction* semigroup if $\|T(t)\| \leq 1$ for all $t \geq 0$.

Given a bounded linear operator B on L, define

(1.1) $$e^{tB} = \sum_{k=0}^{\infty} \frac{1}{k!} t^k B^k, \qquad t \geq 0.$$

A simple calculation gives $e^{(s+t)B} = e^{sB}e^{tB}$ for all $s, t \geq 0$, and hence $\{e^{tB}\}$ is a semigroup, which can easily be seen to be strongly continuous. Furthermore we have

$$\| e^{tB} \| \leq \sum_{k=0}^{\infty} \frac{1}{k!} t^k \| B^k \| \leq \sum_{k=0}^{\infty} \frac{1}{k!} t^k \| B \|^k = e^{t\| B \|}, \qquad t \geq 0. \tag{1.2}$$

An inequality of this type holds in general for strongly continuous semigroups.

1.1 Proposition Let $\{T(t)\}$ be a strongly continuous semigroup on L. Then there exist constants $M \geq 1$ and $\omega \geq 0$ such that

$$\| T(t) \| \leq Me^{\omega t}, \qquad t \geq 0. \tag{1.3}$$

Proof. Note first that there exist constants $M \geq 1$ and $t_0 > 0$ such that $\| T(t) \| \leq M$ for $0 \leq t \leq t_0$. For if not, we could find a sequence $\{t_n\}$ of positive numbers tending to zero such that $\| T(t_n) \| \to \infty$, but then the uniform boundedness principle would imply that $\sup_n \| T(t_n)f \| = \infty$ for some $f \in L$, contradicting the assumption of strong continuity. Now let $\omega = t_0^{-1} \log M$. Given $t \geq 0$, write $t = kt_0 + s$, where k is a nonnegative integer and $0 \leq s < t_0$; then

$$\| T(t) \| = \| T(s)T(t_0)^k \| \leq MM^k \leq MM^{t/t_0} = Me^{\omega t}. \tag{1.4}$$

□

1.2 Corollary Let $\{T(t)\}$ be a strongly continuous semigroup on L. Then, for each $f \in L$, $t \to T(t)f$ is a continuous function from $[0, \infty)$ into L.

Proof. Let $f \in L$. By Proposition 1.1, if $t \geq 0$ and $h \geq 0$, then

$$\begin{aligned} \| T(t + h)f - T(t)f \| &= \| T(t)[T(h)f - f] \| \\ &\leq Me^{\omega t} \| T(h)f - f \|, \end{aligned} \tag{1.5}$$

and if $0 \leq h \leq t$, then

$$\begin{aligned} \| T(t - h)f - T(t)f \| &= \| T(t - h)[T(h)f - f] \| \\ &\leq Me^{\omega t} \| T(h)f - f \|. \end{aligned} \tag{1.6}$$

□

1.3 Remark Let $\{T(t)\}$ be a strongly continuous semigroup on L such that (1.3) holds, and put $S(t) = e^{-\omega t}T(t)$ for each $t \geq 0$. Then $\{S(t)\}$ is a strongly continuous semigroup on L such that

$$\| S(t) \| \leq M, \qquad t \geq 0. \tag{1.7}$$

In particular, if $M = 1$, then $\{S(t)\}$ is a strongly continuous contraction semigroup on L.

Let $\{S(t)\}$ be a strongly continuous semigroup on L such that (1.7) holds, and define the norm $|||\cdot|||$ on L by

$$|||f||| = \sup_{t \geq 0} \|S(t)f\|. \tag{1.8}$$

Then $\|f\| \leq |||f||| \leq M\|f\|$ for each $f \in L$, so the new norm is equivalent to the original norm; also, with respect to $|||\cdot|||$, $\{S(t)\}$ is a strongly continuous contraction semigroup on L.

Most of the results in the subsequent sections of this chapter are stated in terms of strongly continuous contraction semigroups. Using these reductions, however, many of them can be reformulated in terms of noncontraction semigroups. □

A (possibly unbounded) *linear operator* A on L is a linear mapping whose domain $\mathscr{D}(A)$ is a subspace of L and whose range $\mathscr{R}(A)$ lies in L. The *graph* of A is given by

$$\mathscr{G}(A) = \{(f, Af): f \in \mathscr{D}(A)\} \subset L \times L. \tag{1.9}$$

Note that $L \times L$ is itself a Banach space with componentwise addition and scalar multiplication and norm $\|(f, g)\| = \|f\| + \|g\|$. A is said to be *closed* if $\mathscr{G}(A)$ is a closed subspace of $L \times L$.

The (*infinitesimal*) *generator* of a semigroup $\{T(t)\}$ on L is the linear operator A defined by

$$Af = \lim_{t \to 0} \frac{1}{t} \{T(t)f - f\}. \tag{1.10}$$

The domain $\mathscr{D}(A)$ of A is the subspace of all $f \in L$ for which this limit exists.

Before indicating some of the properties of generators, we briefly discuss the calculus of Banach space-valued functions.

Let Δ be a closed interval in $(-\infty, \infty)$, and denote by $C_L(\Delta)$ the space of continuous functions $u: \Delta \to L$. Let $C_L^1(\Delta)$ be the space of continuously differentiable functions $u: \Delta \to L$.

If Δ is the finite interval $[a, b]$, $u: \Delta \to L$ is said to be (*Riemann*) *integrable* over Δ if $\lim_{\delta \to 0} \sum_{k=1}^{n} u(s_k)(t_k - t_{k-1})$ exists, where $a = t_0 \leq s_1 \leq t_1 \leq \cdots \leq t_{n-1} \leq s_n \leq t_n = b$ and $\delta = \max (t_k - t_{k-1})$; the limit is denoted by $\int_\Delta u(t)\,dt$ or $\int_a^b u(t)\,dt$. If $\Delta = [a, \infty)$, $u: \Delta \to L$ is said to be *integrable* over Δ if $u|_{[a, b]}$ is integrable over $[a, b]$ for each $b \geq a$ and $\lim_{b \to \infty} \int_a^b u(t)\,dt$ exists; again, the limit is denoted by $\int_\Delta u(t)\,dt$ or $\int_a^\infty u(t)\,dt$.

We leave the proof of the following lemma to the reader (Problem 3).

1.4 Lemma (a) If $u \in C_L(\Delta)$ and $\int_\Delta \| u(t) \| \, dt < \infty$, then u is integrable over Δ and

$$\left\| \int_\Delta u(t) \, dt \right\| \le \int_\Delta \| u(t) \| \, dt. \tag{1.11}$$

In particular, if Δ is the finite interval $[a, b]$, then every function in $C_L(\Delta)$ is integrable over Δ.

(b) Let B be a closed linear operator on L. Suppose that $u \in C_L(\Delta)$, $u(t) \in \mathscr{D}(B)$ for all $t \in \Delta$, $Bu \in C_L(\Delta)$, and both u and Bu are integrable over Δ. Then $\int_\Delta u(t) \, dt \in \mathscr{D}(B)$ and

$$B \int_\Delta u(t) \, dt = \int_\Delta Bu(t) \, dt. \tag{1.12}$$

(c) If $u \in C_L^1[a, b]$, then

$$\int_a^b \frac{d}{dt} u(t) \, dt = u(b) - u(a). \tag{1.13}$$

1.5 Proposition Let $\{T(t)\}$ be a strongly continuous semigroup on L with generator A.

(a) If $f \in L$ and $t \geq 0$, then $\int_0^t T(s)f \, ds \in \mathscr{D}(A)$ and

$$T(t)f - f = A \int_0^t T(s)f \, ds. \tag{1.14}$$

(b) If $f \in \mathscr{D}(A)$ and $t \geq 0$, then $T(t)f \in \mathscr{D}(A)$ and

$$\frac{d}{dt} T(t)f = AT(t)f = T(t)Af. \tag{1.15}$$

(c) If $f \in \mathscr{D}(A)$ and $t \geq 0$, then

$$T(t)f - f = \int_0^t AT(s)f \, ds = \int_0^t T(s)Af \, ds. \tag{1.16}$$

Proof. (a) Observe that

$$\begin{aligned} \frac{1}{h}[T(h) - I] \int_0^t T(s)f \, ds &= \frac{1}{h} \int_0^t [T(s+h)f - T(s)f] \, ds \\ &= \frac{1}{h} \left\{ \int_h^{t+h} T(s)f \, ds - \int_0^t T(s)f \, ds \right\} \\ &= \frac{1}{h} \int_t^{t+h} T(s)f \, ds - \frac{1}{h} \int_0^h T(s)f \, ds \end{aligned} \tag{1.17}$$

for all $h > 0$, and as $h \to 0$ the right side of (1.17) converges to $T(t)f - f$.

(b) Since

$$\frac{1}{h}[T(t+h)f - T(t)f] = A_h T(t)f = T(t)A_h f \tag{1.18}$$

for all $h > 0$, where $A_h = h^{-1}[T(h) - I]$, it follows that $T(t)f \in \mathscr{D}(A)$ and $(d/dt)^+ T(t)f = AT(t)f = T(t)Af$. Thus, it suffices to check that $(d/dt)^- T(t)f = T(t)Af$ (assuming $t > 0$). But this follows from the identity

$$\begin{aligned}\frac{1}{-h}[T(t-h)f - T(t)f] - T(t)Af \\ = T(t-h)[A_h - A]f + [T(t-h) - T(t)]Af,\end{aligned} \tag{1.19}$$

valid for $0 < h \leq t$.

(c) This is a consequence of (b) and Lemma 1.4(c). □

1.6 Corollary If A is the generator of a strongly continuous semigroup $\{T(t)\}$ on L, then $\mathscr{D}(A)$ is dense in L and A is closed.

Proof. Since $\lim_{t \to 0+} t^{-1} \int_0^t T(s)f\, ds = f$ for every $f \in L$, Proposition 1.5(a) implies that $\mathscr{D}(A)$ is dense in L. To show that A is closed, let $\{f_n\} \subset \mathscr{D}(A)$ satisfy $f_n \to f$ and $Af_n \to g$. Then $T(t)f_n - f_n = \int_0^t T(s)Af_n\, ds$ for each $t > 0$, so, letting $n \to \infty$, we find that $T(t)f - f = \int_0^t T(s)g\, ds$. Dividing by t and letting $t \to 0$, we conclude that $f \in \mathscr{D}(A)$ and $Af = g$. □

2. THE HILLE–YOSIDA THEOREM

Let A be a closed linear operator on L. If, for some real λ, $\lambda - A$ ($\equiv \lambda I - A$) is one-to-one, $\mathscr{R}(\lambda - A) = L$, and $(\lambda - A)^{-1}$ is a bounded linear operator on L, then λ is said to belong to the *resolvent set* $\rho(A)$ of A, and $R_\lambda = (\lambda - A)^{-1}$ is called the *resolvent* (at λ) of A.

2.1 Proposition Let $\{T(t)\}$ be a strongly continuous contraction semigroup on L with generator A. Then $(0, \infty) \subset \rho(A)$ and

$$(\lambda - A)^{-1} g = \int_0^\infty e^{-\lambda t} T(t)g\, dt \tag{2.1}$$

for all $g \in L$ and $\lambda > 0$.

Proof. Let $\lambda > 0$ be arbitrary. Define U_λ on L by $U_\lambda g = \int_0^\infty e^{-\lambda t} T(t)g\, dt$. Since

$$\| U_\lambda g \| \leq \int_0^\infty e^{-\lambda t} \| T(t)g \|\, dt \leq \lambda^{-1} \| g \| \tag{2.2}$$

for each $g \in L$, U_λ is a bounded linear operator on L. Now given $g \in L$,

$$\text{(2.3)} \qquad \frac{1}{h}[T(h) - I]U_\lambda g = \frac{1}{h}\int_0^\infty e^{-\lambda t}[T(t+h)g - T(t)g]\,dt$$
$$= \frac{e^{\lambda h} - 1}{h}\int_0^\infty e^{-\lambda t}T(t)g\,dt - \frac{e^{\lambda h}}{h}\int_0^h e^{-\lambda t}T(t)g\,dt$$

for every $h > 0$, so, letting $h \to 0$, we find that $U_\lambda g \in \mathscr{D}(A)$ and $AU_\lambda g = \lambda U_\lambda g - g$, that is,

$$\text{(2.4)} \qquad (\lambda - A)U_\lambda g = g, \qquad g \in L.$$

In addition, if $g \in \mathscr{D}(A)$, then (using Lemma 1.4(b))

$$\text{(2.5)} \qquad U_\lambda Ag = \int_0^\infty e^{-\lambda t}T(t)Ag\,dt = \int_0^\infty A(e^{-\lambda t}T(t)g)\,dt$$
$$= A\int_0^\infty e^{-\lambda t}T(t)g\,dt = AU_\lambda g,$$

so

$$\text{(2.6)} \qquad U_\lambda(\lambda - A)g = g, \qquad g \in \mathscr{D}(A).$$

By (2.6), $\lambda - A$ is one-to-one, and by (2.4), $\mathscr{R}(\lambda - A) = L$. Also, $(\lambda - A)^{-1} = U_\lambda$ by (2.4) and (2.6), so $\lambda \in \rho(A)$. Since $\lambda > 0$ was arbitrary, the proof is complete. □

Let A be a closed linear operator on L. Since $(\lambda - A)(\mu - A) = (\mu - A)(\lambda - A)$ for all $\lambda, \mu \in \rho(A)$, we have $(\mu - A)^{-1}(\lambda - A)^{-1} = (\lambda - A)^{-1}(\mu - A)^{-1}$, and a simple calculation gives the *resolvent identity*

$$\text{(2.7)} \qquad R_\lambda R_\mu = R_\mu R_\lambda = (\lambda - \mu)^{-1}(R_\mu - R_\lambda), \qquad \lambda, \mu \in \rho(A).$$

If $\lambda \in \rho(A)$ and $|\lambda - \mu| < \|R_\lambda\|^{-1}$, then

$$\text{(2.8)} \qquad \sum_{n=0}^\infty (\lambda - \mu)^n R_\lambda^{n+1}$$

defines a bounded linear operator that is in fact $(\mu - A)^{-1}$. In particular, this implies that $\rho(A)$ is open in $\mathbb{R}$.

A linear operator A on L is said to be *dissipative* if $\|\lambda f - Af\| \geq \lambda\|f\|$ for every $f \in \mathscr{D}(A)$ and $\lambda > 0$.

2.2 Lemma Let A be a dissipative linear operator on L and let $\lambda > 0$. Then A is closed if and only if $\mathscr{R}(\lambda - A)$ is closed.

Proof. Suppose A is closed. If $\{f_n\} \subset \mathscr{D}(A)$ and $(\lambda - A)f_n \to h$, then the dissipativity of A implies that $\{f_n\}$ is Cauchy. Thus, there exists $f \in L$ such that

$f_n \to f$, and hence $Af_n \to \lambda f - h$. Since A is closed, $f \in \mathscr{D}(A)$ and $h = (\lambda - A)f$. It follows that $\mathscr{R}(\lambda - A)$ is closed.

Suppose $\mathscr{R}(\lambda - A)$ is closed. If $\{f_n\} \subset \mathscr{D}(A)$, $f_n \to f$, and $Af_n \to g$, then $(\lambda - A)f_n \to \lambda f - g$, which equals $(\lambda - A)f_0$ for some $f_0 \in \mathscr{D}(A)$. By the dissipativity of A, $f_n \to f_0$, and hence $f = f_0 \in \mathscr{D}(A)$ and $Af = g$. Thus, A is closed. □

2.3 Lemma Let A be a dissipative closed linear operator on L, and put $\rho^+(A) = \rho(A) \cap (0, \infty)$. If $\rho^+(A)$ is nonempty, then $\rho^+(A) = (0, \infty)$.

Proof. It suffices to show that $\rho^+(A)$ is both open and closed in $(0, \infty)$. Since $\rho(A)$ is necessarily open in $\mathbb{R}$, $\rho^+(A)$ is open in $(0, \infty)$. Suppose that $\{\lambda_n\} \subset \rho^+(A)$ and $\lambda_n \to \lambda > 0$. Given $g \in L$, let $g_n = (\lambda - A)(\lambda_n - A)^{-1}g$ for each n, and note that, because A is dissipative,

$$(2.9)\quad \lim_{n\to\infty} \|g_n - g\| = \lim_{n\to\infty} \|(\lambda - \lambda_n)(\lambda_n - A)^{-1}g\| \le \lim_{n\to\infty} \frac{|\lambda - \lambda_n|}{\lambda_n} \|g\| = 0.$$

Hence $\mathscr{R}(\lambda - A)$ is dense in L, but because A is closed and dissipative, $\mathscr{R}(\lambda - A)$ is closed by Lemma 2.2, and therefore $\mathscr{R}(\lambda - A) = L$. Using the dissipativity of A once again, we conclude that $\lambda - A$ is one-to-one and $\|(\lambda - A)^{-1}\| \le \lambda^{-1}$. It follows that $\lambda \in \rho^+(A)$, so $\rho^+(A)$ is closed in $(0, \infty)$, as required. □

2.4 Lemma Let A be a dissipative closed linear operator on L, and suppose that $\mathscr{D}(A)$ is dense in L and $(0, \infty) \subset \rho(A)$. Then the *Yosida approximation* A_λ of A, defined for each $\lambda > 0$ by $A_\lambda = \lambda A(\lambda - A)^{-1}$, has the following properties:

(a) For each $\lambda > 0$, A_λ is a bounded linear operator on L and $\{e^{tA_\lambda}\}$ is a strongly continuous contraction semigroup on L.

(b) $A_\lambda A_\mu = A_\mu A_\lambda$ for all $\lambda, \mu > 0$.

(c) $\lim_{\lambda\to\infty} A_\lambda f = Af$ for every $f \in \mathscr{D}(A)$.

Proof. For each $\lambda > 0$, let $R_\lambda = (\lambda - A)^{-1}$ and note that $\|R_\lambda\| \le \lambda^{-1}$. Since $(\lambda - A)R_\lambda = I$ on L and $R_\lambda(\lambda - A) = I$ on $\mathscr{D}(A)$, it follows that

$$(2.10)\qquad A_\lambda = \lambda^2 R_\lambda - \lambda I \quad \text{on} \quad L, \qquad \lambda > 0,$$

and

$$(2.11)\qquad A_\lambda = \lambda R_\lambda A \quad \text{on} \quad \mathscr{D}(A), \qquad \lambda > 0.$$

By (2.10), we find that, for each $\lambda > 0$, A_λ is bounded and

$$(2.12)\qquad \|e^{tA_\lambda}\| = e^{-t\lambda}\|e^{t\lambda^2 R_\lambda}\| \le e^{-t\lambda}e^{t\lambda^2\|R_\lambda\|} \le 1$$

for all $t \geq 0$, proving (a). Conclusion (b) is a consequence of (2.10) and (2.7). As for (c), we claim first that

$$\lim_{\lambda \to \infty} \lambda R_\lambda f = f, \qquad f \in L. \tag{2.13}$$

Noting that $\| \lambda R_\lambda f - f \| = \| R_\lambda Af \| \leq \lambda^{-1} \| Af \| \to 0$ as $\lambda \to \infty$ for each $f \in \mathscr{D}(A)$, (2.13) follows from the facts that $\mathscr{D}(A)$ is dense in L and $\| \lambda R_\lambda - I \| \leq 2$ for all $\lambda > 0$. Finally, (c) is a consequence of (2.11) and (2.13). □

2.5 Lemma If B and C are bounded linear operators on L such that $BC = CB$ and $\| e^{tB} \| \leq 1$ and $\| e^{tC} \| \leq 1$ for all $t \geq 0$, then

$$\| e^{tB}f - e^{tC}f \| \leq t \| Bf - Cf \| \tag{2.14}$$

for every $f \in L$ and $t \geq 0$.

Proof. The result follows from the identity

$$\begin{aligned} e^{tB}f - e^{tC}f &= \int_0^t \frac{d}{ds} [e^{sB}e^{(t-s)C}]f \, ds = \int_0^t e^{sB}(B - C)e^{(t-s)C}f \, ds \\ &= \int_0^t e^{sB}e^{(t-s)C}(B - C)f \, ds. \end{aligned} \tag{2.15}$$

(Note that the last equality uses the commutivity of B and C.) □

We are now ready to prove the Hille–Yosida theorem.

2.6 Theorem A linear operator A on L is the generator of a strongly continuous contraction semigroup on L if and only if:

- **(a)** $\mathscr{D}(A)$ is dense in L.
- **(b)** A is dissipative.
- **(c)** $\mathscr{R}(\lambda - A) = L$ for some $\lambda > 0$.

Proof. The necessity of the conditions (a)–(c) follows from Corollary 1.6 and Proposition 2.1. We therefore turn to the proof of sufficiency.

By (b), (c), and Lemma 2.2, A is closed and $\rho(A) \cap (0, \infty)$ is nonempty, so by Lemma 2.3, $(0, \infty) \subset \rho(A)$. Using the notation of Lemma 2.4, we define for each $\lambda > 0$ the strongly continuous contraction semigroup $\{T_\lambda(t)\}$ on L by $T_\lambda(t) = e^{tA_\lambda}$. By Lemmas 2.4(b) and 2.5,

$$\| T_\lambda(t)f - T_\mu(t)f \| \leq t \| A_\lambda f - A_\mu f \| \tag{2.16}$$

for all $f \in L$, $t \geq 0$, and $\lambda, \mu > 0$. Thus, by Lemma 2.4(c), $\lim_{\lambda \to \infty} T_\lambda(t)f$ exists for all $t \geq 0$, uniformly on bounded intervals, for all $f \in \mathscr{D}(A)$, hence for every $f \in \overline{\mathscr{D}(A)} = L$. Denoting the limit by $T(t)f$ and using the identity

$$(2.17) \quad T(s+t)f - T(s)T(t)f = [T(s+t) - T_\lambda(s+t)]f + T_\lambda(s)[T_\lambda(t) - T(t)]f + [T_\lambda(s) - T(s)]T(t)f,$$

we conclude that $\{T(t)\}$ is a strongly continuous contraction semigroup on L.

It remains only to show that A is the generator of $\{T(t)\}$. By Proposition 1.5(c),

$$(2.18) \qquad T_\lambda(t)f - f = \int_0^t T_\lambda(s)A_\lambda f\, ds$$

for all $f \in L$, $t \geq 0$, and $\lambda > 0$. For each $f \in \mathscr{D}(A)$ and $t \geq 0$, the identity

$$(2.19) \qquad T_\lambda(s)A_\lambda f - T(s)Af = T_\lambda(s)(A_\lambda f - Af) + [T_\lambda(s) - T(s)]\, Af,$$

together with Lemma 2.4(c), implies that $T_\lambda(s)A_\lambda f \to T(s)Af$ as $\lambda \to \infty$, uniformly in $0 \leq s \leq t$. Consequently, (2.18) yields

$$(2.20) \qquad T(t)f - f = \int_0^t T(s)Af\, ds$$

for all $f \in \mathscr{D}(A)$ and $t \geq 0$. From this we find that the generator B of $\{T(t)\}$ is an extension of A. But, for each $\lambda > 0$, $\lambda - B$ is one-to-one by the necessity of (b), and $\mathscr{R}(\lambda - A) = L$ since $\lambda \in \rho(A)$. We conclude that $B = A$, completing the proof. □

The above proof and Proposition 2.9 below yield the following result as a by-product.

2.7 Proposition Let $\{T(t)\}$ be a strongly continuous contraction semigroup on L with generator A, and let A_λ be the Yosida approximation of A (defined in Lemma 2.4). Then

$$(2.21) \qquad \| e^{tA_\lambda}f - T(t)f \| \leq t \| A_\lambda f - Af \|, \qquad f \in \mathscr{D}(A),\ t \geq 0,\ \lambda > 0,$$

so, for each $f \in L$, $\lim_{\lambda \to \infty} e^{tA_\lambda}f = T(t)f$ for all $t \geq 0$, uniformly on bounded intervals.

2.8 Corollary Let $\{T(t)\}$ be a strongly continuous contraction semigroup on L with generator A. For $M \subset L$, let

$$(2.22) \qquad \Lambda_M = \{\lambda > 0 \colon \lambda(\lambda - A)^{-1} \colon M \to M\}.$$

If either (a) M is a closed convex subset of L and Λ_M is unbounded, or (b) M is a closed subspace of L and Λ_M is nonempty, then

$$(2.23) \qquad T(t) \colon M \to M, \qquad t \geq 0.$$

Proof. If $\lambda, \mu > 0$ and $|1 - \mu/\lambda| < 1$, then (cf. (2.8))

$$\mu(\mu - A)^{-1} = \sum_{n=0}^{\infty} \frac{\mu}{\lambda}\left(1 - \frac{\mu}{\lambda}\right)^n [\lambda(\lambda - A)^{-1}]^{n+1}. \tag{2.24}$$

Consequently, if M is a closed convex subset of L, then $\lambda \in \Lambda_M$ implies $(0, \lambda] \subset \Lambda_M$, and if M is a closed subspace of L, then $\lambda \in \Lambda_M$ implies $(0, 2\lambda) \subset \Lambda_M$. Therefore, under either (a) or (b), we have $\Lambda_M = (0, \infty)$. Finally, by (2.10),

$$\begin{aligned} \exp\{tA_\lambda\} &= \exp\{-t\lambda\} \exp\{t\lambda[\lambda(\lambda - A)^{-1}]\} \\ &= e^{-t\lambda} \sum_{n=0}^{\infty} \frac{(t\lambda)^n}{n!} [\lambda(\lambda - A)^{-1}]^n \end{aligned} \tag{2.25}$$

for all $t \geq 0$ and $\lambda > 0$, so the conclusion follows from Proposition 2.7. □

2.9 Proposition Let $\{T(t)\}$ and $\{S(t)\}$ be strongly continuous contraction semigroups on L with generators A and B, respectively. If $A = B$, then $T(t) = S(t)$ for all $t \geq 0$.

Proof. This result is a consequence of the next proposition. □

2.10 Proposition Let A be a dissipative linear operator on L. Suppose that $u: [0, \infty) \to L$ is continuous, $u(t) \in \mathscr{D}(A)$ for all $t > 0$, $Au: (0, \infty) \to L$ is continuous, and

$$u(t) = u(\varepsilon) + \int_\varepsilon^t Au(s)\, ds, \tag{2.26}$$

for all $t > \varepsilon > 0$. Then $\| u(t) \| \leq \| u(0) \|$ for all $t \geq 0$.

Proof. Let $0 < \varepsilon = t_0 < t_1 < \cdots < t_n = t$. Then

(2.27)

$$\begin{aligned} \| u(t) \| &= \| u(\varepsilon) \| + \sum_{i=1}^{n} [\| u(t_i) \| - \| u(t_{i-1}) \|] \\ &= \| u(\varepsilon) \| + \sum_{i=1}^{n} [\| u(t_i) \| - \| u(t_i) - (t_i - t_{i-1})Au(t_i) \|] \\ &\quad + \sum_{i=1}^{n} [\| u(t_i) - (t_i - t_{i-1})Au(t_i) \| - \| u(t_i) - (u(t_i) - u(t_{i-1})) \|] \\ &\leq \| u(\varepsilon) \| + \sum_{i=1}^{n} \left[\| u(t_i) - (t_i - t_{i-1})Au(t_i) \| - \left\| u(t_i) - \int_{t_{i-1}}^{t_i} Au(s)\, ds \right\| \right] \\ &\leq \| u(\varepsilon) \| + \sum_{i=1}^{n} \int_{t_{i-1}}^{t_i} \| Au(t_i) - Au(s) \|\, ds, \end{aligned}$$

where the first inequality is due to the dissipativity of A. The result follows from the continuity of Au and u by first letting $\max(t_i - t_{i-1}) \to 0$ and then letting $\varepsilon \to 0$. □

In many applications, an alternative form of the Hille–Yosida theorem is more useful. To state it, we need two definitions and a lemma.

A linear operator A on L is said to be *closable* if it has a closed linear extension. If A is closable, then the *closure* $\bar{A}$ of A is the minimal closed linear extension of A; more specifically, it is the closed linear operator B whose graph is the closure (in $L \times L$) of the graph of A.

2.11 Lemma Let A be a dissipative linear operator on L with $\mathscr{D}(A)$ dense in L. Then A is closable and $\overline{\mathscr{R}(\lambda - A)} = \mathscr{R}(\lambda - \bar{A})$ for every $\lambda > 0$.

Proof. For the first assertion, it suffices to show that if $\{f_n\} \subset \mathscr{D}(A)$, $f_n \to 0$, and $Af_n \to g \in L$, then $g = 0$. Choose $\{g_m\} \subset \mathscr{D}(A)$ such that $g_m \to g$. By the dissipativity of A,

$$(2.28) \qquad \|(\lambda - A)g_m - \lambda g\| = \lim_{n\to\infty} \|(\lambda - A)(g_m + \lambda f_n)\| \geq \lim_{n\to\infty} \lambda \|g_m + \lambda f_n\| = \lambda \|g_m\|$$

for every $\lambda > 0$ and each m. Dividing by λ and letting $\lambda \to \infty$, we find that $\|g_m - g\| \geq \|g_m\|$ for each m. Letting $m \to \infty$, we conclude that $g = 0$.

Let $\lambda > 0$. The inclusion $\overline{\mathscr{R}(\lambda - A)} \supset \mathscr{R}(\lambda - \bar{A})$ is obvious, so to prove equality, we need only show that $\mathscr{R}(\lambda - \bar{A})$ is closed. But this is an immediate consequence of Lemma 2.2. □

2.12 Theorem A linear operator A on L is closable and its closure $\bar{A}$ is the generator of a strongly continuous contraction semigroup on L if and only if:

- **(a)** $\mathscr{D}(A)$ is dense in L.
- **(b)** A is dissipative.
- **(c)** $\mathscr{R}(\lambda - A)$ is dense in L for some $\lambda > 0$.

Proof. By Lemma 2.11, A satisfies (a)–(c) above if and only if A is closable and $\bar{A}$ satisfies (a)–(c) of Theorem 2.6. □

3. CORES

In this section we introduce a concept that is of considerable importance in Sections 6 and 7.

Let A be a closed linear operator on L. A subspace D of $\mathscr{D}(A)$ is said to be a *core* for A if the closure of the restriction of A to D is equal to A (i.e., if $\overline{A|_D} = A$).

3.1 Proposition Let A be the generator of a strongly continuous contraction semigroup on L. Then a subspace D of $\mathscr{D}(A)$ is a core for A if and only if D is dense in L and $\mathscr{R}(\lambda - A|_D)$ is dense in L for some $\lambda > 0$.

3.2 Remark A subspace of L is dense in L if and only if it is weakly dense (Rudin (1973), Theorem 3.12). □

Proof. The sufficiency follows from Theorem 2.12 and from the observation that, if A and B generate strongly continuous contraction semigroups on L and if A is an extension of B, then $A = B$. The necessity depends on Lemma 2.11. □

3.3 Proposition Let A be the generator of a strongly continuous contraction semigroup $\{T(t)\}$ on L. Let D_0 and D be dense subspaces of L with $D_0 \subset D \subset \mathscr{D}(A)$. (Usually, $D_0 = D$.) If $T(t)\colon D_0 \to D$ for all $t \geq 0$, then D is a core for A.

Proof. Given $f \in D_0$ and $\lambda > 0$,

$$f_n \equiv \frac{1}{n} \sum_{k=0}^{n^2} e^{-\lambda k/n} T\left(\frac{k}{n}\right) f \in D \tag{3.1}$$

for $n = 1, 2, \ldots$. By the strong continuity of $\{T(t)\}$ and Proposition 2.1,

$$\begin{aligned} \lim_{n\to\infty} (\lambda - A) f_n &= \lim_{n\to\infty} \frac{1}{n} \sum_{k=0}^{n^2} e^{-\lambda k/n} T\left(\frac{k}{n}\right)(\lambda - A) f \\ &= \int_0^\infty e^{-\lambda t} T(t)(\lambda - A) f \, dt \\ &= (\lambda - A)^{-1}(\lambda - A) f = f, \end{aligned} \tag{3.2}$$

so $\overline{\mathscr{R}(\lambda - A|_D)} \supset D_0$. This suffices by Proposition 3.1 since D_0 is dense in L. □

Given a dissipative linear operator A with $\mathscr{D}(A)$ dense in L, one often wants to show that $\bar{A}$ generates a strongly continuous contraction semigroup on L. By Theorem 2.12, a necessary and sufficient condition is that $\mathscr{R}(\lambda - A)$ be dense in L for some $\lambda > 0$. We can view this problem as one of characterizing a core (namely, $\mathscr{D}(A)$) for the generator of a strongly continuous contraction semigroup, except that, unlike the situation in Propositions 3.1 and 3.3, the generator is not provided in advance. Thus, the remainder of this section is primarily concerned with verifying the range condition (condition (c)) of Theorem 2.12.

Observe that the following result generalizes Proposition 3.3.

3.4 Proposition Let A be a dissipative linear operator on L, and D_0 a subspace of $\mathscr{D}(A)$ that is dense in L. Suppose that, for each $f \in D_0$, there exists a continuous function $u_f: [0, \infty) \to L$ such that $u_f(0) = f$, $u_f(t) \in \mathscr{D}(A)$ for all $t > 0$, $Au_f: (0, \infty) \to L$ is continuous, and

$$u_f(t) - u_f(\varepsilon) = \int_\varepsilon^t Au_f(s)\, ds \tag{3.3}$$

for all $t > \varepsilon > 0$. Then A is closable, the closure of A generates a strongly continuous contraction semigroup $\{T(t)\}$ on L, and $T(t)f = u_f(t)$ for all $f \in D_0$ and $t \geq 0$.

Proof. By Lemma 2.11, A is closable. Fix $f \in D_0$ and denote u_f by u. Let $t_0 > \varepsilon > 0$, and note that $\int_\varepsilon^{t_0} e^{-t}u(t)\, dt \in \mathscr{D}(\bar{A})$ and

$$\bar{A} \int_\varepsilon^{t_0} e^{-t}u(t)\, dt = \int_\varepsilon^{t_0} e^{-t}Au(t)\, dt. \tag{3.4}$$

Consequently,

$$\begin{aligned} \int_\varepsilon^{t_0} e^{-t}u(t)\, dt &= (e^{-\varepsilon} - e^{-t_0})u(\varepsilon) + \int_\varepsilon^{t_0} e^{-t} \int_\varepsilon^t Au(s)\, ds\, dt \\ &= (e^{-\varepsilon} - e^{-t_0})u(\varepsilon) + \int_\varepsilon^{t_0} (e^{-s} - e^{-t_0})Au(s)\, ds \\ &= \bar{A} \int_\varepsilon^{t_0} e^{-t}u(t)\, dt + e^{-\varepsilon}u(\varepsilon) - e^{-t_0}u(t_0). \end{aligned} \tag{3.5}$$

Since $\|u(t)\| \leq \|f\|$ for all $t \geq 0$ by Proposition 2.10, we can let $\varepsilon \to 0$ and $t_0 \to \infty$ in (3.5) to obtain $\int_0^\infty e^{-t}u(t)\, dt \in \mathscr{D}(\bar{A})$ and

$$(1 - \bar{A}) \int_0^\infty e^{-t}u(t)\, dt = f. \tag{3.6}$$

We conclude that $\mathscr{R}(1 - \bar{A}) \supset D_0$, which by Theorem 2.6 proves that $\bar{A}$ generates a strongly continuous contraction semigroup $\{T(t)\}$ on L. Now for each $f \in D_0$,

$$T(t)f - T(\varepsilon)f = \int_\varepsilon^t \bar{A}T(s)f\, ds \tag{3.7}$$

for all $t > \varepsilon > 0$. Subtracting (3.3) from this and applying Proposition 2.10 once again, we obtain the second conclusion of the proposition. □

The next result shows that a sufficient condition for $\bar{A}$ to generate is that A be triangulizable. Of course, this is a very restrictive assumption, but it is occasionally satisfied.

3.5 Proposition Let A be a dissipative linear operator on L, and suppose that $L_1, L_2, L_3, \ldots$ is a sequence of finite-dimensional subspaces of $\mathscr{D}(A)$ such that $\bigcup_{n=1}^{\infty} L_n$ is dense in L. If $A: L_n \to L_n$ for $n = 1, 2, \ldots$, then A is closable and the closure of A generates a strongly continuous contraction semigroup on L.

Proof. For $n = 1, 2, \ldots, (\lambda - A)(L_n) = L_n$ for all λ not belonging to the set of eigenvalues of $A|_{L_n}$, hence for all but at most finitely many $\lambda > 0$. Consequently, $(\lambda - A)(\bigcup_{n=1}^{\infty} L_n) = \bigcup_{n=1}^{\infty} L_n$ for all but at most countably many $\lambda > 0$ and in particular for some $\lambda > 0$. Thus, the conditions of Theorem 2.12 are satisfied. □

We turn next to a generalization of Proposition 3.3 in a different direction. The idea is to try to approximate A sufficiently well by a sequence of generators for which the conditions of Proposition 3.3 are satisfied. Before stating the result we record the following simple but frequently useful lemma.

3.6 Lemma Let $A_1, A_2, \ldots$ and A be linear operators on L, D_0 a subspace of L, and $\lambda > 0$. Suppose that, for each $g \in D_0$, there exists $f_n \in \mathscr{D}(A_n) \cap \mathscr{D}(A)$ for $n = 1, 2, \ldots$ such that $g_n \equiv (\lambda - A_n)f_n \to g$ as $n \to \infty$ and

$$\lim_{n\to\infty} \|(A_n - A)f_n\| = 0. \tag{3.8}$$

Then $\overline{\mathscr{R}(\lambda - A)} \supset D_0$.

Proof. Given $g \in D_0$, choose $\{f_n\}$ and $\{g_n\}$ as in the statement of the lemma, and observe that $\lim_{n\to\infty} \|(\lambda - A)f_n - g_n\| = 0$ by (3.8). It follows that $\lim_{n\to\infty} \|(\lambda - A)f_n - g\| = 0$, giving the desired result. □

3.7 Proposition Let A be a linear operator on L and D_0 and D_1 dense subspaces of L satisfying $D_0 \subset \mathscr{D}(A) \subset D_1 \subset L$. Let $|||\cdot|||$ be a norm on D_1. For $n = 1, 2, \ldots$, suppose that A_n generates a strongly continuous contraction semigroup $\{T_n(t)\}$ on L and $\mathscr{D}(A) \subset \mathscr{D}(A_n)$. Suppose further that there exist $\omega \geq 0$ and a sequence $\{\varepsilon_n\} \subset (0, \infty)$ tending to zero such that, for $n = 1, 2, \ldots$,

$$\|(A_n - A)f\| \leq \varepsilon_n |||f|||, \qquad f \in \mathscr{D}(A), \tag{3.9}$$

$$T_n(t): D_1 \to D_1, \qquad |||T_n(t)|_{D_1}||| \leq e^{\omega t}, \qquad t \geq 0, \tag{3.10}$$

and

$$T_n(t): D_0 \to \mathscr{D}(A), \qquad t \geq 0. \tag{3.11}$$

Then A is closable and the closure of A generates a strongly continuous contraction semigroup on L.

Proof. Observe first that $\mathscr{D}(A)$ is dense in L and, by (3.9) and the dissipativity of each A_n, A is dissipative. It therefore suffices to verify condition (c) of Theorem 2.12.

Fix $\lambda > \omega$. Given $g \in D_0$, let

$$f_{m,n} = \frac{1}{m} \sum_{k=0}^{m^2} e^{-\lambda k/m} T_n\left(\frac{k}{m}\right) g \in \mathscr{D}(A) \tag{3.12}$$

for each $m, n \geq 1$ (cf. (3.1)). Then, for $n = 1, 2, \ldots$, $(\lambda - A_n) f_{m,n} \to \int_0^\infty e^{-\lambda t} T_n(t)(\lambda - A_n) g \, dt = g$ as $m \to \infty$, so there exists a sequence $\{m_n\}$ of positive integers such that $(\lambda - A_n) f_{m_n, n} \to g$ as $n \to \infty$. Moreover,

$$\begin{aligned} \|(A_n - A) f_{m_n, n}\| &\leq \varepsilon_n |||f_{m_n, n}||| \\ &\leq \varepsilon_n m_n^{-1} \sum_{k=0}^{m_n 2} e^{-\lambda k/m_n} e^{\omega k/m_n} |||g||| \\ &\to 0 \quad \text{as} \quad n \to \infty \end{aligned} \tag{3.13}$$

by (3.9) and (3.10), so Lemma 3.6 gives the desired conclusion. □

3.8 Corollary Let A be a linear operator on L with $\mathscr{D}(A)$ dense in L, and let $|||\cdot|||$ be a norm on $\mathscr{D}(A)$ with respect to which $\mathscr{D}(A)$ is a Banach space. For $n = 1, 2, \ldots$, let T_n be a linear $\|\cdot\|$-contraction on L such that $T_n: \mathscr{D}(A) \to \mathscr{D}(A)$, and define $A_n = n(T_n - I)$. Suppose there exist $\omega \geq 0$ and a sequence $\{\varepsilon_n\} \subset (0, \infty)$ tending to zero such that, for $n = 1, 2, \ldots$, (3.9) holds and

$$|||T_n|_{\mathscr{D}(A)}||| \leq 1 + \frac{\omega}{n}. \tag{3.14}$$

Then A is closable and the closure of A generates a strongly continuous contraction semigroup on L.

Proof. We apply Proposition 3.7 with $D_0 = D_1 = \mathscr{D}(A)$. For $n = 1, 2, \ldots$, $\exp\{tA_n\}: \mathscr{D}(A) \to \mathscr{D}(A)$ and

$$|||\exp\{tA_n\}|_{\mathscr{D}(A)}||| \leq \exp\{-nt\} \exp\{nt\, |||T_n|_{\mathscr{D}(A)}|||\} \leq \exp\{\omega t\} \tag{3.15}$$

for all $t \geq 0$, so the hypotheses of the proposition are satisfied. □

4. MULTIVALUED OPERATORS

Recall that if A is a linear operator on L, then the graph $\mathscr{G}(A)$ of A is a subspace of $L \times L$ such that $(0, g) \in \mathscr{G}(A)$ implies $g = 0$. More generally, we regard an arbitrary subset A of $L \times L$ as a *multivalued* operator on L with *domain* $\mathscr{D}(A) = \{f: (f, g) \in A \text{ for some } g\}$ and range $\mathscr{R}(A) = \{g: (f, g) \in A \text{ for some } f\}$. $A \subset L \times L$ is said to be *linear* if A is a subspace of $L \times L$. If A is linear, then A is said to be *single-valued* if $(0, g) \in A$ implies $g = 0$; in this case,

A is a graph of a linear operator on L, also denoted by A, so we write $Af = g$ if $(f, g) \in A$. If $A \subset L \times L$ is linear, then A is said to be *dissipative* if $\| \lambda f - g \| \geq \lambda \| f \|$ for all $(f, g) \in A$ and $\lambda > 0$; the closure $\bar{A}$ of A is of course just the closure in $L \times L$ of the subspace A. Finally, we define $\lambda - A = \{(f, \lambda f - g): (f, g) \in A\}$ for each $\lambda > 0$.

Observe that a (single-valued) linear operator A is closable if and only if the closure of A (in the above sense) is single-valued. Consequently, the term "closable" is no longer needed.

We begin by noting that the generator of a strongly continuous contraction semigroup is a maximal dissipative (multivalued) linear operator.

4.1 Proposition Let A be the generator of a strongly continuous contraction semigroup on L. Let $B \subset L \times L$ be linear and dissipative, and suppose that $A \subset B$. Then $A = B$.

Proof. Let $(f, g) \in B$ and $\lambda > 0$. Then $(f, \lambda f - g) \in \lambda - B$. Since $\lambda \in \rho(A)$, there exists $h \in \mathscr{D}(A)$ such that $\lambda h - Ah = \lambda f - g$. Hence $(h, \lambda f - g) \in \lambda - A \subset \lambda - B$. By linearity, $(f - h, 0) \in \lambda - B$, so by dissipativity, $f = h$. Hence $g = Ah$, so $(f, g) \in A$. □

We turn next to an extension of Lemma 2.11.

4.2 Lemma Let $A \subset L \times L$ be linear and dissipative. Then

$$A_0 = \{(f, g) \in \bar{A}: g \in \overline{\mathscr{D}(A)}\} \tag{4.1}$$

is single-valued and $\overline{\mathscr{R}(\lambda - A)} = \mathscr{R}(\lambda - \bar{A})$ for every $\lambda > 0$.

Proof. Given $(0, g) \in A_0$, we must show that $g = 0$. By the definition of A_0, there exists a sequence $\{(g_n, h_n)\} \subset A$ such that $g_n \to g$. For each n, $(g_n, h_n + \lambda g) \in \bar{A}$ by the linearity of $\bar{A}$, so $\| \lambda g_n - h_n - \lambda g \| \geq \lambda \| g_n \|$ for every $\lambda > 0$ by the dissipativity of $\bar{A}$. Dividing by λ and letting $\lambda \to \infty$, we find that $\| g_n - g \| \geq \| g_n \|$ for each n. Letting $n \to \infty$, we conclude that $g = 0$.

The proof of the second assertion is similar to that of the second assertion of Lemma 2.11. □

The main result of this section is the following version of the Hille–Yosida theorem.

4.3 Theorem Let $A \subset L \times L$ be linear and dissipative, and define A_0 by (4.1). Then A_0 is the generator of a strongly continuous contraction semigroup on $\overline{\mathscr{D}(A)}$ if and only if $\overline{\mathscr{R}(\lambda - A)} \supset \overline{\mathscr{D}(A)}$ for some $\lambda > 0$.

Proof. A_0 is single-valued by Lemma 4.2 and is clearly dissipative, so by the Hille–Yosida theorem (Theorem 2.6), A_0 generates a strongly continuous contraction semigroup on $\overline{\mathscr{D}(A)}$ if and only if $\mathscr{D}(A_0)$ is dense in $\overline{\mathscr{D}(A)}$ and $\mathscr{R}(\lambda - A_0) = \overline{\mathscr{D}(A)}$ for some $\lambda > 0$. The latter condition is clearly equivalent to

$\overline{\mathscr{R}(\lambda - \bar{A})} \supset \overline{\mathscr{D}(A)}$ for some $\lambda > 0$, which by Lemma 4.2 is equivalent to $\overline{\mathscr{R}(\lambda - A)} \supset \mathscr{D}(A)$ for some $\lambda > 0$. Thus, to complete the proof, it suffices to show that $\mathscr{D}(A_0)$ is dense in $\overline{\mathscr{D}(A)}$ assuming that $\mathscr{R}(\lambda - A_0) = \overline{\mathscr{D}(A)}$ for some $\lambda > 0$.

By Lemma 2.3, $\mathscr{R}(\lambda - A_0) = \overline{\mathscr{D}(A)}$ for every $\lambda > 0$, so $\overline{\mathscr{R}(\lambda - A)} = \mathscr{R}(\lambda - \bar{A}) \supset \overline{\mathscr{D}(A)}$ for every $\lambda > 0$. By the dissipativity of $\bar{A}$, we may regard $(\lambda - \bar{A})^{-1}$ as a (single-valued) bounded linear operator on $\mathscr{R}(\lambda - \bar{A})$ of norm at most λ^{-1} for each $\lambda > 0$. Given $(f, g) \in \bar{A}$ and $\lambda > 0$, $\lambda f - g \in \mathscr{R}(\lambda - \bar{A})$ and $f \in \mathscr{D}(\bar{A}) \subset \overline{\mathscr{D}(A)} \subset \mathscr{R}(\lambda - \bar{A})$, so $g \in \mathscr{R}(\lambda - \bar{A})$, and therefore $\| \lambda(\lambda - \bar{A})^{-1} f - f \| = \| (\lambda - \bar{A})^{-1} g \| \leq \lambda^{-1} \| g \|$. Since $\mathscr{D}(\bar{A})$ is dense in $\mathscr{D}(A)$, it follows that

$$\lim_{\lambda \to \infty} \lambda(\lambda - \bar{A})^{-1} f = f, \qquad f \in \overline{\mathscr{D}(A)}. \tag{4.2}$$

(Note that this does *not* follow from (2.13).) But clearly, $(\lambda - \bar{A})^{-1}$: $\mathscr{R}(\lambda - A_0) \to \mathscr{D}(A_0)$, that is, $(\lambda - \bar{A})^{-1}$: $\overline{\mathscr{D}(A)} \to \mathscr{D}(A_0)$, for all $\lambda > 0$. In view of (4.2), this completes the proof. □

Multivalued operators arise naturally in several ways. For example, the following concept is crucial in Sections 6 and 7.

For $n = 1, 2, \ldots$, let L_n, in addition to L, be a Banach space with norm also denoted by $\| \cdot \|$, and let π_n: $L \to L_n$ be a bounded linear transformation. Assume that $\sup_n \| \pi_n \| < \infty$. If $A_n \subset L_n \times L_n$ is linear for each $n \geq 1$, the *extended limit* of the sequence $\{A_n\}$ is defined by

$$\operatorname*{ex-lim}_{n \to \infty} A_n = \{(f, g) \in L \times L: \text{there exists } (f_n, g_n) \in A_n \text{ for each } n \geq 1 \text{ such that } \|f_n - \pi_n f\| \to 0 \text{ and } \|g_n - \pi_n g\| \to 0\}. \tag{4.3}$$

We leave it to the reader to show that $\text{ex-lim}_{n \to \infty} A_n$ is necessarily closed in $L \times L$ (Problem 11).

To see that $\text{ex-lim}_{n \to \infty} A_n$ need not be single-valued even if each A_n is, let $L_n = L$, $\pi_n = I$, and $A_n = B + nC$ for each $n \geq 1$, where B and C are bounded linear operators on L. If f belongs to $\mathscr{N}(C)$, the null space of C, and $h \in L$, then $A_n(f + (1/n)h) \to Bf + Ch$, so

$$\{(f, Bf + Ch): f \in \mathscr{N}(C), h \in L\} \subset \operatorname*{ex-lim}_{n \to \infty} A_n. \tag{4.4}$$

Another situation in which multivalued operators arise is described in the next section.

5. SEMIGROUPS ON FUNCTION SPACES

In this section we want to extend the notion of the generator of a semigroup, but to do so we need to be able to integrate functions u: $[0, \infty) \to L$ that are not continuous and to which the Riemann integral of Section 1 does not apply. For our purposes, the most efficient way to get around this difficulty is to restrict the class of Banach spaces L under consideration. We therefore assume in this section that L is a "function space" that arises in the following way.

Let $(M, \mathscr{M})$ be a measurable space, let Γ be a collection of positive measures on $\mathscr{M}$, and let $\mathscr{L}$ be the vector space of $\mathscr{M}$-measurable functions f such that

$$\|f\| \equiv \sup_{\mu \in \Gamma} \int |f| \, d\mu < \infty. \tag{5.1}$$

Note that $\|\cdot\|$ is a seminorm on $\mathscr{L}$ but need not be a norm. Let $\mathscr{N} = \{f \in \mathscr{L}: \|f\| = 0\}$ and let L be the quotient space $\mathscr{L}/\mathscr{N}$, that is, L is the space of equivalence classes of functions in $\mathscr{L}$, where $f \sim g$ if $\|f - g\| = 0$. As is typically the case in discussions of L^p-spaces, we do not distinguish between a function in $\mathscr{L}$ and its equivalence class in L unless necessary.

L is a Banach space, the completeness following as for L^p-spaces. In fact, if v is a σ-finite measure on $\mathscr{M}$, $1 \le q \le \infty$, $p^{-1} + q^{-1} = 1$, and

$$\Gamma = \left\{\mu \ge 0: \mu \ll v, \left\|\frac{d\mu}{dv}\right\|_q \le 1\right\}, \tag{5.2}$$

where $\|\cdot\|_q$ is the norm on $L^q(v)$, then $L = L^p(v)$. Of course, if Γ is the set of probability measures on $\mathscr{M}$, then $L = B(M, \mathscr{M})$, the space of bounded $\mathscr{M}$-measurable functions on M with the sup norm.

Let $(S, \mathscr{S}, v)$ be a σ-finite measure space, let $f: S \times M \to \mathbb{R}$ be $\mathscr{S} \times \mathscr{M}$-measurable, and let $g: S \to [0, \infty)$ be $\mathscr{S}$-measurable. If $\|f(s, \cdot)\| \le g(s)$ for all $s \in S$ and $\int g(s)v(ds) < \infty$, then

$$\begin{aligned} \sup_{\mu \in \Gamma} \int \left| \int f(s, x)v(ds) \right| \mu(dx) &\le \sup_{\mu \in \Gamma} \iint |f(s, x)| \mu(dx)v(ds) \\ &\le \int g(s)v(ds) < \infty, \end{aligned} \tag{5.3}$$

and we can define $\int f(s, \cdot)v(ds) \in L$ to be the equivalence class of functions in $\mathscr{L}$ equivalent to h, where

$$h(x) = \begin{cases} \int f(s, x)v(ds), & \int |f(s, x)| \, v(ds) < \infty, \\ 0, & \text{otherwise.} \end{cases} \tag{5.4}$$

With the above in mind, we say that $u: S \to L$ is *measurable* if there exists an $\mathscr{S} \times \mathscr{M}$-measurable function v such that $v(s, \cdot) \in u(s)$ for each $s \in S$. We define a semigroup $\{T(t)\}$ on L to be *measurable* if $T(\cdot)f$ is measurable as a function on $([0, \infty), \mathscr{B}[0, \infty))$ for each $f \in L$. We define the *full generator* $\hat{A}$ of a measurable contraction semigroup $\{T(t)\}$ on L by

(5.5) $$\hat{A} = \left\{(f, g) \in L \times L:\ T(t)f - f = \int_0^t T(s)g\, ds, \qquad t \geq 0\right\}.$$

We note that $\hat{A}$ is not, in general, single-valued. For example, if $L = B(\mathbb{R})$ with the sup norm and $T(t)f(x) \equiv f(x + t)$, then $(0, g) \in \hat{A}$ for each $g \in B(\mathbb{R})$ that is zero almost everywhere with respect to Lebesgue measure.

5.1 Proposition Let L be as above, and let $\{T(t)\}$ be a measurable contraction semigroup on L. Then the full generator $\hat{A}$ of $\{T(t)\}$ is linear and dissipative and satisfies

(5.6) $$(\lambda - \hat{A})^{-1}h = \int_0^\infty e^{-\lambda t}T(t)h\, dt$$

for all $h \in \mathscr{R}(\lambda - \hat{A})$ and $\lambda > 0$. If

(5.7) $$T(s)\int_0^\infty e^{-\lambda t}T(t)h\, dt = \int_0^\infty e^{-\lambda t}T(s + t)h\, dt$$

for all $h \in L$, $\lambda > 0$, and $s \geq 0$, then $\mathscr{R}(\lambda - \hat{A}) = L$ for every $\lambda > 0$.

Proof. Let $(f, g) \in \hat{A}$, $\lambda > 0$, and $h = \lambda f - g$. Then

(5.8) $$\begin{aligned}\int_0^\infty e^{-\lambda t}T(t)h\, dt &= \lambda\int_0^\infty e^{-\lambda t}T(t)f\, dt - \int_0^\infty e^{-\lambda t}T(t)g\, dt \\ &= \lambda\int_0^\infty e^{-\lambda t}T(t)f\, dt - \lambda\int_0^\infty e^{-\lambda t}\int_0^t T(s)g\, ds\, dt \\ &= f.\end{aligned}$$

Consequently, $\|f\| \leq \lambda^{-1}\|h\|$, proving dissipativity, and (5.6) holds.

Assuming (5.7), let $h \in L$ and $\lambda > 0$, and define $f = \int_0^\infty e^{-\lambda t}T(t)h\, dt$ and $g = \lambda f - h$. Then

(5.9) $$\begin{aligned}\int_0^t T(s)g\, ds &= \lambda\int_0^t\int_0^\infty e^{-\lambda u}T(s + u)h\, du\, ds - \int_0^t T(s)h\, ds \\ &= \lambda\int_0^t e^{\lambda s}\int_s^\infty e^{-\lambda u}T(u)h\, du\, ds - \int_0^t T(s)h\, ds \\ &= e^{\lambda t}\int_t^\infty e^{-\lambda u}T(u)h\, du - \int_0^\infty e^{-\lambda u}T(u)h\, du \\ &\quad + \int_0^t T(s)h\, ds - \int_0^t T(s)h\, ds \\ &= T(t)f - f\end{aligned}$$

for all $t \geq 0$, so $(f, g) \in \hat{A}$ and $h = \lambda f - g \in \mathscr{R}(\lambda - \hat{A})$. □

The following proposition, which is analogous to Proposition 1.5(a), gives a useful description of some elements of $\hat{A}$.

5.2 Proposition Let L and $\{T(t)\}$ be as in Theorem 5.1, let $h \in L$ and $u \geq 0$, and suppose that

$$T(t) \int_0^u T(s)h \, ds = \int_0^u T(t+s)h \, ds \tag{5.10}$$

for all $t \geq 0$. Then

$$\left(\int_0^u T(s)h \, ds, \quad T(u)h - h \right) \in \hat{A}. \tag{5.11}$$

Proof. Put $f = \int_0^u T(s)h \, ds$. Then

$$\begin{aligned} T(t)f - f &= \int_0^u T(t+s)h \, ds - \int_0^u T(s)h \, ds \\ &= \int_u^{t+u} T(s)h \, ds - \int_0^t T(s)h \, ds \\ &= \int_0^t T(s)(T(u)h - h) \, ds \end{aligned} \tag{5.12}$$

for all $t \geq 0$. □

In the present context, given a dissipative closed linear operator $A \subset L \times L$, it may be possible to find measurable functions $u: [0, \infty) \to L$ and $v: [0, \infty) \to L$ such that $(u(t), v(t)) \in A$ for every $t > 0$ and

$$u(t) = u(0) + \int_0^t v(s) \, ds, \qquad t \geq 0. \tag{5.13}$$

One would expect u to be continuous, and since A is closed and linear, it is reasonable to expect that

$$\left(\int_0^t u(s) \, ds, \int_0^t v(s) \, ds \right) = \left(\int_0^t u(s) \, ds, u(t) - u(0) \right) \in A \tag{5.14}$$

for all $t > 0$. With these considerations in mind, we have the following multivalued extension of Proposition 2.10. Note that this result is in fact valid for arbitrary L.

5.3 Proposition Let $A \subset L \times L$ be a dissipative closed linear operator. Suppose u: $[0, \infty) \to L$ is continuous and $(\int_0^t u(s)\, ds,\ u(t) - u(0)) \in A$ for each $t > 0$. Then

$$\| u(t) \| \leq \| u(0) \| \tag{5.15}$$

for all $t \geq 0$. Given $\lambda > 0$, define

$$f = \int_0^\infty e^{-\lambda t} u(t)\, dt, \quad g = \lambda \int_0^\infty e^{-\lambda t}(u(t) - u(0))\, dt. \tag{5.16}$$

Then $(f, g) \in A$ and $\lambda f - g = u(0)$.

Proof. Fix $t \geq 0$, and for each $\varepsilon > 0$, put $u_\varepsilon(t) = \varepsilon^{-1} \int_t^{t+\varepsilon} u(s)\, ds$. Then

$$u_\varepsilon(t) = u_\varepsilon(0) + \int_0^t \varepsilon^{-1}(u(s + \varepsilon) - u(s))\, ds. \tag{5.17}$$

Since $(u_\varepsilon(t),\ \varepsilon^{-1}(u(t + \varepsilon) - u(t))) \in A$, it follows as in Proposition 2.10 that $\| u_\varepsilon(t) \| \leq \| u_\varepsilon(0) \|$. Letting $\varepsilon \to 0$, we obtain (5.15).

Integrating by parts,

$$f = \int_0^\infty e^{-\lambda t} u(t)\, dt = \lambda \int_0^\infty e^{-\lambda t} \int_0^t u(s)\, ds\, dt, \tag{5.18}$$

so $(f, g) \in A$ by the continuity of u and the fact that A is closed and linear. The equation $\lambda f - g = u(0)$ follows immediately from the definition of f and g. □

Heuristically, if $\{S(t)\}$ has generator B and $\{T(t)\}$ has generator $A + B$, then (cf. Lemma 6.2)

$$T(t)f = S(t)f + \int_0^t S(t - s)AT(s)f\, ds \tag{5.19}$$

for all $t \geq 0$. Consequently, a weak form of the equation $u_t = (A + B)u$ is

$$u(t) = S(t)u(0) + \int_0^t S(t - s)Au(s)\, ds. \tag{5.20}$$

We extend Proposition 5.3 to this setting.

5.4 Proposition Let L be as in Proposition 5.1, let $A \subset L \times L$ be a dissipative closed linear operator, and let $\{S(t)\}$ be a strongly continuous, measurable, contraction semigroup on L. Suppose u: $[0, \infty) \to L$ is continuous, v: $[0, \infty) \to L$ is bounded and measurable, and

(5.21)
$$u(t) = S(t)u(0) + \int_0^t S(t-s)v(s)\,ds$$

for all $t \geq 0$. If

(5.22)
$$\left(\int_0^t u(s)\,ds,\ \int_0^t v(s)\,ds\right) \in A$$

for every $t > 0$, and

(5.23)
$$\int_0^t S(q+r)v(s)\,ds = S(q)\int_0^t S(r)v(s)\,ds$$

for all $q, r, t \geq 0$, then (5.15) holds for all $t \geq 0$.

5.5 Remark The above result holds in an arbitrary Banach space under the assumption that v is strongly measurable, that is, v can be uniformly approximated by measurable simple functions. □

Proof. Assume first that $u: [0, \infty) \to L$ is continuously differentiable, $v: [0, \infty) \to L$ is continuous, and $(u(t), v(t)) \in A$ for all $t \geq 0$. Let $0 = t_0 < t_1 < \cdots < t_n = t$. Then, as in the proof of Proposition 2.10,

(5.24)
$$\begin{aligned}
\|u(t)\| &= \|u(0)\| + \sum_{i=1}^n [\|u(t_i)\| - \|u(t_{i-1})\|] \\
&= \|u(0)\| + \sum_{i=1}^n \Bigg[\|u(t_i)\| - \Bigg\|u(t_i) - (S(t_i - t_{i-1}) - I)u(t_{i-1}) \\
&\quad - \int_{t_{i-1}}^{t_i} S(t_i - s)v(s)\,ds\Bigg\|\Bigg] \\
&\leq \|u(0)\| + \sum_{i=1}^n [\|u(t_i)\| - \|u(t_i) - (S(t_i - t_{i-1}) - I)u(t_i) - (t_i - t_{i-1})v(t_i)\|] \\
&\quad + \sum_{i=1}^n \left\|(S(t_i - t_{i-1}) - I)(u(t_i) - u(t_{i-1})) - \int_{t_{i-1}}^{t_i} (S(t_i - s)v(s) - v(t_i))\,ds\right\| \\
&\leq \|u(0)\| + \sum_{i=1}^n [\|u(t_i)\| - \|2u(t_i) - (t_i - t_{i-1})v(t_i)\| + \|S(t_i - t_{i-1})u(t_i)\|] \\
&\quad + \sum_{i=1}^n \left\|\int_{t_{i-1}}^{t_i} [(S(t_i - t_{i-1}) - I)u'(s) - S(t_i - s)v(s) + v(t_i)]\,ds\right\| \\
&\leq \|u(0)\| + \int_0^t \|(S(s'' - s') - I)u'(s) - S(s'' - s)v(s) + v(s'')\|\,ds,
\end{aligned}$$

where $s' = t_{i-1}$ and $s'' = t_i$ for $t_{i-1} \leq s < t_i$. Since the integrand on the right is bounded and tends to zero as $\max(t_i - t_{i-1}) \to 0$, we obtain (5.15) in this case.

In the general case, fix $t \geq 0$, and for each $\varepsilon > 0$, put

$$u_\varepsilon(t) = \varepsilon^{-1} \int_t^{t+\varepsilon} u(s)\, ds, \quad v_\varepsilon(t) = \varepsilon^{-1} \int_t^{t+\varepsilon} v(s)\, ds. \tag{5.25}$$

Then

$$\begin{aligned} u_\varepsilon(t) &= \varepsilon^{-1} \int_0^\varepsilon u(t+s)\, ds \\ &= \varepsilon^{-1} \int_0^\varepsilon S(t+s)u(0)\, ds + \varepsilon^{-1} \int_0^\varepsilon \int_0^{t+s} S(t+s-r)v(r)\, dr\, ds \\ &= \varepsilon^{-1} S(t) \int_0^\varepsilon S(s)u(0)\, ds + \varepsilon^{-1} \int_0^\varepsilon \int_0^s S(t+s-r)v(r)\, dr\, ds \\ &\quad + \varepsilon^{-1} \int_0^\varepsilon \int_0^t S(t-r)v(r+s)\, dr\, ds \\ &= S(t)\left[\varepsilon^{-1} \int_0^\varepsilon S(s)u(0)\, ds + \varepsilon^{-1} \int_0^\varepsilon \int_0^s S(s-r)v(r)\, dr\, ds\right] \\ &\quad + \int_0^t S(t-r)v_\varepsilon(r)\, dr. \end{aligned} \tag{5.26}$$

By the special case already treated,

$$\| u_\varepsilon(t) \| \leq \left\| \varepsilon^{-1} \int_0^\varepsilon S(s)u(0)\, ds + \varepsilon^{-1} \int_0^\varepsilon \int_0^s S(s-r)v(r)\, dr\, ds \right\|, \tag{5.27}$$

and letting $\varepsilon \to 0$, we obtain (5.15) in general. □

6. APPROXIMATION THEOREMS

In this section, we adopt the following conventions. For $n = 1, 2, \ldots, L_n$, in addition to L, is a Banach space (with norm also denoted by $\|\cdot\|$) and $\pi_n: L \to L_n$ is a bounded linear transformation. We assume that $\sup_n \|\pi_n\| < \infty$. We write $f_n \to f$ if $f_n \in L_n$ for each $n \geq 1, f \in L$, and $\lim_{n\to\infty} \|f_n - \pi_n f\| = 0$.

6.1 Theorem For $n = 1, 2, \ldots$, let $\{T_n(t)\}$ and $\{T(t)\}$ be strongly continuous contraction semigroups on L_n and L with generators A_n and A. Let D be a core for A. Then the following are equivalent:

(a) For each $f \in L$, $T_n(t)\pi_n f \to T(t)f$ for all $t \geq 0$, uniformly on bounded intervals.

(b) For each $f \in L$, $T_n(t)\pi_n f \to T(t)f$ for all $t \geq 0$.

(c) For each $f \in D$, there exists $f_n \in \mathscr{D}(A_n)$ for each $n \geq 1$ such that $f_n \to f$ and $A_n f_n \to Af$ (i.e., $\{(f, Af) : f \in D\} \subset \text{ex-lim}_{n\to\infty} A_n$).

The proof of this result depends on the following two lemmas, the first of which generalizes Lemma 2.5.

6.2 Lemma Fix a positive integer n. Let $\{S_n(t)\}$ and $\{S(t)\}$ be strongly continuous contraction semigroups on L_n and L with generators B_n and B. Let $f \in \mathscr{D}(B)$ and assume that $\pi_n S(s)f \in \mathscr{D}(B_n)$ for all $s \geq 0$ and that $B_n \pi_n S(\cdot)f$: $[0, \infty) \to L_n$ is continuous. Then, for each $t \geq 0$,

$$S_n(t)\pi_n f - \pi_n S(t)f = \int_0^t S_n(t-s)(B_n \pi_n - \pi_n B)S(s)f \, ds, \tag{6.1}$$

and therefore

$$\| S_n(t)\pi_n f - \pi_n S(t)f \| \leq \int_0^t \| (B_n \pi_n - \pi_n B)S(s)f \| \, ds. \tag{6.2}$$

Proof. It suffices to note that the integrand in (6.1) is $-(d/ds)S_n(t-s)\pi_n S(s)f$ for $0 \leq s \leq t$. □

6.3 Lemma Suppose that the hypotheses of Theorem 6.1 are satisfied together with condition (c) of that theorem. For $n = 1, 2, \ldots$ and $\lambda > 0$, let A_n^λ and A^λ be the Yosida approximations of A_n and A (cf. Lemma 2.4). Then $A_n^\lambda \pi_n f \to A^\lambda f$ for every $f \in L$ and $\lambda > 0$.

Proof. Fix $\lambda > 0$. Let $f \in D$ and $g = (\lambda - A)f$. By assumption, there exists $f_n \in \mathscr{D}(A_n)$ for each $n \geq 1$ such that $f_n \to f$ and $A_n f_n \to Af$, and therefore $(\lambda - A_n)f_n \to g$. Now observe that

$$\begin{aligned} \| A_n^\lambda \pi_n g - \pi_n A^\lambda g \| &= \| [\lambda^2(\lambda - A_n)^{-1} - \lambda I]\pi_n g - \pi_n[\lambda^2(\lambda - A)^{-1} - \lambda I]g \| \\ &= \lambda^2 \| (\lambda - A_n)^{-1}\pi_n g - \pi_n(\lambda - A)^{-1} g \| \\ &\leq \lambda^2 \| (\lambda - A_n)^{-1}\pi_n g - f_n \| + \lambda^2 \| f_n - \pi_n(\lambda - A)^{-1} g \| \\ &\leq \lambda \| \pi_n g - (\lambda - A_n)f_n \| + \lambda^2 \| f_n - \pi_n f \| \end{aligned} \tag{6.3}$$

for every $n \geq 1$. Consequently, $\| A_n^\lambda \pi_n g - \pi_n A^\lambda g \| \to 0$ for all $g \in \mathscr{R}(\lambda - A|_D)$. But $\mathscr{R}(\lambda - A|_D)$ is dense in L and the linear transformations $A_n^\lambda \pi_n - \pi_n A^\lambda$, $n = 1, 2, \ldots$, are uniformly bounded, so the conclusion of the lemma follows. □

Proof of Theorem 6.1. **(a ⇒ b)** Immediate.

(b ⇒ c) Let $\lambda > 0$, $f \in \mathscr{D}(A)$, and $g = (\lambda - A)f$, so that $f = \int_0^\infty e^{-\lambda t}T(t)g\, dt$. For each $n \geq 1$, put $f_n = \int_0^\infty e^{-\lambda t}T_n(t)\pi_n g\, dt \in \mathscr{D}(A_n)$. By (b) and the dominated convergence theorem, $f_n \to f$, so since $(\lambda - A_n)f_n = \pi_n g \to g = (\lambda - A)f$, we also have $A_n f_n \to Af$.

(c ⇒ a) For $n = 1, 2, \ldots$ and $\lambda > 0$, let $\{T_n^\lambda(t)\}$ and $\{T^\lambda(t)\}$ be the strongly continuous contraction semigroups on L_n and L generated by the Yosida approximations A_n^λ and A^λ. Given $f \in D$, choose $\{f_n\}$ as in (c). Then

$$\begin{aligned} T_n(t)\pi_n f - \pi_n T(t)f = {} & T_n(t)(\pi_n f - f_n) + [T_n(t)f_n - T_n^\lambda(t)f_n] \\ & + T_n^\lambda(t)(f_n - \pi_n f) + [T_n^\lambda(t)\pi_n f - \pi_n T^\lambda(t)f] \\ & + \pi_n[T^\lambda(t)f - T(t)f] \end{aligned} \tag{6.4}$$

for every $n \geq 1$ and $t \geq 0$. Fix $t_0 \geq 0$. By Proposition 2.7 and Lemma 6.3,

$$\begin{aligned} \varlimsup_{n\to\infty} \sup_{0 \leq t \leq t_0} \| T_n(t)f_n - T_n^\lambda(t)f_n \| & \leq \varlimsup_{n\to\infty} t_0 \| A_n f_n - A_n^\lambda f_n \| \\ & \leq \varlimsup_{n\to\infty} t_0 \{ \| A_n f_n - \pi_n Af \| + \| \pi_n(Af - A^\lambda f) \| \\ & \quad + \| \pi_n A^\lambda f - A_n^\lambda \pi_n f \| + \| A_n^\lambda(\pi_n f - f_n) \| \} \\ & \leq Kt_0 \| Af - A^\lambda f \|, \end{aligned} \tag{6.5}$$

where $K = \sup_n \| \pi_n \|$. Using Lemmas 6.2, 6.3, and the dominated convergence theorem, we obtain

$$\begin{aligned} \varlimsup_{n\to\infty} \sup_{0 \leq t \leq t_0} & \| T_n^\lambda(t)\pi_n f - \pi_n T^\lambda(t)f \| \\ & \leq \varlimsup_{n\to\infty} \int_0^{t_0} \| (A_n^\lambda \pi_n - \pi_n A^\lambda)T^\lambda(s)f \| \, ds = 0. \end{aligned} \tag{6.6}$$

Applying (6.5), (6.6), and Proposition 2.7 to (6.4), we find that

$$\varlimsup_{n\to\infty} \sup_{0 \leq t \leq t_0} \| T_n(t)\pi_n f - \pi_n T(t)f \| \leq 2Kt_0 \| A^\lambda f - Af \|. \tag{6.7}$$

Since λ was arbitrary, Lemma 2.4(c) shows that the left side of (6.7) is zero. But this is valid for all $f \in D$, and since D is dense in L, it holds for all $f \in L$. □

There is a discrete-parameter analogue of Theorem 6.1, the proof of which depends on the following lemma.

6.4 Lemma Let B be a linear contraction on L. Then

$$\| B^n f - e^{n(B-I)}f \| \leq \sqrt{n} \| Bf - f \| \tag{6.8}$$

for all $f \in L$ and $n = 0, 1, \ldots$.

Proof. Fix $f \in L$ and $n \geq 0$. For $k = 0, 1, \ldots,$

$$\|B^n f - B^k f\| \leq \|B^{|k-n|}f - f\| = \left\| \sum_{j=0}^{|k-n|-1} B^j(Bf - f) \right\| \leq |k - n| \, \|Bf - f\|. \tag{6.9}$$

Therefore

$$\begin{aligned} \|B^n f - e^{n(B-I)}f\| &= \left\| B^n f - e^{-n} \sum_{k=0}^{\infty} \frac{n^k}{k!} B^k f \right\| \\ &= e^{-n} \left\| \sum_{k=0}^{\infty} (B^n f - B^k f) \frac{n^k}{k!} \right\| \\ &\leq e^{-n} \sum_{k=0}^{\infty} |k - n| \frac{n^k}{k!} \|Bf - f\| \\ &\leq \left\{ e^{-n} \sum_{k=0}^{\infty} (k - n)^2 \frac{n^k}{k!} \right\}^{1/2} \|Bf - f\| \\ &= \sqrt{n} \|Bf - f\|. \end{aligned} \tag{6.10}$$

(Note that the last equality follows from the fact that a Poisson random variable with parameter n has mean n and variance n.) □

6.5 Theorem For $n = 1, 2, \ldots$, let T_n be a linear contraction on L_n, let ε_n be a positive number, and put $A_n = \varepsilon_n^{-1}(T_n - I)$. Assume that $\lim_{n\to\infty} \varepsilon_n = 0$. Let $\{T(t)\}$ be a strongly continuous contraction semigroup on L with generator A, and let D be a core for A. Then the following are equivalent:

(a) For each $f \in L$, $T_n^{[t/\varepsilon_n]}\pi_n f \to T(t)f$ for all $t \geq 0$, uniformly on bounded intervals.

(b) For each $f \in L$, $T_n^{[t/\varepsilon_n]}\pi_n f \to T(t)f$ for all $t \geq 0$.

(c) For each $f \in D$, there exists $f_n \in L_n$ for each $n \geq 1$ such that $f_n \to f$ and $A_n f_n \to Af$ (i.e., $\{(f, Af): f \in D\} \subset \text{ex-lim}_{n\to\infty} A_n$).

Proof. **(a ⇒ b)** Immediate.

(b ⇒ c) Let $\lambda > 0$, $f \in \mathscr{D}(A)$, and $g = (\lambda - A)f$, so that $f = \int_0^\infty e^{-\lambda t}T(t)g \, dt$. For each $n \geq 1$, put

$$f_n = \varepsilon_n \sum_{k=0}^{\infty} e^{-\lambda k \varepsilon_n} T_n^k \pi_n g. \tag{6.11}$$

By (b) and the dominated convergence theorem, $f_n \to f$, and a simple calculation shows that

$$(6.12)\qquad (\lambda - A_n)f_n = \pi_n g + \lambda\varepsilon_n \pi_n g + (\lambda\varepsilon_n e^{-\lambda\varepsilon_n} - 1 + e^{-\lambda\varepsilon_n}) \sum_{k=0}^{\infty} e^{-\lambda k\varepsilon_n} T_n^{k+1} \pi_n g$$

for every $n \geq 1$, so $(\lambda - A_n)f_n \to g = (\lambda - A)f$. It follows that $A_n f_n \to Af$.

(c ⇒ a) Given $f \in D$, choose $\{f_n\}$ as in (c). Then

$$(6.13)\qquad T_n^{[t/\varepsilon_n]}\pi_n f - \pi_n T(t)f = T_n^{[t/\varepsilon_n]}(\pi_n f - f_n) + T_n^{[t/\varepsilon_n]}f_n - \exp\left\{\varepsilon_n\left[\frac{t}{\varepsilon_n}\right]A_n\right\}f_n + \exp\left\{\varepsilon_n\left[\frac{t}{\varepsilon_n}\right]A_n\right\}(f_n - \pi_n f) + \exp\left\{\varepsilon_n\left[\frac{t}{\varepsilon_n}\right]A_n\right\}\pi_n f - \pi_n T(t)f$$

for every $n \geq 1$ and $t \geq 0$. Fix $t_0 \geq 0$. By Lemma 6.4,

$$(6.14)\qquad \overline{\lim_{n\to\infty}} \sup_{0 \leq t \leq t_0} \left\| T_n^{[t/\varepsilon_n]}f_n - \exp\left\{\varepsilon_n\left[\frac{t}{\varepsilon_n}\right]A_n\right\}f_n \right\| \leq \overline{\lim_{n\to\infty}} \left(\left[\frac{t_0}{\varepsilon_n}\right]\right)^{1/2} \varepsilon_n \| A_n f_n \| = 0,$$

and by Theorem 6.1,

$$(6.15)\qquad \lim_{n\to\infty} \sup_{0 \leq t \leq t_0} \left\| \exp\left\{\varepsilon_n\left[\frac{t}{\varepsilon_n}\right]A_n\right\}\pi_n f - \pi_n T(t)f \right\| = 0.$$

Consequently,

$$(6.16)\qquad \lim_{n\to\infty} \sup_{0 \leq t \leq t_0} \| T_n^{[t/\varepsilon_n]}\pi_n f - \pi_n T(t)f \| = 0.$$

But this is valid for all $f \in D$, and since D is dense in L, it holds for all $f \in L$. □

6.6 Corollary Let $\{V(t): t \geq 0\}$ be a family of linear contractions on L with $V(0) = I$, and let $\{T(t)\}$ be a strongly continuous contraction semigroup on L with generator A. Let D be a core for A. If $\lim_{\varepsilon\to 0} \varepsilon^{-1}[V(\varepsilon)f - f] = Af$ for every $f \in D$, then, for each $f \in L$, $V(t/n)^n f \to T(t)f$ for all $t \geq 0$, uniformly on bounded intervals.

Proof. It suffices to show that if $\{t_n\}$ is a sequence of positive numbers such that $t_n \to t \geq 0$, then $V(t_n/n)^n f \to T(t)f$ for every $f \in L$. But this is an immediate consequence of Theorem 6.5 with $T_n = V(t_n/n)$ and $\varepsilon_n = t_n/n$ for each $n \geq 1$. □

6.7 Corollary Let $\{T(t)\}$, $\{S(t)\}$, and $\{U(t)\}$ be strongly continuous contraction semigroups on L with generators A, B, and C, respectively. Let D be a core for A, and assume that $D \subset \mathscr{D}(B) \cap \mathscr{D}(C)$ and that $A = B + C$ on D. Then, for each $f \in L$,

$$\lim_{n \to \infty} \left[S\left(\frac{t}{n}\right) U\left(\frac{t}{n}\right) \right]^n f = T(t)f \tag{6.17}$$

for all $t \geq 0$, uniformly on bounded intervals. Alternatively, if $\{\varepsilon_n\}$ is a sequence of positive numbers tending to zero, then, for each $f \in L$,

$$\lim_{n \to \infty} [S(\varepsilon_n)U(\varepsilon_n)]^{[t/\varepsilon_n]} f = T(t)f \tag{6.18}$$

for all $t \geq 0$, uniformly on bounded intervals.

Proof. The first result follows easily from Corollary 6.6 with $V(t) = S(t)U(t)$ for all $t \geq 0$. The second follows directly from Theorem 6.5. □

6.8 Corollary Let $\{T(t)\}$ be a strongly continuous contraction semigroup on L with generator A. Then, for each $f \in L$, $(I - (t/n)A)^{-n} f \to T(t)f$ for all $t \geq 0$, uniformly on bounded intervals. Alternatively, if $\{\varepsilon_n\}$ is a sequence of positive numbers tending to zero, then, for each $f \in L$, $(I - \varepsilon_n A)^{-[t/\varepsilon_n]} f \to T(t)f$ for all $t \geq 0$, uniformly on bounded intervals.

Proof. The first result is a consequence of Corollary 6.6. Simply take $V(t) = (I - tA)^{-1}$ for each $t \geq 0$, and note that if $\varepsilon > 0$ and $\lambda = \varepsilon^{-1}$, then

$$\varepsilon^{-1}\{V(\varepsilon)f - f\} = \lambda^2(\lambda - A)^{-1} f - \lambda f = A_\lambda f, \tag{6.19}$$

where A_λ is the Yosida approximation of A (cf. Lemma 2.4). The second result follows from (6.19) and Theorem 6.5. □

We would now like to generalize Theorem 6.1 in two ways. First, we would like to be able to use some extension $\hat{A}_n$ of the generator A_n in verifying the conditions for convergence. That is, given $(f, g) \in A$, it may be possible to find $(f_n, g_n) \in \hat{A}_n$ for each $n \geq 1$ such that $f_n \to f$ and $g_n \to g$ when it is not possible (or at least more difficult) to find $(f_n, g_n) \in A_n$ for each $n \geq 1$. Second, we would like to consider notions of convergence other than norm convergence. For example, convergence of bounded sequences of functions pointwise or uniformly on compact sets may be more appropriate than uniform convergence for some applications. An analogous generalization of Theorem 6.5 is also given.

Let LIM denote a notion of convergence of certain sequences $f_n \in L_n$, $n = 1, 2, \ldots$, to elements $f \in L$ satisfying the following conditions:

(6.20) $\quad \text{LIM } f_n = f$ and $\text{LIM } g_n = g$ imply

$$\text{LIM } (\alpha f_n + \beta g_n) = \alpha f + \beta g \quad \text{for all} \quad \alpha, \beta \in \mathbb{R}.$$

(6.21) $\quad \text{LIM } f_n^{(k)} = f^{(k)}$ for each $k \geq 1$ and

$$\lim_{k \to \infty} \sup_{n \geq 1} \|f_n^{(k)} - f_n\| \vee \|f^{(k)} - f\| = 0 \text{ imply } \text{LIM } f_n = f.$$

(6.22) There exists $K > 0$ such that for each $f \in L$, there is a sequence $f_n \in L_n$ with $\|f_n\| \leq K\|f\|$, $n = 1, 2, \ldots$, satisfying $\text{LIM } f_n = f$.

If $A_n \subset L_n \times L_n$ is linear for each $n \geq 1$, then, by analogy with (4.3), we define

(6.23)
$$\text{ex-LIM } A_n = \{(f, g) \in L \times L: \text{there exists } (f_n, g_n) \in A_n \text{ for each } n \geq 1 \text{ such that LIM } f_n = f \text{ and LIM } g_n = g\}.$$

6.9 Theorem For $n = 1, 2, \ldots$, let $A_n \subset L_n \times L_n$ and $A \subset L \times L$ be linear and dissipative with $\mathscr{R}(\lambda - A_n) = L_n$ and $\mathscr{R}(\lambda - A) = L$ for some (hence all) $\lambda > 0$, and let $\{T_n(t)\}$ and $\{T(t)\}$ be the corresponding strongly continuous contraction semigroups on $\overline{\mathscr{D}(A_n)}$ and $\overline{\mathscr{D}(A)}$. Let LIM satisfy (6.20)–(6.22) together with

(6.24) $\quad \text{LIM } f_n = 0$ implies $\text{LIM } (\lambda - A_n)^{-1} f_n = 0$ for all $\lambda > 0$.

(a) If $A \subset \text{ex-LIM } A_n$, then, for each $(f, g) \in A$, there exists $(f_n, g_n) \in A_n$ for each $n \geq 1$ such that $\sup_n \|f_n\| < \infty$, $\sup_n \|g_n\| < \infty$, $\text{LIM } f_n = f$, $\text{LIM } g_n = g$, and $\text{LIM } T_n(t) f_n = T(t) f$ for all $t \geq 0$.

(b) If in addition $\{T_n(t)\}$ extends to a contraction semigroup (also denoted by $\{T_n(t)\}$) on L_n for each $n \geq 1$, and if

(6.25) $\quad \text{LIM } f_n = 0$ implies $\text{LIM } T_n(t) f_n = 0$ for all $t \geq 0$,

then, for each $f \in \overline{\mathscr{D}(A)}$, $\text{LIM } f_n = f$ implies $\text{LIM } T_n(t) f_n = T(t) f$ for all $t \geq 0$.

6.10 Remark Under the hypotheses of the theorem, ex-LIM A_n is closed in $L \times L$ (Problem 16). Consequently, the conclusion of (a) is valid for all $(f, g) \in \bar{A}$. □

Proof. By renorming L_n, $n = 1, 2, \ldots$, if necessary, we can assume $K = 1$ in (6.22).

Let $\mathscr{L}$ denote the Banach space $(\prod_{n \geq 1} L_n) \times L$ with norm given by $\|(\{f_n\}, f)\| = \sup_{n \geq 1} \|f_n\| \vee \|f\|$, and let

(6.26)
$$\mathscr{L}_0 = \{(\{f_n\}, f) \in \mathscr{L}: \text{LIM } f_n = f\}.$$

Conditions (6.20) and (6.21) imply that $\mathscr{L}_0$ is a closed subspace of $\mathscr{L}$, and Condition (6.22) (with $K = 1$) implies that, for each $f \in L$, there is an element $(\{f_n\}, f) \in \mathscr{L}_0$ with $\|(\{f_n\}, f)\| = \|f\|$.

Let

$$(6.27) \quad \mathscr{A} = \{[(\{f_n\}, f), (\{g_n\}, g)] \in \mathscr{L} \times \mathscr{L} : (f_n, g_n) \in A_n \text{ for each } n \geq 1 \text{ and } (f, g) \in A\}.$$

Then $\mathscr{A}$ is linear and dissipative, and $\mathscr{R}(\lambda - \mathscr{A}) = \mathscr{L}$ for all $\lambda > 0$. The corresponding strongly continuous semigroup $\{\mathscr{T}(t)\}$ on $\overline{\mathscr{D}(\mathscr{A})}$ is given by

$$(6.28) \qquad \mathscr{T}(t)(\{f_n\}, f) = (\{T_n(t)f_n\}, T(t)f).$$

We would like to show that

$$(6.29) \qquad \mathscr{T}(t): \mathscr{L}_0 \cap \overline{\mathscr{D}(\mathscr{A})} \to \mathscr{L}_0 \cap \overline{\mathscr{D}(\mathscr{A})}, \qquad t \geq 0.$$

To do so, we need the following observation. If $(f, g) \in A$, $\lambda > 0$, $h = \lambda f - g$, $(\{h_n\}, h) \in \mathscr{L}_0$, and

$$(6.30) \qquad (f_n, g_n) = ((\lambda - A_n)^{-1} h_n, \lambda f_n - h_n)$$

for each $n \geq 1$, then

$$(6.31) \qquad [(\{f_n\}, f), (\{g_n\}, g)] \in (\mathscr{L}_0 \times \mathscr{L}_0) \cap \mathscr{A}.$$

To prove this, since $A \subset \text{ex-LIM } A_n$, choose $(\hat{f}_n, \hat{g}_n) \in A_n$ for each $n \geq 1$ such that $\text{LIM } \hat{f}_n = f$ and $\text{LIM } \hat{g}_n = g$. Then $\text{LIM } (h_n - (\lambda \hat{f}_n - \hat{g}_n)) = 0$, so by (6.24), $\text{LIM } (\lambda - A_n)^{-1} h_n - \hat{f}_n = 0$. It follows that $\text{LIM } f_n = \text{LIM } (\lambda - A_n)^{-1} h_n = \text{LIM } \hat{f}_n = f$ and $\text{LIM } g_n = \text{LIM } (\lambda f_n - h_n) = \lambda f - h = g$. Also, $\sup_n \|f_n\| \leq \lambda^{-1} \sup_n \|h_n\| < \infty$ and $\sup_n \|g_n\| \leq 2 \sup_n \|h_n\| < \infty$. Consequently, $[(\{f_n\}, f), (\{g_n\}, g)]$ belongs to $\mathscr{L}_0 \times \mathscr{L}_0$, and it clearly also belongs to $\mathscr{A}$.

Given $(\{h_n\}, h) \in \mathscr{L}_0$ and $\lambda > 0$, there exists $(f, g) \in A$ such that $\lambda f - g = h$. Define $(f_n, g_n) \in A_n$ for each $n \geq 1$ by (6.30). Then $(\lambda - \mathscr{A})^{-1}(\{h_n\}, h) = (\{f_n\}, f) \in \mathscr{L}_0$ by (6.31), so

$$(6.32) \qquad (\lambda - \mathscr{A})^{-1}: \mathscr{L}_0 \to \mathscr{L}_0, \qquad \lambda > 0.$$

By Corollary 2.8, this proves (6.29).

To prove (a), let $(f, g) \in A$, $\lambda > 0$, and $h = \lambda f - g$. By (6.22), there exists $(\{h_n\}, h) \in \mathscr{L}_0$ with $\|(\{h_n\}, h)\| = \|h\|$. Define $(f_n, g_n) \in A_n$ for each $n \geq 1$ by (6.30). By (6.31), (6.29), and (6.28), $(\{T_n(t)f_n\}, T(t)f) \in \mathscr{L}_0$ for all $t \geq 0$, so the conclusion of (a) is satisfied.

As for (b), observe that, by (a) together with (6.25), $\text{LIM } f_n = f \in \mathscr{D}(A)$ implies $\text{LIM } T_n(t)f_n = T(t)f$ for all $t \geq 0$. Let $f \in \overline{\mathscr{D}(A)}$ and choose $\{f^{(k)}\} \subset \mathscr{D}(A)$ such that $\|f^{(k)} - f\| \leq 2^{-k}$ for each $k \geq 1$. Put $f^{(0)} = 0$, and by (6.22), choose

$(\{u_n^{(k)}\}, f^{(k)} - f^{(k-1)}) \in \mathscr{L}_0$ such that $\|(\{u_n^{(k)}\}, f^{(k)} - f^{(k-1)})\| = \|f^{(k)} - f^{(k-1)}\|$ for each $k \geq 1$. Fix $t \geq 0$. Then

$$\text{(6.33)} \qquad \text{LIM} \sum_1^k u_n^{(l)} = f^{(k)}, \qquad \text{LIM } T_n(t) \sum_1^k u_n^{(l)} = T(t)f^{(k)}$$

for each $k \geq 1$. Since

$$\text{(6.34)} \qquad \left\| \sum_{k+1}^{\infty} u_n^{(l)} \right\| \leq 3 \cdot 2^{-k}, \qquad \|f^{(k)} - f\| \leq 2^{-k},$$

and

$$\text{(6.35)} \qquad \left\| T_n(t) \sum_{k+1}^{\infty} u_n^{(l)} \right\| \leq 3 \cdot 2^{-k}, \qquad \|T(t)f^{(k)} - T(t)f\| \leq 2^{-k},$$

for each $n \geq 1$ and $k \geq 1$, (6.21) implies that

$$\text{(6.36)} \qquad \text{LIM} \sum_1^{\infty} u_n^{(l)} = f, \qquad \text{LIM } T_n(t) \sum_1^{\infty} u_n^{(l)} = T(t)f,$$

so the conclusion of (b) follows from (6.25). □

6.11 Theorem For $n = 1, 2, \ldots$, let T_n be a linear contraction on L_n, let $\varepsilon_n > 0$, and put $A_n = \varepsilon_n^{-1}(T_n - I)$. Assume that $\lim_{n\to\infty} \varepsilon_n = 0$. Let $A \subset L \times L$ be linear and dissipative with $\mathscr{R}(\lambda - A) = L$ for some (hence all) $\lambda > 0$, and let $\{T(t)\}$ be the corresponding strongly continuous contraction semigroup on $\overline{\mathscr{D}(A)}$. Let LIM satisfy (6.20)–(6.22), (6.24), and

$$\text{(6.37)} \qquad \lim \|f_n\| = 0 \quad \text{implies} \quad \text{LIM } f_n = 0.$$

(a) If $A \subset$ ex-LIM A_n, then, for each $(f, g) \in A$, there exists $f_n \in L_n$ for each $n \geq 1$ such that $\sup_n \|f_n\| < \infty$, $\sup_n \|A_n f_n\| < \infty$, LIM $f_n = f$, LIM $A_n f_n = g$, and LIM $T_n^{[t/\varepsilon_n]} f_n = T(t)f$ for all $t \geq 0$.

(b) If in addition

$$\text{(6.38)} \qquad \text{LIM } f_n = 0 \quad \text{implies} \quad \text{LIM } T_n^{[t/\varepsilon_n]} f_n = 0 \quad \text{for all} \quad t \geq 0,$$

then for each $f \in \overline{\mathscr{D}(A)}$, LIM $f_n = f$ implies LIM $T_n^{[t/\varepsilon_n]} f_n = T(t)f$ for all $t \geq 0$.

Proof. Let $(f, g) \in A$. By Theorem 6.9, there exists $f_n \in L_n$ for each $n \geq 1$ such that $\sup_n \|f_n\| < \infty$, $\sup_n \|A_n f_n\| < \infty$, LIM $f_n = f$, LIM $A_n f_n = g$, and LIM $e^{tA_n} f_n = T(t)f$ for all $t \geq 0$. Since

$$\text{(6.39)} \qquad \lim_{n\to\infty} \left\| \exp\left\{\varepsilon_n \left[\frac{t}{\varepsilon_n}\right] A_n\right\} f_n - \exp\{tA_n\} f_n \right\|$$
$$\leq \lim_{n\to\infty} \left| \varepsilon_n \left[\frac{t}{\varepsilon_n}\right] - t \right| \|A_n f_n\| = 0$$

for all $t \geq 0$, we deduce from (6.37) that

$$\text{LIM} \exp\left\{\varepsilon_n\left[\frac{t}{\varepsilon_n}\right]A_n\right\}f_n = T(t)f, \qquad t \geq 0. \tag{6.40}$$

The conclusion of (a) therefore follows from (6.14) and (6.37).

The proof of (b) is completely analogous to that of Theorem 6.9 (b). □

7. PERTURBATION THEOREMS

One of the main results of this section concerns the approximation of semigroups with generators of the form $A + B$, where A and B themselves generate semigroups. (By definition, $\mathscr{D}(A + B) = \mathscr{D}(A) \cap \mathscr{D}(B)$.) First, however, we give some sufficient conditions for $A + B$ to generate a semigroup.

7.1 Theorem Let A be a linear operator on L such that $\bar{A}$ is single-valued and generates a strongly continuous contraction semigroup on L. Let B be a dissipative linear operator on L such that $\mathscr{D}(B) \supset \mathscr{D}(A)$. (In particular, $\bar{B}$ is single-valued by Lemma 4.2.) If

$$\| Bf \| \leq \alpha \| Af \| + \beta \| f \|, \qquad f \in \mathscr{D}(A), \tag{7.1}$$

where $0 \leq \alpha < 1$ and $\beta \geq 0$, then $\overline{A + B}$ is single-valued and generates a strongly continuous contraction semigroup on L. Moreover, $\overline{A + B} = \bar{A} + \bar{B}$.

Proof. Let $\gamma \geq 0$ be arbitrary. Clearly, $\mathscr{D}(A + \gamma B) = \mathscr{D}(A)$ is dense in L. In addition, $A + \gamma B$ is dissipative. To see this, let A_μ be the Yosida approximation of $\bar{A}$ for each $\mu > 0$, so that $A_\mu = \mu[\mu(\mu - \bar{A})^{-1} - I]$. If $f \in \mathscr{D}(A)$ and $\lambda > 0$, then

$$\begin{aligned} \| \lambda f - (A + \gamma B)f \| &= \lim_{\mu \to \infty} \| \lambda f - (A_\mu + \gamma B)f \| \\ &= \lim_{\mu \to \infty} \| (\lambda + \mu)f - \gamma Bf - \mu^2(\mu - \bar{A})^{-1}f \| \\ &\geq \overline{\lim_{\mu \to \infty}} \{ \| (\lambda + \mu)f - \gamma Bf \| - \| \mu^2(\mu - \bar{A})^{-1}f \| \} \\ &\geq \lim_{\mu \to \infty} \{ (\lambda + \mu) \| f \| - \mu \| f \| \} \\ &= \lambda \| f \| \end{aligned} \tag{7.2}$$

by Lemma 2.4(c) and the dissipativity of γB.

If $f \in \mathscr{D}(\bar{A})$, then there exists $\{f_n\} \subset \mathscr{D}(A)$ such that $f_n \to f$ and $Af_n \to \bar{A}f$. By (7.1), $\{Bf_n\}$ is Cauchy, so $f \in \mathscr{D}(\bar{B})$ and $Bf_n \to \bar{B}f$. Hence $\mathscr{D}(\bar{A}) \subset \mathscr{D}(\bar{B})$ and (7.1) extends to

$$\|\bar{B}f\| \leq \alpha\|\bar{A}f\| + \beta\|f\|, \qquad f \in \mathscr{D}(\bar{A}). \tag{7.3}$$

In addition, if $f \in \mathscr{D}(\bar{A})$ and if $\{f_n\}$ is as above, then

$$(\bar{A} + \gamma\bar{B})f = \lim_n Af_n + \gamma \lim_n Bf_n = \lim_n (A + \gamma B)f_n = \overline{(A + \gamma B)}f, \tag{7.4}$$

implying that $\overline{A + \gamma B}$ is a dissipative extension of $\bar{A} + \gamma\bar{B}$.

Let

$$\Gamma = \{\gamma \geq 0: \mathscr{R}(\lambda - \bar{A} - \gamma\bar{B}) = L \text{ for some (hence all) } \lambda > 0\}. \tag{7.5}$$

To complete the proof, it suffices by Theorem 2.6 and Proposition 4.1 to show that $1 \in \Gamma$. Noting that $0 \in \Gamma$ by assumption, it is enough to show that

$$\gamma \in \Gamma \cap [0, 1) \quad \text{implies} \quad \left[\gamma, \gamma + \frac{1 - \alpha\gamma}{2\alpha}\right) \subset \Gamma. \tag{7.6}$$

To prove (7.6), let $\gamma \in \Gamma \cap [0, 1)$, $0 \leq \varepsilon < (2\alpha)^{-1}(1 - \alpha\gamma)$, and $\lambda > 0$. If $g \in \mathscr{D}(\bar{A})$, then $f \equiv (\lambda - \bar{A} - \gamma\bar{B})^{-1}g$ satisfies

$$\|\bar{B}f\| \leq \alpha\|\bar{A}f\| + \beta\|f\| \leq \alpha\|(\bar{A} + \gamma\bar{B})f\| + \alpha\gamma\|\bar{B}f\| + \beta\|f\| \tag{7.7}$$

by (7.3), that is,

$$\|\bar{B}f\| \leq \alpha(1 - \alpha\gamma)^{-1}\|(\bar{A} + \gamma\bar{B})f\| + \beta(1 - \alpha\gamma)^{-1}\|f\|, \tag{7.8}$$

and consequently,

$$\|\bar{B}(\lambda - \bar{A} - \gamma\bar{B})^{-1}g\| \leq [2\alpha(1 - \alpha\gamma)^{-1} + \beta(1 - \alpha\gamma)^{-1}\lambda^{-1}]\|g\|. \tag{7.9}$$

Thus, for λ sufficiently large, $\|\varepsilon\bar{B}(\lambda - \bar{A} - \bar{B})^{-1}\| < 1$, which implies that $I - \varepsilon\bar{B}(\lambda - \bar{A} - \gamma\bar{B})^{-1}$ is invertible. We conclude that

$$\begin{aligned} \mathscr{R}(\lambda - \bar{A} - (\gamma + \varepsilon)\bar{B}) &\supset \mathscr{R}((\lambda - \bar{A} - (\gamma + \varepsilon)\bar{B})(\lambda - \bar{A} - \gamma\bar{B})^{-1}) \\ &= \mathscr{R}(I - \varepsilon\bar{B}(\lambda - \bar{A} - \gamma\bar{B})^{-1}) \\ &= L \end{aligned} \tag{7.10}$$

for such λ, so $\gamma + \varepsilon \in \Gamma$, implying (7.6) and completing the proof. □

7.2 Corollary If A generates a strongly continuous contraction semigroup on L and B is a bounded linear operator on L, then $A + B$ generates a strongly continuous semigroup $\{T(t)\}$ on L such that

$$\|T(t)\| \leq e^{\|B\|t}, \qquad t \geq 0. \tag{7.11}$$

Proof. Apply Theorem 7.1 with $B - \|B\|I$ in place of B. □

Before turning to limit theorems, we state the following lemma, the proof of which is left to the reader (Problem 18). For an operator A, let $\mathcal{N}(A) \equiv \{f \in \mathcal{D}(A): Af = 0\}$ denote the null space of A.

7.3 Lemma Let B generate a strongly continuous contraction semigroup $\{S(t)\}$ on L, and assume that

$$\lim_{\lambda \to 0+} \lambda \int_0^\infty e^{-\lambda t} S(t) f \, dt \equiv Pf \quad \text{exists for all} \quad f \in L. \tag{7.12}$$

Then the following conclusions hold:

(a) P is a linear contraction on L and $P^2 = P$.
(b) $S(t)P = PS(t) = P$ for all $t \geq 0$.
(c) $\mathcal{R}(P) = \mathcal{N}(B)$.
(d) $\mathcal{N}(P) = \overline{\mathcal{R}(B)}$.

7.4 Remark If in the lemma

$$B = \gamma^{-1}(Q - I), \tag{7.13}$$

where Q is a linear contraction on L and $\gamma > 0$, then a simple calculation shows that (7.12) is equivalent to

$$\lim_{\rho \to 1-} (1 - \rho) \sum_{k=0}^{\infty} \rho^k Q^k f \equiv Pf \quad \text{exists for all} \quad f \in L. \tag{7.14}$$

□

7.5 Remark If in the lemma $\lim_{t \to \infty} S(t)f$ exists for every $f \in L$, then (7.12) holds and

$$Pf = \lim_{t \to \infty} S(t)f, \qquad f \in L. \tag{7.15}$$

If B is as in Remark 7.4 and if $\lim_{k \to \infty} Q^k f$ exists for every $f \in L$, then (7.14) holds (in fact, so does (7.15)) and

$$Pf = \lim_{k \to \infty} Q^k f, \qquad f \in L. \tag{7.16}$$

The proofs of these assertions are elementary. □

For the following result, recall the notation introduced in the first paragraph of Section 6, as well as the notion of the extended limit of a sequence of operators (Section 4).

7.6 Theorem Let $A \subset L \times L$ be linear, and let B generate a strongly continuous contraction semigroup $\{S(t)\}$ on L satisfying (7.12). Let D be a subspace of $\mathscr{D}(A)$ and D' a core for B. For $n = 1, 2, \ldots$, let A_n be a linear operator on L_n and let $\alpha_n > 0$. Suppose that $\lim_{n\to\infty} \alpha_n = \infty$ and that

$$\{(f, g) \in A : f \in D\} \subset \underset{n\to\infty}{\text{ex-lim}}\ A_n, \tag{7.17}$$

$$\{(h, Bh) : h \in D'\} \subset \underset{n\to\infty}{\text{ex-lim}}\ \alpha_n^{-1} A_n. \tag{7.18}$$

Define $C = \{(f, Pg) : (f, g) \in A, f \in D\}$ and assume that $\{(f, g) \in \bar{C} : g \in \bar{D}\}$ is single-valued and generates a strongly continuous contraction semigroup $\{T(t)\}$ on $\bar{D}$.

(a) If A_n is the generator of a strongly continuous contraction semigroup $\{T_n(t)\}$ on L_n for each $n \geq 1$, then, for each $f \in \bar{D}$, $T_n(t)\pi_n f \to T(t)f$ for all $t \geq 0$, uniformly on bounded intervals.

(b) If $A_n = \varepsilon_n^{-1}(T_n - I)$ for each $n \geq 1$, where T_n is a linear contraction on L_n and $\varepsilon_n > 0$, and if $\lim_{n\to\infty} \varepsilon_n = 0$, then, for each $f \in \bar{D}$, $T_n^{[t/\varepsilon_n]}\pi_n f \to T(t)f$ for all $t \geq 0$, uniformly on bounded intervals.

Proof. Theorems 6.1 and 6.5 are applicable, provided we can show that

$$\{(f, g) \in \bar{C} : g \in \bar{D}\} \subset \left(\underset{n\to\infty}{\text{ex-lim}}\ A_n\right) \cap (\bar{D} \times \bar{D}). \tag{7.19}$$

Since $\text{ex-lim}_{n\to\infty} A_n$ is closed, it suffices to show that $C \subset \text{ex-lim}_{n\to\infty} A_n$. Given $(f, g) \in A$ with $f \in D$, choose $f_n \in \mathscr{D}(A_n)$ for each $n \geq 1$ such that $f_n \to f$ and $A_n f_n \to g$. Given $h \in D'$, choose $h_n \in \mathscr{D}(A_n)$ for each $n \geq 1$ such that $h_n \to h$ and $\alpha_n^{-1} A_n h_n \to Bh$. Then $f_n + \alpha_n^{-1} h_n \to f$ and $A_n(f_n + \alpha_n^{-1} h_n) \to g + Bh$. Consequently,

$$\{(f, g + Bh) : (f, g) \in A,\quad f \in D,\quad h \in D'\} \subset \underset{n\to\infty}{\text{ex-lim}}\ A_n. \tag{7.20}$$

But since $\text{ex-lim}_{n\to\infty} A_n$ is closed and since, by Lemma 7.3(d),

$$Pg - g \in \mathscr{N}(P) = \overline{\mathscr{R}(B)} = \overline{\mathscr{R}(B|_{D'})} \tag{7.21}$$

for all $g \in L$, we conclude that

$$\{(f, Pg) : (f, g) \in A,\quad f \in D) \subset \underset{n\to\infty}{\text{ex-lim}}\ A_n, \tag{7.22}$$

completing the proof. □

We conclude this section with two corollaries. The first one extends the conclusions of Theorem 7.6, and the other describes an important special case of the theorem.

7.7 Corollary Assume the hypotheses of Theorem 7.6(a) and suppose that (7.15) holds. If $h \in \mathscr{N}(P)$ and if $\{t_n\} \subset [0, \infty)$ satisfies $\lim_{n\to\infty} t_n \alpha_n = \infty$, then $T_n(t_n)\pi_n h \to 0$. Consequently, for each $f \in P^{-1}(\bar{D})$ and $\delta \in (0, 1)$, $T_n(t)\pi_n f \to T(t)Pf$, uniformly in $\delta \le t \le \delta^{-1}$.

Assume the hypotheses of Theorem 7.6(b), and suppose that either (i) $\lim_{n\to\infty} \alpha_n \varepsilon_n = 0$ and (7.15) holds, or (ii) $\lim_{n\to\infty} \alpha_n \varepsilon_n = \gamma > 0$ and (7.16) holds (where Q is as in (7.13)). If $h \in \mathcal{N}(P)$ and if $\{k_n\} \subset \{0, 1, \dots\}$ satisfies $\lim_{n\to\infty} k_n \alpha_n \varepsilon_n = \infty$, then $T_n^{k_n}\pi_n h \to 0$. Consequently, for each $f \in P^{-1}(\bar{D})$ and $\delta \in (0, 1)$, $T_n^{[t/\varepsilon_n]}\pi_n f \to T(t)Pf$, uniformly in $\delta \le t \le \delta^{-1}$.

Proof. We give the proof assuming the hypotheses of Theorem 7.6(a), the other case being similar. Let $h \in \mathcal{N}(P)$, let $\{t_n\}$ be as above, and let $\varepsilon > 0$. Choose $s \ge 0$ such that $\| S(s)h \| \le \varepsilon/2K$, where $K = \sup_{n \ge 1} \| \pi_n \|$, and let $s_n = s \wedge t_n \alpha_n$ for each $n \ge 1$. Then

$$(7.23) \qquad \| T_n(t_n)\pi_n h \| \le \left\| T_n\left(\frac{s_n}{\alpha_n}\right)\pi_n h - \pi_n S(s)h \right\| + \| \pi_n S(s)h \| \le \varepsilon$$

for all n sufficiently large by (7.18) and Theorem 6.1. If $f \in L$, then $f - Pf \in \mathcal{N}(P)$, so $T_n(t_n)\pi_n(f - Pf) \to 0$ whenever $\{t_n\} \subset [0, \infty)$ satisfies $\lim_{n\to\infty} t_n = t \ne 0$. If $f \in P^{-1}(\bar{D})$, this, together with the conclusion of the theorem applied to Pf, completes the proof. □

7.8 Corollary Let Π, A, and B be linear operators on L such that B generates a strongly continuous contraction semigroup $\{S(t)\}$ on L satisfying (7.12). Assume that $\mathcal{D}(\Pi) \cap \mathcal{D}(A) \cap \mathcal{D}(B)$ is a core for B. For each α sufficiently large, suppose that an extension of $\Pi + \alpha A + \alpha^2 B$ generates a strongly continuous contraction semigroup $\{T_\alpha(t)\}$ on L. Let D be a subspace of

$$(7.24) \quad \{f \in \mathcal{D}(\Pi) \cap \mathcal{D}(A) \cap \mathcal{N}(B): \text{ there exists } h \in \mathcal{D}(\Pi) \cap \mathcal{D}(A) \cap \mathcal{D}(B) \text{ with } Bh = -Af\},$$

and define

$$(7.25) \quad C = \{(f, P\Pi f + PAh): f \in D, \quad h \in \mathcal{D}(\Pi) \cap \mathcal{D}(A) \cap \mathcal{D}(B), \quad Bh = -Af\}.$$

Then C is dissipative, and if $\{(f, g) \in \bar{C}: g \in \bar{D}\}$, which is therefore single-valued, generates a strongly continuous contraction semigroup $\{T(t)\}$ on $\bar{D}$, then, for each $f \in \bar{D}$, $\lim_{\alpha\to\infty} T_\alpha(t)f = T(t)f$ for all $t \ge 0$, uniformly on bounded intervals.

Proof. Let $\{\alpha_n\}$ be a sequence of (sufficiently large) positive numbers such that $\lim_{n\to\infty} \alpha_n = \infty$, and apply Theorem 7.6(a) with $L_n = L$, $\pi_n = I$, A replaced by

$$(7.26) \quad \{(f, \Pi f + Ah): f \in D, \quad h \in \mathcal{D}(\Pi) \cap \mathcal{D}(A) \cap \mathcal{D}(B), \quad Bh = -Af\},$$

A_n equal to the generator of $\{T_{\alpha_n}(t)\}$, α_n replaced by α_n^2, and $D' = \mathcal{D}(\Pi) \cap \mathcal{D}(A) \cap \mathcal{D}(B)$. Since $A_n(f + \alpha_n^{-1}h) = \Pi f + Ah + \alpha_n^{-1}\Pi h$ whenever $f \in D$, $h \in \mathcal{D}(\Pi) \cap \mathcal{D}(A) \cap \mathcal{D}(B)$, $Bh = -Af$, and $n \ge 1$, and since $\lim_{n\to\infty}$

$\alpha_n^{-2} A_n h = Bh$ for all $h \in D'$, we find that (7.17) and (7.18) hold, so the theorem is applicable. The dissipativity of C follows from the dissipativity of ex-$\lim_{n\to\infty} A_n$. □

7.9 Remark **(a)** Observe that in Corollary 7.8 it is necessary that $PAf = 0$ for all $f \in D$ by Lemma 7.3(d).

(b) Let $f \in \mathscr{D}(A)$ satisfy $PAf = 0$. To actually solve the equation $Bh = -Af$ for h, suppose that

$$\int_0^\infty \| (S(t) - P)g \| \, dt < \infty, \qquad g \in L. \tag{7.27}$$

Then $h \equiv \lim_{\lambda \to 0+} (\lambda - B)^{-1} Af = \int_0^\infty (S(t) - P)Af \, dt$ belongs to $\mathscr{D}(B)$ (since B is closed) and satisfies $Bh = -Af$. Of course, the requirement that h belong to $\mathscr{D}(\Pi) \cap \mathscr{D}(A)$ must also be satisfied.

(c) When applying Corollary 7.8, it is not necessary to determine C explicitly. Instead, suppose a linear operator C_0 on $\bar{D}$ can be found such that C_0 generates a strongly continuous contraction semigroup on $\bar{D}$ and $C_0 \subset \bar{C}$. Then $\{(f, g) \in \bar{C} : g \in \bar{D}\} = C_0$ by Proposition 4.1.

(d) See Problem 20 for a generalization and Problem 22 for a closely related result. □

8. PROBLEMS

1. Define $\{T(t)\}$ on $\hat{C}(\mathbb{R})$ by $T(t)f(x) = f(x + t)$. Show that $\{T(t)\}$ is a strongly continuous contraction semigroup on L, and determine its generator A. (In particular, this requires that $\mathscr{D}(A)$ be characterized.)

2. Define $\{T(t)\}$ on $\hat{C}(\mathbb{R})$ by

$$T(t)f(x) = \frac{1}{\sqrt{2\pi t}} \int_{-\infty}^{\infty} f(y) \exp\left\{ -\frac{(y - x)^2}{2t} \right\} dy \tag{8.1}$$

for each $t > 0$ and $T(0) = I$. Show that $\{T(t)\}$ is a strongly continuous contraction semigroup on L, and determine its generator A.

3. Prove Lemma 1.4.

4. Let $\{T(t)\}$ be a strongly continuous contraction semigroup on L with generator A, and let $f \in \mathscr{D}(A^2)$.

(a) Prove that

$$T(t)f = f + tAf + \int_0^t (t - s)T(s)A^2 f \, ds, \qquad t \geq 0. \tag{8.2}$$

(b) Show that $\| Af \|^2 \leq 4 \| A^2 f \| \, \| f \|$.

5. Let A generate a strongly continuous semigroup on L. Show that $\bigcap_{n=1}^{\infty} \mathscr{D}(A^n)$ is dense in L.

6. Show directly that the linear operator $A = \frac{1}{2} d^2/dx^2$ on L satisfies conditions (a)–(c) of Theorem 2.6 when $\mathscr{D}(A)$ and L are as follows:

(a) $\mathscr{D}(A) = \{f \in C^2[0, 1]: \alpha_i f''(i) - (-1)^i \beta_i f'(i) = 0, \quad i = 0, 1\}$.
$L = C[0, 1]$, $\alpha_0, \beta_0, \alpha_1, \beta_1 \geq 0$, $\alpha_0 + \beta_0 > 0$, $\alpha_1 + \beta_1 > 0$.

(b) $\mathscr{D}(A) = \{f \in \hat{C}^2[0, \infty): \alpha_0 f''(0) - \beta_0 f'(0) = 0\}$
$L = \hat{C}[0, \infty)$, $\alpha_0, \beta_0 \geq 0$, $\alpha_0 + \beta_0 > 0$.

(c) $\mathscr{D}(A) = \hat{C}^2(\mathbb{R})$, $L = \hat{C}(\mathbb{R})$.

Hint: Look for solutions of $\lambda f - \frac{1}{2} f'' = g$ of the form $f(x) = \exp\{-\sqrt{2\lambda}\, x\} h(x)$.

7. Show that $C_c^{\infty}(\mathbb{R})$ is a core for the generators of the semigroups of Problems 1 and 2.

8. In this problem, every statement involving k, l, or n is assumed to hold for all $k, l, n \geq 1$.

Let $L_1 \subset L_2 \subset L_3 \subset \cdots$ be a sequence of closed subspaces of L. Let U_k, M_k, and $M_k^{(n)}$ be bounded linear operators on L. Assume that U_k and $M_k^{(n)}$ map L_n into L_n, and that for some $\mu_k > 0$, $\| M_k^{(n)} \| \leq \mu_k$ and

$$\lim_{m \to \infty} \| M_k^{(m)} - M_k \| = 0. \tag{8.3}$$

Suppose that the restriction of $A_n \equiv \sum_{j=1}^{n} M_j^{(n)} U_j$ to L_n is dissipative and that there exist nonnegative constants $\alpha_{kl} (= \alpha_{lk})$, β_{kl}, and γ such that

$$\| U_k U_l f - U_l U_k f \| \leq \alpha_{kl} (\| U_k f \| + \| U_l f \|), \qquad f \in L, \tag{8.4}$$

$$\sum_{j=1}^{\infty} \mu_j \alpha_{jl} \leq \gamma, \tag{8.5}$$

$$\| U_k M_l^{(n)} - M_l^{(n)} U_k \| \leq \beta_{kl}, \tag{8.6}$$

and

$$\sum_{j=1}^{\infty} \mu_j \beta_{jl} \leq \gamma \mu_l. \tag{8.7}$$

Define $A = \sum_{j=1}^{\infty} M_j U_j$ on

$$\mathscr{D}(A) = \left\{ f \in \bigcup_{m=1}^{\infty} L_m : \sum_{j=1}^{\infty} \mu_j \| U_j f \| < \infty \right\}. \tag{8.8}$$

If $\mathscr{D}(A)$ is dense in L, show that $\bar{A}$ is single-valued and generates a strongly continuous contraction semigroup on L.

Hint: Fix $\lambda > 3\gamma$ and apply Lemma 3.6. Show first that for $g \in \mathscr{D}(A)$ and $f_n = (\lambda - A_n)^{-1}g$,

$$(8.9) \qquad (\lambda - \gamma)\| U_k f_n \| \leq \| U_k g \| + \sum_{j=1}^{n} (\beta_{kj} + \mu_l \alpha_{kj})\| U_j f_n \|.$$

Denoting by μ the positive measure on the set of positive integers that gives mass μ_k to k, observe that the formula

$$(8.10) \qquad (BF)_k = \sum_{j=1}^{\infty} (\beta_{kj} + \mu_j \alpha_{kj})F_j$$

defines a positive bounded linear operator on $L^1(\mu)$ of norm at most 2γ.

9. As an application of Corollary 3.8, prove the following result, which yields the conclusion of Theorem 7.1 under a different set of hypotheses.

Let A and B generate strongly continuous contraction semigroups $\{T(t)\}$ and $\{S(t)\}$ on L. Let D be a dense subspace of L and $|||\cdot|||$ a norm on D with respect to which D is a Banach space. Assume that $|||f||| \geq \|f\|$ for all $f \in D$. Suppose there exists $\mu \geq 0$ such that

$$(8.11) \qquad D \subset \mathscr{D}(A^2); \quad \|A^2 f\| \leq \mu |||f|||, \qquad f \in D;$$

$$(8.12) \qquad D \subset \mathscr{D}(B^2); \quad \|B^2 f\| \leq \mu |||f|||, \qquad f \in D;$$

$$(8.13) \qquad T(t)\colon D \to D, \qquad S(t)\colon D \to D, \qquad t \geq 0;$$

$$(8.14) \qquad |||T(t)||| \leq e^{\mu t}, \qquad |||S(t)||| \leq e^{\mu t}, \qquad t \geq 0.$$

Then the closure of the restriction of $A + B$ to D is single-valued and generates a strongly continuous contraction semigroup on L.

We remark that only one of the two conditions (8.11) and (8.12) is really needed. See Ethier (1976).

10. Define the bounded linear operator B on $L \equiv C([0, 1] \times [0, 1])$ by $Bf(x, y) = \int_0^1 f(x, z)\,dz$, and define $A \subset L \times L$ by

$$(8.15) \qquad A = \{(f, \tfrac{1}{2}f_{xx} + h) \colon f \in C^2([0, 1] \times [0, 1]) \cap \mathscr{R}(B),$$
$$f_x(0, y) = f_x(1, y) = 0 \quad \text{for all} \quad y \in [0, 1],$$
$$h \in \mathscr{N}(B)\}.$$

Show that A satisfies the conditions of Theorem 4.3.

11. Show that ex-$\lim_{n\to\infty} A_n$, defined by (4.3), is closed in $L \times L$.

12. Does the dissipativity of A_n for each $n \geq 1$ imply the dissipativity of ex-$\lim_{n\to\infty} A_n$?

13. In Theorem 6.1 (and Theorem 6.5), show that (a)–(c) are equivalent to the following:

(d) There exists $\lambda > 0$ such that $(\lambda - A_n)^{-1}\pi_n g \to (\lambda - A)^{-1}g$ for all $g \in L$.

14. Let L, $\{L_n\}$, and $\{\pi_n\}$ be as in Section 6. For each $n \geq 1$, let $\{T_n(t)\}$ be a contraction semigroup on L_n, or, for each $n \geq 1$, let $\{T_n(t)\}$ be defined in terms of a linear contraction T_n on L_n and a number $\varepsilon_n > 0$ by $T_n(t) = T_n^{[t/\varepsilon_n]}$ for all $t \geq 0$; in the latter case assume that $\lim_{n\to\infty} \varepsilon_n = 0$. Let $\{T(t)\}$ be a contraction semigroup on L, let $f, g \in L$, and suppose that $\lim_{t\to\infty} T(t)f = g$ and

$$\lim_{n\to\infty} \sup_{0 \leq t \leq t_0} \| T_n(t)\pi_n f - \pi_n T(t)f \| = 0 \tag{8.16}$$

for every $t_0 > 0$. Show that

$$\lim_{n\to\infty} \sup_{t \geq 0} \| T_n(t)\pi_n f - \pi_n T(t)f \| = 0 \tag{8.17}$$

if and only if

$$\lim_{n\to\infty} \sup_{t \geq 0} \| T_n(t)\pi_n g - \pi_n T(t)g \| = 0. \tag{8.18}$$

15. Using the results of Problem 2 and Theorem 6.5, prove the central limit theorem. That is, if $X_1, X_2, \ldots$ are independent, identically distributed, real-valued random variables with mean 0 and variance 1, show that $n^{-1/2}\sum_{i=1}^n X_i$ converges in distribution to a standard normal random variable as $n \to \infty$. (Define $T_n f(x) = E[f(x + n^{-1/2}X_1)]$ and $\varepsilon_n = n^{-1}$.)

16. Under the hypotheses of Theorem 6.9, show that ex-LIM A_n is closed in $L \times L$.

17. Show that (6.21) implies (6.37) under the following (very reasonable) additional assumption.

(8.19) If $f_n \in L_n$ for each $n \geq 1$ and if, for some $n_0 \geq 1, f_n = 0$ for all $n \geq n_0$, then LIM $f_n = 0$.

18. Prove Lemma 7.3 and the remarks following it.

19. Under the assumptions of Corollary 6.7, prove (6.18) using Theorem 7.6. *Hint*: For each $n \geq 1$, define the contraction operator T_n on $L \times L$ by

$$T_n\begin{pmatrix} f \\ g \end{pmatrix} = \begin{pmatrix} S(\varepsilon_n)g \\ U(\varepsilon_n)f \end{pmatrix}. \tag{8.20}$$

20. Corollary 7.8 has been called a second-order limit theorem. Prove the following kth-order limit theorem as an application of Theorem 7.6.

Let $A_0, A_1, \ldots, A_k$ be linear operators on L such that A_k generates a strongly continuous contraction semigroup $\{S(t)\}$ on L satisfying (7.12). Assume that $\mathscr{D} \equiv \bigcap_{j=0}^k \mathscr{D}(A_j)$ is a core for A_k. For each α sufficiently large, suppose that an extension of $\sum_{j=0}^{k} \alpha^j A_j$ generates a strongly continuous contraction semigroup $\{T_\alpha(t)\}$ on L. Let D be a subspace of

$$(8.21) \qquad \Big\{ f_0 \in \mathscr{D}: \text{there exist } f_1, f_2, \ldots, f_{k-1} \in \mathscr{D} \text{ with } \sum_{j=0}^{m} A_{k-m+j} f_j = 0 \text{ for } m = 0, \ldots, k-1 \Big\},$$

and define

$$(8.22) \qquad C = \Big\{ \Big(f_0, \sum_{j=0}^{k-1} PA_j f_j \Big) : f_0 \in D, f_1, \ldots, f_{k-1} \text{ as above} \Big\}.$$

Then C is dissipative and if $\{(f, g) \in \bar{C}: g \in \bar{D}\}$, which is therefore single-valued, generates a strongly continuous contraction semigroup $\{T(t)\}$ on $\bar{D}$, then, for each $f \in \bar{D}$, $\lim_{\alpha \to \infty} T_\alpha(t)f = T(t)f$ for all $t \geq 0$, uniformly on bounded intervals.

21. Prove the following generalization of Theorem 7.6.

Let M be a closed subspace of L, let $A \subset L \times L$ be linear, and let B_1 and B_2 generate strongly continuous contraction semigroups $\{S_1(t)\}$ and $\{S_2(t)\}$ on M and L, respectively, satisfying

$$(8.23) \qquad \lim_{\lambda \to 0+} \lambda \int_0^\infty e^{-\lambda t} S_1(t) f \, dt \equiv P_1 f \quad \text{exists for all} \quad f \in M,$$

$$(8.24) \qquad \lim_{\lambda \to 0+} \lambda \int_0^\infty e^{-\lambda t} S_2(t) f \, dt \equiv P_2 f \quad \text{exists for all} \quad f \in L.$$

Assume that $\mathscr{R}(P_2) \subset M$. Let D be a subspace of $\mathscr{D}(A)$, D_1 a core for B_1, and D_2 a core for B_2. For $n = 1, 2, \ldots$, let A_n be a linear operator on L_n and let $\alpha_n, \beta_n > 0$. Suppose that $\lim_{n \to \infty} \alpha_n = \infty$, $\lim_{n \to \infty} \beta_n = \infty$, and

$$(8.25) \qquad \{(f, g) \in A: f \in D\} \subset \operatorname*{ex-lim}_{n \to \infty} A_n,$$

$$(8.26) \qquad \{(h, B_1 h): h \in D_1\} \subset \operatorname*{ex-lim}_{n \to \infty} \alpha_n^{-1} A_n,$$

$$(8.27) \qquad \{(k, B_2 k): k \in D_2\} \subset \operatorname*{ex-lim}_{n \to \infty} \beta_n^{-1} A_n.$$

Define $C = \{(f, P_1 P_2 g): (f, g) \in A, f \in D\}$ and assume that $\{(f, g) \in \bar{C}: g \in \bar{D}\}$ generates a strongly continuous contraction semigroup $\{T(t)\}$ on $\bar{D}$. Then conclusions (a) and (b) of Theorem 7.6 hold.

22. Prove the following modification of Corollary 7.8.

Let Π, A, and B be linear operators on L such that B generates a strongly continuous contraction semigroup $\{S(t)\}$ on L satisfying (7.12). Assume that $\mathscr{D}(\Pi) \cap D(A) \cap \mathscr{D}(B)$ is a core for B. For each α sufficiently large, suppose that an extension of $\Pi + \alpha A + \alpha^2 B$ generates a strongly continuous contraction semigroup $\{T_\alpha(t)\}$ on L. Let D be a subspace of $\mathscr{D}(\Pi) \cap \mathscr{D}(A) \cap \mathscr{N}(B)$ with $\mathscr{R}(P) \subset \bar{D}$, and define $C = \{(f, PAf): f \in D\}$. Then C is dissipative. Suppose that $\bar{C}$ generates a strongly continuous contraction semigroup $\{U(t)\}$ on $\bar{D}$, and that

$$(8.28) \qquad \lim_{\lambda \to 0+} \lambda \int_0^\infty e^{-\lambda t} U(t) f \, dt \equiv P_0 f \quad \text{exists for every} \quad f \in \bar{D}.$$

Let D_0 be a subspace of $\{f \in D$: there exists $h \in \mathscr{D}(\Pi) \cap \mathscr{D}(A) \cap \mathscr{D}(B)$ with $Bh = -Af\}$, and define

$$(8.29) \qquad C_0 = \{(f, P_0 P \Pi f + P_0 PAh): f \in D_0, \quad h \in \mathscr{D}(\Pi) \cap \mathscr{D}(A) \cap \mathscr{D}(B), \quad Bh = -Af\}.$$

Then C_0 is dissipative, and if $\{(f, g) \in \bar{C}_0 : g \in \bar{D}_0\}$ generates a strongly continuous contraction semigroup $\{T(t)\}$ on $\bar{D}_0$, then, for each $f \in \bar{D}_0$, $\lim_{\alpha \to \infty} T_\alpha(t) f = T(t) f$ for all $t \geq 0$, uniformly on bounded intervals.

23. Let A generate a strongly continuous semigroup $\{T(t)\}$ on L, let $B(t): L \to L$, $t \geq 0$, be bounded linear operators such that $\{B(t)\}$ is strongly continuous in $t \geq 0$ (i.e., $t \to B(t)f$ is continuous for each $f \in L$).
    (a) Show that for each $f \in L$ there exists a unique $u: [0, \infty) \to L$ satisfying

$$(8.30) \qquad u(t) = T(t)f + \int_0^t T(t-s)B(s)u(s) \, ds.$$

    (b) Show that if $B(t)g$ is continuously differentiable in t for each $g \in L$, and $f \in \mathscr{D}(A)$, then the solution of (8.30) satisfies

$$(8.31) \qquad \frac{\partial}{\partial t} u(t) = Au(t) + B(t)u(t).$$

9. NOTES

Among the best general references on operator semigroups are Hille and Phillips (1957), Dynkin (1965), Davies (1980), Yosida (1980), and Pazy (1983).

Theorem 2.6 is due to Hille (1948) and Yosida (1948).

To the best of our knowledge, Proposition 3.3 first appeared in a paper of Watanabe (1968).

Theorem 4.3 is the linear version of a theorem of Crandall and Liggett (1971). The concept of the extended limit is due to Sova (1967) and Kurtz (1969).

Sufficient conditions for the convergence of semigroups in terms of convergence of their generators were first obtained by Neveu (1958), Skorohod (1958), and Trotter (1958). The necessary and sufficient conditions of Theorems

6.1 and 6.5 were found by Sova (1967) and Kurtz (1969). The proof given here follows Goldstein (1976). Hasegawa (1964) and Kato (1966) found necessary and sufficient conditions of a different sort. Lemma 6.4 and Corollary 6.6 are due to Chernoff (1968). Corollary 6.7 is known as the Trotter (1959) product formula. Corollary 6.8 can be found in Hille (1948). Theorems 6.9 and 6.11 were proved by Kurtz (1970a).

Theorem 7.1 was obtained by Kato (1966) assuming $\alpha < \frac{1}{2}$ and in general by Gustafson (1966). Lemma 7.3 appears in Hille (1948). Theorem 7.6 is due to Ethier and Nagylaki (1980) and Corollary 7.7 to Kurtz (1977). Corollary 7.8 was proved by Kurtz (1973) and Kertz (1974); related results are given in Davies (1980).

Problem 4(b) is due to Kallman and Rota (1970), Problem 8 to Liggett (1972), Problem 9 to Kurtz (see Ethier (1976)), Problem 13 to Kato (1966), and Problem 14 to Norman (1977). Problem 20 is closely related to a theorem of Kertz (1978).

2 STOCHASTIC PROCESSES AND MARTINGALES

This chapter consists primarily of background material that is needed later. Section 1 defines various concepts in the theory of stochastic processes, in particular the notion of a stopping time. Section 2 gives a basic introduction to martingale theory including the optional sampling theorem, and local martingales are discussed in Section 3, in particular the existence of the quadratic variation or square bracket process. Section 4 contains additional technical material on processes and conditional expectations, including a Fubini theorem. The Doob–Meyer decomposition theorem for submartingales is given in Section 5, and some of the special properties of square integrable martingales are noted in Section 6. The semigroup of conditioned shifts on the space of progressive processes is discussed in Section 7. The optional sampling theorem for martingales indexed by a metric lattice is given in Section 8.

1. STOCHASTIC PROCESSES

A *stochastic process* X (or simply a process) with *index set* $\mathscr{I}$ and *state space* $(E, \mathscr{B})$ (a measurable space) defined on a probability space $(\Omega, \mathscr{F}, P)$ is a function defined on $\mathscr{I} \times \Omega$ with values in E such that for each $t \in \mathscr{I}$, $X(t, \cdot): \Omega \to E$ is an E-valued random variable, that is, $\{\omega: X(t, \omega) \in \Gamma\} \in \mathscr{F}$ for every $\Gamma \in \mathscr{B}$. We assume throughout that E is a metric space with metric r

and that $\mathscr{B}$ is the Borel σ-algebra $\mathscr{B}(E)$. As is usually done, we write $X(t)$ and $X(t, \cdot)$ interchangeably.

In this chapter, with the exception of Section 8, we take $\mathscr{I} = [0, \infty)$. We are primarily interested in viewing X as a "random" function of time. Consequently, it is natural to put further restrictions on X. We say that X is *measurable* if $X: [0, \infty) \times \Omega \to E$ is $\mathscr{B}[0, \infty) \times \mathscr{F}$-measurable. We say that X is (almost surely) *continuous* (*right continuous, left continuous*) if for (almost) every $\omega \in \Omega$, $X(\cdot, \omega)$ is continuous (right continuous, left continuous). Note that the statements "X is measurable" and "X is continuous" are not parallel in that "X is measurable" is stronger than the statement that $X(\cdot, \omega)$ is measurable for each $\omega \in \Omega$. The function $X(\cdot, \omega)$ is called the *sample path* of the process at ω.

A collection $\{\mathscr{F}_t\} \equiv \{\mathscr{F}_t, t \in [0, \infty)\}$ of σ-algebras of sets in $\mathscr{F}$ is a *filtration* if $\mathscr{F}_t \subset \mathscr{F}_{t+s}$ for $t, s \in [0, \infty)$. Intuitively $\mathscr{F}_t$ corresponds to the information known to an observer at time t. In particular, for a process X we define $\{\mathscr{F}_t^X\}$ by $\mathscr{F}_t^X = \sigma(X(s): s \le t)$; that is, $\mathscr{F}_t^X$ is the information obtained by observing X up to time t.

We occasionally need additional structure on $\{\mathscr{F}_t\}$. We say $\{\mathscr{F}_t\}$ is *right continuous* if for each $t \ge 0$, $\mathscr{F}_t = \mathscr{F}_{t+} \equiv \bigcap_{\varepsilon > 0} \mathscr{F}_{t+\varepsilon}$. Note the filtration $\{\mathscr{F}_{t+}\}$ is always right continuous (Problem 7). We say $\{\mathscr{F}_t\}$ is *complete* if $(\Omega, \mathscr{F}, P)$ is complete and $\{A \in \mathscr{F}: P(A) = 0\} \subset \mathscr{F}_0$.

A process X is *adapted* to a filtration $\{\mathscr{F}_t\}$ (or simply $\{\mathscr{F}_t\}$-adapted) if $X(t)$ is $\mathscr{F}_t$-measurable for each $t \ge 0$. Since $\mathscr{F}_t$ is increasing in t, X is $\{\mathscr{F}_t\}$-adapted if and only if $\mathscr{F}_t^X \subset \mathscr{F}_t$ for each $t \ge 0$.

A process X is $\{\mathscr{F}_t\}$-*progressive* (or simply progressive if $\{\mathscr{F}_t\} = \{\mathscr{F}_t^X\}$) if for each $t \ge 0$ the restriction of X to $[0, t] \times \Omega$ is $\mathscr{B}[0, t] \times \mathscr{F}_t$-measurable. Note that if X is $\{\mathscr{F}_t\}$-progressive, then X is $\{\mathscr{F}_t\}$-adapted and measurable, but the converse is not necessarily the case (see Section 4 however). However, every right (left) continuous $\{\mathscr{F}_t\}$-adapted process is $\{\mathscr{F}_t\}$-progressive (Problem 1).

There are a variety of notions of equivalence between two stochastic processes. For $0 \le t_1 < t_2 < \cdots < t_m$, let $\mu_{t_1, \ldots, t_m}$ be the probability measure on $\mathscr{B}(E) \times \cdots \times \mathscr{B}(E)$ induced by the mapping $(X(t_1), \ldots, X(t_m)) \to E^m$, that is, $\mu_{t_1, \ldots, t_m}(\Gamma) = P\{(X(t_1), \ldots, X(t_m)) \in \Gamma\}$, $\Gamma \in \mathscr{B}(E) \times \cdots \times \mathscr{B}(E)$. The probability measures $\{\mu_{t_1, \ldots, t_m}: m \ge 1, 0 \le t_1 < \cdots < t_m\}$ are called the *finite-dimensional distributions of* X. If X and Y are stochastic processes with the same finite-dimensional distributions, then we say Y is a *version* of X (and X is a version of Y). Note that X and Y need not be defined on the same probability space. If X and Y are defined on the same probability space and for each $t \ge 0$, $P\{X(t) = Y(t)\} = 1$, then we say Y is a *modification* of X. (We are implicitly assuming that $(X(t), Y(t))$ is an $E \times E$-valued random variable, which is always the case if E is separable.) If Y is a modification of X, then clearly Y is a version of X. Finally if there exists $N \in \mathscr{F}$ such that $P(N) = 0$ and $X(\cdot, \omega) = Y(\cdot, \omega)$ for all $\omega \notin N$, then we say X and Y are *indistinguishable*. If X and Y are indistinguishable, then clearly Y is a modification of X.

A random variable τ with values in $[0, \infty]$ is an $\{\mathscr{F}_t\}$-*stopping time* if $\{\tau \leq t\} \in \mathscr{F}_t$ for every $t \geq 0$. (Note that we allow $\tau = \infty$.) If $\tau < \infty$ a.s., we say τ is *finite* a.s. If $\tau \leq T < \infty$ for some constant T, we say τ is *bounded*. In some sense a stopping time is a random time that is recognizable by an observer whose information at time t is $\mathscr{F}_t$.

If τ is an $\{\mathscr{F}_t\}$-stopping time, then for $s < t$, $\{\tau \leq s\} \in \mathscr{F}_s \subset \mathscr{F}_t$, $\{\tau < t\} = \bigcup_n \{\tau \leq t - 1/n\} \in \mathscr{F}_t$ and $\{\tau = t\} = \{\tau \leq t\} - \{\tau < t\} \in \mathscr{F}_t$. If τ is *discrete* (i.e., if there exists a countable set $D \subset [0, \infty]$ such that $\{\tau \in D\} = \Omega$), then τ is an $\{\mathscr{F}_t\}$-stopping time if and only if $\{\tau = t\} \in \mathscr{F}_t$ for each $t \in D \cap [0, \infty)$.

1.1 Lemma A $[0, \infty]$-valued random variable τ is an $\{\mathscr{F}_{t+}\}$-stopping time if and only if $\{\tau < t\} \in \mathscr{F}_t$ for every $t \geq 0$.

Proof. If $\{\tau < t\} \in \mathscr{F}_t$ for every $t \geq 0$, then $\{\tau < t + n^{-1}\} \in \mathscr{F}_{t+m^{-1}}$ for $n \geq m$ and $\{\tau \leq t\} = \bigcap_n \{\tau < t + n^{-1}\} \in \bigcap_m \mathscr{F}_{t+m^{-1}} = \mathscr{F}_{t+}$. The necessity was observed above. □

1.2 Proposition Let $\tau_1, \tau_2, \ldots$ be $\{\mathscr{F}_t\}$-stopping times and let $c \in [0, \infty)$. Then the following hold.

(a) $\tau_1 + c$ and $\tau_1 \wedge c$ are $\{\mathscr{F}_t\}$-stopping times.

(b) $\sup_n \tau_n$ is an $\{\mathscr{F}_t\}$-stopping time.

(c) $\min_{k \leq n} \tau_k$ is an $\{\mathscr{F}_t\}$-stopping time for each $n \geq 1$.

(d) If $\{\mathscr{F}_t\}$ is right continuous, then $\inf_n \tau_n$, $\underline{\lim}_{n \to \infty} \tau_n$, and $\overline{\lim}_{n \to \infty} \tau_n$ are $\{\mathscr{F}_t\}$-stopping times.

Proof. We prove (b) and (d) and leave (a) and (c) to the reader. Note that $\{\sup_n \tau_n \leq t\} = \bigcap_n \{\tau_n \leq t\} \in \mathscr{F}_t$ so (b) follows. Similarly $\{\inf_n \tau_n < t\} = \bigcup_n \{\tau_n < t\} \in \mathscr{F}_t$, so if $\{\mathscr{F}_t\}$ is right continuous, then $\inf_n \tau_n$ is a stopping time by Lemma 1.1. Since $\underline{\lim}_{n \to \infty} \tau_n = \sup_m \inf_{n \geq m} \tau_n$ and $\overline{\lim}_{n \to \infty} \tau_n = \inf_m \sup_{n \geq m} \tau_n$, (d) follows. □

By Proposition 1.2(a) every stopping time τ can be approximated by a sequence of bounded stopping times, that is, $\lim_{n \to \infty} \tau \wedge n = \tau$. This fact is very useful in proving theorems about stopping times. A second equally useful approximation is the approximation of arbitrary stopping times by a nonincreasing sequence of discrete stopping times.

1.3 Proposition For $n = 1, 2, \ldots$, let $0 = t_0^n < t_1^n < \cdots$ and $\lim_{k \to \infty} t_k^n = \infty$, and suppose that $\lim_{n \to \infty} \sup_k (t_{k+1}^n - t_k^n) = 0$. Let τ be an $\{\mathscr{F}_{t+}\}$-stopping time and define

$$\tau_n = \begin{cases} t_{k+1}^n & \text{if} \quad t_k^n \leq \tau < t_{k+1}^n, \qquad k \geq 0 \\ \infty & \text{if} \quad \tau = \infty. \end{cases} \tag{1.1}$$

Then τ_n is an $\{\mathscr{F}_t\}$-stopping time and $\lim_{n\to\infty} \tau_n = \tau$. If in addition $\{t_k^n\} \subset \{t_k^{n+1}\}$, then $\tau_n \geq \tau_{n+1}$.

Proof. Let $\gamma_n(t) = \max\{t_k^n : t_k^n \leq t\}$. Then

$$\{\tau_n \leq t\} = \{\tau_n \leq \gamma_n(t)\} = \{\tau < \gamma_n(t)\} \in \mathscr{F}_{\gamma_n(t)} \subset \mathscr{F}_t. \tag{1.2}$$

The rest is clear. □

Recall the intuitive description of $\mathscr{F}_t$ as the information known to an observer at time t. For an $\{\mathscr{F}_t\}$-stopping time τ, the σ-algebra $\mathscr{F}_\tau$ should have the same intuitive meaning. For technical reasons $\mathscr{F}_\tau$ is defined by

$$\mathscr{F}_\tau = \{A \in \mathscr{F} : A \cap \{\tau \leq t\} \in \mathscr{F}_t \quad \text{for all} \quad t \geq 0\}. \tag{1.3}$$

Similarly, $\mathscr{F}_{\tau+}$ is defined by replacing $\mathscr{F}_t$ by $\mathscr{F}_{t+}$. See Problem 6 for some motivation as to why the definition is reasonable. Given an E-valued process X, define $X(\infty) \equiv x_0$ for some fixed $x_0 \in E$.

1.4 Proposition Let τ and σ be $\{\mathscr{F}_t\}$-stopping times, let γ be a nonnegative $\mathscr{F}_\tau$-measurable random variable, and let X be an $\{\mathscr{F}_t\}$-progressive E-valued process. Define X^τ and Y by $X^\tau(t) = X(\tau \wedge t)$ and $Y(t) = X(\tau + t)$, and define $\mathscr{G}_t = \mathscr{F}_{\tau \wedge t}$ and $\mathscr{H}_t = \mathscr{F}_{\tau + t}$, $t \geq 0$. (Recall that $\tau \wedge t$ and $\tau + t$ are stopping times.) Then the following hold:

(a) $\mathscr{F}_\tau$ is a σ-algebra.

(b) τ and $\tau \wedge \sigma$ are $\mathscr{F}_\tau$-measurable.

(c) If $\tau \leq \sigma$, then $\mathscr{F}_\tau \subset \mathscr{F}_\sigma$.

(d) $X(\tau)$ is $\mathscr{F}_\tau$-measurable.

(e) $\{\mathscr{G}_t\}$ is a filtration and X^τ is both $\{\mathscr{G}_t\}$-progressive and $\{\mathscr{F}_t\}$-progressive.

(f) $\{\mathscr{H}_t\}$ is a filtration and Y is $\{\mathscr{H}_t\}$-progressive.

(g) $\tau + \gamma$ is an $\{\mathscr{F}_t\}$-stopping time.

Proof. **(a)** Clearly $\varnothing$ and Ω are in $\mathscr{F}_\tau$, since $\mathscr{F}_t$ is a σ-algebra and $\{\tau \leq t\} \in \mathscr{F}_t$. If $A \cap \{\tau \leq t\} \in \mathscr{F}_t$, then $A^c \cap \{\tau \leq t\} = \{\tau \leq t\} - A \cap \{\tau \leq t\} \in \mathscr{F}_t$, and hence $A \in \mathscr{F}_\tau$ implies $A^c \in \mathscr{F}_\tau$. Similarly $A_k \cap \{\tau \leq t\} \in \mathscr{F}_t$, $k = 1, 2, \ldots$, implies $(\bigcup_k A_k) \cap \{\tau \leq t\} = \bigcup_k (A_k \cap \{\tau \leq t\}) \in \mathscr{F}_t$, and hence $\mathscr{F}_\tau$ is closed under countable unions.

(b) For each $c \geq 0$ and $t \geq 0$,

$$\begin{aligned}\{\tau \wedge \sigma \leq c\} \cap \{\tau \leq t\} &= \{\tau \wedge \sigma \leq c \wedge t\} \cap \{\tau \leq t\} \\ &= (\{\tau \leq c \wedge t\} \cup \{\sigma \leq c \wedge t\}) \cap \{\tau \leq t\} \in \mathscr{F}_t.\end{aligned} \tag{1.4}$$

Hence $\{\tau \wedge \sigma \leq c\} \in \mathscr{F}_\tau$ and $\tau \wedge \sigma$ is $\mathscr{F}_\tau$-measurable, as is τ (take $\sigma = \tau$).

(c) If $A \in \mathscr{F}_\tau$, then $A \cap \{\sigma \le t\} = A \cap \{\tau \le t\} \cap \{\sigma \le t\} \in \mathscr{F}_t$ for all $t \ge 0$. Hence $A \in \mathscr{F}_\sigma$.

(d) Fix $t \ge 0$. By (b), $\tau \wedge t$ is $\mathscr{F}_t$-measurable. Consequently the mapping $\omega \to (\tau(\omega) \wedge t, \omega)$ is a measurable mapping of $(\Omega, \mathscr{F}_t)$ into $([0, t] \times \Omega, \mathscr{B}[0, t] \times \mathscr{F}_t)$ and since X is $\{\mathscr{F}_t\}$-progressive, $(s, \omega) \to X(s, \omega)$ is a measurable mapping of $([0, t] \times \Omega, \mathscr{B}[0, t] \times \mathscr{F}_t)$ into $(E, \mathscr{B}(E))$. Since $X(\tau \wedge t)$ is the composition of these two mappings, it is $\mathscr{F}_t$-measurable. Finally, for $\Gamma \in \mathscr{B}(E)$, $\{X(\tau) \in \Gamma\} \cap \{\tau \le t\} = \{X(\tau \wedge t) \in \Gamma\} \cap \{\tau \le t\} \in \mathscr{F}_t$ and hence $\{X(\tau) \in \Gamma\} \in \mathscr{F}_\tau$.

(e) By (a) and (c), $\{\mathscr{G}_t\}$ is a filtration, and since $\mathscr{G}_t \subset \mathscr{F}_t$ by (c), X^τ is $\{\mathscr{F}_t\}$-progressive if it is $\{\mathscr{G}_t\}$-progressive. To see that X^τ is $\{\mathscr{G}_t\}$-progressive, we begin by showing that if $s \le t$ and $H \in \mathscr{B}[0, t] \times \mathscr{F}_s$, then

$$(1.5) \qquad H \cap ([0, t] \times \{\tau \wedge t \ge s\}) \in \mathscr{B}[0, t] \times \mathscr{F}_{\tau \wedge t} = \mathscr{B}[0, t] \times \mathscr{G}_t.$$

To verify this, note that the collection $\mathscr{H}_{s,t}$ of $H \in \mathscr{B}[0, t] \times \mathscr{F}_s$ satisfying (1.5) is a σ-algebra. Since $A \in \mathscr{F}_s$ implies $A \cap \{\tau \wedge t \ge s\} \in \mathscr{F}_{\tau \wedge t}$, it follows that if $B \in \mathscr{B}[0, t]$ and $A \in \mathscr{F}_s$, then

$$(1.6) \qquad (B \times A) \cap ([0, t] \times \{\tau \wedge t \ge s\}) = B \times (A \cap \{\tau \wedge t \ge s\}) \in \mathscr{B}[0, t] \times \mathscr{G}_t,$$

so $B \times A \in \mathscr{H}_{s,t}$. But the collection of $B \times A$ of this form generates $\mathscr{B}[0, t] \times \mathscr{F}_s$.

Finally, for $\Gamma \in \mathscr{B}(E)$ and $t \ge 0$,

$$\begin{aligned}(1.7) \qquad &\{(s, \omega) \in [0, t] \times \Omega : X(\tau(\omega) \wedge s, \omega) \in \Gamma\}\\ &= \{(s, \omega) : X(\tau(\omega) \wedge s, \omega) \in \Gamma, \ \tau(\omega) \wedge t \le s \le t\}\\ &\quad \cup \{(s, \omega) : X(s, \omega) \in \Gamma, \ s < \tau(\omega) \wedge t\}\\ &= (\{(s, \omega) : \tau(\omega) \wedge t \le s \le t\} \cap ([0, t] \times \{X(\tau \wedge t) \in \Gamma\}))\\ &\quad \cup \bigcup_n \bigcup_k \left(\left\{(s, \omega) : X(s, \omega) \in \Gamma, \ s < \frac{k}{n}\right\} \right.\\ &\quad \left. \cap \left\{(s, \omega) : \frac{k}{n} \le \tau(\omega) \wedge t\right\}\right) \in \mathscr{B}[0, t] \times \mathscr{G}_t\end{aligned}$$

since

$$\begin{aligned}(1.8) \qquad &\{(s, \omega) : \tau(\omega) \wedge t \le s \le t\}\\ &= \bigcap_n \bigcup_k \left(\left[\frac{k}{n}, t\right] \times \left\{\frac{k}{n} \le \tau \wedge t < \frac{k+1}{n}\right\}\right) \in \mathscr{B}[0, t] \times \mathscr{G}_t,\end{aligned}$$

and since the last set on the right in (1.7) is in $\mathscr{B}[0, t] \times \mathscr{G}_t$ by (1.5).

(f) Again $\{\mathscr{H}_t\}$ is a filtration by (a) and (c). Fix $t \ge 0$. By part (e) the mapping $(s, \omega) \to X((\tau(\omega) + t) \wedge s, \omega)$ from $([0, \infty] \times \Omega, \mathscr{B}[0, \infty] \times \mathscr{F}_{\tau+t})$

into $(E, \mathscr{B}(E))$ is measurable, as is the mapping $(u, \omega) \to (\tau(\omega) + u, \omega)$ from $([0, t] \times \Omega, \mathscr{B}[0, t] \times \mathscr{F}_{\tau+t})$ into $([0, \infty] \times \Omega, \mathscr{B}[0, \infty] \times \mathscr{F}_{\tau+t})$. The mapping $(u, \omega) \to X(\tau(\omega) + u, \omega)$ from $([0, t] \times \Omega, \mathscr{B}[0, t] \times \mathscr{F}_{\tau+t})$ into $(E, \mathscr{B}(E))$ is a composition of the first two mappings so it too is measurable. Since $\mathscr{H}_t = \mathscr{F}_{\tau+t}$, Y is $\{\mathscr{H}_t\}$-progressive.

(g) Let $\gamma_n = [n\gamma]/n$. Note that $\{\tau + \gamma_n \le t\} \cap \{\gamma_n = k/n\} = \{\tau \le t - k/n\} \cap \{\gamma_n = k/n\} \in \mathscr{F}_{t-k/n}$, since $\{\gamma_n = k/n\} \in \mathscr{F}_\tau$. Consequently, $\{\tau + \gamma_n \le t\} \in \mathscr{F}_t$. Since $\tau + \gamma = \sup_n(\tau + \gamma_n)$, part (g) follows by Proposition 1.2(b). □

Let X be an E-valued process and let $\Gamma \in \mathscr{B}(E)$. The *first entrance time into* Γ is defined by

$$\tau_e(\Gamma) = \inf \{t: X(t) \in \Gamma\} \tag{1.9}$$

(where $\inf \varnothing = \infty$), and for a $[0, \infty]$-valued random variable σ, the *first entrance time into* Γ *after* σ is defined by

$$\tau_e(\Gamma, \sigma) = \inf \{t \ge \sigma: X(t) \in \Gamma\}. \tag{1.10}$$

For each $\omega \in \Omega$ and $0 \le s \le t$, let $F_X(s, t, \omega) \subset E$ be the closure of $\{X(u, \omega): s \le u \le t\}$. The *first contact time with* Γ is defined by

$$\tau_c(\Gamma) = \inf \{t: F_X(0, t) \cap \Gamma \ne \varnothing\} \tag{1.11}$$

and the *first contact time with* Γ *after* σ by

$$\tau_c(\Gamma, \sigma) = \inf \{t \ge \sigma: F_X(\sigma, t) \cap \Gamma \ne \varnothing\}. \tag{1.12}$$

The *first exit time from* Γ (after σ) is the first entrance time of Γ^c (after σ). Although intuitively the above times are "recognizable" to our observer, they are not in general stopping times (or even random variables). We do, however, have the following result, which is sufficient for our purposes.

1.5 Proposition Suppose that X is a right continuous, $\{\mathscr{F}_t\}$-adapted, E-valued process and that σ is an $\{\mathscr{F}_t\}$-stopping time.

(a) If Γ is closed and X has left limits at each $t > 0$ or if Γ is compact, then $\tau_c(\Gamma, \sigma)$ is an $\{\mathscr{F}_t\}$-stopping time.

(b) If Γ is open, then $\tau_e(\Gamma, \sigma)$ is an $\{\mathscr{F}_{t+}\}$-stopping time.

Proof. Using the right continuity of X, if Γ is open,

$$\{\tau_e(\Gamma, \sigma) < t\} = \bigcup_{s \in \mathbb{Q} \cap [0, t)} \{X(s) \in \Gamma\} \cap \{\sigma < s\} \in \mathscr{F}_t, \tag{1.13}$$

implying part (b). For $n = 1, 2, \ldots$ let $\Gamma_n = \{x: r(x, \Gamma) < 1/n\}$. Then, under the conditions of part (a), $\tau_c(\Gamma, \sigma) = \lim_{n\to\infty} \tau_e(\Gamma_n, \sigma)$, and

$$\{\tau_c(\Gamma, \sigma) \le t\} = (\{\sigma \le t\} \cap \{X(t) \in \Gamma\}) \cup \bigcap_n \{\tau_e(\Gamma_n, \sigma) < t\} \in \mathscr{F}_t. \tag{1.14}$$

□

Under slightly more restrictive hypotheses on $\{\mathscr{F}_t\}$, a much more general result than Proposition 1.5 holds. We do not need this generality, so we simply state the result without proof.

1.6 Theorem Let $\{\mathscr{F}_t\}$ be complete and right continuous, and let X be an E-valued $\{\mathscr{F}_t\}$-progressive process. Then for each $\Gamma \in \mathscr{B}(E)$, $\tau_e(\Gamma)$ is an $\{\mathscr{F}_t\}$-stopping time.

Proof. See, for example, Elliott (1982), page 50. □

2. MARTINGALES

A real-valued process X with $E[|X(t)|] < \infty$ for all $t \geq 0$ and adapted to a filtration $\{\mathscr{F}_t\}$ is an $\{\mathscr{F}_t\}$-*martingale* if

(2.1) $$E[X(t+s)|\mathscr{F}_t] = X(t), \qquad t, s \geq 0,$$

is an $\{\mathscr{F}_t\}$-*submartingale* if

(2.2) $$E[X(t+s)|\mathscr{F}_t] \geq X(t), \qquad t, s \geq 0,$$

and is an $\{\mathscr{F}_t\}$-*supermartingale* if the inequality in (2.2) is reversed. Note that X is a supermartingale if $-X$ is a submartingale, and that X is a martingale if both X and $-X$ are submartingales. Consequently, results proved for submartingales immediately give analogous results for martingales and supermartingales. If $\{\mathscr{F}_t\} = \{\mathscr{F}_t^X\}$ we simply say X is a martingale (submartingale, supermartingale).

Jensen's inequality gives the following.

2.1 Proposition **(a)** Suppose X is an $\{\mathscr{F}_t\}$-martingale, φ is convex, and $E[|\varphi(X(t))|] < \infty$ for all $t \geq 0$. Then $\varphi \circ X$ is an $\{\mathscr{F}_t\}$-submartingale.

(b) Suppose X is an $\{\mathscr{F}_t\}$-submartingale, φ is convex and nondecreasing, and $E[|\varphi(X(t))|] < \infty$ for all $t \geq 0$. Then $\varphi \circ X$ is an $\{\mathscr{F}_t\}$-submartingale.

Proof. By Jensen's inequality, for $t, s \geq 0$,

(2.3) $$E[\varphi(X(t+s))|\mathscr{F}_t] \geq \varphi(E[X(t+s)|\mathscr{F}_t]) \geq \varphi(X(t)).$$

Note that for part (a) the last inequality is in fact equality, and in part (b) the last inequality follows from the assumption that φ is nondecreasing. □

2.2 Lemma Let τ_1 and τ_2 be $\{\mathscr{F}_t\}$-stopping times assuming values in $\{t_1, t_2, \ldots, t_m\} \subset [0, \infty)$. If X is an $\{\mathscr{F}_t\}$-submartingale, then

(2.4) $$E[X(\tau_2)|\mathscr{F}_{\tau_1}] \geq X(\tau_1 \wedge \tau_2).$$

Proof. Assume $t_1 < t_2 < \cdots < t_m$. We must show that for every $A \in \mathscr{F}_{\tau_1}$

$$\int_A X(\tau_2)\, dP \geq \int_A X(\tau_2 \wedge \tau_1)\, dP. \tag{2.5}$$

Since $A = \bigcup_{i=1}^m (A \cap \{\tau_1 = t_i\})$, it is sufficient to show that

$$\int_{A \cap \{\tau_1 = t_i\}} X(\tau_2)\, dP \geq \int_{A \cap \{\tau_1 = t_i\}} X(\tau_2 \wedge \tau_1)\, dP = \int_{A \cap \{\tau_1 = t_i\}} X(\tau_2 \wedge t_i)\, dP, \tag{2.6}$$

but since $A \cap \{\tau_1 = t_i\} \in \mathscr{F}_{t_i}$, (2.5) holds if

$$E[X(\tau_2) | \mathscr{F}_{t_i}] \geq X(\tau_2 \wedge t_i). \tag{2.7}$$

Finally, observe that

$$\begin{aligned} E[X(\tau_2 \wedge t_{k+1}) | \mathscr{F}_{t_k}] &= E[X(t_{k+1})\chi_{\{\tau_2 > t_k\}} + X(\tau_2 \wedge t_k)\chi_{\{\tau_2 \leq t_k\}} | \mathscr{F}_{t_k}] \\ &= E[X(t_{k+1}) | \mathscr{F}_{t_k}]\chi_{\{\tau_2 > t_k\}} + X(\tau_2 \wedge t_k)\chi_{\{\tau_2 \leq t_k\}} \\ &\geq X(t_k)\chi_{\{\tau_2 > t_k\}} + X(\tau_2 \wedge t_k)\chi_{\{\tau_2 \leq t_k\}} \\ &= X(\tau_2 \wedge t_k). \end{aligned} \tag{2.8}$$

Starting with $k = m - 1$ and observing that $\tau_2 = \tau_2 \wedge t_m$, (2.8) may be iterated to give (2.7). □

The following is a simple application of Lemma 2.2. Let $x^+ = x \vee 0$.

2.3 Lemma Let X be a submartingale, $T > 0$, and $F \subset [0, T]$ be finite. Then for each $x > 0$,

$$P\left\{\max_{t \in F} X(t) \geq x\right\} \leq x^{-1} E[X^+(T)] \tag{2.9}$$

and

$$P\left\{\min_{t \in F} X(t) \leq -x\right\} \leq x^{-1}(E[X^+(T)] - E[X(0)]). \tag{2.10}$$

Proof. Let $\tau = \min\{t \in F: X(t) \geq x\}$ and set $\tau_1 = \tau \wedge T$ and $\tau_2 = T$ in (2.4). Then

$$E[X(T)] \geq E[X(\tau \wedge T)] = E[X(\tau)\chi_{\{\tau < \infty\}}] + E[X(T)\chi_{\{\tau = \infty\}}], \tag{2.11}$$

and hence

$$E[X(T)\chi_{\{\tau < \infty\}}] \geq E[X(\tau)\chi_{\{\tau < \infty\}}] \geq xP\{\tau < \infty\} = xP\left\{\max_{t \in F} X(t) \geq x\right\}, \tag{2.12}$$

which implies (2.9). The proof of (2.10) is similar. □

2.4 Corollary Let X be a submartingale and let $F \subset [0, \infty)$ be countable. Then for each $x > 0$ and $T > 0$,

$$P\left\{\sup_{t \in F \cap [0, T]} X(t) \geq x\right\} \leq x^{-1}E[X^+(T)] \tag{2.13}$$

and

$$P\left\{\inf_{t \in F \cap [0, T]} X(t) \leq -x\right\} \leq x^{-1}(E[X^+(T)] - E[X(0)]). \tag{2.14}$$

Proof. Let $F_1 \subset F_2 \subset \cdots$ be finite and $F = \bigcup F_n$. Then, for $0 < y < x$,

(2.15)

$$P\left\{\sup_{t \in F \cap [0, T]} X(t) \geq x\right\} \leq \lim_{n \to \infty} P\left\{\max_{t \in F_n \cap [0, T]} X(t) \geq y\right\} \leq y^{-1}E[X^+(T)].$$

Letting $y \to x$ we obtain (2.13), and (2.14) follows similarly. □

Let X be a real-valued process, and let $F \subset [0, \infty)$ be finite. For $a < b$ define $\tau_1 = \min\{t \in F: X(t) \leq a\}$, and for $k = 1, 2, \ldots$ define $\sigma_k = \min\{t > \tau_k: t \in F, X(t) \geq b\}$ and $\tau_{k+1} = \min\{t > \sigma_k: t \in F, X(t) \leq a\}$. Define

$$U(a, b, F) = \max\{k: \sigma_k < \infty\}. \tag{2.16}$$

The quantity $U(a, b, F)$ is called the number of *upcrossings* of the interval (a, b) by X restricted to F.

2.5 Lemma Let X be a submartingale. If $T > 0$ and $F \subset [0, T]$ is finite, then

$$E[U(a, b, F)] \leq \frac{E[(X(T) - a)^+]}{b - a}. \tag{2.17}$$

Proof. Since $\sigma_k \wedge T \leq \tau_{k+1} \wedge T$, Lemma 2.2 implies

$$\begin{aligned} 0 &\leq E\left[\sum_{k=1}^{\infty} (X(\tau_{k+1} \wedge T) - X(\sigma_k \wedge T))\right] \\ &= E\left[\sum_{k=1}^{U(a, b, F)} (X(\tau_{k+1} \wedge T) - X(\sigma_k \wedge T))\right] \\ &= E\left[-\sum_{k=2}^{U(a, b, F)} (X(\sigma_k \wedge T) - X(\tau_k \wedge T)) \right. \\ &\qquad \left. + X(\tau_{U(a, b, F)+1} \wedge T) - (X(\sigma_1 \wedge T) - a) - a\right] \\ &\leq E[-(b - a)U(a, b, F) + X(\tau_{U(a, b, F)+1} \wedge T) - a] \\ &\leq E[-(b - a)U(a, b, F) + (X(T) - a)^+], \end{aligned} \tag{2.18}$$

which gives (2.17). □

2.6 Corollary Let X be a submartingale. Let $T > 0$, let F be a countable subset of $[0, T]$, and let $F_1 \subset F_2 \subset \cdots$ be finite subsets with $F = \bigcup F_n$. Define $U(a, b, F) = \lim_{n\to\infty} U(a, b, F_n)$. Then $U(a, b, F)$ depends only on F (not the particular sequence $\{F_n\}$) and

$$E[U(a, b, F)] \le \frac{E[(X(T) - a)^+]}{b - a}. \tag{2.19}$$

Proof. The existence of the limit defining $U(a, b, F)$ as well as the independence of $U(a, b, F)$ from the choice of $\{F_n\}$ follows from the fact that $G \subset H$ implies $U(a, b, G) \le U(a, b, H)$. Consequently (2.19) follows from (2.17) and the monotone convergence theorem. □

One implication of the upcrossing inequality (2.19) is that submartingales have modifications with "nice" sample paths. To see this we need the following lemmas.

2.7 Lemma Let (E, r) be a metric space and let $x: [0, \infty) \to E$. Suppose $x(t+) \equiv \lim_{s\to t+} x(s)$ exists for all $t \ge 0$ and $x(t-) \equiv \lim_{s\to t-} x(s)$ exists for all $t > 0$. Then there exists a countable set Γ such that for $t \in (0, \infty) - \Gamma$, $x(t-) = x(t) = x(t+)$.

Let $\Gamma_n = \{t: r(x(t-), x(t)) \vee r(x(t-), x(t+)) \vee r(x(t), x(t+)) > n^{-1}\}$. Then $\Gamma_n \cap [0, T]$ is finite for each $T > 0$.

Proof. Since we may take $\Gamma = \bigcup_n \Gamma_n$, it is enough to verify the last statement. If $\Gamma_n \cap [0, T]$ had a limit point t then either $x(t-)$ or $x(t+)$ would fail to exist. Consequently $\Gamma_n \cap [0, T]$ must be finite. □

2.8 Lemma Let (E, r) be a metric space, let F be a dense subset of $[0, \infty)$, and let $x: F \to E$. If for each $t \ge 0$

$$y(t) \equiv \lim_{\substack{s\to t+ \\ s\in F}} x(s) \tag{2.20}$$

exists, then y is right continuous. If for each $t > 0$

$$y^-(t) \equiv \lim_{\substack{s\to t- \\ s\in F}} x(s) \tag{2.21}$$

exists, then y^- is left continuous (on $(0, \infty)$). If for each $t > 0$ both (2.20) and (2.21) exist, then $y^-(t) = y(t-)$ for all $t > 0$.

Proof. Suppose (2.20) exists for all $t \geq 0$. Given $t_0 > 0$ and $\varepsilon > 0$, there exists a $\delta > 0$ such that $r(y(t_0), x(s)) \leq \varepsilon$ for all $s \in F \cap (t_0, t_0 + \delta)$, and hence

$$r(y(t_0), y(s)) = \lim_{\substack{u \to s+ \\ u \in F}} r(y(t_0), x(u)) \leq \varepsilon \tag{2.22}$$

for all $s \in (t_0, t_0 + \delta)$ and the right continuity of y follows. The proof of the other parts is similar. □

Let F be a countable dense subset of $[0, \infty)$. For a submartingale X, Corollary 2.4 implies $P\{\sup_{t \in F \cap [0,T]} X(t) < \infty\} = 1$ and $P\{\inf_{t \in F \cap [0,T]} X(t) > -\infty\} = 1$ for each $T > 0$, and Corollary 2.6 gives $P\{U(a, b, F \cap [0, T]) < \infty\} = 1$ for all $a < b$ and $T > 0$. Let

$$\Omega_0 = \bigcap_{n=1}^{\infty} \left(\left\{ \sup_{t \in F \cap [0, n]} X(t) < \infty \right\} \cap \left\{ \inf_{t \in F \cap [0, n]} X(t) > -\infty \right\} \cap \bigcap_{\substack{a < b \\ a, b \in \mathbb{Q}}} \{U(a, b, F \cap [0, n]) < \infty\} \right). \tag{2.23}$$

Then $P(\Omega_0) = 1$. For $\omega \in \Omega_0$,

$$Y(t, \omega) = \lim_{\substack{s \to t+ \\ s \in F}} X(s, \omega) \tag{2.24}$$

exists for all $t \geq 0$, and

$$Y^-(t, \omega) = \lim_{\substack{s \to t- \\ s \in F}} X(s, \omega) \tag{2.25}$$

exists for all $t > 0$; furthermore, $Y(\cdot, \omega)$ is right continuous and has left limits with $Y(t-, \omega) = Y^-(t, \omega)$ for all $t > 0$ (Problem 9). Define $Y(t, \omega) = 0$ for all $\omega \notin \Omega_0$ and $t \geq 0$.

2.9 Proposition Let X be a submartingale, and let Y be defined by (2.24). Then $\Gamma \equiv \{t: P\{Y(t) \neq Y(t-)\} > 0\}$ is countable, $P\{X(t) = Y(t)\} = 1$ for $t \notin \Gamma$, and

$$\tilde{X}(t) = \begin{cases} Y(t), & t \in [0, \infty) - \Gamma, \\ X(t), & t \in \Gamma, \end{cases} \tag{2.26}$$

defines a modification of X almost all of whose sample paths have right and left limits at all $t \geq 0$ and are right continuous at all $t \notin \Gamma$.

Proof. For real-valued random variables η and ξ (defined on $(\Omega, \mathscr{F}, P)$) define

$$\gamma(\eta, \xi) = \inf\{\varepsilon > 0: P\{|\eta - \xi| > \varepsilon\} < \varepsilon\}. \tag{2.27}$$

Then γ is a metric corresponding to convergence in probability (Problem 8). Since Y has right and left limits in this metric at all $t \geq 0$, Lemma 2.7 implies Γ is countable.

Let $\alpha \in \mathbb{R}$. Then $X \vee \alpha$ is a submartingale by Proposition 2.1 so for any $T > 0$,

$$\alpha \leq X(t) \vee \alpha \leq E[X(T) \vee \alpha \mid \mathscr{F}_t^X], \qquad 0 \leq t \leq T, \tag{2.28}$$

and since $\{E[X(T) \vee \alpha \mid \mathscr{F}_t^X]: 0 \leq t \leq T\}$ is uniformly integrable (Problem 10), it follows that $\{X(t) \vee \alpha: 0 \leq t \leq T\}$ is uniformly integrable. Therefore

$$X(t) \vee \alpha \leq \lim_{\substack{s \to t+ \\ s \in \mathbb{Q}}} E[X(s) \vee \alpha \mid \mathscr{F}_t^X] = E[Y(t) \vee \alpha \mid \mathscr{F}_t^X], \qquad t \geq 0. \tag{2.29}$$

Furthermore if $t \notin \Gamma$, then

$$E[E[Y(t) \vee \alpha \mid \mathscr{F}_t^X] - X(t) \vee \alpha] \leq \lim_{\substack{s \to t- \\ s \in \mathbb{Q}}} E[Y(t) \vee \alpha - X(s) \vee \alpha] = 0, \tag{2.30}$$

and hence, since $Y(t) = Y(t-)$ a.s. and $Y(t-)$ is $\mathscr{F}_t^X$-measurable,

$$X(t) \vee \alpha = E[Y(t) \vee \alpha \mid \mathscr{F}_t^X] = Y(t) \vee \alpha \qquad \text{a.s.} \tag{2.31}$$

Since α is arbitrary, $P\{X(t) = Y(t)\} = 1$ for $t \notin \Gamma$.

To see that almost all sample paths of $\tilde{X}$ have right and left limits at all $t \geq 0$ and are right continuous at all $t \notin \Gamma$, replace F in the construction of Y by $F \cup \Gamma$. Note that this replaces Ω_0 by $\tilde{\Omega}_0 \subset \Omega_0$, but that for $\omega \in \tilde{\Omega}_0$, $Y(\cdot, \omega)$ and $\tilde{X}(\cdot, \omega)$ do not change. Since for $\omega \in \tilde{\Omega}_0$

$$Y(t, \omega) = \lim_{s \to t+} Y(s, \omega) = \lim_{\substack{s \to t+ \\ s \in F \cup \Gamma}} X(s, \omega), \qquad t \geq 0, \tag{2.32}$$

it follows that

$$Y(t, \omega) = \lim_{s \to t+} \tilde{X}(s, \omega), \qquad t \geq 0, \tag{2.33}$$

which gives both the existence of right limits and the right continuity of $\tilde{X}(\cdot, \omega)$ at $t \notin \Gamma$. The existence of left limits follows similarly. □

2.10 Corollary Let Z be a random variable with $E[|Z|] < \infty$. Then for any filtration $\{\mathscr{F}_t\}$ and $t \geq 0$, $E[Z \mid \mathscr{F}_s] \to E[Z \mid \mathscr{F}_{t+}]$ in L^1 as $s \to t+$.

Proof. Let $X(t) = E[Z \mid \mathscr{F}_t]$, $t \geq 0$. Then X is a martingale and by Proposition 2.9 we may assume X has right limits a.s. at each $t \geq 0$. Since $\{X(t)\}$ is uniformly integrable, $X(t+) \equiv \lim_{s \to t+} X(s)$ exists a.s. and in L^1 for all $t \geq 0$.

We need only check that $X(t+) = E[Z \mid \mathscr{F}_{t+}]$. Clearly $X(t+)$ is $\mathscr{F}_{t+}$-measurable and for $A \in \mathscr{F}_{t+}$,

$$\int_A X(t+)\,dP = \lim_{s \to t+} \int_A X(s)\,dP = \int_A Z\,dP, \tag{2.34}$$

hence $X(t+) = E[Z \mid \mathscr{F}_{t+}]$. □

2.11 Corollary If $\{\mathscr{F}_t\}$ is a right continuous filtration and X is an $\{\mathscr{F}_t\}$-martingale, then X has a right continuous modification.

2.12 Remark It follows from the construction of Y in (2.24) that almost all sample paths of a right continuous submartingale have left limits at all $t > 0$. □

Proof. With reference to (2.24) and Corollary 2.10, for $t < T$,

$$\begin{aligned} Y(t) &= \lim_{\substack{s \to t+ \\ s \in F}} X(s) = \lim_{\substack{s \to t+ \\ s \in F}} E[X(T) \mid \mathscr{F}_s] \\ &= E[X(T) \mid \mathscr{F}_{t+}] = E[X(T) \mid \mathscr{F}_t] = X(t) \quad \text{a.s.,} \end{aligned} \tag{2.35}$$

so Y is the desired modification. □

Essentially, Proposition 2.9 says that we may assume every submartingale has well-behaved sample paths, that is, if all that is prescribed about a submartingale is its finite-dimensional distributions, then we may as well assume that the sample paths have the properties given in the proposition. In fact, in virtually all cases of interest, $\Gamma = \emptyset$, so we can assume right continuity at all $t \geq 0$. We do just that in the remainder of this section. Extension of the results to the somewhat more general case is usually straightforward.

Our next result is the optional sampling theorem.

2.13 Theorem Let X be a right continuous $\{\mathscr{F}_t\}$-submartingale, and let τ_1 and τ_2 be $\{\mathscr{F}_t\}$-stopping times. Then for each $T > 0$,

$$E[X(\tau_2 \wedge T) \mid \mathscr{F}_{\tau_1}] \geq X(\tau_1 \wedge \tau_2 \wedge T) \tag{2.36}$$

If, in addition, τ_2 is finite a.s., $E[|X(\tau_2)|] < \infty$, and

$$\lim_{T \to \infty} E[|X(T)| \chi_{\{\tau_2 > T\}}] = 0, \tag{2.37}$$

then

$$E[X(\tau_2) \mid \mathscr{F}_{\tau_1}] \geq X(\tau_1 \wedge \tau_2). \tag{2.38}$$

2.14 Remark Note that if X is a martingale (X and $-X$ are submartingales), then equality holds in (2.36) and (2.38). Note also that any right continuous $\{\mathscr{F}_t\}$-submartingale is an $\{\mathscr{F}_{t+}\}$-submartingale, and hence corresponding inequalities hold for $\{\mathscr{F}_{t+}\}$-stopping times. □

Proof. For $i = 1, 2$, let $\tau_i^{(n)} = \infty$ if $\tau_i = \infty$ and let $\tau_i^{(n)} = (k+1)/2^n$ if $k/2^n \leq \tau_i < (k+1)/2^n$. Then by Proposition 1.3, $\tau_i^{(n)}$ is an $\{\mathscr{F}_t\}$-stopping time, and by Lemma 2.2, for each $\alpha \in \mathbb{R}$ and $T > 0$,

$$E[X(\tau_2^{(n)} \wedge T) \vee \alpha \mid \mathscr{F}_{\tau_1^{(n)}}] \geq X(\tau_1^{(n)} \wedge \tau_2^{(n)} \wedge T) \vee \alpha. \tag{2.39}$$

Since $\mathscr{F}_{\tau_1} \subset \mathscr{F}_{\tau_1^{(n)}}$ by Proposition 1.4(c), (2.39) implies

$$E[X(\tau_2^{(n)} \wedge T) \vee \alpha \mid \mathscr{F}_{\tau_1}] \geq E[X(\tau_1^{(n)} \wedge \tau_2^{(n)} \wedge T) \vee \alpha \mid \mathscr{F}_{\tau_1}]. \tag{2.40}$$

Since Lemma 2.2 implies

$$\alpha \leq X(\tau_2^{(n)} \wedge T) \vee \alpha \leq E[X(T) \vee \alpha \mid \mathscr{F}_{\tau_2^{(n)}}], \tag{2.41}$$

$\{X(\tau_2^{(n)} \wedge T) \vee \alpha\}$ is uniformly integrable as is $\{X(\tau_1^{(n)} \wedge \tau_2^{(n)} \wedge T) \vee \alpha\}$ (Problem 10). Letting $n \to \infty$, the right continuity of X and the uniform integrability of the sequences gives

$$\begin{aligned} E[X(\tau_2 \wedge T) \vee \alpha \mid \mathscr{F}_{\tau_1}] &\geq E[X(\tau_1 \wedge \tau_2 \wedge T) \vee \alpha \mid \mathscr{F}_{\tau_1}] \\ &= X(\tau_1 \wedge \tau_2 \wedge T) \vee \alpha. \end{aligned} \tag{2.42}$$

Letting $\alpha \to -\infty$ gives (2.36), and under the additional hypotheses, letting $T \to \infty$ gives (2.38). □

The following is an application of the optional sampling theorem.

2.15 Proposition Let X be a right continuous nonnegative $\{\mathscr{F}_t\}$-supermartingale, and let $\tau_c(0)$ be the first contact time with 0. Then $X(t) = 0$ for all $t \geq \tau_c(0)$ with probability one.

Proof. For $n = 1, 2, \ldots$, let $\tau_n = \tau_e([0, n^{-1}))$, the first entrance time into $[0, n^{-1})$. (By Proposition 1.5, τ_n is an $\{\mathscr{F}_{t+}\}$-stopping time.) Then $\tau_c(0) = \lim_{n\to\infty} \tau_n$. If $\tau_n < \infty$, then $X(\tau_n) \leq n^{-1}$. Consequently, for every $t \geq 0$,

$$E[X(t) \mid \mathscr{F}_{\tau_n+}] \leq X(t \wedge \tau_n), \tag{2.43}$$

and hence

$$E[X(t) \mid \mathscr{F}_{\tau_n+}]\chi_{\{\tau_n \leq t\}} \leq n^{-1}. \tag{2.44}$$

Taking expectations and letting $n \to \infty$, we have

$$E[X(t)\chi_{\{\tau_c(0) \leq t\}}] = 0. \tag{2.45}$$

The proposition follows by the nonnegativity and right continuity. □

Next we extend Lemma 2.3.

2.16 Proposition **(a)** Let X be a right continuous submartingale. Then for each $x > 0$ and $T > 0$,

(2.46) $$P\left\{\sup_{t \leq T} X(t) \geq x\right\} \leq x^{-1}E[X^+(T)]$$

and

(2.47) $$P\left\{\inf_{t \leq T} X(t) \leq -x\right\} \leq x^{-1}(E[X^+(T)] - E[X(0)]).$$

(b) Let X be a nonnegative right continuous submartingale. Then for $\alpha > 1$ and $T > 0$,

(2.48) $$E\left[\sup_{t \leq T} X(t)^\alpha\right] \leq \left(\frac{\alpha}{\alpha - 1}\right)^\alpha E[X(T)^\alpha].$$

Proof. Corollary 2.4 implies (2.46) and (2.47), but we need to extend (2.46) in order to obtain (2.48). Under the assumptions of part (b) let $x > 0$, and define $\tau = \inf\{t: X(t) > x\}$. Then τ is an $\{\mathscr{F}_{t+}\}$-stopping time by Proposition 1.5(b), and the right continuity of X implies $X(\tau) \geq x$ if $\tau < \infty$. Consequently for $T > 0$,

(2.49) $$\left\{\sup_{t \leq T} X(t) > x\right\} \subset \{\tau \leq T\} \subset \left\{\sup_{t \leq T} X(t) \geq x\right\},$$

and the three events have equal probability for all but countably many $x > 0$. By Theorem 2.13,

(2.50) $$E[X(\tau \wedge T)] \leq E[X(T)],$$

and hence

(2.51) $$xP\{\tau \leq T\} \leq E[X(\tau)\chi_{\{\tau \leq T\}}] \leq E[X(T)\chi_{\{\tau \leq T\}}].$$

Let φ be absolutely continuous on bounded intervals of $[0, \infty)$ with $\varphi' \geq 0$ and $\varphi(0) = 0$. Define $Z = \sup_{t \leq T} X(t)$. Then for $\beta > 0$,

(2.52) $$\begin{aligned} E[\varphi(Z \wedge \beta)] &= \int_0^\beta \varphi'(x)P\{Z > x\}\, dx \\ &\leq \int_0^\beta \varphi'(x)x^{-1}E[X(T)\chi_{\{Z \geq x\}}]\, dx \\ &= E[X(T)\psi(Z \wedge \beta)] \end{aligned}$$

where $\psi(z) = \int_0^z \varphi'(x)x^{-1}\, dx$.

If $\varphi(x) \equiv x^\alpha$ for some $\alpha > 1$, then

$$(2.53) \qquad E[(Z \wedge \beta)^\alpha] \le \frac{\alpha}{\alpha - 1} E[X(T)(Z \wedge \beta)^{\alpha-1}]$$

$$\le \frac{\alpha}{\alpha - 1} E[X(T)^\alpha]^{1/\alpha} E[(Z \wedge \beta)^\alpha]^{(\alpha-1)/\alpha},$$

and hence

$$(2.54) \qquad E[(Z \wedge \beta)^\alpha]^{1/\alpha} \le \frac{\alpha}{\alpha - 1} E[X(T)^\alpha]^{1/\alpha}.$$

Letting $\beta \to \infty$ gives (2.48). □

2.17 Corollary Let X be a right continuous martingale. Then for $x > 0$ and $T > 0$,

$$(2.55) \qquad P\left\{\sup_{t \le T} |X(t)| \ge x\right\} \le x^{-1} E[|X(T)|],$$

and for $\alpha > 1$ and $T > 0$,

$$(2.56) \qquad E\left[\sup_{t \le T} |X(t)|^\alpha\right] \le \left(\frac{\alpha}{\alpha - 1}\right)^\alpha E[|X(T)|^\alpha].$$

Proof. Since $|X|$ is a submartingale by Proposition 2.1, (2.55) and (2.56) follow directly from (2.46) and (2.48). □

3. LOCAL MARTINGALES

A real-valued process X is an $\{\mathscr{F}_t\}$-*local martingale* if there exist $\{\mathscr{F}_t\}$-stopping times $\tau_1 \le \tau_2 \le \cdots$ with $\tau_n \to \infty$ a.s. such that $X^{\tau_n} \equiv X(\cdot \wedge \tau_n)$ is an $\{\mathscr{F}_t\}$-martingale. Local submartingales and local supermartingales are defined similarly. Of course a martingale is a local martingale. In studying stochastic integrals (Chapter 5) and random time changes (Chapter 6), one is led naturally to local martingales that are not martingales.

3.1 Proposition If X is a right continuous $\{\mathscr{F}_t\}$-local martingale and τ is an $\{\mathscr{F}_t\}$-stopping time, then $X^\tau \equiv X(\cdot \wedge \tau)$ is an $\{\mathscr{F}_t\}$-local martingale.

Proof. There exist $\{\mathscr{F}_t\}$-stopping times $\tau_1 \le \tau_2 \le \cdots$ such that $\tau_n \to \infty$ a.s. and X^{τ_n} is an $\{\mathscr{F}_t\}$-martingale. But then $X^\tau(\cdot \wedge \tau_n) = X^{\tau_n}(\cdot \wedge \tau)$ is an $\{\mathscr{F}_t\}$-martingale, and hence X^τ is an $\{\mathscr{F}_t\}$-local martingale. □

In the next result the stochastic integral is just a Stieltjes integral and consequently needs no special definition. As before, when we say a process V is continuous and locally of bounded variation, we mean that for all $\omega \in \Omega$, $V(\cdot, \omega)$ is continuous and of bounded variation on bounded intervals.

3.2 Proposition Suppose X is a right continuous $\{\mathscr{F}_t\}$-local martingale, and V is real-valued, continuous, locally of bounded variation, and $\{\mathscr{F}_t\}$-adapted. Then

$$M(t) \equiv \int_0^t V(s)\, dX(s) = V(t)X(t) - V(0)X(0) - \int_0^t X(s)\, dV(s) \tag{3.1}$$

is an $\{\mathscr{F}_t\}$-local martingale.

Proof. The last equality in (3.1) is just integration by parts. There exist $\{\mathscr{F}_t\}$-stopping times $\tau_1 \le \tau_2 \le \cdots$ such that $\tau_n \to \infty$ a.s. and X^{τ_n} is an $\{\mathscr{F}_t\}$-martingale. Without loss of generality we may assume $\tau_n \le \tau_c((-\infty, -n] \cup [n, \infty))$, the first contact time of $(-\infty, -n] \cup [n, \infty)$ by X. (If not, replace τ_n by the minimum of the two stopping times.) Let R be the total variation process

$$R(t) = \sup \sum_{i=0}^{m-1} |V(s_{i+1}) - V(s_i)|, \tag{3.2}$$

where the supremum is over partitions of $[0, t]$, $0 = s_0 < s_1 < \cdots < s_m = t$. For $n = 1, 2, \ldots$ let $\gamma_n = \inf\{t: R(t) \ge n\}$. Since R is continuous, γ_n is the first contact time of $[n, \infty)$ and is an $\{\mathscr{F}_t\}$-stopping time by Proposition 1.5. The continuity of R also implies $\gamma_n \to \infty$ a.s.

Let $\sigma_n = \gamma_n \wedge \tau_n$. Then $\sigma_n \to \infty$ a.s. and we claim $M(\cdot \wedge \sigma_n)$ is an $\{\mathscr{F}_t\}$-martingale. To verify this we must show

$$E\left[\int_{t \wedge \sigma_n}^{(t+s) \wedge \sigma_n} V(u)\, dX(u) \,\middle|\, \mathscr{F}_t\right] = E\left[\int_t^{t+s} V^{\sigma_n}(u)\, dX^{\sigma_n}(u) \,\middle|\, \mathscr{F}_t\right] = 0, \tag{3.3}$$

for all $t, s \ge 0$.

Let $t = u_0 < u_1 < \cdots < u_m = t + s$. Then

$$E\left[\sum_{k=0}^{m-1} V^{\sigma_n}(u_k)(X^{\sigma_n}(u_{k+1}) - X^{\sigma_n}(u_k)) \,\middle|\, \mathscr{F}_t\right] = 0, \tag{3.4}$$

since X^{σ_n} is an $\{\mathscr{F}_t\}$-martingale and $V^{\sigma_n}(u_k)$ is $\mathscr{F}_{u_k}$-measurable. Letting $\max_k |u_{k+1} - u_k| \to 0$, the sum in (3.4) converges to the second integral in (3.3) a.s. However, to obtain (3.3), we must show that the convergence is in L^1. Observe

$$(3.5)\quad \left|\sum_{k=0}^{m-1} V^{\sigma_n}(u_k)(X^{\sigma_n}(u_{k+1}) - X^{\sigma_n}(u_k))\right|$$

$$= \left| V^{\sigma_n}(t+s)X^{\sigma_n}(t+s) - V^{\sigma_n}(0)X^{\sigma_n}(0) - \sum_{k=0}^{m-1} X^{\sigma_n}(u_{k+1})(V^{\sigma_n}(u_{k+1}) - V^{\sigma_n}(u_k))\right|$$

$$\leq |V^{\sigma_n}(t+s)X^{\sigma_n}(t+s) - V^{\sigma_n}(0)X^{\sigma_n}(0)| + \sum_{k=0}^{m-1} |X^{\sigma_n}(u_{k+1})|\,|V^{\sigma_n}(u_{k+1}) - V^{\sigma_n}(u_k)|$$

$$\leq |V^{\sigma_n}(t+s)X^{\sigma_n}(t+s) - V^{\sigma_n}(0)X^{\sigma_n}(0)| + (n \vee |X^{\sigma_n}(t+s)|)R(\sigma_n).$$

The right side is in L^1, so the desired convergence follows by the dominated convergence theorem. □

3.3 Corollary Let X and Y be real-valued, right continuous, $\{\mathscr{F}_t\}$-adapted processes. Suppose that for each t, $\inf_{s \leq t} X(s) > 0$. Then

$$(3.6)\quad M_1(t) \equiv X(t) - \int_0^t Y(s)\,ds$$

is an $\{\mathscr{F}_t\}$-local martingale if and only if

$$(3.7)\quad M_2(t) \equiv X(t) \exp\left\{-\int_0^t \frac{Y(s)}{X(s)}\,ds\right\}$$

is an $\{\mathscr{F}_t\}$-local martingale.

Proof. Suppose M_1 is an $\{\mathscr{F}_t\}$-local martingale. Then by Proposition 3.2,

$$(3.8)\quad \int_0^t \exp\left\{-\int_0^s \frac{Y(u)}{X(u)}\,du\right\} dM_1(s)$$

$$= \int_0^t \exp\left\{-\int_0^s \frac{Y(u)}{X(u)}\,du\right\} dX(s) - \int_0^t \exp\left\{-\int_0^s \frac{Y(u)}{X(u)}\,du\right\} Y(s)\,ds$$

$$= X(t) \exp\left\{-\int_0^t \frac{Y(u)}{X(u)}\,du\right\} - X(0)$$

is an $\{\mathscr{F}_t\}$-local martingale. Conversely, if M_2 is an $\{\mathscr{F}_t\}$-local martingale, then

(3.9)
$$\int_0^t \exp\left\{\int_0^s \frac{Y(u)}{X(u)}\,du\right\} dM_2(s) = X(t) - X(0) - \int_0^t Y(s)\,ds$$

is an $\{\mathscr{F}_t\}$-local martingale. □

We close this section with a result concerning the quadratic variation of the sample paths of a local martingale. By an "increasing" process, we mean a process whose sample paths are nondecreasing.

3.4 Proposition Let X be a right continuous $\{\mathscr{F}_t\}$-local martingale. Then there exists a right continuous increasing process, denoted by $[X]$, such that for each $t \ge 0$ and each sequence of partitions $\{u_k^{(n)}\}$ of $[0, t]$ with $\max_k(u_{k+1}^{(n)} - u_k^{(n)}) \to 0$,

(3.10)
$$\sum_k (X(u_{k+1}^{(n)}) - X(u_k^{(n)}))^2 \xrightarrow{P} [X](t)$$

as $n \to \infty$. If, in addition, X is a martingale and $E[X(t)^2] < \infty$ for all $t \ge 0$, then the convergence in (3.10) is in L^1.

Proof. Convergence in probability is metrizable (Problem 8); consequently we want to show that $\{\sum_k(X(u_{k+1}^{(n)}) - X(u_k^{(n)}))^2\}$ is a Cauchy sequence. If this were not the case, then there would exist $\varepsilon > 0$ and $\{n_i\}$ and $\{m_i\}$ such that $n_i \to \infty$, $m_i \to \infty$, and

(3.11)
$$P\left\{\left|\sum_k (X(u_{k+1}^{(n_i)}) - X(u_k^{(n_i)}))^2 - \sum_k (X(u_{k+1}^{(m_i)}) - X(u_k^{(m_i)}))^2\right| \ge \varepsilon\right\} \ge \varepsilon$$

for all i.

Since any pair of partitions of $[0, t]$ has a common refinement, that is, there exists $\{v_k\}$ such that $\{u_k^{(n_i)}\} \subset \{v_k\}$ and $\{u_k^{(m_i)}\} \subset \{v_k\}$, the following lemma contradicts (3.11) and hence proves that the left side of (3.10) converges in probability.

3.5 Lemma Let X be a right continuous $\{\mathscr{F}_t\}$-local martingale. Fix $T > 0$. For $n = 1, 2, \ldots$, let $\{u_k^{(n)}\}$ and $\{v_k^{(n)}\}$ be partitions of $[0, T]$ with $\{v_k^{(n)}\} \subset \{u_k^{(n)}\}$ and $\max_k(v_{k+1}^{(n)} - v_k^{(n)}) \to 0$. Then

(3.12)
$$\sum_k (X(v_{k+1}^{(n)}) - X(v_k^{(n)}))^2 - \sum_k (X(u_{k+1}^{(n)}) - X(u_k^{(n)}))^2 \xrightarrow{P} 0.$$

Proof. Without loss of generality we can assume X is a martingale (otherwise consider the stopped process X^{τ_n}), and $X(0) = 0$. Fix $M > 0$, and let $\tau = \inf\{s: |X(s)| \geq M \text{ or } |X(s-)| \geq M\}$. Note that $P\{\tau \leq t\} \leq E[|X(t)|]/M$ by Corollary 2.17.

Let $\{u_k\}$ and $\{v_k\}$ be partitions of $[0, t]$, and suppose $\{v_k\} \subset \{u_k\}$. Let $w_k = \max\{v_l: v_l \leq u_k\}$, and define $\eta_k \equiv X^\tau(u_k) - X^\tau(u_{k-1})$ and

$$Z = \sum (X^\tau(v_l) - X^\tau(v_{l-1}))^2 - \sum (X^\tau(u_k) - X^\tau(u_{k-1}))^2 = \sum \xi_k, \tag{3.13}$$

where $\xi_k = 2(X^\tau(u_k) - X^\tau(u_{k-1}))(X^\tau(u_{k-1}) - X^\tau(w_{k-1}))$. Note that either $\xi_k = 0$ or $|\xi_k| \leq 4M|X^\tau(u_k) - X^\tau(u_{k-1})|$ and that $E[\xi_{k+1} | \mathscr{F}_{u_k}] = 0$. Consequently,

$$Z_m = \sum_{k=1}^{m} \xi_k \tag{3.14}$$

is a discrete-parameter martingale.

Let

$$\varphi(x) = \begin{cases} x^2, & |x| \leq 4M^2, \\ 8M^2|x| - 16M^4, & |x| > 4M^2. \end{cases} \tag{3.15}$$

Let α be an $\{\mathscr{F}_t\}$-stopping time with values in $\{u_k\}$ and let β be $\mathscr{F}_\alpha$-measurable with values in $\{u_k\}$ and $\beta \geq \alpha$. Let k_α and k_β satisfy $\alpha = u_{k_\alpha}$ and $\beta = u_{k_\beta}$ and let $K = \max\{k: u_k < \tau\}$. Then

$$\begin{aligned} &E[\varphi(X^\tau(\beta)) | \mathscr{F}_\alpha] - \varphi(X^\tau(\alpha)) \\ &= \sum_{k=k_\alpha+1}^{k_\beta} E[\varphi(\eta_k + X^\tau(u_{k-1})) - \varphi(X^\tau(u_{k-1})) | \mathscr{F}_\alpha] \\ &= \sum_{k=k_\alpha+1}^{k_\beta} E[\varphi(\eta_k + X^\tau(u_{k-1})) - \varphi(X^\tau(u_{k-1})) - \eta_k \varphi'(X^\tau(u_{k-1})) | \mathscr{F}_\alpha] \\ &\geq E\left[\sum_{k=k_\alpha+1}^{k_\beta \wedge K} \eta_k^2 \,\middle|\, \mathscr{F}_\alpha\right], \end{aligned} \tag{3.16}$$

where the last inequality follows from the convexity of φ and the fact that for $k \leq K$, $|X^\tau(u_k)| \leq M$.

Using the fact that $\{Z_m\}$ is a discrete-parameter martingale,

$$\begin{aligned} E[\varphi(Z)] &= \sum E[\varphi(\xi_k + Z_{k-1}) - \varphi(Z_{k-1}) - \xi_k \varphi'(Z_{k-1})] \\ &= \sum E\left[\int_0^{\xi_k} \int_0^y \varphi''(z + Z_{k-1})\, dz\, dy\right] \\ &\leq \sum 2E\left[\int_0^{\xi_k} \int_0^{y/2} \varphi''(z)\, dz\, dy\right] \end{aligned} \tag{3.17}$$

$$= \sum 4E\left[\varphi\left(\frac{\xi_k}{2}\right)\right]$$

$$= 4E\left[\varphi\left(\frac{\xi_{K+1}}{2}\right)\right]$$

$$+ 4E\left[\sum_{k=1}^{K} \eta_k^2(X^\tau(u_{k-1}) - X^\tau(w_{k-1}))^2\right].$$

Fix $\varepsilon > 0$. Let $\alpha_l = \min\{u_k: |X^\tau(u_k) - X^\tau(v_l)| \geq \varepsilon\} \cup \{v_{l+1}\}$ and $\beta_l = v_{l+1}$. Note that if $v_l = w_{k-1} \leq u_{k-1} < v_{l+1}$ and $\alpha_l > v_{l+1}$, then $(X^\tau(u_{k-1}) - X^\tau(w_{k-1}))^2 \leq \varepsilon^2$. Consequently, by (3.16) and (3.17),

$$(3.18) \qquad E[\varphi(Z)] \leq 4E\left[\varphi\left(\frac{\xi_{K+1}}{2}\right)\right] + 4\varepsilon^2 E\left[\sum_{k=1}^{K} \eta_k^2\right] + \sum_l E[\chi_{\{\alpha_l < v_{l+1}\}} 16M^2(\varphi(X^\tau(\beta_l)) - \varphi(X^\tau(\alpha_l)))].$$

Fix $N \geq 1$, and let $L = \min\{l: \sum_{i=0}^{l} \chi_{\{\alpha_i < v_{i+1}\}} = N\}$. Let $\gamma = \alpha_L$ if $L < \infty$ and $\gamma = T$ otherwise. Then γ is an $\{\mathscr{F}_t\}$-stopping time, and hence by (3.18) and the convexity of φ,

$$(3.19) \qquad E[\varphi(Z)] \leq 4E\left[\varphi\left(\frac{\xi_{K+1}}{2}\right)\right] + 4\varepsilon^2 E[\varphi(X^\tau(T))] + 16M^2 E[\varphi(X^\tau(T)) - \varphi(X^\tau(\gamma))] + 16M^2 E\left[\sum_{l \leq L} \chi_{\{\alpha_l < v_{l+1}\}}(\varphi(X^\tau(\beta_l)) - \varphi(X^\tau(\alpha_l)))\right].$$

Given ε, $\varepsilon' > 0$, let $D = \{s \in [0, T]: |X(s) - X(s-)| > \varepsilon/2\}$. Then there exists a positive random variable δ such that $s \in D$ and $s \leq t \leq s + \delta$ imply $|X(t) - X(s)| \leq \varepsilon'$, and $0 \leq s < t \leq T$, $t - s \leq \delta$, and $|X(t) - X(s)| \geq \varepsilon$ imply $(s, t] \cap D \neq \varnothing$. Let $|D|$ denote the cardinality of D. On $\{\max(v_{i+1} - v_i) \leq \delta\}$,

$$(3.20) \qquad \sum_{l \leq L} \chi_{\{\alpha_l < v_{l+1}\}}(\varphi(X^\tau(\beta_l)) - \varphi(X^\tau(\alpha_l))) \leq (|D| \wedge N) 8M^2 \varepsilon'.$$

Let $S(T)$ be the collection of $\{\mathscr{F}_t\}$-stopping times α with $\alpha \leq T$. Since

$$(3.21) \qquad \varphi(X^\tau(\alpha)) \leq E[\varphi(X^\tau(T)) \mid \mathscr{F}_\alpha]$$

for all $\alpha \in S(T)$, $\{\varphi(X^\tau(\alpha)): \alpha \in S(T)\}$ is uniformly integrable. Consequently, the right side of (3.19) can be made arbitrarily small by taking ε small, N large (so that $P\{N < |D|\}$ is small), ε' small, and $\max(v_{i+1} - v_i)$ small. Note that if $N > |D|$ and $\max(v_{i+1} - v_i) < \delta$, then $\gamma = T$.

Thus, if $Z^{(n)}$ is defined for $\{v_k^{(n)}\}$ and $\{u_k^{(n)}\}$ as in (3.13), the estimate in (3.19) implies

$$\lim_{n \to \infty} E[\varphi(Z^{(n)})] = 0, \tag{3.22}$$

which, since M is arbitrary, implies (3.12). □

Proof of Proposition 3.4—*continued.* Assume X is a martingale and $E[X(T)^2] < \infty$. Let $\{u_k\}$ be a partition of $[0, T]$, and let X^τ be as in the proof of Lemma 3.5. Then

$$\begin{aligned} &E[|\textstyle\sum(X(u_{k+1}) - X(u_k))^2 - \sum(X^\tau(u_{k+1}) - X^\tau(u_k))^2|] \\ &\quad \leq E[(X(T) - X(T \wedge \tau))^2] + E[|(X(u_{K+1}) - X(\tau))(X(\tau) - X(u_K))|\chi_{\{\tau < T\}}], \end{aligned} \tag{3.23}$$

where $K = \max\{k: u_k < \tau\}$. Since for $M \geq 1$, φ defined by (3.15) satisfies $|x| \leq \varepsilon + \varphi(x)/\varepsilon$ for every $\varepsilon > 0$, the estimates in the proof of Lemma 3.5 imply $\{\sum(X^\tau(u_{k+1}^{(n)}) - X^\tau(u_k^{(n)}))^2\}$ is a Cauchy sequence in L^1 (note that we need $E[X(T)^2] < \infty$ in order that this sequence be in L^1). Consequently, since the right side of (3.23) can be made small by taking M large and $\max(u_{k+1} - u_k)$ small, it follows that $\{\sum(X(u_{k+1}^{(n)}) - X(u_k^{(n)}))^2\}$ is a Cauchy sequence in L^1.

Convergence of the left side of (3.10) determines $[X](t)$ a.s. for each $t \geq 0$. We must show that $[X]$ has a right continuous modification. Since $\{Z_m\}$ given by (3.14) is a discrete-parameter martingale, Proposition 2.16 gives

$$P\left\{\sup_m |Z_m| > \varepsilon\right\} \leq \frac{E[\varphi(Z)]}{\varphi(\varepsilon)}. \tag{3.24}$$

Consequently, for $l \geq n \to \infty$ and $T > 0$,

$$\begin{aligned} \sup_{k \leq 2^n T} \Bigg| &\sum_{i=1}^{k} \left(X\left(\frac{i}{2^n}\right) - X\left(\frac{i-1}{2^n}\right)\right)^2 \\ &- \sum_{j=1}^{k2^{l-n}} \left(X\left(\frac{j}{2^l}\right) - X\left(\frac{j-1}{2^l}\right)\right)^2 \Bigg| \xrightarrow{P} 0, \end{aligned} \tag{3.25}$$

and it follows that we can define $[X]$ on the dyadic rationals so that it is nondecreasing and satisfies

$$\sup_{k \leq 2^n T} \left| \sum_{i=1}^{k} \left(X\left(\frac{i}{2^n}\right) - X\left(\frac{i-1}{2^n}\right)\right)^2 - [X]\left(\frac{k}{2^n}\right) \right| \xrightarrow{P} 0. \tag{3.26}$$

The right continuity of $[X]$ on the dyadic rationals follows from the right continuity of X. For arbitrary $t \geq 0$, define

$$[X](t) = \lim_{n \to \infty} [X]\left(\frac{[2^n t] + 1}{2^n}\right). \tag{3.27}$$

Clearly this definition makes $[X]$ right continuous. We must verify that (3.10) is satisfied.

Let $\{u_k^n\} = \{i/2^n: 0 \le i \le [2^n t]\} \cup \{t\}$. Then

$$(3.28)\qquad \left|\sum_k (X(u_{k+1}^n) - X(u_k^n))^2 - [X](t)\right|$$
$$\le \left|[X]\left(\frac{[2^n t] + 1}{2^n}\right) - [X](t)\right|$$
$$+ \left|\left(X(t) - X\left(\frac{[2^n t]}{2^n}\right)\right)^2 - \left(X\left(\frac{[2^n t] + 1}{2^n}\right) - X\left(\frac{[2^n t]}{2^n}\right)\right)^2\right|$$
$$+ \left|\sum_{i=1}^{[2^n t]+1} \left(X\left(\frac{i}{2^n}\right) - X\left(\frac{i-1}{2^n}\right)\right)^2 - [X]\left(\frac{[2^n t] + 1}{2^n}\right)\right|$$
$$\xrightarrow{P} 0,$$

and (3.10) follows. □

3.6 Proposition Let X be a continuous $\{\mathscr{F}_t\}$-local martingale. Then $[X]$ can be taken to be continuous.

Proof. Let $[X]$ be as in Proposition 3.4. Almost sure continuity of $[X]$ restricted to the dyadic rationals follows from (3.26) and the continuity of X. Since $[X]$ is nondecreasing, it must therefore be almost surely continuous. □

4. THE PROJECTION THEOREM

Recall that an E-valued process X is $\{\mathscr{F}_t\}$-progressive if the restriction of X to $[0, t] \times \Omega$ is $\mathscr{B}[0, t] \times \mathscr{F}_t$-measurable for each $t \ge 0$, that is, if

$$(4.1)\qquad \{(s, \omega): X(s, \omega) \in \Gamma\} \cap ([0, t] \times \Omega) \in \mathscr{B}[0, t] \times \mathscr{F}_t$$

for each $t \ge 0$ and $\Gamma \in \mathscr{B}(E)$. Alternatively, we define the σ-algebra of *$\{\mathscr{F}_t\}$-progressive sets* $\mathscr{W}$ by

$$(4.2)\qquad \mathscr{W} = \{A \in \mathscr{B}[0, \infty) \times \mathscr{F}: A \cap ([0, t] \times \Omega) \in \mathscr{B}[0, t] \times \mathscr{F}_t \text{ for all } t \ge 0\}.$$

(The proof that $\mathscr{W}$ is a σ-algebra is the same as for $\mathscr{F}_\tau$ in Proposition 1.4.) Then (4.1) is just the requirement that X is a $\mathscr{W}$-measurable function on $[0, \infty) \times \Omega$.

The σ-algebra of *$\{\mathscr{F}_t\}$-optional sets* $\mathscr{O}$ is the σ-algebra of subsets of $[0, \infty) \times \Omega$ generated by the real-valued, right continuous $\{\mathscr{F}_t\}$-adapted processes. An E-valued process X is *$\{\mathscr{F}_t\}$-optional* if it is an $\mathscr{O}$-measurable function on $[0, \infty) \times \Omega$. Since every right continuous $\{\mathscr{F}_t\}$-adapted process is $\{\mathscr{F}_t\}$-progressive, $\mathscr{O} \subset \mathscr{W}$, and every $\{\mathscr{F}_t\}$-optional process is $\{\mathscr{F}_t\}$-progressive.

Throughout the remainder of this section we fix $\{\mathscr{F}_t\}$ and simply say adapted, optional, progressive, and so on to mean $\{\mathscr{F}_t\}$-adapted, and so on. In addition we assume that $\{\mathscr{F}_t\}$ is complete.

4.1 Lemma Every martingale has an optional modification.

Proof. By Proposition 2.9, every martingale has a modification X whose sample paths have right and left limits at every $t \in [0, \infty)$ and are right continuous at every t except possibly for t in a countable, deterministic set Γ. We show that X is optional. Since we are assuming $\{\mathscr{F}_t\}$ is complete, X is adapted. First define

$$Y_n(t) = X\left(\frac{k-1}{n}\right), \quad \frac{k}{n} \le t < \frac{k+1}{n}, \tag{4.3}$$

(set $X(-1/n) = X(0)$), and note that $\lim_{n\to\infty} Y_n(t) \equiv Y(t) = X(t-)$. Since Y_n is adapted and right continuous, Y is optional.

Fix $\varepsilon > 0$. Define $\tau_0 = 0$ and, for $n = 0, 1, 2, \ldots,$

$$\tau_{n+1} = \inf\{s > \tau_n : |X(s) - X(s-)| > \varepsilon \quad \text{or} \quad |X(s+) - X(s-)| > \varepsilon \quad \text{or} \quad |X(s+) - X(s)| > \varepsilon\}. \tag{4.4}$$

Since $X(s+) = X(s)$ except for $s \in \Gamma$,

$$\{\tau_n < t\} = \bigcup_l \bigcap_m \bigcup_{\{s_i, t_i\}} \bigcap_{i=1}^{n} \left\{|X(t_i) - X(s_i)| > \varepsilon + \frac{1}{l}\right\} \in \mathscr{F}_t \tag{4.5}$$

where $\{s_i, t_i\}$ ranges over all sets of the form $0 \le s_1 < t_1 < s_2 < t_2 < \cdots < s_n < t_n < t$, $|t_i - s_i| < 1/m$, and $t_i, s_i \in \Gamma \cup \mathbb{Q}$. Define

$$\begin{aligned} Z_m^\varepsilon(t) &= \sum_{n=1}^{\infty} \chi_{[\tau_n, \tau_n + 1/m)}(t)\chi_{\{|X(\tau_n) - X(\tau_n-)| > \varepsilon\}}(X(\tau_n) - X(\tau_n-)) \\ &= \sum_{n=1}^{\infty} \chi_{\{\tau_n < t\}}\chi_{[\tau_n, \tau_n + 1/m)}(t)\chi_{\{|X(\tau_n) - X(\tau_n-)| > \varepsilon\}}(X(\tau_n) - X(\tau_n-)) \\ &\quad + \chi_{\{|X(t) - X(t-)| > \varepsilon\}}(X(t) - X(t-)). \end{aligned} \tag{4.6}$$

Since X has right and left limits at each $t \in [0, \infty)$, $\lim_{n\to\infty} \tau_n = \infty$, and hence Z_m^ε is right continuous and has left limits. By (4.5), $\{\tau_n < s\} \in \mathscr{F}_t$ for $s \le t$, and an examination of the right side of (4.6) shows that $Z_m^\varepsilon(t)$ is $\mathscr{F}_t$-measurable. Therefore Z_m^ε is optional. Finally observe that $|Y(t) + \lim_{m\to\infty} Z_m^\varepsilon(t) - X(t)| \le \varepsilon$, and since ε is arbitrary, X is optional. □

4.2 Theorem Let X be a nonnegative real-valued, measurable process. Then there exists a $[0, \infty]$-valued optional process Y such that

$$E[X(\tau) | \mathscr{F}_\tau] = Y(\tau) \tag{4.7}$$

for all stopping times τ with $P\{\tau < \infty\} = 1$. (Note that we allow both sides of (4.7) to be infinite.)

4.3 Remark Y is called the *optional projection* of X. This theorem implies a partial converse to the observation that an optional process is progressive. Every real-valued, progressive process has an optional modification. The optional process Y is unique in the sense that, if Y_1 and Y_2 are optional processes satisfying (4.7), then Y_1 and Y_2 are indistinguishable. (See Dellacherie and Meyer (1982), page 103.) □

Proof. Let $A \in \mathscr{F}$ and $B \in \mathscr{B}[0, \infty)$, and let Z be an optional process satisfying $E[\chi_A | \mathscr{F}_t] = Z(t)$. Z exists, since $E[\chi_A | \mathscr{F}_t]$ is a martingale. The optional sampling theorem implies $E[\chi_A | \mathscr{F}_\tau] = Z(\tau)$. Consequently, $\chi_B(t)Z(t)$ is optional, and

$$E[\chi_B(\tau)\chi_A | \mathscr{F}_\tau] = \chi_B(\tau)Z(\tau). \tag{4.8}$$

Therefore the collection M of bounded nonnegative measurable processes X for which there exists an optional Y satisfying (4.7) contains processes of the form $\chi_B \chi_A$, $B \in \mathscr{B}[0, \infty)$, $A \in \mathscr{F}$. Since M is closed under nondecreasing limits, and $X_1, X_2 \in M$, $X_1 \geq X_2$ implies $X_1 - X_2 \in M$, the Dynkin class theorem implies M contains all indicators of sets in $\mathscr{B}[0, \infty) \times \mathscr{F}$, and hence all bounded nonnegative measurable processes. The general case is proved by approximating X by $X \wedge n$, $n = 1, 2, \dots$. □

4.4 Corollary Let X be a nonnegative real-valued, measurable process. Then there exists $Y: [0, \infty) \times [0, \infty) \times \Omega \to [0, \infty]$, measurable with respect to $\mathscr{B}[0, \infty) \times \mathscr{O}$, such that

$$E[X(\tau + s) | \mathscr{F}_\tau] = Y(s, \tau) \tag{4.9}$$

for all a.s. finite stopping times τ and all $s \geq 0$.

Proof. Replace $\chi_B(t)$ by $\chi_B(t + s)$ in the proof above. □

4.5 Corollary Let $X: E \times [0, \infty) \times \Omega \to [0, \infty)$ be $\mathscr{B}(E) \times \mathscr{B}[0, \infty) \times \mathscr{F}$-measurable. Then there exists $Y: E \times [0, \infty) \times \Omega \to [0, \infty]$, measurable with respect to $\mathscr{B}(E) \times \mathscr{O}$, such that

$$E[X(x, \tau) | \mathscr{F}_\tau] = Y(x, \tau) \tag{4.10}$$

for all a.s. finite stopping times τ and all $x \in E$.

Proof. Replace $\chi_B(t)$ by $\chi_B(x, t)$, $B \in \mathscr{B}(E) \times \mathscr{B}[0, \infty)$, in the proof of Theorem 4.2. □

The argument used in the proof of Theorem 4.2 also gives us a Fubini theorem for conditional expectations.

4.6 Proposition Let $X: E \times \Omega \to \mathbb{R}$ be $\mathscr{B}(E) \times \mathscr{F}$-measurable, and let μ be a σ-finite measure on $\mathscr{B}(E)$. Suppose $\int E[|X(x)|]\mu(dx) < \infty$. Then for every σ-algebra $\mathscr{D} \subset \mathscr{F}$, there exists $Y: E \times \Omega \to \mathbb{R}$ such that Y is $\mathscr{B}(E) \times \mathscr{D}$-measurable, $Y(x) = E[X(x) | \mathscr{D}]$ for all $x \in E$, $\int | Y(x) | \mu(dx) < \infty$ a.s., and

$$\int Y(x)\mu(dx) = E\left[\int X(x)\mu(dx) \middle| \mathscr{D}\right]. \tag{4.11}$$

4.7 Remark With this proposition in mind, we do not hesitate to write

$$\int E[X(x) | \mathscr{D}]\mu(dx) = E\left[\int X(x)\mu(dx) \middle| \mathscr{D}\right]. \tag{4.12}$$ □

Proof. First assume μ is finite, verify the result for $X = \chi_B \chi_A$, $B \in \mathscr{B}(E)$, $A \in \mathscr{F}$, and then apply the Dynkin class theorem. The σ-finite case follows by writing μ as a sum of finite measures. □

5. THE DOOB–MEYER DECOMPOSITION

Let S denote the collection of all $\{\mathscr{F}_t\}$-stopping times. A right continuous $\{\mathscr{F}_t\}$-submartingale is of *class DL* if for each $T > 0$, $\{X(\tau \wedge T): \tau \in S\}$ is uniformly integrable. If X is an $\{\mathscr{F}_t\}$-martingale or if X is bounded below, then X is of class DL (Problem 10).

A process A is *increasing* if $A(\cdot, \omega)$ is nondecreasing for all $\omega \in \Omega$. Every right continuous nondecreasing function a on $[0, \infty)$ with $a(0) = 0$ determines a Borel measure μ_a on $[0, \infty)$ by $\mu_a[0, t] = a(t)$. We define

$$\int_0^t f(s)\, da(s) = \int_{[0, t]} f(s)\mu_a(ds) \tag{5.1}$$

when the integral on the right exists. Note that this is not a Stieltjes integral if f and a have common discontinuities.

5.1 Theorem Let $\{\mathscr{F}_t\}$ be complete and right continuous, and let X be a right continuous $\{\mathscr{F}_t\}$-submartingale of class DL. Then there exists a unique (up to indistinguishability) right continuous $\{\mathscr{F}_t\}$-adapted increasing process A with $A(0) = 0$ and the following properties:

(a) $M = X - A$ is an $\{\mathscr{F}_t\}$-martingale.

(b) For every nonnegative right continuous $\{\mathscr{F}_t\}$-martingale Y and every $t \geq 0$ and $\tau \in S$,

$$E\left[\int_0^{t\wedge\tau} Y(s-)\,dA(s)\right] = E\left[\int_0^{t\wedge\tau} Y(s)\,dA(s)\right] = E[Y(t\wedge\tau)A(t\wedge\tau)]. \tag{5.2}$$

5.2 Remark **(a)** We allow the possibility that all three terms in (5.2) are infinite. If (5.2) holds for all bounded nonnegative right continuous $\{\mathscr{F}_t\}$-martingales Y, then it holds for all nonnegative $\{\mathscr{F}_t\}$-martingales, since on every bounded interval $[0, T]$ a nonnegative martingale Y is the limit of an increasing sequence $\{Y_n\}$ of bounded nonnegative martingales (e.g., take Y_n to be a right continuous modification of $Y_n^0(t) = E[Y(T) \wedge n \mid \mathscr{F}_t]$).

(b) If A is continuous, then the first equality in (5.2) is immediate. The second equality always holds, since (assuming Y is bounded) by the right continuity of Y

(5.3)

$$\begin{aligned}
E\left[\int_0^{t\wedge\tau} Y(s)\,dA(s)\right] &= \lim_{n\to\infty} E\left[\int_0^{t\wedge\tau} Y\left(\frac{[ns]+1}{n}\right) dA(s)\right] \\
&= \lim_{n\to\infty} E\left[\sum_k \chi_{\{t\wedge\tau\geq(k-1)/n\}} Y\left(\frac{k}{n}\right)\left(A\left(\frac{k}{n}-\right) - A\left(\frac{k-1}{n}-\right)\right)\right] \\
&= \lim_{n\to\infty} E\left[\sum_k \chi_{\{t\wedge\tau\geq(k-1)/n\}} Y(t\wedge\tau + n^{-1})\left(A\left(\frac{k}{n}-\right) - A\left(\frac{k-1}{n}-\right)\right)\right] \\
&= \lim_{n\to\infty} E\left[Y(t\wedge\tau+n^{-1})A\left(\frac{[n(t\wedge\tau)]+1}{n}-\right)\right] \\
&= E[Y(t\wedge\tau)A(t\wedge\tau)].
\end{aligned}$$

The third equality in (5.3) follows from the fact that Y is a martingale.

(c) Property (b) is usually replaced by the requirement that A be *predictable*, but we do not need this concept elsewhere and hence do not introduce it here. See Dellacherie and Meyer (1982), page 194, or Elliot (1982), Theorem 8.15. □

Proof. For each $\varepsilon > 0$, let X_ε be the optional projection of $\varepsilon^{-1}\int_0^\varepsilon X(\cdot + s)ds$,

$$X_\varepsilon(t) = E\left[\varepsilon^{-1}\int_0^\varepsilon X(t+s)\,ds \,\middle|\, \mathscr{F}_t\right]. \tag{5.4}$$

Then X_ε is a submartingale and

$$\lim_{\varepsilon\to 0} E[|X_\varepsilon(t) - X(t)|] = 0, \qquad t \geq 0. \tag{5.5}$$

Let Y_ε be the optional projection of $\varepsilon^{-1}(X(\cdot + \varepsilon) - X(\cdot))$, and define

$$A_\varepsilon(t) = \int_0^t Y_\varepsilon(s)\, ds. \tag{5.6}$$

Since X is a submartingale,

$$Y_\varepsilon(t) = \varepsilon^{-1}E[X(t + \varepsilon) - X(t) \mid \mathscr{F}_t] \geq 0, \tag{5.7}$$

and hence A_ε is an increasing process. Furthermore

$$M_\varepsilon = X_\varepsilon - A_\varepsilon \tag{5.8}$$

is a martingale, since for $t, u \geq 0$ and $B \in \mathscr{F}_t$,

$$\begin{aligned} \int_B (M_\varepsilon(t + u) - M_\varepsilon(t))\, dP \\ = \int_B \biggl(\varepsilon^{-1} \int_0^\varepsilon X(t + u + s)\, ds - \varepsilon^{-1} \int_0^\varepsilon X(t + s)\, ds \\ - \int_t^{t+u} \varepsilon^{-1}(X(s + \varepsilon) - X(s))\, ds\biggr)\, dP = 0. \end{aligned} \tag{5.9}$$

We next observe that $\{A_\varepsilon(t)\colon 0 < \varepsilon \leq 1\}$ is uniformly integrable for each $t \geq 0$. To see this, let $\tau_\lambda^\varepsilon = \inf\{s\colon A_\varepsilon(s) \geq \lambda\}$. Then

$$\begin{aligned} E[A_\varepsilon(t) - \lambda \wedge A_\varepsilon(t)] &= E[A_\varepsilon(t) - A_\varepsilon(\tau_\lambda^\varepsilon \wedge t)] \\ &= E[X_\varepsilon(t) - X_\varepsilon(\tau_\lambda^\varepsilon \wedge t)] \\ &= E[\chi_{\{\tau_\lambda^\varepsilon < t\}}(X_\varepsilon(t) - X_\varepsilon(\tau_\lambda^\varepsilon \wedge t))] \\ &= \varepsilon^{-1} \int_0^\varepsilon E[\chi_{\{\tau_\lambda^\varepsilon < t\}}(X(t + s) - X(\tau_\lambda^\varepsilon \wedge t + s))]\, ds. \end{aligned} \tag{5.10}$$

Since $P\{\tau_\lambda^\varepsilon < t\} \leq \lambda^{-1}E[A_\varepsilon(t)] \leq \lambda^{-1}E[X(t + \varepsilon) - X(0)]$, the uniform integrability of $\{X(\tau \wedge (t + 1))\colon \tau \in S\}$ implies the right side of (5.10) goes to zero as $\lambda \to \infty$ uniformly in $0 < \varepsilon \leq 1$. Consequently $\{A_\varepsilon(t)\colon 0 < \varepsilon \leq 1\}$ is uniformly integrable (Appendix 2). For each $t \geq 0$, this uniform integrability implies the existence of a sequence $\{\varepsilon_n\}$ with $\varepsilon_n \to 0$, and a random variable $A(t)$ on $(\Omega, \mathscr{F})$ such that

$$\lim_{n \to \infty} E[A_{\varepsilon_n}(t)\chi_B] = E[A(t)\chi_B], \tag{5.11}$$

for every $B \in \mathscr{F}$ (Appendix 2). By a diagonalization argument we may assume the same sequence $\{\varepsilon_n\}$ works for all $t \in \mathbb{Q} \cap [0, \infty)$.

Let $0 \leq s < t$, $s, t \in \mathbb{Q}$, and $B = \{A(t) < A(s)\}$. Then

$$E[(A(t) - A(s))\chi_B] = \lim_{n \to \infty} E[(A_{\varepsilon_n}(t) - A_{\varepsilon_n}(s))\chi_B] \geq 0, \tag{5.12}$$

so $A(s) \le A(t)$ a.s. For $s, t \ge 0$, $s, t \in \mathbb{Q}$, and $B \in \mathscr{F}_t$,

$$\begin{aligned}(5.13)\qquad E[(X(t+s) - A(t+s) - X(t) + A(t))\chi_B] \\ = \lim_{n\to\infty} E[(M_{\varepsilon_n}(t+s) - M_{\varepsilon_n}(t))\chi_B] = 0,\end{aligned}$$

and defining $M(t) = X(t) - A(t)$ for $t \in \mathbb{Q} \cap [0, \infty)$, we have $E[M(t+s) \mid \mathscr{F}_t] = M(t)$ for all $s, t \in \mathbb{Q} \cap [0, \infty)$. By the right continuity of $\{\mathscr{F}_t\}$ and Corollary 2.11, M extends to a right continuous $\{\mathscr{F}_t\}$-martingale, and it follows that A has a right continuous increasing modification.

To see that (5.2) holds, let Y be a bounded right continuous $\{\mathscr{F}_t\}$-martingale. Then for $t \ge 0$,

$$\begin{aligned}(5.14)\qquad E[Y(t)A(t)] &= \lim_{n\to\infty} E[Y(t)A_{\varepsilon_n}(t)] \\ &= \lim_{n\to\infty} E\left[\int_0^t Y(s-)\, dA_{\varepsilon_n}(s)\right] \\ &= \lim_{n\to\infty} E\left[\int_0^t Y(s-)\varepsilon_n^{-1} E[X(s+\varepsilon_n) - X(s) \mid \mathscr{F}_s]\, ds\right] \\ &= \lim_{n\to\infty} E\left[\int_0^t Y(s-)\varepsilon_n^{-1}(A(s+\varepsilon_n) - A(s))\, ds\right] \\ &= \lim_{n\to\infty} E\left[\int_0^{t+\varepsilon_n}\int_0^t Y(s-)\varepsilon_n^{-1}\chi_{(s, s+\varepsilon_n]}(u)\, ds\, dA(u)\right] \\ &= E\left[\int_0^t Y(u-)\, dA(u)\right],\end{aligned}$$

and the same argument works with t replaced by $t \wedge \tau$.

Finally, to obtain the uniqueness, suppose A_1 and A_2 are processes with the desired properties. Then $A_1 - A_2$ is a martingale, and by Problem 15, if Y is a bounded, right continuous martingale,

$$\begin{aligned}(5.15)\qquad E[Y(t)A_1(t)] &= E\left[\int_0^t Y(s-)\, dA_1(s)\right] \\ &= E\left[\int_0^t Y(s-)\, dA_2(s)\right] \\ &= E[Y(t)A_2(t)].\end{aligned}$$

Let $B = \{A_1(t) > A_2(t)\}$ and $Y(s) = E[\chi_B \mid \mathscr{F}_s]$ (by Corollary 2.11, Y can be taken to be right continuous). Then (5.15) implies

$$(5.16)\qquad E[(A_1(t) - A_2(t))\chi_B] = 0.$$

Similarly take $B = \{A_2(t) > A_1(t)\}$ and it follows that $A_1(t) = A_2(t)$ a.s. for each $t \ge 0$. The fact that A_1 and A_2 are indistinguishable follows from the right continuity. □

5.3 Corollary If, in addition to the assumptions of Theorem 5.1, X is continuous, then A can be taken to be continuous.

Proof. Let A be as in Theorem 5.1. Let $\alpha > 0$ and $\tau = \inf\{t: X(t) - A(t) \notin [-\alpha, \alpha]\}$, and define $Y = A(\cdot \wedge \tau) - X(\cdot \wedge \tau) + \alpha$. Since X is continuous, $Y \geq 0$, and hence by (5.2),

$$E\left[\int_0^{t \wedge \tau} Y(s-)\, dA(s)\right] = E\left[\int_0^{t \wedge \tau} Y(s)\, dA(s)\right], \qquad t \geq 0. \tag{5.17}$$

For $0 \leq s \leq \tau$, $Y(s-) \leq \alpha$, and hence (5.17) is finite, and

$$\begin{aligned} E\left[\int_0^{t \wedge \tau} (Y(s) - Y(s-))\, dA(s)\right] \\ = E\left[\int_0^{t \wedge \tau} (A(s) - A(s-))\, dA(s)\right] = 0, \qquad t \geq 0. \end{aligned} \tag{5.18}$$

Since α is arbitrary, it follows that A is almost surely continuous. □

5.4 Corollary Let X be a right continuous, $\{\mathscr{F}_t\}$-local submartingale. Then there exists a right continuous, $\{\mathscr{F}_t\}$-adapted, increasing process A satisfying Property (b) of Theorem 5.1 such that $M \equiv X - A$ is an $\{\mathscr{F}_t\}$-local martingale.

Proof. Let $\tau_1 \leq \tau_2 \leq \cdots$ be stopping times such that $\tau_n \to \infty$ and X^{τ_n} is a submartingale, and let $\gamma_n = \inf\{t: X(t) \leq -n\}$. Then $X^{\tau_n \wedge \gamma_n}$ is a submartingale of class DL, since for any $\{\mathscr{F}_t\}$-stopping time τ,

$$X^{\tau_n \wedge \gamma_n}(T) \wedge (-n) \leq X^{\tau_n \wedge \gamma_n}(T \wedge \tau) \leq E[X^{\tau_n \wedge \gamma_n}(T) \mid \mathscr{F}_\tau]. \tag{5.19}$$

Let A_n be the increasing process for $X^{\tau_n \wedge \gamma_n}$ given by Theorem 5.1. Then $A = \lim_{n\to\infty} A_n$. □

6. SQUARE INTEGRABLE MARTINGALES

Fix a filtration $\{\mathscr{F}_t\}$, and assume $\{\mathscr{F}_t\}$ is complete and right continuous. In this section all martingales, local martingales, and so on are $\{\mathscr{F}_t\}$-martingales, $\{\mathscr{F}_t\}$-local martingales, and so on.

A martingale M is *square integrable* if $E[|M(t)|^2] < \infty$ for all $t \geq 0$. A right continuous process M is a *local square integrable* martingale if there exist stopping times $\tau_1 \leq \tau_2 \leq \cdots$ such that $\tau_n \to \infty$ a.s. and for each $n \geq 1$, $M^{\tau_n} \equiv M(\cdot \wedge \tau_n)$ is a square integrable martingale. Let $\mathscr{M}$ denote the collection of right continuous, square integrable martingales, and let $\mathscr{M}_{\text{loc}}$ denote the collection of right continuous local square integrable martingales. We also need to define $\mathscr{M}_c$, the collection of continuous square integrable martingales, and $\mathscr{M}_{c,\,\text{loc}}$, the collection of continuous local martingales. (Note that a continuous local martingale is necessarily a local square integrable martingale.)

Each of these collections is a linear space. Let τ be a stopping time. If $M \in \mathscr{M}(\mathscr{M}_{\text{loc}}, \mathscr{M}_c, \mathscr{M}_{c,\,\text{loc}})$, then clearly $M^\tau \equiv M(\cdot \wedge \tau) \in \mathscr{M}$ $(\mathscr{M}_{\text{loc}}, \mathscr{M}_c, \mathscr{M}_{c,\,\text{loc}})$.

6.1 Proposition If $M \in \mathscr{M}(\mathscr{M}_{\text{loc}})$, then $M^2 - [M]$ is a martingale (local martingale).

Proof. Let $M \in \mathscr{M}$. Since for $t, s \geq 0$ and $t = u_0 < u_1 < \cdots < u_m = t + s$,

$$E[M^2(t+s) - M^2(t) \mid \mathscr{F}_t] = E[(M(t+s) - M(t))^2 \mid \mathscr{F}_t] = E\left[\sum_{k=0}^{m-1} (M(u_{k+1}) - M(u_k))^2 \,\middle|\, \mathscr{F}_t\right], \tag{6.1}$$

the result follows by Proposition 3.4. The extension to local martingales is immediate. □

If $M \in \mathscr{M}(\mathscr{M}_{\text{loc}})$, then M^2 satisfies the conditions of Theorem 5.1 (Corollary 5.4). Let $\langle M \rangle$ be the increasing process given by the theorem (corollary) with $X = M^2$. Then $M^2 - \langle M \rangle$ is a martingale (local martingale). If $M \in \mathscr{M}_c(\mathscr{M}_{c,\,\text{loc}})$, then by Proposition 3.6, $[M]$ is continuous, and Proposition 6.1 implies $[M]$ has the properties required for A in Theorem 5.1 (Corollary 5.4). Consequently, by uniqueness, $[M] = \langle M \rangle$ (up to indistinguishability).

For $M, N \in \mathscr{M}_{\text{loc}}$ we define

$$[M, N] = \tfrac{1}{2}([M+N, M+N] - [M, M] - [N, N]) \tag{6.2}$$

and

$$\langle M, N \rangle = \tfrac{1}{2}(\langle M+N, M+N \rangle - \langle M, M \rangle - \langle N, N \rangle). \tag{6.3}$$

Of course, $[M, N]$ is the *cross variation* of M and N, that is (cf. (3.10)),

$$[M, N](t) = \lim_{n \to \infty} \sum_k (M(u_{k+1}^{(n)}) - M(u_k^{(n)}))(N(u_{k+1}^{(n)}) - N(u_k^{(n)})) \tag{6.4}$$

in probability. Note that $[M, M] = [M]$ and $\langle M, M \rangle = \langle M \rangle$. The following proposition indicates the interest in these quantities.

6.2 Proposition If $M, N \in \mathscr{M}$ $(\mathscr{M}_{\text{loc}})$, then $MN - [M, N]$ and $MN - \langle M, N \rangle$ are martingales (local martingales).

Proof. Observe that

$$MN - [M, N] = \tfrac{1}{2}((M + N)^2 - [M + N, M + N] - (M^2 - [M]) - (N^2 - [N])), \tag{6.5}$$

and similarly for $MN - \langle M, N \rangle$. □

If $\langle M, N \rangle = 0$, then M and N are said to be *orthogonal.* Note that $\langle M, N \rangle = 0$ implies MN and $[M, N]$ are martingales (local martingales).

7. SEMIGROUPS OF CONDITIONED SHIFTS

Let $\{\mathscr{F}_t\}$ be a complete filtration. Again all martingales, stopping times, and so on are $\{\mathscr{F}_t\}$-martingales, $\{\mathscr{F}_t\}$-stopping times, and so on. Let $\mathscr{L}$ be the space of progressive (i.e., $\{\mathscr{F}_t\}$-progressive) processes Y such that $\sup_t E[|Y(t)|] < \infty$. Defining

$$\| Y \| = \sup_t E[|Y(t)|] \tag{7.1}$$

and $\mathscr{N} = \{Y \in \mathscr{L}: \| Y \| = 0\}$, then $\mathscr{L}/\mathscr{N}$ (the quotient space) is a Banach space with norm $\| \cdot \|$ satisfying the conditions of Chapter 1, Section 5, that is, (7.1) is of the form (5.1) of Chapter 1 ($\Gamma = \{\delta_t \times P: t \in [0, \infty)\}$). Since there is little chance of confusion, we do not distinguish between $\mathscr{L}$ and $\mathscr{L}/\mathscr{N}$.

We define a semigroup of operators $\{\mathscr{T}(s)\}$ on $\mathscr{L}$ by

$$\mathscr{T}(s)Y(t) = E[Y(t + s) | \mathscr{F}_t]. \tag{7.2}$$

By Corollary 4.4, we can assume $(s, t, \omega) \to \mathscr{T}(s)Y(t, \omega)$ is $\mathscr{B}[0, \infty) \times \mathscr{O}$-measurable. The semigroup property follows by

$$\begin{aligned} \mathscr{T}(u)\mathscr{T}(s)Y(t) &= E[E[Y(t + u + s) | \mathscr{F}_{t+u}] | \mathscr{F}_t] \\ &= E[Y(t + u + s) | \mathscr{F}_t] \\ &= \mathscr{T}(u + s)Y(t). \end{aligned} \tag{7.3}$$

Since

$$\sup_t E[|\mathscr{T}(s)Y(t)|] \le \sup_t E[|Y(t)|], \tag{7.4}$$

$\{\mathscr{T}(s)\}$ is a measurable contraction semigroup on $\mathscr{L}$.

Integrals of the form $W = \int_a^b f(u)\mathscr{T}(u)Z\,du$ are well defined for Borel measurable f with $\int_a^b |f(u)|\,du < \infty$ and $Z \in \mathscr{L}$ by (5.4) of Chapter 1,

$$W(t) = E\left[\int_a^b f(u)Z(t + u)\,du \,\middle|\, \mathscr{F}_t\right], \tag{7.5}$$

and

$$\left\| \int_a^b f(u)\mathscr{T}(u)Z \, du \right\| \leq \int_a^b |f(u)| \, \| \mathscr{T}(u)Z \| \, du \leq \| Z \| \int_a^b |f(u)| \, du. \tag{7.6}$$

Define

$$\hat{\mathscr{A}} = \left\{(Y, Z) \in \mathscr{L} \times \mathscr{L} \colon Y(t) - \int_0^t Z(s) \, ds \quad \text{is a martingale}\right\}. \tag{7.7}$$

Since $(Y, Z) \in \hat{\mathscr{A}}$ if and only if

$$\mathscr{T}(s)Y = Y + \int_0^s \mathscr{T}(u)Z \, du, \qquad s \geq 0, \tag{7.8}$$

$\hat{\mathscr{A}}$ is the full generator for $\{\mathscr{T}(s)\}$ as defined in Chapter 1, Section 5. Note that the "harmonic functions", that is, the solutions of $\hat{\mathscr{A}}Y = 0$, are the martingales in $\mathscr{L}$.

7.1 Theorem The operator $\hat{\mathscr{A}}$ defined in (7.7) is a dissipative linear operator with $\mathscr{R}(\lambda - \hat{\mathscr{A}}) = \mathscr{L}$ for all $\lambda > 0$ and resolvent

$$(\lambda - \hat{\mathscr{A}})^{-1}W = \int_0^\infty e^{-\lambda s}\mathscr{T}(s)W \, ds. \tag{7.9}$$

The largest closed subspace $\mathscr{L}_0$ of $\mathscr{L}$ on which $\{\mathscr{T}(s)\}$ is strongly continuous is the closure of $\mathscr{D}(\hat{\mathscr{A}})$, and for each $Y \in \mathscr{L}_0$ and $s \geq 0$,

$$\mathscr{T}(s)Y = \lim_{n \to \infty} \left(I - \frac{s}{n}\hat{\mathscr{A}}\right)^{-n} Y. \tag{7.10}$$

Proof. (Cf. the proof of Proposition 5.1 of Chapter 1.) Suppose $(Y, Z) \in \hat{\mathscr{A}}$. Then

$$\begin{aligned} &\int_0^\infty e^{-\lambda s}\mathscr{T}(s)(\lambda Y - Z)(t) \, ds \\ &= \int_0^\infty e^{-\lambda s}E[\lambda Y(t+s) - Z(t+s) \mid \mathscr{F}_t] \, ds \\ &= \int_0^\infty e^{-\lambda s}E\left[\lambda Y(t) + \lambda \int_0^s Z(t+u) \, du - Z(t+s) \,\middle|\, \mathscr{F}_t\right] ds \\ &= Y(t) + E\left[\int_0^\infty e^{-\lambda s}\lambda \int_0^s Z(t+u) \, du \, ds \,\middle|\, \mathscr{F}_t\right] \\ &\quad - E\left[\int_0^\infty e^{-\lambda s}Z(t+s) \, ds \,\middle|\, \mathscr{F}_t\right] = Y(t). \end{aligned} \tag{7.11}$$

The last equality follows by interchanging the order of integration in the second term. This identity implies (7.9), which since $\mathscr{T}(s)$ is a contraction, implies $\hat{\mathscr{A}}$ is dissipative.

To see that $\mathscr{R}(\lambda - \hat{\mathscr{A}}) = \mathscr{L}$, let $W \in \mathscr{L}$, $Y = \int_0^\infty e^{-\lambda s}\mathscr{T}(s)W\,ds$, and $Z = \lambda Y - W$. An interchange in the order of integration gives the identity

$$\begin{aligned}(7.12)\qquad \mathscr{T}(r)Y &= \int_0^\infty \lambda e^{-\lambda s}\int_0^s \mathscr{T}(r+u)W\,du\,ds\\ &= \int_0^\infty \lambda e^{-\lambda s}\int_r^{r+s} \mathscr{T}(u)W\,du\,ds,\end{aligned}$$

and we have

$$\begin{aligned}(7.13)\qquad \int_0^r \mathscr{T}(u)Z\,du &= \int_0^\infty \lambda e^{-\lambda s}\int_0^r \mathscr{T}(s+u)W\,du\,ds - \int_0^r \mathscr{T}(u)W\,du\\ &= \int_0^\infty \lambda e^{-\lambda s}\int_s^{r+s} \mathscr{T}(u)W\,du\,ds - \int_0^r \mathscr{T}(u)W\,du.\end{aligned}$$

Subtracting (7.13) from (7.12) gives

$$\begin{aligned}(7.14)\qquad \mathscr{T}(r)Y - \int_0^r \mathscr{T}(u)Z\,du &= \int_r^\infty \lambda e^{-\lambda s}\int_r^s \mathscr{T}(u)W\,du\,ds\\ &\quad - \int_0^r \lambda e^{-\lambda s}\int_s^r \mathscr{T}(u)W\,du\,ds + \int_0^r \mathscr{T}(u)W\,du\\ &= \int_r^\infty e^{-\lambda u}\mathscr{T}(u)W\,du - \int_0^r (1-e^{-\lambda u})\mathscr{T}(u)W\,du\\ &\quad + \int_0^r \mathscr{T}(u)W\,du = Y,\end{aligned}$$

which verifies (7.8) and implies $(Y, Z) \in \hat{\mathscr{A}}$.

If $W \in \mathscr{L}_0$, then $\lambda \int_0^\infty e^{-\lambda s}\mathscr{T}(s)W\,ds \in \mathscr{D}(\hat{\mathscr{A}})$ and $\lim_{\lambda\to\infty} \lambda\int_0^\infty e^{-\lambda s}\mathscr{T}(s)W\,ds = W$ (the limit being the strong limit in the Banach space $\mathscr{L}$). If $(Y, Z) \in \hat{\mathscr{A}}$, then

$$\begin{aligned}(7.15)\qquad \|\mathscr{T}(s)Y - Y\| &= \sup_t E\left[\left|\int_0^s E[Z(t+u)\mid \mathscr{F}_t]\,du\right|\right] \le s \sup_t E[|Z(t)|]\\ &= s\|Z\|,\end{aligned}$$

and hence $\mathscr{D}(\hat{\mathscr{A}}) \subset \mathscr{L}_0$. Therefore $\mathscr{L}_0$ is the closure of $\mathscr{D}(\hat{\mathscr{A}})$. Corollary 6.8 of Chapter 1 gives (7.10). □

The following lemma may be useful in showing that a process is in $\mathscr{D}(\hat{\mathscr{A}})$.

7.2 Lemma Let $Y, Z_1, Z_2 \in \mathscr{L}$ and suppose that Y is right continuous and that $Z_1(t) \le Z_2(t)$ a.s. for all t. If $Y(t) - \int_0^t Z_1(s)\,ds$ is a submartingale, and $Y(t) - \int_0^t Z_2(s)\,ds$ is a supermartingale, then there exists $Z \in \mathscr{L}$ satisfying $Z_1(t) \le Z(t) \le Z_2(t)$ a.s. for all $t \ge 0$, such that $Y(t) - \int_0^t Z(s)\,ds$ is a martingale.

7.3 Remark The assumption that Y is right continuous is for convenience only. There always exists a modification of Y that has right limits (since $Y(t) - \int_0^t Z_2(s)\,ds$ is a supermartingale). The lemma as stated then implies the existence of Z (adapted to $\{\mathscr{F}_{s+}\}$) for which $Y(t+) - \int_0^t Z(s)\,ds$ is an $\{\mathscr{F}_{t+}\}$-martingale. Since $E[Y(t+)\,|\,\mathscr{F}_t] = Y(t)$ and $E[\int_t^{t+u} Z(s)\,ds\,|\,\mathscr{F}_t] = E[\int_t^{t+u} E[Z(s)\,|\,\mathscr{F}_s]\,ds\,|\,\mathscr{F}_t]$, $Y(t) - \int_0^t E[Z(s)\,|\,\mathscr{F}_s]\,ds$ is an $\{\mathscr{F}_t\}$-martingale. □

Proof. Without loss of generality we may assume $Z_1 = 0$. Then Y is a submartingale, and since $Y \vee 0$ and $(\int_0^t Z_2(s)\,ds - Y(t)) \vee 0$ are submartingales of class DL, Y and $\int_0^t Z_2(s)\,ds - Y(t)$ are also (note that $|Y(t)| \le Y(t) \vee 0 + (\int_0^t Z_2(s)\,ds - Y(t)) \vee 0$). Consequently, by Theorem 5.1, there exist right continuous increasing processes A_1 and A_2 with Property (b) of Theorem 5.1 such that $Y - A_1$ and $Y(t) - \int_0^t Z_2(s)\,ds + A_2(t)$ are martingales. Since $Y + A_2$ is a submartingale of class DL, and $Y + A_2 - (A_1 + A_2)$ and $Y(t) + A_2(t) - \int_0^t Z_2(s)\,ds$ are martingales, the uniqueness in Theorem 5.1 implies that with probability one,

$$A_1(t) + A_2(t) = \int_0^t Z_2(s)\,ds, \qquad t \ge 0. \tag{7.16}$$

Since A_2 is increasing,

$$A_1(t+u) - A_1(t) \le \int_t^{t+u} Z_2(s)\,ds, \qquad t, u \ge 0, \tag{7.17}$$

so A_1 is absolutely continuous with derivative Z, where $0 \le Z \le Z_2$. □

7.4 Corollary If $Y \in \mathscr{L}$, Y is right continuous, and there exists a constant M such that

$$|E[Y(t+s) - Y(t)\,|\,\mathscr{F}_t]| \le Ms, \qquad t, s \ge 0, \tag{7.18}$$

then there exists $Z \in \mathscr{L}$ with $|Z| \le M$ a.s. such that $Y(t) - \int_0^t Z(s)\,ds$ is a martingale.

Proof. Take $Z_1(t) = -M$ and $Z_2(t) = M$ in Lemma 7.2. □

7.5 Proposition Let $Y \in \mathscr{L}$ and let ξ be the optional projection of $\int_0^b Y(\cdot + s)\,ds$ and η the optional projection of $Y(\cdot + b) - Y$, that is, $\xi(t) = E[\int_0^b Y(t+s)\,ds\,|\,\mathscr{F}_t]$ and $\eta(t) = E[Y(t+b) - Y(t)\,|\,\mathscr{F}_t]$. Then $(\xi, \eta) \in \hat{\mathscr{A}}$.

Proof. This is just Proposition 5.2 of Chapter 1. □

7.6 Proposition Let $\xi \in \mathscr{L}$. If $\{s^{-1}E[\xi(t+s) - \xi(t) \mid \mathscr{F}_t] : s > 0, t \geq 0\}$ is uniformly integrable and

$$(7.19) \qquad s^{-1}E[\xi(t+s) - \xi(t) \mid \mathscr{F}_t] \xrightarrow{p} \eta(t) \quad \text{as} \quad s \to 0+, \quad \text{a.e.} \quad t,$$

then $(\xi, \eta) \in \hat{\mathscr{A}}$.

Proof. Let $\xi_\varepsilon(t) = \varepsilon^{-1}E[\int_0^\varepsilon \xi(t+s)\, ds \mid \mathscr{F}_t]$ and $\eta_\varepsilon(t) = \varepsilon^{-1}E[\xi(t+\varepsilon) - \xi(t) \mid \mathscr{F}_t]$. Then $(\xi_\varepsilon, \eta_\varepsilon) \in \hat{\mathscr{A}}$ and as $\varepsilon \to 0$, $\xi_\varepsilon(t) \to \xi(t)$ and

$$(7.20) \qquad \int_0^t \eta_\varepsilon(s)\, ds \to \int_0^t \eta(s)\, ds$$

in L^1 for each $t \geq 0$. □

We close this section with some observations about the relationship between the semigroup of conditioned shifts and the semigroup associated with a Markov process.

For an adapted process X with values in a metric space (E, r), let $\mathscr{M}$ be the subspace of $\mathscr{L}$ of processes of the form $\{f(X(t), t)\}$, where $f \in B(E \times [0, \infty))$, and let $\mathscr{M}_0$ be the subspace of processes of the form $\{f(X(t))\}$, $f \in B(E)$. Then X is a Markov process if and only if $\mathscr{T}(s): \mathscr{M} \to \mathscr{M}$ for all $s \geq 0$, and it is natural to call X temporally homogeneous if $\mathscr{T}(s): \mathscr{M}_0 \to \mathscr{M}_0$ for all $s \geq 0$.

Suppose X is a Markov process corresponding to a transition function $P(s, x, \Gamma)$, define the semigroup $\{T(t)\}$ on $B(E)$ by

$$(7.21) \qquad T(s)f(x) = \int f(y)P(s, x, dy),$$

and let $\hat{\mathscr{A}}$ denote its full generator. Then for $Y \equiv f \circ X \in \mathscr{M}_0$,

$$(7.22) \qquad \mathscr{T}(s)Y = T(s)f \circ X, \qquad s \geq 0,$$

and for $(f, h) \in \hat{A}$, $(f \circ X, h \circ X) \in \hat{\mathscr{A}}$.

8. MARTINGALES INDEXED BY DIRECTED SETS

In Chapter 6, we need a generalization of the optional sampling theorem to martingales indexed by directed sets. We give this generalization here because of its close relationship to the other material in this chapter.

A set $\mathscr{I}$ is *partially ordered* if some pairs $(u, v) \in \mathscr{I} \times \mathscr{I}$ are ordered by a relation denoted $u \leq v$ (or $v \geq u$) that has the following properties:

$$\text{For all} \quad u \in \mathscr{I}, \qquad u \leq u. \tag{8.1}$$

$$\text{If} \quad u \leq v \quad \text{and} \quad v \leq u, \qquad \text{then} \quad u = v. \tag{8.2}$$

$$\text{If} \quad u \leq v \quad \text{and} \quad v \leq w, \qquad \text{then} \quad u \leq w. \tag{8.3}$$

A partially ordered set $\mathscr{I}$ (together with a metric ρ on $\mathscr{I}$) is a *metric lattice* if $(\mathscr{I}, \rho)$ is a metric space, if for $u, v \in \mathscr{I}$ there exist unique elements $u \wedge v \in \mathscr{I}$ and $u \vee v \in \mathscr{I}$ such that

$$\{w \in \mathscr{I} : w \leq u\} \cap \{w \in \mathscr{I} : w \leq v\} = \{w \in \mathscr{I} : w \leq u \wedge v\} \tag{8.4}$$

and

$$\{w \in \mathscr{I} : w \geq u\} \cap \{w \in \mathscr{I} : w \geq v\} = \{w \in \mathscr{I} : w \geq u \vee v\}, \tag{8.5}$$

and if $(u, v) \rightarrow u \wedge v$ and $(u, v) \rightarrow u \vee v$ are continuous mappings of $\mathscr{I} \times \mathscr{I}$ onto $\mathscr{I}$. We write $\min\{u_1, \ldots, u_m\}$ for $u_1 \wedge \cdots \wedge u_m$, and $\max\{u_1, \ldots, u_m\}$ for $u_1 \vee \cdots \vee u_m$. We assume throughout this section that $\mathscr{I}$ is a metric lattice.

For $u, v \in \mathscr{I}$ with $u \leq v$, the set $[u, v] \equiv \{w \in \mathscr{I} : u \leq w \leq v\}$ is called an *interval.* Note that $[u, v]$ is a closed subset of $\mathscr{I}$. A subset $F \subset \mathscr{I}$ is *separable from above* if there exists a sequence $\{\alpha_n\} \subset F$ such that $w = \lim_{n \to \infty} \min\{\alpha_i : w \leq \alpha_i, \ i \leq n\}$ for all $w \in F$. We call the sequence $\{\alpha_n\}$ a *separating sequence.* Note that F can be separable without being separable from above. Define $\mathscr{I}_u = \{v \in \mathscr{I} : v \leq u\}$.

Let $(\Omega, \mathscr{F}, P)$ be a probability space. As in the case $\mathscr{I} = [0, \infty)$, a collection $\{\mathscr{F}_u\} = \{\mathscr{F}_u, u \in \mathscr{I}\}$ of sub-σ-algebras of $\mathscr{F}$ is a *filtration* if $u \leq v$ implies $\mathscr{F}_u \subset \mathscr{F}_v$, and an $\mathscr{I}$-valued random variable τ is a *stopping time* if $\{\tau \leq u\} \in \mathscr{F}_u$ for all $u \in \mathscr{I}$. For a stopping time τ,

$$\mathscr{F}_\tau = \{A \in \mathscr{F} : A \cap \{\tau \leq u\} \in \mathscr{F}_u \quad \text{for all} \quad u \in \mathscr{I}\}. \tag{8.6}$$

A filtration $\{\mathscr{F}_u\}$ is complete if $(\Omega, \mathscr{F}, P)$ is complete and $\mathscr{F}_u \supset \{A \in \mathscr{F} : P(A) = 0\}$ for all $u \in \mathscr{I}$.

Let $\Gamma_u^n = \{v : \inf_{w \leq u} \rho(v, w) < n^{-1}\}$. We say that $\{\mathscr{F}_u\}$ is *right continuous* if

$$\mathscr{F}_u = \bigcap_n \bigvee_{v \in \Gamma_u^n} \mathscr{F}_v. \tag{8.7}$$

See Problem 20 for an alternative definition of right continuity.

8.1 Proposition Let $\tau_1, \tau_2, \ldots$ be $\{\mathscr{F}_u\}$-stopping times. Then the following hold:

(a) $\max_{k \leq n} \tau_k$ is an $\{\mathscr{F}_u\}$-stopping time.

(b) Suppose $\{\mathscr{F}_u\}$ is right continuous and complete. If τ is an $\mathscr{I}$-valued random variable and $\tau = \lim_{n \to \infty} \tau_n$ a.s., then τ is an $\{\mathscr{F}_u\}$-stopping time.

Proof. **(a)** As in the case $\mathscr{I} = [0, \infty)$, $\{\max_{k \le n} \tau_k \le u\} = \bigcap_{k \le n} \{\tau_k \le u\} \in \mathscr{F}_u$.

(b) By the right continuity,

(8.8)
$$B = \bigcap_n \bigcup_m \bigcap_{k \ge m} \{\tau_k \in \Gamma_u^n\} \in \mathscr{F}_u,$$

and hence

(8.9)
$$\{\tau \le u\} = \left(\{\tau \le u\} \cap \left\{\tau \ne \lim_{n\to\infty} \tau_n\right\}\right) \cup \left(B \cap \left\{\tau = \lim_{n\to\infty} \tau_n\right\}\right) \in \mathscr{F}_u. \quad \square$$

8.2 Proposition Suppose τ is an $\{\mathscr{F}_u\}$-stopping time and $a \in \mathscr{I}$. Define

(8.10)
$$\tau^a = \begin{cases} \tau & \text{on } \{\tau \le a\} \\ a & \text{otherwise.} \end{cases}$$

Then τ^a is an $\{\mathscr{F}_u\}$-stopping time.

8.3 Remark Note that τ^a is not in general equal to $\tau \wedge a$, which need not be a stopping time. $\square$

Proof. If $u = a$, $\{\tau^a \le u\} = \Omega \in \mathscr{F}_u$. If $u \le a$, but $u \ne a$, then $\{\tau^a \le u\} = \{\tau \le u\} \in \mathscr{F}_u$. In general, $\{\tau^a \le u\} = \{\tau^a \le u \wedge a\} \in \mathscr{F}_{u \wedge a} \subset \mathscr{F}_u$. $\square$

8.4 Proposition Suppose τ is an $\{\mathscr{F}_u\}$-stopping time, $a \in \mathscr{I}$ with $\tau \le a$, and $\mathscr{I}_a$ is separable from above. Let $\{\alpha_i\}$ be a separating sequence for $\mathscr{I}_a$ with $\alpha_1 = a$, and define

(8.11)
$$\tau_n = \min\{\alpha_i : \tau \le \alpha_i, \quad i \le n\}, \qquad n \ge 1.$$

Then τ_n is a stopping time for each $n \ge 1$, $a = \tau_1 \ge \tau_2 \ge \dots$, and $\lim_{n\to\infty} \tau_n = \tau$.

Proof. Let F_n be the finite collection of possible values of τ_n. For $u \in F_n$,

(8.12)
$$\{\tau_n = u\} = \{\tau \le u\} \cap \bigcap_{\substack{v \in F_n \cap \mathscr{I}_u \\ v \ne u}} \{\tau \le v\}^c \in \mathscr{F}_u,$$

and in general

(8.13)
$$\{\tau_n \le u\} = \bigcup_{v \in F_n \cap \mathscr{I}_u} \{\tau_n = v\} \in \mathscr{F}_u.$$

The rest follows from the definition of a separating sequence. $\square$

Let X be an E-valued process indexed by $\mathscr{I}$. Then X is $\{\mathscr{F}_u\}$-*adapted* if $X(u)$ is $\mathscr{F}_u$-measurable for each $u \in \mathscr{I}$, and X is $\{\mathscr{F}_u\}$-progressive if for each $u \in \mathscr{I}$, the restriction of X to $\mathscr{I}_u \times \Omega$ is $\mathscr{B}(\mathscr{I}_u) \times \mathscr{F}_u$-measurable. As in the case $\mathscr{I} = [0, \infty)$, if $\mathscr{I}_u$ is separable from above for each $u \in \mathscr{I}$, and X is right continuous (i.e., $\lim_{v \to u} X(u \vee v, \omega) = X(u, \omega)$ for all $u \in \mathscr{I}$ and $\omega \in \Omega$) and $\{\mathscr{F}_u\}$-adapted, then X is $\{\mathscr{F}_u\}$-progressive.

8.5 Proposition Let τ and σ be $\{\mathscr{F}_u\}$-stopping times with $\tau \leq \sigma$, and let X be $\{\mathscr{F}_u\}$-progressive. Then the following hold:

(a) $\mathscr{F}_\tau$ is a σ-algebra.

(b) $\mathscr{F}_\tau \subset \mathscr{F}_\sigma$.

(c) If $\mathscr{I}_u$ is separable from above for each $u \in \mathscr{I}$, then τ and $X(\tau)$ are $\mathscr{F}_\tau$-measurable.

Proof. The proofs for parts (a) and (b) are the same as for the corresponding results in Proposition 1.4. Fix $a \in \mathscr{I}$. We first want to show that τ^a is $\mathscr{F}_a$-measurable. Let $\{\alpha_n\}$ be a separating sequence for $\mathscr{I}_a$ with $\alpha_1 = a$, and define $\tau_n^a = \min\{\alpha_i : \tau^a \leq \alpha_i,\ i \leq n\}$. Then (8.12) implies τ_n^a is $\mathscr{F}_a$-measurable. Since $\lim_{n\to\infty} \tau_n^a = \tau^a$, τ^a is $\mathscr{F}_a$-measurable, and $X(\tau^a)$ is $\mathscr{F}_a$-measurable by the argument in the proof of Proposition 1.4(d). Finally, $\{\tau \in \Gamma\} \cap \{\tau \leq a\} = \{\tau^a \in \Gamma\} \cap \{\tau \leq a\} \in \mathscr{F}_a$ for all $a \in \mathscr{I}$ and $\Gamma \in \mathscr{B}(\mathscr{I})$, so $\{\tau \in \Gamma\} \in \mathscr{F}_\tau$ and τ is $\mathscr{F}_\tau$-measurable. The same argument implies that $X(\tau)$ is $\mathscr{F}_\tau$-measurable. □

8.6 Proposition Suppose $\{\mathscr{F}_u\}$ is a right continuous filtration indexed by $\mathscr{I}$. For each $t \geq 0$, let $\tau(t)$ be an $\{\mathscr{F}_u\}$-stopping time such that $s \leq t$ implies $\tau(s) \leq \tau(t)$ and $\tau(t)$ is a right continuous function of t. Let $\mathscr{H}_t = \mathscr{F}_{\tau(t)}$, and let η be an $\{\mathscr{H}_t\}$-stopping time. Then $\tau(\eta)$ is an $\{\mathscr{F}_u\}$-stopping time.

Proof. First assume η is discrete. Then

$$\{\tau(\eta) \leq u\} = \bigcup_i (\{\eta = t_i\} \cap \{\tau(t_i) \leq u\}). \tag{8.14}$$

Since $\{\eta = t_i\} \in \mathscr{F}_{\tau(t_i)}$, $\{\eta = t_i\} \cap \{\tau(t_i) \leq u\} \in \mathscr{F}_u$. For general η approximate η by a decreasing sequence of discrete stopping times (cf. Proposition 1.3) and apply Proposition 8.1(b). □

A real-valued process X indexed by $\mathscr{I}$ is an $\{\mathscr{F}_u\}$-*martingale* if $E[|X(u)|] < \infty$ for all $u \in \mathscr{I}$, X is $\{\mathscr{F}_u\}$-adapted, and

$$E[X(v) \mid \mathscr{F}_u] = X(u) \tag{8.15}$$

for all $u, v \in \mathscr{I}$ with $u \leq v$.

8.7 Theorem Let $\mathscr{I}$ be a metric lattice and suppose each interval in $\mathscr{I}$ is separable from above. Let X be a right continuous $\{\mathscr{F}_u\}$-martingale, and let τ_1

and τ_2 be $\{\mathscr{F}_u\}$-stopping times with $\tau_1 \le \tau_2$. Suppose there exist $\{u_m\}, \{v_m\} \subset \mathscr{I}$ such that

$$\lim_{m\to\infty} P\{u_m \le \tau_1 \le \tau_2 \le v_m\} = 1 \tag{8.16}$$

and

$$\lim_{m\to\infty} E[|X(v_m)|\chi_{\{\tau_2 \le v_m\}^c}] = 0, \tag{8.17}$$

and that $E[|X(\tau_2)|] < \infty$. Then

$$E[X(\tau_2)|\mathscr{F}_{\tau_1}] = X(\tau_1). \tag{8.18}$$

Proof. Fix $m \ge 1$ and for $i = 1, 2$ define

$$\tau_{im} = \begin{cases} \tau_i \vee u_m & \text{on } \{\tau_i \le v_m\} \\ v_m & \text{otherwise.} \end{cases} \tag{8.19}$$

Let $\{\alpha_n\}$ be separating for $[u_m, v_m]$ with $\alpha_1 = v_m$, and define $\tau_{im}^n = \min\{\alpha_k : \tau_{im} \le \alpha_k, k \le n\}$. Note τ_{im}^n assumes finitely many values, and $\tau_{1m}^n \le \tau_{2m}^n$. Fix n, and let Γ be the set of values assumed by τ_{1m}^n and τ_{2m}^n. For $\alpha \in \Gamma$ with $\alpha \ne v_m$, $\{\tau_{1m}^n = \alpha\} = \{\tau_1 \le \alpha\} \cap \{\tau_{1m}^n = \alpha\}$, and hence $A \cap \{\tau_{1m}^n = \alpha\} = A \cap \{\tau_1 \le \alpha\} \cap \{\tau_{1m}^n = \alpha\} \in \mathscr{F}_\alpha$. Consequently for $\alpha \in \Gamma$ with $\alpha \ne v_m$,

$$\begin{aligned} &\int_{A \cap \{\tau_{1m}^n = \alpha\}} X(\tau_{2m}^n)\, dP \\ &= \int_{A \cap \{\tau_{1m}^n = \alpha\}} \sum_{\beta \in \Gamma} E[X(v_m)|\mathscr{F}_\beta]\chi_{\{\tau_{2m}^n = \beta\}}\, dP \\ &= \sum_{\beta\in\Gamma} \int_{A \cap \{\tau_{1m}^n = \alpha\} \cap \{\tau_{2m}^n = \beta\}} X(v_m)\, dP \\ &= \int_{A \cap \{\tau_{1m}^n = \alpha\}} X(v_m)\, dP \\ &= \int_{A \cap \{\tau_{1m}^n = \alpha\}} X(\alpha)\, dP. \end{aligned} \tag{8.20}$$

Since $\tau_{1m}^n = v_m$ implies $\tau_{2m}^n = v_m$, (8.20) is immediate for $\alpha = v_m$, and summing over $\alpha \in \Gamma$, (8.20) implies

$$\int_A X(\tau_{2m}^n)\, dP = \int_A X(\tau_{1m}^n)\, dP. \tag{8.21}$$

Letting $n \to \infty$ and then $m \to \infty$ gives

$$\int_A X(\tau_2)\, dP = \int_A X(\tau_1)\, dP, \tag{8.22}$$

which implies (8.18). □

9. PROBLEMS

1. Show that if X is right (left) continuous and $\{\mathscr{F}_t\}$-adapted, then X is $\{\mathscr{F}_t\}$-progressive.

Hint: Fix $t > 0$ and approximate X on $[0, t] \times \Omega$ by X_n given by $X_n(s, \omega) = X(t \wedge (([ns] + 1)/n), \omega)$.

2. (a) Suppose X is E-valued and $\{\mathscr{F}_t\}$-progressive, and $f \in B(E)$. Show that $f \circ X$ is $\{\mathscr{F}_t\}$-progressive and $Y(t) \equiv \int_0^t f(X(s))\, ds$ is $\{\mathscr{F}_t\}$-adapted.

(b) Suppose X is E-valued, measurable, and $\{\mathscr{F}_t\}$-adapted, and $f \in B(E)$. Show that $f \circ X$ is $\{\mathscr{F}_t\}$-adapted and that $Y(t) \equiv \int_0^t f(X(s))\, ds$ has an $\{\mathscr{F}_t\}$-adapted modification.

3. Let Y be a version of X. Suppose X is right continuous. Show that there is a modification of Y that is right continuous.

4. Let (E, r) be a complete, separable metric space.

(a) Let $\xi_1, \xi_2, \ldots$ be E-valued random variables defined on $(\Omega, \mathscr{F}, P)$. Let $A = \{\omega: \lim \xi_n(\omega) \text{ exists}\}$. Show that $A \in \mathscr{F}$, and that for $x \in E$,

$$\xi(\omega) = \begin{cases} \lim \xi_n(\omega), & \omega \in A, \\ x, & \omega \in A^c, \end{cases} \tag{9.1}$$

is a random variable (i.e., is $\mathscr{F}$-measurable).

(b) Let X be an E-valued process that is right continuous in probability, that is, for each $\varepsilon > 0$ and $t \geq 0$,

$$\lim_{s \to t+} P\{r(X(t), X(s)) > \varepsilon\} = 0. \tag{9.2}$$

Show that X has a modification that is progressive.

Hint: Show that for each n there exists a countable collection of disjoint intervals $[t_\alpha^n, s_\alpha^n)$ such that $[0, \infty) = \bigcup [t_\alpha^n, s_\alpha^n)$ and

$$P\{r(X(t_\alpha^n), X(s)) > 2^{-n}\} < 2^{-n}, \qquad t_\alpha^n \leq s < s_\alpha^n. \tag{9.3}$$

5. Suppose X is a modification of Y, and X and Y are right continuous. Show that X and Y are indistinguishable.

6. Let X be a stochastic process, and let τ be a discrete $\{\mathscr{F}_t^X\}$-stopping time. Show that

$$\mathscr{F}_\tau^X = \sigma(X(t \wedge \tau) : t \geq 0). \tag{9.4}$$

7. Let $\{\mathscr{F}_t\}$ be a filtration. Show that $\{\mathscr{F}_{t+}\}$ is right continuous.

8. Let $(\Omega, \mathscr{F}, P)$ be a probability space. Let S be the collection of equivalence classes of real-valued random variables where two random variables are equivalent if they are almost surely equal. Let γ be defined by (2.27). Show that γ is a metric on S corresponding to convergence in probability.

9. (a) Let $\{x_n\}$ satisfy $\sup_n x_n < \infty$ and $\inf_n x_n > -\infty$, and assume that for each $a < b$ either $\{n: x_n \geq b\}$ or $\{n: x_n \leq a\}$ is finite. Show that $\lim_{n\to\infty} x_n$ exists.

(b) Verify the existence of the limits in (2.24) and (2.25) and show that $Y^-(t, \omega) = Y(t-, \omega)$ for all $t > 0$ and $\omega \in \Omega_0$.

10. (a) Suppose X is a real-valued integrable random variable on $(\Omega, \mathscr{F}, P)$. Let Γ be the collection of sub-σ-algebras of $\mathscr{F}$. Show that $\{E[X \mid \mathscr{D}]: \mathscr{D} \in \Gamma\}$ is uniformly integrable.

(b) Let X be a right continuous $\{\mathscr{F}_t\}$-martingale and let S be the collection of $\{\mathscr{F}_t\}$-stopping times. Show that for each $T > 0$, $\{X(T \wedge \tau): \tau \in S\}$ is uniformly integrable.

(c) Let X be a right continuous, nonnegative $\{\mathscr{F}_t\}$-submartingale. Show that for each $T > 0$, $\{X(T \wedge \tau): \tau \in S\}$ is uniformly integrable.

11. (a) Let X be a right continuous $\{\mathscr{F}_t\}$-submartingale and τ a finite $\{\mathscr{F}_t\}$-stopping time. Suppose that for each $t > 0$, $E[\sup_{s \leq t} |X(\tau + s) - X(\tau)|] < \infty$. Show that $Y(t) \equiv X(\tau + t) - X(\tau)$ is an $\{\mathscr{F}_{\tau+t}\}$-submartingale.

(b) Let X be a right continuous $\{\mathscr{F}_t\}$-submartingale and τ_1 and τ_2 be finite $\{\mathscr{F}_t\}$-stopping times. Suppose $\tau_1 \leq \tau_2$ and $E[\sup_s |X((\tau_1 + s) \wedge \tau_2) - X(\tau_1)|] < \infty$. Show that $E[X(\tau_2) - X(\tau_1) \mid \mathscr{F}_{\tau_1}] \geq 0$.

12. Let X be a submartingale. Show that $\sup_t E[|X(t)|] < \infty$ if and only if $\sup_t E[X^+(t)] < \infty$.

13. Let η and ξ be independent random variables with $P\{\eta = 1\} = P\{\eta = -1\} = \frac{1}{2}$ and $E[|\xi|] = \infty$. Define

$$X(t) = \begin{cases} 0, & 0 \leq t < 1, \\ \eta\xi, & t \geq 1, \end{cases} \tag{9.5}$$

and

$$\mathscr{F}_t = \begin{cases} \sigma(\xi), & 0 \leq t < 1, \\ \sigma(\xi, \eta), & t \geq 1. \end{cases} \tag{9.6}$$

Show that X is an $\{\mathscr{F}_t\}$-local martingale, but that X is not an $\{\mathscr{F}_t^X\}$-local martingale.

14. Let E be a separable Banach space with norm $\|\cdot\|$, and let X be an E-valued random variable defined on $(\Omega, \mathscr{F}, P)$.

(a) Show that for every $\varepsilon > 0$ there exist $\{x_i\} \subset E$ and $\{B_i^\varepsilon\} \subset \mathscr{F}$ with $B_i^\varepsilon \cap B_j^\varepsilon = \varnothing$ for $i \neq j$, such that

$$X_\varepsilon = \sum x_i \chi_{B_i^\varepsilon} \tag{9.7}$$

satisfies $\| X - X_\varepsilon \| \leq \varepsilon$.

(b) Suppose $E[\| X \|] < \infty$. Define

$$E[X_\varepsilon \mid \mathscr{D}] = \sum x_i \, P(B_i^\varepsilon \mid \mathscr{D}), \tag{9.8}$$

and show that

$$\| E[X_{\varepsilon_1} \mid \mathscr{D}] - E[X_{\varepsilon_2} \mid \mathscr{D}] \| \leq E[\| X_{\varepsilon_1} - X_{\varepsilon_2} \| \mid \mathscr{D}] \leq \varepsilon_1 + \varepsilon_2 \tag{9.9}$$

so that one can define

$$E[X \mid \mathscr{D}] = \lim_{\varepsilon \to 0} E[X_\varepsilon \mid \mathscr{D}]. \tag{9.10}$$

(c) Extend Theorem 4.2 and Corollary 4.5 to bounded, measurable, E-valued processes.

15. Let A_1 and A_2 be right continuous increasing processes with $A_1(0) = A_2(0) = 0$ and $E[A_i(t)] < \infty$, $i = 1, 2$, $t > 0$. Suppose that $A_1 - A_2$ is an $\{\mathscr{F}_t\}$-martingale and that Y is bounded, right continuous, has left limits at each $t > 0$, and is $\{\mathscr{F}_t\}$-adapted. Show that

$$\int_0^t Y(s-) \, d(A_1(s) - A_2(s)) = \int_0^t Y(s-) \, dA_1(s) - \int_0^t Y(s-) \, dA_2(s) \tag{9.11}$$

is an $\{\mathscr{F}_t\}$-martingale. (The integrals are defined as in (5.1).)
Hint: Let

$$Y_\varepsilon(t) = \varepsilon^{-1} \int_{t-\varepsilon}^t Y(s) \, ds \tag{9.12}$$

and apply Proposition 3.2 to

$$\int_0^t Y_\varepsilon(s) \, d(A_1(s) - A_2(s)). \tag{9.13}$$

16. Let Y be a unit Poisson process, and define $M(t) = Y(t) - t$. Show that M is a martingale and compute $[M]$ and $\langle M \rangle$.

17. Let W be standard Brownian motion. Use the law of large numbers to compute $[W]$. (Recall $[W] = \langle W \rangle$ since W is continuous.)

18. Let $\mathscr{L}^2$ be the space of real-valued $\{\mathscr{F}_t\}$-progressive processes X satisfying $\int_0^\infty E[|X(t)|^2]\,dt < \infty$. Note that $\mathscr{L}^2$ is a Hilbert space with inner product

$$(9.14) \qquad (X, Y) = \int_0^\infty E[X(t)Y(t)]\,dt$$

and norm $\|X\| = \sqrt{(X, X)}$. Let A be a bounded linear operator on $\mathscr{L}^2$. Then A^*, the adjoint of A, is the unique bounded linear operator satisfying $(AX, Y) = (X, A^*Y)$ for all $X, Y \in \mathscr{L}^2$. Fix $s \geq 0$ and let $U(s)$ be the bounded linear operator defined by

$$(9.15) \qquad U(s)X(t) = \begin{cases} X(t-s), & t \geq s, \\ 0, & t < s. \end{cases}$$

What is $U^*(s)$? (Remember that $U^*(s)X$ must be $\{\mathscr{F}_t\}$-progressive.)

19. Let $M_1, M_2, \ldots, M_m$ be independent martingales. Let $\mathscr{I} = [0, \infty)^m$ and define

$$(9.16) \qquad M(u) = \prod_{i=1}^m M_i(u_i), \qquad u \in \mathscr{I}.$$

(a) Show that M is a martingale indexed by $\mathscr{I}$.

(b) Let $\mathscr{F}_u = \sigma(M(v): v \leq u)$, and let $\tau(t)$, $t \geq 0$, be $\{\mathscr{F}_u\}$-stopping times satisfying $\tau(s) \leq \tau(t)$ for $s \leq t$. Suppose that for each t there exists $c_t \in \mathscr{I}$ such that $\tau(t) \leq c_t$ a.s. Let $X_i(t) = M_i(\tau_i(t))$. Show that $X_1, \ldots, X_m$ are orthogonal $\{\mathscr{F}_{\tau(t)}\}$-martingales. More generally, show that for any $I \subset \{1, \ldots, m\}$, $\prod_{i \in I} X_i$ is an $\{\mathscr{F}_{\tau(t)}\}$-martingale.

20. Let $\mathscr{I}$ be a metric lattice. Show that a filtration $\{\mathscr{F}_u, u \in \mathscr{I}\}$ is right continuous if and only if for every u, $\{u_n\} \subset \mathscr{I}$ with $u \leq u_n$, $n = 1, 2, \ldots$, and $u = \lim_{n \to \infty} u_n$, we have $\mathscr{F}_u = \bigcap \mathscr{F}_{u_n}$.

21. (a) Suppose M is a local martingale and $\sup_{s \leq t} |M(s)| \in L^1$ for each $t > 0$. Show that M is a martingale.

(b) Suppose M is a positive local martingale. Show that M is a supermartingale.

22. Let X and Y be measurable and $\{\mathscr{G}_t\}$-adapted. Suppose $E[|X(t)| \int_0^t |Y(s)|\,ds] < \infty$ and $E[\int_0^t |X(s)Y(s)|\,ds] < \infty$ for every $t \geq 0$, and that X is a $\{\mathscr{G}_t\}$-martingale. Show that $X(t)\int_0^t Y(s)\,ds - \int_0^t X(s)Y(s)\,ds$ is a martingale. (Cf. Proposition 3.2 but note we are not assuming X is right continuous.)

23. Let X be a real-valued $\{\mathscr{G}_t\}$-adapted process, with $E[|X(t)|] < \infty$ for every $t \geq 0$. Show that X is a $\{\mathscr{G}_t\}$-martingale if and only if $E[X(\tau)] = E[X(0)]$ for every $\{\mathscr{G}_t\}$-stopping time τ assuming only finitely many values.

24. Let $M_1, \ldots, M_n$ be right continuous $\{\mathscr{G}_t\}$-martingales, and suppose that, for each $I \subset \{1, \ldots, n\}$, $\prod_{i \in I} M_i$ is also a $\{\mathscr{G}_t\}$-martingale. Let $\tau_1, \ldots, \tau_n$ be $\{\mathscr{G}_t\}$-stopping times, and suppose $E[\prod_{i=1}^n \sup_{t \leq \tau_i} |M_i(t)|] < \infty$. Show that $M(t) \equiv \prod_{i=1}^n M_i(t \wedge \tau_i)$ is a $\{\mathscr{G}_t\}$-martingale.

Hint: Use Problem 23 and induction on n.

25. Let X be a real-valued stochastic process, and $\{\mathscr{F}_t\}$ a filtration. (X is not necessarily $\{\mathscr{F}_t\}$-adapted.) Suppose $E[X(t) \mid \mathscr{F}_t] \geq 0$ for each t. Show that $E[X(\tau) \mid \mathscr{F}_\tau] \geq 0$ for each finite, discrete $\{\mathscr{F}_t\}$-stopping time τ.

26. Let $(M, \mathscr{M}, \mu)$ be a probability space, and let $\mathscr{M}_1 \subset \mathscr{M}_2 \subset \cdots$ be an increasing sequence of discrete σ-algebras, that is, for $n = 1, 2, \ldots$, $\mathscr{M}_n = \sigma(A_i^n, i = 1, 2, \ldots)$ where the A_i^n are disjoint, and $M = \bigcup_i A_i^n$. Let $X \in L^1(\mu)$, and define

$$X_n = \sum_i \mu(A_i^n)^{-1} \int_{A_i^n} X \, d\mu \, \chi_{A_i^n} \tag{9.17}$$

(a) Show that $\{X_n\}$ is an $\{\mathscr{M}_n\}$-martingale.

(b) Suppose $\mathscr{M} = \bigvee_n \mathscr{M}_n$. Show that $\lim_{n \to \infty} X_n = X$ μ-a.s. and in $L^1(\mu)$.

27. Let $\{X(t) \colon t \in \mathscr{I}\}$ be a stochastic process. Show that

$$\sigma(X(s) \colon s \in \mathscr{I}) = \bigcup_{I \subset \mathscr{I}} \sigma(X(s) \colon s \in I) \tag{9.18}$$

where the union is over all *countable* subsets of $\mathscr{I}$.

28. Let τ and σ be $\{\mathscr{F}_t\}$-stopping times. Show that $\mathscr{F}_{\tau \wedge \sigma} = \mathscr{F}_\tau \cap \mathscr{F}_\sigma$ and $\mathscr{F}_{\tau \vee \sigma} = \mathscr{F}_\tau \vee \mathscr{F}_\sigma$.

29. Let X be a right continuous, E-valued process adapted to a filtration $\{\mathscr{F}_t\}$. Let $f \in \bar{C}(E)$ and $g, h \in B(E)$, and suppose that

$$M_f(t) \equiv f(X(t)) - f(X(0)) - \int_0^t g(X(s)) \, ds \tag{9.19}$$

and

$$f(X(t))^2 - f(X(0))^2 - \int_0^t h(X(s)) \, ds \tag{9.20}$$

are $\{\mathscr{F}_t\}$-martingales. Show that

$$M_f(t)^2 - \int_0^t (h(X(s)) - 2f(X(s))g(X(s)))\,ds \tag{9.21}$$

is an $\{\mathscr{F}_t\}$-martingale.

10. NOTES

Most of the material in Section 1 is from Doob (1953) and Dynkin (1961), and has been developed and refined by the Strasbourg school. For the most recent presentation see Dellacherie and Meyer (1978, 1982). Section 2 is almost entirely from Doob (1953).

The notion of a local martingale is due to Itô and Watanabe (1965). Proposition 3.2 is of course a special case of much more general results on stochastic integrals. See Dellacherie and Meyer (1982) and Chapter 5. Proposition 3.4 is due to Doleans-Dade (1969). The projection theorem is due to Meyer (1968). Theorem 5.1 is also due to Meyer. See Meyer (1966), page 122.

The semigroup of conditioned shifts appeared first in work of Rishel (1970). His approach is illustrated by Problem 18. The presentation in Section 7 is essentially that of Kurtz (1975). Chow (1960) gave one version of an optional sampling theorem for martingales indexed by directed sets. Section 8 follows Kurtz (1980b).

Problem 4(b) is essentially Theorem II.2.6 of Doob (1953). See Dellacherie and Meyer (1978), page 99, for a more refined version.

3 CONVERGENCE OF PROBABILITY MEASURES

In this chapter we study convergence of sequences of probability measures defined on the Borel subsets of a metric space (S, d) and in particular of $D_E[0, \infty)$, the space of right continuous functions from $[0, \infty)$ into a metric space (E, r) having left limits. Our starting point in Section 1 is the Prohorov metric ρ on $\mathscr{P}(S)$, the set of Borel probability measures on S, and in Section 2 we give Prohorov's characterization of the compact subsets of $\mathscr{P}(S)$. In Section 3 we define weak convergence of a sequence in $\mathscr{P}(S)$ and consider its relationship to convergence in the Prohorov metric (they are equivalent if S is separable). Section 4 concerns the concepts of separating and convergence determining classes of bounded continuous functions on S.

Sections 5 and 6 are devoted to a study of the space $D_E[0, \infty)$ with the Skorohod topology and Section 7 to weak convergence of sequences in $\mathscr{P}(D_E[0, \infty))$. In Section 8 we give necessary and sufficient conditions in terms of conditional expectations of $r^{\beta}(X_{\alpha}(t + u), X_{\alpha}(t)) \wedge 1$ (conditioning on $\mathscr{F}_t^{X_{\alpha}}$) for a family of processes $\{X_{\alpha}\}$ to be relatively compact (that is, for the family of distributions on $D_E[0, \infty)$ to be relatively compact). Criteria for relative compactness that are particularly useful in the study of Markov processes are given in Section 9. Finally, Section 10 contains necessary and sufficient conditions for a limiting process to have sample paths in $C_E[0, \infty)$.

1. THE PROHOROV METRIC

Throughout Sections 1–4, (S, d) is a metric space (d denoting the metric), $\mathscr{B}(S)$ is the σ-algebra of Borel subsets of S, and $\mathscr{P}(S)$ is the family of Borel probability measures on S. We topologize $\mathscr{P}(S)$ with the Prohorov metric

$$\rho(P, Q) = \inf\{\varepsilon > 0: P(F) \le Q(F^\varepsilon) + \varepsilon \quad \text{for all} \quad F \in \mathscr{C}\}, \tag{1.1}$$

where $\mathscr{C}$ is the collection of closed subsets of S and

$$F^\varepsilon = \left\{x \in S: \inf_{y \in F} d(x, y) < \varepsilon\right\}. \tag{1.2}$$

To see that ρ is a metric, we need the following lemma.

1.1 Lemma Let $P, Q \in \mathscr{P}(S)$ and $\alpha, \beta > 0$. If

$$P(F) \le Q(F^\alpha) + \beta \tag{1.3}$$

for all $F \in \mathscr{C}$, then

$$Q(F) \le P(F^\alpha) + \beta \tag{1.4}$$

for all $F \in \mathscr{C}$.

Proof. Given $F_1 \in \mathscr{C}$, let $F_2 = S - F_1^\alpha$, and note that $F_2 \in \mathscr{C}$ and $F_1 \subset S - F_2^\alpha$. Consequently, by (1.3) with $F = F_2$,

$$P(F_1^\alpha) = 1 - P(F_2) \ge 1 - Q(F_2^\alpha) - \beta \ge Q(F_1) - \beta, \tag{1.5}$$

implying (1.4) with $F = F_1$. □

It follows immediately from Lemma 1.1 that $\rho(P, Q) = \rho(Q, P)$ for all $P, Q \in \mathscr{P}(S)$. Also, if $\rho(P, Q) = 0$, then $P(F) = Q(F)$ for all $F \in \mathscr{C}$ and hence for all $F \in \mathscr{B}(S)$; therefore, $\rho(P, Q) = 0$ if and only if $P = Q$. Finally, if $P, Q, R \in \mathscr{P}(S)$, $\rho(P, Q) < \delta$, and $\rho(Q, R) < \varepsilon$, then

$$\begin{aligned} P(F) &\le Q(F^\delta) + \delta \le Q(\overline{F^\delta}) + \delta \\ &\le R((\overline{F^\delta})^\varepsilon) + \delta + \varepsilon \le R(F^{\delta+\varepsilon}) + \delta + \varepsilon \end{aligned} \tag{1.6}$$

for all $F \in \mathscr{C}$, so $\rho(P, R) \le \delta + \varepsilon$, proving the triangle inequality.

The following theorem provides a probabilistic interpretation of the Prohorov metric when S is separable.

1.2 Theorem Let (S, d) be separable, and let $P, Q \in \mathscr{P}(S)$. Define $\mathscr{M}(P, Q)$ to be the set of all $\mu \in \mathscr{P}(S \times S)$ with marginals P and Q (i.e., $\mu(A \times S) = P(A)$ and $\mu(S \times A) = Q(A)$ for all $A \in \mathscr{B}(S)$). Then

$$\rho(P, Q) = \inf_{\mu \in \mathscr{M}(P, Q)} \inf\{\varepsilon > 0: \mu\{(x, y): d(x, y) \ge \varepsilon\} \le \varepsilon\}. \tag{1.7}$$

Proof. If for some $\varepsilon > 0$ and $\mu \in \mathscr{M}(P, Q)$ we have

$$\mu\{(x, y): d(x, y) \geq \varepsilon\} \leq \varepsilon, \tag{1.8}$$

then

$$\begin{aligned} P(F) &= \mu(F \times S) \\ &\leq \mu((F \times S) \cap \{(x, y): d(x, y) < \varepsilon\}) + \varepsilon \\ &\leq \mu(S \times F^{\varepsilon}) + \varepsilon = Q(F^{\varepsilon}) + \varepsilon \end{aligned} \tag{1.9}$$

for all $F \in \mathscr{C}$, so $\rho(P, Q)$ is less than or equal to the right side of (1.7).

The reverse inequality is an immediate consequence of the following lemma. □

1.3 Lemma Let S be separable. Let $P, Q \in \mathscr{P}(S)$, $\rho(P, Q) < \varepsilon$, and $\delta > 0$. Suppose that $E_1, \dots, E_N \in \mathscr{B}(S)$ are disjoint with diameters less than δ and that $P(E_0) \leq \delta$, where $E_0 = S - \bigcup_{i=1}^{N} E_i$. Then there exist constants $c_1, \dots, c_N \in [0, 1]$ and independent random variables $X, Y_0, \dots, Y_N$ (S-valued) and ξ ($[0, 1]$-valued) on some probability space $(\Omega, \mathscr{F}, \nu)$ such that X has distribution P, ξ is uniformly distributed on $[0, 1]$,

$$Y = \begin{cases} Y_i & \text{on} \quad \{X \in E_i, \xi \geq c_i\}, \qquad i = 1, \dots, N, \\ Y_0 & \text{on} \quad \{X \in E_0\} \cup \bigcup_{i=1}^{N} \{X \in E_i, \xi < c_i\} \end{cases} \tag{1.10}$$

has distribution Q,

$$\{d(X, Y) \geq \delta + \varepsilon\} \subset \{X \in E_0\} \cup \left\{\xi < \max\left[\frac{\varepsilon}{P(E_i)}: P(E_i) > 0\right]\right\}, \tag{1.11}$$

and

$$\nu\{d(X, Y) \geq \delta + \varepsilon\} \leq \delta + \varepsilon. \tag{1.12}$$

The proof of this lemma depends on another lemma.

1.4 Lemma Let μ be a finite positive Borel measure on S, and let $p_i \geq 0$ and $A_i \in \mathscr{B}(S)$ for $i = 1, \dots, n$. Suppose that

$$\sum_{i \in I} p_i \leq \mu\left(\bigcup_{i \in I} A_i\right) \quad \text{for all} \quad I \subset \{1, \dots, n\}. \tag{1.13}$$

Then there exist positive Borel measures $\lambda_1, \dots, \lambda_n$ on S such that $\lambda_i(A_i) = \lambda_i(S) = p_i$ for $i = 1, \dots, n$ and $\sum_{i=1}^{n} \lambda_i(A) \leq \mu(A)$ for all $A \in \mathscr{B}(S)$.

Proof. Note first that it involves no loss of generality to assume that each $p_i > 0$.

We proceed by induction on n. For $n = 1$, define λ_1 on $\mathscr{B}(S)$ by $\lambda_1(A) =$

$p_1\mu(A \cap A_1)/\mu(A_1)$. Then $\lambda_1(A_1) = \lambda_1(S) = p_1$, and since $p_1 \le \mu(A_1)$ by (1.13), we have $\lambda_1(A) \le \mu(A \cap A_1) \le \mu(A)$ for all $A \in \mathscr{B}(S)$. Suppose now that the lemma holds with n replaced by m for $m = 1, \ldots, n-1$ and that μ, p_i, and A_i $(1 \le i \le n)$ satisfy (1.13). Define η on $\mathscr{B}(S)$ by $\eta(A) = \mu(A \cap A_n)/\mu(A_n)$, and let ε_0 be the largest ε such that

$$\sum_{i \in I} p_i \le (\mu - \varepsilon\eta)\left(\bigcup_{i \in I} A_i\right) \quad \text{for all} \quad I \subset \{1, \ldots, n-1\}. \tag{1.14}$$

CASE 1. $\varepsilon_0 \ge p_n$. Let $\lambda_n = p_n \eta$ and put $\mu' = \mu - \lambda_n$. Since $p_n \le \mu(A_n)$ by (1.13), μ' is a positive Borel measure on S, so by (1.14) (with $\varepsilon = p_n$) and the induction hypothesis, there exist positive Borel measures $\lambda_1, \ldots, \lambda_{n-1}$ on S such that $\lambda_i(A_i) = \lambda_i(S) = p_i$ for $i = 1, \ldots, n-1$ and $\sum_{i=1}^{n-1} \lambda_i(A) \le \mu'(A)$ for all $A \in \mathscr{B}(S)$. Also, $\lambda_n(A_n) = \lambda_n(S) = p_n$, so $\lambda_1, \ldots, \lambda_n$ have the required properties.

CASE 2. $\varepsilon_0 < p_n$. Put $\mu' = \mu - \varepsilon_0 \eta$, and note that μ' is a positive Borel measure on S. By the definition of ε_0, there exists $I_0 \subset \{1, \ldots, n-1\}$ (nonempty) such that

$$\sum_{i \in I} p_i \le \mu'\left(\bigcup_{i \in I} A_i\right) \quad \text{for all} \quad I \subset I_0 \tag{1.15}$$

with equality holding for $I = I_0$. By the induction hypothesis, there exist positive Borel measures λ_i on S, $i \in I_0$, such that $\lambda_i(A_i) = \lambda_i(S) = p_i$ for each $i \in I_0$ and $\sum_{i \in I_0} \lambda_i(A) \le \mu'(A)$ for all $A \in \mathscr{B}(S)$. Let $p_i' = p_i$ for $i = 1, \ldots, n-1$ and $p_n' = p_n - \varepsilon_0$. Put $B_0 = \bigcup_{i \in I_0} A_i$, define μ'' on $\mathscr{B}(S)$ by $\mu''(A) = \mu'(A) - \mu'(A \cap B_0)$, and let $I_1 = \{1, \ldots, n\} - I_0$. Then, for all $I \subset I_1$,

$$\begin{aligned} \sum_{i \in I} p_i' + \mu'(B_0) &= \sum_{i \in I \cup I_0} p_i' \\ &\le \mu'\left(\bigcup_{i \in I \cup I_0} A_i\right) \\ &= \mu'\left(\bigcup_{i \in I} A_i\right) + \mu'(B_0) - \mu'\left(\bigcup_{i \in I} A_i \cap B_0\right) \\ &= \mu''\left(\bigcup_{i \in I} A_i\right) + \mu'(B_0). \end{aligned} \tag{1.16}$$

Here, equality in the first line holds because equality in (1.15) holds for $I = I_0$, while the inequality in the second line follows from (1.14) if $n \notin I$ and from (1.13) if $n \in I$; more specifically, if $n \in I$, then

$$\sum_{i \in I \cup I_0} p_i' \le \mu\left(\bigcup_{i \in I \cup I_0} A_i\right) - \varepsilon_0 = \mu'\left(\bigcup_{i \in I \cup I_0} A_i\right). \tag{1.17}$$

By (1.16),

$$\sum_{i \in I} p_i' \le \mu''\left(\bigcup_{i \in I} A_i\right) \quad \text{for all} \quad I \subset I_1, \tag{1.18}$$

so by the induction hypothesis, there exist positive Borel measures λ_i' on S, $i \in I_1$, such that $\lambda_i'(A_i) = \lambda_i'(S) = p_i'$ for each $i \in I_1$ and $\sum_{i \in I_1} \lambda_i'(A) \le \mu''(A)$ for all $A \in \mathscr{B}(S)$. Finally, let $\lambda_i = \lambda_i'$ for $i \in I_1 - \{n\}$ and $\lambda_n = \lambda_n' + \varepsilon_0 n$. Then $\lambda_i(A_i) = \lambda_i(S) = p_i$ for each $i \in I_1$, hence for $i = 1, \ldots, n$, and

$$\begin{aligned} \sum_{i=1}^{n} \lambda_i(A) &= \sum_{i \in I_0} \lambda_i(A) + \sum_{i \in I_1} \lambda_i(A) \\ &= \sum_{i \in I_0} \lambda_i(A \cap B_0) + \sum_{i \in I_1} \lambda_i'(A) + \varepsilon_0 \eta(A) \\ &\le \mu'(A \cap B_0) + \mu''(A) + \varepsilon_0 \eta(A) \\ &= \mu'(A) + \varepsilon_0 \eta(A) \\ &= \mu(A) \end{aligned} \tag{1.19}$$

for all $A \in \mathscr{B}(S)$, so $\lambda_1, \ldots, \lambda_n$ again have the required properties. □

1.5 Corollary Let μ be a finite positive Borel measure on S, and let $p_i \ge 0$ and $A_i \in \mathscr{B}(S)$ for $i = 1, \ldots, n$. Let $\varepsilon > 0$, and suppose that

$$\sum_{i \in I} p_i \le \mu\left(\bigcup_{i \in I} A_i\right) + \varepsilon \quad \text{for all} \quad I \subset \{1, \ldots, n\}. \tag{1.20}$$

Then there exist positive Borel measures $\lambda_1, \ldots, \lambda_n$ on S such that $\lambda_i(A_i) = \lambda_i(S) \le p_i$ for $i = 1, \ldots, n$, $\sum_{i=1}^n \lambda_i(S) \ge \sum_{i=1}^n p_i - \varepsilon$, and $\sum_{i=1}^n \lambda_i(A) \le \mu(A)$ for all $A \in \mathscr{B}(S)$.

Proof. Let $S' = S \cup \{\Delta\}$, where Δ is an isolated point not belonging to S. Extend μ to a Borel measure on S' by defining $\mu(\{\Delta\}) = \varepsilon$. Letting $A_i' = A_i \cup \{\Delta\}$ for $i = 1, \ldots, n$, we have

$$\sum_{i \in I} p_i \le \mu\left(\bigcup_{i \in I} A_i'\right) \quad \text{for all} \quad I \subset \{1, \ldots, n\}. \tag{1.21}$$

By Lemma 1.4, there exist positive Borel measures $\lambda_1', \ldots, \lambda_n'$ on S' such that $\lambda_i'(A_i') = \lambda_i'(S') = p_i$ for $i = 1, \ldots, n$ and $\sum_{i=1}^n \lambda_i'(A) \le \mu(A)$ for all $A \in \mathscr{B}(S')$. Let λ_i be the restriction of λ_i' to $\mathscr{B}(S)$ for $i = 1, \ldots, n$. Then $\lambda_i(A_i) = \lambda_i'(A_i) \le \lambda_i'(A_i') = p_i$ and $\lambda_i(S - A_i) = \lambda_i'(S' - A_i') = 0$ for $i = 1, \ldots, n$. Also,

$$\sum_{i=1}^{n} \lambda_i(S) = \sum_{i=1}^{n} [p_i - \lambda_i'(\{\Delta\})] \ge \sum_{i=1}^{n} p_i - \mu(\{\Delta\}) = \sum_{i=1}^{n} p_i - \varepsilon \tag{1.22}$$

and $\sum_{i=1}^n \lambda_i(A) = \sum_{i=1}^n \lambda_i'(A) \le \mu(A)$ for all $A \in \mathscr{B}(S)$. □

Proof of Lemma 1.3 Let $P, Q, \varepsilon, \delta$, and $E_0, \ldots, E_N$ be as in the statement of the lemma. Let $p_i = P(E_i)$ and $A_i = E_i^\varepsilon$ for $i = 1, \ldots, N$. Then

$$\sum_{i \in I} p_i \le P\left(\overline{\bigcup_{i \in I} E_i}\right) \le Q\left(\bigcup_{i \in I} A_i\right) + \varepsilon \quad \text{for all} \quad I \subset \{1, \ldots, N\}, \tag{1.23}$$

so by Corollary 1.5, there exist positive Borel measures $\lambda_1, \ldots, \lambda_N$ on S such that $\lambda_i(A_i) = \lambda_i(S) \le p_i$ for $i = 1, \ldots, N$,

$$\sum_{i=1}^N \lambda_i(S) \ge \sum_{i=1}^N p_i - \varepsilon, \tag{1.24}$$

and $\sum_{i=1}^N \lambda_i(A) \le Q(A)$ for all $A \in \mathscr{B}(S)$. Define $c_1, \ldots, c_N \in [0, 1]$ by $c_i = (p_i - \lambda_i(S))/p_i$, where $0/0 = 0$, and note that $(1 - c_i)P(E_i) = \lambda_i(S)$ for $i = 1, \ldots, N$ and $P(E_0) + \sum_{i=1}^N c_i P(E_i) = 1 - \sum_{i=1}^N \lambda_i(S)$. Consequently, there exist $Q_0, \ldots, Q_N \in \mathscr{P}(S)$ such that

$$Q_i(B)(1 - c_i)P(E_i) = \lambda_i(B), \qquad i = 1, \ldots, N, \tag{1.25}$$

and

$$Q_0(B)\left(P(E_0) + \sum_{i=1}^N c_i P(E_i)\right) = Q(B) - \sum_{i=1}^N \lambda_i(B) \tag{1.26}$$

for all $B \in \mathscr{B}(S)$.

Let $X, Y_0, \ldots, Y_N$, and ξ be independent random variables on some probability space $(\Omega, \mathscr{F}, \nu)$ with $X, Y_0, \ldots, Y_N$ having distributions $P, Q_0, \ldots, Q_N$ and ξ uniformly distributed on $[0, 1]$. We can assume that $Y_1, \ldots, Y_N$ take values in $A_1, \ldots, A_N$, respectively. Defining Y by (1.10), we have by (1.25) and (1.26),

$$\begin{aligned} \nu\{Y \in B\} &= \sum_{i=1}^N Q_i(B)(1 - c_i)P(E_i) \\ &\quad + Q_0(B)\left(P(E_0) + \sum_{i=1}^N c_i P(E_i)\right) \\ &= Q(B) \end{aligned} \tag{1.27}$$

for all $B \in \mathscr{B}(S)$. Noting that $\{X \in E_i, \ \xi \ge c_i\} \subset \{X \in E_i, \ Y \in A_i\} \subset \{d(X, Y) < \delta + \varepsilon\}$ for $i = 1, \ldots, N$, we have

$$\begin{aligned} \{d(X, Y) \ge \delta + \varepsilon\} &\subset \{X \in E_0\} \cup \bigcup_{i=1}^N \{X \in E_i, \xi < c_i\} \\ &\subset \{X \in E_0\} \cup \left\{\xi < \bigvee_{i=1}^N c_i\right\} \\ &\subset \{X \in E_0\} \cup \left\{\xi < \max\left[\frac{\varepsilon}{P(E_i)} : P(E_i) > 0\right]\right\}, \end{aligned} \tag{1.28}$$

where the third containment follows from $p_i - \lambda_i(S) \le \varepsilon$ for $i = 1, \ldots, N$ (see (1.24)). Finally, by the first containment in (1.28) and by (1.24),

$$\begin{aligned} (1.29) \qquad \nu\{d(X, Y) \ge \delta + \varepsilon\} &\le P(E_0) + \sum_{i=1}^{N} c_i P(E_i) \\ &= P(E_0) + \sum_{i=1}^{N} (p_i - \lambda_i(S)) \\ &\le \delta + \varepsilon. \end{aligned}$$

□

1.6 Corollary Let (S, d) be separable. Suppose that X_n, $n = 1, 2, \ldots$, and X are S-valued random variables defined on the same probability space with distributions P_n, $n = 1, 2, \ldots$, and P, respectively. If $d(X_n, X) \to 0$ in probability as $n \to \infty$, then $\lim_{n\to\infty} \rho(P_n, P) = 0$.

Proof. For $n = 1, 2, \ldots$, let μ_n be the joint distribution of X_n and X. Then $\lim_{n\to\infty} \mu_n\{(x, y): d(x, y) \ge \varepsilon\} = 0$ for every $\varepsilon > 0$, so the result follows from Theorem 1.2. □

The next result shows that the metric space $(\mathscr{P}(S), \rho)$ is complete and separable whenever (S, d) is. We note that while separability is a topological property, completeness is a property of the metric.

1.7 Theorem If S is separable, then $\mathscr{P}(S)$ is separable. If in addition (S, d) is complete, then $(\mathscr{P}(S), \rho)$ is complete.

Proof. Let $\{x_n\}$ be a countable dense subset of S, and let δ_x denote the element of $\mathscr{P}(S)$ with unit mass at $x \in S$. We leave it to the reader to show that the probability measures of the form $\sum_{i=1}^{N} a_i \delta_{x_i}$ with N finite, a_i rational, and $\sum_{i=1}^{N} a_i = 1$, comprise a dense subset of $\mathscr{P}(S)$ (Problem 3).

To prove completeness it is enough to consider sequences $\{P_n\} \subset \mathscr{P}(S)$ with $\rho(P_{n-1}, P_n) < 2^{-n}$ for each $n \ge 2$. For $n = 2, 3, \ldots$, choose $E_1^{(n)}, \ldots, E_{N_n}^{(n)} \in \mathscr{B}(S)$ disjoint with diameters less than 2^{-n} and with $P_{n-1}(E_0^{(n)}) \le 2^{-n}$, where $E_0^{(n)} = S - \bigcup_{i=1}^{N_n} E_i^{(n)}$. By Lemma 1.3, there exists a probability space $(\Omega, \mathscr{F}, \nu)$ on which are defined S-valued random variables $Y_0^{(n)}, \ldots, Y_{N_n}^{(n)}$, $n = 2, 3, \ldots$, $[0, 1]$-valued random variables $\xi^{(n)}$, $n = 2, 3, \ldots$, and an S-valued random variable X_1 with distribution P_1, all of which are independent, such that if the constants $c_1^{(n)}, \ldots, c_{N_n}^{(n)} \in [0, 1]$, $n = 2, 3, \ldots$, are appropriately chosen, then the random variable

$$(1.30) \qquad X_n = \begin{cases} Y_i^{(n)} & \text{on } \{X_{n-1} \in E_i^{(n)}, \xi^{(n)} \ge c_i^{(n)}\}, \quad i = 1, \ldots, N_n, \\ Y_0^{(n)} & \text{on } \{X_{n-1} \in E_0^{(n)}\} \cup \bigcup_{i=1}^{N_n} \{X_{n-1} \in E_i^{(n)}, \xi^{(n)} < c_i^{(n)}\} \end{cases}$$

has distribution P_n and

$$\nu\{d(X_{n-1}, X_n) \geq 2^{-n+1}\} \leq 2^{-n+1}, \tag{1.31}$$

successively for $n = 2, 3, \ldots$. By the Borel–Cantelli lemma,

$$\nu\left\{\sum_{n=2}^{\infty} d(X_{n-1}, X_n) < \infty\right\} = 1, \tag{1.32}$$

so by the completeness of (S, d), $\lim_{n\to\infty} X_n \equiv X$ exists a.s. Letting P be the distribution of X, Corollary 1.6 implies that $\lim_{n\to\infty} \rho(P_n, P) = 0$. □

As a further application of Lemma 1.3, we derive the so-called Skorohod representation.

1.8 Theorem Let (S, d) be separable. Suppose P_n, $n = 1, 2, \ldots$, and P in $\mathscr{P}(S)$ satisfy $\lim_{n\to\infty} \rho(P_n, P) = 0$. Then there exists a probability space $(\Omega, \mathscr{F}, \nu)$ on which are defined S-valued random variables X_n, $n = 1, 2, \ldots$, and X with distributions P_n, $n = 1, 2, \ldots$, and P, respectively, such that $\lim_{n\to\infty} X_n = X$ a.s.

Proof. For $k = 1, 2, \ldots$, choose $E_1^{(k)}, \ldots, E_{N_k}^{(k)} \in \mathscr{B}(S)$ disjoint with diameters less than 2^{-k} and with $P(E_0^{(k)}) \leq 2^{-k}$, where $E_0^{(k)} = S - \bigcup_{i=1}^{N_k} E_i^{(k)}$, and assume (without loss of generality) that $\varepsilon_k \equiv \min_{1\leq i\leq N_k} P(E_i^{(k)}) > 0$. Define the sequence $\{k_n\}$ by

$$k_n = \max\,\{1\} \cup \left\{k \geq 1 : \rho(P_n, P) < \frac{\varepsilon_k}{k}\right\}, \tag{1.33}$$

and apply Lemma 1.3 with $Q = P_n$, $\varepsilon = \varepsilon_{k_n}/k_n$ if $k_n > 1$ and $\varepsilon = \rho(P_n, P) + 1/n$ if $k_n = 1$, $\delta = 2^{-k_n}$, $E_i = E_i^{(k_n)}$, and $N = N_{k_n}$ for $n = 1, 2, \ldots$. We conclude that there exists a probability space $(\Omega, \mathscr{F}, \nu)$ on which are defined S-valued random variables $Y_0^{(n)}, \ldots, Y_{N_{k_n}}^{(n)}$, $n = 1, 2, \ldots$, a random variable ξ uniformly distributed on $[0, 1]$, and an S-valued random variable X with distribution P, all of which are independent, such that if the constants $c_1^{(n)}, \ldots, c_{N_{k_n}}^{(n)} \in [0, 1]$, $n = 1, 2, \ldots$, are appropriately chosen, then the random variable

$$X_n = \begin{cases} Y_i^{(n)} & \text{on } \{X \in E_i^{(k_n)}, \xi \geq c_i^{(n)}\}, \quad i = 1, \ldots, N_{k_n}, \\ Y_0^{(n)} & \text{on } \{X \in E_0^{(k_n)}\} \cup \bigcup_{i=1}^{N_{k_n}} \{X \in E_i^{(k_n)}, \xi < c_i^{(n)}\} \end{cases} \tag{1.34}$$

has distribution P_n and

$$\left\{d(X_n, X) \geq 2^{-k_n} + \frac{\varepsilon_{k_n}}{k_n}\right\} \tag{1.35}$$

$$\subset \{X \in E_0^{(k_n)}\} \cup \left\{\xi < \frac{1}{k_n}\right\} \quad \text{if} \quad k_n > 1$$

for $n = 1, 2, \ldots$. If $K_n \equiv \min_{m \geq n} k_m > 1$, then

$$\nu\left(\bigcup_{m=n}^{\infty} \left\{d(X_m, X) \geq 2^{-k_m} + \frac{\varepsilon_{k_m}}{k_m}\right\}\right) \leq \sum_{k=K_n}^{\infty} \nu\{X \in E_0^{(k)}\} + \nu\left\{\xi < \frac{1}{K_n}\right\} \leq 2^{-K_n+1} + \frac{1}{K_n}, \tag{1.36}$$

and since $\lim_{n\to\infty} K_n = \infty$, we have $\lim_{n\to\infty} X_n = X$ a.s. □

We conclude this section by proving the continuous mapping theorem.

1.9 Corollary Let (S, d) and (S', d') be separable metric spaces, and let $h: S \to S'$ be Borel measurable. Suppose that P_n, $n = 1, 2, \ldots$, and P in $\mathscr{P}(S)$ satisfy $\lim_{n\to\infty} \rho(P_n, P) = 0$, and define Q_n, $n = 1, 2, \ldots$, and Q in $\mathscr{P}(S')$ by

$$Q_n = P_n h^{-1}, \quad Q = Ph^{-1}. \tag{1.37}$$

(By definition, $Ph^{-1}(B) = P\{s \in S: h(s) \in B\}$.) Let C_h be the set of points of S at which h is continuous. If $P(C_h) = 1$, then $\lim_{n\to\infty} \rho'(Q_n, Q) = 0$, where ρ' is the Prohorov metric on $\mathscr{P}(S')$.

Proof. By Theorem 1.8, there exists a probability space $(\Omega, \mathscr{F}, \nu)$ on which are defined S-valued random variables X_n, $n = 1, 2, \ldots$, and X with distributions P_n, $n = 1, 2, \ldots$, and P, respectively, such that $\lim_{n\to\infty} X_n = X$ a.s. Since $\nu\{X \in C_h\} = 1$, we have $\lim_{n\to\infty} h(X_n) = h(X)$ a.s., and by Corollary 1.6, this implies that $\lim_{n\to\infty} \rho'(Q_n, Q) = 0$. □

2. PROHOROV'S THEOREM

We are primarily interested in the convergence of sequences of Borel probability measures on the metric space (S, d). A common approach for verifying the convergence of a sequence $\{x_n\}$ in a metric space is to first show that $\{x_n\}$ is contained in some compact set and then to show that every convergent subsequence of $\{x_n\}$ must converge to the same element x. This then implies that $\lim_{n\to\infty} x_n = x$. We use this argument repeatedly in what follows, and, consequently, a characterization of the compact subsets of $\mathscr{P}(S)$ is crucial. This characterization is given by the theorem of Prohorov that relates compactness to the notion of tightness.

A probability measure $P \in \mathscr{P}(S)$ is said to be *tight* if for each $\varepsilon > 0$ there exists a compact set $K \subset S$ such that $P(K) \geq 1 - \varepsilon$. A family of probability measures $\mathscr{M} \subset \mathscr{P}(S)$ is *tight* if for each $\varepsilon > 0$ there exists a compact set $K \subset S$

such that

$$\inf_{P \in \mathscr{M}} P(K) \geq 1 - \varepsilon. \tag{2.1}$$

2.1 Lemma If (S, d) is complete and separable, then each $P \in \mathscr{P}(S)$ is tight.

Proof. Let $\{x_k\}$ be dense in S, and let $P \in \mathscr{P}(S)$. Given $\varepsilon > 0$, choose positive integers $N_1, N_2, \ldots$ such that

$$P\left(\bigcup_{k=1}^{N_n} B\left(x_k, \frac{1}{n}\right)\right) \geq 1 - \frac{\varepsilon}{2^n} \tag{2.2}$$

for $n = 1, 2, \ldots$. Let K be the closure of $\bigcap_{n \geq 1} \bigcup_{k=1}^{N_n} B(x_k, 1/n)$. Then K is totally bounded and hence compact, and

$$P(K) \geq 1 - \sum_{n=1}^{\infty} \frac{\varepsilon}{2^n} = 1 - \varepsilon. \tag{2.3}$$ □

2.2 Theorem Let (S, d) be complete and separable, and let $\mathscr{M} \subset \mathscr{P}(S)$. Then the following are equivalent:

(a) $\mathscr{M}$ is tight.

(b) For each $\varepsilon > 0$, there exists a compact set $K \subset S$ such that

$$\inf_{P \in \mathscr{M}} P(K^{\varepsilon}) \geq 1 - \varepsilon, \tag{2.4}$$

where K^{ε} is as in (1.2).

(c) $\mathscr{M}$ is relatively compact.

Proof. **(a ⇒ b)** Immediate.

(b ⇒ c) Since the closure of $\mathscr{M}$ is complete by Theorem 1.7, it is sufficient to show that $\mathscr{M}$ is totally bounded. That is, given $\delta > 0$, we must construct a finite set $\mathscr{N} \subset \mathscr{P}(S)$ such that $\mathscr{M} \subset \{Q: \rho(P, Q) < \delta$ for some $P \in \mathscr{N}\}$.

Let $0 < \varepsilon < \delta/2$ and choose a compact set $K \subset S$ such that (2.4) holds. By the compactness of K, there exists a finite set $\{x_1, \ldots, x_n\} \subset K$ such that $K^{\varepsilon} \subset \bigcup_{i=1}^{n} B_i$, where $B_i = B(x_i, 2\varepsilon)$. Fix $x_0 \in S$ and $m \geq n/\varepsilon$, and let $\mathscr{N}$ be the collection of probability measures of the form

$$P = \sum_{i=0}^{n} \left(\frac{k_i}{m}\right) \delta_{x_i}, \tag{2.5}$$

where $0 \leq k_i \leq m$ and $\sum_{i=0}^{n} k_i = m$.

Given $Q \in \mathscr{M}$, let $k_i = [mQ(E_i)]$ for $i = 1, \ldots, n$, where $E_i = B_i$

$-\bigcup_{j=1}^{i-1} B_j$, and let $k_0 = m - \sum_{i=1}^{n} k_i$. Then, defining P by (2.5), we have

$$\text{(2.6)} \qquad \begin{aligned} Q(F) &\le Q\Bigl(\bigcup_{F \cap E_i \neq \varnothing} E_i\Bigr) + \varepsilon \\ &\le \sum_{F \cap E_i \neq \varnothing} \frac{[mQ(E_i)]}{m} + \frac{n}{m} + \varepsilon \\ &\le P(F^{2\varepsilon}) + 2\varepsilon \end{aligned}$$

for all closed sets $F \subset S$, so $\rho(P, Q) \le 2\varepsilon < \delta$.

(c ⇒ a) Let $\varepsilon > 0$. Since $\mathscr{M}$ is totally bounded, there exists for $n = 1, 2, \ldots$ a finite subset $\mathscr{N}_n \subset \mathscr{M}$ such that $\mathscr{M} \subset \{Q: \rho(P, Q) < \varepsilon/2^{n+1}$ for some $P \in \mathscr{N}_n\}$. Lemma 2.1 implies that for $n = 1, 2, \ldots$ we can choose a compact set $K_n \subset S$ such that $P(K_n) \ge 1 - \varepsilon/2^{n+1}$ for all $P \in \mathscr{N}_n$. Given $Q \in \mathscr{M}$, it follows that for $n = 1, 2, \ldots$ there exists $P_n \in \mathscr{N}_n$ such that

$$\text{(2.7)} \qquad Q(K_n^{\varepsilon/2^{n+1}}) \ge P_n(K_n) - \varepsilon/2^{n+1} \ge 1 - \varepsilon/2^n.$$

Letting K be the closure of $\bigcap_{n \ge 1} K_n^{\varepsilon/2^{n+1}}$, we conclude that K is compact and

$$\text{(2.8)} \qquad Q(K) \ge 1 - \sum_{n=1}^{\infty} \frac{\varepsilon}{2^n} = 1 - \varepsilon. \qquad \square$$

2.3 Corollary Let (S, d) be arbitrary, and let $\mathscr{M} \subset \mathscr{P}(S)$. If $\mathscr{M}$ is tight, then $\mathscr{M}$ is relatively compact.

Proof. For each $m \ge 1$ there exists a compact set $K_m \subset S$ such that

$$\text{(2.9)} \qquad \inf_{P \in \mathscr{M}} P(K_m) \ge 1 - \frac{1}{m},$$

and we can assume that $K_1 \subset K_2 \subset \cdots$. For every $P \in \mathscr{M}$ and $m \ge 1$, define $P^{(m)} \in \mathscr{P}(S)$ by $P^{(m)}(A) = P(A \cap K_m)/P(K_m)$, and note that $P^{(m)}$ may be regarded as belonging to $\mathscr{P}(K_m)$. Since compact metric spaces are complete and separable, $\mathscr{M}^{(m)} = \{P^{(m)}: P \in \mathscr{M}\}$ is relatively compact in $\mathscr{P}(K_m)$ for each $m \ge 1$ by Theorem 2.2.

We also have

$$P(A) \leq P(A \cap K_m) + \frac{1}{m} \leq P^{(m)}(A) + \frac{1}{m}, \tag{2.10}$$

$$P^{(m')}(A) \leq \frac{(P(A \cap K_m) + 1/m)}{P(K_{m'})} \leq P^{(m)}(A) + \frac{2}{m}, \quad m' > m, \tag{2.11}$$

$$P(A) \geq P(K_m)P^{(m)}(A) \geq \left(1 - \frac{1}{m}\right)P^{(m)}(A), \tag{2.12}$$

$$P^{(m')}(A) \geq \frac{P(K_m)P^{(m)}(A)}{P(K_{m'})} \geq \left(1 - \frac{1}{m}\right)P^{(m)}(A), \qquad m' > m, \tag{2.13}$$

for all $P \in \mathscr{M}$, $A \in \mathscr{B}(S)$, and $m \geq 1$. By (2.10),

$$\rho(P, P^{(m)}) \leq \frac{1}{m} \tag{2.14}$$

for all $P \in \mathscr{M}$ and $m \geq 1$. Given $A_1, A_2, \ldots \in \mathscr{B}(S)$ disjoint, (2.13) and (2.11) imply that

$$\begin{aligned} &\sum_i |P^{(m')}(A_i) - P^{(m)}(A_i)| \\ &\leq \sum_i \left(P^{(m')}(A_i) - \left(1 - \frac{1}{m}\right)P^{(m)}(A_i) + \frac{1}{m} P^{(m)}(A_i)\right) \\ &\leq P^{(m')}\left(\bigcup_i A_i\right) - P^{(m)}\left(\bigcup_i A_i\right) + \frac{2}{m} \\ &\leq \frac{2}{m} + \frac{2}{m} = \frac{4}{m} \end{aligned} \tag{2.15}$$

for every $P \in \mathscr{M}$ and $m' > m \geq 1$.

Let $\{P_n\} \subset \mathscr{M}$. By the relative compactness of $\mathscr{M}^{(m)}$ in $\mathscr{P}(K_m)$, there exists (through a diagonalization argument) a subsequence $\{P_{n_k}\} \subset \{P_n\}$ and $Q^{(m)} \in \mathscr{P}(K_m)$ such that

$$\lim_{k \to \infty} \rho(P_{n_k}^{(m)}, Q^{(m)}) = 0 \tag{2.16}$$

for every $m \geq 1$. It follows that

$$Q^{(m)}(F) = \lim_{\varepsilon \to 0} \overline{\lim_{k \to \infty}} P_{n_k}^{(m)}(F^\varepsilon) \tag{2.17}$$

for all closed sets $F \subset S$ and $m \geq 1$, and therefore the inequalities (2.11) and (2.13) are preserved for the $Q^{(m)}$ for all closed sets $A \subset S$, hence for all $A \in \mathscr{B}(S)$ (using the regularity of the $Q^{(m)}$). Consequently, we have (2.15) for the $Q^{(m)}$, so

$$Q(A) \equiv \lim_{m \to \infty} Q^{(m)}(A) \tag{2.18}$$

exists for every $A \in \mathscr{B}(S)$ and defines a probability measure $Q \in \mathscr{P}(S)$ (the countable additivity following from (2.15)). But

$$\rho(P_{n_k}, Q) \le \rho(P_{n_k}, P_{n_k}^{(m)}) + \rho(P_{n_k}^{(m)}, Q^{(m)}) + \rho(Q^{(m)}, Q) \le \frac{1}{m} + \rho(P_{n_k}^{(m)}, Q^{(m)}) + \frac{2}{m}, \tag{2.19}$$

for each k and $m \ge 1$, implying that $\lim_{k \to \infty} \rho(P_{n_k}, Q) = 0$. □

We conclude this section with a result concerning countably infinite product spaces.

2.4 Proposition Let (S_k, d_k), $k = 1, 2, \ldots$, be metric spaces, and define the metric space (S, d) by letting $S = \prod_{k=1}^{\infty} S_k$ and $d(x, y) = \sum_{k=1}^{\infty} 2^{-k}(d_k(x_k, y_k) \wedge 1)$ for all $x, y \in S$. Let $\{P_\alpha\} \subset \mathscr{P}(S)$ (where α ranges over some index set), and for $k = 1, 2, \ldots$ and each α, define $P_\alpha^k \in \mathscr{P}(S_k)$ to be the kth marginal distribution of P_α (i.e., $P_\alpha^k = P_\alpha \pi_k^{-1}$, where the projection $\pi_k \colon S \to S_k$ is given by $\pi_k(x) = x_k$). Then $\{P_\alpha\}$ is tight if and only if $\{P_\alpha^k\}$ is tight for $k = 1, 2, \ldots$.

Proof. Suppose that $\{P_\alpha^k\}$ is tight for $k = 1, 2, \ldots$, and let $\varepsilon > 0$. For $k = 1, 2, \ldots$, choose a compact set $K_k \subset S_k$ such that $\inf_\alpha P_\alpha^k(K_k) \ge 1 - \varepsilon/2^k$. Then $K \equiv \prod_{k=1}^{\infty} K_k = \bigcap_{k=1}^{\infty} \pi_k^{-1}(K_k)$ is compact in S, and

$$P_\alpha(K) \ge 1 - \sum_{k=1}^{\infty} (1 - P_\alpha^k(K_k)) \ge 1 - \varepsilon \tag{2.20}$$

for all α. Consequently, $\{P_\alpha\}$ is tight.

The converse follows by observing that for each compact set $K \subset S$, $\pi_k(K)$ is compact in S_k and

$$\inf_\alpha P_\alpha^k(\pi_k(K)) \ge \inf_\alpha P_\alpha(K) \tag{2.21}$$

for $k = 1, 2, \ldots$. □

3. WEAK CONVERGENCE

Let $\bar{C}(S)$ be the space of real-valued bounded continuous functions on the metric space (S, d) with norm $\|f\| = \sup_{x \in S} |f(x)|$. A sequence $\{P_n\} \subset \mathscr{P}(S)$ is said to *converge weakly* to $P \in \mathscr{P}(S)$ if

$$\lim_{n \to \infty} \int f\, dP_n = \int f\, dP, \qquad f \in \bar{C}(S). \tag{3.1}$$

The *distribution* of an S-valued random variable X, denoted by PX^{-1}, is the element of $\mathscr{P}(S)$ given by $PX^{-1}(B) = P\{X \in B\}$. A sequence $\{X_n\}$ of S-valued random variables is said to *converge in distribution* to the S-valued random variable X if $\{PX_n^{-1}\}$ converges weakly to PX^{-1}, or equivalently, if

$$\lim_{n\to\infty} E[f(X_n)] = E[f(X)], \qquad f \in \bar{C}(S). \tag{3.2}$$

Weak convergence is denoted by $P_n \Rightarrow P$ and convergence in distribution by $X_n \Rightarrow X$. When it is useful to emphasize which metric space is involved, we write "$P_n \Rightarrow P$ on S" or "$X_n \Rightarrow X$ in S".

If S' is a second metric space and $f: S \to S'$ is continuous, we note that then $X_n \Rightarrow X$ in S implies $f(X_n) \Rightarrow f(X)$ in S' since $g \in \bar{C}(S')$ implies $g \circ f \in \bar{C}(S)$. For example, if $S = C[0, 1]$ and $S' = \mathbb{R}$, then $f(x) \equiv \sup_{0 \le t \le 1} x(t)$ is continuous, so $X_n \Rightarrow X$ in $C[0, 1]$ implies $\sup_{0 \le t \le 1} X_n(t) \Rightarrow \sup_{0 \le t \le 1} X(t)$ in $\mathbb{R}$. Recall that, if $S = \mathbb{R}$, then (3.2) is equivalent to

$$\lim_{n\to\infty} P\{X_n \le x\} = P\{X \le x\} \tag{3.3}$$

for all x at which the right side of (3.3) is continuous.

We now show that weak convergence is equivalent to convergence in the Prohorov metric. The *boundary* of a subset $A \subset S$ is given by $\partial A = \bar{A} \cap \overline{A^c}$ ($\bar{A}$ and A^c denote the closure and complement of A, respectively). A is said to be a *P-continuity set* if $A \in \mathscr{B}(S)$ and $P(\partial A) = 0$.

3.1 Theorem Let (S, d) be arbitrary, and let $\{P_n\} \subset \mathscr{P}(S)$ and $P \in \mathscr{P}(S)$. Of the following conditions, (b) through (f) are equivalent and are implied by (a). If S is separable, then all six conditions are equivalent:

- **(a)** $\lim_{n\to\infty} \rho(P_n, P) = 0$.
- **(b)** $P_n \Rightarrow P$.
- **(c)** $\lim_{n\to\infty} \int f\, dP_n = \int f\, dP$ for all uniformly continuous $f \in \bar{C}(S)$.
- **(d)** $\overline{\lim}_{n\to\infty} P_n(F) \le P(F)$ for all closed sets $F \subset S$.
- **(e)** $\underline{\lim}_{n\to\infty} P_n(G) \ge P(G)$ for all open sets $G \subset S$.
- **(f)** $\lim_{n\to\infty} P_n(A) = P(A)$ for all P-continuity sets $A \subset S$.

Proof. **(a ⇒ b)** For each n, let $\varepsilon_n = \rho(P_n, P) + 1/n$. Given $f \in \bar{C}(S)$ with $f \ge 0$,

$$\int f\, dP_n = \int_0^{\|f\|} P_n\{f \ge t\}\, dt \le \int_0^{\|f\|} P(\{f \ge t\}^{\varepsilon_n})\, dt + \varepsilon_n \|f\| \tag{3.4}$$

for every n, so

$$\begin{aligned} \overline{\lim_{n\to\infty}} \int f\, dP_n &\le \lim_{n\to\infty} \int_0^{\|f\|} P(\{f \ge t\}^{\varepsilon_n})\, dt \\ &= \int_0^{\|f\|} P\{f \ge t\}\, dt = \int f\, dP. \end{aligned} \tag{3.5}$$

Consequently,

$$\overline{\lim_{n\to\infty}} \int (\|f\| + f)\, dP_n \le \int (\|f\| + f)\, dP,$$

(3.6)

$$\overline{\lim_{n\to\infty}} \int (\|f\| - f)\, dP_n \le \int (\|f\| - f)\, dP$$

for all $f \in \bar{C}(S)$, and this implies (3.1).

(b ⇒ c) Immediate.

(c ⇒ d) Let $F \subset S$ be closed. For each $\varepsilon > 0$, define $f_\varepsilon \in \bar{C}(S)$ by

$$f_\varepsilon(x) = \left(1 - \frac{d(x, F)}{\varepsilon}\right) \vee 0, \tag{3.7}$$

where $d(x, F) = \inf_{y \in F} d(x, y)$. Then f is uniformly continuous, so

$$\overline{\lim_{n\to\infty}} P_n(F) \le \lim_{n\to\infty} \int f_\varepsilon\, dP_n = \int f_\varepsilon\, dP, \tag{3.8}$$

for each $\varepsilon > 0$, and therefore

$$\overline{\lim_{n\to\infty}} P_n(F) \le \lim_{\varepsilon\to 0} \int f_\varepsilon\, dP = P(F). \tag{3.9}$$

(d ⇒ e) For every open set $G \subset S$,

$$\underline{\lim_{n\to\infty}} P_n(G) = 1 - \overline{\lim_{n\to\infty}} P_n(G^c) \ge 1 - P(G^c) = P(G). \tag{3.10}$$

(e ⇒ f) Let A be a P-continuity set in S, and let A° denote its interior ($A^\circ = A - \partial A$). Then

$$\overline{\lim_{n\to\infty}} P_n(A) \le \overline{\lim_{n\to\infty}} P_n(\bar{A}) = 1 - \underline{\lim_{n\to\infty}} P_n(\bar{A}^c) \le 1 - P(\bar{A}^c) = P(A) \tag{3.11}$$

and

$$\underline{\lim_{n\to\infty}} P_n(A) \ge \underline{\lim_{n\to\infty}} P_n(A^\circ) \ge P(A^\circ) = P(A). \tag{3.12}$$

(f ⇒ b) Let $f \in \bar{C}(S)$ with $f \ge 0$. Then $\partial\{f \ge t\} \subset \{f = t\}$, so $\{f \ge t\}$ is a P-continuity set for all but at most countably many $t \ge 0$. Therefore,

$$\begin{aligned} \lim_{n\to\infty} \int f\, dP_n &= \lim_{n\to\infty} \int_0^{\|f\|} P_n\{f \ge t\}\, dt \\ &= \int_0^{\|f\|} P\{f \ge t\}\, dt = \int f\, dP \end{aligned} \tag{3.13}$$

for all nonnegative $f \in \bar{C}(S)$, which clearly implies (3.1).

(e ⇒ a, assuming separability) Let $\varepsilon > 0$ be arbitrary, and let $E_1, E_2, \ldots \in \mathscr{B}(S)$ be a partition of S with diameter$(E_i) < \varepsilon/2$ for $i = 1, 2, \ldots$. Let N be the smallest positive integer n such that $P(\bigcup_{i=1}^n E_i) > 1 - \varepsilon/2$, and let $\mathscr{G}$ be the (finite) collection of open sets of the form $(\bigcup_{i \in I} E_i)^{\varepsilon/2}$, where $I \subset \{1, \ldots, N\}$. Since $\mathscr{G}$ is finite, there exists n_0 such that $P(G) \leq P_n(G) + \varepsilon/2$ for all $G \in \mathscr{G}$ and $n \geq n_0$. Given $F \in \mathscr{C}$, let

$$F_0 = \bigcup \{E_i: 1 \leq i \leq N,\ E_i \cap F \neq \varnothing\}. \tag{3.14}$$

Then $F_0^{\varepsilon/2} \in \mathscr{G}$ and

$$\begin{aligned} P(F) &\leq P(F_0^{\varepsilon/2}) + \varepsilon/2 \\ &\leq P_n(F_0^{\varepsilon/2}) + \varepsilon \leq P_n(F^\varepsilon) + \varepsilon \end{aligned} \tag{3.15}$$

for all $n \geq n_0$. Hence $\rho(P_n, P) \leq \varepsilon$ for each $n \geq n_0$. □

3.2 Corollary Let P_n, $n = 1, 2, \ldots$, and P belong to $\mathscr{P}(S)$, and let $S' \in \mathscr{B}(S)$. For $n = 1, 2, \ldots$, suppose that $P_n(S') = P(S') = 1$, and let P_n' and P' be the restrictions of P_n and P to $\mathscr{B}(S')$ (of course, S' has the relative topology). Then $P_n \Rightarrow P$ on S if and only if $P_n' \Rightarrow P'$ on S'.

Proof. If G' is open in S', then $G' = G \cap S'$ for some open set $G \subset S$. Therefore, if $P_n \Rightarrow P$ on S,

$$\varliminf_{n\to\infty} P_n'(G') = \varliminf_{n\to\infty} P_n(G) \geq P(G) = P'(G'), \tag{3.16}$$

so $P_n' \Rightarrow P'$ on S' by Theorem 3.1. The converse is proved similarly. □

3.3 Corollary Let (S, d) be arbitrary, and let (X_n, Y_n), $n = 1, 2, \ldots$, and X be $(S \times S)$- and S-valued random variables. If $X_n \Rightarrow X$ and $d(X_n, Y_n) \to 0$ in probability, then $Y_n \Rightarrow X$.

3.4 Remark If S is separable, then $\mathscr{B}(S \times S) = \mathscr{B}(S) \times \mathscr{B}(S)$, and hence (X, Y) is an $(S \times S)$-valued random variable whenever X and Y are S-valued random variables defined on the same probability space. This observation has already been used implicitly in Section 1, and we use it again (without mention) in later sections of this chapter. □

Proof. If $f \in \bar{C}(S)$ is uniformly continuous, then

$$\lim_{n\to\infty} E[f(X_n) - f(Y_n)] = 0. \tag{3.17}$$

Consequently,

$$\lim_{n\to\infty} E[f(Y_n)] = \lim_{n\to\infty} E[f(X_n)] = E[f(X)], \tag{3.18}$$

and Theorem 3.1 is again applicable. □

4. SEPARATING AND CONVERGENCE DETERMINING SETS

Let (S, d) be a metric space. A sequence $\{f_n\} \subset B(S)$ is said to converge *boundedly and pointwise* to $f \in B(S)$ if $\sup_n \|f_n\| < \infty$ (where $\|\cdot\|$ denotes the sup norm) and $\lim_{n\to\infty} f_n(x) = f(x)$ for every $x \in S$; we denote this by

$$\text{bp-}\lim_{n\to\infty} f_n = f. \tag{4.1}$$

A set $M \subset B(S)$ is called *bp-closed* if whenever $\{f_n\} \subset M$, $f \in B(S)$, and (4.1) holds, we have $f \in M$. The *bp-closure* of $M \subset B(S)$ is the smallest bp-closed subset of $B(S)$ that contains M. Finally, if the bp-closure of $M \subset B(S)$ is equal to $B(S)$, we say that M is *bp-dense* in $B(S)$. We remark that if M is bp-dense in $B(S)$ and $f \in B(S)$, there need not exist a sequence $\{f_n\} \subset M$ such that (4.1) holds.

4.1 Lemma If $M \subset B(S)$ is a subspace, then the bp-closure of M is also a subspace.

Proof. Let H be the bp-closure of M. For each $f \in H$, define

$$H_f = \{g \in H: af + bg \in H \quad \text{for all} \quad a, b \in \mathbb{R}\}, \tag{4.2}$$

and note that H_f is bp-closed because H is. If $f \in M$, then $H_f \supset M$, so $H_f = H$. If $f \in H$, then $f \in H_g$ for every $g \in M$, hence $g \in H_f$ for every $g \in M$, and therefore $H_f = H$. □

4.2 Proposition Let (S, d) be arbitrary. Then $\bar{C}(S)$ is bp-dense in $B(S)$. If S is separable, then there exists a sequence $\{f_n\}$ of nonnegative functions in $\bar{C}(S)$ such that span $\{f_n\}$ is bp-dense in $B(S)$.

Proof. Let H be the bp-closure of $\bar{C}(S)$. H is closed under monotone convergence of uniformly bounded sequences, H is a subspace of $B(S)$ by Lemma 4.1, and $\chi_G \in H$ for every open set $G \subset S$. By the Dynkin class theorem for functions (Theorem 4.3 of the Appendixes), $H = B(S)$.

If S is separable, let $\{x_i\}$ be dense in S. For every open set $G \subset S$ that is a finite intersection of $B(x_i, 1/k)$, $i, k \geq 1$, choose a sequence $\{f_n^G\}$ of nonnegative functions in $\bar{C}(S)$ such that $\text{bp-}\lim_{n\to\infty} f_n^G = \chi_G$. The Dynkin class theorem for functions now applies to span $\{f_n^G: n, G \text{ as above}\}$. □

For future reference, we extend two of the definitions given at the beginning of this section. A set $M \subset B(S) \times B(S)$ is called *bp-closed* if whenever $\{(f_n, g_n)\} \subset M$, $(f, g) \in B(S) \times B(S)$, $\text{bp-}\lim_{n\to\infty} f_n = f$, and $\text{bp-}\lim_{n\to\infty} g_n = g$, we have $(f, g) \in M$. The *bp-closure* of $M \subset B(S) \times B(S)$ is the smallest bp-closed subset of $B(S) \times B(S)$ that contains M.

A set $M \subset \bar{C}(S)$ is called *separating* if whenever $P, Q \in \mathscr{P}(S)$ and

$$\int f\,dP = \int f\,dQ, \qquad f \in M, \tag{4.3}$$

we have $P = Q$. Also, M is called *convergence determining* if whenever $\{P_n\} \subset \mathscr{P}(S)$, $P \in \mathscr{P}(S)$, and

$$\lim_{n\to\infty} \int f\,dP_n = \int f\,dP, \qquad f \in M, \tag{4.4}$$

we have $P_n \Rightarrow P$.

Given P, $Q \in \mathscr{P}(S)$, the set of all $f \in B(S)$ such that $\int f\,dP = \int f\,dQ$ is bp-closed. Consequently, Proposition 4.2 implies that $\bar{C}(S)$ is itself separating. It follows that if $M \subset \bar{C}(S)$ is convergence determining, then M is separating. The converse is false in general, as Problem 8 indicates. However, if S is compact, then $\mathscr{P}(S)$ is compact by Theorem 2.2, and the following lemma implies that the two concepts are equivalent.

4.3 Lemma Let $\{P_n\} \subset \mathscr{P}(S)$ be relatively compact, let $P \in \mathscr{P}(S)$, and let $M \subset \bar{C}(S)$ be separating. If (4.4) holds, then $P_n \Rightarrow P$.

Proof. If Q is the weak limit of a convergent subsequence of $\{P_n\}$, then (4.4) implies (4.3), so $Q = P$. It follows that $P_n \Rightarrow P$. □

4.4 Proposition Let (S, d) be separable. The space of functions $f \in \bar{C}(S)$ that are uniformly continuous and have bounded support is convergence determining. If S is also locally compact, then $C_c(S)$, the space of $f \in \bar{C}(S)$ with compact support, is convergence determining.

Proof. Let $\{x_i\}$ be dense in S, and define $f_{ij} \in \bar{C}(S)$ for $i, j = 1, 2, \ldots$ by

$$f_{ij}(x) = 2(1 - jd(x, x_i)) \vee 0. \tag{4.5}$$

Given an open set $G \subset S$, define $g_m \in M$ for $m = 1, 2, \ldots$ by $g_m(x) = (\sum f_{ij}(x)) \wedge 1$, where the sum extends over those $i, j \leq m$ such that $\overline{B(x_i, 1/j)} \subset G$ (and $\overline{B(x_i, 1/j)}$ is compact if S is locally compact). If (4.4) holds, then

$$\varliminf_{n\to\infty} P_n(G) \geq \lim_{n\to\infty} \int g_m\,dP_n = \int g_m\,dP \tag{4.6}$$

for $m = 1, 2, \ldots$, so by letting $m \to \infty$, we conclude that condition (e) of Theorem 3.1 holds. □

Recall that a collection of functions $M \subset \bar{C}(S)$ is said to *separate points* if for every x, $y \in S$ with $x \neq y$ there exists $h \in M$ such that $h(x) \neq h(y)$. In addition, M is said to *strongly separate points* if for every $x \in S$ and $\delta > 0$ there exists a finite set $\{h_1, \ldots, h_k\} \subset M$ such that

$$\inf_{y:\, d(y,x) \geq \delta} \max_{1 \leq i \leq k} |h_i(y) - h_i(x)| > 0. \tag{4.7}$$

Clearly, if M strongly separates points, then M separates points.

4.5 Theorem Let (S, d) be complete and separable, and let $M \subset \bar{C}(S)$ be an algebra.

(a) If M separates points, then M is separating.

(b) If M strongly separates points, then M is convergence determining.

Proof. **(a)** Let P, $Q \in \mathscr{P}(S)$, and suppose that (4.3) holds. Then $\int h\,dP = \int h\,dQ$ for all h in the algebra $H = \{f + a : f \in M,\ a \in \mathbb{R}\}$, hence for all h in the closure (with respect to $\|\cdot\|$) of H. Let $g \in \bar{C}(S)$ and $\varepsilon > 0$ be arbitrary. By Lemma 2.1, there exists a compact set $K \subset S$ such that $P(K) \geq 1 - \varepsilon$ and $Q(K) \geq 1 - \varepsilon$. By the Stone–Weierstrass theorem, there exists a sequence $\{g_n\} \subset H$ such that $\sup_{x \in K} |g_n(x) - g(x)| \to 0$ as $n \to \infty$. Now observe that

$$\begin{aligned}
&\left| \int g e^{-\varepsilon g^2}\, dP - \int g e^{-\varepsilon g^2}\, dQ \right| \\
&\leq \left| \int_S g e^{-\varepsilon g^2}\, dP - \int_K g e^{-\varepsilon g^2}\, dP \right| \\
&\quad + \left| \int_K g e^{-\varepsilon g^2}\, dP - \int_K g_n e^{-\varepsilon g_n^2}\, dP \right| \\
&\quad + \left| \int_K g_n e^{-\varepsilon g_n^2}\, dP - \int_S g_n e^{-\varepsilon g_n^2}\, dP \right| \\
&\quad + \left| \int_S g_n e^{-\varepsilon g_n^2}\, dP - \int_S g_n e^{-\varepsilon g_n^2}\, dQ \right| \\
&\quad + \left| \int_S g_n e^{-\varepsilon g_n^2}\, dQ - \int_K g_n e^{-\varepsilon g_n^2}\, dQ \right| \\
&\quad + \left| \int_K g_n e^{-\varepsilon g_n^2}\, dQ - \int_K g e^{-\varepsilon g^2}\, dQ \right| \\
&\quad + \left| \int_K g e^{-\varepsilon g^2}\, dQ - \int_S g e^{-\varepsilon g^2}\, dQ \right|
\end{aligned} \tag{4.8}$$

for each n, and the fourth term on the right is zero since $g_n e^{-\varepsilon g_n^2}$ belongs to the closure of H. The second and sixth terms tend to zero as $n \to \infty$, so the left side of (4.8) is bounded by $4\gamma\sqrt{\varepsilon}$, where $\gamma = \sup_{t \geq 0} t e^{-t^2}$. Letting $\varepsilon \to 0$, it follows that $\int g\,dP = \int g\,dQ$. Since $g \in \bar{C}(S)$ was arbitrary and $\bar{C}(S)$ is separating, we conclude that $P = Q$.

(b) Let $\{P_n\} \subset \mathscr{P}(S)$ and $P \in \mathscr{P}(S)$, and suppose that (4.4) holds. By Lemma 4.3 and part (a), it suffices to show that $\{P_n\}$ is relatively compact.

Let $f_1, \dots, f_k \in M$. Then

$$\lim_{n\to\infty} \int g \circ (f_1, \dots, f_k)\, dP_n = \int g \circ (f_1, \dots, f_k)\, dP \tag{4.9}$$

for all polynomials g in k variables by (4.4) and the assumption that M is an algebra. Since $f_1, \dots, f_k$ are bounded, (4.9) holds for all $g \in \bar{C}(\mathbb{R}^k)$. We conclude that

$$P_n(f_1, \dots, f_k)^{-1} \Rightarrow P(f_1, \dots, f_k)^{-1}, \qquad f_1, \dots, f_k \in M. \tag{4.10}$$

Let $K \subset S$ be compact, and let $\delta > 0$. For each $x \in S$, choose $\{h_1^x, \dots, h_{k(x)}^x\} \subset M$ satisfying

$$\varepsilon(x) \equiv \inf_{y: d(y,x) \ge \delta} \max_{1 \le i \le k(x)} |h_i^x(y) - h_i^x(x)| > 0, \tag{4.11}$$

and let $G_x = \{y \in S: \max_{1 \le i \le k(x)} |h_i^x(y) - h_i^x(x)| < \varepsilon(x)\}$. Then $K \subset \bigcup_{x \in K} G_x \subset K^\delta$, so, since K is compact, there exist $x_1, \dots, x_m \in K$ such that $K \subset \bigcup_{l=1}^m G_{x_l} \subset K^\delta$. Define $g_1, \dots, g_m \in \bar{C}(S)$ by

$$g_l(x) = \max_{1 \le i \le k(x_l)} |h_i^{x_l}(x) - h_i^{x_l}(x_l)|, \tag{4.12}$$

and observe that (4.10) implies that

$$P_n(g_1, \dots, g_m)^{-1} \Rightarrow P(g_1, \dots, g_m)^{-1}. \tag{4.13}$$

It follows that

$$\begin{aligned}
\varliminf_{n\to\infty} P_n(K^\delta) &\ge \varliminf_{n\to\infty} P_n\left(\bigcup_{l=1}^m G_{x_l}\right) \\
&= \varliminf_{n\to\infty} P_n\left\{x \in S: \min_{1 \le l \le m} [g_l(x) - \varepsilon(x_l)] < 0\right\} \\
&\ge P\left\{x \in S: \min_{1 \le l \le m} [g_l(x) - \varepsilon(x_l)] < 0\right\} \\
&= P\left(\bigcup_{l=1}^m G_{x_l}\right) \\
&\ge P(K),
\end{aligned} \tag{4.14}$$

where the middle inequality depends on (4.13) and Theorem 3.1. Applying Lemma 2.1 to P and to finitely many terms in the sequence $\{P_n\}$, we conclude that there exists a compact set $K \subset S$ such that $\inf_n P_n(K^\delta) \ge 1 - \delta$. By Theorem 2.2, $\{P_n\}$ is relatively compact. □

We turn next to a result concerning countably infinite product spaces. Let (S_k, d_k), $k = 1, 2, \dots$, be metric spaces, and define $S = \prod_{k=1}^{\infty} S_k$ and $d(x, y) = \sum_{k=1}^{\infty} 2^{-k}(d_k(x_k, y_k) \wedge 1)$ for all $x, y \in S$. Then (S, d) is separable if the S_k are separable and complete if the (S_k, d_k) are complete. If the S_k are separable, then $\mathscr{B}(S) = \prod_{k=1}^{\infty} \mathscr{B}(S_k)$.

4.6 Proposition Let (S_k, d_k), $k = 1, 2, \dots$, and (S, d) be as above. Suppose $M_k \subset \bar{C}(S_k)$, and let

$$M = \{f(x) \equiv \prod_{k=1}^{n} f_k(x_k): n \geq 1, f_k \in M_k \cup \{1\} \quad \text{for} \quad k = 1, \dots, n\}. \tag{4.15}$$

(a) If the S_k are separable and the M_k are separating, then M is separating.

(b) If the (S_k, d_k) are complete and separable and the M_k are convergence determining, then M is convergence determining.

Proof. **(a)** Suppose that $P, Q \in \mathscr{P}(S)$ and

$$\int f_1(x_1) \cdots f_n(x_n) P(dx) = \int f_1(x_1) \cdots f_n(x_n) Q(dx) \tag{4.16}$$

whenever $n \geq 1$ and $f_k \in M_k \cup \{1\}$ for $k = 1, \dots, n$. Given $n \geq 2$ and $f_k \in M_k \cup \{1\}$ for $k = 2, \dots, n$, put

$$\mu(dx) = f_2(x_2) \cdots f_n(x_n) P(dx), \quad \nu(dx) = f_2(x_2) \cdots f_n(x_n) Q(dx), \tag{4.17}$$

and let μ^1 and ν^1 be the first marginals of μ and ν on $\mathscr{B}(S_1)$. Since M_1 is separating (with respect to Borel probability measures), it is separating with respect to finite signed Borel measures as well. Therefore $\mu^1 = \nu^1$ and hence

$$\int \chi_{A_1}(x_1) f_2(x_2) \cdots f_n(x_n) P(dx) = \int \chi_{A_1}(x_1) f_2(x_2) \cdots f_n(x_n) Q(dx) \tag{4.18}$$

whenever $A_1 \in \mathscr{B}(S_1)$, $n \geq 2$, and $f_k \in M_k \cup \{1\}$ for $k = 2, \dots, n$. Proceeding inductively, we conclude that

$$\int \chi_{A_1}(x_1) \cdots \chi_{A_n}(x_n) P(dx) = \int \chi_{A_1}(x_1) \cdots \chi_{A_n}(x_n) Q(dx) \tag{4.19}$$

whenever $n \geq 1$ and $A_k \in \mathscr{B}(S_k)$ for $k = 1, \dots, n$. It follows that $P = Q$ on $\prod_{k=1}^{\infty} \mathscr{B}(S_k) = \mathscr{B}(S)$ and thus that M is separating.

(b) Let $\{P_n\} \subset \mathscr{P}(S)$ and $P \in \mathscr{P}(S)$, and suppose that (4.4) holds. Then, for $k = 1, 2, \dots$, $\lim_{n \to \infty} \int f \, dP_n^k = \int f \, dP^k$ for all $f \in M_k$, where P_n^k and P^k denote the kth marginals of P_n and P, and hence $P_n^k \Rightarrow P^k$. In particular, this implies that $\{P_n^k\}$ is relatively compact for $k = 1, 2, \dots$, and hence, by

Theorem 2.2 and Proposition 2.4, $\{P_n\}$ is relatively compact. By Lemma 4.3, $P_n \Rightarrow P$, so M is convergence determining. □

We conclude this section by generalizing the concept of separating set. A set $M \subset B(S)$ is called *separating* if whenever $P, Q \in \mathscr{P}(S)$ and (4.3) holds, we have $P = Q$. More generally, if $\mathscr{M} \subset \mathscr{P}(S)$, a set $M \subset M(S)$ (the space of real-valued Borel functions on S) is called *separating on* $\mathscr{M}$ if

$$\int |f|\, dP < \infty, \qquad f \in M,\ P \in \mathscr{M}, \tag{4.20}$$

and if whenever $P, Q \in \mathscr{M}$ and (4.3) holds, we have $P = Q$. For example, the set of monomials on $\mathbb{R}$ (i.e., $1, x, x^2, x^3, \ldots$) is separating on

$$\mathscr{M} = \left\{P \in \mathscr{P}(\mathbb{R}) \colon \overline{\lim_{n\to\infty}} \frac{1}{n}\left(\int_{-\infty}^{\infty} |x|^n P(dx)\right)^{1/n} < \infty\right\} \tag{4.21}$$

(Feller (1971), p. 514).

5. THE SPACE $D_E[0, \infty)$

Throughout the remaining sections of this chapter, (E, r) denotes a metric space, and q denotes the metric $r \wedge 1$.

Most stochastic processes arising in applications have the property that they have right and left limits at each time point for almost every sample path. It has become conventional to assume that sample paths are actually right continuous when this can be done (as it usually can) without altering the finite-dimensional distributions. Consequently, the space $D_E[0, \infty)$ of right continuous functions $x\colon [0, \infty) \to E$ with left limits (i.e., for each $t \geq 0$, $\lim_{s\to t+} x(s) = x(t)$ and $\lim_{s\to t-} x(s) \equiv x(t-)$ exists; by convention, $\lim_{s\to 0-} x(s) = x(0-) = x(0)$) is of considerable importance.

We begin by observing that functions in $D_E[0, \infty)$ are better behaved than might initially be suspected.

5.1 Lemma If $x \in D_E[0, \infty)$, then x has at most countably many points of discontinuity.

Proof. For $n = 1, 2, \ldots$, let $A_n = \{t > 0\colon r(x(t), x(t-)) > 1/n\}$, and observe that A_n has no limit points in $[0, \infty)$ since $\lim_{s\to t+} x(s)$ and $\lim_{s\to t-} x(s)$ exist for all $t \geq 0$. Consequently, each A_n is countable. But the set of all discontinuities of x is $\bigcup_{n=1}^{\infty} A_n$, and hence it too is countable. □

The results on convergence of probability measures in Sections 1–4 are best suited for complete separable metric spaces. With this in mind we now define a metric on $D_E[0, \infty)$ under which it is a separable metric space if E is separable, and is complete if (E, r) is complete. Let Λ' be the collection of (strictly) increasing functions λ mapping $[0, \infty)$ onto $[0, \infty)$ (in particular, $\lambda(0) = 0$, $\lim_{t \to \infty} \lambda(t) = \infty$, and λ is continuous). Let Λ be the set of Lipschitz continuous functions $\lambda \in \Lambda'$ such that

$$\gamma(\lambda) \equiv \operatorname*{ess\,sup}_{t \geq 0} |\log \lambda'(t)| = \sup_{s > t \geq 0} \left| \log \frac{\lambda(s) - \lambda(t)}{s - t} \right| < \infty. \tag{5.1}$$

For $x, y \in D_E[0, \infty)$, define

$$d(x, y) = \inf_{\lambda \in \Lambda} \left[\gamma(\lambda) \vee \int_0^\infty e^{-u} d(x, y, \lambda, u)\, du \right], \tag{5.2}$$

where

$$d(x, y, \lambda, u) = \sup_{t \geq 0} q(x(t \wedge u), y(\lambda(t) \wedge u)). \tag{5.3}$$

It follows that, given $\{x_n\}, \{y_n\} \subset D_E[0, \infty)$, $\lim_{n \to \infty} d(x_n, y_n) = 0$ if and only if there exists $\{\lambda_n\} \subset \Lambda$ such that $\lim_{n \to \infty} \gamma(\lambda_n) = 0$ and

$$\lim_{n \to \infty} m\{u \in [0, u_0]: d(x_n, y_n, \lambda_n, u) \geq \varepsilon\} = 0 \tag{5.4}$$

for every $\varepsilon > 0$ and $u_0 > 0$, where m is Lebesgue measure; moreover, since

$$\operatorname*{ess\,sup}_{t \geq 0} |\lambda'(t) - 1| \leq 1 - e^{-\gamma(\lambda)} \leq \gamma(\lambda) \tag{5.5}$$

for every $\lambda \in \Lambda$,

$$\lim_{n \to \infty} \gamma(\lambda_n) = 0 \tag{5.6}$$

implies that

$$\lim_{n \to \infty} \sup_{0 \leq t \leq T} |\lambda_n(t) - t| = 0 \tag{5.7}$$

for all $T > 0$.

Let $x, y \in D_E[0, \infty)$, and observe that

$$\sup_{t \geq 0} q(x(t \wedge u), y(\lambda(t) \wedge u)) = \sup_{t \geq 0} q(x(\lambda^{-1}(t) \wedge u), y(t \wedge u)) \tag{5.8}$$

for all $\lambda \in \Lambda$ and $u \geq 0$, and therefore $d(x, y, \lambda, u) = d(y, x, \lambda^{-1}, u)$. Together with the fact that $\gamma(\lambda) = \gamma(\lambda^{-1})$ for every $\lambda \in \Lambda$, this implies that $d(x, y) = d(y, x)$. If $d(x, y) = 0$, then, by (5.4) and (5.7), $x(t) = y(t)$ for every continuity point t of y, and hence $x = y$ by Lemma 5.1 and the right continuity of x and y. Thus, to show that d is a metric, we need only verify the triangle inequality. Let $x, y, z \in D_E[0, \infty)$, $\lambda_1, \lambda_2 \in \Lambda$, and $u \geq 0$. Then

$$(5.9) \qquad \begin{aligned} \sup_{t\geq 0} q(x(t \wedge u), z(\lambda_2(\lambda_1(t)) \wedge u)) &\leq \sup_{t\geq 0} q(x(t \wedge u), y(\lambda_1(t) \wedge u)) \\ &\quad + \sup_{t\geq 0} q(y(\lambda_1(t) \wedge u), z(\lambda_2(\lambda_1(t)) \wedge u)) \\ &= \sup_{t\geq 0} q(x(t \wedge u), y(\lambda_1(t) \wedge u)) \\ &\quad + \sup_{t\geq 0} q(y(t \wedge u), z(\lambda_2(t) \wedge u)), \end{aligned}$$

that is, $d(x, z, \lambda_2 \circ \lambda_1, u) \leq d(x, y, \lambda_1, u) + d(y, z, \lambda_2, u)$. But since $\lambda_2 \circ \lambda_1 \in \Lambda$ and

$$(5.10) \qquad \gamma(\lambda_2 \circ \lambda_1) \leq \gamma(\lambda_1) + \gamma(\lambda_2),$$

we obtain $d(x, z) \leq d(x, y) + d(y, z)$.

The topology induced on $D_E[0, \infty)$ by the metric d is called the *Skorohod topology*.

5.2 Proposition Let $\{x_n\} \subset D_E[0, \infty)$ and $x \in D_E[0, \infty)$. Then $\lim_{n\to\infty} d(x_n, x) = 0$ if and only if there exists $\{\lambda_n\} \subset \Lambda$ such that (5.6) holds and

$$(5.11) \qquad \lim_{n\to\infty} d(x_n, x, \lambda_n, u) = 0 \quad \text{for all continuity points } u \text{ of } x.$$

In particular, $\lim_{n\to\infty} d(x_n, x) = 0$ implies that $\lim_{n\to\infty} x_n(u) = \lim_{n\to\infty} x_n(u-) = x(u)$ for all continuity points u of x.

Proof. The sufficiency follows from Lemma 5.1. Conversely, suppose that $\lim_{n\to\infty} d(x_n, x) = 0$, and let u be a continuity point of x. Recalling (5.4), there exist $\{\lambda_n\} \subset \Lambda$ and $\{u_n\} \subset (u, \infty)$ such that (5.6) holds and

$$(5.12) \qquad \lim_{n\to\infty} \sup_{t\geq 0} q(x_n(t \wedge u_n), x(\lambda_n(t) \wedge u_n)) = 0.$$

Now

$$(5.13) \qquad \begin{aligned} \sup_{t\geq 0} q(x_n(t \wedge u), x(\lambda_n(t) \wedge u)) &\leq \sup_{t\geq 0} q(x_n(t \wedge u), x(\lambda_n(t \wedge u) \wedge u_n)) \\ &\quad + \sup_{t\geq 0} q(x(\lambda_n(t \wedge u) \wedge u_n), x(\lambda_n(t) \wedge u)) \end{aligned}$$

$$\leq \sup_{0 \leq t \leq u} q(x_n(t \wedge u_n), x(\lambda_n(t) \wedge u_n))$$

$$+ \sup_{u \leq s \leq \lambda_n(u) \vee u} q(x(s), x(u))$$

$$\vee \sup_{\lambda_n(u) \wedge u \leq s \leq u} q(x(\lambda_n(u) \wedge u_n), x(s))$$

for each n, where the second half of the second inequality follows by considering separately the cases $t \leq u$ and $t > u$. Thus, $\lim_{n \to \infty} d(x_n, x, \lambda_n, u) = 0$ by (5.12), (5.7), and the continuity of x at u. □

5.3 Proposition Let $\{x_n\} \subset D_E[0, \infty)$ and $x \in D_E[0, \infty)$. Then the following are equivalent:

(a) $\lim_{n \to \infty} d(x_n, x) = 0$.

(b) There exists $\{\lambda_n\} \subset \Lambda$ such that (5.6) holds and

$$\lim_{n \to \infty} \sup_{0 \leq t \leq T} r(x_n(t), x(\lambda_n(t))) = 0 \tag{5.14}$$

for all $T > 0$.

(c) For each $T > 0$, there exists $\{\lambda_n\} \subset \Lambda'$ (possibly depending on T) such that (5.7) and (5.14) hold.

5.4 Remark In conditions (b) and (c) of Proposition 5.3, (5.14) can be replaced by

$$\lim_{n \to \infty} \sup_{0 \leq t \leq T} r(x_n(\lambda_n(t)), x(t)) = 0. \tag{5.14'}$$

Denoting the resulting conditions by (b′) and (c′), this is easily established by checking that (b) is equivalent to (b′) and (c) is equivalent to (c′). □

Proof. **(a ⇒ b)** Assuming (a) holds, there exist $\{\lambda_n\} \subset \Lambda$ and $\{u_n\} \subset (0, \infty)$ such that (5.6) holds, $u_n \to \infty$, and $d(x_n, x, \lambda_n, u_n) \to 0$; in particular,

$$\lim_{n \to \infty} \sup_{t \geq 0} r(x_n(t \wedge u_n), x(\lambda_n(t) \wedge u_n)) = 0. \tag{5.15}$$

Given $T > 0$, note that $u_n \geq T \vee \lambda_n(T)$ for all n sufficiently large, so (5.15) implies (5.14).

(b ⇒ a) Let $\{\lambda_n\} \subset \Lambda$ satisfy the conditions of (b). Then

$$\lim_{n \to \infty} \sup_{t \geq 0} q(x_n(t \wedge u), x(\lambda_n(t) \wedge u)) = 0 \tag{5.16}$$

for every continuity point u of x by (5.13) with $u_n > \lambda_n(u) \vee u$ for each n. Hence (a) holds.

(b ⇒ c) Immediate.

(c ⇒ b) Let N be a positive integer, and choose $\{\lambda_n^N\} \subset \Lambda'$ satisfying the conditions of (c) with $T = N$, such that for each n, $\lambda_n^N(t) = \lambda_n^N(N) + t - N$ for all $t > N$. Define $\tau_0^N = 0$ and, for $k = 1, 2, \ldots,$

$$\tau_k^N = \inf\left\{t > \tau_{k-1}^N : r(x(t), x(\tau_{k-1}^N)) > \frac{1}{N}\right\} \tag{5.17}$$

if $\tau_{k-1}^N < \infty$, $\tau_k^N = \infty$ if $\tau_{k-1}^N = \infty$. Observe that the sequence $\{\tau_k^N\}$ is (strictly) increasing (as long as its terms remain finite) by the right continuity of x and has no finite limit points since x has left limits. For each n, let $u_{k,n}^N = (\lambda_n^N)^{-1}(\tau_k^N)$ for $k = 0, 1, \ldots,$ where $(\lambda_n^N)^{-1}(\infty) = \infty$, and define $\mu_n^N \in \Lambda$ by

$$\begin{aligned} \mu_n^N(t) &= \tau_k^N + (t - u_{k,n}^N)(u_{k+1,n}^N - u_{k,n}^N)^{-1}(\tau_{k+1}^N - \tau_k^N), \\ &\qquad t \in [u_{k,n}^N, u_{k+1,n}^N) \cap [0, N], \qquad k = 0, 1, \ldots, \\ \mu_n^N(t) &= \mu_n^N(N) + t - N, \qquad t > N, \end{aligned} \tag{5.18}$$

where, by convention, $\infty^{-1}\infty = 1$. With this convention,

$$\gamma(\mu_n^N) = \max_{N_{kn} u_{k,n}^N < N} |\log (u_{k+1,n}^N - u_{k,n}^N)^{-1}(\tau_{k+1}^N - \tau_k^N)| \tag{5.19}$$

and

$$\begin{aligned} &\sup_{0 \le t \le N} r(x_n(t), x(\mu_n^N(t))) \\ &\quad \le \sup_{0 \le t \le N} r(x_n(t), x(\lambda_n^N(t))) + \sup_{0 \le t \le N} r(x(\lambda_n^N(t)), x(\mu_n^N(t))) \\ &\quad \le \sup_{0 \le t \le N} r(x_n(t), x(\lambda_n^N(t))) + \frac{2}{N} \end{aligned} \tag{5.20}$$

for all n. Since $\lim_{n\to\infty} u_{k,n}^N = \tau_k^N$ for $k = 0, 1, \ldots,$ (5.18) implies that $\lim_{n\to\infty} \gamma(\mu_n^N) = 0$, which, together with (5.20) and (5.14) with $T = N$, implies that we can select $1 < n_1 < n_2 < \cdots$ such that $\gamma(\mu_n^N) \le 1/N$ and $\sup_{0 \le t \le N} r(x_n(t), x(\mu_n^N(t))) \le 3/N$ for all $n \ge n_N$. For $1 \le n < n_1$, let $\hat{\lambda}_n \in \Lambda$ be arbitrary. For $n_N \le n < n_{N+1}$, where $N \ge 1$, let $\hat{\lambda}_n = \mu_n^N$. Then $\{\hat{\lambda}_n\} \subset \Lambda$ satisfies the conditions of (b). □

5.5 Corollary For $x, y \in D_E[0, \infty)$, define

$$d'(x, y) = \inf_{\lambda \in \Lambda'} \int_0^\infty e^{-u} \sup_{t \ge 0} [q(x(t \wedge u), y(\lambda(t) \wedge u)) \vee (|\lambda(t) \wedge u - t \wedge u| \wedge 1)] \, du. \tag{5.21}$$

The d' is a metric on $D_E[0, \infty)$ that is equivalent to d. (However, the metric space $(D_E[0, \infty), d')$ is not complete.)

Proof. The proof that d' is a metric is essentially the same as that for d. The equivalence of d and d' follows from the equivalence of (a) and (c) in Proposition 5.3. We leave the verification of the parenthetical remark to the reader. □

5.6 Theorem If E is separable, then $D_E[0, \infty)$ is separable. If (E, r) is complete, then $(D_E[0, \infty), d)$ is complete.

Proof. Let $\{\alpha_i\}$ be a countable dense subset of E and let Γ be the collection of functions of the form

(5.22) $$x(t) = \begin{cases} \alpha_{i_k}, & t_{k-1} \le t < t_k, \quad k = 1, \dots, n, \\ \alpha_{i_n}, & t \ge t_n, \end{cases}$$

where $0 = t_0 < t_1 < \cdots < t_n$ are rationals, $i_1, \dots, i_n$ are positive integers, and $n \ge 1$. We leave to the reader the proof that Γ is dense in $D_E[0, \infty)$ (Problem 14).

To prove completeness, it is enough to show that every Cauchy sequence has a convergent subsequence. If $\{x_n\} \subset D_E[0, \infty)$ is Cauchy, then there exist $1 \le N_1 < N_2 < \cdots$ such that $m, n \ge N_k$ implies

(5.23) $$d(x_n, x_m) \le 2^{-k-1}e^{-k}.$$

For $k = 1, 2, \dots$, let $y_k = x_{N_k}$ and, by (5.23), select $u_k > k$ and $\lambda_k \in \Lambda$ such that

(5.24) $$\gamma(\lambda_k) \vee d(y_k, y_{k+1}, \lambda_k, u_k) \le 2^{-k};$$

then, recalling (5.5),

(5.25) $$\mu_k \equiv \lim_{n \to \infty} \lambda_{k+n} \circ \cdots \circ \lambda_{k+1} \circ \lambda_k$$

exists uniformly on bounded intervals, is Lipschitz continuous, satisfies

(5.26) $$\gamma(\mu_k) \le \sum_{l=k}^{\infty} \gamma(\lambda_l) \le 2^{-k+1},$$

and hence belongs to Λ. Since

(5.27) $$\begin{aligned} &\sup_{t \ge 0} q(y_k(\mu_k^{-1}(t) \wedge u_k), y_{k+1}(\mu_{k+1}^{-1}(t) \wedge u_k)) \\ &\quad = \sup_{t \ge 0} q(y_k(\mu_k^{-1}(t) \wedge u_k), y_{k+1}(\lambda_k(\mu_k^{-1}(t)) \wedge u_k)) \\ &\quad = \sup_{t \ge 0} q(y_k(t \wedge u_k), y_{k+1}(\lambda_k(t) \wedge u_k)) \\ &\quad \le 2^{-k} \end{aligned}$$

for $k = 1, 2, \dots$ by (5.24), it follows from the completeness of (E, r) that $z_k \equiv y_k \circ \mu_k^{-1}$ converges uniformly on bounded intervals to a function $y: [0, \infty) \to E$.

But each $z_k \in D_E[0, \infty)$, so y must also belong to $D_E[0, \infty)$. Since $\lim_{k \to \infty} \gamma(\mu_k^{-1}) = 0$ and

$$\lim_{k \to \infty} \sup_{0 \le t \le T} r(y_k(\mu_k^{-1}(t)), y(t)) = 0 \tag{5.28}$$

for all $T > 0$, we conclude that $\lim_{k \to \infty} d(y_k, y) = 0$ by Proposition 5.3 (see Remark 5.4). □

6. THE COMPACT SETS OF $D_E[0, \infty)$

Again let (E, r) denote a metric space. In order to apply Prohorov's theorem to $\mathscr{P}(D_E[0, \infty))$, we must have a characterization of the compact subsets of $D_E[0, \infty)$. With this in mind we first give conditions under which a collection of step functions is compact. Given a step function $x \in D_E[0, \infty)$, define $s_0(x) = 0$ and, for $k = 1, 2, \ldots$,

$$s_k(x) = \inf\{t > s_{k-1}(x) : x(t) \neq x(t-)\} \tag{6.1}$$

if $s_{k-1}(x) < \infty$, $s_k(x) = \infty$ if $s_{k-1}(x) = \infty$.

6.1 Lemma Let $\Gamma \subset E$ be compact, let $\delta > 0$, and define $A(\Gamma, \delta)$ to be the set of step functions $x \in D_E[0, \infty)$ such that $x(t) \in \Gamma$ for all $t \ge 0$ and $s_k(x) - s_{k-1}(x) > \delta$ for each $k \ge 1$ for which $s_{k-1}(x) < \infty$. Then the closure of $A(\Gamma, \delta)$ is compact.

Proof. It is enough to show that every sequence in $A(\Gamma, \delta)$ has a convergent subsequence. Given $\{x_n\} \subset A(\Gamma, \delta)$, there exists by a diagonalization argument a subsequence $\{y_m\}$ of $\{x_n\}$ such that, for $k = 0, 1, \ldots$, either (a) $s_k(y_m) < \infty$ for each m, $\lim_{m \to \infty} s_k(y_m) \equiv t_k$ exists (possibly ∞), and $\lim_{m \to \infty} y_m(s_k(y_m)) \equiv \alpha_k$ exists, or (b) $s_k(y_m) = \infty$ for each m. Since $s_k(y_m) - s_{k-1}(y_m) > \delta$ for each $k \ge 1$ and m for which $s_{k-1}(y_m) < \infty$, it follows easily that $\{y_m\}$ converges to the function $y \in D_E[0, \infty)$ defined by $y(t) = \alpha_k$, $t_k \le t < t_{k+1}$, $k = 0, 1, \ldots$. □

The conditions for compactness are stated in terms of the following modulus of continuity. For $x \in D_E[0, \infty)$, $\delta > 0$, and $T > 0$, define

$$w'(x, \delta, T) = \inf_{\{t_i\}} \max_i \sup_{s, t \in [t_{i-1}, t_i)} r(x(s), x(t)), \tag{6.2}$$

where $\{t_i\}$ ranges over all partitions of the form $0 = t_0 < t_1 < \cdots < t_{n-1} < T \le t_n$ with $\min_{1 \le i \le n}(t_i - t_{i-1}) > \delta$ and $n \ge 1$. Note that $w'(x, \delta, T)$ is nondecreasing in δ and in T, and that

$$w'(x, \delta, T) \le w'(y, \delta, T) + 2 \sup_{0 \le s < T + \delta} r(x(s), y(s)). \tag{6.3}$$

6.2 Lemma (a) For each $x \in D_E[0, \infty)$ and $T > 0$, $w'(x, \delta, T)$ is right continuous in δ and

$$\lim_{\delta \to 0} w'(x, \delta, T) = 0. \tag{6.4}$$

(b) If $\{x_n\} \subset D_E[0, \infty)$, $x \in D_E[0, \infty)$, and $\lim_{n\to\infty} d(x_n, x) = 0$, then

$$\overline{\lim_{n\to\infty}}\, w'(x_n, \delta, T) \le w'(x, \delta, T + \varepsilon) \tag{6.5}$$

for every $\delta > 0$, $T > 0$, and $\varepsilon > 0$.

(c) For each $\delta > 0$ and $T > 0$, $w'(x, \delta, T)$ is Borel measurable in x.

Proof. (a) The right continuity follows from the fact that any partition that is admissible in the definition of $w'(x, \delta, T)$ is admissible for some $\delta' > \delta$. To obtain (6.4), let $N \ge 1$ and define $\{\tau_k^N\}$ as in (5.17). If $0 < \delta < \min \{\tau_{k+1}^N - \tau_k^N : \tau_k^N < T\}$, then $w'(x, \delta, T) \le 2/N$.

(b) Let $\{x_n\} \subset D_E[0, \infty)$, $x \in D_E[0, \infty)$, $\delta > 0$, and $T > 0$. If $\lim_{n\to\infty} d(x_n, x) = 0$, then by Proposition 5.3, there exists $\{\lambda_n\} \subset \Lambda'$ such that (5.7) and (5.14) hold with T replaced by $T + \delta$. For each n, let $y_n(t) = x(\lambda_n(t))$ for all $t \ge 0$ and $\delta_n = \sup_{0\le t\le T}[\lambda_n(t + \delta) - \lambda_n(t)]$. Then, using (6.3) and part (a),

$$\begin{aligned} \overline{\lim_{n\to\infty}}\, w'(x_n, \delta, T) &= \overline{\lim_{n\to\infty}}\, w'(y_n, \delta, T) \\ &\le \overline{\lim_{n\to\infty}}\, w'(x, \delta_n, \lambda_n(T)) \\ &\le \lim_{n\to\infty} w'(x, \delta_n \vee \delta, T + \varepsilon) \\ &= w'(x, \delta, T + \varepsilon) \end{aligned} \tag{6.6}$$

for all $\varepsilon > 0$.

(c) By part (b) and the monotonicity of $w'(x, \delta, T)$ in T, $w'(x, \delta, T+) \equiv \lim_{\varepsilon\to 0+} w'(x, \delta, T + \varepsilon)$ is upper semicontinuous, hence Borel measurable, in x. Therefore it suffices to observe that $w'(x, \delta, T) = \lim_{\varepsilon\to 0+} w'(x, \delta, (T - \varepsilon)+)$ for every $x \in D_E[0, \infty)$. □

6.3 Theorem Let (E, r) be complete. Then the closure of $A \subset D_E[0, \infty)$ is compact if and only if the following two conditions hold:

(a) For every rational $t \ge 0$, there exists a compact set $\Gamma_t \subset E$ such that $x(t) \in \Gamma_t$ for all $x \in A$.

(b) For each $T > 0$,

$$\lim_{\delta\to 0} \sup_{x\in A} w'(x, \delta, T) = 0. \tag{6.7}$$

6.4 Remark In Theorem 6.3 it is actually necessary that for each $T > 0$ there exist a compact set $\Gamma_T \subset E$ such that $x(t) \in \Gamma_T$ for $0 \leq t \leq T$ and all $x \in A$. See Problem 16. □

Proof. Suppose that A satisfies (a) and (b), and let $l \geq 1$. Choose $\delta_l \in (0, 1)$ such that

$$\sup_{x \in A} w'(x, \delta_l, l) \leq \frac{1}{l} \tag{6.8}$$

and $m_l \geq 2$ such that $1/m_l < \delta_l$. Define $\Gamma^{(l)} = \bigcup_{i=0}^{(l+1)m_l} \Gamma_{i/m_l}$ and, using the notation of Lemma 6.1, let $A_l = A(\Gamma^{(l)}, \delta_l)$.

Given $x \in A$, there is a partition $0 = t_0 < t_1 < \cdots < t_{n-1} < l \leq t_n < l + 1 < t_{n+1} = \infty$ with $\min_{1 \leq i \leq n}(t_i - t_{i-1}) > \delta_l$ such that

$$\max_{1 \leq i \leq n} \sup_{s, t \in [t_{i-1}, t_i)} r(x(s), x(t)) \leq \frac{2}{l}. \tag{6.9}$$

Define $x' \in A_l$ by $x'(t) = x(([m_l t] + 1)/m_l)$ for $t_i \leq t < t_{i+1}$, $i = 0, 1, \ldots, n$. Then $\sup_{0 \leq t < l} r(x'(t), x(t)) \leq 2/l$, so

$$\begin{aligned} d(x', x) &\leq \int_0^\infty e^{-u} \sup_{t \geq 0} [r(x'(t \wedge u), x(t \wedge u)) \wedge 1] \, du \\ &\leq 2/l + e^{-l} < 3/l. \end{aligned} \tag{6.10}$$

It follows that $A \subset A_l^{3/l}$. Now l was arbitrary, so since $\bar{A}_l$ is compact for each $l \geq 1$ by Lemma 6.1, and since $A \subset \bigcap_{l \geq 1} A_l^{3/l}$, A is totally bounded and hence has compact closure.

Conversely, suppose that A has compact closure. We leave the proof of (a) to the reader (Problem 16). To see that (b) holds, suppose there exist $\eta > 0$, $T > 0$, and $\{x_n\} \subset A$ such that $w'(x_n, 1/n, T) \geq \eta$ for all n. Since A has compact closure, we may assume that $\lim_{n \to \infty} d(x_n, x) = 0$ for some $x \in D_E[0, \infty)$. But then Lemma 6.2(b) implies that

$$\eta \leq \overline{\lim_{n \to \infty}} \, w'(x_n, \delta, T) \leq w'(x, \delta, T + 1) \tag{6.11}$$

for all $\delta > 0$. Letting $\delta \to 0$, the right side of (6.11) tends to zero by Lemma 6.2(a), and this results in a contradiction. □

We conclude this section with a further characterization of convergence of sequences in $D_E[0, \infty)$. (This result would have been included in Section 5 were it not for the fact that we need Lemma 6.2 in the proof.) We note that $(C_E[0, \infty), d_U)$ is a metric space, where

$$d_U(x, y) = \int_0^\infty e^{-u} \sup_{0 \leq t \leq u} [r(x(t), y(t)) \wedge 1] \, du. \tag{6.12}$$

Moreover, if $\{x_n\} \subset C_E[0, \infty)$ and $x \in C_E[0, \infty)$, then $\lim_{n\to\infty} d_U(x_n, x) = 0$ if and only if whenever $\{t_n\} \subset [0, \infty)$, $t \geq 0$, and $\lim_{n\to\infty} t_n = t$, we have $\lim_{n\to\infty} r(x_n(t_n), x(t)) = 0$. The following proposition gives an analogue of this result for $(D_E[0, \infty), d)$.

6.5 Proposition Let (E, r) be arbitrary, and let $\{x_n\} \subset D_E[0, \infty)$ and $x \in D_E[0, \infty)$. Then $\lim_{n\to\infty} d(x_n, x) = 0$ if and only if whenever $\{t_n\} \subset [0, \infty)$, $t \geq 0$, and $\lim_{n\to\infty} t_n = t$, the following conditions hold:

(a) $\lim_{n\to\infty} r(x_n(t_n), x(t)) \wedge r(x_n(t_n), x(t-)) = 0$.

(b) If $\lim_{n\to\infty} r(x_n(t_n), x(t)) = 0$, $s_n \geq t_n$ for each n, and $\lim_{n\to\infty} s_n = t$, then $\lim_{n\to\infty} r(x_n(s_n), x(t)) = 0$.

(c) If $\lim_{n\to\infty} r(x_n(t_n), x(t-)) = 0$, $0 \leq s_n \leq t_n$ for each n, and $\lim_{n\to\infty} s_n = t$, then $\lim_{n\to\infty} r(x_n(s_n), x(t-)) = 0$.

Proof. Suppose $\lim_{n\to\infty} d(x_n, x) = 0$, and let $\{t_n\} \subset [0, \infty)$, $t \geq 0$, and $\lim_{n\to\infty} t_n = t$. Choose $T > 0$ such that $\{t_n\} \subset [0, T]$ and $t \leq T$. By Proposition 5.3, there exists $\{\lambda_n\} \subset \Lambda'$ such that (5.7) and (5.14) hold. Therefore, since

$$\begin{aligned} r(x_n(t_n), x(t)) \wedge r(x_n(t_n), x(t-)) &\leq \sup_{0\leq u\leq T} r(x_n(u), x(\lambda_n(u))) \\ &\quad + r(x(\lambda_n(t_n)), x(t)) \wedge r(x(\lambda_n(t_n)), x(t-)) \end{aligned} \tag{6.13}$$

for each n, and since $\lim_{n\to\infty} \lambda_n(t_n) = t$, (a) holds. If, in addition, $t_n \leq s_n \leq T$ for each n and $\lim_{n\to\infty} s_n = t$, then

$$\begin{aligned} r(x_n(s_n), x(t)) \leq &\sup_{0\leq u\leq T} r(x_n(u), x(\lambda_n(u))) \\ &+ r(x(\lambda_n(s_n)), x(t)) \end{aligned} \tag{6.14}$$

and

$$\begin{aligned} r(x(\lambda_n(t_n)), x(t)) \leq &\sup_{0\leq u\leq T} r(x(\lambda_n(u)), x_n(u)) \\ &+ r(x_n(t_n), x(t)) \end{aligned} \tag{6.15}$$

for each n. If also $\lim_{n\to\infty} r(x_n(t_n), x(t)) = 0$, then $\lim_{n\to\infty} r(x(\lambda_n(t_n)), x(t)) = 0$ by (6.15), so since $\lambda_n(s_n) \geq \lambda_n(t_n)$ for each n and $\lim_{n\to\infty} \lambda_n(s_n) = t$, it follows that $\lim_{n\to\infty} r(x(\lambda_n(s_n)), x(t)) = 0$. Thus, (b) follows from (6.14), and the proof of (c) is similar.

We turn to the proof of the sufficiency of (a)–(c). Fix $T > 0$ and for each n define

$$\varepsilon_n = 2 \inf \{\varepsilon > 0: \Gamma(t, n, \varepsilon) \neq \varnothing \quad \text{for} \quad 0 \leq t \leq T\}, \tag{6.16}$$

where

$$(6.17)\quad \Gamma(t, n, \varepsilon) = \{s \in (t-\varepsilon, t+\varepsilon) \cap [0, \infty): r(x_n(s), x(t)) < \varepsilon,\ r(x_n(s-), x(t-)) < \varepsilon\}.$$

We claim that $\lim_{n\to\infty} \varepsilon_n = 0$. Suppose not. Then there exist $\varepsilon > 0$, a sequence $\{n_k\}$ of positive integers, and a sequence $\{t_k\} \subset [0, T]$ such that $\Gamma(t_k, n_k, \varepsilon) = \varnothing$ for all k. By choosing a subsequence if necessary, we can assume that $\lim_{k\to\infty} t_k \equiv t$ exists and that $t_k < t$ for all k, $t_k > t$ for all k, or $t_k = t$ for all k. In the first case, $\lim_{k\to\infty} x(t_k) = \lim_{k\to\infty} x(t_k-) = x(t-)$, and in the second case, $\lim_{k\to\infty} x(t_k) = \lim_{k\to\infty} x(t_k-) = x(t)$. Since (a) implies that $\lim_{n\to\infty} x_n(s) = \lim_{n\to\infty} x_n(s-) = x(s)$ for all continuity points s of x, there exist (by Lemma 5.1 and Proposition 5.2) sequences $\{a_n\}$ and $\{b_n\}$ of continuity points of x such that $a_n < t < b_n$ for each n and $a_n \to t$ and $b_n \to t$ sufficiently slowly that $\lim_{n\to\infty} x_n(a_n) = \lim_{n\to\infty} x_n(a_n-) = \lim_{n\to\infty} x(a_n) = x(t-)$ and $\lim_{n\to\infty} x_n(b_n) = \lim_{n\to\infty} x_n(b_n-) = \lim_{n\to\infty} x(b_n) = x(t)$. If $t_k < t$ (respectively, $t_k > t$) for all k, then a_{n_k} (respectively, b_{n_k}) belongs to $\Gamma(t_k, n_k, \varepsilon)$ for all k sufficiently large, a contradiction. It remains to consider the case $t_k = t$ for all k. If $x(t) = x(t-)$, then $t \in \Gamma(t, n_k, \varepsilon)$ for all k sufficiently large by condition (a). Therefore we suppose that $r(x(t), x(t-)) = \delta > 0$. By the choice of $\{a_n\}$ and $\{b_n\}$ and by condition (a), there exists $n_0 \geq 1$ such that for all $n \geq n_0$,

$$(6.18)\qquad r(x_n(a_n), x(t-)) \vee r(x_n(b_n), x(t)) < \frac{\delta \wedge \varepsilon}{2},$$

$$(6.19)\qquad \sup_{a_n \leq s \leq b_n} r(x_n(s), x(t)) \wedge r(x_n(s), x(t-)) < \frac{\delta \wedge \varepsilon}{2},$$

and $a_n, b_n \in (t-\varepsilon, t+\varepsilon)$. Let $n \geq n_0$ and define

$$(6.20)\qquad s_n = \inf\left\{s > a_n: r(x_n(s), x(t)) < \frac{\delta \wedge \varepsilon}{2}\right\}.$$

By (6.18), $a_n < s_n \leq b_n$, and therefore $s_n \in (t-\varepsilon, t+\varepsilon)$, $r(x_n(s_n), x(t)) \leq (\delta \wedge \varepsilon)/2$, and $r(x_n(s_n-), x(t)) \geq (\delta \wedge \varepsilon)/2$. The latter inequality, together with (6.19), implies that $r(x_n(s_n-), x(t-)) < (\delta \wedge \varepsilon)/2$. We conclude that $s_n \in \Gamma(t, n, \varepsilon)$ for all $n \geq n_0$, and this contradiction establishes the claim that $\lim_{n\to\infty} \varepsilon_n = 0$.

For each n, we construct $\lambda_n \in \Lambda'$ as follows. Choose a partition $0 = t_0^n < t_1^n < \cdots < t_{m_n-1}^n < T \leq t_{m_n}^n$ with $\min_{1\leq i\leq m_n}(t_i^n - t_{i-1}^n) > 3\varepsilon_n$ such that

$$(6.21)\qquad \max_{1\leq i\leq m_n} \sup_{s,t\in[t_{i-1}^n, t_i^n)} r(x(s), x(t)) \leq w'(x, 3\varepsilon_n, T) + \varepsilon_n,$$

and put $m_n^* = \max\{i \geq 0: t_i^n \leq T\}$ (m_n^* is $m_n - 1$ or m_n). Define $\lambda_n(0) = 0$ and $\lambda_n(t_i^n) = \inf \Gamma(t_i^n, n, \varepsilon_n)$ for $i = 1, \ldots, m_n^*$, interpolate linearly on $[0, t_{m_n^*}^n]$, and let $\lambda_n(t) = t - t_{m_n^*}^n + \lambda_n(t_{m_n^*}^n)$ for all $t > t_{m_n^*}^n$. Then $\lambda_n \in \Lambda'$ and $\sup_{t\geq 0} |\lambda_n(t) - t| \leq \varepsilon_n$.

We claim that $\lim_{n\to\infty} \sup_{0\leq t\leq T} r(x_n(\lambda_n(t)), x(t)) = 0$ and hence $\lim_{n\to\infty} d(x_n, x) = 0$ by Proposition 5.3 (see Remark 5.4). To verify the claim, it is enough to show that if $\{t_n\} \subset [0, T]$, $0 \le t \le T$, and $\lim_{n\to\infty} t_n = t$, then $\lim_{n\to\infty} r(x_n(\lambda_n(t_n)), x(t_n)) = 0$. If $x(t) = x(t-)$, the result follows from condition (a) since $\lim_{n\to\infty} \lambda_n(t_n) = t$. Therefore, let us suppose that $x(t) \ne x(t-)$. Then, for each n sufficiently large, $t = t_{i_n}^n$ for some $i_n \in \{1, \ldots, m_n^*\}$ by (6.21) and Lemma 6.2(a). To complete the proof, it suffices to consider two cases, $\{t_n\} \subset [t, T]$ and $\{t_n\} \subset [0, t)$. In the first case, $\lambda_n(t_n) \ge \lambda_n(t) = \lambda_n(t_{i_n}^n)$ and $r(x_n(\lambda_n(t_{i_n}^n)), x(t)) \le \varepsilon_n$ for each n sufficiently large, so $\lim_{n\to\infty} r(x_n(\lambda_n(t_n)), x(t)) = 0$ by condition (b), and the desired result follows. In the second case, $\lambda_n(t_n) < \lambda_n(t) = \lambda_n(t_{i_n}^n)$ and either $r(x_n(\lambda_n(t_{i_n}^n)-), x(t-)) < \varepsilon_n$ or $r(x_n(\lambda_n(t_{i_n}^n)), x(t-)) \le \varepsilon_n$ (depending on whether the infimum in the definition of $\lambda_n(t_{i_n}^n)$ is attained or not) for each n sufficiently large. Consequently, for such n, there exists s_n with $\lambda_n(t_n) < s_n \le \lambda_n(t_{i_n}^n)$ such that $r(x_n(s_n), x(t-)) \le \varepsilon_n$, and therefore $\lim_{n\to\infty} r(x_n(\lambda_n(t_n)), x(t-)) = 0$ by condition (c), from which the desired result follows. This completes the proof. □

7. CONVERGENCE IN DISTRIBUTION IN $D_E[0, \infty)$

As in the previous two sections, (E, r) denotes a metric space. Let $\mathscr{S}_E$ denote the Borel σ-algebra of $D_E[0, \infty)$. We are interested in weak convergence of elements of $\mathscr{P}(D_E[0, \infty))$ and naturally it is important to know more about $\mathscr{S}_E$. The following result states that $\mathscr{S}_E$ is just the σ-algebra generated by the coordinate random variables.

7.1 Proposition For each $t \ge 0$, define $\pi_t: D_E[0, \infty) \to E$ by $\pi_t(x) = x(t)$. Then

$$\mathscr{S}_E \supset \mathscr{S}'_E \equiv \sigma(\pi_t: 0 \le t < \infty) = \sigma(\pi_t: t \in D), \tag{7.1}$$

where D is any dense subset of $[0, \infty)$. If E is separable, then $\mathscr{S}_E = \mathscr{S}'_E$.

Proof. For each $\varepsilon > 0$, $t \ge 0$, and $f \in \bar{C}(E)$,

$$f_t^\varepsilon(x) \equiv \frac{1}{\varepsilon} \int_t^{t+\varepsilon} f(\pi_s(x))\, ds \tag{7.2}$$

defines a continuous function f_t^ε on $D_E[0, \infty)$. Since $\lim_{\varepsilon\to 0} f_t^\varepsilon(x) = f(\pi_t(x))$ for every $x \in D_E[0, \infty)$, we find that $f \circ \pi_t$ is Borel measurable for every $f \in \bar{C}(E)$ and hence for every $f \in B(E)$. Consequently,

$$\pi_t^{-1}(\Gamma) = \{x \in D_E[0, \infty): \chi_\Gamma(\pi_t(x)) = 1\} \in \mathscr{S}_E \tag{7.3}$$

for all $\Gamma \in B(E)$, and hence $\mathscr{S}_E \supset \mathscr{S}'_E$. For each $t \ge 0$, there exists $\{t_n\} \subset D \cap [t, \infty)$ such that $\lim_{n\to\infty} t_n = t$. Therefore, $\pi_t = \lim_{n\to\infty} \pi_{t_n}$ is $\sigma(\pi_s: s \in D)$-measurable, and hence we have (7.1).

Assume now that E is separable. Let $n \geq 1$, let $0 = t_0 < t_1 < \cdots < t_n < t_{n+1} = \infty$, and for $\alpha_0, \alpha_1, \ldots, \alpha_n \in E$ define $\eta(\alpha_0, \alpha_1, \ldots, \alpha_n) \in D_E[0, \infty)$ by

(7.4) $$\eta(\alpha_0, \alpha_1, \ldots, \alpha_n)(t) = \alpha_i, \qquad t_i \leq t < t_{i+1}, \qquad i = 0, 1, \ldots, n.$$

Since

(7.5) $$d(\eta(\alpha_0, \alpha_1, \ldots, \alpha_n), \eta(\alpha'_0, \alpha'_1, \ldots, \alpha'_n)) \leq \max_{0 \leq i \leq n} r(\alpha_i, \alpha'_i),$$

η is a continuous function from E^{n+1} into $D_E[0, \infty)$. Since each π_t is $\mathscr{S}'_E$-measurable and E is separable, we have that for fixed $z \in D_E[0, \infty)$, $d(z, \eta \circ (\pi_{t_0}, \pi_{t_1}, \ldots, \pi_{t_n}))$ is an $\mathscr{S}'_E$-measurable function from $D_E[0, \infty)$ into $\mathbb{R}$. Finally, for $m = 1, 2, \ldots$, let η_m be defined as was η with $t_i = i/m$, $i = 0, 1, \ldots, n \equiv m^2$. Then

(7.6) $$\lim_{m \to \infty} d(z, \eta_m(\pi_{t_0}(x), \ldots, \pi_{t_{m^2}}(x))) = d(z, x)$$

for every $x \in D_E[0, \infty)$ (see Problem 12), so $d(z, x)$ is $\mathscr{S}'_E$-measurable in x for fixed $z \in D_E[0, \infty)$. In particular, every open ball $B(z, \varepsilon) = \{x \in D_E[0, \infty): d(z, x) < \varepsilon\}$ belongs to $\mathscr{S}'_E$, so since E (and, by Theorem 5.6, $D_E[0, \infty)$) is separable, $\mathscr{S}'_E$ contains all open sets in $D_E[0, \infty)$ and hence contains $\mathscr{S}_E$. □

A $D_E[0, \infty)$-valued random variable is a stochastic process with sample paths in $D_E[0, \infty)$, although the converse need not be true if E is not separable. Let $\{X_\alpha\}$ (where α ranges over some index set) be a family of stochastic processes with sample paths in $D_E[0, \infty)$ (if E is not separable, *assume* the X_α are $D_E[0, \infty)$-valued random variables), and let $\{P_\alpha\} \subset \mathscr{P}(D_E[0, \infty))$ be the family of associated probability distributions (i.e., $P_\alpha(B) = P\{X_\alpha \in B\}$ for all $B \in \mathscr{S}_E$). We say that $\{X_\alpha\}$ is *relatively compact* if $\{P_\alpha\}$ is (i.e., if the closure of $\{P_\alpha\}$ in $\mathscr{P}(D_E[0, \infty))$ is compact). Theorem 6.3 gives, through an application of Prohorov's theorem, criteria for $\{X_\alpha\}$ to be relatively compact.

7.2 Theorem Let (E, r) be complete and separable, and let $\{X_\alpha\}$ be a family of processes with sample paths in $D_E[0, \infty)$. Then $\{X_\alpha\}$ is relatively compact if and only if the following two conditions hold:

(a) For every $\eta > 0$ and rational $t \geq 0$, there exists a compact set $\Gamma_{\eta, t} \subset E$ such that

(7.7) $$\inf_\alpha P\{X_\alpha(t) \in \Gamma^\eta_{\eta, t}\} \geq 1 - \eta.$$

(b) For every $\eta > 0$ and $T > 0$, there exists $\delta > 0$ such that

(7.8) $$\sup_\alpha P\{w'(X_\alpha, \delta, T) \geq \eta\} \leq \eta.$$

7.3 Remark In fact, if $\{X_\alpha\}$ is relatively compact, then the stronger *compact containment condition* holds; that is, for every $\eta > 0$ and $T > 0$ there is a compact set $\Gamma_{\eta, T} \subset E$ such that

$$\inf_\alpha P\{X_\alpha(t) \in \Gamma_{\eta, T} \quad \text{for} \quad 0 \le t \le T\} \ge 1 - \eta. \tag{7.9}$$ □

Proof. If $\{X_\alpha\}$ is relatively compact, then Theorems 5.6, 2.2, and 6.3 immediately yield (a) and (b); in fact, $\Gamma_{\eta,t}^\eta$ can be replaced by $\Gamma_{\eta,t}$ in (7.7).

Conversely, let $\varepsilon > 0$, let T be a positive integer such that $e^{-T} < \varepsilon/2$, and choose $\delta > 0$ such that (7.8) holds with $\eta = \varepsilon/4$. Let $m > 1/\delta$, put $\Gamma = \bigcup_{i=0}^{mT} \Gamma_{\varepsilon 2^{-i-2}, i/m}$, and observe that

$$\inf_\alpha P\{X_\alpha(i/m) \in \Gamma^{\varepsilon/4}, \qquad i = 0, 1, \ldots, mT\} \ge 1 - \frac{\varepsilon}{2}. \tag{7.10}$$

Using the notation of Lemma 6.1, let $A = A(\Gamma, \delta)$. By the lemma, A has compact closure.

Given $x \in D_E[0, \infty)$ with $w'(x, \delta, T) < \varepsilon/4$ and $x(i/m) \in \Gamma^{\varepsilon/4}$ for $i = 0, 1, \ldots, mT$, choose $0 = t_0 < t_1 < \cdots < t_{n-1} < T \le t_n$ such that $\min_{1 \le i \le n}(t_i - t_{i-1}) > \delta$ and

$$\max_{1 \le i \le n} \sup_{s, t \in [t_{i-1}, t_i)} r(x(s), x(t)) < \frac{\varepsilon}{4}, \tag{7.11}$$

and select $\{y_i\} \subset \Gamma$ such that $r(x(i/m), y_i) < \varepsilon/4$ for $i = 0, 1, \ldots, mT$. Defining $x' \in A$ by

$$x'(t) = \begin{cases} y_{[mt_{i-1}]+1}, & t_{i-1} \le t < t_i, \qquad i = 1, \ldots, n-1, \\ y_{[mt_{n-1}]+1}, & t \ge t_{n-1}, \end{cases} \tag{7.12}$$

we have $\sup_{0 \le t < T} r(x(t), x'(t)) < \varepsilon/2$ and hence $d(x, x') < \varepsilon/2 + e^{-T} < \varepsilon$, implying that $x \in A^\varepsilon$. Consequently, $\inf_\alpha P\{X_\alpha \in A^\varepsilon\} \ge 1 - \varepsilon$, so the relative compactness of $\{X_\alpha\}$ follows from Theorems 5.6 and 2.2. □

7.4 Corollary Let (E, r) be complete and separable, and let $\{X_n\}$ be a sequence of processes with sample paths in $D_E[0, \infty)$. Then $\{X_n\}$ is relatively compact if and only if the following two conditions hold:

(a) For every $\eta > 0$ and rational $t \ge 0$, there exists a compact set $\Gamma_{\eta, t} \subset E$ such that

$$\varliminf_{n \to \infty} P\{X_n(t) \in \Gamma_{\eta, t}^\eta\} \ge 1 - \eta. \tag{7.13}$$

(b) For every $\eta > 0$ and $T > 0$, there exists $\delta > 0$ such that

$$\varlimsup_{n \to \infty} P\{w'(X_n, \delta, T) \ge \eta\} \le \eta. \tag{7.14}$$

Proof. Fix $\eta > 0$, rational $t \geq 0$, and $T > 0$. For each $n \geq 1$, there exist by Lemmas 2.1 and 6.2(a) a compact set $\Gamma_n \subset E$ and $\delta_n > 0$ such that $P\{X_n(t) \in \Gamma_n^\eta\} \geq 1 - \eta$ and $P\{w'(X_n, \delta_n, T) \geq \eta\} \leq \eta$. By (7.13) and (7.14), there exist a compact set $\Gamma_0 \subset E$, $\delta_0 > 0$, and a positive integer n_0 such that

$$\inf_{n \geq n_0} P\{X_n(t) \in \Gamma_0^\eta\} \geq 1 - \eta \tag{7.15}$$

and

$$\sup_{n \geq n_0} P\{w'(X_n, \delta_0, T) \geq \eta\} \leq \eta. \tag{7.16}$$

We can replace n_0 in (7.15) and (7.16) by 1 if we replace Γ_0 by $\Gamma = \bigcup_{n=0}^{n_0 - 1} \Gamma_n$ and δ_0 by $\delta = \bigwedge_{n=0}^{n_0 - 1} \delta_n$, so the result follows from Theorem 7.2. □

7.5 Lemma Let (E, r) be arbitrary, let $\Gamma_1 \subset \Gamma_2 \subset \cdots$ be a nondecreasing sequence of compact subsets of E, and define

$$S = \{x \in D_E[0, \infty): x(t) \in \Gamma_n \quad \text{for} \quad 0 \leq t \leq n, n = 1, 2, \ldots\}. \tag{7.17}$$

Let $\{X_\alpha\}$ be a family of processes with sample paths in S. Then $\{X_\alpha\}$ is relatively compact if condition (b) of Theorem 7.2 holds.

Proof. The proof is similar to that of Theorem 7.2 Let $\varepsilon > 0$, let T be a positive integer such that $e^{-T} < \varepsilon/2$, choose $\delta > 0$ such that (7.8) holds with $\eta = \varepsilon/2$, and let $A = A(\Gamma_T, \delta)$. Given $x \in S$ with $w'(x, \delta, T) < \varepsilon/2$, it is easy to construct $x' \in A \cap S$ with $d(x, x') < \varepsilon$, and hence $x \in (A \cap S)^\varepsilon$. Consequently, $\inf_\alpha P\{X_\alpha \in (A \cap S)^\varepsilon\} \geq 1 - \varepsilon$, so the relative compactness of $\{X_\alpha\}$ follows from Lemma 6.1 and Theorem 2.2. Here we are using the fact that (S, d) is complete and separable (Problem 15). □

7.6 Theorem Let (E, r) be arbitrary, and let $\{X_\alpha\}$ be a family of processes with sample paths in $D_E[0, \infty)$. If the compact containment condition (Remark 7.3) and condition (b) of Theorem 7.2 hold, then the X_α have modifications $\tilde{X}_\alpha$ that are $D_E[0, \infty)$-valued random variables and $\{\tilde{X}_\alpha\}$ is relatively compact.

Proof. By the compact containment condition there exist compact sets $\Gamma_n \subset E$, $n = 1, 2, \ldots$, such that $\inf_\alpha P\{X_\alpha(t) \in \Gamma_n$ for $0 \leq t \leq n\} \geq 1 - n^{-1}$. Let $E_0 = \bigcup_n \Gamma_n$. Note that E_0 is separable and $P\{X_\alpha(t) \in E_0\} = 1$ so X_α has a modification with sample paths in $D_{E_0}[0, \infty)$. Consequently, we may as well assume E is separable. Given $\eta > 0$, we can assume without loss of generality that $\{\Gamma_{\eta 2^{-n}, n}\}$ is a nondecreasing sequence of compact subsets of E. Define

$$S_\eta = \{x \in D_E[0, \infty): x(t) \in \Gamma_{\eta 2^{-n}, n} \quad \text{for} \quad 0 \leq t \leq n, n = 1, 2, \ldots\}, \tag{7.18}$$

and note that $\inf_\alpha P\{X_\alpha \in S_\eta\} \geq 1 - \eta$. By Lemma 7.4, the family $\{P_\alpha^\eta\} \subset \mathscr{P}(S_\eta)$, defined by

$$P_\alpha^\eta(B) = P\{X_\alpha \in B \mid X_\alpha \in S_\eta\}, \tag{7.19}$$

is relatively compact. The proof proceeds analogously to that of Corollary 2.3. We leave the details to the reader. □

7.7 Lemma If X is a process with sample paths in $D_E[0, \infty)$, then the complement in $[0, \infty)$ of

$$D(X) \equiv \{t \geq 0: P\{X(t) = X(t-)\} = 1\} \tag{7.20}$$

is at most countable.

Proof. Let $\varepsilon > 0$, $\delta > 0$, and $T > 0$. If the set

$$\{0 \leq t \leq T: P\{r(X(t), X(t-)) \geq \varepsilon\} \geq \delta\} \tag{7.21}$$

contains a sequence $\{t_n\}$ of distinct points, then

$$P\{r(X(t_n), X(t_n-)) \geq \varepsilon \quad \text{infinitely often}\} \geq \delta > 0, \tag{7.22}$$

contradicting the fact that, for each $x \in D_E[0, \infty)$, $r(x(t), x(t-)) \geq \varepsilon$ for at most finitely many $t \in [0, T]$. Hence the set in (7.21) is finite, and therefore

$$\{t \geq 0: P\{r(X(t), X(t-)) \geq \varepsilon\} > 0\} \tag{7.23}$$

is at most countable. The conclusion follows by letting $\varepsilon \to 0$. □

7.8 Theorem Let E be separable and let X_n, $n = 1, 2, \ldots$, and X be processes with sample paths in $D_E[0, \infty)$.

(a) If $X_n \Rightarrow X$, then

$$(X_n(t_1), \ldots, X_n(t_k)) \Rightarrow (X(t_1), \ldots, X(t_k)) \tag{7.24}$$

for every finite set $\{t_1, \ldots, t_k\} \subset D(X)$. Moreover, for each finite set $\{t_1, \ldots, t_k\} \subset [0, \infty)$, there exist sequences $\{t_1^n\} \subset [t_1, \infty), \ldots, \{t_k^n\} \subset [t_k, \infty)$ converging to $t_1, \ldots, t_k$, respectively, such that $(X_n(t_1^n), \ldots, X_n(t_k^n)) \Rightarrow (X(t_1), \ldots, X(t_k))$.

(b) If $\{X_n\}$ is relatively compact and there exists a dense set $D \subset [0, \infty)$ such that (7.24) holds for every finite set $\{t_1, \ldots, t_k\} \subset D$, then $X_n \Rightarrow X$.

Proof. **(a)** Suppose that $X_n \Rightarrow X$. By Theorem 1.8, there exists a probability space on which are defined processes Y_n, $n = 1, 2, \ldots$, and Y with sample paths in $D_E[0, \infty)$ and with the same distributions as X_n, $n = 1, 2, \ldots$, and X, such that $\lim_{n\to\infty} d(Y_n, Y) = 0$ a.s. If $t \in D(X) = D(Y)$, then, using the notation of Proposition 7.1, π_t is continuous a.s. with respect to the distribution of Y, so $\lim_{n\to\infty} Y_n(t) = Y(t)$ a.s. by Corollary 1.9, and the first conclusion follows. We leave it to the reader to show that the second conclusion is a consequence of the first, together with Lemma 7.7.

(b) It suffices to show that every convergent (in distribution) subsequence of $\{X_n\}$ converges in distribution to X. Relabeling if necessary, suppose that $X_n \Rightarrow Y$. We must show that X and Y have the same distribution. Let $\{t_1, \ldots, t_k\} \subset D(Y)$ and $f_1, \ldots, f_k \in \bar{C}(E)$, and choose sequences $\{t_1^m\} \subset D \cap [t_1, \infty), \ldots, \{t_k^m\} \subset D \cap [t_k, \infty)$ converging to $t_1, \ldots, t_k$, respectively, and $n_1 < n_2 < n_3 < \cdots$ such that

$$\left| E\left[\prod_{i=1}^{k} f_i(X(t_i^m))\right] - E\left[\prod_{i=1}^{k} f_i(X_{n_m}(t_i^m))\right] \right| < \frac{1}{m}. \tag{7.25}$$

Then

$$\begin{aligned} \left| E\left[\prod_{i=1}^{k} f_i(X(t_i))\right] - E\left[\prod_{i=1}^{k} f_i(Y(t_i))\right] \right| &\le \left| E\left[\prod_{i=1}^{k} f_i(X(t_i))\right] - E\left[\prod_{i=1}^{k} f_i(X(t_i^m))\right] \right| \\ &\quad + \left| E\left[\prod_{i=1}^{k} f_i(X(t_i^m))\right] - E\left[\prod_{i=1}^{k} f_i(X_{n_m}(t_i^m))\right] \right| \\ &\quad + \left| E\left[\prod_{i=1}^{k} f_i(X_{n_m}(t_i^m))\right] - E\left[\prod_{i=1}^{k} f_i(Y(t_i))\right] \right| \end{aligned} \tag{7.26}$$

for each $m \ge 1$. All three terms on the right tend to zero as $m \to \infty$, the first by the right continuity of X, the second by (7.25), and the third by the facts that $X_{n_m} \Rightarrow Y$ and $\{t_1, \ldots, t_k\} \subset D(Y)$. Consequently,

$$E\left[\prod_{i=1}^{k} f_i(X(t_i))\right] = E\left[\prod_{i=1}^{k} f_i(Y(t_i))\right] \tag{7.27}$$

for all $\{t_1, \ldots, t_k\} \subset [0, \infty)$ (by Lemma 7.7 and right continuity) and all $f_1, \ldots, f_k \in \bar{C}(E)$. By Proposition 7.1 and the Dynkin class theorem (Appendix 4), we conclude that X and Y have the same distribution. □

8. CRITERIA FOR RELATIVE COMPACTNESS IN $D_E[0, \infty)$

Let (E, r) denote a metric space and $q = r \wedge 1$. We now consider a systematic way of selecting the partition in the definition of $w'(x, \delta, T)$. Given $\varepsilon > 0$ and $x \in D_E[0, \infty)$, define $\tau_0 = \sigma_0 = 0$ and, for $k = 1, 2, \ldots$,

(8.1) $$\tau_k = \inf\left\{t > \tau_{k-1}: r(x(t), x(\tau_{k-1})) > \frac{\varepsilon}{2}\right\}$$

if $\tau_{k-1} < \infty$, $\tau_k = \infty$ if $\tau_{k-1} = \infty$,

(8.2) $$\sigma_k = \sup\left\{t \le \tau_k: r(x(t), x(\tau_k)) \vee r(x(t-), x(\tau_k)) \ge \frac{\varepsilon}{2}\right\}$$

if $\tau_k < \infty$, and $\sigma_k = \infty$ if $\tau_k = \infty$. Given $\delta > 0$ and $T > 0$, observe that $w'(x, \delta, T) < \varepsilon/2$ implies $\min\{\tau_{k+1} - \sigma_k: \tau_k < T\} > \delta$, for if $\tau_{k+1} - \sigma_k \le \delta$ and $\tau_k < T$ for some $k \ge 0$, then any interval $[a, b)$ containing τ_k with $b - a > \delta$ must also contain σ_k or τ_{k+1} (or both) in its interior and hence must satisfy $\sup_{s, t \in [a, b)} r(x(s), x(t)) \ge \varepsilon/2$; in this case, $w'(x, \delta, T) \ge \varepsilon/2$.

Letting

(8.3) $$s_k = \frac{\sigma_k + \tau_k}{2}$$

for $k = 0, 1, \dots$, we have $\lim_{k\to\infty} s_k = \infty$. Observe that, for each $k \ge 0$,

(8.4) $$\sigma_k \le s_k \le \tau_k \le \sigma_{k+1} \le s_{k+1} \le \tau_{k+1},$$

and

(8.5) $$s_{k+1} - s_k \ge \frac{\tau_k + \tau_{k+1}}{2} - \frac{\sigma_k + \tau_k}{2} = \frac{\tau_{k+1} - \sigma_k}{2}$$

if $s_k < \infty$, where the middle inequality in (8.4) follows from the fact that $r(x(\tau_k), x(\tau_{k+1})) \ge \varepsilon/2$ if $\tau_{k+1} < \infty$. We conclude from (8.5) that $\min\{\tau_{k+1} - \sigma_k: \tau_k < T + \delta/2\} > \delta$ implies

(8.6) $$\min\{s_{k+1} - s_k: s_k < T\} > \frac{\delta}{2};$$

for if not, there would exist $k \ge 0$ with $s_k < T$, $\tau_k \ge T + \delta/2$, and $s_{k+1} - s_k \le \delta/2$, a contradiction by (8.4). Finally, (8.6) implies $w'(x, \delta/2, T) \le \varepsilon$.

Let us now regard τ_k, σ_k, and s_k, $k = 0, 1, \dots$, as functions from $D_E[0, \infty)$ into $[0, \infty]$. Assuming that E is separable (recall Remark 3.4), their $\mathscr{S}_E$-measurability follows easily from the identities

(8.7)
$$\{\tau_k < u\} = \{\tau_{k-1} < \infty\} \cap \bigcup_{t \in [0, u) \cap \mathbb{Q}} \left(\left\{r(x(t), x(\tau_{k-1})) > \frac{\varepsilon}{2}\right\} \cap \{t > \tau_{k-1}\}\right)$$

and

(8.8) $$\{\sigma_k \ge u\} = \{\tau_k = \infty\} \cup \left(\{\tau_k < \infty\} \cap \left[\left\{r(x(u-), x(\tau_k)) \ge \frac{\varepsilon}{2}\right\}\right.\right.$$
$$\left.\left.\cup \bigcap_{n=1}^{\infty} \bigcup_{t \in [u, \infty) \cap \mathbb{Q}} \left(\left\{r(x(t), x(\tau_k)) > \frac{\varepsilon}{2} - \frac{1}{n}\right\} \cap \{t \le \tau_k\}\right)\right]\right),$$

valid for $0 < u < \infty$ and $k = 1, 2, \ldots$. We summarize the implications of the two preceding paragraphs for our purposes in the following lemma.

8.1 Lemma Let (E, r) be separable, and let $\{X_\alpha\}$ be a family of processes with sample paths in $D_E[0, \infty)$. Let $\tau_k^{\alpha, \varepsilon}$, $\sigma_k^{\alpha, \varepsilon}$, and $s_k^{\alpha, \varepsilon}$, $k = 0, 1, \ldots$, be defined for given $\varepsilon > 0$ and X_α as in (8.1)–(8.3). Then the following are equivalent:

(8.9) $$\liminf_{\delta \to 0} \inf_\alpha P\{w'(X_\alpha, \delta, T) < \varepsilon\} = 1, \qquad \varepsilon > 0, \quad T > 0.$$

(8.10) $$\liminf_{\delta \to 0} \inf_\alpha P\{\min\{s_{k+1}^{\alpha, \varepsilon} - s_k^{\alpha, \varepsilon} : s_k^{\alpha, \varepsilon} < T\} \geq \delta\} = 1, \qquad \varepsilon > 0, \quad T > 0.$$

(8.11) $$\liminf_{\delta \to 0} \inf_\alpha P\{\min\{\tau_{k+1}^{\alpha, \varepsilon} - \sigma_k^{\alpha, \varepsilon} : \tau_k^{\alpha, \varepsilon} < T\} \geq \delta\} = 1, \qquad \varepsilon > 0, \quad T > 0.$$

Proof. See the discussion above. □

8.2. Lemma For each α, let $0 = s_0^\alpha < s_1^\alpha < s_2^\alpha < \cdots$ be a sequence of random variables with $\lim_{k \to \infty} s_k^\alpha = \infty$, define $\Delta_k^\alpha = s_{k+1}^\alpha - s_k^\alpha$ for $k = 0, 1, \ldots$, let $T > 0$, and put $K_\alpha(T) = \max\{k \geq 0 : s_k^\alpha < T\}$. Define $F: [0, \infty) \to [0, 1]$ by $F(t) = \sup_\alpha \sup_{k \geq 0} P\{\Delta_k^\alpha < t, s_k^\alpha < T\}$. Then

(8.12) $$F(\delta) \leq \sup_\alpha P\left\{\min_{0 \leq k \leq K_\alpha(T)} \Delta_k^\alpha < \delta\right\} \leq LF(\delta) + e^T \int_0^\infty Le^{-Lt} F(t)\, dt$$

for all $\delta > 0$ and $L = 1, 2, \ldots$. Consequently,

(8.13) $$\lim_{\delta \to 0} \sup_\alpha P\left\{\min_{0 \leq k \leq K_\alpha(T)} \Delta_k^\alpha < \delta\right\} = 0$$

if and only if $F(0+) = 0$.

Proof. The first inequality in (8.12) is immediate. As for the second,

(8.14) $$\begin{aligned} P\left\{\min_{0 \leq k \leq K_\alpha(T)} \Delta_k^\alpha < \delta\right\} &\leq \sum_{k=0}^{L-1} P\{\Delta_k^\alpha < \delta, s_k^\alpha < T\} + P\{K_\alpha(T) \geq L\} \\ &\leq LF(\delta) + e^T E\left[\chi_{\{K_\alpha(T) \geq L\}} \exp\left(-\sum_{k=0}^{L-1} \Delta_k^\alpha\right)\right] \\ &\leq LF(\delta) + e^T \prod_{k=0}^{L-1} \{E[\chi_{\{s_k^\alpha < T\}} \exp(-L\Delta_k^\alpha)]\}^{1/L} \\ &\leq LF(\delta) + e^T \int_0^\infty Le^{-Lt} F(t)\, dt. \end{aligned}$$

Finally, observe that $F(0+) = 0$ implies that the right side of (8.12) approaches zero as $\delta \to 0$ and then $L \to \infty$. □

8.3 Proposition Under the assumptions of Lemma 8.1, (8.9) is equivalent to

(8.15) $$\lim_{\delta \to 0} \sup_{\alpha} \sup_{k \geq 0} P\{\tau_{k+1}^{\alpha, \varepsilon} - \sigma_k^{\alpha, \varepsilon} < \delta, \tau_k^{\alpha, \varepsilon} < T\} = 0, \qquad \varepsilon > 0, \quad T > 0.$$

Proof. The result follows from Lemmas 8.1 and 8.2 together with the inequalities

(8.16) $$\begin{aligned} P\left\{s_{k+1}^{\alpha, \varepsilon} - s_k^{\alpha, \varepsilon} < \frac{\delta}{2}, s_k^{\alpha, \varepsilon} < T\right\} &\leq P\{\tau_{k+1}^{\alpha, \varepsilon} - \sigma_k^{\alpha, \varepsilon} < \delta, \tau_k^{\alpha, \varepsilon} < T + \delta\} \\ &\leq P\{s_{k+1}^{\alpha, \varepsilon} - s_k^{\alpha, \varepsilon} < \delta, s_k^{\alpha, \varepsilon} < T + \delta\}. \end{aligned}$$ □

The following lemma gives us a means of estimating the probabilities in (8.15). Let $S(T)$ denote the collection of all $\{\mathscr{F}_{t+}^X\}$-stopping times bounded by T.

8.4 Lemma Let (E, r) be separable, let X be a process with sample paths in $D_E[0, \infty)$, and fix $T > 0$ and $\beta > 0$. Then, for each $\delta > 0$, $\lambda > 0$, and $\tau \in S(T)$,

(8.17) $$\begin{aligned} P\left\{\sup_{0 \leq u \leq \delta} q(X(\tau + u), X(\tau)) \geq \lambda, \sup_{0 \leq v \leq \delta \wedge \tau} q(X(\tau), X(\tau - v)) \geq \lambda\right\} \\ \leq \lambda^{-2\beta}[a_\beta + 2a_\beta^2(a_\beta + 4a_\beta^2)]C(\delta) \end{aligned}$$

and

(8.18) $$\begin{aligned} P\left\{\sup_{0 \leq u \leq \delta} q(X(u), X(0)) \geq \lambda\right\} \\ \leq \lambda^{-2\beta}\{a_\beta(a_\beta + 4a_\beta^2)C(\delta) + a_\beta E[q^\beta(X(\delta), X(0))]\}, \end{aligned}$$

where

(8.19)

$$C(\delta) = \sup_{\tau \in S(T + 2\delta)} \sup_{0 \leq u \leq 2\delta} E\left[\sup_{0 \leq v \leq 3\delta \wedge \tau} q^\beta(X(\tau + u), X(\tau))q^\beta(X(\tau), X(\tau - v))\right]$$

and $a_\beta = 2^{(\beta - 1) \vee 0}$ (and hence $(c + d)^\beta \leq a_\beta(c^\beta + d^\beta)$ for all $c, d \geq 0$).

8.5 Remark (a) In (8.19), $\sup_{\tau \in S(T+2\delta)}$ can be replaced by $\sup_{\tau \in S_0(T+2\delta)}$, where $S_0(T + 2\delta)$ is the collection of all discrete $\{\mathscr{F}_t^X\}$-stopping times bounded by $T + 2\delta$. This follows from the fact that for each $\tau \in S(T + 2\delta)$ there exists a sequence $\{\tau_n\} \subset S_0(T + 2\delta)$ such that $\tau_n \geq \tau$ for each n and $\lim_{n\to\infty} \tau_n = \tau$; we also need the observation that for fixed $x \in D_E[0, \infty)$, $\sup_{0 \leq v \leq 3\delta \wedge t} q^\beta(x(t), x(t - v))$ is right continuous in t $(0 \leq t < \infty)$.

(b) If we take $\lambda = \varepsilon/2 \in (0, 1]$ and $\tau = \tau_k \wedge T$ (recall (8.1)) in Lemma 8.4, where $k \geq 1$, then the left side of (8.17) bounds

$$P\{\tau_{k+1} - \tau_k \leq \delta,\ \tau_k - \sigma_k < \delta,\ \tau_k < T\}, \tag{8.20}$$

which for each $k \geq 1$ bounds $P\{\tau_{k+1} - \sigma_k < \delta, \tau_k < T, \tau_1 > \delta\}$. The left side of (8.18) bounds $P\{\tau_1 \leq \delta\}$, and hence the sum of the right sides of (8.17) and (8.18) bounds $P\{\tau_{k+1} - \sigma_k < \delta, \tau_k < T\}$ for each $k \geq 0$. □

Proof. Given a $\{\mathscr{F}_{t+}^X\}$-stopping time τ, let $M_\tau(\delta)$ be the collection of $\mathscr{F}_{\tau+}^X$-measurable random variables U satisfying $0 \leq U \leq \delta$. We claim that

$$\sup_{\tau \in S(T+\delta)} \sup_{U \in M_\tau(\delta)} E\left[\sup_{0 \leq v \leq 2\delta \wedge \tau} q^\beta(X(\tau + U), X(\tau))q^\beta(X(\tau), X(\tau - v))\right] \leq (a_\beta + 4a_\beta^2)C(\delta). \tag{8.21}$$

To see this, observe that for each $\tau \in S(T + \delta)$ and $U \in M_\tau(\delta)$,

$$\begin{aligned} q^\beta(X(\tau + U), X(\tau)) &\leq a_\beta \delta^{-1} \int_\delta^{2\delta} [q^\beta(X(\tau + \theta), X(\tau)) + q^\beta(X(\tau + \theta), X(\tau + U))]\, d\theta \\ &\leq a_\beta \delta^{-1}\left[\int_\delta^{2\delta} q^\beta(X(\tau + \theta), X(\tau))\, d\theta \right. \\ &\qquad \left. + \int_0^{2\delta} q^\beta(X(\tau + U + \theta), X(\tau + U))\, d\theta\right], \end{aligned} \tag{8.22}$$

and hence

$$\begin{aligned} &\sup_{0 \leq v \leq 2\delta \wedge \tau} q^\beta(X(\tau + U), X(\tau))q^\beta(X(\tau), X(\tau - v)) \\ &\leq a_\beta \delta^{-1} \int_\delta^{2\delta} \sup_{0 \leq v \leq 2\delta \wedge \tau} q^\beta(X(\tau + \theta), X(\tau))q^\beta(X(\tau), X(\tau - v))\, d\theta \\ &\quad + a_\beta^2 \delta^{-1} \int_0^{2\delta} \sup_{0 \leq v \leq 2\delta \wedge \tau} q^\beta(X(\tau + U + \theta), X(\tau + U)) \\ &\qquad\qquad \times q^\beta(X(\tau + U), X(\tau - v))\, d\theta \\ &\quad + a_\beta^2 \delta^{-1} \int_0^{2\delta} q^\beta(X(\tau + U + \theta), X(\tau + U))q^\beta(X(\tau + U), X(\tau))\, d\theta \end{aligned} \tag{8.23}$$

$$\le a_\beta \delta^{-1} \int_\delta^{2\delta} \sup_{0 \le v \le 2\delta \wedge \tau} q^\beta(X(\tau + \theta), X(\tau)) q^\beta(X(\tau), X(\tau - v))\, d\theta$$

$$+ 2a_\beta^2 \delta^{-1} \int_0^{2\delta} \sup_{0 \le v \le 3\delta \wedge (\tau + U)} q^\beta(X(\tau + U + \theta), X(\tau + U))$$

$$\times q^\beta(X(\tau + U), X(\tau + U - v))\, d\theta;$$

also, $\tau + U \in S(T + 2\delta)$, so (8.21) follows from (8.23).

Given $0 < \eta < \lambda$ and $\tau \in S(T)$, define

$$\Delta = \inf\{t > 0: q(X(\tau + t), X(\tau)) > \lambda - \eta\}, \tag{8.24}$$

and observe that

$$\begin{aligned} &q^\beta(X(\tau + \Delta \wedge \delta), X(\tau)) q^\beta(X(\tau), X(\tau - v)) \\ &\quad \le a_\beta q^\beta(X(\tau + \delta), X(\tau)) q^\beta(X(\tau), X(\tau - v)) \\ &\quad\quad + a_\beta^2 q^\beta(X(\tau + \delta), X(\tau + \Delta \wedge \delta)) q^\beta(X(\tau + \Delta \wedge \delta), X(\tau)) \\ &\quad\quad + a_\beta^2 q^\beta(X(\tau + \delta), X(\tau + \Delta \wedge \delta)) q^\beta(X(\tau + \Delta \wedge \delta), X(\tau - v)) \end{aligned} \tag{8.25}$$

for $0 \le v \le \delta \wedge \tau$. Since $\tau + \Delta \wedge \delta \in S(T + \delta)$, $\delta - \Delta \wedge \delta \in M_{\tau + \Delta \wedge \delta}(\delta)$, and $\Delta \wedge \delta + v \le 2\delta$, (8.21) and (8.25) imply that

$$E\left[\sup_{0 \le v \le \delta \wedge \tau} q^\beta(X(\tau + \Delta \wedge \delta), X(\tau)) q^\beta(X(\tau), X(\tau - v))\right] \le [a_\beta + 2a_\beta^2(a_\beta + 4a_\beta^2)]C(\delta). \tag{8.26}$$

But the left side of (8.17) is bounded by $(\lambda - \eta)^{-\beta}\lambda^{-\beta}$ times the left side of (8.26), so (8.17) follows by letting $\eta \to 0$.

Now define Δ as in (8.24) with $\tau = 0$. Then

$$\begin{aligned} q^{2\beta}(X(\Delta \wedge \delta), X(0)) &\le a_\beta[q^\beta(X(\delta), X(\Delta \wedge \delta)) q^\beta(X(\Delta \wedge \delta), X(0)) \\ &\quad + q^\beta(X(\delta), X(0)) q^\beta(X(\Delta \wedge \delta), X(0))] \\ &\le a_\beta q^\beta(X(\delta), X(\Delta \wedge \delta)) q^\beta(X(\Delta \wedge \delta), X(0)) + a_\beta q^\beta(X(\delta), X(0)), \end{aligned} \tag{8.27}$$

so (8.18) follows as above. □

8.6 Theorem Let (E, r) be complete and separable, and let $\{X_\alpha\}$ be a family of processes with sample paths in $D_E[0, \infty)$. Suppose that condition (a) of Theorem 7.2 holds. Then the following are equivalent:

(a) $\{X_\alpha\}$ is relatively compact.

(b) For each $T > 0$, there exist $\beta > 0$ and a family $\{\gamma_\alpha(\delta): 0 < \delta < 1, \text{ all } \alpha\}$ of nonnegative random variables satisfying

$$E[q^\beta(X_\alpha(t+u), X_\alpha(t)) \mid \mathscr{F}_t^\alpha] q^\beta(X_\alpha(t), X_\alpha(t-v)) \le E[\gamma_\alpha(\delta) \mid \mathscr{F}_t^\alpha] \tag{8.28}$$

for $0 \le t \le T$, $0 \le u \le \delta$, and $0 \le v \le \delta \wedge t$, where $\mathscr{F}_t^\alpha \equiv \mathscr{F}_t^{X_\alpha}$; in addition,

$$\lim_{\delta \to 0} \sup_\alpha E[\gamma_\alpha(\delta)] = 0 \tag{8.29}$$

and

$$\lim_{\delta \to 0} \sup_\alpha E[q^\beta(X_\alpha(\delta), X_\alpha(0))] = 0. \tag{8.30}$$

(c) For each $T > 0$, there exists $\beta > 0$ such that the quantities

$$C_\alpha(\delta) = \sup_{\tau \in S_0^\alpha(T)} \sup_{0 \le u \le \delta} E\left[\sup_{0 \le v \le \delta \wedge \tau} q^\beta(X_\alpha(\tau + u), X_\alpha(\tau)) q^\beta(X_\alpha(\tau), X_\alpha(\tau - v))\right], \tag{8.31}$$

defined for $0 < \delta < 1$ and all α, satisfy

$$\lim_{\delta \to 0} \sup_\alpha C_\alpha(\delta) = 0; \tag{8.32}$$

in addition (8.30) holds. (Here $S_0^\alpha(T)$ is the collection of all discrete $\{\mathscr{F}_t^\alpha\}$-stopping times bounded by T.)

8.7 Remark **(a)** If, as will typically be the case,

$$E[q^\beta(X_\alpha(t+u), X_\alpha(t)) \mid \mathscr{F}_t^\alpha] \le E[\gamma_\alpha(\delta) \mid \mathscr{F}_t^\alpha] \tag{8.33}$$

in place of (8.28), then $E[q^\beta(X_\alpha(\delta), X_\alpha(0))] \le E[\gamma_\alpha(\delta)]$ and we need only verify (8.29) in condition (b).

(b) For sequences $\{X_n\}$, one can replace $\sup_\alpha$ in (8.29), (8.30), and (8.32) by $\overline{\lim}_{n \to \infty}$ as was done in Corollary 7.4. □

Proof. **(a ⇒ b)** In view of Theorem 7.2, this follows from the facts that

$$\begin{aligned} q(X_\alpha(t+u), X_\alpha(t)) q(X_\alpha(t), X_\alpha(t-v)) &\le q(X_\alpha(t+u), X_\alpha(t)) \wedge q(X_\alpha(t), X_\alpha(t-v)) \\ &\le w'(X_\alpha, 2\delta, T+\delta) \wedge 1 \end{aligned} \tag{8.34}$$

for $0 \le t \le T$, $0 \le u \le \delta$, and $0 \le v \le \delta \wedge t$, and

$$q(X_\alpha(\delta), X_\alpha(0)) \le w'(X_\alpha, \delta, T) \wedge 1. \tag{8.35}$$

(b ⇒ c) Observe that t in (8.28) may be replaced by $\tau \in S_0^\alpha(T)$ (Problem 25 of Chapter 2), and that we may replace the right side of (8.28) by its supremum over $v \in [0, \delta \wedge \tau] \cap \mathbb{Q}$ and hence over $v \in [0, \delta \wedge \tau]$. Consequently, (8.29) implies (8.32).

(c ⇒ a) This follows from Lemma 8.4, Remark 8.5, Proposition 8.3, and Theorem 7.2. □

The following result gives sufficient conditions for the existence of $\{\gamma_\alpha(\delta): 0 < \delta < 1, \text{ all } \alpha\}$ with the properties required by condition (b) of Theorem 8.6.

8.8 Theorem Let (E, r) be separable, and let $\{X_\alpha\}$ be a family of processes with sample paths in $D_E[0, \infty)$. Fix $T > 0$ and suppose there exist $\beta > 0$, $C > 0$, and $\theta > 1$ such that for all α

$$E[q^\beta(X_\alpha(t+h), X_\alpha(t)) \wedge q^\beta(X_\alpha(t), X_\alpha(t-h))] \le Ch^\theta, \qquad 0 \le t \le T+1, 0 \le h \le t, \tag{8.36}$$

which is implied by

$$E[q^{\beta/2}(X_\alpha(t+h), X_\alpha(t))q^{\beta/2}(X_\alpha(t), X_\alpha(t-h))] \le Ch^\theta, \qquad 0 \le t \le T+1, 0 \le h \le t. \tag{8.37}$$

Then there exists a family $\{\gamma_\alpha(\delta): 0 < \delta < 1, \text{ all } \alpha\}$ of nonnegative random variables for which (8.29) holds, and

$$q^\beta(X_\alpha(t+u), X_\alpha(t))q^\beta(X_\alpha(t), X_\alpha(t-v)) \le \gamma_\alpha(\delta) \tag{8.38}$$

for $0 \le t \le T, 0 \le u \le \delta$, and $0 \le v \le \delta \wedge t$.

8.9 Remark **(a)** The inequality (8.28) follows by taking conditional expectations on both sides of (8.38).

(b) Let $\varepsilon > 0$, $C > 0$, $\theta > 1$, and $0 < h \le t$, and suppose that

$$P\{r(X_\alpha(t+h), X_\alpha(t)) \ge \lambda, r(X_\alpha(t), X_\alpha(t-h)) \ge \lambda\} \le \lambda^{-\varepsilon}Ch^\theta \tag{8.39}$$

for all $\lambda > 0$. Then, letting $\beta = 1 + \varepsilon$,

$$\begin{aligned} &E[q^\beta(X_\alpha(t+h), X_\alpha(t)) \wedge q^\beta(X_\alpha(t), X_\alpha(t-h))] \\ &\quad = \int_0^1 P\{q^\beta(X_\alpha(t+h), X_\alpha(t)) \ge x, q^\beta(X_\alpha(t), X_\alpha(t-h)) \ge x\}\, dx \\ &\quad \le \int_0^1 x^{-\varepsilon/\beta} Ch^\theta\, dx = \beta Ch^\theta. \end{aligned} \tag{8.40}$$

□

Proof. We prove the theorem in the case $\beta > 1$; the proof for $0 < \beta \le 1$ is similar and in fact simpler since q^β is a metric (satisfies the triangle inequality) in this case. In the calculations that follow we drop the subscript α. Define

(8.41)

$$\eta_m = \sum_{1 \le k \le 2^m(T+1)-1} q^\beta(X((k+1)2^{-m}), X(k2^{-m})) \wedge q^\beta(X(k2^{-m}), X((k-1)2^{-m}))$$

for $m = 0, 1, \ldots$, and fix a nonnegative integer n. We claim that for integers $m \ge n$ and j, k_1, k_2, and k_3 satisfying

$$(8.42) \qquad 0 \le j2^{-n} \le k_1 2^{-m} < k_2 2^{-m} < k_3 2^{-m} \le (j+2)2^{-n} \le T+1,$$

we have

$$(8.43) \qquad q(X(k_3 2^{-m}), X(k_2 2^{-m})) \wedge q(X(k_2 2^{-m}), X(k_1 2^{-m})) \le 2 \sum_{i=n}^{m} \eta_i^{1/\beta}.$$

(If $0 < \beta \le 1$, replace q by q^β and $\eta_i^{1/\beta}$ by η_i in (8.43).)

We prove the claim by induction. For $m = n$, (8.43) is immediate since (8.42) implies that $k_1 = j$, $k_2 = j+1$, and $k_3 = j+2$, and

$$(8.44) \quad q(X((j+2)2^{-n}), X((j+1)2^{-n})) \wedge q(X((j+1)2^{-n}), X(j2^{-n})) \le \eta_n^{1/\beta}.$$

Suppose (8.43) holds for some $m \ge n$ and $0 \le j2^{-n} \le k_1 2^{-m-1} < k_2 2^{-m-1} < k_3 2^{-m-1} \le (j+2)2^{-n} \le T+1$. For $i = 1, 2, 3$, let $\varepsilon_i = q(X(k_i' 2^{-m}), X(k_i 2^{-m-1}))$, where if k_i is even, $k_i' = k_i/2$, and if k_i is odd, $k_i' = (k_i \pm 1)/2$ as determined by

(8.45)

$$\varepsilon_i = q(X((k_i+1)2^{-m-1}), X(k_i 2^{-m-1})) \wedge q(X(k_i 2^{-m-1}), X((k_i-1)2^{-m-1})).$$

Note that $\varepsilon_i = 0$ if k_i is even and $\varepsilon_i \le \eta_{m+1}^{1/\beta}$ otherwise, so the triangle inequality implies that

$$\begin{aligned}
(8.46) \qquad & q(X(k_3 2^{-m-1}), X(k_2 2^{-m-1})) \wedge q(X(k_2 2^{-m-1}), X(k_1 2^{-m-1})) \\
&\le [\varepsilon_3 + q(X(k_3' 2^{-m}), X(k_2' 2^{-m})) + \varepsilon_2] \\
&\quad \wedge [\varepsilon_2 + q(X(k_2' 2^{-m}), X(k_1' 2^{-m})) + \varepsilon_1] \\
&\le 2\eta_{m+1}^{1/\beta} + q(X(k_3' 2^{-m}), X(k_2' 2^{-m})) \wedge q(X(k_2' 2^{-m}), X(k_1' 2^{-m})).
\end{aligned}$$

By the definition of k_i', we still have $0 \le j2^{-n} \le k_1' 2^{-m} \le k_2' 2^{-m} \le k_3' 2^{-m} \le (j+2)2^{-n} \le T+1$, and hence the induction step is verified.

If $0 \le t_1 < t_2 < t_3 \le T + \frac{1}{2}$ and t_1, t_2, and t_3 are dyadic rational with $t_3 - t_1 \le 2^{-n}$ for some $n \ge 1$, then there exist j, m, k_1, k_2, and k_3 satisfying (8.42) and $t_i = k_i 2^{-m}$. Consequently,

$$(8.47) \qquad q(X(t_3), X(t_2)) \wedge q(X(t_2), X(t_1)) \le 2 \sum_{i=n}^{\infty} \eta_i^{1/\beta} \equiv \varphi_n.$$

By right continuity, (8.47) holds for all $0 \le t_1 < t_2 < t_3 < T + \frac{1}{2}$ with $t_3 - t_1 \le 2^{-n}$. If $\delta \ge \frac{1}{4}$, let $\gamma(\delta) = 1$; if $0 < \delta < \frac{1}{4}$, let n_δ be the largest integer n

satisfying $2\delta < 2^{-n}$, and define $\gamma(\delta) = \varphi_{n_\delta}$. Since $ab \le a \wedge b$ for all $a, b \in [0, 1]$, we conclude that (8.38) holds. Also,

$$\text{(8.48)} \qquad E[\gamma(\delta)] = 2 \sum_{i=n_\delta}^{\infty} E[\eta_i^{1/\beta}] \le 2 \sum_{i=n_\delta}^{\infty} E[\eta_i]^{1/\beta}$$

$$\le 2 \sum_{i=n_\delta}^{\infty} [2^i(T + 1)C\, 2^{-i\theta}]^{1/\beta},$$

so $\lim_{\delta\to 0} E[\gamma(\delta)] = 0$ (and the limit is uniform in α). □

8.10 Corollary Let (E, r) be complete and separable, and let X be a process with values in E that is right continuous in probability. Suppose that for each $T > 0$, there exist $\beta > 0$, $C > 0$, and $\theta > 1$ such that

$$\text{(8.49)} \qquad E[q^\beta(X(t + h_2), X(t)) \wedge q^\beta(X(t), X(t - h_1))] \le C(h_1 \vee h_2)^\theta$$

whenever $0 \le t - h_1 \le t \le t + h_2 \le T$. Then X has a version with sample paths in $D_E[0, \infty)$.

Proof. Define the sequence of processes $\{X_n\}$ with sample paths in $D_E[0, \infty)$ by $X_n(t) = X(([nt] + 1)/n)$. It suffices to show that $\{X_n\}$ is relatively compact, for by Theorem 7.8 and the assumed right continuity in probability of X, the limit in distribution of any convergent subsequence of $\{X_n\}$ has the same finite-dimensional distributions as X.

Given $\eta > 0$ and $t \ge 0$, choose by Lemma 2.1 a compact set $\Gamma_{\eta, t} \subset E$ such that $P\{X(t) \in \Gamma_{\eta, t}^\eta\} \ge 1 - \eta$. Then (7.13) holds by Theorem 3.1 and the fact that $X_n(t) \Rightarrow X(t)$ in E. Consequently, it suffices to verify condition (b) of Theorem 8.6, and for this we apply Theorem 8.8. By (8.49) with T replaced by $T + 2$, there exist $\beta > 0$, $C > 0$, and $\theta > 1$ such that for each n

$$\text{(8.50)} \quad E[q^\beta(X_n(t + h), X_n(t)) \wedge q^\beta(X_n(t), X_n(t - h))]$$

$$\le C\left(\frac{[nt] - [n(t - h)]}{n} \vee \frac{[n(t + h)] - [nt]}{n}\right)^\theta,$$

$$0 \le t \le T + 1, \quad 0 \le h \le t.$$

But the left side of (8.50) is zero if $2h \le 1/n$ and is bounded by $C(h + n^{-1})^\theta \le 3^\theta C h^\theta$ if $2h > 1/n$. Thus, Theorem 8.8 implies that (8.29) holds, and the verification of (8.30) is immediate. □

9. FURTHER CRITERIA FOR RELATIVE COMPACTNESS IN $D_E[0, \infty)$

We now consider criteria for relative compactness that are particularly useful in approximating by Markov processes. These criteria are based on the following simple result. As usual, (E, r) denotes a metric space.

9.1 Theorem Let (E, r) be complete and separable, and let $\{X_\alpha\}$ be a family of processes with sample paths in $D_E[0, \infty)$. Suppose that the compact containment condition holds. That is, for every $\eta > 0$ and $T > 0$ there exists a compact set $\Gamma_{\eta, T} \subset E$ for which

$$\inf_\alpha P\{X_\alpha(t) \in \Gamma_{\eta, T} \quad \text{for} \quad 0 \le t \le T\} \ge 1 - \eta. \tag{9.1}$$

Let H be a dense subset of $\bar{C}(E)$ in the topology of uniform convergence on compact sets. Then $\{X_\alpha\}$ is relatively compact if and only if $\{f \circ X_\alpha\}$ is relatively compact (as a family of processes with sample paths in $D_{\mathbb{R}}[0, \infty)$) for each $f \in H$.

Proof. Given $f \in \bar{C}(E)$, the mapping $x \to f \circ x$ from $D_E[0, \infty)$ into $D_{\mathbb{R}}[0, \infty)$ is continuous (Problem 13). Consequently, convergence in distribution of a sequence of processes $\{X_n\}$ with sample paths in $D_E[0, \infty)$ implies convergence in distribution of $\{f \circ X_n\}$, and hence relative compactness of $\{X_\alpha\}$ implies relative compactness of $\{f \circ X_\alpha\}$.

Conversely, suppose that $\{f \circ X_\alpha\}$ is relatively compact for every $f \in H$. It then follows from (9.1), (6.3), and Theorem 7.2 that $\{f \circ X_\alpha\}$ is relatively compact for every $f \in \bar{C}(E)$ and in particular that $\{q(\cdot, z) \circ X_\alpha\}$ is relatively compact for each $z \in E$, where $q = r \wedge 1$. Let $\eta > 0$ and $T > 0$. By the compactness of $\Gamma_{\eta, T}$, there exists for each $\varepsilon > 0$ a finite set $\{z_1, \dots, z_N\} \subset \Gamma_{\eta, T}$ such that $\min_{1 \le i \le N} q(x, z_i) < \varepsilon$ for all $x \in \Gamma_{\eta, T}$. If $y \in \Gamma_{\eta, T}$, then, for some $i \in \{1, \dots, N\}$, $q(y, z_i) < \varepsilon$ and hence

$$q(x, y) < |q(x, z_i) - q(y, z_i)| + 2\varepsilon \tag{9.2}$$

for all $x \in E$. Consequently, for $0 \le t \le T$, $0 \le u \le \delta$, and $0 \le v \le \delta \wedge t$,

$$\begin{aligned} &q(X_\alpha(t + u), X_\alpha(t))q(X_\alpha(t), X_\alpha(t - v)) \\ &\quad \le \bigvee_{i=1}^N |q(X_\alpha(t + u), z_i) - q(X_\alpha(t), z_i)| \, |q(X_\alpha(t), z_i) - q(X_\alpha(t - v), z_i)| \\ &\qquad + 4(\varepsilon + \varepsilon^2) + \chi_{\{X_\alpha(s) \notin \Gamma_{\eta, T} \text{ for some } s \in [0, T]\}} \\ &\quad \le \bigvee_{i=1}^N w'(q(\cdot, z_i) \circ X_\alpha, 2\delta, T + \delta) \wedge 1 \\ &\qquad + 4(\varepsilon + \varepsilon^2) + \chi_{\{X_\alpha(s) \notin \Gamma_{\eta, T} \text{ for some } s \in [0, T]\}} \\ &\quad \equiv \gamma_\alpha(\delta), \end{aligned} \tag{9.3}$$

where $0 < \delta < 1$. Note that N depends on η, T, and ε.

Since $\lim_{\delta \to 0} \sup_\alpha E[w'(q(\cdot, z) \circ X_\alpha, 2\delta, T + \delta) \wedge 1] = 0$ for each $z \in E$ by Theorem 7.2, we may select η and ε depending on δ in such a way that $\lim_{\delta \to 0} \sup_\alpha E[\gamma_\alpha(\delta)] = 0$. Finally, (9.2) implies that

$$(9.4)\quad q(X_\alpha(\delta), X_\alpha(0)) \leq \bigvee_{i=1}^{N} |q(X_\alpha(\delta), z_i) - q(X_\alpha(0), z_i)| + 2\varepsilon + \chi_{\{X_\alpha(0) \notin \Gamma_{\eta, T}\}}$$

for all $\delta > 0$, so $\lim_{\delta \to 0} \sup_\alpha E[q(X_\alpha(\delta), X_\alpha(0))] = 0$ by Theorem 8.6. Thus, the relative compactness of $\{X_\alpha\}$ follows from Theorem 8.6. □

9.2 Corollary Let (E, r) be complete and separable, and let $\{X_n\}$ be a sequence of processes with sample paths in $D_E[0, \infty)$. Let $M \subset \bar{C}(E)$ strongly separate points. Suppose there exists a process X with sample paths in $D_E[0, \infty)$ such that for each finite set $\{g_1, \ldots, g_k\} \subset M$,

$$(9.5)\qquad (g_1, \ldots, g_k) \circ X_n \Rightarrow (g_1, \ldots, g_k) \circ X \quad \text{in} \quad D_{\mathbb{R}^k}[0, \infty).$$

Then $X_n \Rightarrow X$.

Proof. Let H be the smallest algebra containing $M \cup \{1\}$, that is, the algebra of functions of the form $\sum_{i=1}^{l} a_i f_{i1} f_{i2} \cdots f_{im}$, where $l \geq 1$, $m \geq 1$, and $a_i \in \mathbb{R}$ and $f_{ij} \in M \cup \{1\}$ for $i = 1, \ldots, l$ and $j = 1, \ldots, m$. By the Stone–Weierstrass theorem, H is dense in $\bar{C}(E)$ in the topology of uniform convergence on compact sets. Note that (9.5) holds for every finite set $\{g_1, \ldots, g_k\} \subset H$. Thus, by Theorem 9.1, to prove the relative compactness of $\{X_n\}$ we need only verify (9.1).

Let $\Gamma \subset E$ be compact, and let $\delta > 0$. For each $x \in E$, choose $\{h_1^x, \ldots, h_{k(x)}^x\} \subset H$ satisfying

$$(9.6)\qquad \varepsilon(x) \equiv \inf_{y: r(x, y) \geq \delta} \max_{1 \leq i \leq k(x)} |h_i^x(y) - h_i^x(x)| > 0,$$

and let $U_x = \{y \in E: \max_{1 \leq i \leq k(x)} |h_i^x(y) - h_i^x(x)| < \varepsilon(x)\}$. Then $\Gamma \subset \bigcup_{x \in \Gamma} U_x \subset \Gamma^\delta$, so, since Γ is compact, there exist $x_1, \ldots, x_n \in \Gamma$ such that $\Gamma \subset \bigcup_{l=1}^{N} U_{x_l} \subset \Gamma^\delta$. Define $\sigma: D_{\mathbb{R}}[0, \infty) \to D_{\mathbb{R}}[0, \infty)$ by $\sigma(x)(t) = \sup_{0 \leq s \leq t} x(s)$ and observe that σ is continuous (by Proposition 5.3). For each n, let

$$(9.7)\qquad Y_n(t) = \min_{1 \leq l \leq N} \{g_l(X_n(t)) - \varepsilon(x_l)\}, \qquad t \geq 0,$$

where $g_l(x) \equiv \max_{1 \leq i \leq k(x_l)} |h_i^{x_l}(x) - h_i^{x_l}(x_l)|$, and put $Z_n = \sigma(Y_n)$. It follows from (9.5) and the continuity of σ that $Z_n \Rightarrow Z$, where Z is defined in terms of X as Z_n is in terms of X_n. Therefore $Z_n(T) \Rightarrow Z(T)$ for all $T \in D(Z)$, and for such T

$$(9.8)\qquad \begin{aligned} &\varliminf_{n \to \infty} P\{X_n(t) \in \Gamma^\delta \quad \text{for} \quad 0 \leq t \leq T\} \\ &\geq \varliminf_{n \to \infty} P\{Z_n(T) < 0\} \\ &\geq P\{Z(T) < 0\} \\ &\geq P\{X(t) \in \Gamma \quad \text{for} \quad 0 \leq t \leq T\} \end{aligned}$$

by Theorem 3.1, where the last inequality uses the fact that

(9.9) $$\sup_{x \in \Gamma} \min_{1 \le l \le N} \{g_l(x) - \varepsilon(x_l)\} < 0.$$

Let $\eta > 0$, let $T > 0$ be as above, let $m \ge 1$, and choose a compact set $\Gamma_{0,m} \subset E$ such that

(9.10) $$P\{X(t) \in \Gamma_{0,m} \quad \text{for} \quad 0 \le t \le T\} \ge 1 - \eta 2^{-m-1};$$

this is possible by Theorem 5.6, Lemma 2.1, and Remark 6.4. By (9.8), there exists $n_m \ge 1$ such that

(9.11) $$\inf_{n \ge n_m} P\{X_n(t) \in \Gamma_{0,m}^{1/m} \quad \text{for} \quad 0 \le t \le T\} \ge 1 - \eta 2^{-m}.$$

Finally, for $n = 1, \ldots, n_m - 1$, choose a compact set $\Gamma_{n,m} \subset E$ such that

(9.12) $$P\{X_n(t) \in \Gamma_{n,m}^{1/m} \quad \text{for} \quad 0 \le t \le T\} \ge 1 - \eta 2^{-m}.$$

Letting $\Gamma_m = \bigcup_{n=0}^{n_m - 1} \Gamma_{n,m}$, we have

(9.13) $$\inf_{n \ge 1} P\{X_n(t) \in \Gamma_m^{1/m} \quad \text{for} \quad 0 \le t \le T\} \ge 1 - \eta 2^{-m},$$

so if we define $\Gamma_{\eta,T}$ to be the closure of $\bigcap_{m \ge 1} \Gamma_m^{1/m}$, then $\Gamma_{\eta,T}$ is compact (being complete and totally bounded) and

(9.14) $$\inf_{n \ge 1} P\{X_n(t) \in \Gamma_{\eta,T} \quad \text{for} \quad 0 \le t \le T\} \ge 1 - \eta.$$

Finally, we note that

(9.15) $$(g_1 \wedge a_1, \ldots, g_k \wedge a_k) \circ X_n \Rightarrow (g_1 \wedge a_1, \ldots, g_k \wedge a_k) \circ X$$

for all $g_1, \ldots, g_k \in H$ and $a_1, \ldots, a_k \in \mathbb{R}$. This, together with the fact that H is dense in $\bar{C}(E)$ in the topology of uniform convergence on compact sets, allows one to conclude that the finite-dimensional distributions converge. The details are left to the reader. □

9.3 Corollary Let E be locally compact and separable, and let E^{Δ} be its one-point compactification. If $\{X_n\}$ is a sequence of processes with sample paths in $D_E[0, \infty)$ and if $\{f \circ X_n\}$ is relatively compact for every $f \in \hat{C}(E)$ (the space of continuous functions on E vanishing at infinity), then $\{X_n\}$ is relatively compact considered as a sequence of processes with sample paths in $D_{E^{\Delta}}[0, \infty)$. If, in addition, $(X_n(t_1), \ldots, X_n(t_k)) \Rightarrow (X(t_1), \ldots, X(t_k))$ for all finite subsets $\{t_1, \ldots, t_k\}$ of some dense set $D \subset [0, \infty)$, where X has sample paths in $D_E[0, \infty)$, then $X_n \Rightarrow X$ in $D_E[0, \infty)$.

Proof. If $f \in C(E^\Delta)$, then $f(\cdot) - f(\Delta)$ restricted to E belongs to $\hat{C}(E)$. Consequently, $\{f \circ X_n\}$ is relatively compact for every $f \in C(E^\Delta)$, and the relative compactness of $\{X_n\}$ in $D_{E^\Delta}[0, \infty)$ follows from Theorem 9.1. Under the additional assumptions, $X_n \Rightarrow X$ in $D_{E^\Delta}[0, \infty)$ by Theorem 7.7, and hence $X_n \Rightarrow X$ in $D_E[0, \infty)$ by Corollary 3.2. □

We now consider the problem of verifying the relative compactness of $\{f \circ X_\alpha\}$, where $f \in \bar{C}(E)$ is fixed. First, however, recall the notation and terminology of Section 7 of Chapter 2. For each α, let X_α be a process with sample paths in $D_E[0, \infty)$ defined on a probability space $(\Omega_\alpha, \mathscr{F}^\alpha, P_\alpha)$ and adapted to a filtration $\{\mathscr{F}_t^\alpha\}$. Let $\mathscr{L}_\alpha$ be the Banach space of real-valued $\{\mathscr{F}_t^\alpha\}$-progressive processes with norm $\|Y\| = \sup_{t \geq 0} E[|Y(t)|] < \infty$. Let

$$(9.16) \qquad \hat{\mathscr{A}}_\alpha = \left\{(Y, Z) \in \mathscr{L}_\alpha \times \mathscr{L}_\alpha: \quad Y(t) - \int_0^t Z(s)\, ds \text{ is an } \{\mathscr{F}_t^\alpha\}\text{-martingale}\right\},$$

and note that completeness of $\{\mathscr{F}_t^\alpha\}$ is not needed here.

9.4 Theorem Let (E, r) be arbitrary, and let $\{X_\alpha\}$ be a family of processes as above. Let C_a be a subalgebra of $\bar{C}(E)$ (e.g., the space of bounded, uniformly continuous functions with bounded support), and let D be the collection of $f \in \bar{C}(E)$ such that for every $\varepsilon > 0$ and $T > 0$ there exist $(Y_\alpha, Z_\alpha) \in \hat{\mathscr{A}}_\alpha$ with

$$(9.17) \qquad \sup_\alpha E\left[\sup_{t \in [0, T] \cap \mathbb{Q}} |Y_\alpha(t) - f(X_\alpha(t))|\right] < \varepsilon$$

and

$$(9.18) \qquad \sup_\alpha E[\|Z_\alpha\|_{p, T}] < \infty \quad \text{for some} \quad p \in (1, \infty].$$

($\|h\|_{p, T} = [\int_0^T |h(t)|^p\, dt]^{1/p}$ if $p < \infty$; $\|h\|_{\infty, T} = \text{ess sup}_{0 \leq t \leq T} |h(t)|$.) If C_a is contained in the closure of D (in the sup norm), then $\{f \circ X_\alpha\}$ is relatively compact for each $f \in C_a$; more generally, $\{(f_1, \ldots, f_k) \circ X_\alpha\}$ is relatively compact in $D_{\mathbb{R}^k}[0, \infty)$ for all $f_1, f_2, \ldots, f_k \in C_a$, $1 \leq k \leq \infty$.

9.5 Remark **(a)** Taking $p = 1$ in condition (9.18) is not sufficient. For example, let $n \geq 1$ and consider the two-state ($E = \{0, 1\}$) Markov process X_n with infinitesimal matrix

$$(9.19) \qquad \begin{pmatrix} -1 & 1 \\ n & -n \end{pmatrix}$$

and $P\{X_n(0) = 0\} = 1$. Given a function f on $\{0, 1\}$, put $Y_n = f \circ X_n$ and $Z_n = (A_n f) \circ X_n$, so that $(Y_n, Z_n) \in \hat{\mathscr{A}}_n$ and

$$(9.20) \qquad \begin{aligned} &E[|A_n f(X_n(t))|] \\ &\quad = n|f(0) - f(1)|P\{X_n(t) = 1\} + |f(1) - f(0)|P\{X_n(t) = 0\} \\ &\quad = |f(1) - f(0)|(1 + (n-1)(n+1)^{-1}(1 - e^{-(n+1)t})) \\ &\quad \leq 2|f(1) - f(0)|, \end{aligned}$$

for all $t \geq 0$. However, observe that the finite-dimensional distributions of X_n converge to those of a process that is identically zero, but $\{X_n\}$ does not converge in distribution.

(b) For sequences $\{X_n\}$, one can replace $\sup_\alpha$ in (9.17) and (9.18) by $\overline{\lim}_{n \to \infty}$. □

Proof. Since D is actually closed, we may assume that $C_a \subset D$. Let $f \in C_a$, $\varepsilon > 0$, and $T > 0$. Then $f^2 \in C_a$ and there exist (Y_α, Z_α), $(Y'_\alpha, Z'_\alpha) \in \hat{\mathscr{A}}_\alpha$ such that

$$(9.21) \qquad \sup_\alpha E\left[\sup_{t \in [0, T+1] \cap \mathbb{Q}} |f(X_\alpha(t)) - Y_\alpha(t)|\right] < \varepsilon,$$

$$(9.22) \qquad \sup_\alpha E\left[\sup_{t \in [0, T+1] \cap \mathbb{Q}} |f^2(X_\alpha(t)) - Y'_\alpha(t)|\right] < \varepsilon,$$

$$(9.23) \qquad \sup_\alpha E[\| Z_\alpha \|_{p, T+1}] < \infty \quad \text{for some} \quad p \in (1, \infty],$$

$$(9.24) \qquad \sup_\alpha E[\| Z'_\alpha \|_{p', T+1}] < \infty \quad \text{for some} \quad p' \in (1, \infty].$$

Let $0 < \delta < 1$. For each $t \in [0, T] \cap \mathbb{Q}$ and $u \in [0, \delta] \cap \mathbb{Q}$,

$$(9.25) \qquad \begin{aligned} &E[(f(X_\alpha(t+u)) - f(X_\alpha(t)))^2 \mid \mathscr{F}_t^\alpha] \\ &\quad = E[f^2(X_\alpha(t+u)) - f^2(X_\alpha(t)) \mid \mathscr{F}_t^\alpha] \\ &\qquad - 2f(X_\alpha(t))E[f(X_\alpha(t+u)) - f(X_\alpha(t)) \mid \mathscr{F}_t^\alpha] \\ &\quad \leq 2E\left[\sup_{s \in [0, T+1] \cap \mathbb{Q}} |f^2(X_\alpha(s)) - Y'_\alpha(s)| \,\middle|\, \mathscr{F}_t^\alpha\right] \\ &\qquad + 4\|f\| E\left[\sup_{s \in [0, T+1] \cap \mathbb{Q}} |f(X_\alpha(s)) - Y_\alpha(s)| \,\middle|\, \mathscr{F}_t^\alpha\right] \\ &\qquad + E\left[\sup_{0 \leq r \leq T} \int_r^{r+\delta} |Z'_\alpha(s)| \, ds \,\middle|\, \mathscr{F}_t^\alpha\right] \\ &\qquad + 2\|f\| E\left[\sup_{0 \leq r \leq T} \int_r^{r+\delta} |Z_\alpha(s)| \, ds \,\middle|\, \mathscr{F}_t^\alpha\right]. \end{aligned}$$

Let $1/p + 1/q = 1$ and $1/p' + 1/q' = 1$, and note that $\int_r^{r+\delta} |h(s)|\, ds \leq \delta^{1/q} \| h \|_{p, T+1}$ for $0 \leq r \leq T$. Therefore, if we define

$$\begin{aligned}(9.26) \qquad \gamma_\alpha(\delta) = 2 &\sup_{s \in [0, T+1] \cap \mathbb{Q}} |f^2(X_\alpha(s)) - Y'_\alpha(s)| \\ &+ 4\|f\| \sup_{s \in [0, T+1] \cap \mathbb{Q}} |f(X_\alpha(s)) - Y_\alpha(s)| \\ &+ \delta^{1/q'} \| Z'_\alpha \|_{p', T+1} + 2\|f\| \delta^{1/q} \| Z_\alpha \|_{p, T+1},\end{aligned}$$

then

$$(9.27) \qquad E[(f(X_\alpha(t+u)) - f(X_\alpha(t)))^2 \mid \mathscr{F}_t^\alpha] \leq E[\gamma_\alpha(\delta) \mid \mathscr{F}_t^\alpha].$$

Note that this holds for all $0 \leq t \leq T$ and $0 \leq u \leq \delta$ (not just for rational t and u) by the right continuity of X_α. Since

$$\begin{aligned}(9.28) \quad \sup_\alpha E[\gamma_\alpha(\delta)] &\leq (2 + 4\|f\|)\varepsilon \\ &+ \delta^{1/q'} \sup_\alpha E[\| Z'_\alpha \|_{p', T+1}] + 2\|f\| \delta^{1/q} \sup_\alpha E[\| Z_\alpha \|_{p, T+1}],\end{aligned}$$

we may select ε depending on δ in such a way that (8.29) holds. Therefore, $\{f \circ X_\alpha\}$ is relatively compact by Theorem 8.6 (see Remark 8.7(a)).

Let $1 \leq k < \infty$. Given $f_1, \ldots, f_k \in C_a$, define $\gamma_\alpha^j(\delta)$ as in (9.26) and set $\gamma_\alpha(\delta) = \sum_{j=1}^k \gamma_\alpha^j(\delta)$. Then

$$(9.29) \qquad E\left[\sum_{j=1}^k (f_j(X_\alpha(t+u)) - f_j(X_\alpha(t)))^2 \,\middle|\, \mathscr{F}_t^\alpha\right] \leq E[\gamma_\alpha(\delta) \mid \mathscr{F}_t^\alpha]$$

for $0 \leq t \leq T$ and $0 \leq u \leq \delta$, and the $\gamma_\alpha^j(\delta)$ can be selected so that (8.29) holds. Finally, relative compactness for $k = \infty$ follows from relative compactness for all $k < \infty$. (See Problem 23.) □

10. CONVERGENCE TO A PROCESS IN $C_E[0, \infty)$

Let (E, r) be a metric space, and let $C_E[0, \infty)$ be the space of continuous functions $x: [0, \infty) \to E$. For $x \in D_E[0, \infty)$, define

$$(10.1) \qquad J(x) = \int_0^\infty e^{-u} [J(x, u) \wedge 1]\, du,$$

where

$$(10.2) \qquad J(x, u) = \sup_{0 \leq t \leq u} r(x(t), x(t-)).$$

Since the mapping $x \to J(x, \cdot)$ from $D_E[0, \infty)$ into $D_{[0, \infty)}[0, \infty)$ is continuous (by Proposition 5.3), it follows that J is continuous on $D_E[0, \infty)$. For each $x \in D_E[0, \infty)$, $J(x, \cdot)$ is nondecreasing, so $J(x) = 0$ if and only if $x \in C_E[0, \infty)$.

10.1 Lemma Suppose $\{x_n\} \subset D_E[0, \infty)$, $x \in D_E[0, \infty)$, and $\lim_{n\to\infty} d(x_n, x) = 0$. Then

$$\overline{\lim_{n\to\infty}} \sup_{0\le t\le u} r(x_n(t), x(t)) \le J(x, u) \tag{10.3}$$

for all $u \ge 0$.

Proof. By Proposition 5.3, there exists $\{\lambda_n\} \subset \Lambda$ such that $\lim_{n\to\infty} \sup_{0\le t\le u} |\lambda_n(t) - t| = 0$ and $\lim_{n\to\infty} \sup_{0\le t\le u} r(x_n(t), x(\lambda_n(t))) = 0$ for all $u \ge 0$. Therefore

$$\begin{aligned} \overline{\lim_{n\to\infty}} \sup_{0\le t\le u} r(x_n(t), x(t)) &\le \overline{\lim_{n\to\infty}} \sup_{0\le t\le u} r(x_n(t), x(\lambda_n(t))) \\ &\quad + \overline{\lim_{n\to\infty}} \sup_{0\le t\le u} r(x(\lambda_n(t)), x(t)) \\ &\le J(x, u) \end{aligned} \tag{10.4}$$

for all $u \ge 0$. □

Let $\bar{C}(D_E[0, \infty), d_U)$ be the space of bounded, real-valued functions on $D_E[0, \infty)$ that are continuous with respect to the metric

$$d_U(x, y) = \int_0^\infty e^{-u} \sup_{0\le t\le u} [r(x(t), y(t)) \wedge 1]\, du, \tag{10.5}$$

that is, continuous in the topology of uniform convergence on compact subsets of $[0, \infty)$. Since $d \le d_U$, we have $\mathscr{S}_E \equiv \mathscr{B}(D_E[0, \infty), d) \subset \mathscr{B}(D_E[0, \infty), d_U)$.

10.2 Theorem Let X_n, $n = 1, 2, \ldots$, and X be processes with sample paths in $D_E[0, \infty)$, and suppose that $X_n \Rightarrow X$. Then (a) X is a.s. continuous if and only if $J(X_n) \Rightarrow 0$, and (b) if X is a.s. continuous, then $f(X_n) \Rightarrow f(X)$ for every $\mathscr{S}_E$-measurable $f \in \bar{C}(D_E[0, \infty), d_U)$.

Proof. Part (a) follows from the continuity of J on $D_E[0, \infty)$.

By Lemma 10.1, if $\{x_n\} \subset D_E[0, \infty)$, $x \in C_E[0, \infty)$, and $\lim_{n\to\infty} d(x_n, x) = 0$, then $\lim_{n\to\infty} d_U(x_n, x) = 0$. Letting $F \subset \mathbb{R}$ be closed and f be as in the statement of the theorem, $f^{-1}(F)$ is d_U-closed. Denoting its d-closure by $\overline{f^{-1}(F)}$, it follows that $\overline{f^{-1}(F)} \cap C_E[0, \infty) = f^{-1}(F) \cap C_E[0, \infty)$. Therefore, if $P_n \Rightarrow P$ on $(D_E[0, \infty), d)$ and $P(C_E[0, \infty)) = 1$,

$$\begin{aligned} \overline{\lim_{n\to\infty}} P_n f^{-1}(F) &\le \overline{\lim_{n\to\infty}} P_n(\overline{f^{-1}(F)}) \le P(\overline{f^{-1}(F)}) \\ &= P(\overline{f^{-1}(F)} \cap C_E[0, \infty)) = P(f^{-1}(F) \cap C_E[0, \infty)) = Pf^{-1}(F) \end{aligned} \tag{10.6}$$

by Theorem 3.1, so we conclude that $P_n f^{-1} \Rightarrow Pf^{-1}$. This implies (b). □

The next result provides a useful criterion for a process with sample paths in $D_E[0, \infty)$ to have sample paths in $C_E[0, \infty)$. It can also be used in conjunction with Corollary 8.10.

10.3 Proposition Let (E, r) be separable, and let X be a process with sample paths in $D_E[0, \infty)$. Suppose that for each $T > 0$, there exist $\beta > 0$, $C > 0$, and $\theta > 1$ such that

$$E[q^\beta(X(t), X(s))] \le C(t - s)^\theta \tag{10.7}$$

whenever $0 \le s \le t \le T$, where $q = r \wedge 1$. Then almost all sample paths of X belong to $C_E[0, \infty)$.

Proof. Let T be a positive integer and observe that

$$\sum_{0 < t \le T} q^\beta(X(t), X(t-)) \le \varliminf_{n \to \infty} \sum_{k=1}^{2^n T} q^\beta(X(k2^{-n}), X((k-1)2^{-n})). \tag{10.8}$$

By Fatou's lemma and (10.7), the right side of (10.8) has zero expectation, and hence the left side is zero a.s. □

10.4 Proposition For $n = 1, 2, \ldots,$ let $\{Y_n(k), k = 0, 1, \ldots\}$ be a discrete-parameter $\mathbb{R}^d$-valued process, let $\alpha_n > 0$, and define

$$X_n(t) = Y_n([\alpha_n t]) \tag{10.9}$$

and

$$Z_n(t) = Y_n([\alpha_n t]) + (\alpha_n t - [\alpha_n t])(Y_n([\alpha_n t] + 1) - Y_n([\alpha_n t])) \tag{10.10}$$

for all $t \ge 0$. Note that X_n has sample paths in $D_{\mathbb{R}^d}[0, \infty)$ and Z_n has sample paths in $C_{\mathbb{R}^d}[0, \infty)$. If $\lim_{n \to \infty} \alpha_n = \infty$ and X is a process with sample paths in $C_{\mathbb{R}^d}[0, \infty)$, then $X_n \Rightarrow X$ if and only if $Z_n \Rightarrow X$.

Proof. We apply Corollary 3.3. It suffices to show that, if either $X_n \Rightarrow X$ or $Z_n \Rightarrow X$, then $d(X_n, Z_n) \to 0$ in probability. (The two uses of the letter d here should cause no confusion.) For $n = 1, 2, \ldots,$

$$\begin{aligned} d(X_n, Z_n) &\le \int_0^\infty e^{-u} \sup_{0 \le t \le u + \alpha_n^{-1}} (|X_n(t) - X_n(t-)| \wedge 1)\, du \\ &\le e^{\alpha_n^{-1}} J(X_n), \end{aligned} \tag{10.11}$$

and

$$J(X_n) \le \int_0^\infty e^{-u} \sup_{0 \le t \le u} \sup_{t - \varepsilon \wedge t \le s < t} (|Z_n(t) - Z_n(s)| \wedge 1)\, du \tag{10.12}$$

provided $\alpha_n^{-1} \le \varepsilon$. But the function $J_\varepsilon: D_{\mathbb{R}^d}[0, \infty) \to [0, 1]$, defined for each $\varepsilon > 0$ by

$$(10.13) \quad J_\varepsilon(x) = \int_0^\infty e^{-u} \sup_{0 \le t \le u} \sup_{t - \varepsilon \wedge t \le s < t} (|x(t) - x(s)| \wedge 1)\, du \qquad (\ge J(x)),$$

is continuous and satisfies $\lim_{\varepsilon \to 0} J_\varepsilon(x) = J(x)$ for all $x \in D_{\mathbb{R}^d}[0, \infty)$. Consequently, if $Z_n \Rightarrow X$, then (10.12) and Theorem 3.1 imply that

$$(10.14) \qquad \overline{\lim_{n\to\infty}} P\{J(X_n) \ge \delta\} \le \lim_{\varepsilon \to 0} \overline{\lim_{n \to \infty}} P\{J_\varepsilon(Z_n) \ge \delta\} \le \lim_{\varepsilon \to 0} P\{J_\varepsilon(X) \ge \delta\} = 0$$

for all $\delta > 0$, so we conclude that $J(X_n) \to 0$ in probability. The same conclusion follows from Theorem 10.2 if $X_n \Rightarrow X$. In either case, (10.11) implies that $d(X_n, Z_n) \to 0$ in probability, as required for Corollary 3.3. □

11. PROBLEMS

1. Let (S, d) be complete and separable, and let $P, Q \in \mathscr{P}(S)$. Show that there exists $\mu \in \mathscr{M}(P, Q)$ (see Theorem 1.2) such that

$$(11.1) \qquad \rho(P, Q) = \inf\{\varepsilon > 0: \mu\{(x, y): d(x, y) \ge \varepsilon\} \le \varepsilon\}.$$

2. Define $\|f\|_{BL} = \sup_x |f(x)| \vee \sup_{x \ne y} |f(x) - f(y)|/d(x, y)$ for each $f \in \bar{C}(S)$. Given $P, Q \in \mathscr{P}(S)$, let

$$(11.2) \qquad \|P - Q\| = \sup_{\|f\|_{BL} = 1} \left| \int f\, dP - \int f\, dQ \right|,$$

and show that $\rho^2(P, Q) \le \|P - Q\| \le 3\rho(P, Q)$.
Hint: Recall that $\int f\, dP = \int_0^{\|f\|} P\{f \ge t\}\, dt$ if $f \ge 0$, and note that $\|(\varepsilon - d(\cdot, F)) \vee 0\|_{BL} \le 1$ for $0 < \varepsilon < 1$.

3. Show that $\mathscr{P}(S)$ is separable whenever S is.

4. Suppose $\{P_n\} \subset \mathscr{P}(\mathbb{R})$, $P \in \mathscr{P}(\mathbb{R})$, and $P_n \Rightarrow P$. Define

$$(11.3) \qquad G_n(x) = \inf\{y \in \mathbb{R}: P_n((-\infty, y]) \ge x\}$$

and

$$(11.4) \qquad G(x) = \inf\{y \in \mathbb{R}: P((-\infty, y]) \ge x\}$$

for $0 < x < 1$, and let ξ be uniformly distributed on $(0, 1)$. Show that $G_n(\xi)$ has distribution P_n for each n, $G(\xi)$ has distribution P, and $\lim_{n \to \infty} G_n(\xi) = G(\xi)$ a.s.

5. Let (S, d) and (S', d') be separable. Let X_n, $n = 1, 2, \ldots$, and X be S-valued random variables with $X_n \Rightarrow X$. Suppose there exist Borel measurable mappings h_k, $k = 1, 2, \ldots$, and h from S into S' such that:

   (a) For $k = 1, 2, \ldots$, h_k is continuous a.s. with respect to the distribution of X.

   (b) $h_k \to h$ as $k \to \infty$ a.s. with respect to the distribution of X.

   (c) $\lim_{k \to \infty} \overline{\lim}_{n \to \infty} P\{d'(h_k(X_n), h(X_n)) > \varepsilon\} = 0$ for every $\varepsilon > 0$.

   Show that $h(X_n) \Rightarrow h(X)$. (Note that this generalizes Corollary 1.9.)

6. Let X_n, $n = 1, 2, \ldots$, and X be real-valued random variables with finite second moments defined on a common probability space $(\Omega, \mathscr{F}, P)$. Suppose that $\{X_n\}$ converges weakly to X in $L^2(P)$ (i.e., $\lim_{n \to \infty} E[X_n Z] = E[XZ]$, $Z \in L^2(P)$), and $\{X_n\}$ converges in distribution to some real-valued random variable Y. Give necessary and sufficient conditions for X and Y to have the same distribution.

7. Let X and Y be S-valued random variables defined on a probability space $(\Omega, \mathscr{F}, P)$, and let $\mathscr{G}$ be a sub-σ-algebra of $\mathscr{F}$. Suppose that $M \subset \bar{C}(S)$ is separating and

$$E[f(X) \mid \mathscr{G}] = f(Y) \tag{11.5}$$

   for every $f \in M$. Show that $X = Y$ a.s.

8. Let $M = \{f \in \bar{C}(\mathbb{R}): f \text{ has period } N \text{ for some positive integer } N\}$. Show that M is separating but not convergence determining.

9. Let $M \subset \bar{C}(S)$ and suppose that for every open set $G \subset S$ there exists a sequence $\{f_n\} \subset M$ with $0 \le f_n \le \chi_G$ for each n such that $\text{bp-}\lim_{n \to \infty} f_n = \chi_G$. Show that M is convergence determining.

10. Show that the collection of all twice continuously Frechet differentiable functions with bounded support on a separable Hilbert space is convergence determining.

11. Let S be locally compact and separable. Show that $M \subset \hat{C}(S)$ is convergence determining if and only if M is dense in $\hat{C}(S)$ in the supremum norm.

12. Let $x \in D_E[0, \infty)$, and for each $n \ge 1$, define $x_n \in D_E[0, \infty)$ by $x_n(t) = x(([nt]/n) \wedge n)$. Show that $\lim_{n \to \infty} d(x_n, x) = 0$.

13. Let E and F be metric spaces, and let $f: E \to F$ be continuous. Show that the mapping $x \to f \circ x$ from $D_E[0, \infty)$ into $D_F[0, \infty)$ is continuous.

14. Show that $D_E[0, \infty)$ is separable whenever E is.

15. Let (E, r) be arbitrary, let $\Gamma_1 \subset \Gamma_2 \subset \cdots$ be a nondecreasing sequence of compact subsets of E, and define

$$(11.6) \quad S = \{x \in D_E[0, \infty): x(t) \in \Gamma_n \quad \text{for} \quad 0 \le t \le n, n = 1, 2, \ldots\}.$$

Show that (S, d) is complete and separable, where d is defined by (5.2).

16. Let (E, r) be complete. Show that if A is compact in $D_E[0, \infty)$, then for each $T > 0$ there exists a compact set $\Gamma_T \subset E$ such that $x(t) \in \Gamma_T$ for $0 \le t \le T$ and all $x \in A$.

17. Prove the following variation of Proposition 6.5.

Let $\{x_n\} \subset D_E[0, \infty)$ and $x \in D_E[0, \infty)$. Then $\lim_{n\to\infty} d(x_n, x) = 0$ if and only if whenever $t_n \ge s_n \ge 0$ for each n, $t \ge 0$, $\lim_{n\to\infty} s_n = t$, and $\lim_{n\to\infty} t_n = t$, we have

$$(11.7) \qquad \lim_{n\to\infty} [r(x_n(t_n), x(t)) \vee r(x_n(s_n), x(t))] \wedge r(x_n(s_n), x(t-)) = 0$$

and

$$(11.8) \qquad \lim_{n\to\infty} r(x_n(t_n), x(t)) \wedge [r(x_n(t_n), x(t-)) \vee r(x_n(s_n), x(t-))] = 0.$$

18. Let (E, r) be complete and separable. Let $\{X_\alpha\}$ be a family of processes with sample paths in $D_E[0, \infty)$. Suppose that for every $\varepsilon > 0$ and $T > 0$ there exists a family $\{X_\alpha^{\varepsilon, T}\}$ of processes with sample paths in $D_E[0, \infty)$ (with $X_\alpha^{\varepsilon, T}$ and X_α defined on the same probability space) such that

$$(11.9) \qquad \sup_\alpha P\left\{\sup_{0 \le t \le T} r(X_\alpha^{\varepsilon, T}(t), X_\alpha(t)) \ge \varepsilon\right\} < \varepsilon$$

and $\{X_\alpha^{\varepsilon, T}\}$ is relatively compact. Show that $\{X_\alpha\}$ is relatively compact.

19. Let (E, r) be complete and separable. Show that if $\{X_\alpha\}$ is relatively compact in $D_E[0, \infty)$, then the compact containment condition holds.

20. Let $\{N_\alpha\}$ be a family of right continuous counting processes (i.e., $N_\alpha(0) = 0$, $N_\alpha(t) - N_\alpha(t-) = 0$ or 1 for all $t > 0$). For $k = 0, 1, \ldots,$ let $\tau_k^\alpha = \inf\{t \ge 0: N_\alpha(t) \ge k\}$ and $\Delta_k^\alpha = \tau_k^\alpha - \tau_{k-1}^\alpha$ (if $\tau_{k-1}^\alpha < \infty$). Use Lemma 8.2 to give necessary and sufficient conditions for the relative compactness of $\{N_\alpha\}$.

21. Let (E, r) be complete and separable, and let $\{X_\alpha\}$ be a family of processes with sample paths in $D_E[0, \infty)$. Show that $\{X_\alpha\}$ is relatively compact if and only if for every $\varepsilon > 0$ there exists a family $\{X_\alpha^\varepsilon\}$ of pure jump processes (with X_α^ε and X_α defined on the same probability space) such that $\sup_\alpha \sup_{t \ge 0} r(X_\alpha^\varepsilon(t), X_\alpha(t)) < \varepsilon$ a.s., $\{X_\alpha^\varepsilon(t)\}$ is relatively compact for each rational $t \ge 0$, and $\{N_\alpha^\varepsilon\}$ is relatively compact, where $N_\alpha^\varepsilon(t)$ is the number of jumps of X_α^ε in $(0, t]$.

22. (a) Give an example in which $\{X_n\}$ and $\{Y_n\}$ are relatively compact in $D_{\mathbb{R}}[0, \infty)$, but $\{(X_n, Y_n)\}$ is not relatively compact in $D_{\mathbb{R}^2}[0, \infty)$.

(b) Show that if $\{X_n\}$, $\{Y_n\}$, and $\{X_n + Y_n\}$ are relatively compact in $D_{\mathbb{R}}[0, \infty)$, then $\{(X_n, Y_n)\}$ is relatively compact in $D_{\mathbb{R}^2}[0, \infty)$.

(c) More generally, if $2 \le r < \infty$, show that $\{(X_n^1, X_n^2, \ldots, X_n^r)\}$ is relatively compact in $D_{\mathbb{R}^r}[0, \infty)$ if and only if $\{X_n^k\}$ and $\{X_n^k + X_n^l\}$ ($k, l = 1, \ldots, r$) are relatively compact in $D_{\mathbb{R}}[0, \infty)$.

23. Show that $\{(X_n^1, X_n^2, \ldots)\}$ is relatively compact in $D_{\mathbb{R}^\infty}[0, \infty)$ (where $\mathbb{R}^\infty$ has the product topology) if and only if $\{(X_n^1, \ldots, X_n^r)\}$ is relatively compact in $D_{\mathbb{R}^r}[0, \infty)$ for $r = 1, 2, \ldots$.

24. Let (E, r) be complete and separable, and let $\{X_n\}$ be a sequence of processes with sample paths in $D_E[0, \infty)$. Let M be a subspace of $\bar{C}(E)$ that strongly separates points. Show that if the finite-dimensional distributions of X_n converge to those of a process X with sample paths in $D_E[0, \infty)$, and if $\{g \circ X_n\}$ is relatively compact in $D_{\mathbb{R}}[0, \infty)$ for every $g \in M$, then $X_n \Rightarrow X$.

25. Let (E, r) be separable, and consider the metric space $(C_E[0, \infty), d_U)$, where d_U is defined by (10.5). Let $\mathscr{B}$ denote its Borel σ-algebra.

(a) For each $t \ge 0$, define $\pi_t\colon C_E[0, \infty) \to E$ by $\pi_t(x) = x(t)$. Show that $\mathscr{B} = \sigma(\pi_t\colon 0 \le t < \infty)$.

(b) Show that d_U determines the same topology on $C_E[0, \infty)$ as d (the latter defined by (5.2)).

(c) Show that $C_E[0, \infty)$ is a closed subset of $D_E[0, \infty)$, hence it belongs to $\mathscr{S}_E$, and therefore $\mathscr{B} \subset \mathscr{S}_E$.

(d) Suppose that $\{P_n\} \subset \mathscr{P}(D_E[0, \infty))$, $P \in \mathscr{P}(D_E[0, \infty))$, and $P_n(C_E[0, \infty)) = P(C_E[0, \infty)) = 1$ for each n. Define $\{Q_n\} \subset \mathscr{P}(C_E[0, \infty))$ and $Q \in \mathscr{P}(C_E[0, \infty))$ by $Q_n = P_n|_{\mathscr{B}}$ and $Q = P|_{\mathscr{B}}$. Show that $P_n \Rightarrow P$ on $D_E[0, \infty)$ if and only if $Q_n \Rightarrow Q$ on $C_E[0, \infty)$.

26. Show that each of the following functions $f_k\colon D_{\mathbb{R}}[0, \infty) \to D_{\mathbb{R}}[0, \infty)$ is continuous:

$$\begin{aligned} f_1(x)(t) &= \sup_{s \le t} x(s), \\ f_2(x)(t) &= \inf_{s \le t} x(s), \\ f_3(x)(t) &= \int_0^t x(s)\, ds, \\ f_4(x)(t) &= \sup_{s \le t} (x(s) - x(s-)). \end{aligned} \tag{11.10}$$

27. Let $\mathscr{A} \subset \mathscr{B}(S)$ be closed under finite intersections and suppose each open

set in S is a countable union of sets in $\mathscr{A}$. Suppose $P, P_n \in \mathscr{P}(S)$, $n = 1, 2, \ldots$, and $\lim_{n\to\infty} P_n(A) = P(A)$ for every $A \in \mathscr{A}$. Show that $P_n \Rightarrow P$.

28. Let (S, d) be complete and separable and let $\mathscr{A} \subset \mathscr{B}(S)$. Suppose for each closed set F and open set U with $F \subset U$, there exists $A \in \mathscr{A}$ such that $F \subset A \subset U$. Show that if $\{P_n\} \subset \mathscr{P}(S)$ is relatively compact and $\lim_{n\to\infty} P_n(A)$ exists for each $A \in \mathscr{A}$, then there exists $P \in \mathscr{P}(S)$ such that $P_n \Rightarrow P$.

12. NOTES

The standard reference on the topic of convergence of probability measures is Billingsley's (1968) book of that title, where additional historical remarks can be found.

As originally defined, the Prohorov (1956) metric was a symmetrized version of the present ρ. Strassen (1965) noticed that ρ is already symmetric and obtained Theorem 1.2. Lemma 1.3 is essentially due to Dudley (1968). Lemma 1.4 is a modification of the marriage lemma of Hall (1935), and is a special case of a result of Artstein (1983). Prohorov (1956) obtained Theorem 1.7. The Skorohod (1956) representation theorem (Theorem 1.8) originally required that (S, d) be complete; Dudley (1968) removed this assumption. For a recent somewhat stronger result see Blackwell and Dubins (1983). The continuous mapping theorem (Corollary 1.9) can be attributed to Mann and Wald (1943) and Chernoff (1956).

Theorem 2.2 is of course due to Prohorov (1956).

Theorem 3.1 (without (a)) is known as the Portmanteau theorem and goes back to Alexandroff (1940–1943); the equivalence of (a) is due to Prohorov (1956) assuming completeness and to Dudley (1968) in general. Corollary 3.3 is called Slutsky's theorem.

The topology on $D_E[0, \infty)$ is Stone's (1963) analogue of Skorohod's (1956) J_1 topology. Metrizability was first shown by Prohorov (1956). The metric d is analogous to Billingsley's (1968) d_0 on $D[0, 1]$. Theorem 5.6 is essentially due to Kolmogorov (1956).

With a different modulus of continuity, Theorem 6.3 was proved by Prohorov (1956); in its present form, it is due to Billingsley (1968).

Similar remarks apply to Theorem 7.2.

Condition (b) of Theorem 8.6 for relative compactness is due to Kurtz (1975), as are the results preceding it in Section 8; Aldous (1978) is responsible for condition (c). See also Jacod, Mémin, and Métivier (1983). Theorem 8.8 is due to Chenčov (1956).

The results of Section 9 are based on Kurtz (1975).

Proposition 10.4 was proved by Sato (1977).

Problem 5 is due to Lindvall (1974) and can be derived as a consequence of Theorem 4.2 of Billingsley (1968).

4 GENERATORS AND MARKOV PROCESSES

In this chapter we study Markov processes from the standpoint of the generators of their corresponding semigroups. In Section 1 we give the basic definitions of a Markov process, its transition function, and its corresponding semigroup, show that a transition function and initial distribution uniquely determine a Markov process, and verify the important martingale relationship between a Markov process and its generator. Section 2 is devoted to the study of Feller processes and the properties of their sample paths, and Sections 3 through 7 to the martingale problem as a means of characterizing the Markov process corresponding to a given generator. In Section 8 we exploit the characterization of a Markov process by its generator (either through the determination of its semigroup or as the unique solution of a martingale problem) to give general conditions for the weak convergence of a sequence of processes to a Markov process. Stationary distributions are the subject of Section 9. Some conditions under which sums of generators characterize Markov processes are given in Section 10.

Throughout this chapter E is a metric space, $M(E)$ is the collection of all real-valued, Borel measurable functions on E, $B(E) \subset M(E)$ is the Banach space of bounded functions with $\|f\| = \sup_{x \in E} |f(x)|$, and $\bar{C}(E) \subset B(E)$ is the subspace of bounded continuous functions.

1. MARKOV PROCESSES AND TRANSITION FUNCTIONS

Let $\{X(t), t \geq 0\}$ be a stochastic process defined on a probability space $(\Omega, \mathscr{F}, P)$ with values in E, and let $\mathscr{F}_t^X = \sigma(X(s): s \leq t)$. Then X is a *Markov process* if

$$P\{X(t+s) \in \Gamma \mid \mathscr{F}_t^X\} = P\{X(t+s) \in \Gamma \mid X(t)\} \tag{1.1}$$

for all $s, t \geq 0$ and $\Gamma \in \mathscr{B}(E)$. If $\{\mathscr{G}_t\}$ is a filtration with $\mathscr{F}_t^X \subset \mathscr{G}_t$, $t \geq 0$, then X is a *Markov process with respect to* $\{\mathscr{G}_t\}$ if (1.1) holds with $\mathscr{F}_t^X$ replaced by $\mathscr{G}_t$. (Of course if X is a Markov process with respect to $\{\mathscr{G}_t\}$, then it is a Markov process.) Note that (1.1) implies

$$E[f(X(t+s)) \mid \mathscr{F}_t^X] = E[f(X(t+s)) \mid X(t)] \tag{1.2}$$

for all $s, t \geq 0$ and $f \in B(E)$.

A function $P(t, x, \Gamma)$ defined on $[0, \infty) \times E \times \mathscr{B}(E)$ is a time homogeneous *transition function* if

$$P(t, x, \cdot) \in \mathscr{P}(E), \qquad (t, x) \in [0, \infty) \times E, \tag{1.3}$$

$$P(0, x, \cdot) = \delta_x \quad \text{(unit mass at } x\text{)}, \qquad x \in E, \tag{1.4}$$

$$P(\cdot, \cdot, \Gamma) \in B([0, \infty) \times E), \qquad \Gamma \in \mathscr{B}(E), \tag{1.5}$$

and

$$P(t+s, x, \Gamma) = \int P(s, y, \Gamma) P(t, x, dy), \qquad s, t \geq 0,\ x \in E,\ \Gamma \in \mathscr{B}(E). \tag{1.6}$$

A transition function $P(t, x, \Gamma)$ is a *transition function for a time-homogeneous Markov process* X if

$$P\{X(t+s) \in \Gamma \mid \mathscr{F}_t^X\} = P(s, X(t), \Gamma) \tag{1.7}$$

for all $s, t \geq 0$ and $\Gamma \in \mathscr{B}(E)$, or equivalently, if

$$E[f(X(t+s)) \mid \mathscr{F}_t^X] = \int f(y) P(s, X(t), dy) \tag{1.8}$$

for all $s, t \geq 0$ and $f \in B(E)$.

To see that (1.6), called the *Chapman–Kolmogorov property*, is a reasonable assumption given (1.7), observe that (1.7) implies

$$\begin{aligned} P(t+s, X(u), \Gamma) &= P\{X(u+t+s) \in \Gamma \mid \mathscr{F}_u^X\} \\ &= E[P\{X(u+t+s) \in \Gamma \mid \mathscr{F}_{u+t}^X\} \mid \mathscr{F}_u^X] \\ &= E[P(s, X(u+t), \Gamma) \mid \mathscr{F}_u^X] \\ &= \int P(s, y, \Gamma) P(t, X(u), dy) \end{aligned} \tag{1.9}$$

for all $s, t, u \geq 0$ and $\Gamma \in \mathscr{B}(E)$.

The probability measure $\nu \in \mathscr{P}(E)$ given by $\nu(\Gamma) = P\{X(0) \in \Gamma\}$ is called the *initial distribution* of X.

A transition function for X and the initial distribution determine the finite-dimensional distributions of X by

$$(1.10)\quad P\{X(0) \in \Gamma_0, X(t_1) \in \Gamma_1, \dots, X(t_n) \in \Gamma_n\}$$
$$= \int_{\Gamma_0} \cdots \int_{\Gamma_{n-1}} P(t_n - t_{n-1}, y_{n-1}, \Gamma_n) P(t_{n-1} - t_{n-2}, y_{n-2}, dy_{n-1})$$
$$\cdots P(t_1, y_0, dy_1)\nu(dy_0).$$

In particular we have the following theorem.

1.1 Theorem Let $P(t, x, \Gamma)$ satisfy (1.3)–(1.6) and let $\nu \in \mathscr{P}(E)$. If for each $t \geq 0$ the probability measure $\int P(t, x, \cdot)\nu(dx)$ is tight (which will be the case if (E, r) is complete and separable by Lemma 2.1 of Chapter 3), then there exists a Markov process X in E whose finite-dimensional distributions are uniquely determined by (1.10).

Proof. For $I \subset [0, \infty)$, let E^I denote the product space $\prod_{s \in I} E_s$ where for each s, $E_s = E$, and let $\mathscr{P}_I$ denote the collection of probability measures defined on the product σ-algebra $\prod_{s \in I} \mathscr{B}(E_s)$. For $s \in I$, let $X(s)$ denote the coordinate random variable. Note that $\prod_{s \in I} \mathscr{B}(E_s) = \sigma(X(s): s \in I)$.

Let $\{s_i, i = 0, 1, 2, \dots\} \subset [0, \infty)$ satisfy $s_i \neq s_j$ for $i \neq j$ and $s_0 = 0$, and fix $x_0 \in E$. For $n > 1$, let $t_1 < t_2 < \cdots < t_n$ be the increasing rearrangement of $s_1, \dots, s_n$. Then it follows from Tulcea's theorem (Appendix 9) that there exists $P_n \in \mathscr{P}_{\{s_i\}}$ such that $P_n(X(0) \in \Gamma_0, X(t_1) \in \Gamma_1, \dots, X(t_n) \in \Gamma_n\}$ equals the right side of (1.10) and $P_n\{X(s_i) = x_0\} = 1$ for $i > n$. Tulcea's theorem gives a measure Q_n on $E^{\{s_1, \dots, s_n\}}$. Fix $x_0 \in E$ and define $P_n = Q_n \times \delta_{\{x_0, x_0, \dots\}}$ on $E^I = E^{\{s_1, \dots, s_n\}} \times E^{\{s_{n+1}, s_{n+2}, \dots\}}$. The sequence $\{P_n\}$ is tight by Proposition 2.4 of Chapter 3, and any limit point will satisfy (1.10) for $\{t_1, \dots, t_n\} \subset \{s_i\}$. Consequently $\{P_n\}$ converges weakly to $P^{\{s_i\}} \in \mathscr{P}_{\{s_i\}}$.

By Problem 27 of Chapter 2 for $B \in \mathscr{B}(E)^{[0, \infty)}$ there exists a countable subset $\{s_i\} \subset [0, \infty)$ such that $B \in \sigma(X(s_i): i = 1, 2, \dots)$, that is, there exists $\hat{B} \in \mathscr{B}(E)^{\{s_i\}}$ such that $B = \{(X(s_1), X(s_2), \dots) \in \hat{B}\}$. Define $P(B) \equiv P^{\{s_i\}}(\hat{B})$. We leave it to the reader to verify the consistency of this definition and to show that X, defined on $(E^{[0, \infty)}, \mathscr{B}(E)^{[0, \infty)}, P)$, is Markov. □

Let P_x denote the measure on $\mathscr{B}(E)^{[0, \infty)}$ given by Theorem 1.1 with $\nu = \delta_x$, and let X be the corresponding coordinate process, that is, $X(t, \omega) = \omega(t)$. It follows from (1.5) and (1.10) that $P_x(B)$ is a Borel measurable function of x for $B = \{X(0) \in \Gamma_0, \dots, X(t_n) \in \Gamma_n\}$, $0 < t_1 < t_2 < \cdots < t_n$, $\Gamma_0, \Gamma_1, \dots, \Gamma_n \in \mathscr{B}(E)$. In fact this is true for all $B \in \mathscr{B}(E)^{[0, \infty)}$.

1.2 Proposition Let P_x be as above. Then $P_x(B)$ is a Borel measurable function of x for each $B \in \mathscr{B}(E)^{[0, \infty)}$.

Proof. The collection of B for which $P_x(B)$ is a Borel measurable function is a Dynkin class containing all sets of the form $\{X(0) \in \Gamma_0, \ldots, X(t_n) \in \Gamma_n\} \in \mathscr{B}(E)^{[0, \infty)}$, and hence all $B \in \mathscr{B}(E)^{[0, \infty)}$. (See Appendix 4.) □

Let $\{Y(n),\ n = 0,\ 1,\ 2, \ldots\}$ be a discrete-parameter process defined on $(\Omega, \mathscr{F}, P)$ with values in E, and let $\mathscr{F}_n^Y = \sigma(Y(k): k \le n)$. Then Y is a *Markov chain* if

$$(1.11) \qquad P\{Y(n + m) \in \Gamma \mid \mathscr{F}_n^Y\} = P\{Y(n + m) \in \Gamma \mid Y(n)\}$$

for all $m,\ n \ge 0$ and $\Gamma \in \mathscr{B}(E)$. A function $\mu(x, \Gamma)$ defined on $E \times \mathscr{B}(E)$ is a *transition function* if

$$(1.12) \qquad \mu(x, \cdot\,) \in \mathscr{P}(E), \qquad x \in E,$$

and

$$(1.13) \qquad \mu(\cdot, \Gamma) \in B(E), \qquad \Gamma \in \mathscr{B}(E).$$

A transition function $\mu(x, \Gamma)$ is a *transition function for a time-homogeneous Markov chain* Y if

$$(1.14) \qquad P\{Y(n + 1) \in \Gamma \mid \mathscr{F}_n^Y\} = \mu(Y(n), \Gamma), \qquad n \ge 0,\ \Gamma \in \mathscr{B}(E).$$

Note that

$$(1.15) \quad P\{Y(n + m) \in \Gamma \mid \mathscr{F}_n^Y\} = \int \cdots \int \mu(y_{m-1}, \Gamma)\mu(y_{m-2}, dy_{m-1}) \cdots \mu(Y(n), dy_1).$$

As before, the probability measure $\nu \in \mathscr{P}(E)$ given by $\nu(\Gamma) = P\{Y(0) \in \Gamma\}$ is called the initial distribution for Y. The analogues of Theorem 1.1 and Proposition 1.2 are left to the reader.

Let $\{X(t), t \ge 0\}$, defined on $(\Omega, \mathscr{F}, P)$, be an E-valued Markov process with respect to a filtration $\{\mathscr{G}_t\}$ such that X is $\{\mathscr{G}_t\}$-progressive. Suppose $P(t, x, \Gamma)$ is a transition function for X, and let τ be a $\{\mathscr{G}_t\}$-stopping time with $\tau < \infty$ a.s. Then X is *strong Markov at* τ if

$$(1.16) \qquad P\{X(\tau + t) \in \Gamma \mid \mathscr{G}_\tau\} = P(t, X(\tau), \Gamma)$$

for all $t \ge 0$ and $\Gamma \in \mathscr{B}(E)$, or equivalently, if

$$(1.17) \qquad E[f(X(\tau + t)) \mid \mathscr{G}_\tau] = \int f(y)P(t, X(\tau), dy)$$

for all $t \ge 0$ and $f \in B(E)$. X is a *strong Markov process* with respect to $\{\mathscr{G}_t\}$ if X is strong Markov at τ for all $\{\mathscr{G}_t\}$-stopping times τ with $\tau < \infty$ a.s.

1.3 Proposition Let X be E-valued, $\{\mathscr{G}_t\}$-progressive, and $\{\mathscr{G}_t\}$-Markov, and let $P(t, x, \Gamma)$ be a transition function for X. Let τ be a discrete $\{\mathscr{G}_t\}$-stopping time with $\tau < \infty$ a.s. Then X is strong Markov at τ.

Proof. Let $t_1, t_2, \ldots$ be the values assumed by τ, and let $f \in B(E)$. If $B \in \mathscr{G}_\tau$, then $B \cap \{\tau = t_i\} \in \mathscr{G}_{t_i}$ and hence for $t \geq 0$,

$$(1.18)\qquad \begin{aligned}\int_{B \cap \{\tau = t_i\}} f(X(\tau + t))\, dP &= \int_{B \cap \{\tau = t_i\}} f(X(t_i + t))\, dP \\ &= \int_{B \cap \{\tau = t_i\}} \int f(y)P(t, X(t_i), dy)\, dP \\ &= \int_{B \cap \{\tau = t_i\}} \int f(y)P(t, X(\tau), dy)\, dP.\end{aligned}$$

Summing over i we obtain

$$(1.19)\qquad \int_B f(X(\tau + t))\, dP = \int_B \int f(y)P(t, X(\tau), dy)\, dP$$

for all $B \in \mathscr{G}_\tau$, which implies (1.17). □

1.4 Remark Recall (Proposition 1.3 of Chapter 2) that every stopping time is the limit of a decreasing sequence of discrete stopping times. This fact can frequently be exploited to extend Proposition 1.3 to more-general stopping times. See, for example, Theorem 2.7 below.

1.5 Proposition Let X be E-valued, $\{\mathscr{G}_t\}$-progressive, and $\{\mathscr{G}_t\}$-Markov, and let $P(t, x, \Gamma)$ be a transition function for X. Let τ be a $\{\mathscr{G}_t\}$-stopping time, suppose X is strong Markov at $\tau + t$ for all $t \geq 0$, and let $B \in \mathscr{B}(E)^{[0, \infty)}$. Then

$$(1.20)\qquad P\{X(\tau + \cdot) \in B \mid \mathscr{G}_\tau\} = P_{X(\tau)}(B).$$

Proof. First consider B of the form

$$(1.21)\qquad \{x \in E^{[0, \infty)}: x(t_i) \in \Gamma_i,\ i = 1, \ldots, n\}$$

where $0 \leq t_1 \leq t_2 \leq \cdots \leq t_n$, $\Gamma_1, \Gamma_2, \ldots, \Gamma_n \in \mathscr{B}(E)$. Proceeding by induction on n, for $B = \{x \in E^{[0, \infty)}: x(t_1) \in \Gamma_1\}$ we have

$$(1.22)\qquad P\{X(\tau + \cdot) \in B \mid \mathscr{G}_\tau\} = P\{X(\tau + t_1) \in \Gamma_1 \mid \mathscr{G}_\tau\} = P(t_1, X(\tau), \Gamma_1)$$

by (1.16), but this is just (1.20). Suppose now that (1.20) holds for all B of the form (1.21) for some fixed n. Then

$$(1.23)\quad E\left[\prod_{i=1}^{n} f_i(X(\tau + t_i))\middle| \mathscr{G}_\tau\right]$$
$$= \int \cdots \int \prod_{i=1}^{n} f_i(y_i) P(t_n - t_{n-1}, y_{n-1}, dy_n) \cdots P(t_1, X(\tau), dy_1)$$

for $0 \le t_1 \le t_2 \le \cdots \le t_n$ and $f_i \in B(E)$. Let B be of the form (1.21) with $n + 1$ in place of n. Then

$$(1.24)\quad P\{B \mid \mathscr{G}_\tau\} = E\left[\prod_{i=1}^{n+1} \chi_{\Gamma_i}(X(\tau + t_i))\middle| \mathscr{G}_\tau\right]$$
$$= E\left[E[\chi_{\Gamma_{n+1}}(X(\tau + t_{n+1})) \mid \mathscr{G}_{\tau+t_n}] \prod_{i=1}^{n} \chi_{\Gamma_i}(X(\tau + t_i))\middle| \mathscr{G}_\tau\right]$$
$$= E\left[P(t_{n+1} - t_n, X(\tau + t_n), \Gamma_{n+1}) \prod_{i=1}^{n} \chi_{\Gamma_i}(X(\tau + t_i))\middle| \mathscr{G}_\tau\right]$$
$$= \int \cdots \int P(t_{n+1} - t_n, y_n, \Gamma_{n+1}) \chi_{\Gamma_n}(y_n) \prod_{i=1}^{n-1} \chi_{\Gamma_i}(y_i)$$
$$\times P(t_n - t_{n-1}, y_{n-1}, dy_n) \cdots P(t_1, X(\tau), dy_1)$$
$$= \int_{\Gamma_1} \cdots \int_{\Gamma_n} P(t_{n+1} - t_n, y_n, \Gamma_{n+1})$$
$$\times P(t_n - t_{n-1}, y_{n-1}, dy_n) \cdots P(t_1, X(\tau), dy_1)$$
$$= P_{X(\tau)}(B).$$

Therefore (1.20) holds for all B of the form (1.21). The proposition now follows by the Dynkin class theorem. □

Ordinarily, formulas for transition functions cannot be obtained, the most notable exception being that of one-dimensional Brownian motion given by

$$(1.25)\quad P(t, x, \Gamma) = \int_\Gamma \frac{1}{\sqrt{2\pi t}} \exp\left\{-\frac{(y - x)^2}{2t}\right\} dy.$$

Consequently, directly defining a transition function is usually not a useful method of specifying a Markov process. In this chapter and in Chapters 5 and 6, we consider other methods of specifying processes. In particular, in this chapter we exploit the fact that

$$(1.26)\quad T(t)f(x) \equiv \int f(y) P(t, x, dy)$$

defines a measurable contraction semigroup on $B(E)$ (by the Chapman–Kolmogorov property (1.6)).

Let $\{T(t)\}$ be a semigroup on a closed subspace $L \subset B(E)$. With reference to (1.8), we say that an E-valued Markov process X *corresponds to* $\{T(t)\}$ if

$$E[f(X(t+s)) \mid \mathscr{F}_t^X] = T(s)f(X(t)) \tag{1.27}$$

for all $s, t \geq 0$ and $f \in L$. Of course if $\{T(t)\}$ is given by a transition function as in (1.26), then (1.27) is just (1.8).

1.6 Proposition Let E be separable. Let X be an E-valued Markov process with initial distribution ν corresponding to a semigroup $\{T(t)\}$ on a closed subspace $L \subset B(E)$. If L is separating, then $\{T(t)\}$ and ν determine the finite-dimensional distributions of X.

Proof. For $f \in L$ and $t \geq 0$, we have

$$\begin{aligned} \int f(y)P\{X(t) \in dy\} &= E[f(X(t))] \\ &= E[E[f(X(t)) \mid \mathscr{F}_0^X]] \\ &= E[T(t)f(X(0))] = \int T(t)f(x)\nu(dx). \end{aligned} \tag{1.28}$$

Since L is separating, $\nu_t(\Gamma) \equiv P\{X(t) \in \Gamma\}$ is determined. Similarly if $f \in L$ and $g \in B(E)$, then for $0 \leq t_1 < t_2$,

$$\begin{aligned} E[f(X(t_1))g(X(t_2))] &= E[f(X(t_1))T(t_2 - t_1)g(X(t_1))] \\ &= \int f(x)T(t_2 - t_1)g(x)\nu_{t_1}(dx) \end{aligned} \tag{1.29}$$

and the joint distribution of $X(t_1)$ and $X(t_2)$ is determined (cf. Proposition 4.6 of Chapter 3). Proceeding in this manner, the proposition can be proved by induction. □

Since the finite-dimensional distributions of a Markov process are determined by a corresponding semigroup $\{T(t)\}$, they are in turn determined by its full generator $\hat{A}$ or by a sufficiently large set $A \subset \hat{A}$. One of the best approaches for determining when a set is "sufficiently large" is through the martingale problem of Stroock and Varadhan, which is based on the observation in the following proposition.

1.7 Proposition Let X be an E-valued, progressive Markov process with transition function $P(t, x, \Gamma)$ and let $\{T(t)\}$ and $\hat{A}$ be as above. If $(f, g) \in \hat{A}$ then

$$M(t) \equiv f(X(t)) - \int_0^t g(X(s))\, ds \tag{1.30}$$

is an $\{\mathscr{F}_t^X\}$-martingale.

Proof. For each $t, u \geq 0$

$$\begin{aligned} E[M(t+u) \mid \mathscr{F}_t^X] &= \int f(y) P(u, X(t), dy) \\ &\quad - \int_t^{t+u} \int g(y) P(s-t, X(t), dy)\, ds \\ &\quad - \int_0^t g(X(s))\, ds \\ &= T(u) f(X(t)) - \int_0^u T(s) g(X(t))\, ds \\ &\quad - \int_0^t g(X(s))\, ds \\ &= f(X(t)) - \int_0^t g(X(s))\, ds = M(t). \end{aligned} \tag{1.31}$$

□

We study the basic properties of the martingale problem in Sections 3–7.

2. MARKOV JUMP PROCESSES AND FELLER PROCESSES

The simplest Markov process to describe is a Markov jump process with a bounded generator. Let $\mu(x, \Gamma)$ be a transition function on $E \times \mathscr{B}(E)$ and let $\lambda \in B(E)$ be nonnegative. Then

$$Af(x) = \lambda(x) \int (f(y) - f(x)) \mu(x, dy) \tag{2.1}$$

defines a bounded linear operator A on $B(E)$, and A is the generator for a Markov process in E that can be constructed as follows.

Let $\{Y(k), k = 0, 1, \cdots\}$ be a Markov chain in E with initial distribution ν and transition function $\mu(x, \Gamma)$. That is, $P\{Y(0) \in \Gamma\} = \nu(\Gamma)$ and

$$P\{Y(k+1) \in \Gamma \mid Y(0), \cdots, Y(k)\} = \mu(Y(k), \Gamma) \tag{2.2}$$

for all $\Gamma \in \mathscr{B}(E)$ and $k = 0, 1, \ldots$. Let $\Delta_0, \Delta_1, \ldots$ be independent and exponentially distributed with parameter 1 (and independent of $Y(\cdot)$). Then

$$X(t) = \begin{cases} Y(0), & 0 \le t < \dfrac{\Delta_0}{\lambda(Y(0))}, \\ Y(k), & \displaystyle\sum_{j=0}^{k-1} \frac{\Delta_j}{\lambda(Y(j))} \le t < \sum_{j=0}^{k} \frac{\Delta_j}{\lambda(Y(j))}, \end{cases} \tag{2.3}$$

defines a Markov process X in E with initial distribution ν and generator A. (Note that we allow $\lambda(x) = 0$, taking $\Delta/0 = \infty$.)

To see this, we make use of an even simpler representation. Let $\lambda = \sup_{x \in E} \lambda(x)$, and to avoid trivialities, assume that $\lambda > 0$. Define the transition function $\mu'(x, \Gamma)$ on $E \times \mathscr{B}(E)$ by

$$\mu'(x, \Gamma) = \left(1 - \frac{\lambda(x)}{\lambda}\right)\delta_x(\Gamma) + \frac{\lambda(x)}{\lambda}\, \mu(x, \Gamma), \tag{2.4}$$

and note that (2.1) can be rewritten as

$$Af(x) = \lambda \int (f(y) - f(x))\mu'(x, dy). \tag{2.5}$$

Let $\{Y'(k), k = 0, 1, \ldots\}$ be a Markov chain in E with initial distribution ν and transition function $\mu'(x, \Gamma)$, and let V be an independent Poisson process with parameter λ. Define

$$X'(t) = Y'(V(t)), \qquad t \ge 0. \tag{2.6}$$

We leave it to the reader to show that X and X' have the same finite-dimensional distributions (Problem 4).

Observe that

$$Pf(x) = \int f(y)\mu'(x, dy) \tag{2.7}$$

defines a linear contraction P on $B(E)$ and that, by (2.5), $A = \lambda(P - I)$. Consequently, the semigroup $\{T(t)\}$ generated by A is given by

$$T(t) = \sum_{k=0}^{\infty} e^{-\lambda t} \frac{(\lambda t)^k}{k!} P^k. \tag{2.8}$$

Let $f \in B(E)$. By the Markov property of Y' (cf. (1.15)),

$$E[f(Y'(k+l)) \mid Y'(0), \ldots, Y'(l)] = P^k f(Y'(l)) \tag{2.9}$$

for $k, l = 0, 1, \ldots$, and we claim that

$$E[f(Y'(k + V(t))) \mid \mathscr{F}_t] = P^k f(X'(t)) \tag{2.10}$$

for $k = 0, 1, \ldots$ and $t \geq 0$, where

$$\mathscr{F}_t = \mathscr{F}_t^V \vee \mathscr{F}_t^{X'}. \tag{2.11}$$

To see this, let $A \in \mathscr{F}_t^V$ and $B \in \mathscr{F}_l^{Y'}$. Then

$$\begin{aligned} \int_{A \cap B \cap \{V(t) = l\}} & f(Y'(k + V(t)))\, dP \\ &= \int_{A \cap B \cap \{V(t) = l\}} f(Y'(k + l))\, dP \\ &= P(A \cap \{V(t) = l\}) \int_B f(Y'(k + l))\, dP \\ &= P(A \cap \{V(t) = l\}) \int_B P^k f(Y'(l))\, dP \\ &= \int_{A \cap B \cap \{V(t) = l\}} P^k f(X'(t))\, dP. \end{aligned} \tag{2.12}$$

Since $\{A \cap B \cap \{V(t) = l\} : A \in \mathscr{F}_t^V,\ B \in \mathscr{F}_l^{Y'},\ l = 0, 1, 2, \ldots\}$ is closed under finite intersections and generates $\mathscr{F}_t$, by the Dynkin class theorem (Appendix 4) we have

$$\int_A f(Y'(k + V(t)))\, dP = \int_A P^k f(X'(t))\, dP \tag{2.13}$$

for all $A \in \mathscr{F}_t$, and (2.10) follows. Finally, since V has independent increments,

$$\begin{aligned} E[f(X'(t + s)) \mid \mathscr{F}_t] & \\ &= E[f(Y'(V(t + s) - V(t) + V(t))) \mid \mathscr{F}_t] \\ &= \sum_{k=0}^{\infty} e^{-\lambda s} \frac{(\lambda s)^k}{k!} E[f(Y'(k + V(t))) \mid \mathscr{F}_t] \\ &= \sum_{k=0}^{\infty} e^{-\lambda s} \frac{(\lambda s)^k}{k!} P^k f(X'(t)) \\ &= T(s) f(X'(t)) \end{aligned} \tag{2.14}$$

for all $s, t \geq 0$. Hence X' is a Markov process in E with initial distribution ν corresponding to the semigroup $\{T(t)\}$ generated by A.

We now assume E is locally compact and consider Markov processes with semigroups that are strongly continuous on the Banach space $\hat{C}(E)$ of continuous functions vanishing at infinity with norm $\|f\| = \sup_{x \in E} |f(x)|$. Note that $\hat{C}(E) = C(E)$ if E is compact. Let $\Delta \notin E$ be the point at infinity if E is noncompact and an isolated point if E is compact, and put $E^{\Delta} = E \cup \{\Delta\}$; in the noncompact case, E^{Δ} is the one-point compactification of E. We note that if E is also separable, then E^{Δ} is metrizable. (See, for example, pages 201 and 202 of Cohn (1980).)

A semigroup $\{T(t)\}$ on $\hat{C}(E)$ is said to be *positive* if $T(t)$ is a positive operator for each $t \geq 0$. (A *positive operator* is one that maps nonnegative functions to nonnegative functions.)

An operator A on $\hat{C}(E)$ is said to satisfy the *positive maximum principle* if whenever $f \in \mathscr{D}(A)$, $x_0 \in E$, and $\sup_{x \in E} f(x) = f(x_0) \geq 0$, we have $Af(x_0) \leq 0$.

2.1 Lemma Let E be locally compact. A linear operator A on $\hat{C}(E)$ satisfying the positive maximum principle is dissipative.

Proof. Let $f \in \mathscr{D}(A)$ and $\lambda > 0$. There exists $x_0 \in E$ such that $|f(x_0)| = \|f\|$. Suppose $f(x_0) \geq 0$ (otherwise replace f by $-f$). Since $\sup_{x \in E} f(x) = f(x_0) \geq 0$, $Af(x_0) \leq 0$ and hence

(2.15) $$\|\lambda f - Af\| \geq \lambda f(x_0) - Af(x_0) \geq \lambda f(x_0) = \lambda \|f\|. \qquad \square$$

We restate the Hille–Yosida theorem in the present context.

2.2 Theorem Let E be locally compact. The closure $\bar{A}$ of a linear operator A on $\hat{C}(E)$ is single-valued and generates a strongly continuous, positive, contraction semigroup $\{T(t)\}$ on $\hat{C}(E)$ if and only if:

- **(a)** $\mathscr{D}(A)$ is dense in $\hat{C}(E)$.
- **(b)** A satisfies the positive maximum principle.
- **(c)** $\mathscr{R}(\lambda - A)$ is dense in $\hat{C}(E)$ for some $\lambda > 0$.

Proof. The necessity of (a) and (c) follows from Theorem 2.12 of Chapter 1. As for (b), if $f \in \mathscr{D}(A)$, $x_0 \in E$, and $\sup_{x \in E} f(x) = f(x_0) \geq 0$, then

(2.16) $$T(t)f(x_0) \leq T(t)(f^+)(x_0) \leq \|f^+\| = f(x_0)$$

for each $t \geq 0$, so $Af(x_0) \leq 0$.

Conversely, suppose A satisfies (a)–(c). Since (b) implies A is dissipative by Lemma 2.1, $\bar{A}$ is single-valued and generates a strongly continuous contraction semigroup $\{T(t)\}$ by Theorem 2.12 of Chapter 1. To complete the proof, we must show that $\{T(t)\}$ is positive.

Let $f \in \mathscr{D}(\bar{A})$ and $\lambda > 0$, and suppose that $\inf_{x \in E} f(x) < 0$. Choose $\{f_n\} \subset \mathscr{D}(A)$ such that $(\lambda - A)f_n \to (\lambda - \bar{A})f$, and let $x_n \in E$ and $x_0 \in E$ be points at which f_n and f, respectively, take on their minimum values. Then

$$\begin{aligned} \inf_{x \in E}(\lambda - \bar{A})f(x) &\leq \varliminf_{n \to \infty} (\lambda - A)f_n(x_n) \\ &\leq \lim_{n \to \infty} \lambda f_n(x_n) \\ &= \lambda f(x_0) \\ &< 0, \end{aligned} \tag{2.17}$$

where the second inequality is due to the fact that $\inf_{x \in E} f_n(x) = f_n(x_n) \leq 0$ for n sufficiently large. We conclude that if $f \in \mathscr{D}(\bar{A})$ and $\lambda > 0$, then $(\lambda - \bar{A})f \geq 0$ implies $f \geq 0$, so the positivity of $\{T(t)\}$ is a consequence of Corollary 2.8 of Chapter 1. □

An operator $A \subset B(E) \times B(E)$ (possibly multivalued) is said to be conservative if $(1, 0)$ is in the bp-closure of A. For example, if $(1, 0)$ is in the full generator of a measurable contraction semigroup $\{T(t)\}$, then $T(t)1 = 1$ for all $t \geq 0$, and conversely. For semigroups given by transition functions, this property is just the fact that $P(t, x, E) = 1$.

A strongly continuous, positive, contraction semigroup on $\hat{C}(E)$ whose generator is conservative is called a *Feller semigroup.* Our aim in this section is to show (assuming in addition that E is separable) that every Feller semigroup on $\hat{C}(E)$ corresponds to a Markov process with sample paths in $D_E[0, \infty)$. First, however, we require several preliminary results, including our first convergence theorem.

2.3 Lemma Let E be locally compact and separable and let $\{T(t)\}$ be a strongly continuous, positive, contraction semigroup on $\hat{C}(E)$. Define the operator $T^\Delta(t)$ on $C(E^\Delta)$ for each $t \geq 0$ by

$$T^\Delta(t)f = f(\Delta) + T(t)(f - f(\Delta)). \tag{2.18}$$

(We do not distinguish notationally between functions on E^Δ and their restrictions to E.) Then $\{T^\Delta(t)\}$ is a Feller semigroup on $C(E^\Delta)$.

Proof. It is easy to verify that $\{T^\Delta(t)\}$ is a strongly continuous semigroup on $C(E^\Delta)$. Fix $t \geq 0$. To show that $T^\Delta(t)$ is a positive operator, we must show that if $\alpha \in \mathbb{R}$, $f \in \hat{C}(E)$, and $\alpha + f \geq 0$, then $\alpha + T(t)f \geq 0$. By the positivity of $T(t)$, $T(t)(f^+) \geq 0$ and $T(t)(f^-) \geq 0$. Hence $-T(t)f \leq T(t)(f^-)$, and so $(T(t)f)^- \leq T(t)(f^-)$. Since $T(t)$ is a contraction, $\| T(t)(f^-) \| \leq \| f^- \| \leq \alpha$. Therefore $(T(t)f)^- \leq \alpha$, so $\alpha + T(t)f \geq 0$.

Next, the positivity of $T^\Delta(t)$ gives $|T^\Delta(t)f| \leq T^\Delta(t) \| f \| = \| f \|$ for all $f \in C(E^\Delta)$, so $\| T^\Delta(t) \| = 1$. Finally, the generator A^Δ of $\{T^\Delta(t)\}$ clearly contains $(1, 0)$. □

2.4 Proposition Let E be locally compact and separable. Let $\{T(t)\}$ be a strongly continuous, positive, contraction semigroup on $\hat{C}(E)$, and define the semigroup $\{T^{\Delta}(t)\}$ on $C(E^{\Delta})$ as in Lemma 2.3. Let X be a Markov process corresponding to $\{T^{\Delta}(t)\}$ with sample paths in $D_{E^{\Delta}}[0, \infty)$, and let $\tau = \inf\{t \geq 0 : X(t) = \Delta \text{ or } X(t-) = \Delta\}$. Then

$$P\{\tau < \infty, X(\tau + s) = \Delta \quad \text{for all} \quad s \geq 0\} = P\{\tau < \infty\}. \tag{2.19}$$

Let A be the generator of $\{T(t)\}$ and suppose further that A is conservative. If $P\{X(0) \in E\} = 1$, then $P\{X \in D_E[0, \infty)\} = 1$.

Proof. Recalling that E^{Δ} is metrizable, there exists $g \in C(E^{\Delta})$ with $g > 0$ on E and $g(\Delta) = 0$. Put $f = \int_0^\infty e^{-u}T^{\Delta}(u)g\, du$, and note that $f > 0$ on E and $f(\Delta) = 0$. By the Markov property of X,

$$\begin{aligned} E[e^{-t}f(X(t)) \mid \mathscr{F}^X_{s+}] &= e^{-t}T^{\Delta}(t-s)f(X(s)) \\ &= e^{-s}\int_{t-s}^{\infty} e^{-u}T^{\Delta}(u)g(X(s))\, du \\ &\leq e^{-s}f(X(s)), \qquad 0 \leq s < t, \end{aligned} \tag{2.20}$$

so $e^{-t}f(X(t))$ is a nonnegative $\{\mathscr{F}^X_{t+}\}$-supermartingale. Therefore, (2.19) is a consequence of Proposition 2.15 of Chapter 2. It also follows that $P\{X(t) = \Delta\} = P\{\tau \leq t\}$ for all $t \geq 0$.

Let A^{Δ} denote the generator of $\{T^{\Delta}(t)\}$. The assumption that A is conservative (which refers to the bp-closure of A in $B(E) \times B(E)$) implies that $(\chi_E, 0)$ is in the bp-closure of A^{Δ} (considering A^{Δ} as a subspace of $B(E^{\Delta}) \times B(E^{\Delta})$). Since the collection of $(f, g) \in B(E^{\Delta}) \times B(E^{\Delta})$ satisfying

$$E[f(X(t))] = E[f(X(0))] + E\left[\int_0^t g(X(s))\, ds\right] \tag{2.21}$$

is bp-closed and contains A^{Δ}, for all $t \geq 0$ we have

$$P\{\tau > t\} = P\{X(t) \in E\} = P\{X(0) \in E\}, \tag{2.22}$$

and if $P\{X(0) \in E\} = 1$, we conclude that $P\{X \in D_E[0, \infty)\} = P\{\tau = \infty\} = 1$. □

A converse to the second assertion of Proposition 2.4 is provided by Corollary 2.8.

2.5 Theorem Let E be locally compact and separable. For $n = 1, 2, \ldots$ let $\{T_n(t)\}$ be a Feller semigroup on $\hat{C}(E)$, and suppose X_n is a Markov process corresponding to $\{T_n(t)\}$ with sample paths in $D_E[0, \infty)$. Suppose that $\{T(t)\}$ is a Feller semigroup on $\hat{C}(E)$ and that for each $f \in \hat{C}(E)$,

$$\lim_{n \to \infty} T_n(t)f = T(t)f, \qquad t \geq 0. \tag{2.23}$$

If $\{X_n(0)\}$ has limiting distribution $\nu \in \mathscr{P}(E)$, then there is a Markov process X corresponding to $\{T(t)\}$ with initial distribution ν and sample paths in $D_E[0, \infty)$, and $X_n \Rightarrow X$.

Proof. For each $n \geq 1$, let A_n be the generator of $\{T_n(t)\}$. By Theorem 6.1 of Chapter 1, (2.23) implies that for each $f \in \mathscr{D}(A)$, there exist $f_n \in \mathscr{D}(A_n)$ such that $f_n \to f$ and $A_n f_n \to Af$. Since $f_n(X_n(t)) - \int_0^t A_n f_n(X_n(s))\, ds$ is an $\{\mathscr{F}_t^{X_n}\}$-martingale for each $n \geq 1$, and since $\mathscr{D}(A)$ is dense in $\hat{C}(E)$, Chapter 3's Corollary 9.3 and Theorem 9.4 imply that $\{X_n\}$ is relatively compact in $D_{E^\Delta}[0, \infty)$.

We next prove the convergence of the finite-dimensional distributions of $\{X_n\}$. For each $n \geq 1$, let $\{T_n^\Delta(t)\}$ and $\{T^\Delta(t)\}$ be the semigroups on $C(E^\Delta)$ defined in terms of $\{T_n(t)\}$ and $\{T(t)\}$ as in Lemma 2.3. Then, for each $f \in C(E^\Delta)$ and $t \geq 0$,

$$\lim_{n\to\infty} E[f(X_n(t))] = \lim_{n\to\infty} E[T_n^\Delta(t) f(X_n(0))] = \int T^\Delta(t) f(x)\nu(dx) \tag{2.24}$$

by the Markov property, the strong convergence of $\{T_n^\Delta(t)\}$, the continuity of $T^\Delta(t)f$, and the convergence in distribution of $\{X_n(0)\}$. Proceeding by induction, let m be a positive integer, and suppose that

$$\lim_{n\to\infty} E[f_1(X_n(t_1)) \cdots f_m(X_n(t_m))] \tag{2.25}$$

exists for all $f_1, \ldots, f_m \in C(E^\Delta)$ and $0 \leq t_1 < \cdots < t_m$. Then

$$\begin{aligned} &\lim_{n\to\infty} E[f_1(X_n(t_1)) \cdots f_m(X_n(t_m)) f_{m+1}(X_n(t_{m+1}))] \\ &= \lim_{n\to\infty} E[f_1(X_n(t_1)) \cdots f_m(X_n(t_m)) T_n^\Delta(t_{m+1} - t_m) f_{m+1}(X_n(t_m))] \\ &= \lim_{n\to\infty} E[f_1(X_n(t_1)) \cdots f_m(X_n(t_m)) T^\Delta(t_{m+1} - t_m) f_{m+1}(X_n(t_m))] \end{aligned} \tag{2.26}$$

exists for all $f_1, \ldots, f_{m+1} \in C(E^\Delta)$ and $0 \leq t_1 < \cdots < t_{m+1}$.

It follows that every convergent subsequence of $\{X_n\}$ has the same limit, so there exists a process X with initial distribution ν and with sample paths in $D_{E^\Delta}[0, \infty)$ such that $X_n \Rightarrow X$. By (2.26), X is a Markov process corresponding to $\{T^\Delta(t)\}$, so by Proposition 2.4, X can be assumed to have sample paths in $D_E[0, \infty)$. Finally, Corollary 9.3 of Chapter 3 implies $X_n \Rightarrow X$ in $D_E[0, \infty)$. □

2.6 Theorem Let E be locally compact and separable. For $n = 1, 2, \ldots$ let $\mu_n(x, \Gamma)$ be a transition function on $E \times \mathscr{B}(E)$ such that T_n, defined by

$$T_n f(x) = \int f(y)\mu_n(x, dy), \tag{2.27}$$

satisfies $T_n : \hat{C}(E) \to \hat{C}(E)$. Suppose that $\{T(t)\}$ is a Feller semigroup on $\hat{C}(E)$. Let $\varepsilon_n > 0$ satisfy $\lim_{n\to\infty} \varepsilon_n = 0$ and suppose that for every $f \in \hat{C}(E)$,

$$\lim_{n\to\infty} T_n^{[t/\varepsilon_n]} f = T(t)f, \qquad t \geq 0. \tag{2.28}$$

For each $n \geq 1$, let $\{Y_n(k),\ k = 0, 1, 2, \ldots\}$ be a Markov chain in E with transition function $\mu_n(x, \Gamma)$, and suppose $\{Y_n(0)\}$ has limiting distribution $\nu \in \mathscr{P}(E)$. Define X_n by $X_n(t) \equiv Y_n([t/\varepsilon_n])$. Then there is a Markov process X corresponding to $\{T(t)\}$ with initial distribution ν and sample paths in $D_E[0, \infty)$, and $X_n \Rightarrow X$.

Proof. Following the proof of Theorem 2.5, use Theorem 6.5 of Chapter 1 in place of Theorem 6.1. □

2.7 Theorem Let E be locally compact and separable, and let $\{T(t)\}$ be a Feller semigroup on $\hat{C}(E)$. Then for each $\nu \in \mathscr{P}(E)$, there exists a Markov process X corresponding to $\{T(t)\}$ with initial distribution ν and sample paths in $D_E[0, \infty)$. Moreover, X is strong Markov with respect to the filtration $\mathscr{G}_t = \mathscr{F}_{t+}^X = \bigcap_{\varepsilon>0} \mathscr{F}_{t+\varepsilon}^X$.

Proof. Let n be a positive integer, and let

$$A_n = A(I - n^{-1}A)^{-1} = n[(I - n^{-1}A)^{-1} - I] \tag{2.29}$$

be the Yosida approximation of A. Note that since $(I - n^{-1}A)^{-1}$ is a positive contraction on $\hat{C}(E)$, there exists for each $x \in E$ a positive Borel measure $\mu_n(x, \Gamma)$ on E such that

$$(I - n^{-1}A)^{-1} f(x) = \int f(y)\mu_n(x, dy) \tag{2.30}$$

for all $f \in \hat{C}(E)$. It follows that $\mu_n(\cdot, \Gamma)$ is Borel measurable for each $\Gamma \in \mathscr{B}(E)$. For each $(f, g) \in A$, (2.30) implies

$$f(x) = \int (f(y) - n^{-1}g(y))\mu_n(x, dy), \qquad x \in E. \tag{2.31}$$

Since the collection of $(f, g) \in B(E) \times B(E)$ satisfying (2.31) is bp-closed, it includes $(1, 0)$ and hence $\mu_n(x, E) = 1$ for each $x \in E$, implying that $\mu_n(x, \Gamma)$ is a transition function on $E \times \mathscr{B}(E)$. Therefore, by the discussion at the beginning of this section, the semigroup $\{T_n(t)\}$ on $\hat{C}(E)$ with generator A_n corresponds to a jump Markov process X_n with initial distribution ν and with sample paths in $D_E[0, \infty)$.

Now letting $n \to \infty$, Proposition 2.7 of Chapter 1 implies that for each $f \in \hat{C}(E)$ and $t \geq 0$, $\lim_{n\to\infty} T_n(t)f = T(t)f$, so the existence of X follows from Theorem 2.5.

Let τ be a discrete $\{\mathscr{G}_t\}$-stopping time with $\tau < \infty$ a.s. concentrated on $\{t_1, t_2, \ldots\}$. Let $A \in \mathscr{G}_\tau$, $s > 0$, and $f \in \hat{C}(E)$. Then $A \cap \{\tau = t_i\} \in \mathscr{F}^X_{t_i+\varepsilon}$ for every $\varepsilon > 0$, so

$$(2.32) \qquad \int_{A\cap\{\tau=t_i\}} f(X(\tau + s))\, dP = \int_{A\cap\{\tau=t_i\}} f(X(t_i + s))\, dP$$
$$= \int_{A\cap\{\tau=t_i\}} T(s - \varepsilon) f(X(t_i + \varepsilon))\, dP$$

for $0 < \varepsilon \le s$ and $i = 1, 2, \ldots$. Since $\{T(t)\}$ is strongly continuous, $T(s)f$ is continuous on E, and X has right continuous sample paths, we can take $\varepsilon = 0$ in (2.32). This gives

$$(2.33) \qquad E[f(X(\tau + s)) \mid \mathscr{G}_\tau] = T(s)f(X(\tau))$$

for discrete τ.

If τ is an arbitrary $\{\mathscr{G}_t\}$-stopping time, with $\tau < \infty$ a.s., it is the limit of a decreasing sequence $\{\tau_n\}$ of discrete stopping times (Proposition 1.3 of Chapter 2), so (2.33) follows from the continuity of $T(s)f$ on E and the right continuity of the sample paths of X. (Replace τ by τ_n in (2.33), condition on $\mathscr{G}_\tau$, and then let $n \to \infty$.) □

2.8 Corollary Let E be locally compact and separable. Let A be a linear operator on $\hat{C}(E)$ satisfying (a)–(c) of Theorem 2.2, and let $\{T(t)\}$ be the strongly continuous, positive, contraction semigroup on $\hat{C}(E)$ generated by $\bar{A}$. Then there exists for each $x \in E$ a Markov process X_x corresponding to $\{T(t)\}$ with initial distribution δ_x and with sample paths in $D_E[0, \infty)$ if and only if A is conservative.

Proof. The sufficiency follows from Theorem 2.7. As for necessity, let $\{g_n\} \subset \mathscr{R}(I - A)$ satisfy bp-$\lim_{n\to\infty} g_n = 1$, and define $\{f_n\} \subset \mathscr{D}(A)$ by $f_n = (I - A)^{-1} g_n$. Then

$$(2.34) \qquad \lim_{n\to\infty} f_n(x) = \lim_{n\to\infty} E\left[\int_0^\infty e^{-t} g_n(X_x(t))\, dt\right] = 1$$

for all $x \in E$, so bp-$\lim_{n\to\infty} f_n = 1$ and bp-$\lim_{n\to\infty} Af_n =$ bp-$\lim_{n\to\infty} (f_n - g_n) = 0$. □

We next give criteria for the continuity of the sample paths of the process obtained in Theorem 2.7. Since we know the process has sample paths in $D_E[0, \infty)$, to show the sample paths are continuous it is enough to show that they have no jumps.

2.9 Proposition Let (E, r) be locally compact and separable, and let $\{T(t)\}$ be a Feller semigroup on $\hat{C}(E)$. Let $P(t, x, \Gamma)$ be the transition function for $\{T(t)\}$ and suppose for each $x \in E$ and $\varepsilon > 0$,

$$\lim_{t \to 0} t^{-1} P(t, x, B(x, \varepsilon)^c) = 0. \tag{2.35}$$

Then the process X given by Theorem 2.7 satisfies $P\{X \in C_E[0, \infty)\} = 1$.

2.10 Remark Suppose A is the generator of a Feller semigroup $\{T(t)\}$ on $\hat{C}(E)$ with transition function $P(t, x, \Gamma)$, and that for each $x \in E$ and $\varepsilon > 0$ there exists $f \in \mathscr{D}(A)$ with $f(x) = \|f\|$, $\sup_{y \in B(x, \varepsilon)^c} f(y) \equiv M < \|f\|$, and $Af(x) = 0$. Then (2.35) holds. To see this, note that

$$\begin{aligned}(\|f\| - M)P(t, x, B(x, \varepsilon)^c) &\le f(x) - E_x[f(X(t))] \\ &= -\int_0^t T(s)Af(x)\, ds.\end{aligned} \tag{2.36}$$

Divide by t and let $t \to 0$ to obtain (2.35). □

Proof. Note that for each $x \in E$ and $t \ge 0$,

$$\overline{\lim_{y \to x}}\, P(t, y, B(y, \varepsilon)^c) \le \overline{\lim_{y \to x}}\, P\left(t, y, B\left(x, \frac{\varepsilon}{2}\right)^c\right) \le P\left(t, x, B\left(x, \frac{\varepsilon}{2}\right)^c\right). \tag{2.37}$$

For each $\delta > 0$ there is a $t(x, \delta) \le \delta$ such that for $t = t(x, \delta)$ the right side of (2.37) is less than $\delta t(x, \delta)$. Consequently, there is a neighborhood U_x of x such that $y \in U_x$ implies

$$P(t(x, \delta), y, B(y, \varepsilon)^c) \le 2\delta t(x, \delta). \tag{2.38}$$

Since any compact subset of E can be covered by finitely many such U_x, we can define a Borel measurable function $s(y, \delta) \le \delta$ such that

$$P(s(y, \delta), y, B(y, \varepsilon)^c) \le 2\delta s(y, \delta), \tag{2.39}$$

and for each compact $K \subset E$

$$\inf_{y \in K} s(y, \delta) > 0. \tag{2.40}$$

Define $\tau_0 = 0$ and

$$\tau_{k+1} = \tau_k + s(X(\tau_k), \delta). \tag{2.41}$$

Note that $\lim_{k \to \infty} \tau_k = \infty$ since $\{X(s) \colon s \le t\}$ has compact closure for each $t \ge 0$.

Let

$$N_\delta(n) = \sum_{k=0}^{n-1} \chi_{\{r(X(\tau_{k+1}), X(\tau_k)) \ge \varepsilon\}}, \tag{2.42}$$

and observe that

$$M_\delta(n) = N_\delta(n) - \sum_{k=0}^{n-1} P(s(X(\tau_k), \delta), X(\tau_k), B(X(\tau_k), \varepsilon)^c) \tag{2.43}$$

is a martingale. Let $K \subset E$ be compact, let $T > 0$, and define

$$\gamma = \gamma_\delta = \min \{n: N_\delta(n) = 1, \tau_n > T, \quad \text{or} \quad X(\tau_n) \notin K\}. \tag{2.44}$$

Then by the optional sampling theorem

$$\begin{aligned} E[N_\delta(\gamma_\delta)] &= E\left[\sum_{k=0}^{\gamma-1} P(s(X(\tau_k), \delta), X(\tau_k), B(X(\tau_k), \varepsilon)^c)\right] \\ &\le E\left[\sum_{k=0}^{\gamma-1} 2\delta s(X(\tau_k), \delta)\right] \\ &\le E[2\delta\tau_\gamma] \le 2\delta(T + \delta). \end{aligned} \tag{2.45}$$

Finally, observe that $\lim_{\delta\to 0} N_\delta(\gamma_\delta) = 1$ on the set where X has a jump of size larger than ε before T and before leaving K. Consequently, with probability one, no such jump occurs. Since ε, T, and K are arbitrary, we have the desired result. □

We close this section with two theorems generalizing Theorems 2.5 and 2.6. Much more general results are given in Section 8, but these results can be obtained here using essentially the same argument as in the proof of Theorem 2.5.

2.11 Theorem Let $E, E_1, E_2, \ldots$ be metric spaces with E locally compact and separable. For $n = 1, 2, \ldots$, let $\eta_n : E_n \to E$ be measurable, let $\{T_n(t)\}$ be a semigroup on $B(E_n)$ given by a transition function, and suppose Y_n is a Markov process in E_n corresponding to $\{T_n(t)\}$ such that $X_n = \eta_n \circ Y_n$ has sample paths in $D_E[0, \infty)$. Define $\pi_n : B(E) \to B(E_n)$ by $\pi_n f = f \circ \eta_n$ (cf. Section 6 of Chapter 1). Suppose that $\{T(t)\}$ is a Feller semigroup on $\hat{C}(E)$ and that for each $f \in \hat{C}(E)$ and $t \ge 0$, $T_n(t)\pi_n f \to T(t)f$ (i.e., $\| T_n(t)\pi_n f - \pi_n T(t)f \| \to 0$). If $\{X_n(0)\}$ has limiting distribution $\nu \in \mathscr{P}(E)$, then there is a Markov process X corresponding to $\{T(t)\}$ with initial distribution ν and sample paths in $D_E[0, \infty)$, and $X_n \Rightarrow X$.

Proof. Note that

$$\begin{aligned} E[f(X_n(t + s)) \mid \mathscr{F}_t^{Y_n}] &= T_n(s)\pi_n f(Y_n(t)) \\ &\approx \pi_n T(s)f(Y_n(t)) \\ &= T(s)f(X_n(t)). \end{aligned} \tag{2.46}$$

With this observation the proof is essentially the same as for Theorem 2.5. □

Finally, we give a similar extension of Theorem 2.6.

2.12 Theorem Let $E, E_1, E_2, \ldots$ be metric spaces with E locally compact and separable. For $n = 1, 2, \ldots$, let $\eta_n : E_n \to E$ be measurable, let $\mu_n(x, \Gamma)$ be a transition function on $E_n \times \mathscr{B}(E_n)$, and suppose $\{Y_n(k),\ k = 0, 1, 2, \ldots\}$ is a Markov chain in E_n corresponding to $\mu_n(x, \Gamma)$. Let $\varepsilon_n > 0$ satisfy $\lim_{n\to\infty} \varepsilon_n = 0$. Define $X_n(t) = \eta_n(Y_n([t/\varepsilon_n]))$,

$$T_n f(x) = \int f(y)\mu_n(x, dy), \qquad f \in B(E_n), \tag{2.47}$$

and $\pi_n \colon B(E) \to B(E_n)$ by $\pi_n f = f \circ \eta_n$. Suppose that $\{T(t)\}$ is a Feller semigroup on $\hat{C}(E)$ and that for each $f \in \hat{C}(E)$ and $t \geq 0$, $T_n^{[t/\varepsilon_n]}\pi_n f \to T(t)f$. If $\{X_n(0)\}$ has limiting distribution $\nu \in \mathscr{P}(E)$, then there is a Markov process X corresponding to $\{T(t)\}$ with initial distribution ν and sample paths in $D_E[0, \infty)$, and $X_n \Rightarrow X$.

3. THE MARTINGALE PROBLEM: GENERALITIES AND SAMPLE PATH PROPERTIES

In Proposition 1.7 we observed that, if X is a Markov process with full generator $\hat{A}$, then

$$f(X(t)) - \int_0^t g(X(s))\, ds \tag{3.1}$$

is a martingale for all $(f, g) \in \hat{A}$. In the next several sections we develop the idea of Stroock and Varadhan of using this martingale property as a means of characterizing the Markov process associated with a given generator A. As elsewhere in this chapter, E (or more specifically (E, r)) denotes a metric space. Occasionally we want to allow A to be a multivalued operator (cf. Chapter 1, Section 4), and hence think of A as a subset (not necessarily linear) of $B(E) \times B(E)$. By a *solution of the martingale problem for A* we mean a measurable stochastic process X with values in E defined on some probability space $(\Omega, \mathscr{F}, P)$ such that for each $(f, g) \in A$, (3.1) is a martingale with respect to the filtration

$$^*\mathscr{F}_t^X \equiv \mathscr{F}_t^X \vee \sigma\left(\int_0^s h(X(u))\, du \colon s \leq t,\ h \in B(E)\right). \tag{3.2}$$

Note that if X is progressive, in particular if X is right continuous, then $^*\mathscr{F}_t^X = \mathscr{F}_t^X$. In general, every event in $^*\mathscr{F}_t^X$ differs from an event in $\mathscr{F}_t^X$ by an event of probability zero. See Problem 2 of Chapter 2.

If $\{\mathscr{G}_t\}$ is a filtration with $\mathscr{G}_t \supset {}^*\mathscr{F}_t^X$ for all $t \geq 0$, and (3.1) is a $\{\mathscr{G}_t\}$-martingale for all $(f, g) \in A$, we say X is a *solution of the martingale* *problem for A with respect to $\{\mathscr{G}_t\}$.* When an initial distribution $\mu \in \mathscr{P}(E)$ is specified, we say that a solution X of the martingale problem for A is a *solution of the martingale problem for* (A, μ) if $PX(0)^{-1} = \mu$.

Usually X has sample paths in $D_E[0, \infty)$. It is convenient to call a probability measure $P \in \mathscr{P}(D_E[0, \infty))$ a *solution of the martingale problem for A* (or for (A, μ)) if the coordinate process defined on $(D_E[0, \infty), \mathscr{S}_E, P)$ by

$$X(t, \omega) \equiv \omega(t), \qquad \omega \in D_E[0, \infty), \qquad t \geq 0, \tag{3.3}$$

is a solution of the martingale problem for A (or for (A, μ)) as defined above.

Note that a measurable process X is a solution of the martingale problem for A if and only if

$$\begin{aligned} 0 &= E\left[\left(f(X(t_{n+1})) - f(X(t_n)) - \int_{t_n}^{t_{n+1}} g(X(s))\, ds\right) \prod_{k=1}^{n} h_k(X(t_k))\right] \\ &= E\left[f(X(t_{n+1})) \prod_{k=1}^{n} h_k(X(t_k))\right] - E\left[f(X(t_n)) \prod_{k=1}^{n} h_k(X(t_k))\right] \\ &\quad - \int_{t_n}^{t_{n+1}} E\left[g(X(s)) \prod_{k=1}^{n} h_k(X(t_k))\right] ds \end{aligned} \tag{3.4}$$

whenever $0 \leq t_1 < t_2 < \cdots < t_{n+1}$, $(f, g) \in A$, and $h_1, \ldots, h_n \in B(E)$ (or equivalently $\bar{C}(E)$). Consequently the statement that a (measurable) process is a solution of a martingale problem is a statement about its finite-dimensional distributions. In particular, any measurable modification of a solution of the martingale problem for A is also a solution.

Let A_S denote the linear span of A. Then any solution of the martingale problem for A is a solution for A_S. Note also that, if $A^{(1)} \subset A^{(2)}$, then any solution of the martingale problem for $A^{(2)}$ is also a solution for $A^{(1)}$, but not necessarily conversely. Finally, observe that the set of pairs (f, g) for which (3.1) is a $\{\mathscr{G}_t\}$-martingale is bp-closed. Consequently, any solution of the martingale problem for A is a solution for the bp-closure of A_S. (See Appendix 3.)

3.1 Proposition Let $A^{(1)}$ and $A^{(2)}$ be subsets of $B(E) \times B(E)$. If the bp-closures of $(A^{(1)})_S$ and $(A^{(2)})_S$ are equal, then X is a solution of the martingale problem for $A^{(1)}$ if and only if it is a solution for $A^{(2)}$.

Proof. This is immediate from the discussion above. □

The following lemma gives two useful equivalences to (3.1) being a martingale.

3.2 Lemma Let X be a measurable process, $\mathscr{G}_t \supset {}^*\mathscr{F}_t^X$, and let $f, g \in B(E)$. Then for fixed $\lambda \in \mathbb{R}$, (3.1) is a $\{\mathscr{G}_t\}$-martingale if and only if

$$e^{-\lambda t} f(X(t)) + \int_0^t e^{-\lambda s}(\lambda f(X(s)) - g(X(s)))\, ds \tag{3.5}$$

is a $\{\mathscr{G}_t\}$-martingale. If $\inf_x f(x) > 0$, then (3.1) is a $\{\mathscr{G}_t\}$-martingale if and only if

$$f(X(t)) \exp\left\{-\int_0^t \frac{g(X(s))}{f(X(s))}\, ds\right\} \tag{3.6}$$

is a $\{\mathscr{G}_t\}$-martingale.

Proof. If (3.1) is a $\{\mathscr{G}_t\}$-martingale, then by Proposition 3.2 of Chapter 2 (see Problem 22 of the same chapter),

$$\begin{aligned}
&\left[f(X(t)) - \int_0^t g(X(s))\, ds\right] e^{-\lambda t} \\
&\quad + \int_0^t \left[f(X(s)) - \int_0^s g(X(u))\, du\right] \lambda e^{-\lambda s}\, ds \\
&= e^{-\lambda t} f(X(t)) + \int_0^t e^{-\lambda s} \lambda f(X(s))\, ds \\
&\quad - \int_0^t g(X(s))\, ds\, e^{-\lambda t} - \int_0^t \int_0^s g(X(u))\, du\, \lambda e^{-\lambda s}\, ds \\
&= e^{-\lambda t} f(X(t)) + \int_0^t e^{-\lambda s}[\lambda f(X(s)) - g(X(s))]\, ds
\end{aligned} \tag{3.7}$$

is a $\{\mathscr{G}_t\}$-martingale. (The last equality follows by Fubini's theorem.) If $\inf_x f(x) > 0$ and (3.1) is a $\{\mathscr{G}_t\}$-martingale, then

$$\begin{aligned}
&\left[f(X(t)) - \int_0^t g(X(s))\, ds\right] \exp\left\{-\int_0^t \frac{g(X(s))}{f(X(s))}\, ds\right\} \\
&\quad + \int_0^t \left[f(X(s)) - \int_0^s g(X(u))\, du\right] \frac{g(X(s))}{f(X(s))} \exp\left\{-\int_0^s \frac{g(X(u))}{f(X(u))}\, du\right\} ds \\
&= f(X(t)) \exp\left\{-\int_0^t \frac{g(X(s))}{f(X(s))}\, ds\right\} \\
&\quad - \int_0^t g(X(s))\, ds \exp\left\{-\int_0^t \frac{g(X(s))}{f(X(s))}\, ds\right\} \\
&\quad + \int_0^t g(X(s)) \exp\left\{-\int_0^s \frac{g(X(u))}{f(X(u))}\, du\right\} ds \\
&\quad - \int_0^t \int_0^s g(X(u))\, du \frac{g(X(s))}{f(X(s))} \exp\left\{-\int_0^s \frac{g(X(u))}{f(X(u))}\, du\right\} ds \\
&= f(X(t)) \exp\left\{-\int_0^t \frac{g(X(s))}{f(X(s))}\, ds\right\}
\end{aligned} \tag{3.8}$$

is a $\{\mathscr{G}_t\}$-martingale. The converses follow by similar calculations (Problem 14). □

The above lemma gives the following equivalent formulations of the martingale problem.

3.3 Proposition Let A be a linear subset of $B(E) \times B(E)$ containing $(1, 0)$ and define

$$A^+ = \{(f, g) \in A: \inf_x f(x) > 0\}. \tag{3.9}$$

Let X be a measurable E-valued process and let $\mathscr{G}_t \supset {}^*\mathscr{F}_t^X$. Then the following are equivalent:

(a) X is a solution of the martingale problem for A with respect to $\{\mathscr{G}_t\}$.

(b) X is a solution of the martingale problem for A^+ with respect to $\{\mathscr{G}_t\}$.

(c) For each $(f, g) \in A$, (3.5) is a $\{\mathscr{G}_t\}$-martingale.

(d) For each $(f, g) \in A^+$, (3.6) is a $\{\mathscr{G}_t\}$-martingale.

Proof. Since $(A^+)_S = A$, (a) and (b) are equivalent. The other equivalences follow by Lemma 3.2. □

For right continuous X, the fact that (3.5) is a martingale whenever (3.1) is, is a special case of the following lemma.

3.4 Lemma Let X be a measurable stochastic process on $(\Omega, \mathscr{F}, P)$ with values in E. Let u, v: $[0, \infty) \times E \times \Omega \to \mathbb{R}$ be bounded and $\mathscr{B}[0, \infty) \times \mathscr{B}(E) \times \mathscr{F}$-measurable, and let w: $[0, \infty) \times [0, \infty) \times E \times \Omega \to \mathbb{R}$ be bounded and $\mathscr{B}[0, \infty) \times \mathscr{B}[0, \infty) \times \mathscr{B}(E) \times \mathscr{F}$-measurable. Assume that $v(t, x, \omega)$ is continuous in x for fixed t and ω, that $u(t, X(t))$ is adapted to a filtration $\{\mathscr{G}_t\}$, and that $v(t, X(t))$ and $w(t, t, X(t))$ are $\{\mathscr{G}_t\}$-progressive. Suppose further that the conditions in either (a) or (b) hold:

(a) For every $t_2 > t_1 \geq 0$,

$$E[u(t_2, X(t_2)) - u(t_1, X(t_2)) \mid \mathscr{G}_{t_1}] = E\left[\int_{t_1}^{t_2} v(s, X(t_2))\, ds \,\middle|\, \mathscr{G}_{t_1}\right] \tag{3.10}$$

and

$$E[u(t_1, X(t_2)) - u(t_1, X(t_1)) \mid \mathscr{G}_{t_1}] = E\left[\int_{t_1}^{t_2} w(t_1, s, X(s))\, ds \,\middle|\, \mathscr{G}_{t_1}\right]. \tag{3.11}$$

Moreover, X is right continuous and

$$\lim_{\delta \to 0+} E[|w(t-\delta, t, X(t)) - w(t, t, X(t))|] = 0, \qquad t > 0. \tag{3.12}$$

(b) For every $t_2 > t_1 \geq 0$,

$$E[u(t_2, X(t_2)) - u(t_2, X(t_1)) | \mathscr{G}_{t_1}] = E\left[\int_{t_1}^{t_2} w(t_2, s, X(s))\, ds \,\middle|\, \mathscr{G}_{t_1}\right] \tag{3.13}$$

and

$$E[u(t_2, X(t_1)) - u(t_1, X(t_1)) | \mathscr{G}_{t_1}] = E\left[\int_{t_1}^{t_2} v(s, X(t_1))\, ds \,\middle|\, \mathscr{G}_{t_1}\right]. \tag{3.14}$$

Moreover, X is left continuous and

$$\lim_{\delta \to 0+} E[|w(t+\delta, t, X(t)) - w(t, t, X(t))|] = 0, \qquad t \geq 0. \tag{3.15}$$

Under the above assumptions,

$$u(t, X(t)) - \int_0^t \{v(s, X(s)) + w(s, s, X(s))\}\, ds \tag{3.16}$$

is a $\{\mathscr{G}_t\}$-martingale.

Proof. Fix $t_2 > t_1 \geq 0$. For any partition $t_1 = s_0 < s_1 < s_2 < \cdots < s_n = t_2$, we have

$$\begin{aligned} &E[u(t_2, X(t_2)) - u(t_1, X(t_1)) | \mathscr{G}_{t_1}] \\ &\qquad = E\left[\int_{t_1}^{t_2} \{v(s, X(s'')) + w(s', s, X(s))\}\, ds \,\middle|\, \mathscr{G}_{t_1}\right] \end{aligned} \tag{3.17}$$

under the assumptions in (a), and

$$\begin{aligned} &E[u(t_2, X(t_2)) - u(t_1, X(t_1)) | \mathscr{G}_{t_1}] \\ &\qquad = E\left[\int_{t_1}^{t_2} \{v(s, X(s')) + w(s'', s, X(s))\}\, ds \,\middle|\, \mathscr{G}_{t_1}\right] \end{aligned} \tag{3.18}$$

under the assumptions in (b), where $s' = s_k$ and $s'' = s_{k+1}$ for $s_k \leq s < s_{k+1}$. Letting $\max |s_{k+1} - s_k| \to 0$, we obtain

$$\begin{aligned} &E[u(t_2, X(t_2)) - u(t_1, X(t_1)) | \mathscr{G}_{t_1}] \\ &\qquad = E\left[\int_{t_1}^{t_2} \{v(s, X(s)) + w(s, s, X(s))\}\, ds \,\middle|\, \mathscr{G}_{t_1}\right]. \end{aligned} \tag{3.19}$$

□

Clearly, only dissipative operators arise as generators of Markov processes. One consequence of Lemma 3.2 is that we must still restrict our attention to dissipative operators in order to have solutions of the martingale problem.

3.5 Proposition Let A be a linear subset of $B(E) \times B(E)$. If there exists a solution X_x of the martingale problem for (A, δ_x) for each $x \in E$, then A is dissipative.

Proof. Given $(f, g) \in A$ and $\lambda > 0$, (3.5) is a martingale and hence

$$f(x) = E\left[\int_0^\infty e^{-\lambda s}(\lambda f(X_x(s)) - g(X_x(s)))\, ds\right], \qquad x \in E. \tag{3.20}$$

Therefore

$$|f(x)| \le \int_0^\infty e^{-\lambda s} \|\lambda f - g\|\, ds \le \lambda^{-1} \|\lambda f - g\| \tag{3.21}$$

and $\lambda \|f\| \le \|\lambda f - g\|$. □

As stated above, we usually are interested in solutions with sample paths in $D_E[0, \infty)$. The following theorem demonstrates that in most cases this is not a restriction.

3.6 Theorem Let E be separable. Let $A \subset \bar{C}(E) \times B(E)$ and suppose that $\mathscr{D}(A)$ is separating and contains a countable subset that separates points. Let X be a solution of the martingale problem for A and assume that for every $\varepsilon > 0$ and $T > 0$, there exists a compact set $K_{\varepsilon, T}$ such that

$$P\{X(t) \in K_{\varepsilon, T} \quad \text{for all} \quad t \in [0, T] \cap \mathbb{Q}\} > 1 - \varepsilon. \tag{3.22}$$

Then there is a modification of X with sample paths in $D_E[0, \infty)$.

Proof. Let X be defined on $(\Omega, \mathscr{F}, P)$. By assumption, there exists a sequence $\{(f_i, g_i)\} \subset A$ such that $\{f_i\}$ separates points in E. By Proposition 2.9 of Chapter 2, there exists $\Omega' \subset \Omega$ with $P(\Omega') = 1$ such that

$$f_i(X(t)) - \int_0^t g_i(X(s))\, ds \tag{3.23}$$

has limits through the rationals from above and below for all $t \ge 0$, all i, and all $\omega \in \Omega'$. By (3.22) there exists $\Omega'' \subset \Omega'$ with $P(\Omega'') = 1$ such that $\{X(t, \omega): t \in [0, T] \cap \mathbb{Q}\}$ has compact closure for all $T > 0$ and $\omega \in \Omega''$. Suppose $\omega \in \Omega''$. Then for each $t \ge 0$ there exist $s_n \in \mathbb{Q}$ such that $s_n > t$, $\lim_{n\to\infty} s_n = t$, and $\lim_{n\to\infty} X(s_n, \omega)$ exists, and hence

$$f_i(\lim_{n\to\infty} X(s_n, \omega)) = \lim_{\substack{s \to t+ \\ s \in \mathbb{Q}}} f_i(X(s, \omega)), \tag{3.24}$$

where the limit on the right exists since $\omega \in \Omega'$. Since $\{f_i\}$ separates points we have

$$\lim_{\substack{s \to t+ \\ s \in \mathbb{Q}}} X(s) \equiv Y(t) \tag{3.25}$$

exists for all $t \geq 0$ and $\omega \in \Omega''$. Similarly

$$\lim_{\substack{s \to t- \\ s \in \mathbb{Q}}} X(s) \equiv Y^-(t) \tag{3.26}$$

exists for all $t > 0$ and $\omega \in \Omega''$, so Y has sample paths in $D_E[0, \infty)$ by Lemma 2.8 of Chapter 2.

Since X is a solution of the martingale problem, if follows that

$$E[f(Y(t)) \mid \mathscr{F}_t^X] = \lim_{\substack{s \to t+ \\ s \in \mathbb{Q}}} E[f(X(s)) \mid \mathscr{F}_t^X] = f(X(t)) \tag{3.27}$$

for every $f \in \mathscr{D}(A)$ and $t \geq 0$. Since $\mathscr{D}(A)$ is separating, $P\{Y(t) = X(t)\} = 1$ for all $t \geq 0$. (See Problem 7 of Chapter 3.) □

3.7 Corollary Let E be locally compact and separable. Let $A \subset \hat{C}(E) \times B(E)$ and suppose that $\mathscr{D}(A)$ is dense in $\hat{C}(E)$ in the norm topology. Then any solution of the martingale problem for A has a modification with sample paths in $D_{E^\Delta}[0, \infty)$ where E^Δ is the one-point compactification of E.

Proof. Let $A' \subset C(E^\Delta) \times B(E^\Delta)$ be given by

$$A' = \{(f, g) : f(\Delta) = g(\Delta) = 0, (f|_E, g|_E) \in A\} \cup \{(1, 0)\} \tag{3.28}$$

and $A'' = (A')_S$. Then any solution of the martingale problem for A considered as a process with values in E^Δ is a solution of the martingale problem for A''. Since A'' satisfies the conditions of Theorem 3.6, the corollary follows. □

In the light of condition (3.22) and in particular Corollary 3.7, it is sometimes useful to first prove the existence of a modification with sample paths in $D_{\hat{E}}[0, \infty)$ (where $\hat{E}$ is some compactification of E) and then to prove that the modification actually has sample paths in $D_E[0, \infty)$. With this in mind we prove the following theorem.

3.8 Theorem Let $(\hat{E}, r)$ be a metric space and let $A \subset B(\hat{E}) \times B(\hat{E})$. Let $E \subset \hat{E}$ be open, and suppose that X is a solution of the martingale problem for A with sample paths in $D_{\hat{E}}[0, \infty)$. Suppose $(\chi_E, 0)$ is in the bp-closure of $A \cap (\bar{C}(\hat{E}) \times B(\hat{E}))$. If $P\{X(0) \in E\} = 1$, then $P\{X \in D_E[0, \infty)\} = 1$.

Proof. For $m = 1, 2, \ldots$, define the $\{\mathscr{F}_{t+}^X\}$-stopping time

$$\tau_m = \inf\left\{t\colon \inf_{y \in \hat{E} - E} r(y, X(t)) < \frac{1}{m}\right\}. \tag{3.29}$$

Then $\tau_1 \le \tau_2 \le \cdots$ and $\lim_{m\to\infty} X(\tau_m \wedge t) \equiv Y(t)$ exists. Note that $Y(t)$ is in $\hat{E} - E$ if and only if $\lim_{m\to\infty} \tau_m \equiv \tau \le t$. For $(f, g) \in A \cap (\bar{C}(\hat{E}) \times B(\hat{E}))$,

$$f(X(t)) - \int_0^t g(X(s))\, ds \tag{3.30}$$

is a right continuous $\{\mathscr{F}_t^X\}$-martingale, and hence the optional sampling theorem implies that for each $t \ge 0$,

$$E[f(X(\tau_m \wedge t))] = E[f(X(0))] + E\left[\int_0^{\tau_m \wedge t} g(X(s))\, ds\right]. \tag{3.31}$$

Letting $m \to \infty$, we have

$$E[f(Y(t))] = E[f(X(0))] + E\left[\int_0^{\tau \wedge t} g(X(s))\, ds\right], \tag{3.32}$$

and this holds for all (f, g) in the bp-closure of $A \cap (\bar{C}(\hat{E}) \times B(\hat{E}))$. Taking $(f, g) = (\chi_E, 0)$, we have

$$P\{\tau > t\} = P\{Y(t) \in E\} = 1, \qquad t \ge 0. \tag{3.33}$$

Consequently, with probability 1, X has no limit points in $\hat{E} - E$ on any bounded time interval and therefore has almost all sample paths in $D_E[0, \infty)$. □

3.9 Proposition Let $\hat{E}$, A, and X be as above. Let $E \subset \hat{E}$ be open. Suppose there exists $\{(f_n, g_n)\} \subset A \cap (\bar{C}(\hat{E}) \times B(\hat{E}))$ such that

$$\operatorname*{bp-lim}_{n\to\infty} f_n = \chi_E, \tag{3.34}$$

$$\inf_n \inf_x g_n(x) > -\infty \tag{3.35}$$

and $\{g_n\}$ converges pointwise to zero. If $P\{X(0) \in E\} = 1$, then $P\{X \in D_E[0, \infty)\} = 1$.

Proof. Substituting (f_n, g_n) in (3.32) and letting $n \to \infty$, Fatou's lemma gives

$$P\{Y(t) \in E\} \ge P\{X(0) \in E\} = 1. \tag{3.36}$$

□

3.10 Proposition Let $\hat{E}$, A, and X be as above. Let $E_1, E_2, \ldots$ be open subsets of $\hat{E}$ and let $E = \bigcap_k E_k$. Suppose $(\chi_E, 0)$ is in the bp-closure of $A \cap (\bar{C}(\hat{E}) \times B(\hat{E}))$. If $P\{X(0) \in E\} = 1$, then $P\{X \in D_E[0, \infty)\} = 1$.

Proof. Let τ_m^k be defined as in (3.29) with E replaced by E_k. Then the analogue of (3.32) gives

$$P\{\lim_{m\to\infty} X(\tau_m^k \wedge t) \in E_k\} \geq P\{\lim_{m\to\infty} X(\tau_m^k \wedge t) \in E\} = 1. \tag{3.37}$$

Therefore almost all sample paths of X are in $D_{E_k}[0, \infty)$ for every k, and hence in $D_E[0, \infty)$. □

3.11 Remark In the application of Theorem 3.8 and Propositions 3.9 and 3.10, E might be locally compact and $\hat{E} = E^\Delta$, or $E = \prod_k F_k$, where the F_k are locally compact, $\hat{E} = \prod_k F_k^\Delta$, and $E_k = \prod_{l \leq k} F_l \times \prod_{l > k} F_l^\Delta$. □

We close this section by showing, under the conditions of Theorem 3.6, that any solution of the martingale problem for A with sample paths in $D_E[0, \infty)$ is *quasi-left continuous*, that is, for every nondecreasing sequence of stopping times τ_n with $\lim_{n\to\infty} \tau_n = \tau < \infty$ a.s., we have $\lim_{n\to\infty} X(\tau_n) = X(\tau)$ a.s.

3.12 Theorem Let E be separable. Let $A \subset \bar{C}(E) \times B(E)$ and suppose $\mathscr{D}(A)$ is separating. Let X be a solution of the martingale problem for A with respect to $\{\mathscr{G}_t\}$, having sample paths in $D_E[0, \infty)$. Let $\tau_1 \leq \tau_2 \leq \cdots$ be a sequence of $\{\mathscr{G}_t\}$-stopping times and let $\tau = \lim_{n\to\infty} \tau_n$. Then

$$P\left\{\lim_{n\to\infty} X(\tau_n) = X(\tau),\ \tau < \infty\right\} = P\{\tau < \infty\}. \tag{3.38}$$

In particular, $P\{X(t) = X(t-)\} = 1$ for each $t > 0$.

Proof. Clearly the limit in (3.38) exists. For $(f, g) \in A$ and $t \geq 0$,

$$\begin{aligned} \lim_{n\to\infty} f(X(\tau_n \wedge t)) &= \lim_{n\to\infty} E\left[f(X(\tau \wedge t)) - \int_{\tau_n \wedge t}^{\tau \wedge t} g(X(s))\, ds \,\middle|\, \mathscr{G}_{\tau_n}\right] \\ &= E\Big[f(X(\tau \wedge t)) \,\Big|\, \bigvee_n \mathscr{G}_{\tau_n}\Big], \end{aligned} \tag{3.39}$$

and (3.38) follows. (See Problem 7 of Chapter 3.) □

3.13 Corollary Let (E, r) be separable, and let A and X satisfy the conditions of Theorem 3.12. Let $F \subset E$ be closed and define $\tau = \inf\{t\colon X(t) \in F$ or $X(t-) \in F\}$ and $\sigma = \inf\{t\colon X(t) \in F\}$. (Note that σ need not be measurable.) Then $\tau = \sigma$ a.s.

Proof. Note that $\{\tau = \sigma\} = \{\tau < \infty, X(\tau) \in F\} \cup \{\tau = \infty\}$. Note that by the right continuity of the martingales, X is a solution of the martingale problem for A with respect to $\{\mathscr{F}_{t+}^X\}$. Let $U_n = \{y\colon \inf_{x \in F} r(x, y) < 1/n\}$, and define

$\tau_n = \inf\{t: X(t) \in U_n\}$. Then τ_n is an $\{\mathscr{F}_{t+}^X\}$-stopping time, $\tau_1 \le \tau_2 \le \cdots$ and $\lim_{n\to\infty} \tau_n = \tau$. Since $X(\tau_n) \in \bar{U}_n$, Theorem 3.12 implies

$$P\left\{\tau < \infty, X(\tau) = \lim_{n\to\infty} X(\tau_n) \in F\right\} = P\{\tau < \infty\}. \tag{3.40}$$ □

4. THE MARTINGALE PROBLEM: UNIQUENESS, THE MARKOV PROPERTY, AND DUALITY

As was observed above, the statement that a measurable process X is a solution of the martingale problem for (A, μ) is a statement about the finite-dimensional distributions of X. Consequently, we say that *uniqueness* holds for solutions of the martingale problem for (A, μ) if any two solutions have the same finite-dimensional distributions. If there exists a solution of the martingale problem for (A, μ) and uniqueness holds, we say that the martingale problem for (A, μ) is *well-posed*. If this is true for all $\mu \in \mathscr{P}(E)$, then the martingale problem for A is said to be well-posed. (Typically, if the martingale problem for (A, δ_x) is well-posed for each $x \in E$, then the martingale problem for (A, μ) is well-posed for each $\mu \in \mathscr{P}(E)$. See Problems 49 and 50.) We say that the martingale problem for (A, μ) is well-posed in $D_E[0, \infty)$ ($C_E[0, \infty)$) if there is a unique solution $P \in \mathscr{P}(D_E[0, \infty))$ ($P \in \mathscr{P}(C_E[0, \infty))$). Note that a martingale problem may be well-posed in $D_E[0, \infty)$ without being well-posed, that is, uniqueness may hold under the restriction that the solution have sample paths in $D_E[0, \infty)$ but not in general. See Problem 21. However, Theorem 3.6 shows that this difficulty is rare. The following theorem says essentially that a Markov process is the unique solution of the martingale problem for its generator.

4.1 Theorem Let E be separable, and let $A \subset B(E) \times B(E)$ be linear and dissipative. Suppose there exists $A' \subset A$, A' linear, such that $\overline{\mathscr{R}(\lambda - A')} = \overline{\mathscr{D}(A')} \equiv L$ for some $\lambda > 0$, and L is separating. Let $\mu \in \mathscr{P}(E)$ and suppose X is a solution of the martingale problem for (A, μ). Then X is a Markov process corresponding to the semigroup on L generated by the closure of A', and uniqueness holds for the martingale problem for (A, μ).

Proof. Without loss of generality we can assume A' is closed (it is single-valued by Lemma 4.2 of Chapter 1) and hence, by Theorem 2.6 of Chapter 1, it generates a strongly continuous contraction semigroup $\{T(t)\}$ on L. In particular, by Corollary 6.8 of Chapter 1,

$$T(t)f = \lim_{n\to\infty} (I - n^{-1}A')^{-[nt]}f, \qquad f \in L,\ t \ge 0. \tag{4.1}$$

We want to show that

$$E[f(X(t+u)) \mid \mathscr{F}_t^X] = T(u)f(X(t)) \tag{4.2}$$

for all $f \in L$, which implies the Markov property, and the uniqueness follows by Proposition 1.6.

If $(f, g) \in A'$ and $\lambda > 0$, then (3.5) in Lemma 3.2 is a martingale and hence

$$f(X(t)) = E\left[\int_0^\infty e^{-\lambda s}(\lambda f(X(t+s)) - g(X(t+s)))\, ds \,\middle|\, \mathscr{F}_t^X\right], \tag{4.3}$$

which gives

$$\begin{aligned}(I - n^{-1}A')^{-1}h(X(t)) &= E\left[n \int_0^\infty e^{-ns}h(X(t+s))\, ds \,\middle|\, \mathscr{F}_t^X\right] \\ &= E\left[\int_0^\infty e^{-s}h(X(t+n^{-1}s))\, ds \,\middle|\, \mathscr{F}_t^X\right]\end{aligned} \tag{4.4}$$

for all $h \in L$. Iterating (4.4) gives

$$\begin{aligned}(I - n^{-1}A')^{-k}h(X(t)) &= E\left[\int_0^\infty \cdots \int_0^\infty \exp\{-(s_1 + s_2 + \cdots + s_k)\}\right. \\ &\quad \left. \times\, h(X(t + n^{-1}(s_1 + s_2 + \cdots + s_k)))\, ds_1 \cdots ds_k \,\middle|\, \mathscr{F}_t^X\right] \\ &= E\left[\int_0^\infty \Gamma(k)^{-1}s^{k-1}e^{-s}h(X(t+n^{-1}s))\, ds \,\middle|\, \mathscr{F}_t^X\right].\end{aligned} \tag{4.5}$$

Suppose $h \in \mathscr{D}(A')$. Then

$$\begin{aligned}(I - n^{-1}A')^{-[nu]}h(X(t)) &= E[h(X(t+u)) \mid \mathscr{F}_t^X] \\ &\quad + E\left[\int_0^\infty \Gamma([nu])^{-1}s^{[nu]-1}e^{-s}\int_u^{s/n} A'h(X(t+v))\, dv\, ds \,\middle|\, \mathscr{F}_t^X\right].\end{aligned} \tag{4.6}$$

The second term on the right is bounded by

$$\begin{aligned}&\| A'h \| \int_0^\infty \left|\frac{s}{n} - u\right| \Gamma([nu])^{-1}s^{[nu]-1}e^{-s}\, ds \\ &\qquad = \| A'h \| \, E\left[\left| n^{-1}\sum_{k=1}^{[nu]} \Delta_k - u\right|\right]\end{aligned} \tag{4.7}$$

where the Δ_k are independent and exponentially distributed with mean 1. Consequently (4.7) goes to zero as $n \to \infty$, and we have by (4.1)

$$(4.8) \qquad T(u)h(X(t)) = \lim_{n\to\infty} (I - n^{-1}A')^{-[nu]}h(X(t)) = E[h(X(t+u)) \mid \mathscr{F}_t^X].$$

Since $\overline{\mathscr{D}(A')} = L$, (4.2) holds for all $f \in L$. □

Under the conditions of Theorem 4.1, every solution of the martingale problem for A is Markovian. We now show that uniqueness of the solution of the martingale problem always implies the Markov property.

4.2 Theorem Let E be separable, and let $A \subset B(E) \times B(E)$. Suppose that for each $\mu \in \mathscr{P}(E)$ any two solutions X, Y of the martingale problem for (A, μ) have the same one-dimensional distributions, that is, for each $t > 0$,

$$(4.9) \qquad P\{X(t) \in \Gamma\} = P\{Y(t) \in \Gamma\}, \qquad \Gamma \in \mathscr{B}(E).$$

Then the following hold.

(a) Any solution of the martingale problem for A with respect to a filtration $\{\mathscr{G}_t\}$ is a Markov process with respect to $\{\mathscr{G}_t\}$, and any two solutions of the martingale problem for (A, μ) have the same finite-dimensional distributions (i.e., (4.9) implies uniqueness).

(b) If $A \subset \bar{C}(E) \times B(E)$, X is a solution of the martingale problem for A with respect to a filtration $\{\mathscr{G}_t\}$, and X has sample paths in $D_E[0, \infty)$, then for each a.s. finite $\{\mathscr{G}_t\}$-stopping time τ,

$$(4.10) \qquad E[f(X(\tau + t)) \mid \mathscr{G}_\tau] = E[f(X(\tau + t)) \mid X(\tau)]$$

for all $f \in B(E)$ and $t \geq 0$.

(c) If, in addition to the conditions of part (b), for each $x \in E$ there exists a solution $P_x \in \mathscr{P}(D_E[0, \infty))$ of the martingale problem for (A, δ_x) such that $P_x(B)$ is a Borel measurable function of x for each $B \in \mathscr{S}_E$ (cf. Theorem 4.6), then, defining $T(t)f(x) = \int f(\omega(t))P_x(d\omega)$,

$$(4.11) \qquad E[f(X(\tau + t)) \mid \mathscr{G}_\tau] = T(t)f(X(\tau))$$

for all $f \in B(E)$, $t \geq 0$, and a.s. finite $\{\mathscr{G}_t\}$-stopping times τ (i.e., X is strong Markov).

Proof. Let X, defined on $(\Omega, \mathscr{F}, P)$, be a solution of the martingale problem for A with respect to a filtration $\{\mathscr{G}_t\}$, fix $r \geq 0$, and let $F \in \mathscr{G}_r$ satisfy $P(F) > 0$. For $B \in \mathscr{F}$ define

$$(4.12) \qquad P_1(B) = \frac{E[\chi_F E[\chi_B \mid \mathscr{G}_r]]}{P(F)}$$

and

$$P_2(B) = \frac{E[\chi_F E[\chi_B \mid X(r)]]}{P(F)}, \tag{4.13}$$

and set $Y(\cdot) = X(r + \cdot)$. Note that

$$P_1\{Y(0) \in \Gamma\} = P_2\{Y(0) \in \Gamma\} = P\{X(r) \in \Gamma \mid F\}. \tag{4.14}$$

With (3.4) in mind we set

$$\eta(Y) = \left[f(Y(t_{n+1})) - f(Y(t_n)) - \int_{t_n}^{t_{n+1}} g(Y(s))\, ds \right] \prod_{k=1}^{n} h_k(Y(t_k)) \tag{4.15}$$

where $0 \le t_1 < t_2 < \cdots < t_n < t_{n+1}$, $(f, g) \in A$, and $h_k \in B(E)$. Since $E[\eta(X(r + \cdot)) \mid \mathscr{G}_r] = 0$,

$$E_2[\eta(Y)] = \frac{E[\chi_F E[\eta(X(r + \cdot)) \mid X(r)]]}{P(F)} = 0 \tag{4.16}$$

and similarly for $E_1[\eta(Y)]$. Consequently, Y is a solution of the martingale problem for A on $(\Omega, \mathscr{F}, P_1)$ and $(\Omega, \mathscr{F}, P_2)$. By (4.9), $E_1[f(Y(t))] = E_2[f(Y(t))]$ for each $f \in B(E)$ and $t \ge 0$, and hence

$$E[\chi_F E[f(X(r + t)) \mid \mathscr{G}_r]] = E[\chi_F E[f(X(r + t)) \mid X(r)]]. \tag{4.17}$$

Since $F \in \mathscr{G}_r$ is arbitrary, (4.17) implies

$$E[f(X(r + t)) \mid \mathscr{G}_r] = E[f(X(r + t)) \mid X(r)], \tag{4.18}$$

which is the Markov property.

Uniqueness is proved in much the same way. Let X and Y be solutions of the martingale problem for (A, μ) defined on $(\Omega, \mathscr{F}, P)$ and $(\Gamma, \mathscr{G}, Q)$ respectively. We want to show

$$E^P\left[\prod_{k=1}^{m} f_k(X(t_k))\right] = E^Q\left[\prod_{k=1}^{m} f_k(Y(t_k))\right] \tag{4.19}$$

for all choices of $t_k \in [0, \infty)$ and $f_k \in B(E)$ (cf. Proposition 4.6 of Chapter 3). It is sufficient to consider only $f_k > 0$. For $m = 1$, (4.19) holds by (4.9). Proceeding by induction, assume (4.19) holds for all $m \le n$, and fix $0 \le t_1 < t_2 < \cdots < t_n$ and $f_1, \ldots, f_n \in B(E)$, $f_k > 0$. Define

$$\tilde{P}(B) = \frac{E^P[\chi_B \prod_{k=1}^{n} f_k(X(t_k))]}{E[\prod_{k=1}^{n} f_k(X(t_k))]}, \qquad B \in \mathscr{F}, \tag{4.20}$$

$$\tilde{Q}(B) = \frac{E^Q[\chi_B \prod_{k=1}^{n} f_k(Y(t_k))]}{E[\prod_{k=1}^{n} f_k(Y(t_k))]}, \qquad B \in \mathscr{G}, \tag{4.21}$$

and set $\tilde{X}(t) = X(t_n + t)$ and $\tilde{Y}(t) = Y(t_n + t)$. By the argument used above, $\tilde{X}$ on $(\Omega, \mathscr{F}, \tilde{P})$ and $\tilde{Y}$ on $(\Gamma, \mathscr{G}, \tilde{Q})$ are solutions of the martingale problem for A. Furthermore, by (4.19) with $m = n$,

$$(4.22) \qquad E^{\tilde{P}}[f(\tilde{X}(0))] = \frac{E^P[f(X(t_n)) \prod_{k=1}^n f_k(X(t_k))]}{E^P[\prod_{k=1}^n f_k(X(t_k))]}$$

$$= \frac{E^Q[f(Y(t_n)) \prod_{k=1}^n f_k(Y(t_k))]}{E^Q[\prod_{k=1}^n f_k(Y(t_k))]}$$

$$= E^{\tilde{Q}}[f(\tilde{Y}(0))], \qquad f \in B(E),$$

so $\tilde{X}$ and $\tilde{Y}$ have the same initial distributions. Consequently, (4.9) applies and

$$(4.23) \qquad E^{\tilde{P}}[f(\tilde{X}(t))] = E^{\tilde{Q}}[f(\tilde{Y}(t))], \qquad t \geq 0, \ f \in B(E).$$

As in (4.22), this implies

$$(4.24) \quad E^P\left[f(X(t_n + t)) \prod_{k=1}^n f_k(X(t_k))\right] = E^Q\left[f(Y(t_n + t)) \prod_{k=1}^n f_k(Y(t_k))\right],$$

and, setting $t_{n+1} = t_n + t$, we have (4.19) for $m = n + 1$.

For part (b), assume that $A \subset \bar{C}(E) \times B(E)$ and that X has sample paths in $D_E[0, \infty)$. Then (3.1) is a right continuous martingale with bounded increments for all $(f, g) \in A$ and the optional sampling theorem (see Problem 11 of Chapter 2) implies

$$(4.25) \qquad E[\eta(X(\tau + \cdot)) \mid \mathscr{G}_\tau] = 0,$$

so (4.10) follows in the same way as (4.18).

Similarly for part (c), if $F \in \mathscr{G}_\tau$ and $P(F) > 0$, then

$$(4.26) \qquad P_1(B) = \frac{E[\chi_F P_{X(\tau)}(B)]}{P(F)}$$

and

$$(4.27) \qquad P_2(B) = \frac{E[\chi_F E[\chi_B(X(\tau + \cdot)) \mid \mathscr{G}_\tau]]}{P(F)}$$

define solutions of the martingale problem with the same initial distribution and (4.11) follows as before. □

Since it is possible to have uniqueness among solutions of a martingale problem with sample paths in $D_E[0, \infty)$ without having uniqueness among solutions that are only required to be measurable, it is useful to introduce the terminology $D_E[0, \infty)$ *martingale problem* and $C_E[0, \infty)$ *martingale problem* to indicate when we are requiring the designated sample path behavior.

4.3 Corollary Let E be separable, and let $A \subset B(E) \times B(E)$. Suppose that for each $\mu \in \mathscr{P}(E)$, any two solutions X, Y of the martingale problem for (A, μ)

with sample paths in $D_E[0,\infty)$ (respectively, $C_E[0, \infty)$) satisfy (4.9) for each $t \geq 0$. Then for each $\mu \in \mathscr{P}(E)$, any two solutions of the martingale problem for (A, μ) with sample paths in $D_E[0, \infty)$ ($C_E[0, \infty)$) have the same distribution on $D_E[0, \infty)$ ($C_E[0, \infty)$).

Proof. Note that $\tilde{X}$ and $\tilde{Y}$ defined in the proof of Theorem 4.2 have sample paths in $D_E[0, \infty)$ ($C_E[0, \infty)$) if X and Y do. Consequently, the proof that X and Y have the same finite-dimensional distributions is the same as before. Since E is separable, by Proposition 7.1 of Chapter 3, the finite-dimensional distributions of X and Y determine their distributions on $D_E[0, \infty)$ ($C_E[0, \infty)$). □

4.4 Corollary Let E be separable, and let $A \subset B(E) \times B(E)$ be linear and dissipative. Suppose that for some (hence all) $\lambda > 0$, $\overline{\mathscr{R}(\lambda - A)} \supset \mathscr{D}(A)$, and that there exists $M \subset B(E)$ such that M is separating and $M \subset \overline{\mathscr{R}(\lambda - A)}$ for every $\lambda > 0$. Then for each $\mu \in \mathscr{P}(E)$ any two solutions of the martingale problem for (A, μ) with sample paths in $D_E[0, \infty)$ have the same distribution on $D_E[0, \infty)$.

4.5 Remark Note that the significance of this result, in contrast with Theorem 4.1, is that we do not require $\mathscr{D}(A)$ to be separating. See Problem 22 for an example in which $\mathscr{D}(A)$ is not separating. □

Proof. If X and Y are solutions of the martingale problem for (A, μ) with sample paths in $D_E[0, \infty)$, and if $h \in M$, then by (4.3),

$$E\left[\int_0^\infty e^{-\lambda t}h(X(t))\,dt\right] = \int (\lambda - A)^{-1}h\,d\mu = E\left[\int_0^\infty e^{-\lambda t}h(Y(t))\,dt\right] \tag{4.28}$$

for every $\lambda > 0$. Since M is separating, the identity

$$\int_0^\infty e^{-\lambda t}E[h(X(t))]\,dt = \int_0^\infty e^{-\lambda t}E[h(Y(t))]\,dt \tag{4.29}$$

holds for all $h \in B(E)$ (think of $\int_0^\infty e^{-\lambda t}E[h(X(t))]\,dt \equiv \int h\,d\nu_{\lambda, X}$). By the uniqueness of the Laplace transform, for almost every $t \geq 0$,

$$E[h(X(t))] = E[h(Y(t))], \tag{4.30}$$

and if h is continuous, the right continuity of X and Y imply (4.30) holds for all $t \geq 0$. This in turn implies (4.9) and the uniqueness follows from Corollary 4.3. □

The following theorem shows that the measurability condition in Theorem 4.2(c) typically holds.

4.6 Theorem Let (E, r) be complete and separable, and let $A \subset \bar{C}(E) \times B(E)$. Suppose there exists a countable subset $A_0 \subset A$ such that A is contained in the bp-closure of A_0 (for example, suppose $A \subset L \times L$ where L is a separable subspace of $\bar{C}(E)$). Suppose that the $D_E[0, \infty)$ martingale problem for A is well-posed. Then, denoting the solution for (A, δ_x) by P_x, $P_x(B)$ is Borel measurable in x for each $B \in \mathscr{S}_E$.

Proof. By Theorems 5.6 and 1.7 of Chapter 3, $(\mathscr{P}(D_E[0, \infty)), \rho)$, where ρ is the Prohorov metric, is complete and separable. By the separability of E and Proposition 4.2 of Chapter 3, there is a countable set $M \subset \bar{C}(E)$ such that M is bp-dense in $B(E)$.

Let H be the collection of functions on $D_E[0, \infty)$ of the form

$$\eta = \left(f(X(t_{n+1})) - f(X(t_n)) - \int_{t_n}^{t_{n+1}} g(X(s))\, ds \right) \prod_{k=1}^{n} h_k(X(t_k)) \tag{4.31}$$

where X is the coordinate process, $(f, g) \in A_0$, $h_1, \dots, h_n \in M$, $0 \le t_1 < t_2 < \cdots < t_{n+1}$, and $t_k \in \mathbb{Q}$. Note that since A_0 and M are countable, H is countable, and since f and the h_k are continuous, $P \in \mathscr{P}(D_E[0, \infty))$ is a solution of the martingale problem for A if and only if

$$\int \eta\, dP = 0, \qquad \eta \in H. \tag{4.32}$$

Let $\mathscr{M}_A \subset \mathscr{P}(D_E[0, \infty))$ be the collection of all such solutions. Then $\mathscr{M}_A = \bigcap_{\eta \in H}\{P: \int \eta\, dP = 0\}$, and $\mathscr{M}_A$ is a Borel set since H is countable and $\{P: \int \eta\, dP = 0\}$ is a Borel set. (Note that if $\eta \in \bar{C}(D_E[0, \infty))$, then $F_\eta(P) \equiv \int \eta\, dP$ is continuous, hence Borel measurable, and the collection of $\eta \in B(D_E[0, \infty))$ for which F_η is Borel measurable is bp-closed.)

Let $G: \mathscr{P}(D_E[0, \infty)) \to \mathscr{P}(E)$ be given by $G(P) = PX(0)^{-1}$. Note that G is continuous. The fact that the martingale problem for A is well-posed implies that the restriction of G to $\mathscr{M}_A$ is one-to-one and onto. But a one-to-one Borel measurable mapping of a Borel subset of a complete, separable metric space onto a Borel subset of a complete, separable metric space has a Borel measurable inverse (see Appendix 10), that is, letting P_μ denote the solution of the martingale problem for (A, μ), the mapping of $\mathscr{P}(E)$ into $\mathscr{P}(D_E[0, \infty))$ given by $\mu \to P_\mu$ is Borel measurable and it follows that the mapping of E into $\mathscr{P}(D_E[0, \infty))$ given by $x \to P_x \equiv P_{\delta_x}$ is also Borel measurable. □

Theorem 4.2 is the basic tool for proving uniqueness for solutions of a martingale problem. The problem, of course, is to verify (4.9). One approach to doing this which, despite its strange, ad hoc appearance, has found widespread applicability involves the notion of *duality*.

Let (E_1, r_1) and (E_2, r_2) be separable metric spaces. Let $A_1 \subset B(E_1) \times B(E_1)$, $A_2 \subset B(E_2) \times B(E_2)$, $f \in M(E_1 \times E_2)$, $\alpha \in M(E_1)$, $\beta \in M(E_2)$, $\mu_1 \in \mathscr{P}(E_1)$, and $\mu_2 \in \mathscr{P}(E_2)$. Then the martingale problems for (A_1, μ_1) and (A_2, μ_2) are *dual* with respect to (f, α, β) if for each solution X of the martingale problem for (A_1, μ_1) and each solution Y for (A_2, μ_2), $\int_0^t |\alpha(X(s))|\, ds < \infty$ a.s., $\int_0^t |\beta(Y(s))|\, ds < \infty$ a.s.,

(4.33)
$$\int E\left[\left| f(X(t), y) \exp\left\{\int_0^t \alpha(X(s))\, ds\right\}\right|\right] \mu_2(dy) < \infty,$$

(4.34)
$$\int E\left[\left| f(x, Y(t)) \exp\left\{\int_0^t \beta(Y(s))\, ds\right\}\right|\right] \mu_1(dx) < \infty,$$

and

(4.35)
$$\int E\left[f(X(t), y) \exp\left\{\int_0^t \alpha(X(s))\, ds\right\}\right] \mu_2(dy) = \int E\left[f(x, Y(t)) \exp\left\{\int_0^t \beta(Y(s))\, ds\right\}\right] \mu_1(dx)$$

for every $t \geq 0$. Note that if X and Y are defined on the same sample space and are independent, then (4.35) can be written

(4.36)
$$E\left[f(X(t), Y(0)) \exp\left\{\int_0^t \alpha(X(s))\, ds\right\}\right] = E\left[f(X(0), Y(t)) \exp\left\{\int_0^t \beta(Y(s))\, ds\right\}\right].$$

4.7 Proposition Let (E_1, r_1) be complete and separable and let E_2 be separable. Let $A_1 \subset B(E_1) \times B(E_1)$, $A_2 \subset B(E_2) \times B(E_2)$, $f \in M(E_1 \times E_2)$, and $\beta \in M(E_2)$. Let $\mathscr{M} \subset \mathscr{P}(E_1)$ contain $PX(t)^{-1}$ for all solutions X of the martingale problem for A_1 with $PX(0)^{-1}$ having compact support, and all $t \geq 0$. Suppose that (A_1, μ) and (A_2, δ_y) are dual with respect to $(f, 0, \beta)$ (i.e., $\alpha = 0$ in (4.35)) for every $\mu \in \mathscr{P}(E_1)$ with compact support and every $y \in E_2$, and that $\{f(\cdot, y) : y \in E_2\}$ is separating on $\mathscr{M}$. If for every $y \in E_2$ there exists a solution of the martingale problem for (A_2, δ_y), then for each $\mu \in \mathscr{P}(E_1)$ uniqueness holds for the martingale problem for (A_1, μ).

4.8 Remark **(a)** The restriction to μ with compact support in the hypotheses is important since we are not assuming boundedness for f and β. Completeness is needed only so that arbitrary $\mu \in \mathscr{P}(E_1)$ can be approximated by μ with compact support.

(b) The proposition transforms the uniqueness problem for A_1 into an existence problem for A_2. Existence problems, of course, are typically simpler to handle. □

Proof. Let Y_y be a solution of the martingale problem for (A_2, δ_y). If $\mu \in \mathscr{P}(E_1)$ has compact support and X and $\tilde{X}$ are solutions of the martingale problem for (A_1, μ), then

$$(4.37) \qquad E[f(X(t), y)] = \int E\left[f(x, Y_y(t)) \exp\left\{\int_0^t \beta(Y_y(s))\, ds\right\}\right]\mu(dx)$$
$$= E[f(\tilde{X}(t), y)].$$

Since $\{f(\cdot, y): y \in E_2\}$ is separating on $\mathscr{M}$, (4.9) holds for X and $\tilde{X}$.

Now let $\mu \in \mathscr{P}(E_1)$ be arbitrary. If X and $\tilde{X}$ are solutions of the martingale problem for (A_1, μ) and K is compact with $\mu(K) > 0$, then X conditioned on $\{X(0) \in K\}$ and $\tilde{X}$ conditioned on $\{\tilde{X}(0) \in K\}$ are solutions of the martingale problem for $(A_1, \mu(\cdot \cap K)/\mu(K))$. Consequently,

$$(4.38) \quad P\{X(t) \in \Gamma \mid X(0) \in K\} = P\{\tilde{X}(t) \in \Gamma \mid \tilde{X}(0) \in K\}, \qquad \Gamma \in \mathscr{B}(E_1).$$

Since K is arbitrary and μ is tight, (4.9) follows, and Theorem 4.2 gives the uniqueness. □

The next step is to give conditions under which (4.35) holds. For the moment proceeding heuristically, suppose X and Y are independent E_1- and E_2-valued processes, $g, h \in M(E_1 \times E_2)$,

$$(4.39) \qquad f(X(t), y) - \int_0^t g(X(s), y)\, ds$$

is an $\{\mathscr{F}_t^X\}$-martingale for every $y \in E_2$, and

$$(4.40) \qquad f(x, Y(t)) - \int_0^t h(x, Y(s))\, ds$$

is an $\{\mathscr{F}_t^Y\}$-martingale for every $x \in E_1$. Then

$$(4.41) \quad \frac{d}{ds} E\left[f(X(s), Y(t-s)) \exp\left\{\int_0^s \alpha(X(u))\, du + \int_0^{t-s} \beta(Y(u))\, du\right\}\right]$$
$$= E\Big[(g(X(s), Y(t-s)) - h(X(s), Y(t-s)) + (\alpha(X(s))$$
$$-\beta(Y(t-s)))f(X(s), Y(t-s)))$$
$$\times \exp\left\{\int_0^s \alpha(X(u))\, du + \int_0^{t-s} \beta(Y(u))\, du\right\}\Big],$$

which is zero if

$$(4.42) \qquad g(x, y) + \alpha(x)f(x, y) = h(x, y) + \beta(y)f(x, y).$$

(Compare this calculation with (2.15) of Chapter 1.) Integrating gives (4.36).

4.9 Example To see that there is some possibility of the above working, suppose $E_1 = (-\infty, \infty)$, $E_2 = \{0, 1, 2, \dots\}$, $A_1 f(x) = f''(x) - xf'(x)$, and $A_2 f(y) = y(y-1)(f(y-2) - f(y))$. Of course A_1 corresponds to an Ornstein–Uhlenbeck process and A_2 to a jump process that jumps down by two until it absorbs in 0 or 1. Let $f(x, y) = x^y$. Let X be a solution of the martingale problem for A_1. Then

$$X(t)^y - \int_0^t (y(y-1)X(s)^{y-2} - yX(s)^y)\, ds \tag{4.43}$$

is a martingale provided the appropriate expectations exist; they will if the distribution of $X(0)$ has compact support. Let $g(x, y) = y(y-1)x^{y-2} - yx^y$ and $\alpha(x) = 0$. Then $g(x, y) = A_2 f(x, y) + (y^2 - 2y)x^y$, and we have (4.42) if we set $\beta(y) = y^2 - 2y$. Then, assuming the calculation in (4.41) is justified (and it is in this case), we have

$$E[X(t)^{Y(0)}] = E\left[X(0)^{Y(t)} \exp\left\{\int_0^t (Y^2(u) - 2Y(u))\, du\right\}\right], \tag{4.44}$$

and the moments of $X(t)$ are determined. In general, of course, this is not enough to determine the distribution of $X(t)$. However, in this case, (4.44) can be used to estimate the growth rate of the moments and the distribution is in fact determined. (See (4.21) of Chapter 3.)

Note that (4.44) suggests another use for duality. If $Y(0) = y$ is odd, then Y absorbs at 1 and

$$\begin{aligned} \lim_{t\to\infty} E[X(t)^y] &= \lim_{t\to\infty} E\left[X(0)^{Y(t)} \exp\left\{\int_0^t (Y^2(u) - 2Y(u))\, du\right\}\right] \\ &= 0, \end{aligned} \tag{4.45}$$

since the integrand in the exponent is -1 after absorption. Note that in order to justify this limit one needs to check that

$$E\left[\exp\left\{\int_0^{\tau_1} (Y^2(u) - 2Y(u))\, du\right\}\right] < \infty,$$

where $\tau_1 = \inf\{t: Y(t) = 1\}$. Similarly, if $Y(0) = y$ is even, then Y absorbs at 0 and setting $\tau_0 = \inf\{t: Y(t) = 0\}$,

$$\lim_{t\to\infty} E[X(t)^y] = E\left[\exp\left\{\int_0^{\tau_0} (Y^2(u) - 2Y(u))\, du\right\}\right]. \tag{4.46}$$

This identity can be used to determine the moments of the limiting distribution (which is Gaussian). See Problem 23. □

The next lemma gives the first step in justifying the calculation in (4.41).

4.10 Lemma Suppose $f(s, t)$ on $[0, \infty) \times [0, \infty)$ is absolutely continuous in s for each fixed t and absolutely continuous in t for each fixed s, and, setting $(f_1, f_2) \equiv \nabla f$, suppose

$$\int_0^T \int_0^T |f_i(s, t)| \, ds \, dt < \infty \qquad i = 1, 2, \qquad T > 0. \tag{4.47}$$

Then for almost every $t \geq 0$,

$$f(t, 0) - f(0, t) = \int_0^t (f_1(s, t - s) - f_2(s, t - s)) \, ds. \tag{4.48}$$

Proof.

$$\begin{aligned}
&\int_0^T \int_0^t (f_1(s, t - s) - f_2(s, t - s)) \, ds \, dt \\
&= \int_0^T \int_0^t f_1(t - s, s) \, ds \, dt - \int_0^T \int_0^t f_2(s, t - s) \, ds \, dt \\
&= \int_0^T \int_s^T f_1(t - s, s) \, dt \, ds - \int_0^T \int_s^T f_2(s, t - s) \, dt \, ds \\
&= \int_0^T (f(T - s, s) - f(0, s)) \, ds - \int_0^T (f(s, T - s) - f(s, 0)) \, ds \\
&= \int_0^T (f(s, 0) - f(0, s)) \, ds.
\end{aligned} \tag{4.49}$$

Differentiating with respect to T gives the desired result. □

The following theorem gives conditions under which the calculation in (4.41) is valid.

4.11 Theorem Let X and Y be independent measurable processes in E_1 and E_2, respectively. Let $f, g, h \in M(E_1 \times E_2)$, $\alpha \in M(E_1)$, and $\beta \in M(E_2)$. Suppose that for each $T > 0$ there exist an integrable random variable Γ_T and a constant C_T such that

$$\begin{aligned}
&\sup_{r, s, t \leq T} (|\alpha(X(r))| + 1) |f(X(s), Y(t))| \leq \Gamma_T, \\
&\sup_{r, s, t \leq T} (|\beta(Y(r))| + 1) |f(X(s), Y(t))| \leq \Gamma_T, \\
&\sup_{r, s, t \leq T} (|\alpha(X(r))| + 1) |g(X(s), Y(t))| \leq \Gamma_T, \\
&\sup_{r, s, t \leq T} (|\beta(Y(r))| + 1) |h(X(s), Y(t))| \leq \Gamma_T,
\end{aligned} \tag{4.50}$$

and

$$\int_0^T |\alpha(X(u))|\,du + \int_0^T |\beta(Y(u))|\,du \le C_T. \tag{4.51}$$

Suppose that

$$f(X(t), y) - \int_0^t g(X(s), y)\,ds \tag{4.52}$$

is an $\{{}^*\mathscr{F}_t^X\}$-martingale for each y, and

$$f(x, Y(t)) - \int_0^t h(x, Y(s))\,ds \tag{4.53}$$

is an $\{{}^*\mathscr{F}_t^Y\}$-martingale for each x. (The integrals in (4.52) and (4.53) are assumed to exist.) Then for almost every $t \ge 0$,

$$\begin{aligned} &E\left[f(X(t), Y(0)) \exp\left\{\int_0^t \alpha(X(u))\,du\right\}\right] \\ &\quad - E\left[f(X(0), Y(t)) \exp\left\{\int_0^t \beta(Y(u))\,du\right\}\right] \\ &\quad = E\left[\int_0^t \{g(X(s), Y(t-s)) - h(X(s), Y(t-s)) \right. \\ &\quad + (\alpha(X(s)) - \beta(Y(t-s)))f(X(s), Y(t-s))\} \\ &\quad \left. \times \exp\left\{\int_0^s \alpha(X(u))\,du + \int_0^{t-s} \beta(Y(u))\,du\right\} ds\right]. \end{aligned} \tag{4.54}$$

4.12 Remark Note that (4.54) can frequently be extended to more-general α and β by approximating by bounded α and β. □

Proof. Since (4.52) is a martingale and Y is independent of X, for $h \ge 0$,

(4.55)

$$\begin{aligned} &E\left[f(X(s+h), Y(t)) \exp\left\{\int_0^{s+h} \alpha(X(u))\,du\right\} \exp\left\{\int_0^t \beta(Y(u))\,du\right\}\right] \\ &\quad - E\left[f(X(s), Y(t)) \exp\left\{\int_0^s \alpha(X(u))\,du\right\} \exp\left\{\int_0^t \beta(Y(u))\,du\right\}\right] \\ &= E\left[f(X(s+h), Y(t)) \int_s^{s+h} \alpha(X(r)) \right. \\ &\quad \left. \times \exp\left\{\int_0^r \alpha(X(u))\,du\right\} dr \exp\left\{\int_0^t \beta(Y(u))\,du\right\}\right] \\ &\quad + E\left[\int_s^{s+h} g(X(r), Y(t))\,dr \exp\left\{\int_0^s \alpha(X(u))\,du\right\} \exp\left\{\int_0^t \beta(Y(u))\,du\right\}\right] \end{aligned}$$

$$= \int_s^{s+h} E\left[f(X(r), Y(t))\alpha(X(r)) \exp\left\{\int_0^r \alpha(X(u))\, du\right\} \exp\left\{\int_0^t \beta(Y(u))\, du\right\}\right] dr$$

$$+ \int_s^{s+h} E\left[\int_r^{s+h} g(X(v), Y(t))\, dv\, \alpha(X(r))\right.$$

$$\left.\times \exp\left\{\int_0^r \alpha(X(u))\, du\right\} \exp\left\{\int_0^t \beta(Y(u))\, du\right\}\right] dr$$

$$+ \int_s^{s+h} E\left[g(X(r), Y(t)) \exp\left\{\int_0^r \alpha(X(u))\, du\right\} \exp\left\{\int_0^t \beta(Y(u))\, du\right\}\right] dr$$

$$+ \int_s^{s+h} E\left[g(X(r), Y(t))\left(\exp\left\{\int_0^s \alpha(X(u))\, du\right\}\right.\right.$$

$$\left.\left. - \exp\left\{\int_0^r \alpha(X(u))\, du\right\}\right) \exp\left\{\int_0^t \beta(Y(u))\, du\right\}\right] dr.$$

We use (4.50) and (4.51) to ensure the integrability of the random variables above.

Note that for $t,\ s + h \le T$, the absolute values of the second and fourth terms are bounded by

$$\tag{4.56} \tfrac{1}{2}h^2 E[\Gamma_T] e^{C_T}.$$

Set

$$\tag{4.57} F(s, t) = E\left[f(X(s), Y(t)) \exp\left\{\int_0^s \alpha(X(u))\, du\right\} \exp\left\{\int_0^t \beta(Y(u))\, du\right\}\right].$$

For $0 = s_0 < s_1 < \cdots < s_m = s$, write

$$\tag{4.58} F(s, t) - F(0, t) = \sum_{i=1}^{m} (F(s_i, t) - F(s_{i-1}, t)).$$

Letting $\max_i (s_i - s_{i-1}) \to 0$, (4.55) and the fact that (4.56) is $O(h^2)$ imply

$$\tag{4.59} F(s, t) - F(0, t)$$

$$= \int_0^s E\left[(f(X(r), Y(t))\alpha(X(r)) + g(X(r), Y(t)))\right.$$

$$\left.\times \exp\left\{\int_0^r \alpha(X(u))\, du\right\} \exp\left\{\int_0^t \beta(Y(u))\, du\right\}\right] dr.$$

A similar identity holds for $F(s, t) - F(s, 0)$ and (4.54) follows from Lemma 4.10. □

4.13 Corollary If, in addition to the conditions of Theorem 4.11, $g(x, y) + \alpha(x)f(x, y) \equiv h(x, y) + \beta(y)f(x, y)$, then for all $t \geq 0$,

$$E\left[f(X(t), Y(0)) \exp\left\{\int_0^t \alpha(X(u))\, du\right\}\right] = E\left[f(X(0), Y(t)) \exp\left\{\int_0^t \beta(Y(u))\, du\right\}\right]. \tag{4.60}$$

Proof. By (4.54), (4.60) holds for almost every t and extends to all t since $F(t, 0)$ and $F(0, t)$ are continuous (see (4.55)). □

The estimates in (4.50) may be difficult to obtain, and it may be simpler to work first with the processes stopped at exit times from certain sets and then to take limits through expanding sequences of sets to obtain the desired result.

4.14 Corollary Let $\{\mathscr{F}_t\}$ and $\{\mathscr{G}_t\}$ be independent filtrations. Let X be $\{\mathscr{F}_t\}$-progressive and let τ be an $\{\mathscr{F}_t\}$-stopping time. Let Y be $\{\mathscr{G}_t\}$-progressive and let σ be a $\{\mathscr{G}_t\}$-stopping time. Suppose that (4.50) and (4.51) hold with X and Y replaced by $X(\cdot \wedge \tau)$ and $Y(\cdot \wedge \sigma)$ and that

$$f(X(t \wedge \tau), y) - \int_0^{t \wedge \tau} g(X(s), y)\, ds \tag{4.61}$$

is an $\{\mathscr{F}_t\}$-martingale for each y, and

$$f(x, Y(t \wedge \sigma)) - \int_0^{t \wedge \sigma} h(x, Y(s))\, ds \tag{4.62}$$

is a $\{\mathscr{G}_t\}$-martingale for each x. (The integrals in (4.61) and (4.62) are assumed to exist.) Then for almost every $t \geq 0$,

$$\begin{aligned}
&E\left[f(X(t \wedge \tau), Y(0)) \exp\left\{\int_0^{t \wedge \tau} \alpha(X(u))\, du\right\}\right] \\
&\quad - E\left[f(X(0), Y(t \wedge \sigma)) \exp\left\{\int_0^{t \wedge \sigma} \beta(Y(u))\, du\right\}\right] \\
&= \int_0^t E\Big[(\chi_{\{s \leq \tau\}}\, g(X(s), Y((t-s) \wedge \sigma)) - \chi_{\{t-s \leq \sigma\}}\, h(X(s \wedge \tau), Y(t-s)) \\
&\quad + (\chi_{\{s \leq \tau\}}\, \alpha(X(s)) - \chi_{\{t-s \leq \sigma\}}\, \beta(Y(t-s)))f(X(s \wedge \tau), Y((t-s) \wedge \sigma))) \\
&\quad \times \exp\left\{\int_0^{s \wedge \tau} \alpha(X(u))\, du + \int_0^{(t-s) \wedge \sigma} \beta(Y(u))\, du\right\}\Big]\, ds.
\end{aligned} \tag{4.63}$$

Proof. Note that (4.61), for example, can be rewritten

$$f(X(t \wedge \tau), y) - \int_0^t \chi_{\{s \le \tau\}} g(X(s), y)\, ds. \tag{4.64}$$

The proof of (4.63) is then essentially the same as the proof of (4.54). □

4.15 Corollary Under the conditions of Corollary 4.14, if $g(x, y) + \alpha(x)f(x, y) \equiv h(x, y) + \beta(y)f(x, y)$, then for all $t \ge 0$,

$$\begin{aligned} &E\left[f(X(t \wedge \tau), Y(0)) \exp\left\{\int_0^{t \wedge \tau} \alpha(X(u))\, du\right\}\right] \\ &\quad - E\left[f(X(0), Y(t \wedge \sigma)) \exp\left\{\int_0^{t \wedge \sigma} \beta(Y(u))\, du\right\}\right] \\ &= \int_0^t E\left[(\chi_{\{s \le \tau\}} - \chi_{\{t-s \le \sigma\}})(g(X(s \wedge \tau), Y((t - s) \wedge \sigma)) \right. \\ &\quad + \alpha(X(s \wedge \tau))f(X(s \wedge \tau), Y((t - s) \wedge \sigma))) \\ &\quad \left. \times \exp\left\{\int_0^{s \wedge \tau} \alpha(X(u))\, du + \int_0^{(t-s) \wedge \sigma} \beta(Y(u))\, du\right\}\right] ds. \end{aligned} \tag{4.65}$$

4.16 Remark As $\tau, \sigma \to \infty$ in (4.65), the integrand on the right goes to zero. The difficulty in practice is to justify the interchange of limits and expectations. □

5. THE MARTINGALE PROBLEM: EXISTENCE

In this section we are concerned with the existence of solutions of a martingale problem, in particular with the existence of solutions that are Markov or strong Markov. As a part of this discussion, we also examine the structure of the set $\mathcal{M}_A$ of all solutions of the $D_E[0, \infty)$ martingale problem for a given A, considered as a subset of $\mathcal{P}(D_E[0, \infty))$. One of the simplest ways of obtaining solutions is as weak limits of solutions of approximating martingale problems, as indicated by the following lemma.

5.1 Lemma Let $A \subset \bar{C}(E) \times \bar{C}(E)$ and let $A_n \subset B(E) \times B(E)$, $n = 1, 2, \ldots$. Suppose that for each $(f, g) \in A$, there exist $(f_n, g_n) \in A_n$ such that

$$\lim_{n \to \infty} \| f_n - f \| = 0, \quad \lim_{n \to \infty} \| g_n - g \| = 0. \tag{5.1}$$

If for each n, X_n is a solution of the martingale problem for A_n with sample paths in $D_E[0, \infty)$, and if $X_n \Rightarrow X$, then X is a solution of the martingale problem for A.

5.2 Remark Suppose that (E, r) is complete and separable and that $\overline{\mathscr{D}(A)}$ contains an algebra that separates points and vanishes nowhere (and hence is dense in $\bar{C}(E)$ in the topology of uniform convergence on compact sets). Then $\{X_n\}$ is relatively compact if (5.1) holds for each $(f, g) \in A$ and if for every $\varepsilon, T > 0$, there exists a compact $K_{\varepsilon, T} \subset E$ such that

$$\inf_n P\{X_n(t) \in K_{\varepsilon, T} \quad \text{for} \quad 0 \le t \le T\} \ge 1 - \varepsilon. \tag{5.2}$$

See Theorems 9.1 and 9.4 of Chapter 3. □

Proof. Let $0 \le t_i \le t < s$, $t_i, t, s \in D(X) \equiv \{u: P\{X(u) = X(u-)\} = 1\}$, and $h_i \in \bar{C}(E)$, $i = 1, \ldots, k$. Then for $(f, g) \in A$ and $(f_n, g_n) \in A_n$ satisfying (5.1),

$$\begin{aligned} E\left[\left(f(X(s)) - f(X(t)) - \int_t^s g(X(u))\, du\right) \prod_{i=1}^k h_i(X(t_i))\right] \\ = \lim_{n \to \infty} E\left[\left(f_n(X_n(s)) - f_n(X_n(t)) - \int_t^s g_n(X_n(u))\, du\right) \prod_{i=1}^k h_i(X_n(t_i))\right] \\ = 0. \end{aligned} \tag{5.3}$$

By Lemma 7.7 of Chapter 3 and the right continuity of X, the equality holds for all $0 \le t_i \le t < s$, and hence X is a solution of the martingale problem for A. □

We now give conditions under which we can approximate A as in Lemma 5.1.

5.3 Lemma Let E be compact and let A be a dissipative linear operator on $C(E)$ such that $\mathscr{D}(A)$ is dense in $C(E)$ and $(1, 0) \in A$. Then there exists a sequence $\{T_n\}$ of positive contraction operators on $B(E)$ given by transition functions such that

$$\lim_{n \to \infty} n(T_n - I)f = Af, \qquad f \in \mathscr{D}(A). \tag{5.4}$$

Proof. Note that

$$\Lambda f \equiv n(n - A)^{-1} f(x) \tag{5.5}$$

defines a bounded linear functional on $\mathscr{R}(n - A)$ for each $n \ge 1$ and $x \in E$. Since $\Lambda 1 = 1$ and $|\Lambda f| \le \|f\|$, for $f \ge 0$,

$$\|f\| - \Lambda f = \Lambda(\|f\| - f) \le \|f\|, \tag{5.6}$$

and hence $\Lambda f \ge 0$. Consequently, Λ is a positive linear functional on $\mathscr{R}(n - A)$ with $\|\Lambda\| = 1$. By the Hahn–Banach theorem Λ extends to a positive linear functional of norm 1 on all of $C(E)$ and hence by the Riesz representation theorem there exists a measure $\mu \in \mathscr{P}(E)$ (not necessarily unique) such that

$$(5.7) \qquad \Lambda f = \int f \, d\mu \quad \text{for all} \quad f \in \mathscr{R}(n - A).$$

Consequently, the set

$$(5.8) \qquad M_x^n \equiv \left\{\mu \in \mathscr{P}(E): n(n - A)^{-1}f(x) = \int f \, d\mu \quad \text{for all} \quad f \in \mathscr{R}(n - A)\right\}$$

is nonempty for each $n \geq 1$ and $x \in E$.

If $\lim_{k \to \infty} x_k = x$ and $\mu_k \in M_{x_k}^n$, then by the compactness of $\mathscr{P}(E)$ (Theorem 2.2 of Chapter 3), there exists a subsequence of $\{\mu_k\}$ that converges in the Prohorov metric to a measure $\mu_\infty \in \mathscr{P}(E)$. Since for all $f \in \mathscr{R}(n - A)$,

$$(5.9) \qquad \int f \, d\mu_\infty = \lim_{k \to \infty} \int f \, d\mu_k = \lim_{k \to \infty} n(n - A)^{-1}f(x_k) = n(n - A)^{-1}f(x),$$

$\mu_\infty \in M_x^n$ and the conditions of the measurable selection theorem (Appendix 10) hold for the mapping $x \to M_x^n$. Consequently, there exist $\mu_x^n \in M_x^n$ such that the mapping $x \to \mu_x^n$ is a measurable function from E into $\mathscr{P}(E)$. It follows that $\mu_n(x, \Gamma) \equiv \mu_x^n(\Gamma)$ is a transition function, and hence

$$(5.10) \qquad T_n f(x) \equiv \int f(y)\mu_n(x, dy)$$

is a positive contraction operator on $B(E)$.

It remains to verify (5.4). For $f \in \mathscr{D}(A)$,

$$(5.11) \qquad \begin{aligned} T_n f &= T_n\left(f - \frac{1}{n} Af\right) + \frac{1}{n} T_n Af \\ &= f + \frac{1}{n} T_n Af \end{aligned}$$

and hence

$$(5.12) \qquad \lim_{n \to \infty} T_n f = f.$$

Since $\mathscr{D}(A)$ is dense in $C(E)$ it follows that (5.12) holds for all $f \in C(E)$. Therefore for $f \in \mathscr{D}(A)$,

$$(5.13) \qquad \lim_{n \to \infty} n(T_n - I)f = \lim_{n \to \infty} T_n Af = Af,$$

since $Af \in C(E)$. □

If E is locally compact and separable, then we can apply Lemma 5.1 to obtain existence of solutions of the martingale problem for a large class of operators.

5.4 Theorem Let E be locally compact and separable, and let A be a linear operator on $\hat{C}(E)$. Suppose $\mathscr{D}(A)$ is dense in $\hat{C}(E)$ and A satisfies the positive maximum principle (i.e., conditions (a) and (b) of Theorem 2.2 are satisfied). Define the linear operator A^{Δ} on $C(E^{\Delta})$ by

$$(A^{\Delta}f)|_E = A((f - f(\Delta))|_E), \qquad A^{\Delta}f(\Delta) = 0, \tag{5.14}$$

for all $f \in C(E^{\Delta})$ such that $(f - f(\Delta))|_E \in \mathscr{D}(A)$. Then for each $v \in \mathscr{P}(E^{\Delta})$, there exists a solution of the martingale problem for (A^{Δ}, v) with sample paths in $D_{E^{\Delta}}[0, \infty)$.

5.5 Remark If A^{Δ} satisfies the conditions of Theorem 3.8 (with $\hat{E} = E^{\Delta}$) and $v(E) = 1$, then the above solution of the martingale problem for (A^{Δ}, v) will have sample paths in $D_E[0, \infty)$. In particular, this will be the case if E is compact and $(1, 0) \in A$. □

Proof. Note that $\overline{\mathscr{D}(A^{\Delta})} = C(E^{\Delta})$ and that $f(x_0) = \sup_y f(y) \geq 0$ implies $f(x_0) - f(\Delta) = \sup_y f(y) - f(\Delta) \geq 0$ so $A^{\Delta}f(x_0) = A(f - f(\Delta))(x_0) \leq 0$. (If $x_0 = \Delta$ then $A^{\Delta}f(x_0) = 0$ by definition.) Since $A^{\Delta}1 = A0 = 0$, A^{Δ} satisfies the conditions of Lemma 5.3 and there exists a sequence of transition functions $\mu_n(x, \Gamma)$ on $E^{\Delta} \times \mathscr{B}(E^{\Delta})$ such that

$$A_n f \equiv n \int (f(y) - f(\cdot))\mu_n(\cdot, dy), \qquad f \in B(E^{\Delta}), \tag{5.15}$$

satisfies

$$\lim_{n \to \infty} A_n f = A^{\Delta}f \quad \text{for all} \quad f \in \mathscr{D}(A^{\Delta}). \tag{5.16}$$

For every $v \in \mathscr{P}(E^{\Delta})$ the martingale problem for (A_n, v) has a solution (a Markov jump process) and hence by Lemma 5.1 and Remark 5.2 there exists a solution of the martingale problem for (A, v). □

We now consider the question of the existence of a Markov process that solves a given martingale problem.

Throughout the remainder of this section, X denotes the coordinate process on $D_E[0, \infty)$, $\mathscr{T}$ a collection of nonnegative, bounded, Borel measurable functions on $D_E[0, \infty)$ containing all nonnegative constants, and

$$\mathscr{F}_t = \mathscr{F}_t^X \vee \sigma(\{\tau \leq s\} : s \leq t, \tau \in \mathscr{T}). \tag{5.17}$$

Note that all $\tau \in \mathscr{T}$ are $\{\mathscr{F}_t\}$-stopping times.

Let $\Gamma \subset \mathscr{P}(D_E[0, \infty))$ and for each $v \in \mathscr{P}(E)$ let $\Gamma_v = \{P \in \Gamma : PX(0)^{-1} = v\}$. Assume $\Gamma_v \neq \varnothing$ for each v and for $f \in B(E)$ define

$$\gamma(\Gamma_v, f) = \sup_{P \in \Gamma_v} E^P\left[\int_0^{\infty} e^{-t} f(X(t))\, dt\right]. \tag{5.18}$$

The following lemma gives an important relationship between γ and the martingale problem.

5.6 Lemma Suppose $f, g \in B(E)$ and

$$f(X(t)) - \int_0^t g(X(s))\, ds \tag{5.19}$$

is an $\{\mathscr{F}_t\}$-martingale for all $P \in \Gamma_\nu$. Then

$$\gamma(\Gamma_\nu, f - g) = \int f\, d\nu = E^P\left[\int_0^\infty e^{-t}(f(X(t)) - g(X(t)))\, dt\right] \tag{5.20}$$

for all $P \in \Gamma_\nu$.

Proof. This is immediate from Lemma 3.2. □

We are interested in the following possible conditions on Γ and $\mathscr{T}$.

5.7 Conditions

C_1: For $P \in \Gamma$, $\tau \in \mathscr{T}$, $\mu = PX(\tau)^{-1}$, and $P' \in \Gamma_\mu$, there exists $\hat{Q} \in \mathscr{P}(D_E[0, \infty) \times [0, \infty))$ with marginal $Q \in \Gamma$ such that

$$\begin{aligned} &E^{\hat{Q}}[\chi_B(X(\cdot \wedge \eta), \eta)\chi_C(X(\eta + \cdot))] \\ &\quad = \int E^P[\chi_B(X(\cdot \wedge \tau), \tau) \mid X(\tau) = x]E^{P'}[\chi_C(X(\cdot)) \mid X(0) = x]\mu(dx) \end{aligned} \tag{5.21}$$

for all $B \in \mathscr{S}_E \times \mathscr{B}[0, \infty)$ and $C \in \mathscr{S}_E$, where (X, η) denotes the coordinate random variable on $D_E[0, \infty) \times [0, \infty)$. (Note there can be at most one such $\hat{Q}$.)

C_2: For $P \in \Gamma$, $\tau \in \mathscr{T}$, and $H \geq 0$ $\mathscr{F}_\tau$-measurable with $0 < E^P[H] < \infty$, the measure $Q \in \mathscr{P}(D_E[0, \infty))$ defined by

$$Q(C) = \frac{E^P[H\chi_C(X(\tau + \cdot))]}{E^P[H]} \tag{5.22}$$

is in Γ.

C_3: Γ is convex.

C_4: For each $h \in \bar{C}(E)$ such that $h \geq 0$, there is a $u \in B(E)$ such that $\gamma(\Gamma_\nu, h) = \int u\, d\nu$ for all $\nu \in \mathscr{P}(E)$.

C_5: Γ_ν is compact for all $\nu \in \mathscr{P}(E)$.

We also use a stronger version of C_5 and a condition that is implied by C_2.

C_2': For $\nu, \mu_1, \mu_2 \in \mathscr{P}(E)$ such that $\nu = \alpha\mu_1 + (1 - \alpha)\mu_2$ for some $\alpha \in (0, 1)$ and for $P \in \Gamma_\nu$, there exist $Q_1 \in \Gamma_{\mu_1}$ and $Q_2 \in \Gamma_{\mu_2}$ such that $P = \alpha Q_1 + (1 - \alpha)Q_2$.

C$'_5$: $\bigcup_{\nu \in V} \Gamma_\nu$ is compact for all compact $V \subset \mathscr{P}(E)$.

5.8 Lemma Condition C_2 implies C'_2.

Proof. Let $h_1 = d\mu_1/d\nu$ and $h_2 = d\mu_2/d\nu$, and note that $\alpha h_1 + (1 - \alpha)h_2 = 1$. Then setting $H_i = h_i(X(0))$, $i = 1, 2$,

$$Q_i(C) = \frac{E^P[H_i \chi_C(X)]}{E^P[H_i]} = E^P[H_i \chi_C(X)] \tag{5.23}$$

is in Γ_{μ_i}, and $P = \alpha Q_1 + (1 - \alpha)Q_2$. □

Condition C'_5 is important because of its relationship to C_4. To see this we make use of the following lemma.

5.9 Lemma Let E be separable. Let $\varphi: \mathscr{P}(E) \to [0, c]$ for some $c > 0$. Suppose φ satisfies

$$\varphi(\alpha\mu_1 + (1 - \alpha)\mu_2) = \alpha\varphi(\mu_1) + (1 - \alpha)\varphi(\mu_2) \tag{5.24}$$

for $\alpha \in (0, 1)$ and $\mu_1, \mu_2 \in \mathscr{P}(E)$ and that φ is upper semicontinuous in the sense that $\nu_n, \nu \in \mathscr{P}(E)$, $\nu_n \Rightarrow \nu$ implies

$$\overline{\lim_{n\to\infty}}\, \varphi(\nu_n) \leq \varphi(\nu). \tag{5.25}$$

Then there exists $u \in B(E)$ such that

$$\varphi(\nu) = \int u \, d\nu, \qquad \nu \in \mathscr{P}(E). \tag{5.26}$$

Proof. By (5.25), $u(x) \equiv \varphi(\delta_x)$ is upper semicontinuous and hence measurable ($\{x: u(x) < a\}$ is open). Let E_i^n, $i = 1, 2, \ldots$, be disjoint with diameter less than $1/n$ and $E = \bigcup_i E_i^n$; let $x_i^n \in E_i^n$ satisfy $u(x_i^n) \geq \sup_{x \in E_i^n} u(x) - 1/n$. Fix ν and define $u_n \in B(E)$ by

$$u_n(x) = \sum_i u(x_i^n)\chi_{E_i^n}(x) \tag{5.27}$$

and $\nu_n \in \mathscr{P}(E)$ by

$$\nu_n = \sum_i \nu(E_i^n)\delta_{x_i^n}. \tag{5.28}$$

Then bp-$\lim_{n\to\infty} u_n = u$ and $\nu_n \Rightarrow \nu$. Consequently,

$$\int u \, d\nu = \lim_{n\to\infty} \int u_n \, d\nu = \lim_{n\to\infty} \int u \, d\nu_n = \lim_{n\to\infty} \varphi(\nu_n) \leq \varphi(\nu). \tag{5.29}$$

To obtain the inequality in the other direction (and hence (5.26)) let $\mu_i^n(B) = \nu(B \cap E_i^n)/\nu(E_i^n)$ when $\nu(E_i^n) > 0$ and $v_n(x) = \sum \varphi(\mu_i^n)\chi_{E_i^n}(x)$. Note that $\lim_{n\to\infty} v_n(x) \le u(x)$ by (5.25), and hence

(5.30)
$$\varphi(\nu) = \sum_i \varphi(\mu_i^n)\nu(E_i^n) = \int v_n\, d\nu$$
$$= \lim_{n\to\infty} \int v_n\, d\nu \le \int \overline{\lim_{n\to\infty}}\, v_n\, d\nu \le \int u\, d\nu.$$

(Note $v_n \le c$.) □

5.10 Lemma Let (E, r) be complete and separable. Suppose conditions C_2' and C_3 hold. Then for $h \in B(E)$ with $h \ge 0$,

(5.31)
$$\gamma(\Gamma_{\alpha\mu_1 + (1-\alpha)\mu_2}, h) = \alpha\gamma(\Gamma_{\mu_1}, h) + (1 - \alpha)\gamma(\Gamma_{\mu_2}, h)$$

for all $\alpha \in (0, 1)$ and $\mu_1, \mu_2 \in \mathscr{P}(E)$. If, in addition, C_5' holds, then C_4 holds.

Proof. Condition C_2' implies the right side of (5.31) is greater than or equal to the left while C_3 implies the reverse inequality. If C_5' holds, then for ν_n, $\nu \in \mathscr{P}(E)$, $\nu_n \Rightarrow \nu$, we have $\Gamma_\nu \cup \bigcup_n \Gamma_{\nu_n}$ compact. Consequently, every sequence $P_n \in \Gamma_{\nu_n}$ has a subsequence that converges weakly to some $P \in \Gamma_\nu$. Since, for $h \in \bar{C}(E)$, $\int_0^\infty e^{-t}h(X(t))\, dt$ is continuous on $D_E[0, \infty)$, it follows that

(5.32)
$$\overline{\lim_{n\to\infty}}\, \gamma(\Gamma_{\nu_n}, h) \le \gamma(\Gamma_\nu, h).$$

C_4 now follows by Lemma 5.9. □

Let A_0 be the collection of all pairs $(f, g) \in B(E) \times B(E)$ for which (5.19) is an $\{\mathscr{F}_t\}$- martingale for all $P \in \Gamma$. Our goal is to produce, under conditions C_1–C_5, an extension A of A_0 satisfying the conditions of Theorem 4.1 such that for each $\nu \in \mathscr{P}(E)$, there exists in Γ_ν a solution (necessarily unique) of the martingale problem for (A, ν). The solution will then be a Markov process by Theorem 4.2. Of course typically one begins with an operator A_0 and seeks a set of solutions Γ rather than the reverse. Therefore, to motivate our consideration of C_1–C_5 we first prove the following theorem.

5.11 Theorem Let (E, r) be complete and separable.

(a) Let $A \subset B(E) \times B(E)$, let $\Gamma = \mathscr{M}_A$ (recall that $\mathscr{M}_A$ is the collection of all solutions of the $D_E[0, \infty)$ martingale problem for A), and let $\mathscr{T}$ be the collection of nonnegative constants. Suppose $\Gamma_\nu \ne \emptyset$ for all $\nu \in \mathscr{P}(E)$. Then C_1–C_3 hold.

(b) Let $A \subset \bar{C}(E) \times \bar{C}(E)$, and let $\Gamma = \mathscr{M}_A$ and $\mathscr{T} = \{\tau\colon \{\tau < t\} \in \mathscr{F}_t^X$ for all $t \ge 0$, τ bounded$\}$. Suppose $\mathscr{D}(A)$ contains an algebra that separates points and vanishes nowhere, and suppose for each compact $K \subset E$, $\varepsilon > 0$, and $T > 0$ there exists a compact $K' \subset E$ such that

$$(5.33) \qquad P\{X(t) \in K' \quad \text{for all} \quad t < T,\ X(0) \in K\} \geq (1 - \varepsilon)P\{X(0) \in K\} \quad \text{for all} \quad P \in \Gamma.$$

Then C_1–C_5 and C_5' hold.

(c) In addition to the assumptions of part (b), suppose the $D_E[0, \infty)$ martingale problem for A is well-posed. Then the solutions are Markov processes corresponding to a semigroup that maps $\bar{C}(E)$ into $\bar{C}(E)$.

5.12 Remark Part (b) is the result of primary interest. Before proving the theorem we give a lemma that may be useful in verifying condition (5.33). Of course if E is compact, (5.33) is immediate, and if E is locally compact with $A \subset \hat{C}(E) \times \hat{C}(E)$, one can replace E by its one-point compactification E^Δ and consider the corresponding martingale problem in $D_{E^\Delta}[0, \infty)$. □

5.13 Lemma Let (E, r) be complete, and let $A \subset \bar{C}(E) \times B(E)$. Suppose for each compact $K \subset E$ and $\eta > 0$ there exists a sequence of compact $K_n \subset E$, $K \subset K_n$, and $(f_n, g_n) \in A$ such that for $F_n \equiv \{z: \inf_{x \in K_n} r(x, z) \leq \eta\}$,

$$(5.34) \qquad \beta_{n,\eta} \equiv \inf_{y \in K} f_n(y) - \sup_{y \in E - F_n} f_n(y) > 0,$$

$$(5.35) \qquad \lim_{n \to \infty} \beta_{n,\eta}^{-1} \sup_{y \in F_n} g_n^-(y) = 0,$$

and

$$(5.36) \qquad \lim_{n \to \infty} \beta_{n,\eta}^{-1}(\|f_n\| - \inf_{y \in K} f_n(y)) = 0.$$

Then for each compact $K \subset E$, $\varepsilon > 0$, and $T > 0$, there exists a compact $K' \subset E$ such that

$$(5.37) \qquad P\{X(t) \in K' \quad \text{for all} \quad t < T,\ X(0) \in K\} \geq (1 - \varepsilon)P\{X(0) \in K\},$$

for all $P \in \mathcal{M}_A$.

5.14 Example Let $E = \mathbb{R}^d$ and $A = \{(f, Gf): f \in C_c^\infty(\mathbb{R}^d)\}$ where

$$(5.38) \qquad Gf = \frac{1}{2} \sum_{i,j} a_{ij}\, \partial_i\, \partial_j f + \sum_i b_i\, \partial_i f$$

and the a_{ij} and b_i are measurable functions satisfying $|a_{ij}(x)| \leq M(1 + |x|^2)$ and $|b_i(x)| \leq M(1 + |x|)$ for some $M > 0$. For compact $K \subset B(0, k) \equiv \{z \in \mathbb{R}^d: |z| < k\}$ and $\eta > 0$, let $K_n = \overline{B(0, k + n)}$ and let $f_n \in C_c^\infty(\mathbb{R}^d)$ satisfy

$$f_n(x) = 1 + \log(1 + (k + n + \eta)^2) - \log(1 + |x|^2)$$

for $|x| \leq k + n + \eta$ and $0 \leq f_n(x) \leq 1$ for $|x| > k + n + \eta$. The calculations are left to the reader. □

Proof. Given $T > 0$, a compact $K \subset E$, and $\eta > 0$, let F_n be as hypothesized, and define $\tau_n = 0$ if $X(0) \notin K$ and $\tau_n = \inf\{t: X(t) \notin F_n\}$ otherwise. Then for $P \in \mathcal{M}_A$,

$$E^P[f_n(X(0)) - f_n(X(\tau_n \wedge T))] = E^P\left[-\int_0^{\tau_n \wedge T} g_n(X(s))\, ds\right], \tag{5.39}$$

and hence

$$\begin{aligned}\beta_{n,\eta} P\{0 < \tau_n \le T\} &\le E^P[(f_n(X(0)) - f_n(X(\tau_n)))\chi_{\{0 < \tau_n \le T\}}] \\ &= E^P\left[-\int_0^{\tau_n \wedge T} g_n(X(s))\, ds\right] \\ &\quad + E^P[(f_n(X(T)) - f_n(X(0)))\chi_{\{\tau_n > T\}}] \\ &\le TP\{X(0) \in K\} \sup_{y \in F_n} g_n^-(y) \\ &\quad + (\| f_n\| - \inf_{y \in K} f_n(y))P\{X(0) \in K\},\end{aligned} \tag{5.40}$$

which gives

$$\begin{aligned}P\{X(t) \in F_n \quad &\text{for all} \quad t \le T,\ X(0) \in K\} \\ &= P\{X(0) \in K\} - P\{0 < \tau_n \le T\} \\ &\ge P\{X(0) \in K\}\left(1 - \beta_{n,\eta}^{-1}\left(T \sup_{y \in F_n} g_n^-(y) + \|f_n\| - \inf_{y \in K} f_n(y)\right)\right).\end{aligned} \tag{5.41}$$

From (5.41), (5.35), and (5.36), it follows that for each $m > 0$ there exists a compact $\hat{K}_m \subset E$ such that

$$\begin{aligned}P\{X(t) \in \hat{K}_m^{1/m} \quad \text{for all} \quad t \le T,\ X(0) \in K\} \\ \ge P\{X(0) \in K\}(1 - \varepsilon 2^{-m}).\end{aligned} \tag{5.42}$$

Hence taking K' to be the closure of $\bigcap_m \hat{K}_m^{1/m}$, we have (5.37). □

In order to be able to verify C_1 we need the following technical lemmas.

5.15 Lemma Let (E, r), (S_1, ρ_1), and (S_2, ρ_2) be complete, separable metric spaces, let $P_1 \in \mathcal{P}(S_1)$ and $P_2 \in \mathcal{P}(S_2)$ and suppose that $X_1: S_1 \to E$ and $X_2: S_2 \to E$ are Borel measurable and that $\mu \in \mathcal{P}(E)$ satisfies $\mu = P_1 X_1^{-1} =$

$P_2 X_2^{-1}$. Let $\{B_i^m\} \subset \mathscr{B}(E)$, $m = 1, 2, \ldots$, be a sequence of countable partitions of E with $\{B_i^{m+1}\}$ a refinement of $\{B_i^m\}$ and $\lim_{m \to \infty} \sup_i$ diameter $(B_i^m) = 0$. Define $P^m \in \mathscr{P}(S_1 \times S_2)$ by

$$P^m(C) = \sum_i \frac{E^{P_1 \times P_2}[\chi_C \chi_{B_i^m}(X_1)\chi_{B_i^m}(X_2)]}{\mu(B_i^m)} \tag{5.43}$$

for $C \in \mathscr{B}(S_1 \times S_2)$. Then $\{P^m\}$ converges weakly to a probability measure $P \in \mathscr{P}(S_1 \times S_2)$ satisfying

$$P(A_1 \times A_2) = \int E^{P_1}[\chi_{A_1} | X_1 = x] E^{P_2}[\chi_{A_2} | X_2 = x] \mu(dx) \tag{5.44}$$

for $A_1 \in \mathscr{B}(S_1)$ and $A_2 \in \mathscr{B}(S_2)$. In particular $P(A_1 \times S_2) = P_1(A_1)$ and $P(S_1 \times A_2) = P_2(A_2)$. More generally, if $Z_k \in B(S_k)$, $k - 1, 2$, then

$$E^P[Z_1 Z_2] = \int E^{P_1}[Z_1 | X_1 = x] E^{P_2}[Z_2 | X_2 = x] \mu(dx). \tag{5.45}$$

Proof. For $k = 1, 2$, let $A_k \in \mathscr{B}(S_k)$. Note that $E^{P_k}[\chi_{A_k} | X_k = x]$ is the unique (μ-a.s.) $\mathscr{B}(E)$-measurable function satisfying

$$\int_B E^{P_k}[\chi_{A_k} | X_k = x] \mu(dx) = E^{P_k}[\chi_{A_k} \chi_B(X_k)] \tag{5.46}$$

for all $B \in \mathscr{B}(E)$.

By the martingale convergence theorem (Problem 26 of Chapter 2),

$$\lim_{m \to \infty} \sum_i \frac{E^{P_k}[\chi_{A_k} \chi_{B_i^m}(X_k)] \chi_{B_i^m}(x)}{\mu(B_i^m)} = E^{P_k}[\chi_{A_k} | X_k = x] \tag{5.47}$$

μ-a.s. and in $L^2(\mu)$. Consequently,

$$\lim_{m \to \infty} P^m(A_1 \times A_2) = P(A_1 \times A_2) \tag{5.48}$$

($P(A_1 \times A_2)$ given by (5.44)), and since at most one $P \in \mathscr{P}(S_1 \times S_2)$ can satisfy (5.44), it suffices to show that $\{P^m\}$ is tight (cf. Lemma 4.3 of Chapter 3). Let $\varepsilon > 0$, and let K_1 and K_2 be compact subsets of S_1 and S_2 such that $P_k(K_k) \geq 1 - \varepsilon^2$. Then, since $P_k(K_k) = \int E[\chi_{K_k} | X_k = x] \mu(dx)$,

$$\mu\{x: E[\chi_{K_k} | X_k = x] \leq 1 - \varepsilon\} \leq \varepsilon \tag{5.49}$$

and

$$P(K_1 \times K_2) \geq (1 - \varepsilon)^2 (1 - 2\varepsilon). \tag{5.50}$$

Tightness for $\{P^m\}$ now follows easily from (5.48). □

5.16 Lemma Let (E, r) be complete and separable, and let $A \subset B(E) \times B(E)$. Suppose for each $\nu \in \mathscr{P}(E)$ there exists a solution of the martingale problem for (A, ν) with sample paths in $D_E[0, \infty)$. Let Z be a process with sample paths in $D_E[0, \infty)$ and let τ be a $[0, \infty]$-valued random variable. Suppose, for $(f, g) \in A$, that

$$(5.51) \qquad f(Z(t \wedge \tau)) - \int_0^{t \wedge \tau} g(Z(s))\, ds$$

is a martingale with respect to $\mathscr{G}_t = \sigma(Z(s \wedge \tau), s \wedge \tau \colon s \le t)$. If τ is discrete or if $\mathscr{D}(A) \subset \bar{C}(E)$, then there exists a solution Y of the martingale problem for A with sample paths in $D_E[0, \infty)$ and a $[0, \infty]$-valued random variable η such that $(Y(\cdot \wedge \eta), \eta)$ has the same distribution as $(Z(\cdot \wedge \tau), \tau)$.

Proof. Let $P_1 \in \mathscr{P}(D_E[0, \infty) \times [0, \infty])$ denote the distribution of (Z, τ) and $\mu \in \mathscr{P}(E)$ the distribution of $Z(\tau)$ (fix $x_0 \in E$ and set $Z(\tau) = x_0$ on $\{\tau = \infty\}$). Let $P_2 \in \mathscr{P}(D_E[0, \infty))$ be a solution of the martingale problem for (A, μ). By Lemma 5.15 there exists $Q \in \mathscr{P}(D_E[0, \infty) \times [0, \infty] \times D_E[0, \infty))$ such that, for $B \in \mathscr{S}_E \times \mathscr{B}[0, \infty]$ and $C \in \mathscr{S}_E$,

$$(5.52) \qquad \begin{aligned} Q(B \times C) &= \int E^{P_1}[\chi_B(X, \eta) \mid X(\eta) = x] E^{P_2}[\chi_C(X) \mid X(0) = x] \mu(dx) \\ &= \int E[\chi_B(Z, \tau) \mid Z(\tau) = x] E^{P_2}[\chi_C(X) \mid X(0) = x] \mu(dx) \end{aligned}$$

where (X, η) denotes the coordinate random variable on $D_E[0, \infty) \times [0, \infty]$. Let (X_1, η, X_2) denote the coordinate random variable on $\Omega = D_E[0, \infty) \times [0, \infty] \times D_E[0, \infty)$ and define

$$(5.53) \qquad Y(t) = \begin{cases} X_1(t) & \text{for } t < \eta \\ X_2(t - \eta) & \text{for } t \ge \eta. \end{cases}$$

Note that on $(\Omega, \mathscr{B}(\Omega), Q)$, $Y(\cdot \wedge \eta) = X_1(\cdot \wedge \eta)$ has the same distribution as $Z(\cdot \wedge \tau)$. It remains to show that Y is a solution of the martingale problem for A.

With reference to (3.4), let $(f, g) \in A$, $h_k \in \bar{C}(E)$, $t_1 < t_2 < t_3 < \cdots < t_{n+1}$ and define

$$(5.54) \qquad R = \left(f(Y(t_{n+1})) - f(Y(t_n)) - \int_{t_n}^{t_{n+1}} g(Y(s))\, ds \right) \prod_{k=1}^{n} h_k(Y(t_k)).$$

We must show $E^Q[R] = 0$. Note that

(5.55)

$$R = \left(f(Y(t_{n+1} \wedge \eta)) - f(Y(t_n \wedge \eta)) - \int_{t_n \wedge \eta}^{t_{n+1} \wedge \eta} g(Y(s))\, ds \right) \prod_{k=1}^{n} h_k(Y(t_k))$$

$$+ \left(f(Y(t_{n+1} \vee \eta)) - f(Y(t_n \vee \eta)) - \int_{t_n \vee \eta}^{t_{n+1} \vee \eta} g(Y(s))\, ds \right) \prod_{k=1}^{n} h_k(Y(t_k))$$

$$\equiv R_1 + R_2 .$$

Since R_1 is zero unless $t_n < \eta$, we have

(5.56)
$$E^Q[R_1] = E^Q\left[\left(f(Y(t_{n+1} \wedge \eta)) - f(Y(t_n \wedge \eta)) - \int_{t_n \wedge \eta}^{t_{n+1} \wedge \eta} g(Y(s))\, ds \right) \prod_{k=1}^{n} h_k(Y(t_k \wedge \eta)) \right]$$

$$= E^{P_1}\left[\left(f(X(t_{n+1} \wedge \eta)) - f(X(t_n \wedge \eta)) - \int_{t_n \wedge \eta}^{t_{n+1} \wedge \eta} g(X(s))\, ds \right) \prod_{k=1}^{n} h_k(X(t_k \wedge \eta)) \right]$$

$$= E\left[\left(f(Z(t_{n+1} \wedge \tau)) - f(Z(t_n \wedge \tau)) - \int_{t_n \wedge \tau}^{t_{n+1} \wedge \tau} g(Z(s))\, ds \right) \prod_{k=1}^{n} h_k(Z(t_k \wedge \tau)) \right]$$

$$= 0.$$

It remains to show that $E^Q[R_2] = 0$. Suppose first that $\mathscr{D}(A) \subset \bar{C}(E)$. Define

(5.57)
$$\eta_m = \begin{cases} \dfrac{[m\eta]}{m} & \text{for } \eta < \infty \\ \infty & \text{for } \eta = \infty \end{cases}$$

and

(5.58)
$$R_2^m = \left(f(X_2(t_{n+1} \vee \eta_m - \eta_m)) - f(X_2(t_n \vee \eta_m - \eta_m)) - \int_{t_n \vee \eta_m}^{t_{n+1} \vee \eta_m} g(X_2(s - \eta_m))\, ds \right)$$

$$\times \prod_{t_k \geq \eta_m} h_k(X_2(t_k - \eta_m)) \prod_{t_k < \eta_m} h_k(X_1(t_k)).$$

By the right continuity of X_2 and the continuity of f, as $m \to \infty$ R_2^m converges a.s. to R_2. Noting that $R_2^n = 0$ unless $\eta_m < t_{n+1}$, we have

$$(5.59)\quad E^Q[R_2^m] = \sum_{l < mt_{n+1}} E^Q[R_2^m \chi_{\{\eta_m = l/m\}}]$$

$$= \sum_{l < mt_{n+1}} \int E^{P_2}\left[\left(f\left(X\left(t_{n+1} - \frac{l}{m}\right)\right) - f\left(X\left(t_n \vee \frac{l}{m} - \frac{l}{m}\right)\right)\right.\right.$$

$$\left.\left. - \int_{t_n \vee l/m}^{t_{n+1}} g\left(X\left(s - \frac{l}{m}\right)\right) ds\right) \prod_{t_k \geq l/m} h_k\left(X\left(t_k - \frac{l}{m}\right)\right) \middle| X(0) = x\right]$$

$$\times E^{P_1}\left[\chi_{\{\eta_m = l/m\}} \prod_{t_k < l/m} h_k(X(t_k)) \middle| X(\eta) = x\right] \mu(dx)$$

$$= 0,$$

since P_2 is a solution of the martingale problem for (A, μ). Letting $m \to \infty$, we see that $E^Q[R_2] = 0$.

If $\mathscr{D}(A) \not\subset \bar{C}(E)$ but η is discrete, then $E^Q[R_2] = 0$ by the same argument as in (5.59). □

Proof of Theorem 5.11 (a). **(C_1)** Let $P \in \Gamma$, $\tau \in \mathscr{T}$, $\mu = PX(\tau)^{-1}$, and $P' \in \Gamma_\mu$. In the construction of Q in the proof of Lemma 5.16 take $P_1(B) = P\{(X, \tau) \in B\}$ for $B \in \mathscr{S}_E \times \mathscr{B}[0, \infty]$ and $P_2 = P'$. Then the desired $\hat{Q}$ is the distribution of (Y, η) defined by (5.53) on $(\Omega, \mathscr{B}(\Omega), Q)$. Note that Lemma 5.16 applies under either the conditions of part (a) or of part (b).

(C_2) Let $P \in \Gamma$, $\tau \in \mathscr{T}$, and $H \geq 0$ and $\mathscr{F}_\tau$-measurable with $0 < E^P[H] < \infty$. Define Q by (5.22). Then for $(f, g) \in A$, $h_k \in B(E)$, and $t_1 < t_2 < \cdots < t_{n+1}$,

$$(5.60)\quad E^Q\left[\left(f(X(t_{n+1})) - f(X(t_n)) - \int_{t_n}^{t_{n+1}} g(X(s))\, ds\right) \prod_{k=1}^{n} h_k(X(t_k))\right]$$

$$= \frac{E^P\left[\left(f(X(\tau + t_{n+1})) - f(X(\tau + t_n)) - \int_{\tau + t_n}^{\tau + t_{n+1}} g(X(s))\, ds\right) \prod_{k=1}^{n} h_k(X(\tau + t_k))H\right]}{E^P[H]}$$

$$= 0,$$

since H is $\mathscr{F}_\tau$-measurable and $P \in \mathscr{M}_A$. (Under the assumptions of part (b), the continuity of f allows the application of the optional sampling theorem.)

(C_3) The set of $P \in \mathscr{P}(D_E[0, \infty))$ for which (3.4) holds is clearly convex.

Proof of Theorem 5.11 (b). To complete the proof of part (b) we need only verify C_5' (which implies C_5), since C_4 will then follow from Lemma 5.10.

(C_5') Let $V \subset \mathscr{P}(E)$ be compact. Then for $0 < \varepsilon < 1$ and $T > 0$, by Theorem 2.2 of Chapter 3, there exist compact $K \subset E$ such that $\nu(K) \geq 1 - \varepsilon/2$ for all $\nu \in V$ and (by (5.33)) compact $K_{\varepsilon, T} \subset E$ such that

(5.61)
$$\begin{aligned} P\{X(t) \in K_{\varepsilon, T} \quad &\text{for all} \quad t < T\} \\ &\geq P\{X(t) \in K_{\varepsilon, T} \quad \text{for all} \quad t < T, X(0) \in K\} \\ &\geq \left(1 - \frac{\varepsilon}{2}\right) P\{X(0) \in K\} \geq \left(1 - \frac{\varepsilon}{2}\right)^2 \geq 1 - \varepsilon \end{aligned}$$

for all $P \in \bigcup_{\nu \in V} \Gamma_\nu$. The following lemma completes the proof of C_5' and hence of part (b).

5.17 Lemma Let (E, r) be complete and separable. For $\varepsilon, T > 0$, let $K_{\varepsilon, T} \subset E$ be compact and define $K_{\varepsilon, T}^* = \{x \in D_E[0, \infty): x(t) \in K_{\varepsilon, T}$ for all $t < T\}$. If $A \subset \bar{C}(E) \times B(E)$ and $\mathscr{D}(A)$ contains an algebra that separates points and vanishes nowhere, then

(5.62)
$$\{P \in \mathscr{M}_A: P(K_{\varepsilon, T}^*) \geq 1 - \varepsilon \qquad \text{for all} \quad \varepsilon, T > 0\}$$

is relatively compact. If, in addition, $A \subset \bar{C}(E) \times \bar{C}(E)$, then (5.62) is compact.

Proof. The relative compactness follows from Theorems 9.1 and 9.4 of Chapter 3. If $A \subset \bar{C}(E) \times \bar{C}(E)$, then compactness follows from Lemma 5.1 with $A_n = A$ for all n. Note that $K_{\varepsilon, T}^*$ is closed, and hence $P_n \Rightarrow P$ implies $P(K_{\varepsilon, T}^*) \geq \overline{\lim}_{n \to \infty} P_n(K_{\varepsilon, T}^*)$. □

Proof of Theorem 5.11 (c). Let P_x denote the solution of the $D_E[0, \infty)$ martingale problem for (A, δ_x). By C_5' and uniqueness, P_x is weakly continuous as a function of x, and hence by Theorem 4.2 the solutions are Markov and correspond to a semigroup $\{T(t)\}$. By Theorem 3.12, $P_x\{X(t) = X(t-)\} = 1$ for all t and the weak continuity of P_x implies $T(t): \bar{C}(E) \to \bar{C}(E)$. □

We now give a partial converse to Lemma 5.6, which demonstrates the importance of condition C_4.

5.18 Lemma Let $\Gamma \subset \mathscr{P}(D_E[0, \infty))$ and $\mathscr{T}$ satisfy C_2. Suppose $u, h \in B(E)$ and

(5.63)
$$\gamma(\Gamma_\nu, h) = \int u(x)\, d\nu = E^P\left[\int_0^\infty e^{-t} h(X(t))\, dt\right]$$

for all $P \in \Gamma_\nu$ and $\nu \in \mathscr{P}(E)$. Then for each $P \in \Gamma$,

(5.64)
$$u(X(t)) - \int_0^t (u(X(s)) - h(X(s)))\, ds$$

is an $\{\mathscr{F}_t\}$-martingale.

Proof. Let $P \in \Gamma$, $t \geq 0$, $B \in \mathscr{F}_t$ with $P(B) > 0$, and $\nu(C) = P\{X(t) \in C \mid B\}$ for all $C \in \mathscr{B}(E)$. Then with Q given by (5.22) for $H = \chi_B$,

$$\begin{aligned}(5.65)\qquad e^t \int_B \int_t^\infty e^{-u} h(X(u))\, du\, dP &= \int_B \int_0^\infty e^{-s} h(X(t+s))\, ds\, dP \\ &= P(B) E^Q\left[\int_0^\infty e^{-s} h(X(s))\, ds\right] \\ &= P(B)\gamma(\Gamma_\nu, h) \\ &= P(B) \int u(x)\, d\nu = \int_B u(X(t))\, dP.\end{aligned}$$

Hence

$$\begin{aligned}(5.66)\qquad E^P\left[\int_0^\infty e^{-s} h(X(s))\, ds \,\middle|\, \mathscr{F}_t\right] \\ = e^{-t} u(X(t)) + \int_0^t e^{-s} h(X(s))\, ds.\end{aligned}$$

Since (5.66) is clearly a martingale, the lemma follows from Lemma 3.2. □

5.19 Theorem Let (E, r) be complete and separable. Let $\mathscr{T}$ be a collection of nonnegative, bounded Borel measurable functions on $D_E[0, \infty)$ containing all nonnegative constants, and let $\{\mathscr{F}_t\}$ be given by (5.17). Let $\Gamma \subset \mathscr{P}(D_E[0, \infty))$ and suppose $\Gamma_\nu \neq \varnothing$ for all $\nu \in \mathscr{P}(E)$. Let A_0 be the set of $(f, g) \in B(E) \times B(E)$ such that

$$(5.67)\qquad f(X(t)) - \int_0^t g(X(s))\, ds$$

is an $\{\mathscr{F}_t\}$-martingale for all $P \in \Gamma$. Assuming C_1–C_5, the following hold:

(a) There exists a linear dissipative operator $A \supset A_0$ such that $\mathscr{R}(I - A) = B(E)$ (hence by Lemma 2.3 of Chapter 1, $\mathscr{R}(\lambda - A) = B(E)$ for all $\lambda > 0$) and $\mathscr{D}(A)$ is bp-dense in $B(E)$.

(b) Either Γ_ν is a singleton for all ν or there exists more than one such extension of A_0.

(c) For each $\nu \in \mathscr{P}(E)$ there exists $P_\nu \in \Gamma_\nu$, which is the unique (hence Markovian) solution of the martingale problem for (A, ν), and if P_x is the unique solution for (A, δ_x), then P_x is a measurable function of x and $P_\nu = \int P_x\, \nu(dx)$.

(d) Every solution P of the martingale problem for A satisfies

$$P\{X(\tau + \cdot) \in C \mid \mathscr{F}_\tau\} = P_{X(\tau)}(C) \tag{5.68}$$

for all $C \in \mathscr{S}_E$ and $\tau \in \mathscr{T}$.

Proof. Let $f_1, f_2, \ldots \in \bar{C}(E)$ be nonnegative and suppose the span of $\{f_k\}$ is bp-dense in $B(E)$ (such a sequence always exists by Proposition 4.2 of Chapter 3). Let $\Gamma^{(0)} = \Gamma$ and $\Gamma_\nu^{(0)} = \Gamma_\nu$. Define

$$\Gamma_\nu^{(k+1)} = \left\{P \in \Gamma_\nu^{(k)}: E^P\left[\int_0^\infty e^{-t} f_{k+1}(X(t))\, dt\right] = \gamma(\Gamma_\nu^{(k)}, f_{k+1})\right\} \tag{5.69}$$

for all $\nu \in \mathscr{P}(E)$ and set $\Gamma^{(k)} = \bigcup \Gamma_\nu^{(k)}$. Since $\Gamma_\nu^{(0)}$ is nonempty and compact and $P \to E^P[\int_0^\infty e^{-t} f_1(X(t))\, dt]$ is a continuous function from $\mathscr{P}(D_E[0, \infty))$ to $\mathbb{R}$, it follows that $\Gamma_\nu^{(1)}$ is nonempty and compact and similarly that $\Gamma_\nu^{(k)}$ is nonempty and compact for each k. The key to our proof is the following induction lemma.

5.20 Lemma Fix $k \geq 0$, and let $\Gamma^{(k)}$ be as above. If $\Gamma^{(k)}$ and $\mathscr{T}$ satisfy C_1–C_5, then $\Gamma^{(k+1)}$ and $\mathscr{T}$ satisfy C_1–C_5. (We denote these conditions by $C_i^{(k)}$ and $C_i^{(k+1)}$ as they apply to $\Gamma^{(k)}$ and $\Gamma^{(k+1)}$.)

Proof. Let $\mu \in \mathscr{P}(E)$ and $P \in \Gamma_\mu^{(k+1)}$. For $B \in \mathscr{B}(E)$ with $0 < \mu(B) < 1$, $C_2^{(k)}$ implies

$$\begin{aligned} \gamma(\Gamma_\mu^{(k)}, f_{k+1}) &= E^P\left[\chi_B(X(0)) \int_0^\infty e^{-t} f_{k+1}(X(t))\, dt\right] \\ &\quad + E^P\left[\chi_{B^c}(X(0)) \int_0^\infty e^{-t} f_{k+1}(X(t))\, dt\right] \\ &\leq \mu(B)\gamma(\Gamma_{\mu_1}^{(k)}, f_{k+1}) + \mu(B^c)\gamma(\Gamma_{\mu_2}^{(k)}, f_{k+1}), \end{aligned} \tag{5.70}$$

where $\mu_1(C) = \mu(B \cap C)/\mu(B)$ and $\mu_2(C) = \mu(B^c \cap C)/\mu(B^c)$ for all $C \in \mathscr{B}(E)$, and the inequality holds term by term. But $C_2^{(k)}$, $C_3^{(k)}$ and Lemma 5.10 imply equality holds in (5.70), so by $C_4^{(k)}$ there exists a $u_{k+1} \in B(E)$ such that

$$E^P\left[\chi_B(X(0)) \int_0^\infty e^{-t} f_{k+1}(X(t))\, dt\right] = \int_B u_{k+1}(x)\mu(dx). \tag{5.71}$$

Hence

$$E^P\left[\int_0^\infty e^{-t} f_{k+1}(X(t))\, dt \,\middle|\, X(0) = x\right] = u_{k+1}(x) \quad \mu\text{-a.s.} \tag{5.72}$$

We now verify $C_1^{(k+1)}$–$C_5^{(k+1)}$.

($C_1^{(k+1)}$) For $P \in \Gamma_\nu^{(k+1)} \subset \Gamma_\nu^{(k)}$, $\tau \in \mathscr{T}$, $\mu = PX(\tau)^{-1}$, and $P' \in \Gamma_\mu^{(k+1)} \subset \Gamma_\mu^{(k)}$, there exists $\hat{Q}$ with marginal $Q \in \Gamma_\nu^{(k)}$ such that (5.21) holds. We must show $Q \in \Gamma_\nu^{(k+1)}$. Let $\mu^*(dx) = (E^P[e^{-\tau} \mid X(\tau) = x]/E^P[e^{-\tau}])\mu(dx)$. Then, using (5.21), (5.72), and $C_4^{(k)}$,

$$(5.73)\quad E^Q\left[\int_0^\infty e^{-t} f_{k+1}(X(t))\,dt\right] = E^{\hat{Q}}\left[\int_0^\eta e^{-t} f_{k+1}(X(t))\,dt\right]$$

$$+ E^{\hat{Q}}\left[e^{-\eta}\int_0^\infty e^{-t} f_{k+1}(X(\eta + t))\,dt\right]$$

$$= E^P\left[\int_0^\tau e^{-t} f_{k+1}(X(t))\,dt\right]$$

$$+ \int E^P[e^{-\tau} \mid X(\tau) = x] E^{P'}\left[\int_0^\infty e^{-t} f_{k+1}(X(t))\,dt \,\middle|\, X(0) = x\right]\mu(dx)$$

$$= E^P\left[\int_0^\tau e^{-t} f_{k+1}(X(t))\,dt\right] + E^P[e^{-\tau}]\int u_{k+1}(x)\mu^*(dx)$$

$$= E^P\left[\int_0^\tau e^{-t} f_{k+1}(X(t))\,dt\right] + E^P[e^{-\tau}]\gamma(\Gamma_{\mu^*}^{(k)}, f_{k+1}).$$

By $C_2^{(k)}$ there exists $P'' \in \Gamma_{\mu^*}^{(k)}$ such that

$$(5.74)\quad \gamma(\Gamma_\nu^{(k)}, f_{k+1}) = E^P\left[\int_0^\infty e^{-t} f_{k+1}(X(t))\,dt\right]$$

$$= E^P\left[\int_0^\tau e^{-t} f_{k+1}(X(t))\,dt\right]$$

$$+ E^P[e^{-\tau}]E^{P''}\left[\int_0^\infty e^{-t} f_{k+1}(X(t))\,dt\right]$$

$$\leq E^P\left[\int_0^\tau e^{-t} f_{k+1}(X(t))\,dt\right]$$

$$+ E^P[e^{-\tau}]\gamma(\Gamma_{\mu^*}^{(k)}, f_{k+1})$$

$$= E^Q\left[\int_0^\infty e^{-t} f_{k+1}(X(t))\,dt\right].$$

Hence equality must hold in (5.74), so $Q \in \Gamma_\nu^{(k+1)}$.

($C_2^{(k+1)}$) Let $P \in \Gamma^{(k+1)}$ and $\tau \in \mathscr{T}$, and let μ^* be as above. Then for $B \in \mathscr{F}_\tau$ with $0 < P(B) < 1$, $C_2^{(k)}$ and the fact that equality holds in (5.74) imply

(5.75)
$$\begin{aligned}
E^P[e^{-\tau}]\gamma(\Gamma_{\mu^*}^{(k)}, f_{k+1})
&= E^P\left[\int_\tau^\infty e^{-t} f_{k+1}(X(t))\, dt\right] \\
&= E^P\left[\chi_B e^{-\tau} \int_0^\infty e^{-t} f_{k+1}(X(\tau + t))\, dt\right] \\
&\quad + E^P\left[\chi_{B^c} e^{-\tau} \int_0^\infty e^{-t} f_{k+1}(X(\tau + t))\, dt\right] \\
&\le E^P[\chi_B e^{-\tau}]\gamma(\Gamma_{\mu_1^*}^{(k)}, f_{k+1}) + E^P[\chi_{B^c} e^{-\tau}]\gamma(\Gamma_{\mu_2^*}^{(k)}, f_{k+1}),
\end{aligned}$$

where $\mu_1^*(C) = E^P[\chi_B e^{-\tau}\chi_C(X(\tau))]/E^P[\chi_B e^{-\tau}]$ and $\mu_2^*(C) = E^P[\chi_{B^c} e^{-\tau} \chi_C(X(\tau))]/E^P[\chi_{B^c} e^{-\tau}]$ for all $C \in \mathscr{B}(E)$. As before $C_2^{(k)}$, $C_3^{(k)}$, and Lemma 5.10 imply equality in (5.75), and since the inequality is term by term we must have

(5.76)
$$\begin{aligned}
E^P\left[\chi_B e^{-\tau} \int_0^\infty e^{-t} f_{k+1}(X(\tau + t))\, dt\right]
&= E^P[\chi_B e^{-\tau}] \int u_{k+1}(x)\mu_1^*(dx) \\
&= E^P[\chi_B e^{-\tau} u_{k+1}(X(\tau))],
\end{aligned}$$

which implies

(5.77)
$$E^P\left[\int_0^\infty e^{-t} f_{k+1}(X(\tau + t)) \,\middle|\, \mathscr{F}_\tau\right] = u_{k+1}(X(\tau)).$$

Now let $H \ge 0$ be $\mathscr{F}_\tau$-measurable with $0 < E^P[H] < \infty$. Then Q, given by (5.22), is in $\Gamma^{(k)}$, and setting $\nu = QX(0)^{-1}$, (5.77) implies

(5.78)
$$\begin{aligned}
E^Q\left[\int_0^\infty e^{-t} f_{k+1}(X(t))\, dt\right]
&= \frac{E^P\left[H \int_0^\infty e^{-t} f_{k+1}(X(\tau + t))\, dt\right]}{E^P[H]} \\
&= \frac{E^P[H u_{k+1}(X(\tau))]}{E^P[H]} \\
&= \int u_{k+1}(x)\nu(dx) = \gamma(\Gamma_\nu^{(k)}, f_{k+1}),
\end{aligned}$$

and $Q \in \Gamma^{(k+1)}$.

($C_3^{(k+1)}$) $C_2^{(k)}$ and $C_3^{(k)}$ imply, by Lemma 5.10,

(5.79)
$$\gamma(\Gamma_{\alpha\mu_1 + (1-\alpha)\mu_2}^{(k)}, f_{k+1}) = \alpha\gamma(\Gamma_{\mu_1}^{(k)}, f_{k+1}) + (1 - \alpha)\gamma(\Gamma_{\mu_2}^{(k)}, f_{k+1})$$

for $\mu_1, \mu_2 \in \mathscr{P}(E)$ and $0 < \alpha < 1$, which in turn implies the convexity of $\Gamma^{(k+1)}$.

($C_4^{(k+1)}$) Let u_{k+1} be as above. By $C_4^{(k)}$, for $h \in \bar{C}(E)$ with $h \geq 0$ and $\varepsilon > 0$, there exists $v_\varepsilon \in B(E)$ such that

$$\gamma(\Gamma_\nu^{(k)}, f_{k+1} + \varepsilon h) = \int v_\varepsilon \, d\nu, \qquad \nu \in \mathscr{P}(E). \tag{5.80}$$

We claim that for each $x \in E$

$$v \equiv \lim_{\varepsilon \to 0} \varepsilon^{-1}(v_\varepsilon(x) - u_{k+1}(x)) = \gamma(\Gamma_{\delta_x}^{(k+1)}, h) \tag{5.81}$$

and

$$\int v \, d\nu = \gamma(\Gamma_\nu^{(k+1)}, h). \tag{5.82}$$

First observe

$$\int v_\varepsilon \, d\nu \geq \int u_{k+1} \, d\nu + \varepsilon \gamma(\Gamma_\nu^{(k+1)}, h), \qquad \nu \in \mathscr{P}(E), \tag{5.83}$$

and in particular $v_\varepsilon \geq u_{k+1}$ and

$$\underline{\lim}_{\varepsilon \to 0} \varepsilon^{-1}(v_\varepsilon(x) - u_{k+1}(x)) \geq \gamma(\Gamma_{\delta_x}^{(k+1)}, h). \tag{5.84}$$

For each $\varepsilon > 0$, let $P_\varepsilon \in \Gamma_\nu^{(k)}$ satisfy

$$\int v_\varepsilon \, d\nu = E^{P_\varepsilon}\left[\int_0^\infty e^{-t}(f_{k+1}(X(t)) + \varepsilon h(X(t))) \, dt\right]. \tag{5.85}$$

Clearly $\lim_{\varepsilon \to 0} \int v_\varepsilon \, d\nu = \gamma(\Gamma_\nu^{(k)}, f_{k+1})$, and by the continuity of $\int_0^\infty e^{-\lambda t} f_{k+1}(X(t)) \, dt$, all limit points of $\{P_\varepsilon\}$ as $\varepsilon \to 0$ are in $\Gamma_\nu^{(k+1)}$. Consequently,

$$\begin{aligned} \overline{\lim}_{\varepsilon \to 0} \int \varepsilon^{-1}(v_\varepsilon - u_{k+1}) \, d\nu &\leq \overline{\lim}_{\varepsilon \to 0} E^{P_\varepsilon}\left[\int_0^\infty e^{-t} h(X(t)) \, dt\right] \\ &\leq \gamma(\Gamma_\nu^{(k+1)}, h). \end{aligned} \tag{5.86}$$

In particular,

$$\overline{\lim}_{\varepsilon \to 0} \varepsilon^{-1}(v_\varepsilon(x) - u_{k+1}(x)) \leq \gamma(\Gamma_{\delta_x}^{(k+1)}, h). \tag{5.87}$$

Therefore (5.81) holds and since $0 \leq \varepsilon^{-1}(v_\varepsilon - u_{k+1}) \leq \|h\|$, (5.82) follows by the dominated convergence theorem.

($C_5^{(k+1)}$) This was verified above. □

Proof of Theorem 5.19 continued. Let $\Gamma^{(\infty)} = \bigcap_k \Gamma^{(k)}$ and for each $k \geq 0$ let u_{k+1} be as above. Then, by Lemma 5.18, for each $k \geq 0$,

(5.88) $$u_{k+1}(X(t)) - \int_0^t (u_{k+1}(X(s)) - f_{k+1}(X(s)))\, ds$$

is an $\{\mathscr{F}_t\}$-martingale for all $P \in \Gamma^{(k+1)}$, hence for all $P \in \Gamma^{(\infty)}$. Note that $\Gamma_\nu^{(\infty)} \neq \varnothing$ for all $\nu \in \mathscr{P}(E)$ since $\Gamma_\nu^{(k+1)} \subset \Gamma_\nu^{(k)}$, $\Gamma_\nu^{(k)} \neq \varnothing$ for all k, and $\Gamma_\nu^{(k)}$ is compact. Let A be the collection of $(f, g) \in B(E) \times B(E)$ such that (5.67) is an $\{\mathscr{F}_t\}$-martingale for all $P \in \Gamma^{(\infty)}$. Then $A \supset A_0$ and since $(u_{k+1}, u_{k+1} - f_{k+1}) \in A$ for $k = 1, 2, \ldots$, $\mathscr{R}(I - A)$ contains the linear span of $\{f_k\}$ and hence equals $B(E)$. By Proposition 3.5, A is dissipative, hence by Lemma 2.3 of Chapter 1 $\mathscr{R}(\lambda - A) = B(E)$ for all $\lambda > 0$. Lemma 3.2 implies

(5.89) $$(\lambda - A)^{-1} f(x) = E^P\left[\int_0^\infty e^{-\lambda t} f(X(t))\, dt\right]$$

for every $P \in \Gamma_{\delta_x}^{(\infty)}$. Therefore if $f \in \bar{C}(E)$, then

(5.90) $$\operatorname*{bp-lim}_{\lambda \to \infty} \lambda(\lambda - A)^{-1} f = f,$$

and it follows that $\mathscr{D}(A)$ is bp-dense in $B(E)$, which gives part (a).

Since A satisfies the conditions of Theorem 4.1, the martingale problem for (A, ν), $\nu \in \mathscr{P}(E)$, has at most one solution, and $\Gamma_\nu^{(\infty)} \neq \varnothing$ implies it has exactly one solution.

If Γ_ν is not a singleton for some $\nu \in \mathscr{P}(E)$, then there exist $P, P' \in \Gamma_\nu$ and $k > 0$ such that

(5.91) $$E^P\left[\int_0^\infty e^{-t} f_k(X(t))\, dt\right] \neq E^{P'}\left[\int_0^\infty e^{-t} f_k(X(t))\, dt\right].$$

(Otherwise $\Gamma_\nu = \Gamma_\nu^{(k)}$ for all k and $\Gamma_\nu = \Gamma_\nu^{(\infty)}$.) Therefore replacing f_k by $\|f_k\| - f_k$ for all k in the above procedure would produce a different sequence

(5.92) $$\hat{\Gamma}_\nu^{(k+1)} = \left\{P \in \hat{\Gamma}_\nu^{(k)} : E^P\left[\int_0^\infty e^{-t}(\|f_{k+1}\| - f_{k+1}(X(t)))\, dt\right] = \gamma(\hat{\Gamma}_\nu^{(k)}, \|f_{k+1}\| - f_{k+1})\right\}.$$

Let k_0 be the smallest k for which there exist ν_0 and $P, P' \in \Gamma_{\nu_0}$ such that (5.91) holds. Then $\Gamma_{\nu_0}^{(k)} \neq \hat{\Gamma}_{\nu_0}^{(k)}$, in fact $\Gamma_{\nu_0}^{(k)} \cap \hat{\Gamma}_{\nu_0}^{(k)} = \varnothing$, for $k > k_0$. Consequently $\Gamma_{\nu_0}^{(\infty)} \neq \hat{\Gamma}_{\nu_0}^{(\infty)}$ and the extension of A_0 corresponding to $\|f_k\| - f_k$ differs from A.

Let P_x denote the unique probability measure in $\Gamma_{\delta_x}^{(\infty)}$. The semigroup $\{T(t)\}$ corresponding to A (defined on $\mathscr{D}(A)$) can be represented by

(5.93) $$T(t)f(x) = E^{P_x}[f(X(t))].$$

Since $T(t)\colon \overline{\mathscr{D}(A)} \to \overline{\mathscr{D}(A)}$ and $\mathscr{D}(A)$ is bp-dense in $B(E)$, $\{T(t)\}$ can be extended to a semigroup on all of $B(E)$ satisfying (5.93). Consequently,

$$P(t, x, \Gamma) \equiv E^{P_x}[\chi_\Gamma(X(t))] \tag{5.94}$$

is a transition function, and by Proposition 1.2 $P_x(B)$ is a Borel measurable function of x for all $B \in \mathscr{S}_E$. For each $v \in \mathscr{P}(E)$,

$$P_v \equiv \int P_x\, v(dx) \tag{5.95}$$

is a solution of the martingale problem for (A, v) and hence is the unique element of $\Gamma_v^{(\infty)}$. This completes the proof of part (c).

Since $\Gamma^{(k)}$ satisfies C_2 for all k, $\Gamma^{(\infty)}$ satisfies C_2. For $P \in \Gamma^{(\infty)}$, $\tau \in \mathscr{T}$, and $B \in \mathscr{F}_\tau$ with $P(B) > 0$, uniqueness implies

$$\frac{E^P[\chi_B\, \chi_C(X(\tau + \cdot))]}{P(B)} = \int P_x(C)\mu(dx) \tag{5.96}$$

where $\mu(D) = E^P[\chi_B\, \chi_D(X(\tau))]/P(B)$ for all $D \in \mathscr{B}(E)$. Since B is arbitrary in $\mathscr{F}_\tau$, (5.68) follows. □

6. THE MARTINGALE PROBLEM: LOCALIZATION

Let $A \subset B(E) \times B(E)$, let U be an open subset of E, and let X be a process with initial distribution $v \in \mathscr{P}(E)$ and sample paths in $D_E[0, \infty)$. Define the $\{\mathscr{F}_t^X\}$-stopping time

$$\tau = \inf\{t \geq 0\colon X(t) \notin U \quad \text{or} \quad X(t-) \notin U\}. \tag{6.1}$$

Then X is a solution of the *stopped martingale problem* for (A, v, U) if $X(\cdot) = X(\cdot \wedge \tau)$ a.s. and

$$f(X(t)) - \int_0^{t \wedge \tau} g(X(s))\, ds \tag{6.2}$$

is an $\{\mathscr{F}_t^X\}$-martingale for all $(f, g) \in A$. (Note that the stopped martingale problem requires sample paths in $D_E[0, \infty)$.)

6.1 Theorem Let (E, r) be complete and separable, and let $A \subset \bar{C}(E) \times B(E)$. If the $D_E[0, \infty)$ martingale problem for A is well-posed, then for each $v \in \mathscr{P}(E)$ and open $U \subset E$ there exists a unique solution of the stopped martingale problem for (A, v, U).

Proof. Let X be the solution of the $D_E[0, \infty)$ martingale problem for (A, v), define τ by (6.1), and define $\tilde{X}(\cdot) = X(\cdot \wedge \tau)$. Then $\tilde{X}$ is a solution of the stopped martingale problem for (A, ν, U) by the optional sampling theorem (Theorem 2.13 of Chapter 2).

For uniqueness, fix ν and U and let X be a solution of the stopped martingale problem for (A, ν, U). By Lemma 5.16 there exists a solution Y of the $D_E[0, \infty)$ martingale problem for (A, ν) and a nonnegative random variable η such that X $(= X(\cdot \wedge \tau))$ has the same distribution as $Y(\cdot \wedge \eta)$. Note that in this case the η constructed in the proof of Lemma 5.16 is $\inf \{t \geq 0: Y(t) \notin U$ or $Y(t-) \notin U\}$, and since the distribution of Y is uniquely determined, it follows that the distribution of $Y(\cdot \wedge \eta)$ (and hence of X) is uniquely determined. □

Our primary interest in this section is in possible converses for the above theorem. That is, we are interested in conditions under which existence and, more importantly, uniqueness of solutions of the stopped martingale problem imply existence and uniqueness for the (unstopped) $D_E[0, \infty)$ martingale problem. Recall that uniqueness for the $D_E[0, \infty)$ martingale problem is typically equivalent to the general uniqueness question (cf. Theorem 3.6) but not necessarily (Problem 21).

6.2 Theorem Let E be separable, and let $A \subset \bar{C}(E) \times B(E)$. Suppose that for each $\nu \in \mathscr{P}(E)$ there exists a solution of the $D_E[0, \infty)$ martingale problem for (A, ν). If there exist open subsets U_k, $k = 1, 2, \ldots$, with $E = \bigcup_{k=1}^{\infty} U_k$ such that for each $\nu \in \mathscr{P}(E)$ and $k = 1, 2, \ldots$ the solution of the stopped martingale problem for (A, ν, U_k) is unique, then for each $\nu \in \mathscr{P}(E)$ the solution of the $D_E[0, \infty)$ martingale problem for (A, ν) is unique.

Proof. Let $V_1, V_2, \ldots$ be a sequence of open subsets of E such that for each i there exists a k with $V_i = U_k$ and that for each k there exist infinitely many i with $V_i = U_k$. Fix $\nu \in \mathscr{P}(E)$, and let X be a solution of the martingale problem for (A, ν). Let $\tau_0 = 0$ and for $i \geq 1$

$$\tau_i = \inf \{t \geq \tau_{i-1}: X(t) \notin V_i \quad \text{or} \quad X(t-) \notin V_i\}. \tag{6.3}$$

We note that $\lim_{i \to \infty} \tau_i = \infty$. (See Problem 27.) For $f \in \bar{C}(E)$ and $\lambda > 0$,

$$\begin{aligned} E\left[\int_0^\infty e^{-\lambda t} f(X(t))\, dt\right] &= \sum_{i=1}^{\infty} E\left[\int_{\tau_{i-1}}^{\tau_i} e^{-\lambda t} f(X(t))\, dt\right] \\ &= \sum_{i=1}^{\infty} E\left[e^{-\lambda \tau_{i-1}} \chi_{\{\tau_{i-1} < \infty\}} \int_0^{\eta_i} e^{-\lambda t} f(X(\tau_{i-1} + t))\, dt\right], \end{aligned} \tag{6.4}$$

where on $\{\tau_{i-1} < \infty\}$

$$\eta_i = \tau_i - \tau_{i-1} = \inf \{t \geq 0: X(\tau_{i-1} + t) \notin V_i \quad \text{or} \quad X((\tau_{i-1} + t)-) \notin V_i\}. \tag{6.5}$$

For $i \geq 1$ such that $P\{\tau_{i-1} < \infty\} > 0$, define

$$\mu_i(B) = \frac{E[e^{-\lambda\tau_{i-1}}\chi_{\{\tau_{i-1}<\infty\}}\chi_B(X(\tau_{i-1}))]}{E[e^{-\lambda\tau_{i-1}}\chi_{\{\tau_{i-1}<\infty\}}]} \tag{6.6}$$

for $B \in \mathscr{B}(E)$, and

$$P_i(C) = \frac{E[e^{-\lambda\tau_{i-1}}\chi_{\{\tau_{i-1}<\infty\}}\chi_C(X(\tau_{i-1} + \cdot \wedge \eta_i))]}{E[e^{-\lambda\tau_{i-1}}\chi_{\{\tau_{i-1}<\infty\}}]} \tag{6.7}$$

for $C \in \mathscr{S}_E$.

Let Y_i be the coordinate process on $(D_E[0, \infty), \mathscr{S}_E, P_i)$. Then Y_i is a solution of the stopped martingale problem for (A, μ_i, V_i), and hence, given μ_i, its distribution P_i is uniquely determined. Set

$$\gamma_i = \inf\{t: Y_i(t) \notin V_i \quad \text{or} \quad Y_i(t-) \notin V_i\}. \tag{6.8}$$

Then for $i \geq 1$ with $P\{\tau_i < \infty\} > 0$

$$\begin{aligned} \mu_{i+1}(B) &= \frac{E[e^{-\lambda\tau_{i-1}}\chi_{\{\tau_{i-1}<\infty\}}e^{-\lambda\eta_i}\chi_{\{\eta_i<\infty\}}\chi_B(X(\tau_i))]}{\alpha_i} \\ &= \frac{E^{P_i}[e^{-\lambda\gamma_i}\chi_{\{\gamma_i<\infty\}}\chi_B(Y_i(\gamma_i))]\alpha_{i-1}}{\alpha_i}, \end{aligned} \tag{6.9}$$

where

$$\begin{aligned} \alpha_i = E[e^{-\lambda\tau_i}\chi_{\{\tau_i<\infty\}}] &= E[e^{-\lambda\tau_{i-1}}\chi_{\{\tau_{i-1}<\infty\}}e^{-\lambda\eta_i}\chi_{\{\eta_i<\infty\}}] \\ &= E^{P_i}[e^{-\lambda\gamma_i}\chi_{\{\gamma_i<\infty\}}]\alpha_{i-1} \\ &= \prod_{k=1}^{i} E^{P_k}[e^{-\lambda\gamma_k}\chi_{\{\gamma_k<\infty\}}]. \end{aligned} \tag{6.10}$$

Consequently, $P_1, \ldots, P_i$ determine μ_{i+1}, which in turn determines P_{i+1}. Since $\mu_1 = \nu$, it follows that the P_i are the same for all solutions of the martingale problem for (A, ν) with sample paths in $D_E[0, \infty)$. But the right side of (6.4) can be written

$$\sum_{i=1}^{\infty} E^{P_i}\left[\int_0^{\gamma_i} e^{-\lambda t} f(Y_i(t))\, dt\right]\alpha_{i-1} \tag{6.11}$$

so that (6.4) is the same for all solutions of the $D_E[0, \infty)$ martingale problem for (A, ν), and since λ is arbitrary, the uniqueness of the Laplace transform implies $E[f(X(t))]$ is the same for all solutions. (Note $E[f(X(t))]$ is right continuous as a function of t.) Since $f \in \bar{C}(E)$ is arbitrary, the one-dimensional distributions are determined and uniqueness follows by Corollary 4.3. □

Note that the proof of Theorem 6.2 uses the uniqueness of the solution of the stopped martingale problem for (A, μ, U_k) for *every* choice of μ. The next result does not require this.

6.3 Theorem Let (E, r) be complete and separable, and let $A \subset \bar{C}(E) \times B(E)$. Let $U_1 \subset U_2 \subset \cdots$ be open subsets of E. Fix $\nu \in \mathscr{P}(E)$, and suppose that for each k there exists a unique solution X_k of the stopped martingale problem for (A, ν, U_k) with sample paths in $D_E[0, \infty)$. Setting

$$\tau_k = \inf \{t: X_k(t) \notin U_k \quad \text{or} \quad X_k(t-) \notin U_k\}, \tag{6.12}$$

suppose that for each $t > 0$,

$$\lim_{k \to \infty} P\{\tau_k \leq t\} = 0. \tag{6.13}$$

Then there exists a unique solution of the $D_E[0, \infty)$ martingale problem for (A, ν).

Proof. Let

$$\tau_m^k = \inf \{t: X_m(t) \notin U_k \quad \text{or} \quad X_m(t-) \notin U_k\}. \tag{6.14}$$

For $k < m$, $X_m(\cdot \wedge \tau_m^k)$ is a solution of the stopped martingale problem for (A, ν, U_k) and hence has the same distribution as X_k. It follows from (6.13) that there exists a process X_∞ such that $X_k \Rightarrow X_\infty$. (In particular, for the metric on $D_E[0, \infty)$ given by (5.2) of Chapter 3, the Prohorov distance between the distributions of X_k and X_m is less than $E[e^{-\tau_k \wedge m}]$.) In fact, for any bounded measurable function G on $D_E[0, \infty)$ and any $T > 0$,

$$| E[G(X_k(\cdot \wedge T))] - E[G(X_\infty(\cdot \wedge T))] | \leq 2\|G\| P\{\tau_k \leq T\}. \tag{6.15}$$

Let τ_∞^k be defined as in (6.14). Since the distribution of $X_m(\cdot \wedge \tau_m^k)$ does not depend on $m \geq k$, it follows that the distribution of $X_\infty(\cdot \wedge \tau_\infty^k)$ is the same as that of X_k. Hence

$$f(X_\infty(t \wedge \tau_\infty^k)) - \int_0^{t \wedge \tau_\infty^k} g(X_\infty(s))\, ds \tag{6.16}$$

is an $\{\mathscr{F}_t^{X_\infty}\}$-martingale for each k. Since $\lim_{k \to \infty} \tau_\infty^k = \infty$ a.s. ($P\{\tau_\infty^k \leq t\} = P\{\tau_k \leq t\}$), we see that X_∞ is a solution of the martingale problem for (A, ν) with sample paths in $D_E[0, \infty)$. If X is a solution of the $D_E[0, \infty)$ martingale problem for (A, ν) and

$$\gamma_k = \inf \{t: X(t) \notin U_k \quad \text{or} \quad X(t-) \notin U_k\}, \tag{6.17}$$

then $X(\cdot \wedge \gamma_k)$ has the same distribution as X_k, and hence X has the same distribution as X_∞. □

6.4 Corollary Let (E, r) be complete and separable. Let A_k, $k = 1, 2, \ldots$, $A \subset \bar{C}(E) \times B(E)$ and suppose there exist open subsets $U_1 \subset U_2 \subset \cdots$ with $\bigcup_k U_k = E$ such that

$$\{(f, \chi_{U_k} g): (f, g) \in A_k\} = \{(f, \chi_{U_k} g): (f, g) \in A\}. \tag{6.18}$$

(In the single-valued case this is just $\mathscr{D}(A_k) = \mathscr{D}(A)$ and $A_k f|_{U_k} = Af|_{U_k}$ for all $f \in \mathscr{D}(A)$.) If for each k, the $D_E[0, \infty)$ martingale problem for A_k is well-posed, and if for each $\nu \in \mathscr{P}(E)$, the sequence of solutions $\{X_k\}$ of the $D_E[0, \infty)$ martingale problems for (A_k, ν), $k = 1, 2, \ldots$, is relatively compact, then the $D_E[0, \infty)$ martingale problem for A is well-posed.

Proof. Any solution of the stopped martingale problem for (A, ν, U_k) is a solution for the stopped martingale problem for (A_k, ν, U_k) and hence is unique by Theorem 6.1. Set $\tau_k = \inf\{t: X_k(t) \notin U_k \text{ or } X_k(t-) \notin U_k\}$. Then $\tilde{X}_k = X_k(\cdot \wedge \tau_k)$ is the solution of the stopped martingale problem and (6.13) follows from the relative compactness of $\{X_k\}$. Theorem 6.3 then gives the desired result. □

The following lemma is useful in obtaining the monotone sequence $\{U_k\}$ in Theorem 6.3.

6.5 Lemma Let E be locally compact and separable, and let U_1, U_2 be open subsets of E with compact closure. Let $A \subset \bar{C}(E) \times B(E)$, and suppose $\mathscr{D}(A)$ separates points. If for each $\nu \in \mathscr{P}(E)$ and $k = 1, 2$, there exists a solution of the stopped martingale problem for (A, ν, U_k), then for each $\nu \in \mathscr{P}(E)$ there exists a solution of the stopped martingale problem for $(A, \nu, U_1 \cup U_2)$.

Proof. Let $V_k = U_1$ for k odd and $V_k = U_2$ for k even, and fix $x_0 \in E$. Lemma 5.15 can be used to construct a process X and stopping times τ_i such that $\tau_0 = 0$, τ_i is given by (6.3), and $X(t) = X_i(t - \tau_i)$, $\tau_i \le t < \tau_{i+1}$, where X_i is a solution of the stopped martingale problem for (A, μ_i, V_i), $\mu_0 = \nu$, and $\mu_i(\Gamma) = P\{X(\tau_i) \in \Gamma, \tau_i < \infty\} + \delta_{x_0}(\Gamma)P\{\tau_i = \infty\}$. Let $\tau_\infty = \lim_{i\to\infty} \tau_i$ and note that

$$f(X(t \wedge \tau_\infty)) - \int_0^{t \wedge \tau_\infty} g(X(s))\, ds \tag{6.19}$$

is an $\{\mathscr{F}_t^X\}$-martingale for every $(f, g) \in A$. Either $\tau_\infty = \infty$, $\tau_i = \tau_\infty < \infty$ for $i \ge i_0$ (some i_0) or $\tau_i < \tau_\infty < \infty$ for all $i \ge 0$.

In the second case $\tau_{i+1} = \tau_i$ implies $X(\tau_i) \notin V_i$ and hence $X(\tau_\infty) \notin U_1 \cup U_2$. In the third case, the fact that (6.19) is a martingale implies $\lim_{t\to\tau_\infty -} f(X(t))$ exists for $f \in \mathscr{D}(A)$, and the compactness of $\overline{U_1 \cup U_2}$ and the fact that $\mathscr{D}(A)$ is separating imply $\lim_{t\to\tau_\infty -} X(t)$ exists. But either $X(\tau_i)$ or $X(\tau_i -) \notin V_i$, and hence $\lim_{t\to\tau_\infty -} X(t) \notin U_1 \cup U_2$. Consequently, $\tau \equiv \inf\{t: X(t) \notin U_1 \cup U_2 \text{ or } X(t-) \notin U_1 \cup U_2\} \le \tau_\infty$, and $X(\cdot \wedge \tau)$ is a solution of the stopped martingale problem for $(A, \nu, U_1 \cup U_2)$. □

6.6 Theorem Let E be compact, $A \subset \bar{C}(E) \times B(E)$, and suppose that $\mathscr{D}(A)$ separates points. Suppose that for each $x \in E$ there exists an open set $U \subset E$ with $x \in U$ such that for each $\nu \in \mathscr{P}(E)$ there exists a solution of the stopped martingale problem for (A, ν, U). Then for each $\nu \in \mathscr{P}(E)$ there exists a solution of the $D_E[0, \infty)$ martingale problem for (A, ν).

Proof. This is an immediate consequence of Lemma 6.5. □

7. THE MARTINGALE PROBLEM: GENERALIZATIONS

A. The Time-dependent Martingale Problem

It is natural to consider processes whose parameters vary in time. With this in mind let $A \subset B(E) \times B(E \times [0, \infty))$. Then a measurable E-valued process X is a solution of the martingale problem for A if, for each $(f, g) \in A$,

$$f(X(t)) - \int_0^t g(X(s), s)\, ds \tag{7.1}$$

is an $\{{}^*\mathscr{F}_t^X\}$- martingale. As before, X is a solution of the martingale problem for A with respect to $\{\mathscr{G}_t\}$, where $\mathscr{G}_t \supset {}^*\mathscr{F}_t^X$, if (7.1) is a $\{\mathscr{G}_t\}$-martingale for each $(f, g) \in A$. Most of the basic results concerning martingale problems can be extended to the time-dependent case by considering the space-time process $X^0(t) = (X(t), t)$.

7.1 Theorem Let $A \subset B(E) \times B(E \times [0, \infty))$ and define $A^0 \subset B(E \times [0, \infty)) \times B(E \times [0, \infty))$ by

$$A^0 = \{(f\gamma, g\gamma + f\gamma') : (f, g) \in A, \gamma \in C_c^1[0, \infty)\}. \tag{7.2}$$

Then X is a solution of the martingale problem for A with respect to $\{\mathscr{G}_t\}$ if and only if the space-time process X^0 is a solution of the martingale problem for A^0 with respect to $\{\mathscr{G}_t\}$.

Proof. If X is a solution of the martingale problem for A with respect to $\{\mathscr{G}_t\}$, then for $(f, g) \in A$ and $\gamma \in C_c^1[0, \infty)$,

$$f(X(t))\gamma(t) - \int_0^t (g(X(s), s)\gamma(s) + f(X(s))\gamma'(s))\, ds \tag{7.3}$$

is a $\{\mathscr{G}_t\}$-martingale by the argument used to prove Lemma 3.4. The converse follows by considering $\gamma \equiv 1$ on $[0, T]$, $T > 0$. □

Suppose (7.1) can be written

$$f(X(t)) - \int_0^t A(s) f(X(s))\, ds \tag{7.4}$$

where $\{A(t): t \geq 0\}$ is a family of bounded operators. In this case in considering the space-time process the boundedness of the operators is lost. Fortunately, the bounded case is easy to treat directly. Consider generators of the form

$$A(t)f(x) = \lambda(t, x) \int (f(y) - f(x))\mu(t, x, dy) \tag{7.5}$$

where $\lambda \in M([0, \infty) \times E)$ is nonnegative and bounded in x for each fixed t, $\mu(t, x, \cdot) \in \mathscr{P}(E)$ for every $(t, x) \in [0, \infty) \times E$, and $\mu(\cdot, \cdot, \Gamma) \in B([0, \infty) \times E)$ for every $\Gamma \in \mathscr{B}(E)$. We can obtain a (time inhomogeneous) transition function for a Markov process as a solution of the equation

$$\begin{aligned} P(s, t, x, \Gamma) = \delta_x(\Gamma) + \int_s^t \lambda(u, x) \int (P(u, t, y, \Gamma) \\ - P(u, t, x, \Gamma))\mu(u, x, dy)\, du. \end{aligned} \tag{7.6}$$

7.2 Lemma Suppose there exists a measurable function γ on $[0, \infty)$ such that $\lambda(t, x) \leq \gamma(t)$ for all t and x and that for each $T > 0$

$$\int_0^T \gamma(t)\, dt < \infty. \tag{7.7}$$

Then (7.6) has a unique solution.

Proof. We first obtain uniqueness. Suppose P and $\tilde{P}$ are solutions of (7.6) and set

$$M(s, t) = \sup_{x, \Gamma} \sup_{s \leq u \leq t} |P(u, t, x, \Gamma) - \tilde{P}(u, t, x, \Gamma)|. \tag{7.8}$$

Since M is nonincreasing in s, it is measurable in s and

$$M(s, t) \leq \int_s^t \gamma(u) 2M(u, t)\, du. \tag{7.9}$$

A slight modification of Gronwall's inequality (Appendix 5) implies $M(s, t) = 0$.

Existence is obtained by iteration. To see that the solution is a transition function it is simplest to first transform the equation to

$$\begin{aligned} P(s, t, x, \Gamma) = \delta_x(\Gamma) \exp\left\{-\int_s^t \lambda(u, x)\, du\right\} \\ + \int_s^t \lambda(u, x) \exp\left\{-\int_s^u \lambda(r, x)\, dr\right\} \int P(u, t, y, \Gamma)\mu(u, x, dy)\, du. \end{aligned} \tag{7.10}$$

(To see that (7.10) is equivalent to (7.6), differentiate both sides with respect to s.) Fix t and set $P^0(s, t, x, \Gamma) = \delta_x(\Gamma)$ and

$$(7.11)\quad P^{n+1}(s, t, x, \Gamma) = \delta_x(\Gamma) \exp\left\{-\int_s^t \lambda(u, x)\, du\right\} + \int_s^t \lambda(u, x) \exp\left\{-\int_s^u \lambda(r, x)\, dr\right\} \int P^n(u, t, y, \Gamma)\mu(u, x, dy)\, du.$$

Note that for each n, $P^n(s, t, x, \cdot) \in \mathscr{P}(E)$, and for each $\Gamma \in \mathscr{B}(E)$, $P^n(\cdot, \cdot, \cdot, \Gamma)$ is a Borel measurable function. Let

$$M_n(s, t) = \sup_{x, \Gamma} \sup_{s \le u \le t} |P^{n+1}(u, t, x, \Gamma) - P^n(u, t, x, \Gamma)|.$$

Then

$$(7.12)\quad M_n(s, t) \le \sup_x \int_s^t \lambda(u, x) \exp\left\{-\int_s^u \lambda(r, x)\, dr\right\} M_{n-1}(u, t)\, du \le \int_s^t \gamma(u) M_{n-1}(u, t)\, du.$$

Consequently,

$$(7.13)\quad \sum_{k=0}^n M_k(s, t) \le M_0(s, t) + \int_s^t \gamma(u) \sum_{k=0}^n M_k(u, t)\, du \le 2 + \int_s^t \gamma(u) \sum_{k=0}^n M_k(u, t)\, du,$$

and by Gronwall's inequality,

$$(7.14)\quad \sum_{k=0}^n M_k(s, t) \le 2 \exp\left\{\int_s^t \gamma(u)\, du\right\}.$$

From (7.14) we conclude that $\{P^n(s, t, x, \Gamma)\}$ is a Cauchy sequence whose limit must satisfy (7.6). □

7.3 Theorem Let λ and μ be as above, and define

$$(7.15)\quad A = \left\{\left(f, \lambda(\cdot, \cdot) \int (f(y) - f(\cdot))\mu(\cdot, \cdot, dy)\right) : f \in B(E)\right\}.$$

If λ satisfies the conditions of Lemma 7.2, then the martingale problem for A is well-posed.

Proof. The proof is left as Problem 29. □

B. The Local-Martingale Problem

If we relax the condition that the functions in A be bounded, the natural requirement to place on X is that

$$f(X(t)) - \int_0^t g(X(s))\, ds \tag{7.16}$$

be a local martingale. (In particular if we drop the boundedness assumption, (7.16) need not be in L^1.) Consequently, for $A \subset M(E) \times M(E)$ we say that a measurable E-valued process X is a solution of the local-martingale problem for A if for each $(f, g) \in A$

$$\int_0^t |g(X(s))|\, ds < \infty \qquad \text{a.s.} \tag{7.17}$$

and (7.16) is an $\{{}^*\mathscr{F}_t^X\}$-local martingale. For example, let

$$A = \{(f, \tfrac{1}{2}f''): f \in C^2(\mathbb{R})\}. \tag{7.18}$$

Then the unique solution of the local-martingale problem for (A, ν), $\nu \in \mathscr{P}(\mathbb{R})$, is just Brownian motion with mean zero, variance parameter 1, and initial distribution ν. Itô's formula (Theorem 2.9 of Chapter 5) ensures that (7.16) is a local martingale.

Let $A_1 \subset B(E) \times B(E)$. Let β be nonnegative and measurable (but not necessarily bounded) and set

$$A_2 = \{(f, \beta g): (f, g) \in A_1\}. \tag{7.19}$$

If Y is a solution of the martingale problem for A_1 and τ satisfies

$$\int_0^{\tau(t)} \frac{1}{\beta(Y(s))}\, ds = t, \qquad t \geq 0, \tag{7.20}$$

then $X \equiv Y(\tau(\cdot))$ is a solution of the local-martingale problem for A_2. (See Chapter 6.)

Many of the results in the previous sections extend easily to local-martingale problems.

C. The Martingale Problem Corresponding to a Given Semigroup

Let $\{T(t)\}$ be a semigroup of operators defined on a closed subspace $L \subset B(E)$. Then X is a solution of the martingale problem for $\{T(t)\}$ if, for each $u > 0$ and $f \in L$,

$$T(u - t)f(X(t)) \tag{7.21}$$

is a martingale on the time interval $[0, u]$ with respect to $\{\mathscr{F}_t^X\}$. Of course if L is separating, then X is a Markov process corresponding to $\{T(t)\}$.

8. CONVERGENCE THEOREMS

In Theorem 2.5 we related the weak convergence of a sequence of Feller processes to the convergence of the corresponding semigroups. In this section we give conditions for more general sequences of processes to converge to Markov processes and allow the limiting Markov process to be determined either by its semigroup or as a solution of a martingale problem.

If a sequence of processes $\{X_n\}$ approximates a Markov process, it is reasonable to expect the processes to be approximately Markovian in some sense. One way of expressing this would be to require

$$\lim_{n\to\infty} E[\,|E[f(X_n(t+s))\,|\,\mathscr{F}_t^{X_n}] - T(s)f(X_n(t))\,|\,] = 0 \tag{8.1}$$

where $\{T(s)\}$ is the semigroup corresponding to the limiting process. The following lemma shows that a condition weaker than (8.1) is in fact sufficient.

8.1 Lemma Let (E, r) be complete and separable. Let $\{X_n\}$ be a sequence of processes with values in E, and let $\{T(t)\}$ be a semigroup on a closed subspace $L \subset \bar{C}(E)$. Let $M \subset \bar{C}(E)$ be separating, and suppose either L is separating and $\{X_n(t)\}$ is relatively compact for each $t \geq 0$, or L is convergence determining. If X is a Markov process corresponding to $\{T(t)\}$, $X_n(0) \Rightarrow X(0)$, and

$$\lim_{n\to\infty} E\left[(f(X_n(t+s)) - T(s)f(X_n(t)))\prod_{i=1}^{k} g_i(X_n(t_i))\right] = 0 \tag{8.2}$$

for all $k \geq 0$, $0 \leq t_1 < t_2 < \cdots < t_k \leq t < t+s$, $f \in L$, and $g_1, \ldots, g_k \in M$, then the finite-dimensional distributions of X_n converge weakly to those of X.

Proof. Let $f \in L$ and $t > 0$. Then, since $X_n(0) \Rightarrow X(0)$ and $T(t)f$ is continuous, (8.2) implies

$$\begin{aligned} \lim_{n\to\infty} E[f(X_n(t))] &= \lim_{n\to\infty} E[T(t)f(X_n(0))] \\ &= E[T(t)f(X(0))] = E[f(X(t))], \end{aligned} \tag{8.3}$$

and hence $X_n(t) \Rightarrow X(t)$, using Lemma 4.3 of Chapter 3 under the first conditions on L. Fix $m > 0$ and suppose $(X_n(t_1), \ldots, X_n(t_m)) \Rightarrow (X(t_1), \ldots, X(t_m))$ for all $0 \leq t_1 < t_2 < \cdots < t_m$. Then (8.2) again implies

$$\begin{aligned} \lim_{n\to\infty} E\left[f(X_n(t_{m+1}))\prod_{i=1}^{m} g_i(X_n(t_i))\right] &= \lim_{n\to\infty} E\left[T(t_{m+1}-t_m)f(X_n(t_m))\prod_{i=1}^{m} g_i(X_n(t_i))\right] \\ &= E\left[T(t_{m+1}-t_m)f(X(t_m))\prod_{i=1}^{m} g_i(X(t_i))\right] \\ &= E\left[f(X(t_{m+1}))\prod_{i=1}^{m} g_i(X(t_i))\right] \end{aligned} \tag{8.4}$$

for all $0 \le t_1 < t_2 < \cdots < t_{m+1}$, $f \in L$, and $g_1, \ldots, g_m \in M$. Since relative compactness of $\{X_n(t)\}$, $t \ge 0$, implies relative compactness of $\{(X_n(t_1), \ldots, X_n(t_{m+1})\}$, we may apply Lemma 4.3 and Proposition 4.6, both from Chapter 3, to conclude $(X_n(t_1), \ldots, X_n(t_{m+1})) \Rightarrow (X(t_1), \ldots, X(t_{m+1}))$. □

The convergence in (8.1) (or (8.2)) can be viewed as a type of semigroup convergence (cf. Chapter 2, Section 7). For $n = 1, 2, \ldots$ let $\{\mathscr{G}_t^n\}$ be a complete filtration, and let $\mathscr{L}_n$ be the space of real-valued $\{\mathscr{G}_t^n\}$-progressive processes ξ satisfying

$$\sup_{t \le T} E[|\xi(t)|] < \infty \tag{8.5}$$

for each $T > 0$. Let $\hat{\mathscr{A}}_n$ be the collection of pairs $(\xi, \varphi) \in \mathscr{L}_n \times \mathscr{L}_n$ such that

$$\xi(t) - \int_0^t \varphi(s)\, ds \tag{8.6}$$

is a $\{\mathscr{G}_t^n\}$-martingale. Note that if X_n is a $\{\mathscr{G}_t^n\}$-progressive solution of the martingale problem for $A_n \subset B(E) \times B(E)$ with respect to $\{\mathscr{G}_t^n\}$, then $(f \circ X_n, g \circ X_n) \in \hat{\mathscr{A}}_n$ for each $(f, g) \in A_n$.

8.2 Theorem Let (E, r) be complete and separable. Let $A \subset \bar{C}(E) \times \bar{C}(E)$ be linear, and suppose the closure of A generates a strongly continuous contraction semigroup $\{T(t)\}$ on $L \equiv \overline{\mathscr{D}(A)}$. Suppose X_n, $n = 1, 2, \ldots$, is a $\{\mathscr{G}_t^n\}$-progressive E-valued process, X is a Markov process corresponding to $\{T(t)\}$, and $X_n(0) \Rightarrow X(0)$. Let $M \subset \bar{C}(E)$ be separating and suppose either L is separating and $\{X_n(t)\}$ is relatively compact for each $t \ge 0$, or L is convergence determining. Then the following are equivalent:

(a) The finite-dimensional distributions of X_n converge weakly to those of X.

(b) For each $(f, g) \in A$,

(8.7)

$$\lim_{n \to \infty} E\left[\left(f(X_n(t+s)) - f(X_n(t)) - \int_t^{t+s} g(X_n(u))\, du\right) \prod_{i=1}^k h_i(X_n(t_i))\right] = 0$$

for all $k \ge 0$, $0 \le t_1 < t_2 < \cdots < t_k \le t < t + s$, and $h_1, \ldots, h_k \in M$.

(c) For each $(f, g) \in A$ and $T > 0$, there exist $(\xi_n, \varphi_n) \in \hat{\mathscr{A}}_n$ such that

$$\sup_n \sup_{s \le T} E[|\xi_n(s)|] < \infty, \tag{8.8}$$

$$\sup_n \sup_{s \le T} E[|\varphi_n(s)|] < \infty, \tag{8.9}$$

$$\lim_{n \to \infty} E\left[(\xi_n(t) - f(X_n(t))) \prod_{i=1}^{k} h_i(X_n(t_i))\right] = 0, \tag{8.10}$$

and

$$\lim_{n \to \infty} E\left[(\varphi_n(t) - g(X_n(t))) \prod_{i=1}^{k} h_i(X_n(t_i))\right] = 0, \tag{8.11}$$

for all $k \ge 0$, $0 \le t_1 < t_2 < \cdots < t_k \le t \le T$, and $h_1, \ldots, h_k \in M$.

8.3 Remark (a) Note that (8.10) and (8.11) are implied by

$$\lim_{n \to \infty} E[|\xi_n(t) - f(X_n(t))|] = \lim_{n \to \infty} E[|\varphi_n(t) - g(X_n(t))|] = 0. \tag{8.12}$$

If this stronger condition holds, then (8.1) will hold.

(b) Frequently a good choice for (ξ_n, φ_n) will be

$$\xi_n(t) = \varepsilon_n^{-1} \int_0^{\varepsilon_n} E[f(X_n(t+s)) \mid \mathscr{G}_t^n] \, ds \tag{8.13}$$

and

$$\varphi_n(t) = \varepsilon_n^{-1} E[f(X_n(t + \varepsilon_n)) - f(X_n(t)) \mid \mathscr{G}_t^n] \tag{8.14}$$

for some positive sequence $\{\varepsilon_n\}$ tending to zero. See Proposition 7.5 of Chapter 2.

(c) Conditions (8.9) and (8.11) can be relaxed. In fact it is sufficient to have

$$\lim_{n \to \infty} \int_0^{T-t} \left| E\left[(\varphi_n(t+s) - g(X_n(t+s))) \prod_{i=1}^{k} h_i(X_n(t_i))\right]\right| ds = 0 \tag{8.15}$$

for all $k \ge 0$, $0 \le t_1 < t_2 < \cdots < t_k \le t \le T$, and $h_1, \ldots, h_k \in M$. See (8.23) below.

(d) For the implication $(c \Rightarrow a)$ we may drop the assumption that $\mathscr{D}(A) \subset \bar{C}(E)$ provided $f \in L$ and $h \in M$ imply $f \cdot h \in L$ and

$$\lim_{n \to \infty} E[f(X_n(0))] = E[f(X(0))], \qquad f \in L. \tag{8.16}$$

Note that (8.16) may be stronger than convergence in distribution since f need not be continuous. □

Proof. **(a ⇒ b)** This is immediate.

(b ⇒ c) Let $(f, g) \in A$, and define ξ_n and φ_n by

$$\xi_n(t) = \int_0^\infty e^{-s} E[f(X_n(t+s)) - g(X_n(t+s)) \mid \mathscr{G}_t^n] \, ds \tag{8.17}$$

and

$$\varphi_n(t) = \xi_n(t) - f(X_n(t)) + g(X_n(t)). \tag{8.18}$$

Then $(\xi_n, \varphi_n) \in \hat{\mathscr{A}}_n$ by Theorem 7.1 of Chapter 2. Clearly (8.8) and (8.9) hold, and since $\xi_n - f \circ X_n = \varphi_n - g \circ X_n$, it is enough to verify (8.10).

Integrating by parts gives

$$\begin{aligned} \xi_n(t) &= \int_0^\infty e^{-s} E[f(X_n(t+s)) - g(X_n(t+s)) \mid \mathscr{G}_t^n] \, ds \\ &= \int_0^\infty e^{-s} E[f(X_n(t+s)) \mid \mathscr{G}_t^n] \, ds \\ &\quad - \int_0^\infty e^{-s} E\left[\int_0^s g(X_n(t+u)) \, du \,\middle|\, \mathscr{G}_t^n \right] ds \\ &= f(X_n(t)) + \int_0^\infty e^{-s} E\Bigg[f(X_n(t+s)) \\ &\quad - f(X_n(t)) - \int_0^s g(X_n(t+u)) \, du \,\Bigg|\, \mathscr{G}_t^n \Bigg] ds \end{aligned} \tag{8.19}$$

and (8.10) follows from (8.7) and the dominated convergence theorem.

(c ⇒ a) Let $(f, g) \in A$, and let (ξ_n, φ_n) be defined by (8.17) and (8.18). We claim that $\{(\xi_n, \varphi_n)\}$ satisfies (8.8)–(8.11) with T replaced by ∞. As above, it is enough to consider (8.10). Let $T > 0$ and let $(\hat{\xi}_n, \hat{\varphi}_n) \in \hat{\mathscr{A}}_n$ satisfy (8.8)–(8.11) for all $k \geq 0$, $0 \leq t_1 < \cdots < t_k \leq t \leq T$, and $h_1, \ldots, h_k \in M$. Then

$$e^{-t}\hat{\xi}_n(t) + \int_0^t e^{-u}(\hat{\xi}_n(u) - \hat{\varphi}_n(u)) \, du \tag{8.20}$$

is a $\{\mathscr{G}_t^n\}$-martingale (by the same argument as used in the proof of Lemma 3.2), and for $0 \leq t \leq T$,

$$\hat{\xi}_n(t) = e^{-(T-t)} E[\hat{\xi}_n(T) \mid \mathscr{G}_t^n] + \int_t^T e^{-(s-t)} E[\hat{\xi}_n(s) - \hat{\varphi}_n(s) \mid \mathscr{G}_t^n] \, ds. \tag{8.21}$$

Let $k \geq 0$, $0 \leq t_1 < \cdots < t_k \leq t \leq T$, and $h_1, \ldots, h_k \in M$. Then

$$\begin{aligned} &E\left[(\xi_n(t) - f(X_n(t))) \prod_{i=1}^k h_i(X_n(t_i)) \right] \\ &= E\left[(\xi_n(t) - \hat{\xi}_n(t)) \prod_{i=1}^k h_i(X_n(t_i)) \right] \\ &\quad + E\left[(\hat{\xi}_n(t) - f(X_n(t))) \prod_{i=1}^k h_i(X_n(t_i)) \right]. \end{aligned} \tag{8.22}$$

The first term on the right of (8.22) can be written, by (8.21), as

$$(8.23)\quad E\left[\int_{T-t}^{\infty} e^{-s}E[f(X_n(t+s)) - g(X_n(t+s)) \mid \mathscr{G}_t^n]\, ds \prod_{i=1}^{k} h_i(X_n(t_i))\right]$$
$$- E\left[e^{-(T-t)}(E[\hat{\xi}_n(T) \mid \mathscr{G}_t^n] - f(X_n(T))) \prod_{i=1}^{k} h_i(X_n(t_i))\right]$$
$$- E\left[e^{-(T-t)} f(X_n(T)) \prod_{i=1}^{k} h_i(X_n(t_i))\right]$$
$$+ E\left[\int_0^{T-t} e^{-s}(E[f(X_n(t+s)) - g(X_n(t+s)) \mid \mathscr{G}_t^n]\right.$$
$$\left. - E[\hat{\xi}_n(t+s) - \hat{\varphi}_n(t+s) \mid \mathscr{G}_t^n])\, ds \prod_{i=1}^{k} h_i(X_n(t_i))\right].$$

As $n \to \infty$ the second term on the right of (8.22) goes to zero by (8.10), as do the second and fourth terms in (8.23). (Note that the conditioning may be dropped, and the dominated convergence theorem justifies the interchange of limits and integration in the fourth term. Observe that (8.15) can be used here in place of (8.9) and (8.11).) Consequently,

$$(8.24)\quad \overline{\lim_{n\to\infty}} \left| E\left[(\xi_n(t) - f(X_n(t))) \prod_{i=1}^{k} h_i(X_n(t_i))\right]\right|$$
$$\le e^{-(T-t)}(\|f-g\| + \|f\|) \prod_{i=1}^{k} \|h_i\|.$$

Since T is arbitrary the limit is in fact zero.

Let $\mathscr{L}_n^0$ be the Banach space of real-valued $\{\mathscr{G}_t^n\}$-progressive processes ξ with norm $\|\xi\| = \sup_t E[|\xi(t)|]$. Define $\pi_n: L \to \mathscr{L}_n^0$ by $\pi_n f(t) = f(X_n(t))$, and for $\xi_n \in \mathscr{L}_n^0$, $n = 1, 2, \ldots$, and $f \in L$, define LIM $\xi_n = f$ if $\sup_n \|\xi_n\| < \infty$ and

$$(8.25)\qquad \lim_{n\to\infty} E\left[(\xi_n(t) - \pi_n f(t)) \prod_{i=1}^{k} h_i(X_n(t_i))\right] = 0$$

for all $k \ge 0$, $0 \le t_1 < \cdots < t_k \le t$, and $h_1, \ldots, h_k \in M$.

Let $\{\mathscr{T}_n(s)\}$ denote the semigroup of conditioned shifts on $\mathscr{L}_n^0$. Clearly LIM $\xi_n = 0$ implies LIM $\mathscr{T}_n(s)\xi_n = 0$ for all $s \ge 0$, and LIM satisfies the conditions of Theorem 6.9 of Chapter 1. For each $(f, g) \in A$, we have shown there exist $(\xi_n, \varphi_n) \in \hat{\mathscr{A}}_n$ such that LIM $\xi_n = f$ and LIM $\varphi_n = g$. Consequently, Theorem 6.9 of Chapter 1 implies

$$(8.26)\qquad \text{LIM } \mathscr{T}_n(s)\pi_n f = T(s)f$$

for all $f \in L$ and $s \geq 0$. But (8.26) is just (8.2), and hence Lemma 8.1 implies (a). □

8.4 Corollary Suppose in Theorem 8.2 that $X_n = \eta_n \circ Y_n$ and $\{\mathscr{G}_t^n\} = \{\mathscr{F}_t^{Y_n}\}$, where Y_n is a progressive Markov process in a metric space E_n corresponding to a measurable contraction semigroup $\{T_n(t)\}$ with full generator $\hat{A}_n$, and η_n: $E_n \to E$ is Borel measurable. Then (a), (b), and (c) are equivalent to the following:

(d) For each $(f, g) \in A$ and $T > 0$, there exist $(f_n, g_n) \in \hat{A}_n$ such that $\{(\xi_n, \varphi_n)\} \equiv \{(f_n \circ Y_n, g_n \circ Y_n)\}$ satisfies (8.8)–(8.11) for all $k \geq 0$, $0 \leq t_1 < \cdots < t_k \leq t \leq T$, and $h_1, \ldots, h_k \in M$.

Proof. **(d ⇒ c)** It only needs to be observed that $(f_n, g_n) \in \hat{A}_n$ implies $(f_n \circ Y_n, g_n \circ Y_n) \in \hat{\mathscr{A}}_n$.

(c ⇒ d) By the Markov property, (ξ_n, φ_n) defined by (8.17) and (8.18) is of the form $(f_n \circ Y_n, g_n \circ Y_n)$ for some $(f_n, g_n) \in \hat{A}_n$, and (d) follows by (8.24). □

8.5 Corollary Suppose in Theorem 8.2 that $X_n = \eta_n(Y_n([\alpha_n \cdot]))$ and $\{\mathscr{G}_t^n\} = \{\mathscr{F}_{[\alpha_n t]}^{Y_n}\}$, where $\{Y_n(k), k = 0, 1, 2, \ldots\}$ is a Markov chain in a metric space E_n with transition function $\mu_n(x, \Gamma)$, η_n: $E_n \to E$ is Borel measurable, and $\alpha_n \to \infty$ as $n \to \infty$. Define T_n: $B(E_n) \to B(E_n)$ by

$$T_n f(x) = \int f(y)\mu_n(x, dy), \tag{8.27}$$

and let $A_n = \alpha_n(T_n - I)$. Then (a), (b), and (c) are equivalent to the following:

(e) For each $(f, g) \in A$ and $T > 0$, there exist $f_n \in B(E_n)$ such that for $g_n = A_n f_n$, $\{(\xi_n, \varphi_n)\} \equiv \{(f_n(Y_n([\alpha_n \cdot])), g_n(Y_n([\alpha_n \cdot])))\}$ satisfies (8.8)–(8.11) for all $k \geq 0$, $0 \leq t_1 < \cdots < t_k \leq t \leq T$, and $h_1, \ldots, h_k \in M$.

Proof. **(e ⇒ c)** In this case $(\xi_n, \varphi_n) \notin \hat{\mathscr{A}}_n$. Consequently, we define $(\hat{\xi}_n, \hat{\varphi}_n) \in \hat{\mathscr{A}}_n$ by

$$\begin{aligned}\hat{\xi}_n(t) &= \alpha_n \int_0^{1/\alpha_n} E[f_n(Y_n([\alpha_n(t + s)])) \mid \mathscr{G}_t^n]\, ds \\ &= f_n(Y_n([\alpha_n t])) + \left(\frac{t - [\alpha_n t]}{\alpha_n}\right) g_n(Y_n([\alpha_n t]))\end{aligned} \tag{8.28}$$

and

$$\hat{\varphi}_n(t) = \alpha_n E[f_n(Y_n([\alpha_n t] + 1)) - f_n(Y_n([\alpha_n t])) \mid \mathscr{G}_t^n] = g_n(Y_n([\alpha_n t])) = \varphi_n(t) \tag{8.29}$$

(see Proposition 7.5 of Chapter 2) and note that $E[|\xi_n(t) - \hat{\xi}_n(t)|] \le E[|\varphi_n(t)|]/\alpha_n$.

(b ⇒ e) Define $\pi_n: B(E) \to B(E_n)$ by $\pi_n f = f \circ \eta_n$. For $(f, g) \in A$, set

$$f_n = (1 + \alpha_n)^{-1} \sum_{k=0}^{\infty} \left(\frac{\alpha_n}{1 + \alpha_n}\right)^k T_n^k \pi_n(f - g) \tag{8.30}$$

and note that

$$g_n = A_n f_n = f_n + \pi_n g - \pi_n f. \tag{8.31}$$

For $t = k/\alpha_n$ ($k \in \mathbb{Z}_+$), (8.30) gives

(8.32)

$$\begin{aligned} f_n(Y_n([\alpha_n t])) &= E\left[(1 + \alpha_n)^{-1}\alpha_n \int_0^\infty \left(\frac{\alpha_n}{1 + \alpha_n}\right)^{[\alpha_n s]} (f(X_n(t + s)) - g(X_n(t + s)))\, ds \,\middle|\, \mathscr{G}_t^n\right] \\ &= E\left[(1 + \alpha_n)^{-1}\alpha_n \int_0^\infty \left(\left(\frac{\alpha_n}{1 + \alpha_n}\right)^{[\alpha_n s]} - e^{-s}\right)\right. \\ &\quad \left. \times (f(X_n(t + s)) - g(X_n(t + s)))\, ds \,\middle|\, \mathscr{G}_t^n\right] \\ &\quad + (1 + \alpha_n)^{-1}\alpha_n \int_0^\infty e^{-s} E[f(X_n(t + s)) - g(X_n(t + s)) \mid \mathscr{G}_t^n]\, ds. \end{aligned}$$

Since $\lim_{n\to\infty} (\alpha_n/(1 + \alpha_n))^{[\alpha_n s]} = e^{-s}$ uniformly in $s \ge 0$, the result now follows as in the proof of ($b \Rightarrow c$). □

8.6 Corollary Suppose in Theorem 8.2 that the X_n and X have sample paths in $D_E[0, \infty)$, and there is an algebra $C_a \subset L$ that separates points. Suppose either that the compact containment condition (7.9) of Chapter 3 holds for $\{X_n\}$ or that C_a strongly separates points. If $\{(\xi_n, \varphi_n)\}$ in condition (c) can be selected so that

$$\lim_{n\to\infty} E\left[\sup_{t \in \mathbb{Q} \cap [0, T]} |\xi_n(t) - f(X_n(t))|\right] = 0 \tag{8.33}$$

and

$$\sup_n E[\|\varphi_n\|_{p, T}] < \infty \quad \text{for some} \quad p \in (1, \infty], \tag{8.34}$$

then $X_n \Rightarrow X$.

Proof. This follows from Theorems 9.4, 9.1, 7.8 and Corollary 9.2, all of Chapter 3. Note that D in that chapter's Theorem 9.4 contains $\mathscr{D}(A)$. □

8.7 Corollary Suppose in Theorem 8.2 that the X_n and X have sample paths in $D_E[0, \infty)$, and that there is an algebra $C_a \subset L$ that separates points. Suppose either that the compact containment condition ((7.9) of Chapter 3) holds for $\{X_n\}$ or that C_a strongly separates points. If X_n has the form in Corollary 8.4 and $\pi_n f = f \circ \eta_n$, then either of the following conditions implies $X_n \Rightarrow X$.

(f) For each $(f, g) \in A$ and $T > 0$, there exist $(f_n, g_n) \in \hat{A}_n$ and $G_n \subset E_n$ such that $\{Y_n(t) \in G_n, 0 \leq t \leq T\}$ are events satisfying

$$\lim_{n\to\infty} P\{Y_n(t) \in G_n, 0 \leq t \leq T\} = 1, \tag{8.35}$$

$\sup_n \|f_n\| < \infty$, and

$$\lim_{n\to\infty} \sup_{y \in G_n} |\pi_n f(y) - f_n(y)| = \lim_{n\to\infty} \sup_{y \in G_n} |\pi_n g(y) - g_n(y)| = 0. \tag{8.36}$$

(g) For each $f \in L$ and $T > 0$, there exist $G_n \subset E_n$ such that (8.35) holds and

$$\lim_{n\to\infty} \sup_{y \in G_n} |T_n(t)\pi_n f(y) - \pi_n T(t)f(y)| = 0, \qquad 0 \leq t \leq T. \tag{8.37}$$

8.8 Remark **(a)** If $G_n = E_n$, then (8.35) is immediate and (f) and (g) are equivalent by Theorem 6.1 of Chapter 1.

(b) If the Y_n are continuous and the G_n are compact, then the assumption that $\sup_n \|f_n\| < \infty$ can be dropped. In general it is needed. For example, with $E_n = E = \{0, 1, 2, \ldots\}$ let

$$A_n f(k) = n^{-1}\delta_{0k}(f(n) - f(0)). \tag{8.38}$$

(Clearly, if X_n has generator A_n with $X_n(0) = 0$, then $X_n \Rightarrow X$ where $X \equiv 0$.) Let

$$Af(k) = \delta_{0k}(f(1) - f(0)). \tag{8.39}$$

Set $G_n = \{0, 1, 2, \ldots, n-1\}$ and

$$f_n(k) = f(k) + \delta_{nk}\, n(f(1) - f(0)). \tag{8.40}$$

Then

$$A_n f_n(k) = \delta_{0k}(f(1) - f(0) + n^{-1}(f(n) - f(0))), \tag{8.41}$$

and hence

$$\lim_{n\to\infty} \sup_{k \in G_n} |f_n(k) - f(k)| = \lim_{n\to\infty} \sup_{k \in G_n} |A_n f_n(k) - Af(k)| = 0 \tag{8.42}$$

suggesting (but not implying!) that $X_n \Rightarrow \tilde{X}$, where $\tilde{X}$ has generator A. Of course $\sup_n \|f_n\| = \infty$. □

Proof. Assume (g) holds. For $(f, g) \in A$, let

$$f_n = \int_0^\infty e^{-t} T_n(t)\pi_n(f - g)\, dt, \qquad g_n = f_n - \pi_n(f - g). \tag{8.43}$$

Then $(f_n, g_n) \in \hat{A}_n$ and

$$\begin{aligned} |\pi_n f(y) - f_n(y)| &= |\pi_n g(y) - g_n(y)| \\ &= \left| \int_0^\infty e^{-t} T_n(t)\pi_n(f - g)(y)\, dt - \int_0^\infty e^{-t}\pi_n T(t)(f - g)(y)\, dt \right| \\ &\le \int_0^T e^{-t} |T_n(t)\pi_n(f - g)(y) - \pi_n T(t)(f - g)(y)|\, dt \\ &\quad + 2e^{-T}\|f - g\|. \end{aligned} \tag{8.44}$$

Using (8.37), the dominated convergence theorem, and the arbitrariness of T, a sequence G_n can be constructed satisfying (8.35) (for each $T > 0$) and (8.36). Consequently (g) implies (f).

To see that (f) implies $X_n \Rightarrow X$, fix $(f, g) \in A$ and $T > 0$. Assuming $(f_n, g_n) \in \hat{A}_n$ and $G_n \subset E_n$ are as in (f), define

$$\tau_n = \inf\left\{t > 0: \int_0^t |g_n(Y_n(s))|^2\, ds \ge t(\|g\|^2 + 1)\right\}. \tag{8.45}$$

Note that (8.35) and (8.36) imply $\lim_{n\to\infty} P\{\tau_n < T\} = 0$. Set

$$\xi_n(t) = f_n(Y_n(t \wedge \tau_n)), \qquad \varphi_n(t) = g_n(Y_n(t))\chi_{\{\tau_n > t\}}. \tag{8.46}$$

Then ξ_n and φ_n satisfy (8.33) and (8.34) with $p = 2$ as well as (8.8), (8.10), and (8.15). □

8.9 Corollary Suppose in Theorem 8.2 that the X_n and X have sample paths in $D_E[0, \infty)$ and that there is an algebra $C_a \subset L$ that separates points. Suppose either that the compact containment condition ((7.9) of Chapter 3) holds for $\{X_n\}$ or that C_a strongly separates points. If X_n has the form in Corollary 8.5 and $\pi_n f = f \circ \eta_n$, then either of the following conditions implies $X_n \Rightarrow X$:

(h) For each $(f, g) \in A$ and $T > 0$, there exist $f_n \in B(E_n)$ and $G_n \subset E_n$ such that $\{Y_n([\alpha_n t]) \in G_n, 0 \le t \le T\}$ are events satisfying

$$\lim_{n\to\infty} P\{Y_n([\alpha_n t]) \in G_n, 0 \le t \le T\} = 1, \tag{8.47}$$

$\sup_n \|f_n\| < \infty$, and

$$(8.48) \qquad \lim_{n\to\infty} \sup_{y \in G_n} |\pi_n f(y) - f_n(y)| = \lim_{n\to\infty} \sup_{y \in G_n} |\pi_n g(y) - A_n f_n(y)| = 0.$$

(i) For each $f \in L$ and $T > 0$, there exist $G_n \subset E_n$ such that (8.47) holds and

$$(8.49) \qquad \lim_{n\to\infty} \sup_{y \in G_n} | T_n^{[\alpha_n t]} \pi_n f(y) - \pi_n T(t) f(y)| = 0, \qquad 0 \le t \le T.$$

Proof. The proof is essentially the same as that of Corollary 8.7 using (8.30) in place of (8.43), and (8.28) (appropriately stopped) in place of (8.45). □

We now give an analogue of Theorem 8.2 in which the assumption that the closure of A generates a strongly continuous semigroup is relaxed to the assumption of uniqueness of solutions of the martingale problem. We must now, however, assume a priori that the sequence $\{X_n\}$ is relatively compact. Note that $\{\mathscr{G}_t^n\}$ and $\hat{\mathscr{A}}_n$ are as in Theorem 8.2.

8.10 Theorem Let (E, r) be complete and separable. Let $A \subset \bar{C}(E) \times \bar{C}(E)$ and $\nu \in \mathscr{P}(E)$, and suppose that the $D_E[0, \infty)$ martingale problem for (A, ν) has at most one solution. Suppose X_n, $n = 1, 2, \ldots$, is a $\{\mathscr{G}_t^n\}$-adapted process with sample paths in $D_E[0, \infty)$, $\{X_n\}$ is relatively compact, $PX_n(0)^{-1} \Rightarrow \nu$, and $M \subset \bar{C}(E)$ is separating. Then the following are equivalent:

(a′) There exists a solution X of the $D_E[0, \infty)$ martingale problem for (A, ν), and $X_n \Rightarrow X$.

(b′) There exists a countable set $\Gamma \subset [0, \infty)$ such that for each $(f, g) \in A$

(8.50)

$$\lim_{n\to\infty} E\left[\left(f(X_n(t+s)) - f(X_n(t)) - \int_t^{t+s} g(X_n(u))\, du\right) \prod_{i=1}^k h_i(X_n(t_i))\right] = 0$$

for all $k \ge 0$, $0 \le t_1 < t_2 < \cdots < t_k \le t < t + s$ with $t_i, t, t + s \notin \Gamma$, and $h_1, \ldots, h_k \in M$.

(c′) There exists a countable set $\Gamma \subset [0, \infty)$ such that for each $(f, g) \in A$ and $T > 0$, there exist $(\xi_n, \varphi_n) \in \hat{\mathscr{A}}_n$ such that

$$\sup_n \sup_{s \le T} E[|\xi_n(s)|] < \infty, \tag{8.51}$$

$$\sup_n \sup_{s \le T} E[|\varphi_n(s)|] < \infty, \tag{8.52}$$

$$\lim_{n \to \infty} E\left[(\xi_n(t) - f(X_n(t))) \prod_{i=1}^{k} h_i(X_n(t_i))\right] = 0, \tag{8.53}$$

and

$$\lim_{n \to \infty} E\left[(\varphi_n(t) - g(X_n(t))) \prod_{i=1}^{k} h_i(X_n(t_i))\right] = 0 \tag{8.54}$$

for all $k \ge 0$, $0 \le t_1 < t_2 < \cdots < t_k \le t \le T$ with t_i, $t \notin \Gamma$, and $h_1, \ldots, h_k \in M$.

8.11 Remark As in Theorem 8.2, (8.52) and (8.54) can be replaced by (8.15). □

Proof. **(a′ ⇒ b′)** Take $\Gamma = [0, \infty) - D(X)$. ($D(X) = \{t \ge 0\colon P\{X(t) = X(t-)\} = 1\}$.) By Theorem 7.8 of Chapter 3, $(X_n(t_1), \ldots, X_n(t_k)) \Rightarrow (X(t_1), \ldots, X(t_k))$ for all finite sets $\{t_1, t_2, \ldots, t_k\} \subset D(X)$, and this implies (8.50).

(b′ ⇒ c′) The proof is essentially the same as in Theorem 8.2.

(c′ ⇒ a′) Let Y be a limit point of $\{X_n\}$. Let $(f, g) \in A$ and $T > 0$, and let $\{(\xi_n, \varphi_n)\}$ satisfy the assertions of condition (c′). Let $k \ge 0$, $0 \le t_1 < \cdots < t_k \le t < t + s \le T$ with t_i, t, $t + s \in D(Y)$ and $h_1, \ldots, k_k \in M$. Since

$$E\left[\left(\xi_n(t + s) - \xi_n(t) - \int_t^{t+s} \varphi_n(u)\, du\right) \prod_{i=1}^{k} h_i(X_n(t_i))\right] = 0, \tag{8.55}$$

it follows that

(8.56)

$$\lim_{n \to \infty} E\left[\left(f(X_n(t + s)) - f(X_n(t)) - \int_t^{t+s} g(X_n(u))\, du\right) \prod_{i=1}^{k} h_i(X_n(t_i))\right] = 0,$$

and hence

$$E\left[\left(f(Y(t + s)) - f(Y(t)) - \int_t^{t+s} g(Y(u))\, du\right) \prod_{i=1}^{k} h_i(Y(t_i))\right] = 0. \tag{8.57}$$

By the right continuity of Y, (8.55) holds for all $0 \le t_1 < \cdots < t_k \le t < t + s$, and hence Y is a solution of the martingale problem for (A, ν). Therefore (a′) follows by the assumption of uniqueness. □

We state analogues of Corollaries 8.4–8.7 and 8.9. Their proofs are similar and are omitted.

8.12 Corollary Suppose in Theorem 8.10 that $X_n = \eta_n \circ Y_n$ and $\{\mathscr{G}_t^n\} = \{\mathscr{F}_t^{Y_n}\}$, where Y_n is a progressive Markov process in a metric space E_n corresponding to a measurable contraction semigroup $\{T_n(t)\}$ with full generator $\hat{A}_n$, and $\eta_n: E_n \to E$ is Borel measurable. Then condition (f) of Corollary 8.7 implies (a′), and (a′), (b′), and (c′) are equivalent to the following:

(d′) There exists a countable set $\Gamma \subset [0, \infty)$ such that for every $(f, g) \in A$ and $T > 0$, there exist $(f_n, g_n) \in \hat{A}_n$ such that $\{(\xi_n, \varphi_n)\} = \{(f_n \circ Y_n, g_n \circ Y_n)\}$ satisfies (8.51)–(8.54) for all $k \geq 0$, $0 \leq t_1 < \cdots < t_k \leq t \leq T$ with $t_i, t \notin \Gamma$, and $h_1, \ldots, h_k \in M$.

8.13 Corollary Suppose in Theorem 8.10 that $X_n = \eta_n \circ Y_n([\alpha_n \cdot])$ and $\{\mathscr{G}_t^n\} = \{\mathscr{F}_{[\alpha_n t]}^{Y_n}\}$, where $\{Y_n(k), k = 0, 1, \ldots\}$ is a Markov chain in a metric space E_n with transition function $\mu_n(x, \Gamma)$, $\eta_n: E_n \to E$ is Borel measurable, and $\alpha_n \to \infty$ as $n \to \infty$. Define $T_n: B(E_n) \to B(E_n)$ by (8.27) and let $A_n = \alpha_n(T_n - I)$. Then condition (h) of Corollary 8.9 implies (a′), and (a′), (b′), and (c′) are equivalent to the following:

(e′) There exists a countable set $\Gamma \subset [0, \infty)$ such that for every $(f, g) \in A$ and $T > 0$, there exist $f_n \in B(E_n)$ such that for $g_n = A_n f_n$, $\{(\xi_n, \varphi_n)\} = \{(f_n(Y_n([\alpha_n \cdot])), g_n(Y_n([\alpha_n \cdot])))\}$ satisfies (8.51)–(8.54) for all $k \geq 0$, $0 \leq t_1 < \cdots < t_k \leq t \leq T$ with $t_i, t \notin \Gamma$, and $h_1, \ldots, h_k \in M$. (Note that we are not claiming $(\xi_n, \varphi_n) \in \mathscr{A}_n$.)

8.14 Remark In the following three corollaries we do not assume a priori that $\{X_n\}$ is relatively compact. We do assume the compact containment condition. The assumption that C_a strongly separates points used in the analogous corollaries to Theorem 8.2 does not suffice. □

8.15 Corollary Let (E, r) be complete and separable and let $\{X_n\}$ be a sequence of processes with sample paths in $D_E[0, \infty)$. Suppose $PX_n(0)^{-1} \Rightarrow \nu \in \mathscr{P}(E)$ and the compact containment condition ((7.9) of Chapter 3) holds. Suppose $A \subset \bar{C}(E) \times \bar{C}(E)$, the closure of the linear span of $\mathscr{D}(A)$ contains an algebra that separates points, and the $D_E[0, \infty)$ martingale problem for (A, ν) has at most one solution. If $\{(\xi_n, \varphi_n)\}$ in condition (c′) of Theorem 8.10 can be selected so that (8.33) and (8.34) hold, then (a′) holds.

8.16 Corollary Instead of assuming in Corollary 8.12 that $\{X_n\}$ is relatively compact, suppose that $\{X_n\}$ satisfies the compact containment condition ((7.9) of Chapter 3) and the closure of the linear span of $\mathscr{D}(A)$ contains an algebra that separates points. Then condition (f) of Corollary 8.7 implies condition (a′) of Theorem 8.10.

8.17 Corollary Instead of assuming in Corollary 8.13 that $\{X_n\}$ is relatively compact, suppose that $\{X_n\}$ satisfies the compact containment condition ((7.9) of Chapter 3) and that the closure of the linear span of $\mathscr{D}(A)$ contains an algebra that separates points. Then condition (h) of Corollary 8.9 implies condition (a′) of Theorem 8.10.

The following proposition may be useful in verifying (8.7) or (8.50) and as a result gives an alternative convergence criterion.

8.18 Proposition Let X_n, A, and $\{\mathscr{G}_t^n\}$ be as in Theorem 8.2. Let $(f, g) \in A$. For $n = 1, 2, \ldots$, let $0 = \tau_0^n < \tau_1^n < \cdots$ be an increasing sequence of $\{\mathscr{G}_t^n\}$-stopping times with $\tau_k^n < \infty$ a.s. and $\lim_{k\to\infty} \tau_k^n = \infty$ a.s. Define

$$\tau_n^+(t) = \min \{\tau_k^n : \tau_k^n > t\}, \tag{8.58}$$

$$\tau_n^-(t) = \max \{\tau_k^n : \tau_k^n \le t\}, \tag{8.59}$$

and

$$H_n(t) = \sum_{k=0}^{\infty} (E[f(X_n(\tau_{k+1}^n)) - f(X_n(\tau_k^n)) \mid \mathscr{G}_{\tau_k^n}^n] - g(X_n(\tau_k^n))E[\tau_{k+1}^n - \tau_k^n \mid \mathscr{G}_{\tau_k^n}^n])\chi_{\{\tau_k^n \le t\}}. \tag{8.60}$$

If

$$\lim_{n\to\infty} E[\,|f(X_n(\tau_n^+(t))) - f(X_n(t))|\,] = \lim_{n\to\infty} E[\tau_n^+(t) - \tau_n^-(t)] = 0, \qquad t \ge 0, \tag{8.61}$$

$$\lim_{n\to\infty} E[\,|g(X_n(\tau_n^-(t))) - g(X_n(t))|\,] = 0, \qquad \text{a.e. } t \ge 0, \tag{8.62}$$

and

$$\lim_{n\to\infty} E[\,|H_n(t)|\,] = 0, \qquad t \ge 0, \tag{8.63}$$

then

$$\lim_{n\to\infty} E\left[\left|E\left[f(X_n(t+s)) - f(X_n(t)) - \int_t^{t+s} g(X_n(u))\,du \,\middle|\, \mathscr{G}_t^n\right]\right|\right] = 0 \tag{8.64}$$

for all $s, t \ge 0$, which in turn implies, by (8.19),

$$\lim_{n\to\infty} E[\,|\xi_n(t) - f(X_n(t))|\,] = \lim_{n\to\infty} E[\,|\varphi_n(t) - g(X_n(t))|\,] = 0, \qquad t \ge 0, \tag{8.65}$$

where ξ_n and φ_n are given by (8.17) and (8.18).

Proof. For any $\{\mathscr{G}_t^n\}$-stopping time τ and any Z such that $E[|Z|] < \infty$,

$$E[E[Z \mid \mathscr{G}_\tau^n] \mid \mathscr{G}_t^n]\chi_{\{\tau > t\}} = E[Z \mid \mathscr{G}_t^n]\chi_{\{\tau > t\}} = E[Z\chi_{\{\tau > t\}} \mid \mathscr{G}_t^n]. \tag{8.66}$$

Consequently,

$$\begin{aligned}
&E\left[f(X_n(t+s)) - f(X_n(t)) - \int_t^{t+s} g(X_n(u))\,du - (H_n(t+s) - H_n(t)) \middle| \mathscr{G}_t^n \right] \\
&\quad = E\left[f(X_n(t+s)) - f(X_n(t)) - \int_t^{t+s} g(X_n(u))\,du \middle| \mathscr{G}_t^n \right] \\
&\qquad - \sum E[(f(X_n(\tau_{k+1}^n)) - f(X_n(\tau_k^n)))\chi_{\{t < \tau_k^n \le t+s\}} \mid \mathscr{G}_t^n] \\
&\qquad + \sum E[g(X_n(\tau_k^n))(\tau_{k+1}^n - \tau_k^n)\chi_{\{t < \tau_k^n \le t+s\}} \mid \mathscr{G}_t^n] \\
&\quad = E[f(X_n(t+s)) - f(X_n(\tau_n^+(t+s))) \mid \mathscr{G}_t^n] \\
&\qquad - E[f(X_n(t)) - f(X_n(\tau_n^+(t))) \mid \mathscr{G}_t^n] \\
&\qquad - E\left[\int_t^{t+s} g(X_n(u))\,du - \int_{\tau_n^+(t)}^{\tau_n^+(t+s)} g(X_n(\tau_n^-(u)))\,du \middle| \mathscr{G}_t^n \right] \\
&\quad \equiv I_n(t).
\end{aligned} \tag{8.67}$$

By (8.61), (8.62), and the dominated convergence theorem, $I_n(t)$ converges to zero in L^1. The quantity in (8.64) is bounded by $E(|H_n(t+s)|) + E(|H_n(t)|) + E(|I_n(t)|)$ and the limit follows by (8.63). □

9. STATIONARY DISTRIBUTIONS

Let $A \subset B(E) \times B(E)$ and suppose the martingale problem for A is well-posed. Then $\mu \in \mathscr{P}(E)$ is a *stationary distribution for A* if every solution X of the martingale problem for (A, μ) is a *stationary process*, that is, if $P\{X(t+s_1) \in \Gamma_1, X(t+s_2) \in \Gamma_2, \ldots, X(t+s_k) \in \Gamma_k\}$ is independent of $t \ge 0$ for all $k \ge 1$, $0 \le s_1 < \cdots < s_k$, and $\Gamma_1, \ldots, \Gamma_k \in \mathscr{B}(E)$.

The following lemma shows that to check that μ is a stationary distribution it is sufficient to consider only the one-dimensional distributions.

9.1 Lemma Let $A \subset B(E) \times B(E)$ and suppose the martingale problem for A is well-posed. Let $\mu \in \mathscr{P}(E)$ and let X be a solution of the martingale problem for (A, μ). Then μ is a stationary distribution for A if and only if $X(t)$ has distribution μ for all $t \ge 0$.

Proof. The necessity is immediate. For sufficiency observe that $X_t \equiv X(t + \cdot)$ is a solution of the martingale problem for (A, μ), and hence, by uniqueness, has the same finite-dimensional distributions as X. □

9.2 Proposition Suppose A generates a strongly continuous contraction semigroup $\{T(t)\}$ on a closed subspace $L \subset B(E)$, L is separating , and the martingale problem for A is well-posed. If D is a core for A and $\mu \in \mathscr{P}(E)$, then the following are equivalent:

(a) μ is a stationary distribution for A.

(b) $\int T(t)f \, d\mu = \int f \, d\mu, \quad f \in L, t \geq 0$.

(c) $\int Af \, d\mu = 0, \quad f \in D$.

Proof. **(a ⇒ b)** If X is a solution of the martingale problem for (A, μ), then by (4.2) and (a),

$$E[T(t)f(X(0))] = E[f(X(t))] = E[f(X(0))], \qquad f \in L, \; t \geq 0, \tag{9.1}$$

which is (b).

(b ⇒ a) Let X be a solution of the martingale problem for (A, μ). By (4.2) and (b),

$$E[f(X(t))] = E[T(t)f(X(0))] = E[f(X(0))], \qquad f \in L, \; t \geq 0. \tag{9.2}$$

Since L is separating, $X(t)$ has distribution μ for each $t \geq 0$, and (a) follows by Lemma 9.1.

(b ⇒ c) This is immediate from the definition of A.

(c ⇒ b) Since A is the closure of A restricted to D, we may as well take $D = \mathscr{D}(A)$. Then, by (c),

$$\begin{aligned} \int (T(t)f - f) \, d\mu &= \int\int_0^t AT(s)f \, ds \, d\mu \\ &= \int_0^t \int AT(s)f \, d\mu \, ds = 0, \end{aligned} \tag{9.3}$$

for each $f \in \mathscr{D}(A)$ and $t \geq 0$. Since $\mathscr{D}(A)$ is dense in L, (b) follows. □

If $\{T(t)\}$ is a semigroup on $B(E)$ given by a transition function and condition (b) of Proposition 9.2 holds (with $L = B(E)$), we say that μ is a *stationary distribution for* $\{T(t)\}$.

An immediate question is that of the existence of a stationary distribution. Compactness of the state space is usually sufficient for existence. This observation is a special case of the following theorem.

9.3 Theorem Suppose A generates a strongly continuous contraction semigroup $\{T(t)\}$ on a closed subspace $L \subset \bar{C}(E)$, L is separating, and the martingale problem for A is well-posed. Let X be a solution of the martingale problem for A, and for some sequence $t_n \to \infty$, define $\{\mu_n\} \subset \mathscr{P}(E)$ by

$$\mu_n(\Gamma) = t_n^{-1} \int_0^{t_n} P\{X(s) \in \Gamma\}\, ds, \qquad \Gamma \in \mathscr{B}(E). \tag{9.4}$$

If μ is the weak limit of a subsequence of $\{\mu_n\}$, then μ is a stationary distribution for A.

9.4 Remark The theorem avoids the question of existence of a weak limit point for $\{\mu_n\}$. Of course, if E is compact, then $\{\mu_n\}$ is relatively compact by Theorem 2.2 of Chapter 3, and existence follows. □

Proof. Since the sequence $\{t_n\}$ was arbitrary, we may as well assume $\mu_n \Rightarrow \mu$. For $f \in L$ and $t \geq 0$, $T(t)f \in L \subset \bar{C}(E)$, so

$$\begin{aligned}
\int T(t)f\, d\mu &= \lim_{n\to\infty} \int T(t)f\, d\mu_n \\
&= \lim_{n\to\infty} t_n^{-1} \int_0^{t_n} E[T(t)f(X(s))]\, ds \\
&= \lim_{n\to\infty} t_n^{-1} \int_0^{t_n} \int T(t+s)f\, dv\, ds \\
&= \lim_{n\to\infty} t_n^{-1} \int_t^{t+t_n} \int T(s)f\, dv\, ds \\
&= \lim_{n\to\infty} t_n^{-1} \int_0^{t_n} \int T(s)f\, dv\, ds \\
&= \lim_{n\to\infty} t_n^{-1} \int_0^{t_n} E[f(X(s))]\, ds \\
&= \int f\, d\mu,
\end{aligned} \tag{9.5}$$

where $v = PX(0)^{-1}$, and hence μ is a stationary distribution for A by Proposition 9.2. □

We now turn to the problem of verifying the relative compactness of $\{\mu_n\}$ in Theorem 9.3. Probably the most useful general approach involves the construction of what is called a Lyapunov function. The following lemmas provide one approach to the construction. For diffusion processes, Itô's formula (Theorem 2.9 of Chapter 5) provides a more direct approach. See also Problem 35.

9.5 Lemma Let $A \subset B(E) \times B(E)$ and let $\varphi, \psi \in M(E)$. Suppose there exist $\{(f_k, g_k)\} \subset A$ and a constant K such that

(9.6) $$0 \le f_k \le \varphi, \qquad k = 1, 2, \ldots,$$

(9.7) $$\lim_{k \to \infty} f_k(x) = \varphi(x), \qquad x \in E,$$

(9.8) $$g_k \le K, \qquad k = 1, 2, \ldots,$$

and

(9.9) $$\lim_{k \to \infty} g_k(x) = \psi(x), \qquad x \in E.$$

If X is a solution of the martingale problem for A with $E[\varphi(X(0))] < \infty$, then

(9.10) $$\varphi(X(t)) - \int_0^t \psi(X(s))\, ds$$

is an $\{{}^*\mathscr{F}_t^X\}$-supermartingale.

9.6 Remark Note that we only require that the g_k be bounded above. □

Proof. Since

(9.11) $$E[f_k(X(t))] = E[f_k(X(0))] + E\left[\int_0^t g_k(X(s))\, ds\right] \le E[f_k(X(0))] + Kt,$$

it follows that

(9.12) $$E[\varphi(X(t))] = \lim_{k \to \infty} E[f_k(X(t))] \le E[\varphi(X(0))] + Kt$$

and

(9.13) $$\lim_{k \to \infty} E[\,|\varphi(X(t)) - f_k(X(t))|\,] = 0.$$

Since g_k is bounded above uniformly in k, Fatou's lemma implies

(9.14) $$\overline{\lim_{k \to \infty}}\, E\left[\int_t^{t+s} g_k(X(u))\, du \,\middle|\, {}^*\mathscr{F}_t^X\right] \le E\left[\int_t^{t+s} \psi(X(u))\, du \,\middle|\, {}^*\mathscr{F}_t^X\right]$$

Putting (9.13) and (9.14) together we have

(9.15) $$\begin{aligned} 0 &= \lim_{k \to \infty} E\left[f_k(X(t+s)) - f_k(X(t)) - \int_t^{t+s} g_k(X(u))\, du \,\middle|\, {}^*\mathscr{F}_t^X\right] \\ &\ge E\left[\varphi(X(t+s)) - \varphi(X(t)) - \int_t^{t+s} \psi(X(u))\, du \,\middle|\, {}^*\mathscr{F}_t^X\right], \end{aligned}$$

and it follows that (9.10) is a supermartingale. □

9.7 Lemma Let E be locally compact and separable but not compact, and let $E^\Delta = E \cup \{\Delta\}$ be its one-point compactification. Let $\varphi, \psi \in M(E)$ and let X be a measurable E-valued process. Suppose

$$\varphi(X(t)) - \int_0^t \psi(X(s))\, ds \tag{9.16}$$

is a supermartingale, $\varphi \geq 0$, $\psi \leq C$ for some constant C, and $\lim_{x \to \Delta} \psi(x) = -\infty$. Then $\{\mu_t : t \geq 1\} \subset \mathscr{P}(E)$, defined by

$$\mu_t(\Gamma) = t^{-1} \int_0^t P\{X(s) \in \Gamma\}\, ds, \tag{9.17}$$

is relatively compact.

Proof. Given $m \geq 1$, select K_m compact so that

$$\sup_{x \notin K_m} \psi(x) \leq -m. \tag{9.18}$$

Then, for each $t \geq 1$,

$$\begin{aligned} 0 \leq E[\varphi(X(t))] &\leq E[\varphi(X(0))] + E\left[\int_0^t \psi(X(s))\, ds\right] \\ &\leq E[\varphi(X(0))] + CE\left[\int_0^t \chi_{K_m}(X(s))\, ds\right] \\ &\quad - mE\left[\int_0^t \chi_{K_m^c}(X(s))\, ds\right] \\ &= E[\varphi(X(0))] + (C + m) \int_0^t P\{X(s) \in K_m\}\, ds - mt, \end{aligned} \tag{9.19}$$

and therefore

$$\begin{aligned} \frac{m}{C+m} - \frac{t^{-1}}{C+m} E[\varphi(X(0))] &\leq t^{-1} \int_0^t P\{X(s) \in K_m\}\, ds \\ &= \mu_t(K_m). \end{aligned} \tag{9.20}$$

Consequently, the relative compactness follows by Prohorov's theorem (Theorem 2.2 of Chapter 3). □

9.8 Corollary Let E be locally compact and separable. Let $A \subset B(E) \times B(E)$ and $\varphi, \psi \in M(E)$. Suppose that $\varphi \geq 0$, that $\psi \leq C$ for some constant C and $\lim_{x \to \Delta} \psi(x) = -\infty$, and that for every solution X of the martingale problem for A satisfying $E[\varphi(X(0))] < \infty$, (9.16) is a supermartingale. If X is a stationary solution of the martingale problem for A and $\mu = PX(0)^{-1}$, then

$$\mu(K_m) \geq \frac{m}{C+m}, \qquad m \geq 1, \tag{9.21}$$

where $K_m = \{x: \psi(x) \geq -m\}$.

Proof. Let $\alpha > 0$ satisfy $P\{\varphi(X(0)) \leq \alpha\} > 0$, and let Y be a solution of the martingale problem for A with $P\{Y \in B\} = P\{X \in B \mid \varphi(X(0)) \leq \alpha\}$ (cf. (4.12)). By (9.20)

$$\begin{aligned} \frac{m}{C+m} P\{\varphi(X(0)) \leq \alpha\} &\leq \underline{\lim}_{t\to\infty} t^{-1} \int_0^t P\{Y(s) \in K_m\}\, ds\, P\{\varphi(X(0)) \leq \alpha\} \\ &= \underline{\lim}_{t\to\infty} t^{-1} \int_0^t P\{X(s) \in K_m, \varphi(X(0)) \leq \alpha\}\, ds \\ &\leq \mu(K_m). \end{aligned} \tag{9.22}$$

Since α is arbitrary, (9.21) follows. □

If X is a Markov process corresponding to a semigroup $T(t): \bar{C}(E) \to \bar{C}(E)$, then we can relax the conditions on the Lyapunov function φ and still show existence of a stationary distribution even though we do not get relative compactness for $\{\mu_t: t \geq 1\}$.

9.9 Theorem Let E be locally compact and separable. Let $\{T(t)\}$ be a semigroup on $B(E)$ given by a transition function $P(t, x, \Gamma)$ such that $T(t): \bar{C}(E) \to \bar{C}(E)$ for all $t \geq 0$, and let X be a measurable Markov process corresponding to $\{T(t)\}$. Let $\varphi, \psi \in M(E)$, $\varphi \geq 0$, $\psi \leq C$ for some constant C, and $\overline{\lim}_{x\to\Delta} \psi(x) < 0$, and suppose (9.16) is a supermartingale. Then there is a stationary distribution for $\{T(t)\}$.

Proof. Select $\varepsilon > 0$ and K compact so that $\sup_{x\notin K} \psi(x) \leq -\varepsilon$. Then, as in the proof of Lemma 9.7,

$$\mu_t(K) \geq \frac{\varepsilon}{C+\varepsilon} - \frac{t^{-1}}{C+\varepsilon} E[\varphi(X(0))], \tag{9.23}$$

for all $t \geq 1$, where μ_t is given by (9.17). By Theorem 2.2 of Chapter 3, $\{\mu_t\}$ is relatively compact in $\mathscr{P}(E^\Delta)$. Let $\nu \in \mathscr{P}(E^\Delta)$ be a weak limit point of $\{\mu_t\}$ as $t \to \infty$, and let ν_E be its restriction to E. It follows as in (9.5) that for nonnegative $f \in \hat{C}(E)$,

$$\int f\, d\nu_E = \int f\, d\nu = \lim_{n\to\infty} \int T(t) f\, d\mu_{t_n} \geq \int T(t) f\, d\nu_E. \tag{9.24}$$

Note that if $T(t)f \in \hat{C}(E)$, then we have equality in (9.24). By (9.23), $\nu_E(E) > 0$ so $\mu \equiv \nu_E/\nu_E(E) \in \mathscr{P}(E)$ and

$$\int f\,d\mu \geq \int T(t)f\,d\mu \tag{9.25}$$

for all nonnegative $f \in \hat{C}(E)$ and $t \geq 0$. By the Dynkin class theorem (Appendix 4), (9.25) holds for all nonnegative $f \in B(E)$, in particular for indicators, so

$$\mu(\Gamma) \geq \int P(t, x, \Gamma)\mu(dx), \tag{9.26}$$

for all $\Gamma \in \mathscr{B}(E)$ and $t \geq 0$. But both sides of (9.26) are probability measures, so we must in fact have equality in (9.26) and hence in (9.25). □

The results in Section 8 give conditions under which a sequence of processes converge in some sense to a limiting process. If $\{X_n\}$ is a sequence of stationary processes and $X_n \Rightarrow X$ or, more generally, the finite-dimensional distributions of X_n converge weakly to those of X, then X is stationary. Given this observation, if $\{A_n\}$ is a sequence of generators determining Markov processes (i.e., the martingale problem for A_n is well-posed) and if, for each n, μ_n is a stationary distribution for A_n, then, under the hypotheses of one of the convergence theorems of Section 8, one would expect that the sequence $\{\mu_n\}$ would converge weakly to a stationary distribution for the limiting generator A. This need not be the case in general, but the following theorem is frequently applicable.

9.10 Theorem Let (E, r) be complete and separable. Let $\{T_n(t)\}$, $\{T(t)\}$ be contraction semigroups corresponding to Markov processes in E, and suppose that for each n, $\mu_n \in \mathscr{P}(E)$ is a stationary distribution for $\{T_n(t)\}$. Let $L \subset \bar{C}(E)$ be separating and $T(t)\colon L \to L$ for all $t \geq 0$. Suppose that for each $f \in L$ and compact $K \subset E$,

$$\lim_{n\to\infty} \sup_{x \in K} |T_n(t)f(x) - T(t)f(x)| = 0, \qquad t \geq 0. \tag{9.27}$$

Then every weak limit point of $\{\mu_n\}$ is a stationary distribution for $\{T(t)\}$.

9.11 Remark Note that if $x_n \to x$ implies $T_n(t)f(x_n) \to T(t)f(x)$ for all $t \geq 0$, then (9.27) holds. □

Proof. For simplicity, assume $\mu_n \Rightarrow \mu$. Then for each $f \in L$, $t \geq 0$, and compact $K \subset E$,

$$\begin{aligned}(9.28)\qquad & \left|\int T(t)f\,d\mu - \int f\,d\mu\right| \\ &\leq \overline{\lim_{n\to\infty}} \left|\int T(t)f\,d\mu - \int T(t)f\,d\mu_n\right| \\ &\quad + \overline{\lim_{n\to\infty}} \left|\int (T(t)f - T_n(t)f)\,d\mu_n\right| \\ &\quad + \overline{\lim_{n\to\infty}} \left|\int f\,d\mu_n - \int f\,d\mu\right| \\ &\leq \overline{\lim_{n\to\infty}}\, 2\|f\|\mu_n(K^c).\end{aligned}$$

But by Prohorov's theorem (Theorem 2.2 of Chapter 3), for every $\varepsilon > 0$ there is a compact $K \subset E$ such that $\mu_n(K^c) < \varepsilon$ for all n. Consequently, the left side of (9.28) is zero. □

Theorem 9.10 can be generalized considerably.

9.12 Theorem Let (E_n, r_n), $n = 1, 2, \ldots$, and (E, r) be complete, separable metric spaces, let $A_n \subset B(E_n) \times B(E_n)$ and $A \subset B(E) \times B(E)$, and suppose that the martingale problems for the A_n and A are well-posed. For $\nu_n \in \mathscr{P}(E_n)$ (respectively, $\nu \in \mathscr{P}(E)$), let $X_n^{\nu_n}$ (respectively, X^ν) denote a solution of the martingale problem for (A_n, ν_n) (respectively, (A, ν)). Let $\eta_n: E_n \to E$ be Borel measurable, and suppose that for each choice of $\nu_n \in \mathscr{P}(E_n)$, $n = 1, 2, \ldots$, and any subsequence of $\{\nu_n \eta_n^{-1}\}$ that converges weakly to $\nu \in \mathscr{P}(E)$, the finite-dimensional distributions of the corresponding subsequence of $\{\eta_n \circ X_n^{\nu_n}\}$ converge weakly to those of X^ν. If for $n = 1, 2, \ldots$, μ_n is a stationary distribution for A_n, then any weak limit point of $\{\mu_n \eta_n^{-1}\}$ is a stationary distribution for A.

Proof. If a subsequence of $\{\mu_n \eta_n^{-1}\}$ converges weakly to μ, then the finite dimensional distributions of the corresponding subsequence of $\{\eta_n \circ X_n^{\mu_n}\}$ converge weakly to those of X^μ. But $\eta_n \circ X_n^{\mu_n}$ is a stationary process so X^μ must also be stationary. □

The convergence of the finite-dimensional distributions required in the above theorem can be proved using the results of Section 8. The hidden difficulty is that there is no guarantee that $\{\mu_n \eta_n^{-1}\}$ has any convergent subsequences; thus, we need conditions under which $\{\mu_n \eta_n^{-1}\}$ is relatively compact. Corollary 9.8 suggests one set of conditions.

9.13 Lemma Let E_n, E, A_n, A, and η_n be as in Theorem 9.12, and in addition assume that E is locally compact. Let $\varphi_n, \psi_n \in M(E_n)$ and $\psi \in M(E)$. Suppose that $\varphi_n \geq 0$, $\psi_n \leq \psi \circ \eta_n$, $\psi \leq C$ for some constant C, that $\lim_{x \to \Delta} \psi(x) = -\infty$, and that for every solution X_n of the martingale problem for A_n with $E[\varphi_n(X_n(0))] < \infty$,

$$\varphi_n(X_n(t)) - \int_0^t \psi_n(X_n(s))\, ds \tag{9.29}$$

is a supermartingale. For $n = 1, 2, \ldots$, let μ_n be a stationary distribution for A_n. Then, for each $m, n \geq 1$,

$$\mu_n \eta_n^{-1}(K_m) \geq \frac{m}{m + C}, \tag{9.30}$$

where $K_m = \{x: \psi(x) \geq -m\}$, and hence $\{\mu_n \eta_n^{-1}\}$ is relatively compact.

Proof. Let X_n be a solution of the martingale problem for A_n with $E[\varphi_n(X_n(0))] < \infty$. Then, as in (9.19),

$$\begin{aligned} 0 &\leq E[\varphi_n(X_n(0))] + E\left[\int_0^t \psi_n(X_n(s))\, ds\right] \\ &\leq E[\varphi_n(X_n(0))] + E\left[\int_0^t \psi(\eta_n \circ X_n(s))\, ds\right] \\ &\leq E[\varphi_n(X_n(0))] + CE\left[\int_0^t \chi_{K_m}(\eta_n \circ X_n(s))\, ds\right] \\ &\quad - mE\left[\int_0^t \chi_{K_m^c}(\eta_n \circ X_n(s))\, ds\right], \end{aligned} \tag{9.31}$$

and the estimate follows by the same argument used in the proof of Corollary 9.8. □

Analogues of Theorem 9.12 and Lemma 9.13 hold for sequences of discrete-parameter processes. See Problems 46 and 47.

We give one additional theorem that, while it is not stated in terms of convergence of stationary distributions, typically implies this.

9.14 Theorem Let $\{T_n(t)\}$, $n = 1, 2, \ldots$, and $\{T(t)\}$ be strongly continuous semigroups on a closed subspace $L \subset \bar{C}(E)$ corresponding to transition functions $P_n(t, x, \Gamma)$, $n = 1, 2, \ldots$, and $P(t, x, \Gamma)$. Suppose for each compact $K_1 \subset E$ and $\varepsilon > 0$ there exists a compact $K_2 \subset E$ such that

$$\inf_{x \in K_1} \inf_n \inf_t P_n(t, x, K_2) \geq 1 - \varepsilon. \tag{9.32}$$

Suppose that for each $f \in L$, $t_0 > 0$, and compact $K \subset E$,

$$\lim_{n\to\infty} \sup_{x\in K} \sup_{t\le t_0} |T_n(t)f(x) - T(t)f(x)| = 0, \tag{9.33}$$

and suppose that there exists an operator $\pi: L \to L$ such that for each $f \in L$ and compact $K \subset E$,

$$\lim_{t\to\infty} \sup_{x\in K} |T(t)f(x) - \pi f(x)| = 0 \tag{9.34}$$

and

$$\lim_{n\to\infty} \sup_{x\in K} \sup_{0\le t<\infty} |T_n(t)\pi f(x) - \pi f(x)| = 0. \tag{9.35}$$

Then for each $f \in L$ and compact $K \subset E$,

$$\lim_{n\to\infty} \sup_{x\in K} \sup_{0\le t<\infty} |T_n(t)f(x) - T(t)f(x)| = 0. \tag{9.36}$$

9.15 Remark Note that frequently $\pi f(x)$ is a constant independent of x, that is, $\pi f(x) = \int f\,d\mu$ where μ is the unique stationary distribution for $\{T(t)\}$. This result can be generalized in various ways, for example, to discrete time or to semigroups on different spaces. The proofs are straightforward. □

Proof. For $t > t_0$,

$$\begin{aligned} T_n(t)f - T(t)f &= T_n(t - t_0)(T_n(t_0)f - T(t_0)f) \\ &\quad + T_n(t - t_0)(T(t_0)f - \pi f) \\ &\quad + T_n(t - t_0)\pi f - \pi f \\ &\quad + \pi f - T(t)f. \end{aligned} \tag{9.37}$$

Each term on the right can be made small uniformly on compact sets by first taking t_0 sufficiently large and then letting $n \to \infty$. The details are left to the reader. □

We now reconsider condition (c) of Proposition 9.2. Note that D is required to be a core for the generator of the semigroup. Consequently, if one only knows that the martingale problem for A is well-posed and not that the closure of A generates a semigroup, then Proposition 9.2 is not applicable. To see that this is more than a technical difficulty, see Problem 40. The next theorem gives conditions under which condition (c) of Proposition 9.2 implies μ is a stationary distribution without requiring that A (or the closure of A) generate a semigroup.

We need the following lemma.

9.16 Lemma Let $A \subset \bar{C}(E) \times \bar{C}(E)$. Suppose for each $\nu \in \mathscr{P}(E)$ that the martingale problem for (A, ν) has a solution with sample paths in $D_E[0, \infty)$. Suppose that φ is continuously differentiable and convex on $G \subset \mathbb{R}^m$, that $(f_1, g_1), \ldots, (f_m, g_m) \in A$, $(f_1, \ldots, f_m)\colon E \to G$, and that $(\varphi(f_1, f_2, \ldots, f_m), h) \in A$. Then

$$h \geq \nabla\varphi(f_1, f_2, \ldots, f_m) \cdot (g_1, g_2, \ldots, g_m). \tag{9.38}$$

Proof. Let $x \in E$ and let X be a solution of the martingale problem for (A, δ_x). Then by convexity,

(9.39)

$$\begin{aligned} E\left[\int_0^t h(X(s))\, ds\right] &= E[\varphi(f_1(X(t)), \ldots, f_m(X(t)))] - \varphi(f_1(x), \ldots, f_m(x)) \\ &\geq \nabla\varphi(f_1(x), \ldots, f_m(x)) \cdot E[(f_1(X(t)) - f_1(x), \ldots, f_m(X(t)) - f_m(x))] \\ &= \nabla\varphi(f_1(x), \ldots, f_m(x)) \cdot E\left[\left(\int_0^t g_1(X(s))\, ds, \ldots, \int_0^t g_m(X(s))\, ds\right)\right], \end{aligned}$$

for all $t > 0$. Dividing by t and letting $t \to 0$ gives (9.38). □

9.17 Theorem Let E be locally compact and separable, and let A be a linear operator on $\hat{C}(E)$ satisfying the positive maximum principle such that $\mathscr{D}(A)$ is an algebra and dense in $\hat{C}(E)$. If $\mu \in \mathscr{P}(E)$ satisfies

$$\int Af\, d\mu = 0, \qquad f \in \mathscr{D}(A), \tag{9.40}$$

then there exists a stationary solution of the martingale problem for (A, μ).

Proof. Without loss of generality we may assume E is compact and $(1, 0) \in A$. If not, construct A^Δ as in Theorem 5.4 and extend μ to $\mathscr{B}(E^\Delta)$ by setting $\mu(\{\Delta\}) = 0$. Then A^Δ and μ satisfy the hypotheses of the theorem. If X is a stationary solution of the martingale problem for (A^Δ, μ), then $P\{X(t) \in E\} = \mu(E) = 1$, and hence X has a modification taking values in E, which is then a stationary solution of the martingale problem for (A, μ).

Existence of a solution of the martingale problem for (A, μ) is assured by Theorem 5.4, but there is no guarantee that the solution constructed there will be stationary. For $n = 1, 2, \ldots$, let

$$A_n = \{(f, n[(I - n^{-1}A)^{-1}f - f]) : f \in \mathscr{R}(I - n^{-1}A)\}. \tag{9.41}$$

For $f \in \mathscr{D}(A)$ and $f_n = (I - n^{-1}A)f$, we see that $\|f_n - f\| \to 0$ and $\|A_n f_n - Af\| \to 0$ (in fact $A_n f_n = Af$). It follows from Theorem 9.1 of Chapter 3 that if X_n is a solution of the martingale problem for A_n, $n = 1, 2, 3, \ldots$, with sample paths in $D_E[0, \infty)$, then $\{X_n\}$ is relatively compact.

By Lemma 5.1, any limit point of $\{X_n\}$ is a solution of the martingale problem for A. Consequently, to complete the proof of the theorem it suffices to construct, for each $n \geq 1$, a stationary solution of the $D_E[0, \infty)$ martingale problem for (A_n, μ). Note that for $f \in \mathscr{D}(A_n)$, $f = (I - n^{-1}A)g$ for some $g \in \mathscr{D}(A)$ and

$$\int A_n f \, d\mu = \int Ag \, d\mu = 0. \tag{9.42}$$

The key step of the proof is the construction of a measure $\nu \in \mathscr{P}(E \times E)$ such that

$$\nu(\Gamma \times E) = \nu(E \times \Gamma) = \mu(\Gamma), \qquad \Gamma \in \mathscr{B}(E), \tag{9.43}$$

and

$$\int h(x)g(y)\nu(dx \times dy) = \int h(x)(I - n^{-1}A)^{-1}g(x)\mu(dx) \tag{9.44}$$

for all $h \in C(E)$ and $g \in \mathscr{R}(I - n^{-1}A)$. Let $M \subset C(E \times E)$ be the linear space of functions of the form

$$F(x, y) = \sum_{i=1}^{m} h_i(x)g_i(y) + f(y) \tag{9.45}$$

for $h_1, \ldots, h_m, f \in C(E)$, and $g_1, \ldots, g_m \in \mathscr{R}(I - n^{-1}A)$, and define a linear functional Λ on M by

$$\Lambda F = \int \left[\sum_{i=1}^{m} h_i(y)(I - n^{-1}A)^{-1}g_i(y) + f(y) \right] \mu(dy). \tag{9.46}$$

Since $\Lambda 1 = 1$, the Hahn–Banach theorem and the Riesz representation theorem give the existence of a measure $\nu \in \mathscr{P}(E \times E)$ such that $\Lambda F = \int F \, d\nu$ for all $F \in M$ (and hence (9.43) and (9.44)) if we show that $|\Lambda F| \leq \|F\|$. This inequality also implies that if $F \geq 0$, then $\|F\| - \Lambda F = \Lambda(\|F\| - F) \leq \| \|F\| - F\| \leq \|F\|$, so $\Lambda F \geq 0$.

Let $f_1, f_2, \ldots, f_m \in \mathscr{D}(A)$, let $\alpha_k = \|f_k - n^{-1}Af_k\|$, and let φ be a polynomial on $\mathbb{R}^m$ that is convex on $[-\alpha_1, \alpha_1] \times [-\alpha_2, \alpha_2] \times \cdots \times [-\alpha_m, \alpha_m]$. Since $\mathscr{D}(A)$ is an algebra, $\varphi(f_1, \ldots, f_m) \in \mathscr{D}(A)$, and by Lemma 9.16,

$$A\varphi(f_1, \ldots, f_m) \geq \nabla\varphi(f_1, \ldots, f_m) \cdot (Af_1, \ldots, Af_m). \tag{9.47}$$

Consequently,

$$\begin{aligned} &\varphi((I - n^{-1}A)f_1, \ldots, (I - n^{-1}A)f_m) \\ &\geq \varphi(f_1, \ldots, f_m) - \frac{1}{n}\nabla\varphi(f_1, \ldots, f_m) \cdot (Af_1, \ldots, Af_m) \\ &\geq \varphi(f_1, \ldots, f_m) - \frac{1}{n} A\varphi(f_1, \ldots, f_m), \end{aligned} \tag{9.48}$$

and hence

$$\int \varphi((I - n^{-1}A)f_1, \ldots, (I - n^{-1}A)f_m)\, d\mu \geq \int \varphi(f_1, \ldots, f_m)\, d\mu, \tag{9.49}$$

or equivalently

$$\int \varphi(g_1, \ldots, g_m)\, d\mu \geq \int \varphi((I - n^{-1}A)^{-1}g_1, \ldots, (I - n^{-1}A)^{-1}g_m)\, d\mu \tag{9.50}$$

for $g_1, \ldots, g_m \in \mathscr{R}(I - n^{-1}A)$. Since all convex functions on $\mathbb{R}^m$ can be approximated uniformly on any compact set $K \subset \mathbb{R}^m$ by a polynomial that is convex on K, (9.50) holds for all φ convex on $\mathbb{R}^m$.

Let F be given by (9.45), and define $\varphi\colon \mathbb{R}^m \to \mathbb{R}$ by

$$\varphi(u) = \sup_x \sum_{i=1}^m h_i(x)u_i. \tag{9.51}$$

Note that φ is convex. Then

$$\begin{aligned} \Lambda F &= \int \sum_{i=1}^m h_i(y)\left(I - \frac{1}{n}A\right)^{-1} g_i(y)\mu(dy) + \int f(y)\mu(dy) \\ &\leq \int \varphi((I - n^{-1}A)^{-1}g_1, \ldots, (I - n^{-1}A)^{-1}g_m)\, d\mu + \int f\, d\mu \\ &\leq \int \varphi(g_1, \ldots, g_m)\, d\mu + \int f\, d\mu \\ &= \int \sup_x \left[\sum_{i=1}^m h_i(x)g_i(y) + f(y)\right] d\mu \\ &\leq \|F\|. \end{aligned} \tag{9.52}$$

Similarly $-\Lambda F = \Lambda(-F) \leq \|-F\| = \|F\|$, and the existence of the desired ν follows.

There exists a transition function $\eta(x, \Gamma)$ such that

$$\nu(A \times B) = \int_A \eta(x, B)\mu(dx), \qquad A, B \in \mathscr{B}(E), \tag{9.53}$$

(see Appendix 8), and hence

$$\int_E \eta(x, B)\mu(dx) = \nu(E \times B) = \mu(B), \qquad B \in \mathscr{B}(E). \tag{9.54}$$

Let $Y(0)$, $Y(1)$, $Y(2)$, ... be a Markov chain with transition function η and initial distribution μ. By (9.54), $\{Y(k)\}$ is stationary. Since (9.44) holds for all $h \in C(E)$ and $g \in \mathscr{R}(I - n^{-1}A)$, it follows that

$$\int g(y)\eta(x, dy) = (I - n^{-1}A)^{-1}g(x) \qquad \mu\text{-a.s.} \tag{9.55}$$

for all $g \in \mathscr{R}(I - n^{-1}A)$. Therefore

$$g(Y_k) - \sum_{i=0}^{k-1} n^{-1}A_n g(Y_i) \tag{9.56}$$

is a martingale with respect to $\{\mathscr{F}_k^Y\}$. Let V be a Poisson process with parameter n and define $X = Y(V(\cdot))$. Then

$$g(X(t)) - \int_0^t A_n g(X(s))\, ds \tag{9.57}$$

is an $\{\mathscr{F}_t^X\}$-martingale for each $g \in \mathscr{R}(I - n^{-1}A)$ (cf. (2.6)). We leave it to the reader (Problem 41) to show that X is stationary. □

Proposition 9.2 and Theorem 9.17 are special cases of more-general results. Let $A \subset B(E) \times B(E)$. If X is a solution of the martingale problem for A and ν_t is the distribution of $X(t)$, then $\{\nu_t\}$ satisfies

$$\nu_t f + \nu_0 f + \int_0^t \nu_s g\, ds, \qquad (f, g) \in A, \tag{9.58}$$

where $\nu_t f = \int f\, d\nu_t$. Of course (9.40) is a special case of (9.58) with $\nu_t = \mu$ for all $t \geq 0$. We are interested in conditions under which, given ν_0, (9.58) determines ν_t for all $t \geq 0$. The first result gives a generalization of Proposition 9.2 (c ⇒ a).

9.18 Proposition Suppose $\mathscr{R}(\lambda - A)$ is separating for each $\lambda > 0$. If $\{\nu_t\}$ and $\{\mu_t\}$ satisfy (9.58), are weakly right continuous, and $\nu_0 = \mu_0$, then $\nu_t = \mu_t$ for all $t \geq 0$.

Proof. By (9.58), for $(f, g) \in A$,

$$\begin{aligned} \lambda \int_0^\infty e^{-\lambda t}\nu_t f\, dt &= \nu_0 f + \lambda \int_0^\infty e^{-\lambda t} \int_0^t \nu_s g\, ds\, dt \\ &= \nu_0 f + \lambda \int_0^\infty \int_s^\infty e^{-\lambda t}\, dt\, \nu_s g\, ds \\ &= \nu_0 f + \int_0^\infty e^{-\lambda s}\nu_s g\, ds. \end{aligned} \tag{9.59}$$

Consequently,

$$\int_0^\infty e^{-\lambda t}\nu_t(\lambda f - g)\,dt = \nu_0 f, \qquad (f, g) \in A. \tag{9.60}$$

Since $\mathscr{R}(\lambda - A)$ is separating, (9.60) implies that ν_0 uniquely determines the measure $\int_0^\infty e^{-\lambda t}\nu_t\,dt$. Since this holds for each $\lambda > 0$ and $\{\nu_t\}$ is weakly right continuous, the uniqueness of the Laplace transform implies ν_0 determines ν_t, $t \geq 0$. □

We next consider the generalization of Theorem 9.17

9.19 Proposition Let E be locally compact and separable, and let A be a linear operator on $\hat{C}(E)$ satisfying the positive maximum principle such that $\mathscr{D}(A)$ is an algebra and dense in $\hat{C}(E)$. Suppose the martingale problem for A is well-posed. If $\{\nu_t\} \subset \mathscr{P}(E)$ and $\{\mu_t\} \subset \mathscr{P}(E)$ satisfy (9.58) and $\nu_0 = \mu_0$, then $\nu_t = \mu_t$ for all $t \geq 0$.

Proof. Since $\mathscr{D}(A)$ is dense in $\hat{C}(E)$, weak continuity of $\{\nu_t\}$ and $\{\mu_t\}$ follows from (9.58). We reduce the proof to the case considered in Theorem 9.17. As in the proof of Theorem 9.17, without loss of generality we can assume that E is compact. Let $E_0 = E \times \{-1, 1\}$. Fix $\lambda > 0$. For $f_1 \in \mathscr{D}(A)$ and $f_2 \in B(\{-1, 1\})$, let $f = f_1 f_2$ and define

$$Bf(x, n) = f_2(n)Af_1(x) + \lambda\left(f_2(-n)\int f\,d\nu_0 - f_1(x)f_2(n)\right). \tag{9.61}$$

By Theorem 10.3 of the next section, if the martingale problem for A is well-posed, then the martingale problem for B is well-posed. There the new component is a counting process, but essentially the same proof will give uniqueness here.

Define

$$\mu = \left(\lambda\int_0^\infty e^{-\lambda t}\nu_t\,dt\right) \times \left(\frac{1}{2}\,\delta_1 + \frac{1}{2}\,\delta_{-1}\right) \tag{9.62}$$

Then μ satisfies $\int Bf\,d\mu = 0$ for all $f \in \mathscr{D}(B)$, and, since the linear extension of B satisfies the conditions of Theorem 9.17, μ is a stationary distribution for B.

We claim there is only one stationary distribution for B. To see this, we observe that any solution of the $D_{E_0}[0, \infty)$ martingale problem for B is a strong Markov process (Theorem 4.2 and Corollary 4.3). Let $\{\eta_t\}$ be the one-dimensional distributions for the solution of the $D_{E_0}[0, \infty)$ martingale problem for $(B, \nu_0 \times \delta_1)$. Let (Z, N) be any solution of the $D_{E_0}[0, \infty)$ martingale problem for B, and define $\tau_0 = \inf\{t > 0 : N(t) = -1\}$ and $\tau = \inf\{t > \tau_0 : N(t) = 1\}$. Then

$$P\{(Z(t), N(t)) \in \Gamma\} = P\{(Z(t), N(t)) \in \Gamma,\ t < \tau\} + E[\eta_{t-\tau}(\Gamma)\chi_{\{\tau \leq t\}}]. \tag{9.63}$$

Consequently

$$(9.64)\qquad \lim_{t\to\infty} t^{-1}\int_0^t P\{(Z(s), N(s)) \in \Gamma\}\, ds = \lim_{t\to\infty} t^{-1}\int_0^t E[\eta_{s-\tau}(\Gamma)\chi_{\{\tau \le s\}}]\, ds$$
$$= \lim_{t\to\infty} t^{-1}\int_0^t \eta_s(\Gamma)\, ds,$$

and uniqueness of the stationary distribution follows.

If $\tilde{\mu}$ is defined by (9.62) with $\{v_t\}$ replaced by $\{\mu_t\}$, it is a stationary distribution for B and uniqueness gives

$$(9.65)\qquad \int_0^\infty e^{-\lambda t} v_t\, dt = \int_0^\infty e^{-\lambda t} \mu_t\, dt.$$

Since $\lambda > 0$ is arbitrary and $\{v_t\}$ and $\{\mu_t\}$ are weakly continuous, it follows that $v_t = \mu_t$ for all $t \ge 0$. □

10. PERTURBATION RESULTS

Suppose that X_1 is a solution of the martingale problem for $A_1 \subset B(E_1) \times B(E_1)$ and that X_2 is a solution of the martingale problem for $A_2 \subset B(E_2) \times B(E_2)$. If X_1 and X_2 are independent, then (X_1, X_2) is a solution of the martingale problem for $A \subset B(E_1 \times E_2) \times B(E_1 \times E_2)$ given by

$$(10.1)\qquad A = (f_1 f_2, g_1 f_2 + f_1 g_2): (f_1, g_1) \in A_1, (f_2, g_2) \in A_2\}.$$

If uniqueness holds for A_1 and A_2, and if $(1, 0) \in A_i$, $i = 1, 2$, then we also have uniqueness for A.

10.1 Theorem Let (E_1, r_1), (E_2, r_2) be complete, separable metric spaces. For $i = 1, 2$, let $A_i \subset B(E_i) \times B(E_i)$, $(1, 0) \in A_i$, and suppose that uniqueness holds for the martingale problem for A_i. Then uniqueness holds for the martingale problem for A given by (10.1). In particular, if $X = (X_1, X_2)$ is a solution of the martingale problem for A and $X_1(0)$ and $X_2(0)$ are independent, then X_1 and X_2 are independent.

Proof. Note that $X = (X_1, X_2)$ is a solution of the martingale problem for A if and only if it is a solution for

$$(10.2)\qquad \tilde{A} = \{((f_1 + \|f_1\| + 1)(f_2 + \|f_2\| + 1),$$
$$g_1(f_2 + \|f_2\| + 1) + (f_1 + \|f_1\| + 1)g_2): (f_i, g_i) \in A_i, i = 1, 2\},$$

so we may as well assume $f_i \ge 1$ for all $(f_i, g_i) \in A_i$, $i = 1, 2$.

For each $\nu \in \mathscr{P}(E_i)$ for which a solution Y of the martingale problem for (A_i, ν) exists, define $\eta_i(\nu, t, \Gamma) = P\{Y(t) \in \Gamma\}$, $\Gamma \in \mathscr{B}(E_i)$. By uniqueness, η_i is well-defined.

By Lemma 3.2 and by Problem 23 of Chapter 2, Y is a solution of the martingale problem for A_i if and only if for every bounded, discrete $\{{}^*\mathscr{F}_t^Y\}$-stopping time τ,

$$E\left[f(Y(\tau)) \exp\left\{-\int_0^\tau \frac{g(Y(s))}{f(Y(s))}\,ds\right\}\right] = E[f(Y(0))] \tag{10.3}$$

for all $(f, g) \in A_i$.

Let $X = (X_1, X_2)$ be a solution of the martingale problem for A with respect to $\{\mathscr{G}_t\}$ defined on $(\Omega, \mathscr{F}, P)$. For $\Gamma_2 \in \mathscr{B}(E_2)$ with $P\{X_2(0) \in \Gamma_2\} > 0$, define

$$Q(B) = \frac{E[\chi_B \chi_{\Gamma_2}(X_2(0))]}{E[\chi_{\Gamma_2}(X_2(0))]}, \qquad B \in \mathscr{F}. \tag{10.4}$$

Then X_1 on $(\Omega, \mathscr{F}, Q)$ is a solution of the martingale problem for A_1, and hence

$$E[\chi_{\Gamma_1}(X_1(t))\chi_{\Gamma_2}(X_2(0))] = \eta_1(\nu_{\Gamma_2}, t, \Gamma_1)P\{X_2(0) \in \Gamma_2\} \tag{10.5}$$

where $\nu_{\Gamma_2}(\Gamma_1) = P\{X_1(0) \in \Gamma_1 \mid X_2(0) \in \Gamma_2\}$.

For $(f_i, g_i) \in A_i$, $i = 1, 2$, define

$$M_i(t) = f_i(X_i(t)) \exp\left\{-\int_0^t \frac{g_i(X_i(s))}{f_i(X_i(s))}\,ds\right\}. \tag{10.6}$$

Note M_1, M_2, and $M_1 M_2$ are $\{\mathscr{G}_t\}$-martingales. If τ_1 and τ_2 are bounded, discrete $\{\mathscr{G}_t\}$-stopping times, then

$$\begin{aligned} E[M_1(\tau_1)M_2(\tau_2)] &= E[E[M_1(\tau_1)M_2(\tau_2) \mid \mathscr{F}_{\tau_1}]] = E[M_1(\tau_1)M_2(\tau_2 \wedge \tau_1)] \\ &= E[M_1(\tau_2 \wedge \tau_1)M_2(\tau_2 \wedge \tau_1)] \\ &= E[M_1(0)M_2(0)]. \end{aligned} \tag{10.7}$$

Fix τ_2 and $(f_2, g_2) \in A_2$, and for M_2 as above, define

$$\tilde{Q}(B) = \frac{E[\chi_B M_2(\tau_2)]}{E[M_2(\tau_2)]}, \qquad B \in \mathscr{F}. \tag{10.8}$$

For $(f, g) \in A_1$ and any discrete $\{\mathscr{G}_t\}$-stopping time τ, (10.7) implies

$$\begin{aligned} E^{\tilde{Q}}\left[f(X_1(\tau)) \exp\left\{-\int_0^\tau \frac{g(X_1(s))}{f(X_1(s))}\,ds\right\}\right] &= \frac{E[f(X_1(0))M_2(0)]}{E[M_2(\tau_2)]} \\ &= \frac{E[f(X_1(0))M_2(\tau_2)]}{E[M_2(\tau_2)]} \\ &= E^{\tilde{Q}}[f(X_1(0))]. \end{aligned} \tag{10.9}$$

Consequently X_1 on $(\Omega, \mathscr{F}, \tilde{Q})$ is a solution of the martingale problem for A_1, and uniqueness implies that $\tilde{Q}\{X_1(t) \in \Gamma_1\} = \eta_1(v, t, \Gamma_1)$ where $v(\Gamma) = E[\chi_\Gamma(X_1(0)) f_2(X_2(0))]/E[f_2(X_2(0))]$.

Note that v does not depend on τ_2, so

$$E\left[\chi_{\Gamma_1}(X_1(t)) f_2(X_2(\tau_2)) \exp\left\{-\int_0^{\tau_2} \frac{g_2(X_2(s))}{f_2(X_2(s))}\,ds\right\}\right] \tag{10.10}$$

$$= \eta_1(v, t, \Gamma_1) E[f_2(X_2(0))]$$

$$= E[\chi_{\Gamma_1}(X_1(t)) f_2(X_2(0))].$$

Next, defining

$$\hat{Q}(B) = \frac{E[\chi_{\Gamma_1}(X_1(t))\chi_B]}{E[\chi_{\Gamma_1}(X_1(t))]}, \qquad B \in \mathscr{F}, \tag{10.11}$$

(10.10) implies $\hat{Q}\{X_2(t) \in \Gamma_2\} = \eta_2(v_2, t, \Gamma_2)$, where

$$v_2(\Gamma_2) = \frac{E[\chi_{\Gamma_1}(X_1(t))\chi_{\Gamma_2}(X_2(0))]}{E[\chi_{\Gamma_1}(X_1(t))]} \tag{10.12}$$

$$= \frac{\eta_1(v_{\Gamma_2}, t, \Gamma_1) P\{X_2(0) \in \Gamma_2\}}{P\{X_1(t) \in \Gamma_1\}}, \qquad \Gamma_2 \in \mathscr{B}(E_2).$$

Consequently,

$$P\{X_1(t) \in \Gamma_1, X_2(t) \in \Gamma_2\} = \eta_2(v_2, t, \Gamma_2) P\{X_1(t) \in \Gamma_1\}. \tag{10.13}$$

Since by uniqueness the distribution of $X_1(t)$ is determined by the distribution of $X_1(0)$, v_2 is uniquely determined by the distribution of $(X_1(0), X_2(0))$. Consequently, the right side of (10.13) is uniquely determined by the distribution of $(X_1(0), X_2(0))$. The theorem now follows by Theorem 4.2. □

Let $\lambda \in B(E)$ be nonnegative and let $\mu(x, \Gamma)$ be a transition function on $E \times \mathscr{B}(E)$. Define B on $B(E)$ by

$$Bf(x) = \lambda(x) \int (f(y) - f(x))\mu(x, dy). \tag{10.14}$$

Let $A \subset B(E) \times B(E)$ be linear and dissipative. If for some $\lambda > 0$, $B(E)$ is the bp-closure of $\mathscr{R}(\lambda - A)$, then $B(E)$ is the bp-closure of $\mathscr{R}(\lambda - (A + B))$ where $A + B = \{(f, g + Bf) : (f, g) \in A\}$. Consequently, Theorem 4.1 and Corollary 4.4 give uniqueness results for $A + B$. Also see Problem 3.

We now want to give existence and uniqueness results without assuming the range condition.

10.2 Proposition Let (E, r) be complete and separable, let $A \subset B(E) \times B(E)$, and let B be given as in (10.14). Suppose that for every $\nu \in \mathscr{P}(E)$ there exists a solution of the $D_E[0, \infty)$ martingale problem for (A, ν). Then for every $\nu \in \mathscr{P}(E)$ there exists a solution of the $D_E[0, \infty)$ martingale problem for $(A + B, \nu)$.

Proof. By the construction in (2.4) we may assume λ is constant.

Let $\Omega = \prod_{k=1}^{\infty} (D_E[0, \infty) \times [0, \infty))$. Let (X_k, Δ_k) denote the coordinate random variables. Define $\mathscr{G}_k = \sigma(X_l, \Delta_l : l \le k)$ and $\mathscr{G}^k = \sigma(X_l, \Delta_l : l \ge k)$. By an argument similar to the proof of Lemma 5.15, there is a probability distribution on Ω such that for each k, X_k is a solution of the martingale problem for A, Δ_k is independent of $\sigma(X_1, \ldots, X_k, \Delta_1, \ldots, \Delta_{k-1})$ and exponentially distributed with parameter λ, and for $A_1 \in \mathscr{G}_k$ and $A_2 \in \mathscr{G}^{k+1}$,

$$(10.15) \qquad P(A_1 \cap A_2) = E\left[\int P(A_2 \mid X_{k+1}(0) = x)\mu(X_k(\Delta_k), dx)\chi_{A_1}\right].$$

Define $\tau_0 = 0$, $\tau_k = \sum_{i=1}^{k} \Delta_i$, and $N(t) = k$ for $\tau_k \le t < \tau_{k+1}$. Note that N is a Poisson process with parameter λ. Define

$$(10.16) \qquad X(t) = X_{k+1}(t - \tau_k), \qquad \tau_k \le t < \tau_{k+1},$$

and $\mathscr{F}_t = \mathscr{F}_t^X \vee \mathscr{F}_t^N$. We claim that X is a solution of the martingale problem for $A + B$ with respect to $\{\mathscr{F}_t\}$. First note that for $(f, g) \in A$,

(10.17)

$$f(X_{k+1}((t \vee \tau_k) \wedge \tau_{k+1} - \tau_k)) - f(X_{k+1}(0)) - \int_{\tau_k}^{(t \vee \tau_k) \wedge \tau_{k+1}} g(X_{k+1}(s - \tau_k))\, ds$$

is an $\{\mathscr{F}_t\}$-martingale. This follows from the fact that

$$(10.18) \quad E\Bigg[\Bigg(f(X_{k+1}((t_{m+1} \vee \tau_k) \wedge \tau_{k+1} - \tau_k)) - f(X_{k+1}((t_m \vee \tau_k) \wedge \tau_{k+1} - \tau_k)) - \int_{(t_m \vee \tau_k) \wedge \tau_{k+1}}^{(t_{m+1} \vee \tau_k) \wedge \tau_{k+1}} g(X_{k+1}(s - \tau_k))\, ds\Bigg) \times \prod_{i=1}^{m} h_i(X_{k+1}(t_i \vee \tau_k - \tau_k)) \,\Big|\, \mathscr{G}_k \vee \sigma(\Delta_{k+1})\Bigg] = 0,$$

which in turn follows from (10.15) and the fact that Δ_{k+1} is independent of X_{k+1}. (See Appendix 4.)

Consequently, summing (10.17) over k,

$$(10.19) \qquad f(X(t)) - f(X(0)) - \int_0^t g(X(s))\, ds - \sum_{k=1}^{N(t)} (f(X_{k+1}(0)) - f(X_k(\Delta_k)))$$

is an $\{\mathscr{F}_t\}$-martingale, as are

$$\sum_{k=1}^{N(t)} \left(f(X_{k+1}(0)) - \int f(y)\mu(X_k(\Delta_k), dy) \right) \tag{10.20}$$

and

$$\int_0^t \int (f(y) - f(X(s-)))\mu(X(s-), dy)\, d(N(s) - \lambda s). \tag{10.21}$$

Adding (10.20) and (10.21) to (10.19) gives

$$f(X(t)) - f(X(0)) - \int_0^t (g(X(s)) + Bf(X(s)))\, ds, \tag{10.22}$$

which is therefore an $\{\mathscr{F}_t\}$-martingale. □

10.3 Theorem Let $A \subset \bar{C}(E) \times B(E)$, suppose $\mathscr{D}(A)$ is separating, and let B be given by (10.14). Suppose the $D_E[0, \infty)$ martingale problem for A is well-posed, let $P_x \in \mathscr{P}(D_E[0, \infty))$ denote the distribution of the solution for (A, δ_x), and suppose $x \to P_x$ is a Borel measurable function from E into $\mathscr{P}(D_E[0, \infty))$ (cf. Theorem 4.6). Then the $D_{E \times \mathbb{Z}_+}[0, \infty)$ martingale problem for $C \subset B(E \times \mathbb{Z}_+) \times B(E \times \mathbb{Z}_+)$, defined by

$$C = \left\{ \left(fh,\ gh + \lambda(\cdot) \int (f(y)h(\cdot + 1) - f(\cdot)h(\cdot))\mu(\cdot, dy) \right) : (f, g) \in A,\ h \in B(\mathbb{Z}_+) \right\}, \tag{10.23}$$

is well-posed.

10.4 Remark Note that if (X, N) is a solution of the martingale problem for C, then X is a solution of the martingale problem for $A + B$. The component N simply counts the "new" jumps. □

Proof. If the martingale problem for A is well-posed, then by Theorem 10.1 the martingale problem for $A_0 = \{(fh, gh): (f, g) \in A,\ h \in B(\mathbb{Z}_+)\}$ is well-posed (for the second component, $N(t) \equiv N(0)$). For $f \in B(E \times \mathbb{Z}_+)$ define

$$B_0 f(x, k) = \lambda(x) \int (f(y, k+1) - f(x, k))\mu(x, dy). \tag{10.24}$$

Then $C = A_0 + B_0$, and the existence of solutions follows by Proposition 10.2.
For $f \in B(E)$ define

$$T(t)f(x) = E^{P_x}[f(X(t))], \tag{10.25}$$

and note that $\{T(t)\}$ is the semigroup corresponding to the solutions of the martingale problem for A. Let (Y, N) be a solution of the $D_{E \times \mathbb{Z}_+}[0, \infty)$ martingale problem for C.

Note that

$$\chi_{\{N(t)=N(0)\}} \exp\left\{\int_0^t \lambda(Y(s))\, ds\right\} \tag{10.26}$$

is a nonnegative mean one martingale, and let $Q \in \mathscr{P}(D_E[0, \infty))$ be determined by

$$\begin{aligned} (10.27)\quad & Q\{X(t_1) \in \Gamma_1, \ldots, X(t_m) \in \Gamma_m\} \\ &= E\left[\prod_{i=1}^m \chi_{\Gamma_i}(Y(t_i))\chi_{\{N(t_m)=N(0)\}} \exp\left\{\int_0^{t_m} \lambda(Y(s))\, ds\right\}\right], \end{aligned}$$

for $0 \le t_1 < t_2 < \cdots < t_m$, $\Gamma_1, \ldots, \Gamma_m \in \mathscr{B}(E)$. (Here X is the coordinate process.) Since (Y, N) is a solution of the martingale problem for C, it follows by Lemma 3.4 that

$$\begin{aligned} (10.28)\quad & f(Y(t))\chi_{\{N(t)=N(0)\}} \exp\left\{\int_0^t \lambda(Y(s))\, ds\right\} \\ &\qquad - \int_0^t g(Y(s))\chi_{\{N(s)=N(0)\}} \exp\left\{\int_0^s \lambda(Y(u))\, du\right\} ds \end{aligned}$$

is an $\{\mathscr{F}_t^{(Y, N)}\}$-martingale for $(f, g) \in A$. Since (10.26) and (10.28) are martingales,

$$\begin{aligned} (10.29)\quad & E^Q\left[\left(f(X(t_{n+1})) - f(X(t_n)) - \int_{t_n}^{t_{n+1}} g(X(s))\, ds\right) \prod_{k=1}^n h_k(X(t_k))\right] \\ &= E\left[\left(f(Y(t_{n+1})) - f(Y(t_n)) - \int_{t_n}^{t_{n+1}} g(Y(s))\, ds\right) \prod_{k=1}^n h_k(Y(t_k))\right. \\ &\qquad \left. \times\, \chi_{\{N(t_{n+1})=N(0)\}} \exp\left\{\int_0^{t_{n+1}} \lambda(Y(s))\, ds\right\}\right] \\ &= 0 \end{aligned}$$

for $t_1 < t_2 < \cdots < t_{n+1}$, $(f, g) \in A$, and $h_k \in B(E)$, and it follows that Q is a solution of the martingale problem for A. In particular,

$$E[T(t)f(Y(0))] = E\left[f(Y(t)) \exp\left\{\int_0^t \lambda(Y(s))\, ds\right\} \chi_{\{N(t)=N(0)\}}\right]. \tag{10.30}$$

More generally, for $t \ge s$,

$$E[T(t-s)f(Y(s))] = E\left[f(Y(t)) \exp\left\{\int_s^t \lambda(Y(u))\, du\right\} \chi_{\{N(t)=N(s)\}}\right]. \tag{10.31}$$

To complete the proof we need the following two lemmas.

10.5 Lemma For $f \in B(E)$ and $t \geq 0$,

$$(10.32)\quad E\left[\int_0^t f(Y(t)) \exp\left\{\int_0^t \lambda(Y(u))\, du\right\} \chi_{\{N(t)=N(s)\}}\, dN(s)\right] = E\left[\int_0^t T(t-s)f(Y(s))\, dN(s)\right].$$

Proof. Proceeding as above, for $0 \leq a < b$ assume $E[N(b) - N(a)] > 0$, and let Q be determined by

$$(10.33)\quad Q\{X(t_1) \in \Gamma_1, \ldots, X(t_m) \in \Gamma_m\} = \frac{E\left[\int_a^b \prod_{i=1}^m \chi_{\Gamma_i}(Y(s+t_i)) \exp\left\{\int_s^{s+t_m} \lambda(Y(u))\, du\right\} \chi_{\{N(s+t_m)=N(s)\}}\, dN(s)\right]}{E[N(b)-N(a)]}.$$

Then Q is a solution of the martingale problem for A with initial distribution given by

$$(10.34)\quad \nu(\Gamma) = \frac{E\left[\int_a^b \chi_\Gamma(Y(s))\, dN(s)\right]}{E[N(b)-N(a)]}.$$

Consequently,

$$(10.35)\quad E\left[\int_a^b f(Y(s+t)) \exp\left\{\int_s^{s+t} \lambda(Y(u))\, du\right\} \chi_{\{N(s+t)=N(s)\}}\, dN(s)\right]$$
$$= \int T(t)f\, d\nu\, E[N(b)-N(a)]$$
$$= E\left[\int_a^b T(t)f(Y(s))\, dN(s)\right].$$

(Note that if $E[N(b) - N(a)] = 0$, then (10.35) is immediate.)

Since $T(u)f(x)$ is right continuous in u for each $x \in E$ and $f \in \bar{C}(E)$, we have, for $0 = t_0 < \cdots < t_{n+1} = t$,

$$(10.36)\quad E\left[\int_0^t T(t-s)f(Y(s))\, dN(s)\right]$$
$$= \lim_{\max(t_{i+1}-t_i)\to 0} \sum_{i=0}^n E\left[\int_{t_i}^{t_{i+1}} T(t-t_i)f(Y(s))\, dN(s)\right]$$
$$= \lim_{\max(t_{i+1}-t_i)\to 0} \sum_{i=0}^n E\left[\int_{t_i}^{t_{i+1}} f(Y(s+t-t_i))\right.$$
$$\left.\times \exp\left\{\int_s^{s+t-t_i} \lambda(Y(u))\, du\right\} \chi_{\{N(s+t-t_i)=N(s)\}}\, dN(s)\right]$$
$$= E\left[\int_0^t f(Y(t)) \exp\left\{\int_s^t \lambda(Y(u))\, du\right\} \chi_{\{N(t)=N(s)\}}\, dN(s)\right]. \qquad \square$$

10.6 Lemma For $h \in B(E)$ and $t \geq 0$,

$$E\left[\int_0^t h(Y(s))\,dN(s)\right] = E\left[\int_0^t \lambda(Y(s)) \int h(y)\mu(Y(s), dy)\,ds\right]. \tag{10.37}$$

Proof. For $(f, g) \in A$,

$$f(Y(t))\chi_{\{N(t)=k\}} - \int_0^t \Big(g(Y(s))\chi_{\{N(s)=k\}} + \lambda(Y(s)) \int (f(y)\chi_{\{N(s)+1=k\}} - f(Y(s))\chi_{\{N(s)=k\}}\Big)\mu(Y(s), dy))\,ds \tag{10.38}$$

is a right continuous martingale. Consequently, if $\tau_k = \inf\{t: N(t) = k\}$,

$$E[f(Y(\tau_k))\chi_{\{\tau_k \leq t\}}] = E\left[\int_{\tau_{k-1}\wedge t}^{\tau_k \wedge t} \lambda(Y(s)) \int f(y)\mu(Y(s), dy)\,ds\right]. \tag{10.39}$$

Summing over k gives (10.37) with $h = f$. For general h, the result follows from the fact that $\mathscr{D}(A)$ is separating. □

Proof of Theorem 10.3 continued. From (10.30) and Lemmas 10.5 and 10.6,

$$\begin{aligned}
&E[f(Y(t))] - E[T(t)f(Y(0))] \\
&= E\left[f(Y(t))\left(1 - \exp\left\{\int_0^t \lambda(Y(u))\,du\right\}\chi_{\{N(t)=N(0)\}}\right)\right] \\
&= \lim_{\max(s_{i+1}-s_i)\to 0} E\left[\sum f(Y(t))\left(\exp\left\{\int_{s_{i+1}}^t \lambda(Y(u))\,du\right\}\chi_{\{N(t)=N(s_{i+1})\}}\right.\right. \\
&\qquad \left.\left. - \exp\left\{\int_{s_i}^t \lambda(Y(u))\,du\right\}\chi_{\{N(t)=N(s_i)\}}\right)\right] \\
&= E\left[\int_0^t f(Y(t)) \exp\left\{\int_s^t \lambda(Y(u))\,du\right\}\chi_{\{N(t)=N(s)\}}\,dN(s)\right] \\
&\quad - E\left[\int_0^t f(Y(t))\lambda(Y(s)) \exp\left\{\int_s^t \lambda(Y(u))\,du\right\}\chi_{\{N(t)=N(s)\}}\,ds\right] \\
&= E\left[\int_0^t \lambda(Y(s)) \int T(t-s)f(y)\mu(Y(s), dy)\,ds\right] \\
&\quad - E\left[\int_0^t \lambda(Y(s))T(t-s)f(Y(s))\,ds\right] \\
&= \int_0^t E[BT(t-s)f(Y(s))]\,ds,
\end{aligned} \tag{10.40}$$

that is,

(10.41)
$$E[f(Y(t))] = E[T(t)f(Y(0))] + \int_0^t E[BT(t-s)f(Y(s))]\, ds$$

for every $f \in B(E)$. Iterating this identity gives

(10.42)
$$E[f(Y(t))] = E[T(t)f(Y(0))] + \int_0^t E[T(s)BT(t-s)f(Y(0))]\, ds + \int_0^t \int_0^s E[BT(s-u)BT(t-s)f(Y(u))]\, du\, ds,$$

and we see that by repeated iteration the left side is uniquely determined in terms of $Y(0)$, $\{T(t)\}$, and B. Consequently, uniqueness for the martingale problem follows by Theorem 4.2. □

11. PROBLEMS

1. (a) Show that to verify (1.1) it is enough to show that

(11.1)
$$P\{X(u) \in \Gamma \mid X(t_n), X(t_{n-1}), \ldots, X(t_1)\} = P\{X(u) \in \Gamma \mid X(t)\}$$

for every finite collection $0 \le t_1 < t_2 < \cdots < t_n = t < u$.

(b) Show that the process constructed in the proof of Theorem 1.1 is Markov.

2. Let X be a progressive Markov process corresponding to a measurable contraction semigroup $\{T(t)\}$ on $B(E)$ with full generator $\hat{A}$. Let $\Delta_1, \Delta_2, \ldots$ be independent random variables with $P\{\Delta_k > t\} = e^{-t}$, $t \ge 0$, and let V be an independent Poisson process with parameter 1. Show that $X(n^{-1} \sum_{k=1}^{V(nt)} \Delta_k)$ is a Markov process whose full generator is $A_n = \hat{A}(I - n^{-1}\hat{A})^{-1}$, the Yosida approximation of $\hat{A}$.

3. Suppose $\{T(t)\}$ is a semigroup on $B(E)$ given by a transition function and has full generator $\hat{A}$. Let

(11.2)
$$Bf(x) = \lambda(x) \int (f(y) - f(x))\mu(x, dy)$$

where $\lambda \in B(E)$ is nonnegative and $\mu(x, \Gamma)$ is a transition function. Show that $\hat{A} + B$ is the full generator of a semigroup on $B(E)$ given by a transition function.

4. Show that X defined by (2.3) has the same finite-dimensional distributions as X' defined by (2.6).

5. Dropping the assumption that λ is bounded in (2.3), show that $X(t)$ is defined for all $t \geq 0$ with probability 1 if and only if $P\{\sum_{k=0}^{\infty} 1/\lambda(Y(k)) = \infty\} = 1$. In particular, show that $P(\{\sum_{k=0}^{\infty} \Delta_k/\lambda(Y(k)) = \infty\} \Delta \{\sum_{k=0}^{\infty} 1/\lambda(Y(k)) = \infty\}) = 0$.

6. Show that X given by (2.3) is strong Markov.

7. Let X be a Markov process corresponding to a semigroup $\{T(t)\}$ on $B(E)$. Let V be an independent Poisson process with parameter 1. Show that $X(V(nt)/n)$ is a Markov process. What is its generator?

8. Let $E = \{0, 1, 2, \ldots\}$. Let $q_{ij} \geq 0$, $i \neq j$, and let $\sum_{j \neq i} q_{ij} = -q_{ii} < \infty$. Suppose for each i there exists a Markov process X^i with sample paths in $D_E[0, \infty)$ such that $X^i(0) = i$ and

$$\lim_{\varepsilon \to 0+} \varepsilon^{-1}(P\{X^i(t+\varepsilon) = j \mid X^i(t)\} - \chi_{\{j\}}(X^i(t))) = q_{X^i(t)j}, \qquad j \in E,\ t \geq 0. \tag{11.3}$$

   (a) Show that X^i is the unique such process.

   (b) For $i \in E$ and $n = 1, 2, \ldots$, let X_n^i be a Markov process with sample paths in $D_E[0, \infty)$ satisfying $X_n^i(0) = i$ and

$$\lim_{\varepsilon \to 0+} \varepsilon^{-1}(P\{X_n^i(t+\varepsilon) = j \mid X_n^i(t)\} - \chi_{\{j\}}(X_n^i(t))) = q^{(n)}_{X_n^i(t)j}, \qquad j \in E,\ t \geq 0. \tag{11.4}$$

   Show that $X_n^i \Rightarrow X^i$ for all $i \in E$ if and only if

$$\lim_{n \to \infty} q_{ij}^{(n)} = q_{ij}, \qquad i, j \in E \tag{11.5}$$

   (cf. Problem 31).

9. Prove Theorem 2.6.

10. Let $\xi_1, \xi_2, \ldots$ be independent, identically distributed random variables with mean zero and variance one. Let

$$X_n(t) = \frac{1}{\sqrt{n}} \sum_{k=1}^{[nt]} \xi_k. \tag{11.6}$$

Show that $X_n \Rightarrow X$ where X is standard Brownian motion. (Apply Theorem 6.5 of Chapter 1 and Theorem 2.6 of this chapter, using the fact that $C_c^{\infty}(\mathbb{R})$ is a core for the generator for X.)

11. Let Y be a Poisson process with parameter λ and define

$$X_n(t) = n^{-1}(Y(n^2 t) - \lambda n^2 t). \tag{11.7}$$

Use Theorem 6.1 of Chapter 1 and Theorem 2.5 of this chapter to show that $\{X_n\}$ converges in distribution and identify the limit.

12. Let $E = \mathbb{R}$ and $Af(x) = a(x)f''(x) + b(x)f'(x)$ for $f \in C_c^\infty(\mathbb{R})$, where a and b are locally bounded Borel functions satisfying $0 \le a(x) \le K(1 + |x|^2)$ and $xb(x) \le K(1 + |x|^2)$ for some $K > 0$. Show that A is conservative. Extend this result to higher dimensions.

13. Let $E = \mathbb{R}$ and $Af(x) = x^2(\sin^2 x)f'(x)$ and $Bf = f'$ for $f \in C_c^\infty(\mathbb{R})$.

(a) Show that A, B, and $A + B$ satisfy the conditions of Theorem 2.2.

(b) Show that A and B are conservative but $A + B$ is not.

14. Complete the proof of Lemma 3.2.

15. Let E be locally compact and separable with one-point compactification E^Δ. Suppose that $\lambda \in M(E)$ is nonnegative and bounded on compact sets, and that $\mu(x, \Gamma)$ is a transition function on $E \times \mathscr{B}(E)$. Define X as in (2.3), setting $X(t) = \Delta$ for $t \ge \sum_{k=0}^\infty \Delta_k/\lambda(Y(k))$, and let

$$Af(x) = \begin{cases} 0 & x = \Delta \\ \lambda(x) \displaystyle\int (f(y) - f(x))\mu(x, dy) & x \in E \end{cases} \tag{11.8}$$

for each $f \in B(E^\Delta)$ such that

$$\sup_{x \in E} \lambda(x) \int |f(y) - f(x)| \mu(x, dy) < \infty. \tag{11.9}$$

(a) Show that X is a solution of the martingale problem for A.

(b) Suppose $B(E^\Delta)$ is the bp-closure of the collection of f satisfying (11.9). Show that X is the unique solution of the martingale problem for (A, ν), where ν is the initial distribution for Y, if and only if $P\{\sum_{k=0}^\infty 1/\lambda(Y(k)) = \infty\} = 1$.

(c) Let $E = \mathbb{R}^d$. Suppose $\sup_x \lambda(x)\mu(x, \Gamma) < \infty$ for every compact $\Gamma \subset \mathbb{R}^d$, and

$$\lambda(x) \int |y - x| \mu(x, dy) \le K(1 + |x|), \qquad x \in \mathbb{R}^d, \tag{11.10}$$

for some constant K. Use Theorem 3.8 to show that X has sample paths in $D_E[0, \infty)$, and show that X is the unique solution of the martingale problem for (A, ν).

16. (Discrete-time martingale problem)

(a) Let $\mu(x, \Gamma)$ be a transition function on $E \times \mathscr{B}(E)$ and let $X(n)$, $n = 0, 1, 2, \ldots$ be a sequence of E-valued random variables. Define $A: B(E) \to B(E)$ by

$$Af(x) = \int f(y)\mu(x, dy) - f(x), \tag{11.11}$$

and suppose

$$f(X(n)) - \sum_{k=0}^{n-1} Af(X(k)) \tag{11.12}$$

is an $\{\mathscr{F}_n^X\}$-martingale for each $f \in B(E)$. Show that X is a Markov chain with transition function $\mu(x, \Gamma)$.

(b) Let $X(n)$, $n = 0, 1, 2, \ldots$, be a sequence of $\mathbb{Z}$-valued random variables such that for each $n \geq 0$, $|X(n+1) - X(n)| = 1$. Let $g: \mathbb{Z} \to [-1, 1]$ and suppose that

$$X(n) - \sum_{k=0}^{n-1} g(X(k))$$

is an $\{\mathscr{F}_n^X\}$-martingale. Show that X is a Markov chain and calculate its transition probabilities in terms of g.

17. Suppose that (E, r) is complete and separable, and that $P(t, x, \Gamma)$ is a transition function satisfying

$$\lim_{n \to \infty} \sup_x P\left(\frac{1}{n}, x, B(x, \varepsilon)^c\right) = 0 \tag{11.13}$$

for each $\varepsilon > 0$.

(a) Show that each Markov process X corresponding to $P(t, x, \Gamma)$ has a version with sample paths in $D_E[0, \infty)$. (Apply Theorem 8.6 of Chapter 3.)

(b) Suppose

$$\lim_{n \to \infty} \sup_x nP\left(\frac{1}{n}, x, B(x, \varepsilon)^c\right) = 0 \tag{11.14}$$

for each $\varepsilon > 0$. Show that the version obtained in (a) has sample paths a.s. in $C_E[0, \infty)$ (cf. Proposition 2.9.).

18. Let E be compact, and let A be a linear operator on $C(E)$ with $\mathscr{D}(A)$ dense in $C(E)$. Suppose there exist transition functions $\mu_n(x, \Gamma)$ such that for each $f \in \mathscr{D}(A)$

$$Af(x) = \lim_{n \to \infty} n \int (f(y) - f(x))\mu_n(x, dy) \tag{11.15}$$

uniformly in x and that

$$\lim_{n\to\infty} \sup_x n\, \mu_n(x, B(x, \varepsilon)^c) = 0 \tag{11.16}$$

for each $\varepsilon > 0$. Show that for every $\nu \in \mathscr{P}(E)$ there exists a solution of the martingale problem for (A, ν) with continuous sample paths.

19. Let (E, r) be separable, $A \subset B(E) \times B(E)$, $f, g \in M(E \times E)$, and suppose that for each $y \in E$, $(f(\cdot, y), g(\cdot, y)) \in A$. If for each $\varepsilon > 0$ and compact $K \subset E$, $\inf\{f(x, y) - f(y, y): x, y \in K, r(x, y) \geq \varepsilon\} > 0$ and if for each $x \in E$, $\lim_{y\to x} g(x, y) = g(x, x) = 0$, then every solution of the martingale problem for A with sample paths in $D_E[0, \infty)$ has almost all sample paths in $C_E[0, \infty)$ (cf. Proposition 2.9 and Remark 2.10).

20. For $i = 1, 2, \ldots$, let E_i be locally compact and separable, and let A_i be the generator of a Feller semigroup on $\hat{C}(E_i)$. Let $E = \prod_{i=1}^\infty E_i$. For each i, let $\beta_i \in B(E)$ be nonnegative. For $g(x) = \prod_{i=1}^n f_i(x_i)$, $n \geq 1$, $f_i \in \mathscr{D}(A_i)$, define

$$Ag(x) = \sum_{i=1}^n \beta_i(x)\left(\prod_{j\neq i} f_j(x_j)\right) A_i f_i(x_i). \tag{11.17}$$

Show that every solution of the martingale problem for A has a modification with sample paths in $D_E[0, \infty)$.

21. Let E be the set of finite nonnegative integer-valued measures on $\{0, 1, 2, \ldots\}$ with the weak topology (which in this case is equivalent to the discrete topology). For $f \in B(E)$ with compact support, define

$$Af(\alpha) = \int k^2(f(\alpha + \delta_{k-1} - \delta_k) - f(\alpha))\alpha(dk). \tag{11.18}$$

(a) Interpret a solution of the martingale problem for A in terms of particles moving in $\{0, 1, 2, \ldots\}$.

(b) Show that for each $\nu \in \mathscr{P}(E)$, the $D_E[0, \infty)$ martingale problem for (A, ν) has a unique solution, but that uniqueness is lost if the requirement on the sample paths is dropped.

22. Let $E = [0, 1]$ and $A = \{(f, -f'): f \in C^1(E), f(0) = f(1)\}$.

(a) Show that A satisfies the conditions of Corollary 4.4.

(b) Show that the martingale problem for $(A, \delta_{1/2})$ has more than one solution if the requirement that the sample paths be in $D_E[0, \infty)$ is dropped.

23. Use (4.44) to compute the moments for the limit distribution for X.

Hint: Write the integral as a sum over the intervals on which Y is constant.

24. Let $E_1 = [0, 1]$ and $A_1 = \{(f, x(1-x)f'' + (a - bx)f'): f \in C^2(E_1)\}$, where $0 < a < b$. Use duality to show uniqueness for the martingale problem and to show that, if X is a solution, then $X(t)$ converges in distribution as $t \to \infty$ with a limiting distribution that does not depend on $X(0)$.

Hint: Let $E_2 = \{0, 1, 2, \ldots\}$ and $f(x, y) = x^y$.

25. Let $E_1 = E_2 = [0, \infty)$, $A_1 = \{(f, \frac{1}{2}f''): f \in \bar{C}^2(E_1), f''(0) = 0\}$, and $A_2 = \{(f, \frac{1}{2}f''): f \in \bar{C}^2(E_2), f'(0) = 0\}$, that is, let A_1 correspond to absorbing Brownian motion and A_2 to reflecting Brownian motion.

(a) Let $g \in \bar{C}^2(-\infty, \infty)$ satisfy $g(z) = -g(-z)$. Show that the martingale problems for A_1 and A_2 are dual with respect to $(f, 0, 0)$ where $f(x, y) = g(x + y) + g(x - y)$.

(b) Use the result in part (a) to show that $P\{X(t) > y \mid X(0) = x\} = P\{Y(t) < x \mid Y(0) = y\}$, where X is absorbing Brownian motion and Y is reflecting Brownian motion.

26. Let $E = [0, 1]$, $A = \{(f, \frac{1}{2}f''): f \in C^2(E), f'(0) = f'(1) = f''(0) = f''(1) = 0\}$, and let $\Gamma \subset \mathscr{P}(D_E[0, \infty))$ be $\mathscr{M}_A$, the collection of solutions of the martingale problem for A.

(a) Show that Γ satisfies the conditions of Theorem 5.19.

(b) Find a sequence $\{f_k\}$, as in the proof of Theorem 5.19, for which $\Gamma^{(\infty)} = \mathscr{M}_{A_1}$, where

$$A_1 = \{(f, \tfrac{1}{2}f''): f \in C^2(E), f''(0) = f''(1) = 0\}.$$

(c) Find a sequence $\{f_k\}$ for which $\Gamma^{(\infty)} = \mathscr{M}_{A_2}$, where

$$A_2 = \{(f, \tfrac{1}{2}f''): f \in C^2(E), f'(0) = f'(1) = 0\}.$$

27. Let U_k, $k = 1, 2, \ldots$, be open with $E = \bigcup_{i=1}^{\infty} U_k$. Given $x \in D_E[0, \infty)$, let $S_k^t = \inf\{u \geq t: x(u) \notin U_k \text{ or } x(u-) \notin U_k\}$. Show that there exists a sequence of positive integers $k_1, k_2, \ldots$ and a sequence $0 = t_1 < t_2 < \cdots$ such that $t_{i+1} = S_{k_i}^{t_i}$ for each $i \geq 1$ and $\lim_{i \to \infty} t_i = \infty$. In particular every bounded interval $[0, T]$ is contained in a finite union $\bigcup_{i=1}^{n_T} [t_i, S_{k_i}^{t_i})$.

Hint: Select $k^-(t)$ so that $x(t-) \in U_{k^-(t)}$ and $k^+(t)$ so that $x(t) \in U_{k^+(t)}$, and note that there is an open interval I_t with $t \in I_t$ such that $\{x(s-), x(s): s \in I_t\} \subset U_{k^-(t)} \cup U_{k^+(t)}$.

28. Let (E_k, r_k), $k = 0, 1, 2, \ldots$, be complete, separable metric spaces. Let $E = \bigcup_k E_k$ (think of the E_k as distinct even if they are all the same), and define $r(x, y) = r_k(x, y)$ if $x, y \in E_k$ for some k, and $r(x, y) = 1$ otherwise. (Then (E, r) is complete and separable.) For $k = 0, 1, 2, \ldots$, suppose that $A_k \subset \bar{C}(E_k) \times \bar{C}(E_k)$, that the closure of A_k generates a strongly continuous contraction semigroup on $L_k \equiv \overline{\mathscr{D}(A_k)}$, that L_k is separating, and that for each $\nu \in \mathscr{P}(E_k)$ there exists a solution of the martingale problem for (A_k, ν) with sample paths in $D_{E_k}[0, \infty)$. Let $A \subset \bar{C}(E) \times \bar{C}(E)$ be given by

$$(11.19) \quad A = \left\{\left(\sum_{k=0}^{n} \chi_{E_k} f_k, \sum_{k=0}^{n} \chi_{E_k} A_k f_k\right): n \geq 0, f_k \in \mathscr{D}(A_k)\right\}.$$

(a) Show that the closure of A generates a strongly continuous contraction semigroup on $L \equiv \overline{\mathscr{D}(A)}$.

(b) Show that the martingale problem for A is well-posed.

(c) Let $\lambda \in C(E)$, $\lambda \geq 0$, and $\sup_{x \in E_k} \lambda(x) < \infty$ for each k. Let $\mu(x, \Gamma)$ be a transition function on $E \times \mathscr{B}(E)$ and define

$$(11.20) \qquad Bf(x) = \lambda(x) \int (f(y) - f(x))\mu(x, dy)$$

for $f \in \bar{C}(E)$. Suppose $B \subset \bar{C}(E) \times C(E)$.
Let

$$(11.21) \qquad \lambda_{kl} = \sup_{x \in E_k} \lambda(x)\mu(x, E_l)$$

and suppose for some $a, b \geq 0$,

$$(11.22) \qquad \sum_{l>k} l\lambda_{kl} \leq a + bk, \qquad k \geq 0.$$

Show that for each $\nu \in \mathscr{P}(E)$ there exists a unique solution of the local-martingale problem for $(A + B, \nu)$ with sample paths in $D_E[0, \infty)$.

Remark. The primary examples of the type described above are population models. Let S be the space in which the particles live. Then $E_k = S^k$ corresponds to a population of k particles, A_k describes the motion of the k particles in S, and λ and μ describe the reproduction.

29. Let A be given by (7.15), let λ satisfy the conditions of Lemma 7.2, and define

$$(11.23) \qquad U(s, t)f(x) = \int f(y)P(s, t, x, dy),$$

where P is the solution of (7.6).

(a) Let X be a solution of the martingale problem for A. Show that

$$(11.24) \qquad U(s, t)f(X(s)), \qquad 0 \leq s \leq t$$

is a martingale.

(b) Show that the martingale problem for A is well-posed.

30. Let $\xi_1, \xi_2, \ldots$ be a stationary process with $E[\xi_{k+1} \mid \xi_1, \xi_2, \ldots, \xi_k] = 0$. Suppose $E[\xi_k^2] = \sigma^2$ and $\lim_{n\to\infty} n^{-1} \sum_{k=1}^n \xi_k^2 = \sigma^2$ a.s. Apply Theorem 8.2 to show that $\{X_n\}$ given by

$$X_n(t) = \frac{1}{\sqrt{n}} \sum_{k=1}^{[nt]} \xi_k \tag{11.25}$$

converges in distribution.

Hint: Verify relative compactness by estimating $E[(X_n(t+u) - X_n(t))^2 \mid \mathscr{F}_t^n]$ and applying Theorem 8.6 of Chapter 3.

31. (a) Let (E, r) be complete and separable. Let $\{T_n(t)\}$, $n = 1, 2, \ldots$, and $\{T(t)\}$ be semigroups corresponding to Markov processes with state space E. Suppose that $T(t)\colon L \subset \bar{C}(E) \to L$, where L is convergence determining, and that for each $f \in L$ and compact $K \subset E$,

$$\lim_{n\to\infty} \sup_{x \in K} |T_n(t)f(x) - T(t)f(x)|, \qquad t \geq 0. \tag{11.26}$$

Suppose that for each n, X_n is a Markov process corresponding to $\{T_n(t)\}$, X is a Markov process corresponding to $\{T(t)\}$, and $X_n(0) \Rightarrow X(0)$. Show that the finite-dimensional distributions of X_n converge weakly to those of X.

(b) Let $E = \{0, 1, 2, \ldots\}$. For $f \in B(E)$, define

$$A_n f(k) = \begin{cases} 0 & k = 0 \\ f(n) - f(k) & k \neq 0, n \\ n(f(0) - f(n)) & k = n, \end{cases}$$

$$Af(k) = \begin{cases} 0 & k = 0 \\ f(0) - f(k) & k \neq 0. \end{cases}$$

Show that $T_n(t) \equiv e^{tA_n}$ and $T(t) \equiv e^{tA}$ satisfy (11.26).

(c) Fix $k > 0$. For each $n \geq 1$, let X_n be a Markov process corresponding to A_n defined in (b) with $X_n(0) = k$. Show that the finite-dimensional distributions of X_n converge weakly but that X_n does not converge in distribution in $D_E[0, \infty)$.

32. Let E be locally compact and separable, and let $\{T(t)\}$ and $\{S(t)\}$ be Feller semigroups on $\hat{C}(E)$ with generators A and B, respectively. Let $\{Y_n(k), k = 0, 1, \ldots\}$ be the Markov chain satisfying

$$E[f(Y_n(2k+1)) \mid Y_n(2k)] = T\left(\frac{1}{n}\right) f(Y_n(2k))$$

and

$$E[f(Y_n(2k)) \mid Y_n(2k-1)] = S\left(\frac{1}{n}\right) f(Y_n(2k-1)),$$

and set $X_n(t) = Y_n([nt])$. Suppose that $\mathscr{D}(A) \cap \mathscr{D}(B)$ is dense in $\hat{C}(E)$. Show that $\{X_n\}$ is relatively compact in $D_{E^\Delta}[0, \infty)$ (E^Δ is the one-point compactification of E), and that any limit point of $\{X_n\}$ is a solution of the martingale problem for $A^\Delta + B^\Delta$ (A^Δ and B^Δ as in Theorem 5.4).

33. Consider a sequence of single server queueing processes. For the nth process, the customer arrivals form a Poisson process with intensity λ_n, and the service time is exponentially distributed with parameter μ_n (only one customer is served at a time).

(a) Let Y_n be the number of customers waiting to be served. What is the generator corresponding to Y_n?

(b) Let $X_n(t) = n^{-1/2} Y_n(nt)$. What is the generator for X_n?

(c) Show that if $\{X_n(0)\}$ converges in distribution, $\lim_{n \to \infty} \lambda_n = \lambda$, and $\lim_{n \to \infty} n(\lambda_n - \mu_n) = a$, then $\{X_n\}$ converges in distribution. What is the limit?

(d) What is the result analogous to (c) for two servers serving a single queue?

(e) What if there are two servers serving separate queues and new arrivals join the shortest queue?

Hint: Make a change of variable. Consider the sum of the two queue lengths and the difference.

34. (a) Let $\xi_1, \xi_2, \ldots$ be independent, identically distributed real random variables. For $x_0, a \in \mathbb{R}$, let $Y_n(0) = x_0$ and

$$Y_n(k+1) = (1 + n^{-1}a)Y_n(k) + n^{-1/2}\xi_{k+1}, \qquad k = 0, 1, \ldots. \tag{11.27}$$

For $f \in C_c^2(\mathbb{R})$, calculate

$$\lim_{n \to \infty} nE[f(Y_n(k+1)) - f(Y_n(k)) \mid Y_n(k) = x], \tag{11.28}$$

and use this calculation to show that X_n, given by $X_n(t) = Y_n([nt])$, converges in distribution.

(b) Generalize (a) to d dimensions.

35. Let E be locally compact and separable, let $\lambda: E \to [0, \infty)$ be measurable and bounded on compact subsets of E, and let $\mu(x, \Gamma)$ be a transition function on $E \times \mathscr{B}(E)$. For $f \in C_c(E)$, define

$$Af(x) = \lambda(x) \int (f(y) - f(x))\mu(x, dy). \tag{11.29}$$

(a) Let $\nu \in \mathscr{P}(E)$, and suppose that the local-martingale problem for (A, ν) has a solution X with sample paths in $D_E[0, \infty)$ (i.e., does not reach infinity in finite time). Show that the solution is unique.

(b) Suppose that φ and $\int |\varphi(y)| \mu(\cdot, dy)$ are bounded on compact sets. Suppose that X is a solution of the local-martingale problem for (A, ν) with sample paths in $D_E[0, \infty)$. Show that

$$\varphi(X(t)) - \int_0^t \lambda(X(s)) \int (\varphi(y) - \varphi(X(s)))\mu(X(s), dy)\, ds \tag{11.30}$$

is a local martingale.

(c) In addition to the assumptions in (b), suppose that $\varphi \geq 0$ and that there exists a constant K such that

$$\lambda(x) \int (\varphi(y) - \varphi(x))\mu(x, dy) \leq K \tag{11.31}$$

for all x. Show that (11.30) is a supermartingale and that

$$\varphi(X(t)) - Kt \tag{11.32}$$

is a supermartingale.

36. Let $A \subset \bar{C}(E) \times \bar{C}(E)$. Suppose that the martingale problem for A is well-posed, that every solution has a modification with sample paths in $D_E[0, \infty)$, and that there exists $x_0 \in E$ such that every solution (with sample paths in $D_E[0, \infty)$) satisfies $\tau \equiv \inf\{t: X(t) = x_0\} < \infty$ a.s. Show that there is at most one stationary distribution for A.

37. Let $E = \mathbb{R}$, $a, b \in C^2(E)$, $a > 0$, and $A = \{(f, af'' + bf'): f \in C_c^\infty(E)\}$. Suppose there exists $g \in C^2(E)$, $g \geq 0$, satisfying

$$\frac{d}{dx}\left[\frac{d}{dx}(ag) - bg\right] = 0 \tag{11.33}$$

and $\int_{-\infty}^{\infty} g\, dx = 1$. Show that if the martingale problem for A is well-posed, then g is the density for the unique stationary distribution for A.

38. Let $E = \mathbb{R}$ and $A = \{(f, f'' + x^4 f'): f \in C_c^\infty(E)\}$. Show that there exists a stationary solution of the martingale problem for A and describe the behavior of this process.

39. Let $E = [0, 1]$, $a, b \in C(E)$, $a > 0$, and $A = \{(f, af'' + bf'): f \in C^\infty(E), f'(0) = f'(1) = 0\}$. Find the stationary distribution for A.

40. Let $E = [0, 1]$, and $A = \{(f, \frac{1}{2}f''): f \in C^2(E), f'(0) = f'(1) = 0$, and $f'(\frac{1}{3}) = f'(\frac{2}{3})\}$. Show the following:

(a) $\overline{\mathscr{D}(A)} = C(E)$.

(b) $\mathscr{R}(\lambda - A) \neq C(E)$ for some (hence all) $\lambda > 0$.

(c) The martingale problem for A is well-posed.

(d) $\mu(\Gamma) = 3m(\Gamma \cap [\frac{1}{3}, \frac{2}{3}])$ (where m is Lebesgue measure) satisfies

$$\int \tfrac{1}{2} f'' \, d\mu = 0, \qquad f \in \mathscr{D}(A), \tag{11.34}$$

but μ is not a stationary distribution for A.

41. Show that X defined in the proof of Theorem 9.17 is stationary.

42. Let (E, r) be complete and separable. If X is a stationary E-valued process, then the ergodic theorem (see, e.g., Lamperti (1977), page 102) ensures that for $h \in B(E)$,

$$\lim_{t \to \infty} t^{-1} \int_0^t h(X(s)) \, ds \tag{11.35}$$

exists a.s.

(a) Let $\nu \in \mathscr{P}(E)$. Show that if (11.35) equals $\int h \, d\nu$ for all $h \in \bar{C}(E)$, then this equality holds for all $h \in B(E)$.

(b) Let $P(t, x, \Gamma)$ be a transition function such that for some $\nu \in \mathscr{P}(E)$

$$\lim_{t \to \infty} t^{-1} \int_0^t \int h(y) P(s, x, dy) \, ds = \int h \, d\nu, \quad x \in E, \ h \in \bar{C}(E). \tag{11.36}$$

Show that there is at most one such ν.

(c) Let ν and $P(t, x, \Gamma)$ be as in (b). Let X be a measurable Markov process corresponding to $P(t, x, \Gamma)$ with initial distribution ν (hence X is stationary). Suppose that (11.35) equals $\int h \, d\nu$ for all $h \in B(E)$. Show that X is ergodic. (See Lamperti (1977), page 95.)

43. Let (E, r) be complete and separable. Suppose $P(t, x, \Gamma)$ is a transition function with a unique stationary distribution $\nu \in P(E)$. Show that if X is a measurable Markov process corresponding to $P(t, x, \Gamma)$ with initial distribution ν, then X is ergodic.

44. For $n = 1, 2, \ldots$, let X_n be a solution of the martingale problem for $A_n = \{(f, f'' + nb(n\cdot)f') : f \in C_c^2(\mathbb{R})\}$, where b is continuous and in L^1, and let $\alpha = \int_{-\infty}^{\infty} b(x) \, dx$. Let X be a solution of the martingale problem for A with

$$\mathscr{D}(A) = \{f \in C_c(\mathbb{R}) : f' \text{ and } f'' \text{ exist and are continuous except at zero, } f'(0+) = e^{-\alpha} f'(0-), \text{ and } f''(0+) = f''(0-)\},$$

and Af is the continuous extension of f''.

(a) Show that uniqueness holds for the martingale problem for A.

(b) Show that if $X_n(0) \Rightarrow X(0)$, then $X_n \Rightarrow X$.

Hint: For $f \in \mathscr{D}(A)$, let f_n satisfy $f_n''(x) + nb(nx) f_n'(x) = Af(x)$, and apply the results of Section 8.

(c) What happens if $\int_{-\infty}^{\infty} b^+(x)\,dx = \infty$ and $\int_{-\infty}^{\infty} b^-(x)\,dx < \infty$?

45. Let (E, r) be complete and separable and let $A \subset \bar{C}(E) \times B(E)$. Suppose $\mathscr{D}(A)$ is separating. Let X be a solution of the $D_E[0, \infty)$ martingale problem for A, let $\Gamma \in \mathscr{B}(E)$, and suppose $g(x) = 0$ for every $x \in \Gamma$ and $(f, g) \in A$. For $t \geq 0$, define $\gamma_t = \inf \{u > t: \int_t^u \chi_{\Gamma^c}(X(s))\,ds > 0\}$. Show that $X(u \wedge \gamma_t) = X(t)$ a.s. for all $u > t$, and that with probability one, X is constant on any interval $[t, u]$ for which $\int_t^u \chi_{\Gamma^c}(X(s))\,ds = 0$.

46. Let E be separable. For $n = 1, 2, \ldots,$ let $\{Y_n(k), k = 0, 1, 2, \ldots\}$ be an E-valued discrete-time stationary process, let $\varepsilon_n > 0$, and assume $\varepsilon_n \to 0$. Define $X_n(t) = Y_n([t/\varepsilon_n])$, and suppose $X_n \Rightarrow X$. Show that X is stationary.

47. Let E be locally compact and separable but not compact, and let $E^\Delta = E \cup \{\Delta\}$ be the one-point compactification. Let $v(x, \Gamma)$ be a transition function on $E \times \mathscr{B}(E)$, and let $\varphi, \psi \in M(E)$. Suppose that $\varphi \geq 0$, that $\psi \leq C$ for some constant C and $\lim_{x \to \Delta} \psi(x) = -\infty$, and that for every Markov chain $\{Y(k), k = 0, 1, 2, \ldots\}$ with transition function $v(x, \Gamma)$ satisfying $E[\varphi(Y(0))] < \infty$,

$$\varphi(Y(k)) - \sum_{l=0}^{k-1} \psi(Y(l)) \tag{11.37}$$

is a supermartingale. Suppose Y is stationary. Show that

$$P\{Y(0) \in K_m\} \geq \frac{m}{C + m},$$

where $K_m = \{x: \psi(x) \geq -m\}$.

48. For $i = 1, 2$, let E_i be a locally compact (but not compact) separable metric space and let $E_i^\Delta = E_i \cup \{\Delta_i\}$ be its one-point compactification. Let X be a measurable E_1-valued process, let Y be a measurable E_2-valued process, and let $\varphi \in M(E_1)$ and $\psi \in M(E_1 \times E_2)$. For $t > 0$, define

$$\mu_t(\Gamma_1) = \frac{1}{t} \int_0^t P\{X(s) \in \Gamma_1\}\,ds, \qquad \Gamma_1 \in \mathscr{B}(E_1), \tag{11.38}$$

and

$$v_t(\Gamma_2) = \frac{1}{t} \int_0^t P\{X(s) \in \Gamma_2\}\,ds, \qquad \Gamma_2 \in \mathscr{B}(E_2). \tag{11.39}$$

Suppose that

$$\varphi(X(t)) - \int_0^t \psi(X(s), Y(s))\,ds \tag{11.40}$$

is a supermartingale, $\varphi \geq 0$, $\psi \leq C$ for some constant C, and that for each compact $K_2 \subset E_2$, $\lim_{x \to \Delta_1} \sup_{y \in K_2} \psi(x, y) = -\infty$. Show that if $\{\nu_t: t \geq 1\}$ is relatively compact in $\mathscr{P}(E_2)$, then $\{\mu_t: t \geq 1\}$ is relatively compact in $\mathscr{P}(E_1)$. (See Chapter 12.)

49. (a) Let E be compact and $A \subset C(E) \times C(E)$ with $\mathscr{D}(A)$ dense in $C(E)$. Suppose the martingale problem for (A, δ_x) is well-posed for each $x \in E$. Show that the martingale problem for A is well-posed (i.e., the martingale problem for (A, μ) is well-posed for each $\mu \in \mathscr{P}(E)$).

(b) Extend the result in (a) to E and A satisfying the conditions of Theorem 5.11(b).

50. Let $E_1 = E_2 = [0, 1]$, and set $E = E_1 \times E_2$. Let

$$
\begin{aligned}
A_1 &= \{(f_1 f_2, f_1'' f_2): f_2 \in C(E_2), f_1 \in C^2(E_1), f_1'(0) = \\
&\qquad f_1''(0) = f_1'(1) = f_1''(1) = 0\}, \\
A_2 &= \{(f_1 \chi_{\{\alpha\}}, f_1'' \chi_{\{\alpha\}}): \alpha \in E_2, f_1 \in C^2(E_1), f_1'(0) = \\
&\qquad f_1'(1) = 0\},
\end{aligned}
\tag{11.41}
$$

and $A = A_1 \cup A_2$. Show that the martingale problem for $(A, \delta_{(x, y)})$ is well-posed for each $(x, y) \in E$ but that the martingale problem for (A, μ) has more than one solution if μ is absolutely continuous (cf. Problem 26).

12. NOTES

The basic reference for the material in Sections 1 and 2 is Dynkin (1965). Theorem 2.5 originally appeared in Mackevicius (1974) and Kurtz (1975).

Levy (1948) (see Doob (1953)) characterized standard Brownian motion as the unique continuous process W such that $W(t)$ and $W(t)^2 - t$ are martingales. Watanabe (1964) characterized the unit Poisson process as the unique counting process N such that $N(t) - t$ is a martingale. The systematic development of the martingale problem began with Stroock and Varadhan (1969) (see Stroock and Varadhan (1979)) for diffusion processes and was extended to other classes of processes in Stroock and Varadhan (1971), Stroock (1975), Anderson (1976), Holley and Stroock (1976, 1978).

The primary significance of Corollary 4.4 is its applicability to Ray processes. See Williams (1979) for a discussion of this class of processes. Theorem 4.6 is essentially Exercise 6.7.4 of Stroock and Varadhan (1979).

The notion of duality given by (4.36) was developed first in the context of infinite particle systems by Vasershtein (1969), Vasershtein and Leontovitch (1970), Spitzer (1970), Holley and Liggett (1975), Harris (1976), Liggett (1977), Holley and Stroock (1979). It has also found application to birth and death processes (Siegmund (1976)), to diffusion processes, particularly those arising in genetics (Holley, Stroock, and Williams (1977), Shiga (1980, 1981), Cox and Rösler (1982)) (see Problem 25), and to measure-valued processes (Dawson and Kurtz (1982), Ethier and Kurtz (1986)).

Lemma 5.3 is due to Roth (1976). Theorem 5.19 is a refinement of a result of Krylov (1973). The presentation here is in part motivated by an unpublished approach of Gray and Griffeath (1977b). See also the presentation in Stroock and Varadhan (1979).

The use of semigroup approximation theorems to prove convergence to Markov processes began with Trotter (1958) and Skorohod (1958), although work on diffusion approximations by Khintchine (1933) is very much in this spirit. These techniques were refined in Kurtz (1969, 1975). Use of the martingale problem to prove limit theorems began with the work of Stroock and Varadhan (1969) and was developed further in Morkvenas (1974), Papanicolaou, Stroock, and Varadhan (1977), Kushner (1980), and Rebolledo (1979) (cf. Theorem 4.1 of Chapter 7). Proposition 8.18 abstracts an approach of Helland (1981). The recent book of Kushner (1984) gives another development of the convergence theory with many applications.

The results on existence of stationary distributions are due to Khasminskii (1960, 1980), Wonham (1966), Benes (1968), and Zakai (1969). Similar convergence results can be found in Blankenship and Papanicolaou (1978), Costantini, Gerardi, and Nappo (1982), and Kushner (1982). Theorem 9.14 is due to Norman (1977). Theorem 9.17 is due to Echeverria (1982) and has been extended by Weiss (1981).

Problem 25 is from Cox and Rösler (1982). Problem 40 gives an example of a well-posed martingale problem with a compact state space for which the closure of A is not a generator. The first such example was given by Gray and Griffeath (1977a). Problem 44 is due to Rosenkrantz (1975).

5 STOCHASTIC INTEGRAL EQUATIONS

The emphasis in this chapter is on existence and uniqueness of solutions of stochastic integral equations, and the relationship to existence and uniqueness of solutions of the corresponding martingale problems. These results comprise Section 3. Section 1 introduces d-dimensional Brownian motion, while Section 2 defines stochastic integrals with respect to continuous, local martingales and includes Itô's formula.

1. BROWNIAN MOTION

Let $\xi_1, \xi_2, \ldots$ be a sequence of independent, identically distributed, $\mathbb{R}^d$-valued random variables with mean vector 0 and covariance matrix I_d, the $d \times d$ identity matrix. Think of the process

$$X_n(t) = \frac{1}{\sqrt{n}} \sum_{k=1}^{[nt]} \xi_k, \qquad t \geq 0, \tag{1.1}$$

as specifying for fixed $n \geq 1$ the position at time t of a particle subjected to independent, identically distributed, random displacements of order $1/\sqrt{n}$ at times $1/n, 2/n, 3/n, \ldots$. Now let $n \to \infty$. In view of the (multivariate) central limit theorem, the existence of a limiting process (specified in terms of its finite-dimensional distributions) is clear. If such a process also has continuous sample paths, it is called a d-dimensional Brownian motion.

More precisely, a process $W = \{W(t), t \geq 0\}$ with values in $\mathbb{R}^d$ is said to be a (*standard*) *d-dimensional* $\{\mathscr{F}_t\}$-*Brownian motion* if:

(a) $W(0) = 0$ a.s.

(b) W is adapted to the filtration $\{\mathscr{F}_t\}$, and $\mathscr{F}_t$ is independent of $\sigma(W(u) - W(t): u \geq t)$ for each $t \geq 0$.

(c) $W(t) - W(s)$ is $N(0, (t-s)I_d)$ (i.e., normal with mean vector 0 and covariance matrix $(t-s)I_d$) for every $t > s \geq 0$.

(d) W has sample paths in $C_{\mathbb{R}^d}[0, \infty)$.

When $\{\mathscr{F}_t\} = \{\mathscr{F}_t^W\}$ in the above definition, W is said to be a (*standard*) *d-dimensional Brownian motion.*

Note that if W is a d-dimensional $\{\mathscr{F}_t\}$-Brownian motion defined on a probability space $(\Omega, \mathscr{F}, P)$, then W is a d-dimensional $\{\bar{\mathscr{F}}_t\}$-Brownian motion on $(\Omega, \bar{\mathscr{F}}, P)$, where $\bar{\mathscr{F}}_t$ and $\bar{\mathscr{F}}$ denote the P-completions of $\mathscr{F}_t$ and $\mathscr{F}$, and P denotes its own extension to $\bar{\mathscr{F}}$. (If $\mathscr{G}$ is a sub-σ-algebra of $\mathscr{F}$, the P-completion of $\mathscr{G}$ is defined to be the smallest σ-algebra containing $\mathscr{G} \cup \{A \subset \Omega: A \subset N \text{ for some } N \in \mathscr{F} \text{ with } P(N) = 0\}$.)

The existence of a d-dimensional Brownian motion can be proved in a number of ways. The approach taken here, while perhaps not as efficient as others, provides an application of the results of Chapter 4, Section 2.

We begin by constructing the Feller semigroup $\{T(t)\}$ on $\hat{C}(\mathbb{R}^d)$ corresponding to W. The interpretation above suggests that $\{T(t)\}$ should satisfy

$$T(t)f(x) = \lim_{n \to \infty} E\left[f\left(x + \frac{1}{\sqrt{n}} \sum_{k=1}^{[nt]} \xi_k\right)\right] \tag{1.2}$$

for all $f \in \hat{C}(\mathbb{R}^d)$, $x \in \mathbb{R}^d$, and $t \geq 0$. By the central limit theorem, (1.2) is equivalent to

$$T(t)f(x) = E[f(x + \sqrt{t}Z)], \tag{1.3}$$

where Z is $N(0, I_d)$. We take (1.3) as our definition of the semigroup $\{T(t)\}$ on $\hat{C}(\mathbb{R}^d)$.

1.1 Proposition Equation (1.3) defines a Feller semigroup $\{T(t)\}$ on $\hat{C}(\mathbb{R}^d)$. Its generator A is an extension of

$$\{(f, \tfrac{1}{2}\Delta_d f): f \in \hat{C}^2(\mathbb{R}^d)\}, \tag{1.4}$$

where $\Delta_d \equiv \sum_{i=1}^d \partial_i^2$. Moreover, $C_c^\infty(\mathbb{R}^d)$ is a core for A.

Proof. For each $t \geq 0$, $T(t)$: $\hat{C}(\mathbb{R}^d) \to \hat{C}(\mathbb{R}^d)$ by the dominated convergence theorem, so $T(t)$ is a positive linear contraction on $\hat{C}(\mathbb{R}^d)$. Let Z' be an independent copy of Z. Then, by Fubini's theorem,

$$
\begin{aligned}
(1.5) \qquad T(s)T(t)f(x) &= E[T(t)f(x + \sqrt{s}Z)] \\
&= E[f(x + \sqrt{s}Z + \sqrt{t}Z')] \\
&= E[f(x + \sqrt{s + t}Z)] \\
&= T(s + t)f(x)
\end{aligned}
$$

for all $f \in \hat{C}(\mathbb{R}^d)$, $x \in \mathbb{R}^d$, and $s, t \geq 0$. Since $T(0) = I$, this implies that $\{T(t)\}$ is a semigroup. Observe that each $f \in \hat{C}(\mathbb{R}^d)$ is uniformly continuous (with respect to the Euclidean metric), and let $w(f, \delta)$ denote its modulus of continuity, defined for $\delta \geq 0$ by

$$
(1.6) \qquad w(f, \delta) = \sup \{|f(y) - f(x)|: x, y \in \mathbb{R}^d, |y - x| \leq \delta\}.
$$

Then $\|T(t)f - f\| \leq E[w(f, \sqrt{t}Z)]$ for all $t \geq 0$, so by the dominated convergence theorem, $\{T(t)\}$ is strongly continuous.

To show that the generator A of $\{T(t)\}$ extends (1.4), fix $f \in \hat{C}^2(\mathbb{R}^d)$. By Taylor's theorem,

$$
\begin{aligned}
(1.7) \quad T(t)f(x) - f(x) &= E[f(x + \sqrt{t}Z) - f(x)] \\
&= E\Bigg[\sum_{i=1}^{d} \sqrt{t}Z_i\, \partial_i f(x) + \frac{1}{2} \sum_{i,j=1}^{d} tZ_i Z_j\, \partial_i \partial_j f(x) \\
&\qquad + \int_0^1 (1 - u) \sum_{i,j=1}^{d} tZ_i Z_j\{\partial_i \partial_j f(x + u\sqrt{t}Z) - \partial_i \partial_j f(x)\}\, du\Bigg] \\
&= t\{\tfrac{1}{2}\Delta_d f(x) + \varepsilon(t, x)\},
\end{aligned}
$$

for all $x \in \mathbb{R}^d$ and $t \geq 0$, where

$$
\begin{aligned}
(1.8) \quad |\varepsilon(t, x)| &\leq \int_0^1 (1 - u) \sum_{i,j=1}^{d} \{E[Z_i^2]E[Z_j^2 w(\partial_i \partial_j f, u\sqrt{t}Z)^2]\}^{1/2}\, du \\
&\leq \frac{1}{2} \sum_{i,j=1}^{d} \{E[Z_j^2 w(\partial_i \partial_j f, \sqrt{t}Z)^2]\}^{1/2}.
\end{aligned}
$$

We conclude that $f \in \mathscr{D}(A)$ and $Af = \frac{1}{2}\Delta_d f$.

Observe next that (1.3) can be rewritten as

$$
(1.9) \qquad T(t)f(x) = \int_{\mathbb{R}^d} f(y)(2\pi t)^{-d/2} \exp\{-|y - x|^2/2t\}\, dy,
$$

provided $t > 0$. It follows easily that

$$(1.10)\qquad T(t)\colon \hat{C}(\mathbb{R}^d) \to \hat{C}^\infty(\mathbb{R}^d), \qquad t > 0,$$

where $\hat{C}^\infty(\mathbb{R}^d) = \bigcap_{k \ge 1} \hat{C}^k(\mathbb{R}^d)$. By Proposition 3.3 of Chapter 1, $\hat{C}^\infty(\mathbb{R}^d)$ is a core for A. Now choose $h \in C_c^\infty(\mathbb{R}^d)$ such that $\chi_{\{x:\,|x| \le 1\}} \le h \le \chi_{\{x:\,|x| \le 2\}}$, and define $\{h_n\} \subset C_c^\infty(\mathbb{R}^d)$ by $h_n(x) = h(x/n)$. Given $f \in \hat{C}^\infty(\mathbb{R}^d)$, observe that $fh_n \to f$ and $A(fh_n) = (Af)h_n + fAh_n + \nabla f \cdot \nabla h_n \to Af$ uniformly as $n \to \infty$, implying that $C_c^\infty(\mathbb{R}^d)$ is a core for A. Finally, since $\text{bp-lim}_{n\to\infty} (h_n, Ah_n) = (1, 0)$, A is conservative. □

The main result of this section proves the existence of a d-dimensional Brownian motion and describes several of its properties.

1.2 Theorem A d-dimensional Brownian motion exists. Let W be a d-dimensional $\{\mathscr{F}_t\}$-Brownian motion. Then the following hold:

(a) W is a strong Markov process with respect to $\{\mathscr{F}_t\}$ and corresponds to the semigroup $\{T(t)\}$ of Proposition 1.1.

(b) Writing $W = (W_1, \ldots, W_d)$, each W_i is a continuous, square-integrable, $\{\mathscr{F}_t\}$-martingale, and $\langle W_i, W_j\rangle_t = \delta_{ij} t$ for $i, j = 1, \ldots, d$ and all $t \ge 0$.

(c) With $\{X_n\}$ defined as in the first paragraph of this section, $X_n \Rightarrow W$ in $D_{\mathbb{R}^d}[0, \infty)$ as $n \to \infty$.

Proof. By Proposition 1.1 of this chapter and Theorem 2.7 of Chapter 4, there exists a probability space $(\Omega, \mathscr{F}, P)$ on which is defined a process $W = \{W(t), t \ge 0\}$ with $W(0) = 0$ a.s. and sample paths in $D_{\mathbb{R}^d}[0, \infty)$ satisfying part (a) of the theorem with $\{\mathscr{F}_t\} = \{\mathscr{F}_t^W\}$. In fact, we may assume that W has sample paths in $C_{\mathbb{R}^d}[0, \infty)$ by Proposition 2.9 of Chapter 4. For $t > s \ge 0$ and $f \in \hat{C}(\mathbb{R}^d)$, the Markov property and (1.3) give

$$(1.11)\qquad E[f(W(t) - z) \mid \mathscr{F}_s^W] = E[f(\sqrt{t - s}Z + x - z)]\,|_{x = W(s)}.$$

Consequently, since $W(s)$ is $\mathscr{F}_s^W$-measurable,

$$(1.12)\qquad E[f(W(t) - W(s)) \mid \mathscr{F}_s^W] = E[f(\sqrt{t - s}Z)].$$

It follows that $\mathscr{F}_s^W$ is independent of $\sigma(W(u) - W(s)\colon u \ge s)$ and that $W(t) - W(s)$ is $N(0, (t - s)I_d)$ for all $t > s \ge 0$. In particular, W is a d-dimensional Brownian motion.

Let W be a d-dimensional $\{\mathscr{F}_t\}$-Brownian motion where $\{\mathscr{F}_t\}$ need not be $\{\mathscr{F}_t^W\}$. To prove part (a), let τ be an $\{\mathscr{F}_t\}$-stopping time concentrated on $\{t_1,$

$t_2, \ldots\} \subset [0, \infty)$. Let $A \in \mathscr{F}_\tau$, $s > 0$, and $f \in \hat{C}(\mathbb{R}^d)$. Then $A \cap \{\tau = t_i\} \in \mathscr{F}_{t_i}$, so

$$\begin{aligned}(1.13)\qquad \int_{A \cap \{\tau = t_i\}} & f(W(\tau + s))\, dP \\ &= \int_{A \cap \{\tau = t_i\}} E[f(W(t_i + s)) - W(t_i) + W(t_i)) \mid \mathscr{F}_{t_i}]\, dP \\ &= \int_{A \cap \{\tau = t_i\}} E[f(\sqrt{s}Z + x)]\,|_{x = W(t_i)}\, dP \\ &= \int_{A \cap \{\tau = t_i\}} T(s)f(W(t_i))\, dP.\end{aligned}$$

The verification of (a) is completed as in the proof of Theorem 2.7 of Chapter 4.

Applying the Markov property (with respect to $\{\mathscr{F}_t\}$), we have

$$(1.14)\qquad E[f(W(t)) \mid \mathscr{F}_s] = E[f(x + \sqrt{t - s}Z)]\,|_{x = W(s)}$$

for all $f \in \hat{C}(\mathbb{R}^d)$ and $t > s \geq 0$, hence for all $f \in C(\mathbb{R}^d)$ with polynomial growth. Taking $f(x) = x_i$ and then $f(x) = x_i x_j$, we conclude that W_i is a continuous, square-integrable, $\{\mathscr{F}_t\}$-martingale, and

$$(1.15)\qquad E[W_i(t)W_j(t) \mid \mathscr{F}_s] = W_i(s)W_j(s) + \delta_{ij}(t - s), \qquad t > s \geq 0,$$

for $i, j = 1, \ldots, d$. This implies (b).

Part (c) follows from Theorems 6.5 of Chapter 1 and 2.6 of Chapter 4, provided we can show that, for every $f \in C_c^\infty(\mathbb{R}^d)$,

$$(1.16)\qquad nE\left[f\left(x + \frac{1}{\sqrt{n}}\xi_1\right) - f(x)\right] \to \tfrac{1}{2}\Delta_d f(x)$$

as $n \to \infty$, uniformly in $x \in \mathbb{R}^d$. Observe, however, that this follows immediately from (1.7) and (1.8) if we replace t and Z by $1/n$ and ξ_1. □

2. STOCHASTIC INTEGRALS

Let $(\Omega, \mathscr{F}, P)$ be a complete probability space with a filtration $\{\mathscr{F}_t\}$ such that $\mathscr{F}_0$ contains all P-null sets of $\mathscr{F}$. Throughout this section, $\{\mathscr{F}_t\}$ implicitly prefixes each of the following terms: martingale, progressive, adapted, stopping time, local martingale, and Brownian motion.

Let $\mathscr{M}_c$ be the space of continuous, square-integrable martingales M with $M(0) = 0$ a.s. Given $M \in \mathscr{M}_c$, denote its increasing process (see Chapter 2,

Section 6) by $\langle M \rangle$, and let $L^2(\langle M \rangle)$ be the space of all real-valued, progressive processes X such that

$$E\left[\int_0^t X^2 \, d\langle M \rangle\right] < \infty, \qquad t \geq 0. \tag{2.1}$$

In this section we define the stochastic integral

$$\int_0^\cdot X \, dM \tag{2.2}$$

for each $X \in L^2(\langle M \rangle)$ as an element of $\mathscr{M}_c$ itself. Actually, (2.2) is uniquely determined only up to indistinguishability. As in the case with conditional expectations and increasing processes, this indeterminacy is inherent. Therefore we adopt the convention of suppressing the otherwise pervasive phrase "almost surely" whenever it is needed only because of this indeterminacy.

Since the sample paths of M are typically of unbounded variation on every nondegenerate interval, we cannot in general define (2.2) in a pathwise sense. However, if the sample paths of X are of bounded variation on bounded intervals, then we can define (2.2) pathwise as a Stieltjes integral, and integrating by parts gives

$$\int_0^t X \, dM = X(t)M(t) - \int_0^t M \, dX, \qquad t \geq 0. \tag{2.3}$$

In particular, when X belongs to the space S of real-valued, bounded, adapted, right-continuous step functions, that is, when X is a real-valued, bounded process for which there exist $0 = t_0 < t_1 < t_2 < \cdots$ with $t_n \to \infty$ such that

$$X(t) = \sum_{i=0}^{\infty} X(t_i)\chi_{[t_i, t_{i+1})}(t), \qquad t \geq 0, \tag{2.4}$$

and $X(t_i)$ is $\mathscr{F}_{t_i}$-measurable for each $i \geq 0$, we have

$$\int_0^t X \, dM = \sum_{\substack{i \geq 0 \\ t_{i+1} \leq t}} X(t_i)(M(t_{i+1}) - M(t_i)) + X(t_{m(t)})(M(t) - M(t_{m(t)})) \tag{2.5}$$

for all $t \geq 0$, where $m(t) = \max\{i \geq 0 : t_i \leq t\}$. Observe that (2.5) is linear in X and in M.

2.1 Lemma If $M \in \mathscr{M}_c$ and $X \in S$, then (2.5) defines a process $\int_0^\cdot X \, dM \in \mathscr{M}_c$ and

$$\left\langle \int_0^\cdot X \, dM \right\rangle_t = \int_0^t X^2 \, d\langle M \rangle, \qquad t \geq 0. \tag{2.6}$$

If, in addition, $N \in \mathscr{M}_c$ and $Y \in S$, then

$$\left\langle \int_0^\cdot X \, dM, \int_0^\cdot Y \, dN \right\rangle_t = \int_0^t XY \, d\langle M, N \rangle, \qquad t \geq 0. \tag{2.7}$$

Proof. Clearly, (2.5) is continuous and adapted, and it is square-integrable because X is bounded and $M \in \mathcal{M}_c$. Fix $t \geq s \geq 0$. We can assume that the partition $0 = t_0 < t_1 < \cdots$ associated with X as in (2.4) is also associated with Y and that s and t belong to it. Letting

$$\int_s^t X\, dM = \int_0^t X\, dM - \int_0^s X\, dM, \tag{2.8}$$

we have

$$\begin{aligned} E\left[\int_s^t X\, dM \,\middle|\, \mathcal{F}_s\right] &= E\left[\sum_i X(t_i)(M(t_{i+1}) - M(t_i)) \,\middle|\, \mathcal{F}_s\right] \\ &= \sum_i E[X(t_i)E[M(t_{i+1}) - M(t_i) \mid \mathcal{F}_{t_i}] \mid \mathcal{F}_s] = 0 \end{aligned} \tag{2.9}$$

and

$$\begin{aligned} E\Bigg[\int_0^t X\, dM &\int_0^t Y\, dN - \int_0^t XY\, d\langle M, N\rangle \\ &- \left\{\int_0^s X\, dM \int_0^s Y\, dN - \int_0^s XY\, d\langle M, N\rangle\right\} \Bigg| \mathcal{F}_s\Bigg] \\ &= E\left[\left(\int_s^t X\, dM\right)\left(\int_s^t Y\, dN\right) - \int_s^t XY\, d\langle M, N\rangle \,\middle|\, \mathcal{F}_s\right] \\ &= E\Bigg[\sum_i \sum_j X(t_i)Y(t_j)(M(t_{i+1}) - M(t_i))(N(t_{j+1}) - N(t_j)) \\ &\qquad - \sum_i X(t_i)Y(t_i)(\langle M, N\rangle_{t_{i+1}} - \langle M, N\rangle_{t_i}) \Bigg| \mathcal{F}_s\Bigg] \\ &= 0, \end{aligned} \tag{2.10}$$

where sums over i range over $\{i \geq 0\colon t_i \geq s,\ t_{i+1} \leq t\}$, and similarly for sums over j. The final equality in (2.10) follows by conditioning the (i, j)th term in the first sum on $\mathcal{F}_{t_i \vee t_j}$ and the ith term in the second sum on $\mathcal{F}_{t_i}$ (as in (2.9)). This gives (2.7) and, as a special case, (2.6). □

To define (2.2) more generally, we need the following approximation result.

2.2 Lemma If $M \in \mathcal{M}_c$ and $X \in L^2(\langle M\rangle)$, then there exists a sequence $\{X_n\} \subset S$ such that

$$\lim_{n \to \infty} E\left[\int_0^t (X_n - X)^2\, d\langle M\rangle\right] = 0, \qquad t \geq 0. \tag{2.11}$$

Proof. By the dominated convergence theorem, (2.11) holds with

$$X_n(t) = X(t)\chi_{[-n, n]}(X(t)), \tag{2.12}$$

which for each n is bounded and progressive.

Thus we can assume that X is bounded. We claim that (2.11) then holds by the dominated convergence theorem with

$$X_n(t) = \{\langle M\rangle_t - \langle M\rangle_{t-n^{-1}\wedge t} + n^{-1}\}^{-1} \int_{t-n^{-1}\wedge t}^{t} X(u)\, d(\langle M\rangle_u + u), \tag{2.13}$$

which for each n is bounded, adapted, and continuous. Here we use the fact that if $h \in B[0, \infty)$ and μ is a positive Borel measure on $[0, \infty)$ without atoms such that $0 < \mu((s, t]) < \infty$ whenever $0 \leq s < t < \infty$, then

$$\lim_{\varepsilon \to 0+} \mu((t - \varepsilon \wedge t, t])^{-1} \int_{t-\varepsilon\wedge t}^{t} h\, d\mu = h(t) \qquad \mu\text{-a.e.} \tag{2.14}$$

Of course, this is well known when μ is Lebesgue measure, in which case ε is allowed to depend on t. In the general case, it suffices to write the left side of (2.14) as

$$\lim_{\varepsilon \to 0+} (F(t) - F(t - \varepsilon \wedge t))^{-1} \int_{F(t-\varepsilon\wedge t)}^{F(t)} h(F^{-1}(u))\, du, \tag{2.15}$$

where $F(t) \equiv \mu((0, t])$, and to apply the Lebesgue case.

Thus, we can assume that X is bounded and continuous. It then follows that (2.11) holds with

$$X_n(t) = X\left(\frac{[nt]}{n}\right), \tag{2.16}$$

which for each n belongs to S. □

The following result defines the stochastic integral (2.2) for each $M \in \mathcal{M}_c$ and $X \in L^2(\langle M\rangle)$.

2.3 Theorem Let $M \in \mathcal{M}_c$ and $X \in L^2(\langle M\rangle)$. Then there exists a unique (up to indistinguishability) process $\int_0^\cdot X\, dM \in \mathcal{M}_c$ such that whenever $\{X_n\} \subset S$ satisfies

$$\sum_n \left\{E\left[\int_0^n (X_n - X)^2\, d\langle M\rangle\right]\right\}^{1/2} < \infty, \tag{2.17}$$

we have

$$\sup_{0 \leq t \leq T} \left|\int_0^t X_n\, dM - \int_0^t X\, dM\right| \to 0, \qquad T > 0, \tag{2.18}$$

a.s. and in $L^2(P)$ as $n \to \infty$. Moreover, (2.6) holds, and

$$E\left[\left(\int_0^t X\, dM\right)^2\right] = E\left[\int_0^t X^2\, d\langle M\rangle\right]. \tag{2.19}$$

If, in addition, $N \in \mathcal{M}_c$ and $Y \in L^2(\langle N\rangle)$, then

$$\int_0^t |XY\|d\langle M, N\rangle| \leq \left\{\int_0^t X^2\, d\langle M\rangle \int_0^t Y^2\, d\langle N\rangle\right\}^{1/2} \tag{2.20}$$

for all $t \geq 0$, and (2.7) holds.

Proof. Choose $\{X_n\} \subset S$ satisfying (2.17); such a sequence exists by Lemma 2.2. Then, for each $T > 0$,

$$\begin{aligned} &E\left[\sum_n \sup_{0\leq t\leq T}\left|\int_0^t X_{n+1}\, dM - \int_0^t X_n\, dM\right|\right] \\ &\leq \sum_n \left\{E\left[\sup_{0\leq t\leq T}\left|\int_0^t (X_{n+1} - X_n)\, dM\right|^2\right]\right\}^{1/2} \\ &\leq \sum_n \left\{4E\left[\left|\int_0^T (X_{n+1} - X_n)\, dM\right|^2\right]\right\}^{1/2} \\ &= 2\sum_n \left\{E\left[\int_0^T (X_{n+1} - X_n)^2\, d\langle M\rangle\right]\right\}^{1/2} < \infty \end{aligned} \tag{2.21}$$

by Proposition 2.16 of Chapter 2 and by Lemma 2.1 of this Chapter. In particular, the sum inside the expectation on the left side of (2.21) is finite a.s. for every $T > 0$, implying that there exists $A \in \mathscr{F}$ with $P(A) = 0$ such that, for every $\omega \in \Omega$, $\{\chi_{A^c} \int_0^{\cdot} X_n\, dM\}$ converges uniformly on bounded time intervals as $n \to \infty$. By Lemma 2.1, the limiting process, which we denote by $\int_0^{\cdot} X\, dM$, is continuous, square-integrable, and adapted. (Note that $A \in \mathscr{F}_0$ by the assumption on $\{\mathscr{F}_t\}$ made at the beginning of this section.) Clearly, (2.18) holds a.s.

Moreover,

$$\begin{aligned} &E\left[\sup_{0\leq t\leq T}\left|\int_0^t X_n\, dM - \int_0^t X\, dM\right|^2\right] \\ &= E\left[\lim_{m\to\infty} \sup_{0\leq t\leq T}\left|\int_0^t X_n\, dM - \int_0^t X_m\, dM\right|^2\right] \\ &\leq \varliminf_{m\to\infty} E\left[\sup_{0\leq t\leq T}\left|\int_0^t X_n\, dM - \int_0^t X_m\, dM\right|^2\right] \\ &\leq \varliminf_{m\to\infty} 4E\left[\left|\int_0^T (X_n - X_m)\, dM\right|^2\right] \\ &= \varliminf_{m\to\infty} 4E\left[\int_0^T (X_n - X_m)^2\, d\langle M\rangle\right] \\ &= 4E\left[\int_0^T (X_n - X)^2\, d\langle M\rangle\right] \end{aligned} \tag{2.22}$$

for each $T > 0$ and each n, so (2.18) holds in $L^2(P)$. If $\{X'_n\} \subset S$ also satisfies (2.7) (with X_n replaced by X'_n), then

$$E\left[\sup_{0 \le t \le T}\left|\int_0^t X_n\,dM - \int_0^t X'_n\,dM\right|^2\right] \le 4E\left[\int_0^T (X_n - X'_n)^2\,d\langle M\rangle\right] \tag{2.23}$$

for each $T > 0$ and each n. Together with (2.22) this implies the uniqueness (up to indistinguishability) of $\int_0^\cdot X\,dM$.

To show that $\int_0^\cdot X\,dM$ belongs to $\mathcal{M}_c$ and satisfies (2.7), it is enough to check that

$$E\left[\int_s^t X\,dM \,\middle|\, \mathcal{F}_s\right] = 0 \tag{2.24a}$$

and

$$E\left[\left(\int_s^t X\,dM\right)^2 \,\middle|\, \mathcal{F}_s\right] = E\left[\int_s^t X^2\,d\langle M\rangle \,\middle|\, \mathcal{F}_s\right] \tag{2.24b}$$

whenever $t \ge s \ge 0$, where we use the notation (2.8). But these follow immediately from the fact that they hold with X replaced by X_n and the fact that (2.18) holds in $L^2(P)$.

Suppose, in addition, that $N \in \mathcal{M}_c$ and $Y \in L^2(\langle N\rangle)$, fix $t \ge 0$, and let $\Gamma \in \mathcal{B}[0, t]$. Since $\langle M + \alpha N\rangle = \langle M\rangle + 2\alpha\langle M, N\rangle + \alpha^2\langle N\rangle$,

$$\int_s^t \chi_\Gamma\,d\langle M\rangle + 2\alpha\int_s^t \chi_\Gamma\,d\langle M, N\rangle + \alpha^2\int_s^t \chi_\Gamma\,d\langle N\rangle \ge 0, \qquad 0 \le s \le t, \tag{2.25}$$

for all $\alpha \in \mathbb{R}$, and hence

$$\left|\int_s^t \chi_\Gamma\,d\langle M, N\rangle\right| \le \left\{\int_s^t \chi_\Gamma\,d\langle M\rangle \int_s^t \chi_\Gamma\,d\langle N\rangle\right\}^{1/2}, \qquad 0 \le s \le t. \tag{2.26}$$

From (2.26) and the Schwarz inequality, we readily obtain (2.20) in the case in which X and Y are simple functions (that is, linear combinations of indicator functions). A standard approximation procedure then gives (2.20) in general.

To complete the proof, we must check that

$$E\left[\left(\int_s^t X\,dM\right)\left(\int_s^t Y\,dN\right) - \int_s^t XY\,d\langle M, N\rangle \,\middle|\, \mathcal{F}_s\right] = 0 \tag{2.27}$$

for all $t \geq s \geq 0$. Let $\{Y_n\} \subset S$ be chosen by analogy with (2.17). Then by (2.10), (2.27) holds with X and Y replaced by X_n and Y_n. Applying (2.18) (in $L^2(P)$) and its analogue for Y as well as (2.20) with XY replaced by $(X_n - X)Y_n$ and by $X(Y_n - Y)$ (which sum to $X_n Y_n - XY$), we obtain the desired result by passing to the limit. □

Before considering further generalizations, we state two simple lemmas. For the first one, given $M \in \mathscr{M}_c$ and a stopping time τ, define M^τ and $\langle M \rangle^\tau$ by

$$M^\tau(t) = M(t \wedge \tau), \ \langle M \rangle_t^\tau = \langle M \rangle_{t \wedge \tau}, \qquad t \geq 0, \tag{2.28}$$

and observe that $M^\tau \in \mathscr{M}_c$ and

$$\langle M^\tau \rangle = \langle M \rangle^\tau. \tag{2.29}$$

2.4 Lemma If $M \in \mathscr{M}_c$, $X \in L^2(\langle M \rangle)$, and τ is a stopping time, then

$$\int_0^t X \, dM^\tau = \int_0^{t \wedge \tau} X \, dM, \qquad t \geq 0. \tag{2.30}$$

Proof. Fix $t \geq 0$. Observe first that

$$E\left[\left(\int_0^t X \, dM^\tau\right)^2\right] = E\left[\int_0^t X^2 \, d\langle M^\tau \rangle\right] = E\left[\int_0^{t \wedge \tau} X^2 \, d\langle M \rangle\right] \tag{2.31}$$

by Theorem 2.3 and (2.29). Second,

$$E\left[\left(\int_0^{t \wedge \tau} X \, dM\right)^2\right] = E\left[\int_0^{t \wedge \tau} X^2 \, d\langle M \rangle\right], \tag{2.32}$$

also by Theorem 2.3. Finally,

$$\begin{aligned} E\left[\left(\int_0^t X \, dM^\tau\right)\left(\int_0^{t \wedge \tau} X \, dM\right)\right] &= E\left[\left(\int_0^{t \wedge \tau} X \, dM^\tau\right)\left(\int_0^{t \wedge \tau} X \, dM\right)\right] \\ &= E\left[\int_0^{t \wedge \tau} X^2 \, d\langle M^\tau, M \rangle\right] = E\left[\int_0^{t \wedge \tau} X^2 \, d\langle M \rangle\right], \end{aligned} \tag{2.33}$$

where the first equality is obtained by conditioning on $\mathscr{F}_{t \wedge \tau}$, the second depends on (2.7), and the third follows from the fact that $\langle M^\tau, M \rangle_{t \wedge \tau} = \langle M \rangle_{t \wedge \tau}$. We conclude that

$$E\left[\left(\int_0^t X \, dM^\tau - \int_0^{t \wedge \tau} X \, dM\right)^2\right] = 0, \tag{2.34}$$

which suffices for the proof. □

2.5 Lemma If $M \in \mathscr{M}_c$, X is progressive, $Y \in L^2(\langle M \rangle)$, and $XY \in L^2(\langle M \rangle)$, then $X \in L^2(\langle \int_0^\cdot Y\, dM \rangle)$ and

$$\int_0^t X\, d\left(\int_0^\cdot Y\, dM\right) = \int_0^t XY\, dM, \qquad t \geq 0. \tag{2.35}$$

Proof. See Problem 11. □

The integrability assumptions of the preceding results can be removed by extending the definition of the stochastic integral (2.2). Let $\mathscr{M}_{c,\,loc}$ be the space of continuous local martingales M with $M(0) = 0$ a.s. Given $M \in \mathscr{M}_{c,\,loc}$, denote its increasing process (see Chapter 2, Section 6) by $\langle M \rangle$, and let $L^2_{loc}(\langle M \rangle)$ be the space of all real-valued progressive processes X such that

$$\int_0^t X^2\, d\langle M \rangle < \infty \text{ a.s.}, \qquad t \geq 0. \tag{2.36}$$

If τ is a stopping time, define M^τ and $\langle M \rangle^\tau$ by (2.28), and observe that $M^\tau \in \mathscr{M}_{c,\,loc}$ and (2.29) holds.

2.6 Theorem Let $M \in \mathscr{M}_{c,\,loc}$ and $X \in L^2_{loc}(\langle M \rangle)$. Then there exists a unique (up to indistinguishability) process $\int_0^\cdot X\, dM \in \mathscr{M}_{c,\,loc}$ such that whenever τ is a stopping time satisfying $M^\tau \in \mathscr{M}_c$ and $X \in L^2(\langle M^\tau \rangle)$, we have

$$\int_0^{t \wedge \tau} X\, dM = \int_0^t X\, dM^\tau, \qquad t \geq 0. \tag{2.37}$$

(The right side of (2.37) is defined in Theorem 2.3.) Moreover, (2.6) holds. If, in addition, $N \in \mathscr{M}_{c,\,loc}$ and $Y \in L^2_{loc}(\langle N \rangle)$, then (2.20) and (2.7) hold.

Proof. Given $M \in \mathscr{M}_{c,\,loc}$ and $X \in L^2_{loc}(\langle M \rangle)$, there exist stopping times $\tau_1 \leq \tau_2 \leq \cdots$ with $\tau_n \to \infty$ such that for each $n \geq 1$, $M^{\tau_n} \in \mathscr{M}_c$ and $\int_0^{\tau_n} X^2\, d\langle M \rangle \leq n$, implying

$$E\left[\int_0^t X^2\, d\langle M^{\tau_n} \rangle\right] = E\left[\int_0^{t \wedge \tau_n} X^2\, d\langle M \rangle\right] \leq n, \qquad t \geq 0, \tag{2.38}$$

and hence $X \in L^2(\langle M^{\tau_n} \rangle)$. By Lemma 2.4,

$$\int_0^{t \wedge \tau_n} X\, dM^{\tau_m} = \int_0^t X\, dM^{\tau_m \wedge \tau_n} = \int_0^{t \wedge \tau_m} X\, dM^{\tau_n}, \qquad t \geq 0, \tag{2.39}$$

for all $m, n \geq 1$, and existence and uniqueness of $\int_0^\cdot X\, dM$ follow. The conclusions (2.6), (2.20), and (2.7) follow easily from Theorem 2.3 and (2.37). □

We need the analogues of Lemmas 2.4 and 2.5. The extended lemmas are immediate consequences of the earlier ones and (2.37).

2.7 Lemma If $M \in \mathcal{M}_{c,\,loc}$, $X \in L^2_{loc}(\langle M \rangle)$, and τ is a stopping time, then (2.30) holds.

2.8 Lemma If $M \in \mathcal{M}_{c,\,loc}$, X is progressive, $Y \in L^2_{loc}(\langle M \rangle)$, and $XY \in L^2_{loc}(\langle M \rangle)$, then $X \in L^2_{loc}(\langle \int_0^\cdot Y\, dM \rangle)$ and (2.35) holds.

The next result is known as Itô's formula.

2.9 Theorem For $i = 1, \ldots, d$, let V_i be a real-valued, continuous, adapted process of bounded variation on bounded intervals with $V_i(0) = 0$, let $M_i \in \mathcal{M}_{c,\,loc}$, and suppose that X_i is a real-valued process such that $X_i(0)$ is $\mathcal{F}_0$-measurable and

$$X_i(t) = X_i(0) + V_i(t) + M_i(t), \qquad t \geq 0. \tag{2.40}$$

Put $X = (X_1, \ldots, X_d)$ and let $f \in C^{1,2}([0, \infty) \times \mathbb{R}^d)$, that is, f, f_t, f_{x_i}, and $f_{x_i x_j}$ exist and belong to $C([0, \infty) \times \mathbb{R}^d)$ for $i, j = 1, \ldots, d$. Then

$$\begin{aligned} f(t, X(t)) - f(0, X(0)) &= \int_0^t f_t(s, X(s))\, ds + \sum_{i=1}^d \int_0^t f_{x_i}(s, X(s))\, dV_i(s) \\ &\quad + \sum_{i=1}^d \int_0^t f_{x_i}(s, X(s))\, dM_i(s) \\ &\quad + \frac{1}{2} \sum_{i,j=1}^d \int_0^t f_{x_i x_j}(s, X(s))\, d\langle M_i, M_j \rangle_s, \qquad t \geq 0. \end{aligned} \tag{2.41}$$

2.10 Remark Itô's formula (2.41) is often written in the easy-to-remember form

$$\begin{aligned} df(t, X(t)) &= f_t(t, X(t))\, dt + \sum_{i=1}^d f_{x_i}(t, X(t))\, dX_i(t) \\ &\quad + \frac{1}{2} \sum_{i,j=1}^d f_{x_i x_j}(t, X(t))\, dX_i(t)\, dX_j(t), \end{aligned} \tag{2.42}$$

where $dX_i(t) = dV_i(t) + dM_i(t)$ and $dX_i(t)\, dX_j(t)$ is evaluated using the "multiplication table"

(2.43)

	$dV_j(t)$	$dM_j(t)$
$dV_i(t)$	0	0
$dM_i(t)$	0	$d\langle M_i, M_j \rangle_t$.

□

Proof. Denoting by $|V_i|(t)$ the total variation of V_i on $[0, t]$, let

$$\tau_n = \inf\left\{t \geq 0: \max_{1 \leq i \leq d} (|X_i(0)| + |V_i|(t) + |M_i(t)|) \geq n\right\} \tag{2.44}$$

for each n, and note that $\tau_n \to \infty$. Thus it suffices to verify (2.41) with t replaced by $t \wedge \tau_n$ for each n. But this is equivalent (by Lemma 2.7) to proving (2.41) with X_i, V_i, and M_i replaced by $X_i^{\tau_n}$, $V_i^{\tau_n}$, and $M_i^{\tau_n}$ for each n. We conclude therefore that it involves no loss of generality to assume that $X_i(t)$, $|V_i|(t)$, $M_i(t)$, and $\langle M_i \rangle_t$ are uniformly bounded in $t \geq 0$, $\omega \in \Omega$, and $i = 1, \ldots, d$.

With this assumption, we can require that f have compact support. Fix $t \geq 0$, and let $0 = t_0 < t_1 < \cdots < t_m = t$ be a partition of $[0, t]$. For the remainder of the proof, we use the notation that for a given process Y (real- or $\mathbb{R}^d$-valued), $\Delta_k Y = Y(t_{k+1}) - Y(t_k)$ for $k = 0, \ldots, m-1$. By Taylor's theorem,

$$
\begin{aligned}
(2.45) \qquad & f(t, X(t)) - f(0, X(0)) \\
&= \sum_{k=0}^{m-1} \{f(t_{k+1}, X(t_{k+1})) - f(t_k, X(t_{k+1}))\} \\
&\quad + \sum_{k=0}^{m-1} \{f(t_k, X(t_{k+1})) - f(t_k, X(t_k))\} \\
&= \sum_{k=0}^{m-1} \int_{t_k}^{t_{k+1}} f_t(u, X(t_{k+1}))\, du \\
&\quad + \sum_{i=1}^{d} \sum_{k=0}^{m-1} f_{x_i}(t_k, X(t_k))\, \Delta_k X_i \\
&\quad + \frac{1}{2} \sum_{i,j=1}^{d} \sum_{k=0}^{m-1} f_{x_i x_j}(t_k, \xi_k)\, \Delta_k X_i\, \Delta_k X_j,
\end{aligned}
$$

where $|\xi_k - X(t_k)| \leq |X(t_{k+1}) - X(t_k)|$.

The proof now consists of showing that, as the mesh of the partition tends to zero, the right side of (2.45) converges in probability to the right side of (2.41).

Convergence of the sum of the first two terms in (2.45) to the sum of the first three terms in (2.41) is straightforward (see Problem 12).

Note that by Proposition 3.4 of Chapter 2 and the continuity and bounded variation of the V_i,

$$
\begin{aligned}
(2.46) \qquad & \lim_{\max (t_{k+1} - t_k) \to 0} \sum_k \Delta_k X_i\, \Delta_k X_j \\
&= \lim_{\max (t_{k+1} - t_k) \to 0} \sum_k \Delta_k M_i\, \Delta_k M_j \\
&= \langle M_i, M_j \rangle_t
\end{aligned}
$$

in probability, and that

$$
(2.47) \qquad \lim_{\max (t_{k+1} - t_k) \to 0} \max_k |f_{x_i x_j}(t_k, \xi_k) - f_{x_i x_j}(t_k, X(t_k))| = 0.
$$

Observing that $2\, \Delta_k X_i\, \Delta_k X_j = (\Delta_k X_i + \Delta_k X_j)^2 - (\Delta_k X_i)^2 - (\Delta_k X_j)^2$ (i.e., $\Delta_k X_i\, \Delta_k X_j$ is a linear combination of positive quantities), the convergence of the last term in (2.45) to the last term in (2.41) is a consequence of the following lemma. □

2.11 Lemma Let f be continuous, and let F be nonnegative, nondecreasing, and right continuous on $[0, \infty)$. For $n = 1, 2, \ldots$, let $0 = t_0^n < t_1^n < t_2^n < \cdots$, with $t_k^n \to \infty$. Suppose for each $t > 0$ that $\max_{t_k^n < t} (t_{k+1}^n - t_k^n) \to 0$ as $n \to \infty$, and suppose that f_n and a_n satisfy $a_n \geq 0$,

(2.48) $$\lim_{n\to\infty} \max_{t_k^n \leq t} |f_n(t_k^n) - f(t_k^n)| = 0, \qquad t \geq 0,$$

and

(2.49) $$\lim_{n\to\infty} \sum_{t_k^n \leq t} a_n(t_k^n) = F(t)$$

for each t at which F is continuous. Then

(2.50) $$\lim_{n\to\infty} \sum_{t_k^n \leq t} f_n(t_k^n) a_n(t_k^n) = \int_0^t f(s)\, dF(s)$$

for each t at which F is continuous.

Proof. Clearly

(2.51) $$\lim_{n\to\infty} \left[\sum_{t_k^n \leq t} f_n(t_k^n) a_n(t_k^n) - \sum_{t_k^n \leq t} f(t_k^n) a_n(t_k^n) \right] = 0.$$

Suppose t is a continuity point of F and $F(t) > 0$. Let μ_n and μ be the probability measures on $[0, t]$ given by

(2.52) $$\mu_n[0, s] = \frac{\sum_{t_k^n \leq s} a_n(t_k^n)}{\sum_{t_k^n \leq t} a_n(t_k^n)}, \qquad 0 \leq s \leq t,$$

and $\mu[0, s] = F(s)/F(t)$, $0 \leq s \leq t$. Then $\mu_n \Rightarrow \mu$, and hence

(2.53) $$\begin{aligned} \lim_{n\to\infty} \sum_{t_k^n \leq t} f(t_k^n) a_n(t_k^n) &= \lim_{n\to\infty} \left[\sum_{t_k^n \leq t} a_n(t_k^n) \right] \int_0^t f\, d\mu_n \\ &= F(t) \int_0^t f\, d\mu \\ &= \int_0^t f\, dF. \end{aligned}$$

□

We conclude this section by applying Itô's formula to give an important characterization of Brownian motion, which is essentially the converse of Theorem 1.2(b).

2.12 Theorem Suppose that $X_1, \ldots, X_d \in \mathcal{M}_{c,\,loc}$ satisfy $\langle X_i, X_j \rangle_t = \delta_{ij} t$ for $i, j = 1, \ldots, d$ and all $t \geq 0$. Then $X = (X_1, \ldots, X_d)$ is a d-dimensional Brownian motion.

Proof. Let $\theta \in \mathbb{R}^d$ be arbitrary, and define $f: [0, \infty) \times \mathbb{R}^d \to \mathbb{C}$ by

$$f(t, x) = \exp\{i\theta \cdot x + \tfrac{1}{2}|\theta|^2 t\}, \tag{2.54}$$

where $i = \sqrt{-1}$. By Theorem 2.9, $\{f(t, X(t)),\ t \geq 0\}$ is a complex-valued, continuous, local martingale, bounded on bounded time intervals, so

$$E[f(t, X(t)) \mid \mathcal{F}_s] = f(s, X(s)) \tag{2.55}$$

for all $t \geq s \geq 0$, that is,

$$E[\exp\{i\theta \cdot (X(t) - X(s))\} \mid \mathcal{F}_s] = \exp\{-\tfrac{1}{2}|\theta|^2 (t - s)\}. \tag{2.56}$$

Consequently, X is a d-dimensional Brownian motion. □

3. STOCHASTIC INTEGRAL EQUATIONS

Let $\sigma: [0, \infty) \times \mathbb{R}^d \to \mathbb{R}^d \otimes \mathbb{R}^d$ (the space of real, $d \times d$ matrices) and $b: [0, \infty) \times \mathbb{R}^d \to \mathbb{R}^d$ be locally bounded (i.e., bounded on each compact set) and Borel measurable. In this section we consider the stochastic integral equation

$$X(t) = X(0) + \int_0^t \sigma(s, X(s))\, dW(s) + \int_0^t b(s, X(s))\, ds, \qquad t \geq 0, \tag{3.1}$$

where W is a d-dimensional Brownian motion independent of $X(0)$ and $\int_0^{\cdot} \sigma(s, X(s))\, dW(s)$ denotes the $\mathbb{R}^d$-valued process whose ith component is given by

$$\sum_{j=1}^{d} \int_0^{\cdot} \sigma_{ij}(s, X(s))\, dW_j(s). \tag{3.2}$$

Observe that (3.2) is well-defined (and is a continuous, local martingale) if X is a continuous, $\mathbb{R}^d$-valued, $\{\bar{\mathcal{F}}_t\}$-adapted process, where $\mathcal{F}_t = \mathcal{F}_t^W \vee \sigma(X(0))$. In the classical approach of Itô, W and $X(0)$ are given, and one seeks such a solution X. For our purposes, however, it is convenient to regard W as part of the solution and to allow $\{\mathcal{F}_t\}$ to be an arbitrary filtration with respect to which W is a d-dimensional Brownian motion.

Let $\mu \in \mathscr{P}(\mathbb{R}^d)$. We say that $(\Omega, \mathscr{F}, P, \{\mathscr{F}_t\}, W, X)$ is a *solution of the stochastic integral equation corresponding to* (σ, b, μ) (respectively, (σ, b)) if:

(a) $(\Omega, \mathscr{F}, P)$ is a probability space with a filtration $\{\mathscr{F}_t\}$, and W and X are $\mathbb{R}^d$-valued processes on $(\Omega, \mathscr{F}, P)$.

(b) W is a d-dimensional $\{\mathscr{F}_t\}$-Brownian motion.

(c) X is $\{\bar{\mathscr{F}}_t\}$-adapted, where $\bar{\mathscr{F}}_t$ denotes the P-completion of $\mathscr{F}_t$, that is, the smallest σ-algebra containing $\mathscr{F}_t \cup \{A \subset \Omega: A \subset N \text{ for some } N \in \mathscr{F} \text{ with } P(N) = 0\}$.

(d) $PX(0)^{-1} = \mu$ and X has sample paths in $C_{\mathbb{R}^d}[0, \infty)$ (respectively, X has sample paths in $C_{\mathbb{R}^d}[0, \infty)$).

(e) (3.1) holds a.s.

The definition of the stochastic integral is as in Theorem 2.6, the roles of $(\Omega, \mathscr{F}, P)$ and $\{\mathscr{F}_t\}$ being played by $(\Omega, \bar{\mathscr{F}}, P)$ and $\{\bar{\mathscr{F}}_t\}$. Because it is defined only up to indistinguishability, we continue our convention of suppressing the phrase "almost surely" whenever it is needed only because of this indeterminacy.

There are two types of uniqueness of solutions of (3.1) that are considered:

Pathwise uniqueness is said to hold for solutions of the stochastic integral equation corresponding to (σ, b, μ) if, whenever $(\Omega, \mathscr{F}, P, \{\mathscr{F}_t\}, W, X)$ and $(\Omega, \mathscr{F}, P, \{\mathscr{F}_t\}, W, X')$ are solutions (with the same probability space, filtration, and Brownian motion), $P\{X(0) = X'(0)\} = 1$ implies $P\{X(t) = X'(t) \text{ for all } t \geq 0\} = 1$.

Distribution uniqueness is said to hold for solutions of the stochastic integral equation corresponding to (σ, b, μ) if, whenever $(\Omega, \mathscr{F}, P, \{\mathscr{F}_t\}, W, X)$ and $(\Omega', \mathscr{F}', P', \{\mathscr{F}'_t\}, W', X')$ are solutions (not necessarily with the same probability space), we have $PX^{-1} = P'(X')^{-1}$ (as elements of $\mathscr{P}(C_{\mathbb{R}^d}[0, \infty))$).

Let

$$A = \{(f, Gf) : f \in C_c^\infty(\mathbb{R}^d)\}, \tag{3.3}$$

where

$$G = \frac{1}{2} \sum_{i,j=1}^{d} a_{ij}(t, x)\, \partial_i\, \partial_j + \sum_{i=1}^{d} b_i(t, x)\, \partial_i \tag{3.4}$$

and

$$a = \sigma\sigma^T. \tag{3.5}$$

Observe that if $(\Omega, \mathscr{F}, P, \{\mathscr{F}_t\}, W, X)$ is a solution of the stochastic integral equation corresponding to (σ, b, μ), then, by Itô's formula (Theorem 2.9),

(3.6)

$$f(X(t)) - \int_0^t Gf(s, X(s))\, ds = f(X(0)) + \sum_{i,j=1}^{d} \int_0^t f_{x_i}(X(s))\sigma_{ij}(s, X(s))\, dW_j(s)$$

for all $t \geq 0$ and $f \in C_c^\infty(\mathbb{R}^d)$, so X is a solution of the $C_{\mathbb{R}^d}[0, \infty)$ martingale problem for (A, μ) with respect to $\{\bar{\mathscr{F}}_t\}$.

In what sense is the converse of this result true? We consider first the nondegenerate case.

3.1 Proposition Let σ: $[0, \infty) \times \mathbb{R}^d \to \mathbb{R}^d \otimes \mathbb{R}^d$ and b: $[0, \infty) \times \mathbb{R}^d \to \mathbb{R}^d$ be locally bounded and Borel measurable, and let $\mu \in \mathscr{P}(\mathbb{R}^d)$. Suppose $\sigma(t, x)$ is nonsingular for each $(t, x) \in [0, \infty) \times \mathbb{R}^d$ and σ^{-1} is locally bounded. Define A by (3.3)–(3.5). If X is a solution of the $C_{\mathbb{R}^d}[0, \infty)$ martingale problem for (A, μ) with respect to a filtration $\{\mathscr{F}_t\}$ on a probability space $(\Omega, \mathscr{F}, P)$, then there exists a d-dimensional $\{\bar{\mathscr{F}}_t\}$-Brownian motion W such that $(\Omega, \bar{\mathscr{F}}, P, \{\bar{\mathscr{F}}_t\}, W, X)$ is a solution of the stochastic integral equation corresponding to (σ, b, μ).

Proof. Since

$$f(X(t)) - f(X(0)) - \int_0^t Gf(s, X(s))\, ds \tag{3.7}$$

belongs to $\mathscr{M}_c$ for every $f \in C_c^\infty(\mathbb{R}^d)$, it follows easily that (3.7) belongs to $\mathscr{M}_{c, loc}$ for every $f \in C^\infty(\mathbb{R}^d)$. In particular,

$$M_i(t) \equiv X_i(t) - X_i(0) - \int_0^t b_i(s, X(s))\, ds \tag{3.8}$$

belongs to $\mathscr{M}_{c, loc}$ and, by (3.6) and Theorem 2.6 (see (2.7)),

$$\langle M_i, M_j \rangle_t = \int_0^t a_{ij}(s, X(s))\, ds, \qquad t \geq 0. \tag{3.9}$$

We claim that

$$W(t) \equiv \int_0^t \sigma^{-1}(s, X(s))\, dM(s) \tag{3.10}$$

defines a d-dimensional $\{\bar{\mathscr{F}}_t\}$-Brownian motion. This follows from Theorem 2.12 since

$$\begin{aligned} \langle W_i, W_j \rangle_t &= \sum_{k,l=1}^{d} \left\langle \int_0^\cdot (\sigma^{-1})_{ik}(s, X(s))\, dM_k(s), \int_0^\cdot (\sigma^{-1})_{jl}(s, X(s))\, dM_l(s) \right\rangle_t \\ &= \sum_{k,l=1}^{d} \int_0^t [(\sigma^{-1})_{ik}\, a_{kl}((\sigma^T)^{-1})_{lj}](s, X(s))\, ds \\ &= \delta_{ij} t, \qquad t \geq 0, \end{aligned} \tag{3.11}$$

where the second equality depends on Theorem 2.6. Consequently, by Lemma 2.8,

$$\int_0^t \sigma(s, X(s))\, dW(s) = \int_0^t dM(t) = X(t) - X(0) - \int_0^t b(s, X(s))\, ds \tag{3.12}$$

for all $t \geq 0$, which is to say that (3.1) holds. □

We turn to the general case, in which σ may be singular. Here the conclusion of Proposition 3.1 need not hold because X may not have enough randomness in terms of which to construct a Brownian motion. However, by suitably enlarging the probability space, we can obtain a solution of (3.1).

It will be convenient to separate out a preliminary result concerning matrix-valued functions.

3.2 Lemma Let $\sigma\colon [0, \infty) \times \mathbb{R}^d \to \mathbb{R}^d \otimes \mathbb{R}^d$ be Borel measurable, and put $a = \sigma\sigma^T$. Then there exist Borel measurable functions $\rho, \eta\colon [0, \infty) \times \mathbb{R}^d \to \mathbb{R}^d \otimes \mathbb{R}^d$ such that

$$\rho a \rho^T + \eta\eta^T = I_d, \tag{3.13}$$

$$\sigma\eta = 0, \tag{3.14}$$

$$(I_d - \sigma\rho)a(I_d - \sigma\rho)^T = 0. \tag{3.15}$$

Proof. Suppose first that σ is a constant function. Since $a \in S_d$ (the set of real, symmetric, nonnegative-definite, $d \times d$ matrices), there exists a real, orthogonal matrix U (i.e., $UU^T = U^TU = I_d$) and a diagonal matrix Λ with nonnegative entries such that $a = U^T\Lambda U$. Moreover, a has a unique S_d-valued square root $a^{1/2}$, and $a^{1/2} = U^T\Lambda^{1/2}U$. Let Γ be a diagonal matrix with diagonal entries 1 or 0 depending on whether the corresponding entry of Λ is positive or 0, and let Λ_1 be diagonal with $\Lambda_1\Lambda = \Gamma$. Then, since $\sigma\sigma^T = U^T\Lambda U$, we have $(\Lambda_1^{1/2}U\sigma)(\Lambda_1^{1/2}U\sigma)^T = \Gamma$. By the Gram–Schmidt orthogonalization procedure, we can therefore construct a real, orthogonal matrix V such that $\Lambda_1^{1/2}U\sigma = \Gamma V$, and hence $\Gamma U\sigma = \Lambda^{1/2}V$. It follows that $U\sigma = \Lambda^{1/2}V$, from which we conclude $\sigma = a^{1/2}U^TV$. It is now easily checked that (3.13)–(3.15) hold with $\rho = V^T\Lambda_1^{1/2}U$ and $\eta = V^T(I_d - \Gamma)U$.

To complete the proof, it suffices to show that the measurable selection theorem (Appendix 10) is applicable to the multivalued function taking $\sigma \in \mathbb{R}^d \otimes \mathbb{R}^d$ to the set of pairs (U, V) as above. The details of this step are left to the reader. □

3.3 Theorem Let $\sigma\colon [0, \infty) \times \mathbb{R}^d \to \mathbb{R}^d \otimes \mathbb{R}^d$ and $b\colon [0, \infty) \times \mathbb{R}^d \to \mathbb{R}^d$ be locally bounded and Borel measurable, and let $\mu \in \mathcal{P}(\mathbb{R}^d)$. Define A by (3.3)–(3.5), and suppose X is a solution of the $C_{\mathbb{R}^d}[0, \infty)$ martingale problem for (A, μ) with respect to a filtration $\{\mathcal{F}_t\}$ on a probability space $(\Omega, \mathcal{F}, P)$. Let W' be a d-dimensional $\{\mathcal{F}'_t\}$-Brownian motion on a probability space $(\Omega', \mathcal{F}',$

P') and define $\tilde{\Omega} = \Omega \times \Omega'$, $\tilde{\mathscr{F}} = \overline{\mathscr{F} \times \mathscr{F}'}$, $\tilde{P} = P \times P'$, $\tilde{\mathscr{F}}_t = \overline{\mathscr{F}_t \times \mathscr{F}'_t}$, and $\tilde{X}(t, \omega, \omega') = X(t, \omega)$. Then there exists a d-dimensional $\{\tilde{\mathscr{F}}_t\}$-Brownian motion $\tilde{W}$ such that $(\tilde{\Omega}, \tilde{\mathscr{F}}, \tilde{P}, \{\tilde{\mathscr{F}}_t\}, \tilde{W}, \tilde{X})$ is a solution of the stochastic integral equation corresponding to (σ, b, μ).

Proof. Define M by (3.8), $\tilde{M}(t, \omega, \omega') \equiv M(t, \omega)$, and $\tilde{W}'(t, \omega, \omega') \equiv W'(t, \omega')$. Using the notation of Lemma 3.2, we claim that

$$\tilde{W}(t) \equiv \int_0^t \rho(s, \tilde{X}(s))\, d\tilde{M}(s) + \int_0^t \eta(s, \tilde{X}(s))\, d\tilde{W}'(s) \tag{3.16}$$

defines a d-dimensional $\{\tilde{\mathscr{F}}_t\}$-Brownian motion. Again, this is a consequence of Theorem 2.12 since

$$\begin{aligned} \langle \tilde{W}_i, \tilde{W}_j \rangle_t &= \sum_{k,l=1}^{d} \left\langle \int_0^{\cdot} \rho_{ik}(s, \tilde{X}(s))\, d\tilde{M}_k(s), \int_0^{\cdot} \rho_{jl}(s, \tilde{X}(s))\, d\tilde{M}_l(s) \right\rangle_t \\ &\quad + \sum_{k,l=1}^{d} \left\langle \int_0^{\cdot} \eta_{ik}(s, \tilde{X}(s))\, d\tilde{W}'_k(s), \int_0^{\cdot} \eta_{jl}(s, \tilde{X}(s))\, d\tilde{W}'_l(s) \right\rangle_t \\ &= \sum_{k,l=1}^{d} \int_0^t (\rho_{ik}\, a_{kl}\, \rho_{jl})(s, \tilde{X}(s))\, ds \\ &\quad + \sum_{k,l=1}^{d} \int_0^t (\eta_{ik}\, \delta_{kl}\, \eta_{jl})(s, \tilde{X}(s))\, ds \\ &= \int_0^t (\rho a \rho^T + \eta\eta^T)_{ij}(s, \tilde{X}(s))\, ds \\ &= \delta_{ij}\, t, \qquad t \geq 0, \end{aligned} \tag{3.17}$$

where the first equality uses the fact that $\langle \tilde{M}_k, \tilde{W}_l \rangle = 0$ for $k, l = 1, \ldots, d$, the second depends on Theorem 2.6, and the fourth on (3.13). By Lemma 2.8, (3.14), and (3.15),

$$\begin{aligned} \int_0^t \sigma(s, \tilde{X}(s))\, d\tilde{W}(s) &= \int_0^t (\sigma\rho)(s, \tilde{X}(s))\, d\tilde{M}(s) \\ &\quad + \int_0^t (\sigma\eta)(s, \tilde{X}(s))\, d\tilde{W}'(s) \\ &= \int_0^t (\sigma\rho)(s, \tilde{X}(s))\, d\tilde{M}(s) \\ &= \tilde{M}(t) - \int_0^t (I_d - \sigma\rho)(s, \tilde{X}(s))\, d\tilde{M}(s) \\ &= \tilde{M}(t) \\ &= \tilde{X}(t) - \tilde{X}(0) - \int_0^t b(s, \tilde{X}(s))\, ds, \qquad t \geq 0, \end{aligned} \tag{3.18}$$

where the next-to-last equality is a consequence of (3.15) and

(3.19)
$$\left\langle \sum_{j=1}^{d} \int_0^{\cdot} (I_d - \sigma\rho)_{ij}(s, \tilde{X}(s))\, d\tilde{M}_j(s) \right\rangle_t$$
$$= \sum_{k,l=1}^{d} \int_0^t [(I_d - \sigma\rho)_{ik}\, a_{kl}(I_d - \sigma\rho)_{jl}](s, X(s))\, ds$$
$$= 0,$$

and the desired result follows. □

3.4 Corollary Let $\sigma: [0, \infty) \times \mathbb{R}^d \to \mathbb{R}^d \otimes \mathbb{R}^d$ and $b: [0, \infty) \times \mathbb{R}^d \to \mathbb{R}^d$ be locally bounded and Borel measurable, and let $\mu \in \mathcal{P}(\mathbb{R}^d)$. Define A by (3.3)–(3.5). Then there exists a solution of the stochastic integral equation corresponding to (σ, b, μ) if and only if there exists a solution of the $C_{\mathbb{R}^d}[0, \infty)$ martingale problem for (A, μ). Moreover, distribution uniqueness holds for solutions of the stochastic integral equation corresponding to (σ, b, μ) if and only if uniqueness holds for solutions of the $C_{\mathbb{R}^d}[0, \infty)$ martingale problem for (A, μ).

3.5 Proposition Suppose, in addition to the hypotheses of Corollary 3.4, that there exists a constant K such that

(3.20) $$|a(t, x)| \le K(1 + |x|^2), \qquad x \cdot b(t, x) \le K(1 + |x|^2), \qquad t \ge 0,\ x \in \mathbb{R}^d.$$

Then every solution of the martingale problem for (A, μ) has a modification with sample paths in $C_{\mathbb{R}^d}[0, \infty)$. Consequently, the phrase "$C_{\mathbb{R}^d}[0, \infty)$ martingale problem" in the conclusions of Corollary 3.4 can be replaced by the phrase "martingale problem."

Proof. Let X be a solution of the martingale problem for (A, μ). By Theorem 7.1 of Chapter 4, $X^0(t) \equiv (t, X(t))$ is a solution of the martingale problem for A^0, where

(3.21) $$A^0 = \{(\gamma f, \gamma G f + \gamma' f) : f \in C_c^\infty(\mathbb{R}^d),\ \gamma \in C_c^1[0, \infty)\}.$$

By Corollary 3.7 of Chapter 4, X^0 has a modification Y^0 with sample paths in $D_{([0,\infty)\times\mathbb{R}^d)^\Delta}[0, \infty)$. Letting

(3.22) $$(A^0)^\Delta = \{(f, g) \in C(([0, \infty) \times \mathbb{R}^d)^\Delta) \times B(([0, \infty) \times \mathbb{R}^d)^\Delta):$$
$$(f, g)|_{[0,\infty)\times\mathbb{R}^d} \in A^0, \qquad f(\Delta) = g(\Delta) = 0\},$$

it follows that Y^0 is a solution of the martingale problem for $(A^0)^\Delta$. Choose $\varphi \in C_c^\infty[0, \infty)$ with $\chi_{[0,1]} \le \varphi \le \chi_{[0,2]}$ and $\varphi' \le 0$. Then the sequence

$\{(f_n, g_n)\} \subset (A^0)^\Delta$ given by $f_n(t, x) = \varphi(t/n)\varphi(|x|^2/n^2)$, $f_n(\Delta) = 0$, and $g_n = (A^0)^\Delta f_n$ satisfies $\text{bp-lim}_{n\to\infty} f_n = \chi_{[0,\infty)\times\mathbb{R}^d}$, $g_n \to 0$ pointwise, and $\sup_n \|g_n^-\| < \infty$. By Proposition 3.9 of Chapter 4, Y^0 has almost all sample paths in $D_{[0,\infty)\times\mathbb{R}^d}[0, \infty)$, and therefore, by Problem 19 of Chapter 4, in $C_{[0,\infty)\times\mathbb{R}^d}[0, \infty)$. Define η: $[0, \infty) \times \mathbb{R}^d \to \mathbb{R}^d$ by $\eta(t, y) = y$. Then $\eta \circ Y^0$ is a solution of the martingale problem for (A, μ) with almost all sample paths in $C_{\mathbb{R}^d}[0, \infty)$, and the first conclusion follows.

The second conclusion is an immediate consequence of this. □

3.6 Theorem Let σ: $[0, \infty) \times \mathbb{R}^d \to \mathbb{R}^d \otimes \mathbb{R}^d$ and b: $[0, \infty) \times \mathbb{R}^d \to \mathbb{R}^d$ be locally bounded and Borel measurable, and let $\mu \in \mathscr{P}(\mathbb{R}^d)$. Then pathwise uniqueness of solutions of the stochastic integral equation corresponding to (σ, b, μ) implies distribution uniqueness.

Proof. Let $(\Omega, \mathscr{F}, P, \{\mathscr{F}_t\}, W, X)$ and $(\Omega', \mathscr{F}', P', \{\mathscr{F}'_t\}, W', X')$ be two solutions. We apply Lemma 5.15 of Chapter 4 with $E = C_{\mathbb{R}^d}[0, \infty) \times \mathbb{R}^d$, $S_1 = S_2 = C_{\mathbb{R}^d}[0, \infty) \times C_{\mathbb{R}^d}[0, \infty)$, $P_1 = P(W, X)^{-1}$, $P_2 = P'(W', X')^{-1}$, and $X_1(\omega_1, \omega_2) = X_2(\omega_1, \omega_2) \equiv (\omega_1, \omega_2(0))$. Letting $\lambda = PW^{-1} = P'(W')^{-1}$, we have $P(W, X(0))^{-1} = P'(W', X'(0))^{-1} = \lambda \times \mu$. We conclude that there exists a probability space $(\tilde{\Omega}, \tilde{\mathscr{F}}, \tilde{P})$ on which are defined $C_{\mathbb{R}^d}[0, \infty)$-valued random variables $\tilde{W}$, $\tilde{X}$, and $\tilde{X}'$ satisfying $\tilde{P}(\tilde{W}, \tilde{X})^{-1} = P(W, X)^{-1}$, $\tilde{P}(\tilde{W}, \tilde{X}')^{-1} = P'(W', X')^{-1}$, and $\tilde{P}\{\tilde{X}(0) = \tilde{X}'(0)\} = 1$. Moreover, for all $f, g \in B(C_{\mathbb{R}^d}[0, \infty) \times C_{\mathbb{R}^d}[0, \infty))$,

$$(3.23)\quad E^{\tilde{P}}[f(\tilde{W}, \tilde{X})g(\tilde{W}, \tilde{X}')]$$
$$= \iint E^P[f(W, X) \mid (W, X(0)) = (\beta, x)]E^{P'}[g(W', X') \mid (W', X'(0))$$
$$= (\beta, x)]\lambda(d\beta)\mu(dx).$$

Let $0 \le s_1 \le \cdots \le s_k \le s \le t_1 \le \cdots \le t_k$ and $f_{ij} \in B(E)$ for $i = 1, \ldots, k$ and $j = 0, 1, 2, 3$. Then

(3.24)

$$E^{\tilde{P}}\left[\prod_{i=1}^{k} f_{i0}(\tilde{W}(t_i) - \tilde{W}(s))f_{i1}(\tilde{W}(s_i))f_{i2}(\tilde{X}(s_i))f_{i3}(\tilde{X}'(s_i))\right]$$
$$= \iint \prod_{i=1}^{k} f_{i0}(\beta(t_i) - \beta(s))E^P\left[\prod_{i=1}^{k} f_{i1}(W(s_i))f_{i2}(X(s_i)) \,\middle|\, (W, X(0)) = (\beta, x)\right]$$
$$\cdot E^{P'}\left[\prod_{i=1}^{k} f_{i3}(X'(s_i)) \,\middle|\, (W', X'(0)) = (\beta, x)\right]\lambda(d\beta)\mu(dx)$$

$$= \int \prod_{i=1}^{k} f_{i0}(\beta(t_i) - \beta(s))\lambda(d\beta)$$

$$\cdot \iint E^P\left[\prod_{i=1}^{k} f_{i1}(W(s_i))f_{i2}(X(s_i)) \,\middle|\, (W, X(0)) = (\beta, x)\right]$$

$$\cdot E^{P'}\left[\prod_{i=1}^{k} f_{i3}(X'(s_i)) \,\middle|\, (W', X'(0)) = (\beta, x)\right]\lambda(d\beta)\mu(dx)$$

$$= E^{\tilde{P}}\left[\prod_{i=1}^{k} f_{i0}(\tilde{W}(t_i) - \tilde{W}(s))\right] E^{\tilde{P}}\left[\prod_{i=1}^{k} f_{i1}(\tilde{W}(s_i))f_{i2}(\tilde{X}(s_i))f_{i3}(\tilde{X}'(s_i))\right],$$

where the second equality depends on the fact that the two conditional expectations on its left side are functions only of $(\beta(\cdot \wedge s), x)$. It follows that $\tilde{W}$ is a d-dimensional $\{\tilde{\mathscr{F}}_t\}$-Brownian motion, where

$$\tilde{\mathscr{F}}_t = \sigma(\tilde{W}(s), \tilde{X}(s), \tilde{X}'(s): 0 \leq s \leq t), \tag{3.25}$$

and hence $(\tilde{\Omega}, \tilde{\mathscr{F}}, \tilde{P}, \{\tilde{\mathscr{F}}_t\}, \tilde{W}, \tilde{X})$ and $(\tilde{\Omega}, \tilde{\mathscr{F}}, \tilde{P}, \{\tilde{\mathscr{F}}_t\}, \tilde{W}, \tilde{X}')$ are solutions of the stochastic integral equation corresponding to (σ, b, μ). By pathwise uniqueness, $\tilde{P}\{\tilde{X}(t) = \tilde{X}'(t) \text{ for all } t \geq 0\} = 1$, so $PX^{-1} = \tilde{P}\tilde{X}^{-1} = \tilde{P}(\tilde{X}')^{-1} = P'(X')^{-1}$. Thus distribution uniqueness holds. □

The next result gives sufficient conditions for pathwise uniqueness of solutions of (3.1).

3.7 Theorem Let $\sigma: [0, \infty) \times \mathbb{R}^d \to \mathbb{R}^d \otimes \mathbb{R}^d$ and $b: [0, \infty) \times \mathbb{R}^d \to \mathbb{R}^d$ be locally bounded and Borel measurable. Let $U \subset \mathbb{R}^d$ be open, let $T > 0$, and suppose that there exists a constant K such that

$$|\sigma(t, x) - \sigma(t, y)| \vee |b(t, x) - b(t, y)| \leq K|x - y|, \qquad 0 \leq t \leq T, \quad x, y \in U. \tag{3.26}$$

Given two solutions $(\Omega, \mathscr{F}, P, \{\mathscr{F}_t\}, W, X)$ and $(\Omega, \mathscr{F}, P, \{\mathscr{F}_t\}, W, Y)$ of the stochastic integral equation corresponding to (σ, b), let

$$\tau = \inf\{t \geq 0: X(t) \notin U \quad \text{or} \quad Y(t) \notin U\}. \tag{3.27}$$

Then $P\{X(0) = Y(0)\} = 1$ implies $P\{X(t \wedge \tau) = Y(t \wedge \tau) \text{ for } 0 \leq t \leq T\} = 1$.

Proof. For $0 \le t \le T$,

(3.28)
$$
\begin{aligned}
E[|X(t\wedge\tau) - Y(t\wedge\tau)|^2]
&\le 2E\left[\left|\int_0^{t\wedge\tau} (\sigma(s, X(s)) - \sigma(s, Y(s)))\, dW(s)\right|^2\right] \\
&\quad + 2E\left[\left|\int_0^{t\wedge\tau} (b(s, X(s)) - b(s, Y(s)))\, ds\right|^2\right] \\
&\le 2E\left[\int_0^{t\wedge\tau} |\sigma(s, X(s)) - \sigma(s, Y(s))|^2\, ds\right] \\
&\quad + 2tE\left[\int_0^{t\wedge\tau} |b(s, X(s)) - b(s, Y(s))|^2\, ds\right] \\
&\le 2K^2(1 + t)E\left[\int_0^{t\wedge\tau} |X(s) - Y(s)|^2\, ds\right] \\
&\le 2K^2(1 + T)E\left[\int_0^{t} |X(s\wedge\tau) - Y(s\wedge\tau)|^2\, ds\right],
\end{aligned}
$$

and hence the desired result follows from Gronwall's inequality. □

In particular, if $\sigma(t, x)$ and $b(t, x)$ are locally Lipschitz continuous in x, uniformly in t in bounded intervals (i.e., for every bounded open set $U \subset \mathbb{R}^d$ and $T > 0$, (3.26) holds for some K), then we have pathwise uniqueness. This condition suffices for many applications. However, in some cases, $a = \sigma\sigma^T$ is a smooth function but σ is not. In general this causes serious difficulties, but not when $d = 1$.

3.8 Theorem In the case $d = 1$, Theorem 3.7 is valid with (3.26) replaced by

(3.29)
$$
|\sigma(t, x) - \sigma(t, y)|^2 \vee |b(t, x) - b(t, y)| \le K|x - y|, \qquad 0 \le t \le T, \quad x, y \in U.
$$

3.9 Remark If $\sigma \ge 0$, (3.29) is implied by

(3.30)
$$
|\sigma^2(t, x) - \sigma^2(t, y)| \vee |b(t, x) - b(t, y)| \le K|x - y|, \qquad 0 \le t \le T, \quad x, y \in U.
$$
□

Proof. For each $\varepsilon > 0$, define $\varphi_\varepsilon \in C^2(\mathbb{R})$ by $\varphi_\varepsilon(u) = (u^2 + \varepsilon)^{1/2}$ and $\psi_\varepsilon \in C(\mathbb{R})$ by $\psi_\varepsilon(u) = \varepsilon|u|/(u^2 + \varepsilon)^{3/2}$. For $0 \le t \le T$, we have, by Itô's formula,

$$(3.31) \qquad E[\varphi_\varepsilon(X(t \wedge \tau) - Y(t \wedge \tau))]$$

$$= \varphi_\varepsilon(0) + E\left[\int_0^{t \wedge \tau} \{\varphi_\varepsilon'(X(s) - Y(s))(b(s, X(s)) - b(s, Y(s))) + \tfrac{1}{2}\varphi_\varepsilon''(X(s) - Y(s))(\sigma(s, X(s)) - \sigma(s, Y(s)))^2\} \, ds\right]$$

$$\le \varphi_\varepsilon(0) + E\left[\int_0^{t \wedge \tau} \{K|X(s) - Y(s)| + \tfrac{1}{2}K\psi_\varepsilon(X(s) - Y(s))\} \, ds\right].$$

Noting that $\psi_\varepsilon(u) \le \sup_{y \in \mathbb{R}} |y|/(y^2 + 1)^{3/2}$ for all $u \in \mathbb{R}$ and $\varepsilon > 0$, we let $\varepsilon \to 0$ and conclude from the dominated convergence theorem that

$$(3.32) \qquad E[|X(t \wedge \tau) - Y(t \wedge \tau)|] \le KE\left[\int_0^{t \wedge \tau} |X(s) - Y(s)| \, ds\right] \le K \int_0^t E[|X(s \wedge \tau) - Y(s \wedge \tau)|] \, ds$$

for $0 \le t \le T$, and the result again follows from Gronwall's inequality. □

We turn finally to the question of existence of solutions of (3.1). We take two approaches. The first is based on Corollary 3.4 and results in Chapter 4. The second is the classical iteration method.

3.10 Theorem Let $\sigma: [0, \infty) \times \mathbb{R}^d \to \mathbb{R}^d \otimes \mathbb{R}^d$ and $b: [0, \infty) \times \mathbb{R}^d \to \mathbb{R}^d$ be continuous and satisfy

$$(3.33) \qquad |\sigma(t, x)|^2 \le K(1 + |x|^2), \quad x \cdot b(t,x) \le K(1 + |x|^2), \qquad t \ge 0, \quad x \in \mathbb{R}^d,$$

for some constant K, and let $\mu \in \mathscr{P}(\mathbb{R}^d)$. Then there exists a solution of the stochastic integral equation corresponding to (σ, b, μ).

Proof. It suffices by Corollary 3.4 and Proposition 3.5 to prove the existence of a solution of the martingale problem for (A, μ), where A is defined by (3.3)–(3.5). By Theorem 7.1 of Chapter 4 it suffices to prove the existence of a solution of the martingale problem for $(A^0, \delta_0 \times \mu)$, where A^0 is defined by (3.21). Noting that $A^0 \subset \hat{C}([0, \infty) \times \mathbb{R}^d) \times \hat{C}([0, \infty) \times \mathbb{R}^d)$ and A^0 satisfies the positive maximum principle, Theorem 5.4 of Chapter 4 guarantees a solution of the $D_{([0, \infty) \times \mathbb{R}^d)^\Delta}[0, \infty)$ martingale problem for $((A^0)^\Delta, \delta_0 \times \mu)$, where $(A^0)^\Delta$ is defined by (3.22). Arguing as in the proof of Proposition 3.5, we complete the proof using Proposition 3.9 and Problem 19, both of Chapter 4. □

3.11 Theorem Let $\sigma: [0, \infty) \times \mathbb{R}^d \to \mathbb{R}^d \otimes \mathbb{R}^d$ and $b: [0, \infty) \times \mathbb{R}^d \to \mathbb{R}^d$ be locally bounded and Borel measurable. Suppose that for each $T > 0$ and $n \geq 1$ there exist constants K_T and $K_{T,n}$ such that

$$|\sigma(t, x)|^2 \leq K_T(1 + |x|^2), \qquad x \cdot b(t, x) \leq K_T(1 + |x|^2),$$
$$0 \leq t \leq T, \qquad x \in \mathbb{R}^d, \tag{3.34}$$

and

$$|\sigma(t, x) - \sigma(t, y)| \vee |b(t, x) - b(t, y)| \leq K_{T,n}|x - y|,$$
$$0 \leq t \leq T, \qquad |x| \vee |y| \leq n. \tag{3.35}$$

Given a d-dimensional Brownian motion W and an independent $\mathbb{R}^d$-valued random variable ξ on a probability space $(\Omega, \mathscr{F}, P)$ such that $E[|\xi|^2] < \infty$, there exists a process X with $X(0) = \xi$ a.s. such that $(\Omega, \mathscr{F}, P, \{\mathscr{F}_t\}, W, X)$ is a solution of the stochastic integral equation corresponding to (σ, b), where $\mathscr{F}_t = \mathscr{F}_t^W \vee \sigma(\xi)$.

Proof. We first give the proof in the case that (3.34) and (3.35) are replaced by

$$|\sigma(t, x)| \vee |b(t, x)| \leq K_T, \qquad 0 \leq t \leq T, \qquad x \in \mathbb{R}^d \tag{3.36}$$

and

$$|\sigma(t, x) - \sigma(t, y)| \vee |b(t, x) - b(t, y)| \leq K_T|x - y|,$$
$$0 \leq t \leq T, \qquad x, y \in \mathbb{R}^d. \tag{3.37}$$

Let $X_0(t) \equiv \xi$. Having defined $X_0, \ldots, X_k$, let

$$X_{k+1}(t) \equiv \xi + \int_0^t \sigma(s, X_k(s))\, dW(s) + \int_0^t b(s, X_k(s))\, ds, \tag{3.38}$$

and note that $E[|X_{k+1}(t)|^2] < \infty$ for each $t \geq 0$ by (3.36). For $k = 0, 1, \ldots$, let $\varphi_k(t) \equiv E[|X_{k+1}(t) - X_k(t)|^2]$. Given $T > 0$, (3.37) implies that

$$\varphi_k(t) \leq 2K_T^2(1 + T) \int_0^t \varphi_{k-1}(s)\, ds, \qquad 0 \leq t \leq T. \tag{3.39}$$

Since $\varphi_0(t) \leq 2K_T^2(1 + T)t$ for $0 \leq t \leq T$ by (3.36), we have by induction that, for $k = 0, 1, \ldots$,

$$\varphi_k(t) \leq \frac{[2K_T^2(1 + T)]^{k+1} t^{k+1}}{(k + 1)!}, \qquad 0 \leq t \leq T. \tag{3.40}$$

It follows that

$$(3.41)\qquad E\left[\sup_{0\le t\le T}|X_{k+1}(t)-X_k(t)|^2\right]$$

$$\le 2E\left[\sup_{0\le t\le T}\left|\int_0^t(\sigma(s, X_k(s))-\sigma(s, X_{k-1}(s)))\,dW(s)\right|^2\right]$$

$$+\,2E\left[\sup_{0\le t\le T}\left|\int_0^t(b(s, X_k(s))-b(s, X_{k-1}(s)))\,ds\right|^2\right]$$

$$\le 8E\left[\left|\int_0^T(\sigma(s, X_k(s))-\sigma(s, X_{k-1}(s)))\,dW(s)\right|^2\right]$$

$$+\,2TE\left[\int_0^T|b(s, X_k(s))-b(s, X_{k-1}(s))|^2\,ds\right]$$

$$\le 2K_T^2(4+T)\int_0^T E[|X_k(s)-X_{k-1}(s)|^2]\,ds$$

$$\le\frac{2K_T^2(4+T)[2K_T^2(1+T)]^kT^{k+1}}{(k+1)!}\equiv\varepsilon_k(T),$$

and therefore

$$(3.42)\qquad \sum_{k=0}^{\infty}P\left\{\sup_{0\le t\le T}|X_{k+1}(t)-X_k(t)|\ge 2^{-k}\right\}\le\sum_{k=0}^{\infty}4^k\varepsilon_k(T)<\infty.$$

By the Borel–Cantelli lemma, $\sup_{0\le t\le T}|X_{k+1}(t)-X_k(t)|<2^{-k}$ for all $k\ge k(\omega)$ for almost all ω. Now T was arbitrary, so there exists $A\in\mathscr{F}$ with $P(A)=0$ such that, for every $\omega\in\Omega$, $\{\chi_{A^c}X_k\}$ converges uniformly on bounded time intervals. Letting X be the limiting process, we conclude from (3.42) that $X(0)=\xi$ a.s. and $(\Omega, \mathscr{F}, P, \{\mathscr{F}_t\}, W, X)$ is a solution of the stochastic integral equation corresponding to (σ, b).

We now want to obtain the conclusion of the theorem under the original hypotheses ((3.34) and (3.35) instead of (3.36) and (3.37)). For each $n\ge 1$, define $\rho_n\colon[0,\infty)\times\mathbb{R}^d\to[0,\infty)\times\mathbb{R}^d$ by $\rho_n(t, x)=(t, (1\wedge(n/|x|))x)$, and let $\sigma_n=\sigma\circ\rho_n$ and $b_n=b\circ\rho_n$. By the first part of the proof there exists a solution $(\Omega, \mathscr{F}, P, \{\mathscr{F}_t\}, W, X_n)$ of the stochastic integral equation corresponding to (σ_n, b_n). Letting $\tau_n=\inf\{t\ge 0\colon |X_n(t)|\ge n\}$, Theorem 3.7 guarantees that $X_n(t)=X_m(t)$ whenever $0\le t\le\tau_n\wedge\tau_m$ and $m, n\ge 1$. Thus, we can define $X(t)=X_n(t)$ for $0\le t\le\tau_n$, $n\ge 1$.

To complete the proof, it suffices to show that $\tau_n\to\infty$ a.s. By Itô's formula and (3.34),

$$(3.43)\qquad E[\log(1+|X_n(t\wedge\tau_n)|^2)]$$

is bounded above in n for fixed $t\ge 0$. The same is therefore true of $\log(1+n^2)P\{\tau_n\le t\}$, so $P\{\tau_n\le t\}\to 0$ for each $t\ge 0$. Since $\tau_1\le\tau_2\cdots$, the desired conclusion follows. □

4. PROBLEMS

1. Let W be a d-dimensional $\{\mathscr{F}_t\}$-Brownian motion, and let τ be an $\{\mathscr{F}_t\}$-stopping time with $\tau < \infty$ a.s. Show that $W^*(\cdot) \equiv W(\tau + \cdot) - W(\tau)$ ($\equiv 0$ if $\tau = \infty$) is a d-dimensional Brownian motion, and that $\mathscr{F}_t^{W^*}$ is independent of $\mathscr{F}_\tau$ for each $t \geq 0$.

2. Let W be a d-dimensional $\{\mathscr{F}_t\}$-Brownian motion. Show that

$$X_\theta(t) \equiv \exp\left\{\theta \cdot W(t) - \frac{|\theta|^2 t}{2}\right\} \tag{4.1}$$

is an $\{\mathscr{F}_t\}$-martingale. For $d = 1$, $\alpha > 0$, and $\beta > 0$, show that $P\{\sup_{0 \leq s \leq t} (W(s) - \alpha s/2) > \beta\} \leq e^{-\alpha\beta}$.

3. Let W be a one-dimensional Brownian motion. Evaluate the stochastic integral $\int_0^\cdot W^2\, dW$ directly from its definition (Theorem 2.3). Check your result using Itô's formula.

4. Let $M \in \mathscr{M}_{c,\,loc}$ and $X, Y, X_1, X_2, \ldots \in L^2_{loc}(\langle M \rangle)$. Suppose that $|X_n| \leq Y$ for each $n \geq 1$ and $X_n(t) \to X(t)$ a.s. for each $t \geq 0$. Show that for every $T > 0$,

$$\sup_{0 \leq t \leq T} \left| \int_0^t X_n\, dM - \int_0^t X\, dM \right| \xrightarrow{p} 0. \tag{4.2}$$

5. Let W be a d-dimensional $\{\mathscr{F}_t\}$-Brownian motion (with $\{\mathscr{F}_t\}$ a complete filtration), and let $\sigma\colon [0, \infty) \times \Omega \to \mathbb{R}^d \otimes \mathbb{R}^d$ be $\{\mathscr{F}_t\}$-progressive and satisfy $\sigma\sigma^T = I_d$. Show that $\tilde{W} \equiv \int_0^\cdot \sigma(s)\, dW(s)$ is a d-dimensional $\{\mathscr{F}_t\}$-Brownian motion.

6. Show that the spherical coordinates

$$\begin{aligned} \rho &= |B| = (B_1^2 + B_2^2 + B_3^2)^{1/2}, \\ \varphi &= \cos^{-1}(B_3/\rho) = \text{colatitude}, \\ \theta &= \tan^{-1}(B_2/B_1) = \text{longitude} \end{aligned} \tag{4.3}$$

of a three-dimensional Brownian motion $B = (B_1, B_2, B_3)$ evolve according to the stochastic differential equations

$$\begin{aligned} d\rho &= dW_1 + \rho^{-1}\, dt, \\ d\varphi &= \rho^{-1}\, dW_2 + \tfrac{1}{2}\rho^{-2} \cot\varphi\, dt, \\ d\theta &= \rho^{-1} \csc\varphi\, dW_3 \end{aligned} \tag{4.4}$$

with a new three-dimensional Brownian motion $W = (W_1, W_2, W_3)$:

$$W_1 = \int_0^t \rho^{-1}(B_1\, dB_1 + B_2\, dB_2 + B_3\, dB_3),$$

$$(4.5)\qquad W_2 = \int_0^t \rho^{-2}(\csc\varphi) B_3(B_1\, dB_1 + B_2\, dB_2) - \int_0^t \sin\varphi\, dB_3,$$

$$W_3 = \int_0^t \rho^{-1}(\csc\varphi)(B_1\, dB_2 - B_2\, dB_1).$$

7. Let σ: $[0, \infty) \times \mathbb{R}^d \to \mathbb{R}^d \otimes \mathbb{R}^d$ and b: $[0, \infty) \times \mathbb{R}^d \to \mathbb{R}^d$ be locally bounded and Borel measurable and suppose that $(\Omega, \mathscr{F}, P, \{\mathscr{F}_t\}, W, X)$ is a solution of the stochastic integral equation corresponding to (σ, b). Let c: $[0. \infty) \times \mathbb{R}^d \to \mathbb{R}$ be bounded and Borel measurable. Show that if $f \in C_c^{1,2}([0, \infty) \times \mathbb{R}^d)$, then

$$(4.6)\quad E\left[f(t, X(t)) \exp\left\{\int_0^t c(s, X(s))\, ds\right\}\right]$$

$$= E[f(0, X(0))] + E\left[\int_0^t (Gf + cf)(s, X(s)) \exp\left\{\int_0^s c(r, X(r))\, dr\right\} ds\right]$$

for all $t \geq 0$, where G is defined by (3.4) and (3.5).

8. Let Φ: $\mathbb{R}^d \to \mathbb{R}^d$ be a C^2-diffeomorphism (that is, Φ is one-to-one, onto, and twice continuously differentiable, as is its inverse Φ^{-1}). Let σ: $\mathbb{R}^d \to \mathbb{R}^d \otimes \mathbb{R}^d$ and b: $\mathbb{R}^d \to \mathbb{R}^d$ be locally bounded and Borel measurable, and suppose the stochastic integral equation corresponding to (σ, b) has a solution $(\Omega, \mathscr{F}, P, \{\mathscr{F}_t\}, W, X)$. Observe that then there exist $\hat{\sigma}$: $\mathbb{R}^d \to \mathbb{R}^d \otimes \mathbb{R}^d$ and $\hat{b}$: $\mathbb{R}^d \to \mathbb{R}^d$ locally bounded and Borel measurable such that $(\Omega, \mathscr{F}, P, \{\mathscr{F}_t\}, W, \Phi \circ X)$ is a solution of the stochastic integral equation corresponding to $(\hat{\sigma}, \hat{b})$. Define G in terms of σ and b and $\hat{G}$ in terms of $\hat{\sigma}$ and $\hat{b}$ as in (3.4) and (3.5). Show that $Gf = [\hat{G}(f \circ \Phi^{-1})] \circ \Phi$ for all $f \in C_c^2(\mathbb{R}^d)$. Thus the relationship between a stochastic integral equation and its associated differential operator is invariant under diffeomorphism.

9. Let a: $\mathbb{R}^d \to S_d$ and b: $\mathbb{R}^d \to \mathbb{R}^d$ be locally bounded and Borel measurable, and define A and G by (3.3) and (3.4). Let $\varphi \in C^2(\mathbb{R}^d)$ and suppose that for each $n \geq 1$ there exists a constant $K_n \geq 0$ such that

$$(4.7)\qquad \max\{\nabla\varphi \cdot a\nabla\varphi, G\varphi\}\chi_{\{x \in \mathbb{R}^d:\, \varphi(x) > 0,\, |x| \leq n\}} \leq K_n \varphi.$$

Show that if X is a solution of the $C_{\mathbb{R}^d}[0, \infty)$ martingale problem for A, then $P\{\varphi(X(0)) \leq 0\} = 1$ implies $P\{\varphi(X(t)) \leq 0 \text{ for all } t \geq 0\} = 1$.

Hint: Show that Gronwall's inequality applies to $E[\varphi^+(X(t \wedge \tau_n))]$, where $\tau_n = \inf\{t \geq 0: |X(t)| \geq n\}$, by approximating φ^+ by a sequence of the form $\{h_n \circ \varphi\}$.

10. Define $\int_0^{\cdot} \sigma(s, X(s))\, dW(s)$ for $\sigma\colon [0, \infty) \times \mathbb{R}^d \to \mathbb{R}^d \otimes \mathbb{R}^m$ (the space of real $d \times m$ matrices) locally bounded and Borel measurable with W and X as before by defining $\tilde{\sigma}\colon [0, \infty) \times \mathbb{R}^{d \vee m} \to \mathbb{R}^{d \vee m} \otimes \mathbb{R}^{d \vee m}$ in terms of σ in the obvious way. Check to see which of the results of Section 3 extend to nonsquare σ.

11. (a) Let $M \in \mathcal{M}_c$ and $X \in L^2(\langle M \rangle)$, and let $s \geq 0$ and Z be a bounded $\mathcal{F}_s$-measurable random variable. Show that

$$\int_s^t ZX\, dM = Z \int_s^t X\, dM, \qquad t > s. \tag{4.8}$$

(b) Prove Lemma 2.5.

Hint: First consider $X \in S$.

12. Let $M \in \mathcal{M}_{c,\,loc}$ and let X be continuous and adapted. Show that for $0 = t_0 < t_1 < \cdots < t_m = t$,

$$\int_0^t X\, dM = \lim_{\max (t_{k+1} - t_k) \to 0} \sum_k X(t_k)(M(t_{k+1}) - M(t_k)). \tag{4.9}$$

13. Let W be a one-dimensional Brownian motion, and let $X(t) = W(t) + t$. Find a function φ such that $\varphi(X(t))$ is a martingale. (Use Itô's formula.) Let $\tau = \inf\{t\colon X(t) = -a \text{ or } b\}$. Use $\varphi(X(t))$ to find $P\{X(\tau) = b\}$. What is $E[\tau]$?

14. Let X be a solution in $\mathbb{R}$ of

$$X(t) = x + \int_0^t bX(s)\, ds + \int_0^t \sigma X(s)\, dW(s)$$

and let $Y = X^2$.

(a) Use Itô's formula to find the stochastic integral equation satisfied by Y.

(b) Use the equation in (a) to find $E[X^2]$.

(c) Extend the above argument to find $E[X^k]$, $k = 1, 2, 3, \ldots$.

15. Let W be a one-dimensional Brownian motion.

(a) Let $X = (X_1, X_2)$ satisfy

$$X_1(t) = x_1 + \int_0^t X_2(s)\, ds$$

$$X_2(t) = x_2 - \int_0^t X_1(s)\, ds + \int_0^t cX_1(s)\, dW(s).$$

Define $m_1(t) = E[X^2(t)]$, $m_2(t) = E[X(t)Y(t)]$, and $m_3(t) = E[Y^2(t)]$. Find a system of three linear differential equations satisfied by m_1,

m_2, and m_3. Show that the expected "total energy" $(E[X^2(t) + Y^2(t)])$ is asymptotic to $ke^{\lambda t}$ for some $\lambda > 0$ and $k > 0$.

(b) Let $X = (X_1, X_2)$ satisfy

$$X_1(t) = X_1(0) + \int_0^t X_2(s)\, dW(s)$$

$$X_2(t) = X_2(0) - \int_0^t X_1(s)\, dW(s).$$

Show that $X_1^2(t) + X_2^2(t) = (X_1^2(0) + X_2^2(0))\, e^t$.

16. Let W be a one-dimensional Brownian motion.

(a) For $x \geq 0$, let $X(t, x) = x + \int_0^t \lambda X(s, x)\, ds + \int_0^t \sqrt{X(s, x)}\, dW(s)$ and $\tau_x = \inf\{t: X(t, x) = 0\}$. Calculate $P\{\tau_x < \infty\}$ as a function of λ.

(b) For $x > 0$, let $X(t, x) = x - \int_0^t \lambda X(s, x)\, ds + \int_0^t X(s, x)\, dW(s)$ with $\lambda > 0$, and let τ_x be defined as above. Show that $P\{\tau_x < \infty\} = 0$, but that $P\{\lim_{t \to \infty} X(t, x) = 0\} = 1$.

(c) For $x > 0$, let $X(t, x) = x + \int_0^t \sigma(X(s, x))\, dW(s)$, and let τ_x be defined as above. Give conditions on σ that imply $E[\tau_x] < \infty$.

(d) For $x > 0$, let $X(t, x) = x + \int_0^t \lambda\, ds + \int_0^t \sqrt{X(s, x)}\, dW(s)$, and let τ_x be defined as above. For what values of $\lambda > 0$ is $P\{\tau_x < \infty\} > 0$? For these values, show that $P\{\tau_x < \infty\} = 1$, but that $E[\tau_x] = \infty$.

5. NOTES

There are many general references on stochastic integration and stochastic integral equations. These include McKean (1969), Gihman and Skorohod (1972), Friedman (1975), Ikeda and Watanabe (1981), Elliot (1982), Métivier (1982), and Chung and Williams (1983). Our treatment is heavily influenced by Priouret (1974).

Stochastic integrals with respect to square integrable martingales go back to Doob (1953), page 437, and were developed by Courrege (1963) and Kunita and Watanabe (1967). The extension to local martingales is due to Meyer (1967) and Doleans-Dade and Meyer (1970). Itô's formula goes back, of course, to Itô (1951).

Theorem 3.3 is due to Stroock and Varadhan (1972), Theorems 3.6 and 3.8 to Yamada and Watanabe (1971), and Theorem 3.10 to Skorohod (1965). Theorems 3.7 and 3.11 are the classical uniqueness and existence theorems of Itô (1951).

Problems 6 and 8 were borrowed from McKean (1969) and Friedman (1975), respectively.

6 RANDOM TIME CHANGES

In this chapter we continue the study of stochastic equations that determine Markov processes. These equations involve random time changes of other Markov processes and frequently reduce stochastic problems to problems in analysis. Section 1 considers random time changes of a single process. The multiparameter analogue is developed in Section 2. Section 3 gives convergence results based on the random time changes. Sections 4 and 5 give time change equations for large classes of Markov chains and diffusion processes.

1. ONE-PARAMETER RANDOM TIME CHANGES

Let Y be a process with sample paths in $D_E[0, \infty)$, and let β be a nonnegative Borel measurable function on E. Suppose that $\beta \circ Y$ is a.s. bounded on bounded time intervals. We are interested in solutions of

$$Z(t) = Y\left(\int_0^t \beta(Z(s))\, ds\right). \tag{1.1}$$

Observe that if Z is a solution and we set

$$\tau(t) = \int_0^t \beta(Z(s))\, ds = \int_0^t \beta(Y(\tau(s)))\, ds, \tag{1.2}$$

then (see Problem 11)

$$t \geq \lim_{\varepsilon \to 0+} \int_0^t \frac{\beta(Z(s))}{\beta(Z(s)) \vee \varepsilon} ds = \lim_{\varepsilon \to 0+} \int_0^{\tau(t)} \frac{1}{\beta(Y(u)) \vee \varepsilon} du \tag{1.3}$$

$$= \int_0^{\tau(t)} \frac{1}{\beta(Y(u))} du,$$

and equality holds if and only if the Lebesgue measure of $\{s \leq t: \beta(Z(s)) = 0\}$ is zero. Conversely if

$$t = \int_0^{\tau(t)} \frac{1}{\beta(Y(u))} du \tag{1.4}$$

has a solution for all t (of necessity unique), then $\tau(t)$ is locally absolutely continuous (in fact locally Lipschitz) with

$$\dot{\tau}(t) = \beta(Y(\tau(t))), \quad \text{a.e. } t, \tag{1.5}$$

(differentiate both sides of (1.4)), and hence $Z(t) \equiv Y(\tau(t))$ is a solution of (1.1). More generally, let

$$\tau_1 = \inf\left\{s: \int_0^s \frac{1}{\beta(Y(u))} du = \infty\right\} \tag{1.6}$$

and suppose $\int_0^\infty (1/\beta(Y(u)))\, du = \infty$. If $\tau \leq \tau_1$ and $\beta(Y(\tau)) = 0$ when $\tau < \infty$, let $\tau(t)$ satisfy (1.4) for

$$t \leq \int_0^\tau \frac{1}{\beta(Y(u))} du = t_0 \tag{1.7}$$

and

$$\tau(t) = \tau, \qquad t > t_0. \tag{1.8}$$

Then $Z(t) \equiv Y(\tau(t))$ is a solution of (1.1).

1.1 Theorem Let Y, β, and τ_1 be as above. Define

$$\tau_0 = \inf\{s: \beta(Y(s)) = 0\} \tag{1.9}$$

and

$$\tau_2 = \lim_{\varepsilon \to 0+} \inf\{s: \beta(Y(s)) < \varepsilon\}. \tag{1.10}$$

(a) If $\tau_0 = \tau_1$ and $\beta(Y(\tau_0)) = 0$ when $\tau_0 < \infty$, then (1.1) has a unique solution $Z(t)$.

(b) Suppose β is continuous. If $\tau_0 = \tau_1 = \tau_2$, then there is a unique locally absolutely continuous function $\gamma(t)$ satisfying

$$\begin{aligned}(1.11)\qquad & \beta(Y(\gamma(t))) \wedge \beta(Y^-(\gamma(t))) \\ & \le \gamma'(t) \le \beta(Y(\gamma(t))) \vee \beta(Y^-(\gamma(t))), \qquad \text{a.e. } t,\end{aligned}$$

where $Y^-(u) = \lim_{v \to u-} Y(v)$, $u > 0$, and $Y^-(0) = Y(0)$, and $Z(t) = Y(\gamma(t))$, that is, $\tau(t) = \gamma(t)$.

1.2 Remark Note that τ_0, τ_1, and τ_2 may be infinite. □

Proof. Existence follows from the construction in (1.7) and (1.8). By (1.3), any solution, $Z(t)$, with $\tau(t)$ defined by (1.2), must satisfy $\tau(t) \le \tau_1$. If $\tau(t) < \tau_0$, then $\beta(Z(s)) \ne 0$ for all $s \le t$, and (1.4) uniquely determines $\tau(t)$. If $\tau_1 = \tau_0$, $\tau(t)$ is uniquely determined for all t.

If $\gamma(t)$ satisfies (1.11), then as above

$$\begin{aligned}(1.12)\qquad t &\ge \lim_{\varepsilon \to 0+} \int_0^t \frac{\beta(Y(\gamma(s))) \vee \beta(Y^-(\gamma(s)))}{\beta(Y(\gamma(s))) \vee \beta(Y^-(\gamma(s))) \vee \varepsilon}\, ds \\ &\ge \int_0^{\gamma(t)} \frac{1}{\beta(Y(u)) \vee \beta(Y^-(u))}\, du \\ &= \int_0^{\gamma(t)} \frac{1}{\beta(Y(u))}\, du,\end{aligned}$$

since $Y(u) = Y^-(u)$ for almost every u, and $\gamma(t) \le \tau_1$. Since $\tau_2 = \tau_1$, $\beta(Y(u)) \wedge \beta(Y^-(u)) \ne 0$ for $u < \tau_1$, and hence for $\gamma(t) < \tau_1$,

$$(1.13)\qquad t = \int_0^t \frac{\beta(Y(\gamma(s))) \wedge \beta(Y^-(\gamma(s)))}{\beta(Y(\gamma(s))) \wedge \beta(Y^-(\gamma(s)))}\, ds \le \int_0^{\gamma(t)} \frac{1}{\beta(Y(u))}\, du,$$

and (1.12) and (1.13) imply $Z(\gamma(t))$ is the unique solution of (1.1). □

We now relate solutions of (1.1) to solutions of a martingale problem.

1.3 Theorem Let $A \subset \bar{C}(E) \times B(E)$ and suppose Y is a solution of the $D_E[0, \infty)$ martingale problem for A.

If the conditions of Theorem 1.1(a) hold for almost every sample path, then the solution of (1.1) is a solution of the martingale problem for $\beta A \cap (B(E) \times B(E))$, where

$$(1.14)\qquad \beta A \equiv \{(f, \beta g) : (f, g) \in A\}.$$

Proof. Note that $\{\tau(s) \le t\} = \{\int_0^t (1/\beta(Y(u)))\, du \ge s\} \cup \{\tau_1 \le t\} \in \mathscr{F}_{t+}^Y$, so $\tau(s)$ is an $\{\mathscr{F}_{t+}^Y\}$-stopping time. For $(f, g) \in A$

$$f(Y(t)) - \int_0^t g(Y(s))\, ds \tag{1.15}$$

is an $\{\mathscr{F}_{t+}^Y\}$-martingale, hence the optional sampling theorem implies

$$\begin{aligned} f(Y(\tau(t))) &- \int_0^{\tau(t)} g(Y(s))\, ds \\ &= f(Z(t)) - \int_0^t \beta(Z(s))g(Z(s))\, ds \end{aligned} \tag{1.16}$$

is an $\{\mathscr{F}_{\tau(t)+}^Y\}$-martingale. □

We have the following converse to Theorem 1.3.

1.4 Theorem Let (E, r) be complete and separable, and let $A \subset \bar{C}(E) \times B(E)$. Suppose $\mathscr{D}(A)$ is separating, the $D_E[0, \infty)$ martingale problem for A is well-posed, and $\beta \in M(E)$, $\beta \ge 0$, is such that $\beta A \subset B(E) \times B(E)$. If Z is a solution of the $D_E[0, \infty)$ martingale problem for βA, then there is a version of Z satisfying (1.1) for a process Y that is a solution of the martingale problem for A.

Proof. First suppose

$$\tau(\infty) \equiv \int_0^\infty \beta(Z(s))\, ds = \infty \qquad \text{a.s.} \tag{1.17}$$

and define

$$\gamma(t) = \inf\left\{u: \int_0^u \beta(Z(s))\, ds > t\right\}. \tag{1.18}$$

Then $Y(t) \equiv Z(\gamma(t))$ is a solution of the martingale problem for A, that is, for $(f, g) \in A$,

$$f(Z(\gamma(t))) - \int_0^{\gamma(t)} \beta(Z(s))g(Z(s))\, ds = f(Y(t)) - \int_0^t g(Y(u))\, du \tag{1.19}$$

is a martingale by the optional sampling theorem. (See Problem 12 for the equality in (1.19).) Let $\gamma_+(t) = \lim_{s \to t+} \gamma(s)$. We claim that Z is constant on the interval $[\gamma(t), \gamma_+(t)]$ (see Problem 45 of Chapter 4), and since $\gamma(\tau(t)) \le t \le \gamma_+(\tau(t))$,

$$Z(t) = Z(\gamma(\tau(t))) = Y\left(\int_0^t \beta(Z(s))\, ds\right). \tag{1.20}$$

If $\tau(\infty) < \infty$, then $Y(t) = Z(\gamma(t))$ for $t \le \tau(\infty)$ and Y must be extended past $\tau(\infty)$ (on an enlarged sample space using Lemma 5.16 of Chapter 4). □

1.5 Theorem Suppose Y is as above, β is continuous, $\{Y_n\}$ is a sequence of processes with sample paths in $D_E[0, \infty)$ such that $Y_n \Rightarrow Y$, and $\{\beta_n\}$ is a sequence of nonnegative, Borel measurable functions on E such that

$$\lim_{n\to\infty} \sup_{x \in K} |\beta_n(x) - \beta(x)| = 0 \tag{1.21}$$

for every compact K. Let $h_n > 0$ and $\lim_{n\to\infty} h_n = 0$. Suppose Z, Z_n, and W_n satisfy

$$Z(t) = Y\left(\int_0^t \beta(Z(s))\, ds\right), \tag{1.22}$$

$$Z_n(t) = Y_n\left(\int_0^t \beta_n(Z_n(s))\, ds\right), \tag{1.23}$$

and

$$W_n(t) = Y_n\left(\int_0^{[t/h_n]h_n} \beta_n(W_n(s))\, ds\right) \tag{1.24}$$

for all $t \geq 0$. If $\tau_0 = \tau_1 = \tau_2$ a.s., then $Z_n \Rightarrow Z$ and $W_n \Rightarrow Z$.

1.6 Remark If Y is quasi-left continuous (see Theorem 3.12 of Chapter 4), in particular if Y is a Feller process, then $\tau_0 = \tau_2$ a.s. Observe that, for $\varepsilon > 0$, $\tau^{(\varepsilon)} = \inf \{s: \beta(Y(s)) \leq \varepsilon\} \leq \tau_0$, and by quasi-left continuity, $Y(\tau_2) = \lim_{\varepsilon\to 0} Y(\tau^{(\varepsilon)})$ on the set $\{\tau_2 < \infty\}$. Consequently, on $\{\tau_2 < \infty\}$, $\lim_{\varepsilon\to 0} \beta(Y(\tau^{(\varepsilon)})) = \beta(Y(\tau_2)) = 0$ a.s., and hence $\tau_2 = \tau_0$. Of course if $\inf_x \beta(x) > 0$, then $\tau_0 = \tau_1 = \tau_2 = \infty$.

Proof. By Theorem 1.8 of Chapter 3 we can, without loss of generality, assume Y_n and Y are defined on the same sample space $(\Omega, \mathscr{F}, P)$ and that $\lim_{n\to\infty} d(Y_n, Y) = 0$ a.s. Fix $\omega \in \Omega$ for which $\lim_{n\to\infty} d(Y_n, Y) = 0$ and $\tau_0 = \tau_1 = \tau_2$. We first assume $\bar{\beta} \equiv \sup_{x,n} \beta_n(x)$ is finite. Then

$$w'(Z_n, \delta, T) \leq w'(Y_n, \delta\bar{\beta}, T\bar{\beta}), \tag{1.25}$$

$$w'(W_n, \delta, T) \leq w'(Y_n, (\delta + h_n)\bar{\beta}, T\bar{\beta}), \tag{1.26}$$

$$\{Z_n(s): s \leq T\} \subset \{Y_n(u): u \leq T\bar{\beta}\}, \tag{1.27}$$

and

$$\{W_n(s): s \leq T\} \subset \{Y_n(u): u \leq T\bar{\beta}\}. \tag{1.28}$$

Since $\{Y_n\}$ is convergent in $D_E[0, \infty)$, it follows that $\{Z_n\}$ and $\{W_n\}$ are relatively compact in $D_E[0, \infty)$. We show that any convergent subsequence of $\{Z_n\}$ (or $\{W_n\}$) converges to Z, and hence $\lim_{n\to\infty} d(Z_n, Z) = 0$ (and $\lim_{n\to\infty} d(W_n,$

$Z) = 0$). If $\lim_{k\to\infty} d(Z_{n_k}, \hat{Z}) = 0$, then

$$\lim_{k\to\infty} \int_0^t \beta_{n_k}(Z_{n_k}(s))\, ds = \int_0^t \beta(\hat{Z}(s))\, ds \equiv \gamma(t) \tag{1.29}$$

and

$$\gamma'(t) = \beta(\hat{Z}(t)) = \lim_{k\to\infty} \beta\left(Y_{n_k}\left(\int_0^t \beta_{n_k}(Z_{n_k}(s))\, ds\right)\right) \tag{1.30}$$

for almost every t. The right side is either $\beta(Y(\gamma(t)))$ or $\beta(Y^-(\gamma(t)))$. Therefore $\gamma(t)$ satisfies (1.11) and hence $\hat{Z} = Z$. Similarly, if $\lim_{k\to\infty} d(W_{n_k}, \hat{Z}) = 0$, then

$$\lim_{k\to\infty} \int_0^{[t/h_{n_k}]h_{n_k}} \beta_{n_k}(W_{n_k}(s))\, ds = \int_0^t \beta(\hat{Z}(s))\, ds \tag{1.31}$$

and the proof follows as for $\{Z_n\}$.

Now dropping the assumption that the β_n are bounded, let Z^M, Z_n^M, and W_n^M be as above, but with β and β_n replaced by $M \wedge \beta$ and $M \wedge \beta_n$, $M > 0$. Since we are not assuming the solution of (1.23) is unique, take $Z_n^M = Z_n(\theta_n^M(t))$ where

$$\int_0^{\theta_n^M(t)} 1 \vee \left(\frac{\beta_n(Z_n(s))}{M}\right) ds = t. \tag{1.32}$$

As above, fix $\omega \in \Omega$ such that $\lim_{n\to\infty} d(Y_n, Y) = 0$ and $\tau_0 = \tau_1 = \tau_2$. Then $\lim_{n\to\infty} d(Z_n^M, Z^M) = 0$, and $\lim_{n\to\infty} d(W_n^M, Z^M) = 0$. Fix $t > 0$, and let $M > \sup_{s\le t} \beta(Z(s))$. Note that $Z^M(s) = Z(s)$ for $s \le t$. We claim that for n sufficiently large, $M > \sup_{s\le t} \beta_n(Z_n^M(s))$ and hence $Z_n^M(s) = Z_n(s)$ for $s \le t$. (Similarly for W_n^M.) To see this, suppose not. Then there exist $0 \le s_n \le t$ (which we can assume satisfy $s_n \to s_0$) such that $\overline{\lim}_{n\to\infty} \beta_n(Z_n^M(s_n)) \ge M$. But $\{Z_n^M(s_n)\}$ is relatively compact with limit points in $\{Z^M(s_0-), Z^M(s_0)\} = \{Z(s_0-), Z(s_0)\}$. Consequently,

$$\overline{\lim_{n\to\infty}}\, \beta_n(Z_n^M(s_n)) = \overline{\lim_{n\to\infty}}\, \beta(Z_n^M(s_n)) \le \beta(Z(s_0-)) \vee \beta(Z(s_0)) < M. \tag{1.33}$$

Recall that if $Z^M(s) = Z(s)$ for $s \le t$, then $d(Z^M, Z) \le e^{-t}$. Since t is arbitrary, it follows that $d(Z_n, Z) \to 0$. □

2. MULTIPARAMETER RANDOM TIME CHANGES

We now consider a system of random time changes analogous to (1.1). For $k = 1, 2, \ldots$, let (E_k, r_k) be a complete, separable metric space, and let Y_k be a process with sample paths in $D_{E_k}[0, \infty)$ defined on a complete probability space $(\Omega, \mathscr{F}, P)$. Let $\beta_k\colon \prod_l E_l \to [0, \infty)$ be nonnegative Borel measurable functions. We are interested in solutions of the system

$$Z_k(t) = Y_k\left(\int_0^t \beta_k(Z(s))\, ds\right), \tag{2.1}$$

where $Z = (Z_1, Z_2, \ldots)$. (Similarly we set $Y = (Y_1, Y_2, \ldots)$.)

We begin with the following technical lemma.

2.1 Lemma If for almost every $\omega \in \Omega$ a solution Z of (2.1) exists and is unique, then Z is a stochastic process.

Proof. Let $S = \prod_l D_{E_l}[0, \infty)$ and define $\gamma\colon S \times S \to S$ by

$$\gamma_k(y, z) = y_k\left(\int_0^t \beta_k(z(s))\, ds\right). \tag{2.2}$$

Then γ_k is Borel measurable and hence

$$\Gamma = \{(y, z)\colon z = \gamma(y, z)\} \tag{2.3}$$

is a Borel measurable subset of $S \times S$, as is

$$\Gamma_{k, t, B} = \{(y, z)\colon z = \gamma(y, z),\ z_k(t) \in B\} \tag{2.4}$$

for $B \in \mathscr{B}(E_k)$. Therefore $\pi\Gamma_{k, t, B} = \{y\colon (y, z) \in \Gamma_{k, t, B}\}$ is an analytic subset of S and

$$\{Z_k(t) \in B\} = \{Y \in \pi\Gamma_{k, t, B}\} \in \mathscr{F} \tag{2.5}$$

by the completeness of $(\Omega, \mathscr{F}, P)$. (See Appendix 11.) □

In the one-dimensional case we noted that $\tau(t)$ was a stopping time with respect to $\mathscr{F}^Y_{t+}$, at least in the case $\tau_0 = \tau_1$, $\beta(Y(\tau_0)) = 0$ for $\tau_0 < \infty$.

To determine the analogue of this observation in the multiparameter case we define

$$\mathscr{F}^0_u = \sigma(Y_k(s_k);\ s_k \le u_k), \qquad u \in [0, \infty]^\infty, \tag{2.6}$$

and

$$\mathscr{F}_u = \bigcap_n \sigma(\mathscr{N}) \vee \mathscr{F}^0_{u^{(n)}} \tag{2.7}$$

where $\mathscr{N} \subset \mathscr{F}$ is the collection of all sets of probability zero, and $u^{(n)}$ is defined by $u^{(n)}_k = u_k + 1/n$, $k \le n$, and $u^{(n)}_k = \infty$, $k > n$. A random variable $\tau = (\tau_1, \tau_2, \ldots)$ with values in $[0, \infty)^\infty$ is an $\{\mathscr{F}_u\}$-stopping time if $\{\tau \le u\} \equiv \{\tau_1 \le u_1, \tau_2 \le u_2, \ldots\} \in \mathscr{F}_u$, for all $u \in [0, \infty)^\infty$. (See Chapter 2, Section 8, for details concerning multiparameter stopping times.)

2.2 Theorem **(a)** For $u \in [0, \infty]^\infty$, $t > 0$, let $H_{u,t}$ be the set of $\omega \in \Omega$ such that there exists $z \in S \equiv \prod_l D_{E_l}[0, \infty)$ satisfying

$$z_k(r) = Y_k\left(\int_0^r \beta_k(z(s))\, ds, \omega\right), \qquad r \le t, \qquad k = 1, 2, \ldots, \tag{2.8}$$

and

$$\int_0^t \beta_k(z(s))\, ds \le u_k, \qquad k = 1, 2, \ldots. \tag{2.9}$$

Then $H_{u,t} \in \mathscr{F}_u$.

(b) Suppose a solution of (2.1) exists and is unique in the sense that for each $t > 0$ and almost every $\omega \in \Omega$, if z^1 and z^2 satisfy (2.8), then $z^1(r) = z^2(r)$, $r \le t$. Then for all $t \ge 0$, $\tau(t) \equiv (\tau_1(t), \tau_2(t), \ldots)$, with

$$\tau_k(t) = \int_0^t \beta_k(Z(s))\, ds, \tag{2.10}$$

is an $\{\mathscr{F}_u\}$-stopping time.

Proof. **(a)** Proceeding as in the proof of Lemma 2.1, let $\Gamma_{u,t} \subset S \times S$ be the set of (y, z) such that $z_k(r) = y_k(\int_0^r \beta_k(z(s))\, ds)$, $r \le t$, and (2.9) is satisfied. Then

$$H_{u,t} = \{Y \in \pi\Gamma_{u,t}\} \in \mathscr{F}_u.$$

(b) By the uniqueness assumption $P(\{\tau(t) \le u\} \,\Delta\, H_{u,t}) = 0$, and hence by the completeness of $\mathscr{F}_u$, $\{\tau(t) \le u\} \in \mathscr{F}_u$. □

2.3 Remark If we drop the assumption of uniqueness, then there will in general (even in the one-dimensional case) be solutions for which $\tau(t)$ is not an $\{\mathscr{F}_u\}$-stopping time. See Problem 1. □

Given $Y = (Y_1, Y_2, \ldots)$ on $(\Omega, \mathscr{F}, P)$, we say (2.1) has a *weak solution* if there exists a probability space $(\hat{\Omega}, \hat{\mathscr{F}}, \hat{P})$ on which are defined stochastic processes $\hat{Y} = (\hat{Y}_1, \hat{Y}_2, \ldots)$ and $\hat{Z} = (\hat{Z}_1, \hat{Z}_2, \ldots)$ such that $\hat{Y}$ is a version of Y and

$$\hat{Z}_k(t) = \hat{Y}_k\left(\int_0^t \beta_k(\hat{Z}(s))\, ds\right), \qquad \hat{P}\text{-a.s.} \tag{2.11}$$

2.4 Proposition If (2.1) has a weak solution, then for almost every $\omega \in \Omega$ (2.1) has a solution.

Proof. As in the proof of Lemma 2.1, let $S = \prod_k D_{E_k}[0, \infty)$ and let $\Gamma \subset S \times S$ be given by (2.3). Let $\hat{Y}$ be as above, and let $\pi\Gamma = \{y: (y, z) \in \Gamma\}$. Then

$$P\{Y \in \pi\Gamma\} = \hat{P}\{\hat{Y} \in \pi\Gamma\} = 1, \tag{2.12}$$

that is, (2.1) has a solution for almost every $\omega \in \Omega$. □

2.5 Remark In general it may not be possible to define a version of $\hat{Z}$ on $(\Omega, \mathscr{F}, P)$. For example, let Ω consist of a single point, let $Y(t) = t$, and let $\beta(z) = \sqrt{z}$. Let ξ, defined on $(\hat{\Omega}, \hat{\mathscr{F}}, \hat{P})$, be uniformly distributed on [0, 1] and define

$$\hat{Z}(t) = \begin{cases} 0, & t < \xi, \\ \frac{1}{4}(t - \xi)^2, & t \geq \xi. \end{cases} \tag{2.13}$$

Then for $\hat{Y}(t) = t$,

$$\hat{Z}(t) = \hat{Y}\left(\int_0^t \sqrt{\hat{Z}(s)}\, ds\right) = \int_0^t \sqrt{\hat{Z}(s)}\, ds, \tag{2.14}$$

but a version of $\hat{Z}$ cannot be defined on $(\Omega, \mathscr{F}, P)$. □

The condition that $\tau(t)$ is in some sense a stopping time plays an important role as we examine the relationship between random time changes and corresponding martingale problems. With this in mind, we say that a stochastic process Z defined on $(\Omega, \mathscr{F}, P)$ and satisfying (2.1) is a *nonanticipating solution* of (2.1) if there exists a filtration $\{\mathscr{G}_u\}$ indexed by $u \in [0, \infty)^\infty$ such that $\mathscr{F}_u \subset \mathscr{G}_u \subset \mathscr{F}$ ($\mathscr{F}_u$ given by (2.7)),

$$\begin{aligned} P\{(Y_1(u_1 + \cdot), Y_2(u_2 + \cdot), \ldots) \in B \mid \mathscr{G}_u\} \\ = P\{(Y_1(u_1 + \cdot), Y_2(u_2 + \cdot), \ldots) \in B \mid \mathscr{F}_u\} \end{aligned} \tag{2.15}$$

for all Borel subsets B of $\prod_k D_{E_k}[0, \infty)$, and if $\tau(t)$, given by (2.10), is a $\{\mathscr{G}_u\}$-stopping time for each $t \geq 0$.

We have three notions of solution, and hence three different notions of uniqueness. We say that *strong uniqueness* holds if for almost every $\omega \in \Omega$, (2.1) has at most one solution; we say that *weak uniqueness* holds if any two weak solutions have the same finite-dimensional distributions; and we say that we have *weak uniqueness for nonanticipating solutions* if any two weak, nonanticipating solutions have the same finite-dimensional distributions.

We turn now to the analogue of Theorem 1.3. Let Y_k, $k = 1, 2, \ldots$, be independent Markov processes corresponding to semigroups $\{T_k(t)\}$. Suppose $\{T_k(t)\}$ is strongly continuous on a closed subspace $L_k \subset \bar{C}(E_k)$, and let A_k be the (strong) generator for $\{T_k(t)\}$. We assume that L_k is separating, contains the constants, and is an algebra, and that the $D_{E_k}[0, \infty)$ martingale problem for A is well-posed. By analogy with the one-dimensional case, a solution of (2.1) should be a solution of the martingale problem for

$$(2.16)\qquad A \equiv \left\{\left(\prod_{i \in I} f_i,\ \sum_{j \in I} \beta_j A_j f_j \prod_{i \in I - \{j\}} f_i\right) : I \subset \{1, 2, \ldots\},\ I \text{ finite},\ f_i \in \mathscr{D}(A_i)\right\}.$$

2.6 Lemma Let $Y_1, Y_2, \ldots$ be independent Markov processes (as above) defined on $(\Omega, \mathscr{F}, P)$. Then a stochastic process Z satisfying (2.1) is a nonanticipating solution if and only if for every $t \geq 0$, $\tau(t)$ is a $\{\mathscr{G}_u\}$-stopping time for some $\{\mathscr{G}_u\}$ satisfying

$$(2.17)\qquad E\left[\prod_{k \in I} f_k(Y_k(u_k + v_k)) \,\middle|\, \mathscr{G}_u\right] = \prod_{k \in I} T_k(v_k) f_k(Y_k(u_k))$$

for all finite $I \subset \{1, 2, \ldots\}$, $f_k \in L_k$, and $u_k, v_k \geq 0$, or, setting $H_k f_k = A_k f_k / f_k$,

$$(2.18)\qquad E\left[\prod_{k \in I} f_k(Y_k(u_k + v_k)) \exp\left\{-\int_{u_k}^{u_k + v_k} H_k f_k(Y_k(s))\, ds\right\} \,\middle|\, \mathscr{G}_u\right]$$
$$= \prod_{k \in I} f_k(Y_k(u_k))$$

for all finite $I \subset \{1, 2, \ldots\}$, $f_k \in \mathscr{D}^+(A_k)$, and $u_k, v_k \geq 0$. ($\mathscr{D}^+(A_k) = \{f \in \mathscr{D}(A_k): \inf_{x \in E_k} f(x) > 0\}$.)

Proof. The equivalence of (2.17) and (2.15) follows from the Markov property and the independence of the Y_k. The equivalence of (2.18) and (2.15) follows from the uniqueness for the martingale problem for A_k and the independence of the Y_k. If $\mathscr{G}_u$ does not contain $\mathscr{F}_u$, then $\mathscr{G}_u \vee \mathscr{F}_u$ still satisfies (2.17) and (2.18). In particular, $\mathscr{G}_u$ can be replaced by $\tilde{\mathscr{G}}_u$ where $\tilde{\mathscr{G}}_u$ is obtained from $\mathscr{G}_u$ as $\mathscr{F}_u$ is obtained from $\mathscr{F}_u^0$. See (2.7). □

2.7 Lemma Let $Y_1, Y_2, \ldots$ be independent Markov processes (as above). A stochastic process Z satisfying (2.1) is a nonanticipating solution if and only if

$$(2.19)\qquad E\left[\prod_{k \in I} f_k(Y_k(u_k + v_k)) \prod_i g_{ik}(Y_k(s_{ik})) \prod_j h_j\left(\min_{l \in I} (u_l - \tau_l(t_j)) \vee 0\right)\right]$$
$$= E\left[\prod_{k \in I} T_k(v_k) f_k(Y_k(u_k)) \prod_i g_{ik}(Y_k(s_{ik})) \prod_j h_j\left(\min_{l \in I} (u_l - \tau_l(t_j)) \vee 0\right)\right]$$

for all $u_k, v_k \geq 0$, $0 \leq s_{ik} \leq u_k$, $t_j \geq 0$, finite $I \subset \{1, 2, \ldots\}$, $f_k \in L_k$, $g_{ik} \in \bar{C}(E_k)$, and $h_j \in \hat{C}[0, \infty)$, or

$$(2.20)\qquad E\left[\prod_{k \in I} f_k(Y_k(u_k + v_k)) \exp\left\{-\int_{u_k}^{u_k + v_k} H_k f_k(Y_k(s))\, ds\right\} \prod_i g_{ik}(Y_k(s_{ik}))\right.$$
$$\left. \times \prod_j h_j\left(\min_{l \in I} (u_l - \tau_l(t_j)) \vee 0\right)\right]$$
$$= E\left[\prod_{k \in I} f_k(Y_k(u_k)) \prod_i g_{ik}(Y_k(s_{ik})) \prod_j h_j\left(\min_{l \in I} (u_l - \tau_l(t_j)) \vee 0\right)\right]$$

for all $u_k, v_k \geq 0$, $0 \leq s_{ik} \leq u_k$, $t_j \geq 0$, finite $I \subset \{1, 2, \ldots\}$, $f_k \in \mathscr{D}^+(A_k)$, $g_{ik} \in \bar{C}(E_k)$, and $h_j \in \hat{C}[0, \infty)$.

Proof. The necessity of (2.19) and (2.20) is immediate from Lemma 2.6. Define

$$\mathscr{G}^0_{u,I} = \sigma(Y_k(s_k): s_k \leq u_k, k \in I) \vee \sigma\left(\min_{k \in I}(u_k - \tau_k(t)) \vee 0: t \geq 0\right). \tag{2.21}$$

Then (2.19) implies

$$E\left[\prod_{k \in I} f_k(Y_k(u_k + v_k)) \middle| \mathscr{G}^0_{u,I}\right] = \prod_{k \in I} T_k(v_k) f_k(Y_k(u_k)). \tag{2.22}$$

Fix I. If $I' \supset I$, then, by taking $f_k \equiv 1$ for $k \in I' - I$, we can replace $\mathscr{G}^0_{u,I}$ in (2.22) by $\mathscr{G}^0_{u,I'}$. If $u_k = \infty$ for $k \notin I$, then, for $I' \supset I$, $\min_{k \in I'}(u_k - \tau_k(t)) \vee 0 = \min_{k \in I}(u_k - \tau_k(t)) \vee 0$, and hence $\mathscr{G}^0_{u,I'}$, with $I' \supset I$, is increasing in I'. For u satisfying $u_k = \infty$, $k \notin I$, we define

$$\mathscr{G}^0_u = \bigvee_{I' \supset I} \mathscr{G}^0_{u,I'}, \tag{2.23}$$

and we note that we can replace $\mathscr{G}^0_{u,I}$ in (2.22) by $\mathscr{G}^0_u$. For arbitrary u define

$$\mathscr{G}_u = \bigcap_n \mathscr{G}^0_{u^{(n)}} \tag{2.24}$$

where $u_k^{(n)} = u_k + 1/n$ for $k \leq n$, $u_k = \infty$ for $k > n$. If $I \subset \{1, 2, \ldots, n\}$, we have

$$E\left[\prod_{k \in I} f_k\left(Y_k\left(u_k + \frac{1}{n} + v_k\right)\right) \middle| \mathscr{G}^0_{u^{(n)}}\right] = \prod_{k \in I} T_k(v_k) f_k\left(Y_k\left(u_k + \frac{1}{n}\right)\right). \tag{2.25}$$

The right continuity of Y_k, the continuity of f_k and $T_k(v_k)f_k$, and the fact that $\mathscr{G}^0_{u^{(n)}}$ is decreasing in n imply (2.17). A similar argument shows that (2.20) implies (2.18). Finally $\{\tau(t) \leq u\} = \bigcap_I \bigcap_n \{\min_{k \in I}(u_k^{(n)} - \tau_k(t)) > 0\} \in \mathscr{G}_u$. □

2.8 Theorem Let Y_k, $k = 1, 2, \ldots$, be independent Markov processes (as above). Let β_k, $k = 1, 2, \ldots$ be nonnegative bounded Borel functions on $E \equiv \prod_l E_l$, and let A be given by (2.16).

(a) If Z is a nonanticipating solution of (2.1), then Z is a solution of the martingale problem for A.

(b) If Z is a solution of the $D_E[0, \infty)$ martingale problem for A, then there is a version of Z that is a (weak) nonanticipating solution of (2.1).

2.9 Remark **(a)** If $\inf_z \beta_k(z) > 0$, $k = 1, 2, \ldots$, in (b), then Z itself is a nonanticipating solution of (2.1).

(b) The hypothesis that $\sup_z \beta_k(z) < \infty$ is used to ensure that $A \subset \bar{C}(E) \times B(E)$. There are two approaches toward eliminating this hypothesis. One would be to restrict the domain of A so that $A \subset \bar{C}(E) \times B(E)$. (See Problems 2, 3.) The other would be to develop the notion of a "local-martingale problem." (See Chapter 4, Section 7, and Proposition 2.10.)

Proof. **(a)** Let $I \subset \{1, 2, \ldots\}$ be finite. For $k \in I$, let $f_k \in \mathscr{D}(A_k)$, $\inf_{x \in E_k} f_k(x) > 0$, and set $H_k f_k = A_k f_k / f_k$. Define $f = \prod_{k \in I} f_k$, $g = \sum_{k \in I} \beta_k A_k f_k \prod_{j \in I - \{k\}} f_j$, and

(2.26) $$M(u) = \prod_{k \in I} f_k(Y_k(u_k)) \exp\left\{-\int_0^{u_k} H_k f_k(Y_k(s))\, ds\right\}.$$

By (2.17), for fixed u, $Y_k^{(u)}(v_k) = Y_k(u_k + v_k)$, $k = 1, 2, \ldots$, are Markov processes corresponding to A_k and are conditionally independent given $\mathscr{G}_u$. Therefore

(2.27) $$\begin{aligned} E[M(u + v) \mid \mathscr{G}_u] &= \prod_{k \in I} E\left[f_k(Y_k(u_k + v_k)) \exp\left\{-\int_0^{u_k + v_k} H_k f_k(Y_k(s))\, ds\right\} \middle| \mathscr{G}_u \right] \\ &= \prod_{k \in I} f_k(Y_k(u_k)) \exp\left\{-\int_0^{u_k} H_k f_k(Y_k(s))\, ds\right\} \\ &= M(u), \end{aligned}$$

and hence $M(u)$ is a $\{\mathscr{G}_u\}$-martingale.

By assumption, $\tau(t)$ is a $\{\mathscr{G}_u\}$-stopping time and the optional sampling theorem (Theorem 8.7 of Chapter 2) implies $M(\tau(t))$ is a $\mathscr{G}_{\tau(t)}$-martingale. But

(2.28) $$M(\tau(t)) = \prod_{k \in I} f_k(Z_k(t)) \exp\left\{-\int_0^t \beta_k(Z(s)) H_k f_k(Z_k(s))\, ds\right\}$$

and

(2.29) $$f(Z(t)) - \int_0^t g(Z(s))\, ds$$

is a martingale by Lemma 3.2 of Chapter 4.

(b) The basic idea of the proof is to let

(2.30) $$\gamma_k(u) = \inf\left\{t : \int_0^t \beta_k(Z(s))\, ds > u\right\}$$

and define

(2.31) $$Y_k(u) = Z_k(\gamma_k(u)).$$

Note that then, as in the one-dimensional case, the fact that $\mathscr{D}(A_k)$ is separating implies Z satisfies (2.1).

Two difficulties arise. First, $\gamma_k(u)$ need not be defined for all $u \geq 0$, and second, even if γ_k is defined for all $u \geq 0$, it is not immediately clear that the Y_k are independent. Note that for each $t > 0$ and $(f, g) \in A_k$,

$$(2.32)\qquad M(u) = f(Z_k(\gamma_k(u) \wedge t)) - \int_0^{\gamma_k(u) \wedge t} \beta_k(Z(s))g(Z_k(s))\, ds$$

$$= f(Y_k(u \wedge \tau_k(t))) - \int_0^{u \wedge \tau_k(t)} g(Y_k(s))\, ds$$

is an $\{\mathscr{F}^Z_{\gamma_k(u)}\}$-martingale. By Lemma 5.16 of Chapter 4, there exists a solution $\tilde{Y}_{k,t}$ of the martingale problem for A_k and a nonnegative random variable $\eta_k(t)$ such that $Y_k(\cdot \wedge \tau_k(t))$ has the same distribution as $\tilde{Y}_{k,t}(\cdot \wedge \eta_k(t))$. Letting $t \to \infty$, $(\tilde{Y}_{k,t}, \eta_k(t))$ converges in distribution in $D_{E_k}[0, \infty) \times [0, \infty]$ (at least through a sequence of t's) to $(\tilde{Y}_{k,\infty}, \eta_k(\infty))$ and $Y_k(\cdot \wedge \tau_k(\infty))$ has the same distribution as $\tilde{Y}_{k,\infty}(\cdot \wedge \eta_k(\infty))$. In particular,

$$(2.33)\qquad Z_k(\infty) \equiv \lim_{t\to\infty} Z_k(t) = \lim_{u\to\infty} Y_k(u \wedge \tau_k(\infty))$$

exists on $\{\tau_k(\infty) < \infty\}$. Fix $y_k \in E_k$ and set $Z_k(\infty) = y_k$ on $\{\tau_k(\infty) = \infty\}$.

Let $\Omega' = \Omega \times \prod_k D_{E_k}[0, \infty)$ and define

$$(2.34)\qquad Q(C \times B_1 \times B_2 \times B_3 \times \cdots)$$

$$= \int_C \prod_k P^{(k)}_{Z_k(\infty)}(B_k)\, dP$$

for $C \in \mathscr{F}$ and $B_k \in \mathscr{B}(D_{E_k}[0, \infty))$, where $P^{(k)}_y$ is the distribution of the Markov process with generator A_k starting from y. Then Q extends to a measure on $\mathscr{F} \times \prod_k \mathscr{B}(D_{E_k}[0, \infty))$.

Defining Z on Ω' by $Z(t, (\omega, \omega_1, \omega_2, \ldots)) \equiv Z(t, \omega)$, we see that Z on $(\Omega', \mathscr{F} \times \prod_k \mathscr{B}(D_{E_k}[0, \infty)), Q)$ is a version of Z on $(\Omega, \mathscr{F}, P)$. Let W_k denote the coordinate process in $D_{E_k}[0, \infty)$, that is,

$$(2.35)\qquad W_k(t, (\omega, \omega_1, \omega_2, \omega_3, \ldots)) = \omega_k(t).$$

Set

$$(2.36)\qquad \tau_k(t) = \int_0^t \beta_k(Z(s))\, ds,$$

allowing $t = \infty$, and define

$$(2.37)\qquad Y_k(t) = \begin{cases} Z_k(\gamma_k(t)), & t < \tau_k(\infty), \\ W_k(t - \tau_k(\infty)), & t \geq \tau_k(\infty). \end{cases}$$

We must show that there is a family of σ-algebras $\{\mathscr{G}_u\}$ such that $\tau(t)$ is a $\{\mathscr{G}_u\}$-stopping time and the Y_k satisfy (2.17).

Let $f_k \in \mathscr{D}(A_k)$, $f_k > 0$. Let $\psi_k(x_k, t)$, $(\partial/\partial t)\psi_k(x_k, t) \in \bar{C}(E_k \times [0, \infty))$ with $\psi_k \geq 0$, and suppose $q_k(x_k, t)$ satisfies

$$\frac{\partial}{\partial t} q_k(x_k, t) = A_k q_k(x_k, t) - \psi_k(x_k, t) q_k(x_k, t), \tag{2.38}$$

and $q_k(x_k, 0) = f_k(x_k)$. For $u \in [0, \infty)^\infty$ define

$$h_k(x_k, t) = \begin{cases} q_k(x_k, u_k - t), & t \leq u_k, \\ f_k(x_k), & t > u_k, \end{cases} \tag{2.39}$$

and set $Kh_k = (\partial/\partial t)h_k/h_k$. Setting

$$M_k(t) = h_k(Z_k(t), \tau_k(t)) \exp\left\{-\int_0^t \beta_k(Z(s))[Kh_k(Z_k(s), \tau_k(s)) + H_k h_k(Z_k(s), \tau_k(s))]\, ds\right\}, \tag{2.40}$$

Lemmas 3.2 and 3.4 of Chapter 4 imply that

$$\prod_{k \in I} M_k(t) \tag{2.41}$$

is a martingale for any finite $I \subset \{1, 2, \ldots\}$, with respect to $\{\mathscr{F}_t^Z\}$. Defining $\gamma_k(u)$ by (2.30) for $u < \tau_k(\infty)$ and setting $\gamma_k(u) = \infty$ for $u \geq \tau_k(\infty)$, Problem 24 of Chapter 2 implies

$$E\left[\prod_{k \in I} M_k(\gamma_k(v_k)) \,\middle|\, \mathscr{F}_{\gamma_I(u)}^Z\right] = \prod_{k \in I} M_k(\gamma_I(u)) \tag{2.42}$$

for $u \leq v$, where $\gamma_I(u) = \bigwedge_{k \in I} \gamma_k(u_k)$. In particular, from the definition of h_k and q_k,

(2.43)

$$\begin{aligned} E\Bigg[\prod_{k \in I} & h_k(Y_k(v_k \wedge \tau_k(\infty)), v_k \wedge \tau_k(\infty)) \exp\left\{-\int_{u_k \wedge \tau_k(\infty)}^{v_k \wedge \tau_k(\infty)} H_k f_k(Y_k(s))\, ds\right\} \\ & \times \exp\left\{-\int_0^{u_k \wedge \tau_k(\infty)} \psi_k(Y_k(s), u_k \wedge \tau_k(\infty) - s)\, ds\right\} \,\Bigg|\, \mathscr{F}_{\gamma_I(u)}^Z\Bigg] \\ = E\Bigg[\prod_{k \in I} & h_k(Y_k(u_k \wedge \tau_k(\infty)), u_k \wedge \tau_k(\infty)) \\ & \times \exp\left\{-\int_0^{u_k \wedge \tau_k(\infty)} \psi_k(Y_k(s), u_k \wedge \tau_k(\infty) - s)\, ds\right\} \,\Bigg|\, \mathscr{F}_{\gamma_I(u)}^Z\Bigg]. \end{aligned}$$

Observing that

$$(2.44)\quad E\left[\prod_{k\in I} f_k(Y_k(v_k))\exp\left\{-\int_{u_k}^{v_k} H_k f_k(Y_k(s))\,ds\right\}\right.$$
$$\left.\times\exp\left\{-\int_0^{u_k}\psi_k(Y_k(s),u_k-s)\,ds\right\}\middle|\mathscr{F}_\infty^Z\right]$$
$$=E\left[\prod_{k\in I} h_k(Y_k(v_k\wedge\tau_k(\infty)),v_k\wedge\tau_k(\infty))\exp\left\{-\int_{u_k\wedge\tau_k(\infty)}^{v_k\wedge\tau_k(\infty)} H_k f_k(Y_k(s))\,ds\right\}\right.$$
$$\left.\times\exp\left\{-\int_0^{u_k\wedge\tau_k(\infty)}\psi_k(Y_k(s),u_k\wedge\tau_k(\infty)-s)\,ds\right\}\middle|\mathscr{F}_\infty^Z\right],$$

we see that we can drop the "$\wedge\tau_k(\infty)$" on both sides of (2.43).

Let $\varphi_{ik}(x_k)\in L_k$ satisfy $0<\varphi_{ik}\le 1$, and let $0\le s_{ik}\le u_k$. Let $\rho\ge 0$ be continuously differentiable with compact support in $(0,\infty)$ and $\int_0^\infty\rho(s)\,ds=1$. Replace ψ_k in (2.38) by

$$(2.45)\qquad \psi_k^{(n)}(x_k,t)=-\sum_{i=1}^m n\rho((u_k-t-s_{ik})n)\varphi_{ik}(x_k).$$

Since L_k is an algebra, $B_n(t)f\equiv\psi_k^{(n)}(t,\cdot)f$ defines a bounded linear operator on L_k, and the differentiability of ρ ensures the existence of q_k (see Problem 23 of Chapter 1). Letting $n\to\infty$ in (2.43) gives

$$(2.46)\quad E\left[\prod_{k\in I} f_k(Y_k(v_k))\exp\left\{-\int_{u_k}^{v_k} H_k f_k(Y_k(s))\,ds-\sum_i\varphi_{ik}(Y_k(s_{ik}))\right\}\middle|\mathscr{F}_{\gamma_I(u)}^Z\right]$$
$$=E\left[\prod_{k\in I} f_k(Y_k(u_k))\exp\left\{-\sum_i\varphi_{ik}(Y_k(s_{ik}))\right\}\middle|\mathscr{F}_{\gamma_I(u)}^Z\right].$$

Setting

$$(2.47)\qquad \mathscr{G}_u=\sigma(Y_k(s_k)\colon s_k\le u_k)\vee\bigcap_k\mathscr{F}_{\gamma_k(u_k)}^Z,$$

we note that (2.46) implies (2.18) and that

$$(2.48)\qquad \{\tau(t)\le u\}=\left\{\bigwedge_k\gamma_k(u_k)\ge t\right\}\in\bigcap_k\mathscr{F}_{\gamma_k(u_k)}^Z\subset\mathscr{G}_u.$$

Part (b) now follows by Lemma 2.6. □

The following proposition is useful in reducing the study of (2.1) with unbounded β_k to the bounded case.

2.10 Proposition Let α be measurable, and suppose $\inf_z \alpha(z) > 0$. Let Z be an E-valued stochastic process, let η satisfy

$$\int_0^{\eta(t)} \alpha(Z(s))\, ds = t, \tag{2.49}$$

$\lim_{t\to\infty} \eta(t) = \infty$ a.s., and define

$$Z^\alpha(t) = Z(\eta(t)). \tag{2.50}$$

Then Z is a nonanticipating solution of (2.1) if and only if Z^α is a non-anticipating solution of

$$Z_k^\alpha(t) = Y_k\left(\int_0^t \frac{\beta_k(Z^\alpha(s))}{\alpha(Z^\alpha(s))}\, ds\right). \tag{2.51}$$

Proof. If Z satisfies (2.1), a simple change of variable verifies that Z^α satisfies (2.51). Assume Z is a nonanticipating solution, and let $\{\mathscr{G}_u\}$ be the family of σ-algebras in Lemma 2.6. Since the $\tau(t)$ form an increasing family of $\{\mathscr{G}_u\}$-stopping times, and for each s, $\eta(s)$ is a $\{\mathscr{G}_{\tau(t)}\}$-stopping time, Proposition 8.6 of Chapter 2 gives that $\tau^\alpha(s) = \tau(\eta(s))$ is a $\{\mathscr{G}_u\}$-stopping time. Consequently, by Lemma 2.6, Z^α is a nonanticipating solution of (2.51).

The converse is proved similarly. □

3. CONVERGENCE

We now consider criteria for convergence of a sequence of processes $Z^{(n)}$ satisfying

$$Z_k^{(n)}(t) = Y_k^{(n)}\left(\int_0^t \beta_k^{(n)}(Z^{(n)}(s))\, ds\right), \qquad k = 1, 2, \ldots \tag{3.1}$$

where $Y_k^{(n)}$ is a process with sample paths in $D_{E_k}[0, \infty)$. We continue to assume that the (E_k, r_k) are complete and separable. Relative compactness for sequences of this form is frequently quite simple.

3.1 Proposition Let $Z^{(n)}$ satisfy (3.1). If $\{Y_k^{(n)}\}$ is relatively compact in $D_{E_k}[0, \infty)$ and $\bar{\beta}_k \equiv \sup_n \sup_z \beta_k^{(n)}(z) < \infty$, then $\{Z_k^{(n)}\}$ is relatively compact in $D_{E_k}[0, \infty)$, and hence if $\{Y^{(n)}\}$ is relatively compact in $\prod_k D_{E_k}[0, \infty)$ and $\sup_n \sup_z \beta_k^{(n)}(z) < \infty$ for each k, then $\{Z^{(n)}\}$ is relatively compact in $\prod_k D_{E_k}[0, \infty)$.

Proof. The proposition follows immediately from the fact that

$$w'(Z_k^{(n)}, \delta, T) \le w'(Y_k^{(n)}, \bar{\beta}_k \delta, \bar{\beta}_k T) \tag{3.2}$$

and

$$\{Z_k^{(n)}(t) \in K \text{ for all } t \le T\} \supset \{Y_k^{(n)}(t) \in K \text{ for all } t \le \bar{\beta}T\}. \tag{3.3}$$

(Recall that we are assuming the (E_k, r_k) are complete and separable.) □

We would prefer, of course, to have relative compactness in $D_E[0, \infty)$ where $E = \prod_k E_k$, but relative compactness of $\{Y^{(n)}\}$ in $D_E[0, \infty)$ and the boundedness of the $\beta_k^{(n)}$ do not necessarily imply the relative compactness of $\{Z^{(n)}\}$ in $D_E[0, \infty)$. We do note the following.

3.2 Proposition Let $\{Z^{(n)}\}$ be a sequence of processes with sample paths in $D_E[0, \infty)$, $E = \prod_k E_k$. If $Z^{(n)} \Rightarrow Z$ in $\prod_k D_{E_k}[0, \infty)$ and if no two components of Z have simultaneous jumps (i.e., if $P\{Z_k(t) \neq Z_k(t-)$ and $Z_l(t) \neq Z_l(t-)$ for some $t \ge 0\} = 0$ for all $k \neq l$), then $Z^{(n)} \Rightarrow Z$ in $D_E[0, \infty)$.

Proof. The result follows from Proposition 6.5 of Chapter 3. Details are left to the reader (Problem 5). □

We next give the analogue of Theorem 1.5.

3.3 Theorem Suppose that for $k = 1, 2, \ldots, Y_k$, defined on $(\Omega, \mathscr{F}, P)$, has sample paths in $D_{E_k}[0, \infty)$, β_k is nonnegative, bounded, and continuous on $E = \prod_l E_l$, and either Y_k is continuous or $\beta_k(z) > 0$ for all $z \in E$. Suppose that for almost every $\omega \in \Omega$,

$$Z_k(t) = Y_k\left(\int_0^t \beta_k(Z(s))\, ds\right) \tag{3.4}$$

has a unique solution. Let $\{Y^{(n)}\}$ satisfy $Y^{(n)} \Rightarrow Y$ in $\prod_k D_{E_k}[0, \infty)$, and for $k = 1, 2, \ldots,$ let $\beta_k^{(n)}$ be nonnegative Borel measurable functions satisfying $\sup_n \sup_{z \in K} \beta_k^{(n)}(z) < \infty$ and

$$\lim_{n \to \infty} \sup_{z \in K} |\beta_k^{(n)}(z) - \beta_k(z)| = 0 \tag{3.5}$$

for each compact $K \subset E$. Suppose that $Z^{(n)}$ satisfies

$$Z_k^{(n)}(t) = Y_k^{(n)}\left(\int_0^t \beta_k^{(n)}(Z^{(n)}(s))\, ds\right) \tag{3.6}$$

and that $W^{(n)}$ satisfies

$$W_k^{(n)}(t) = Y_k^{(n)}\left(\int_0^{[t/h_n]h_n} \beta_n(W^{(n)}(s))\, ds\right), \tag{3.7}$$

where $h_n > 0$ and $\lim_{n \to \infty} h_n = 0$. Then $Z^{(n)} \Rightarrow Z$ and $W^{(n)} \Rightarrow Z$ in $\prod_k D_{E_k}[0, \infty)$.

Proof. The proof is essentially the same as for Theorem 1.3, so we only give a sketch. We may assume $\lim_{n\to\infty} Y^{(n)} = Y$ a.s. The estimates in (3.2) and (3.3) imply that if $\{Y^{(n)}(\omega)\}$ is convergent in $\prod_k D_{E_k}[0, \infty)$ for some $\omega \in \Omega$, then $\{Z^{(n)}(\omega)\}$ and $\{W^{(n)}(\omega)\}$ are relatively compact in $\prod_k D_{E_k}[0, \infty)$. The continuity and positivity of β_k imply that any limit point $\hat{Z}(\omega)$ of $\{Z^{(n)}(\omega)\}$ or $\{W^{(n)}(\omega)\}$ must satisfy

(3.8)
$$\hat{Z}_k(\omega, t) = Y_k\left(\omega, \int_0^t \beta_k(\hat{Z}(\omega, s))\, ds\right).$$

(If Y_k is not continuous, then the positivity of β_k implies $Y_k(\omega, \int_0^t \beta_k(\hat{Z}(\omega, s))\, ds) = Y_k^-(\omega, \int_0^t \beta_k(\hat{Z}(\omega, s))\, ds)$ for almost every $t \geq 0$. See Problem 6.) Since the solution of (3.4) is almost surely unique, it follows that $\lim_{n\to\infty} Z^{(n)} = Z$ and $\lim_{n\to\infty} W^{(n)} = Z$ in $\prod_k D_{E_k}[0, \infty)$ a.s. □

The proof of Theorem 3.3 is typical of proofs of weak convergence: compactness is verified, it is shown that any possible limit must possess certain properties, and finally it is shown (or in this case assumed) that those properties uniquely determine the possible limit. The uniqueness used above was strong uniqueness. Unfortunately, there are many situations in which weak uniqueness for nonanticipating solutions is known but not strong uniqueness. Consequently we turn now to convergence criteria in which the limiting process is characterized as the unique weak, nonanticipating solution.

We want to cover not only sequences of the form (3.6) and (3.7) but also solutions of equations of the form

(3.9)
$$Z_k^{(n)}(t) = Y_k^{(n)}\left(\int_0^t \beta_k^{(n)}(Z^{(n)}(s), \xi^{(n)}(s))\, ds\right)$$

where $\xi^{(n)}$ is a rapidly fluctuating process that "averages" $\beta_k^{(n)}$ in the sense that

(3.10)
$$\left|\int_0^t \beta_k^{(n)}(Z^{(n)}(s), \xi^{(n)}(s))\, ds - \int_0^t \beta_k(Z^{(n)}(s))\, ds\right| \to 0.$$

The following theorem provides conditions for convergence that apply to all three of these situations.

3.4 Theorem Let $Y^{(n)}$, $n = 1, 2, \ldots$, have values in $\prod_k D_{E_k}[0, \infty)$, let $\{\mathscr{G}_u^{(n)}\}$ be a filtration indexed by $[0, \infty)^\infty$ satisfying $\mathscr{G}_u^{(n)} \supset \sigma(Y_k^{(n)}(s_k): s_k \leq u_k, k = 1, 2, \ldots)$, and let $\tau^{(n)}(t)$, $t \geq 0$, be a nondecreasing (componentwise) family of $\{\mathscr{G}_u^{(n)}\}$-stopping times that is right continuous in t. Define

(3.11)
$$Z_k^{(n)}(t) = Y_k^{(n)}(\tau_k^{(n)}(t)).$$

Suppose for $k = 1, 2, \ldots$ that $\{T_k(t)\}$ is a strongly continuous semigroup on $L_k \subset \bar{C}(E_k)$ corresponding to a Markov process Y_k, and L_k is convergence determining, that $\beta_k: E \to [0, \infty)$ is continuous, and that either $\beta_k > 0$ or Y_k is continuous.

Assume

$$(3.12)\quad \lim_{n\to\infty} E\left[\left|E\left[\prod_{k\in I} f_k(Y_k^{(n)}(u_k+v_k))\,\middle|\,\mathscr{G}_u^{(n)}\right]-\prod_{k\in I} T_k(v_k)f_k(Y_k^{(n)}(u_k))\right|\right]=0$$

for $f_k \in L_k$, finite $I \subset \{1, 2, \ldots\}$, and $u, v \in [0, \infty)^\infty$, and assume

$$(3.13)\quad \left|\tau_k^{(n)}(t)-\int_0^t \beta_k(Z^{(n)}(s))\,ds\right| \xrightarrow{p} 0$$

for each $k = 1, 2, \ldots$ and $t \geq 0$.

(a) If $(Y^{(n)}, Z^{(n)}) \Rightarrow (Y, Z)$ in $\prod_k D_{E_k}[0, \infty) \times \prod_k D_{E_k}[0, \infty)$, then Z is a nonanticipating solution of (2.1).

(b) Suppose that for each $\varepsilon, T > 0$ and $k = 1, 2, \ldots$ there exists a compact $K_{\varepsilon, T}^k \subset E_k$ such that

$$(3.14)\quad \inf_n P\{Z_k^{(n)}(t) \in K_{\varepsilon, T}^k \quad \text{for all} \quad t \leq T\} \geq 1 - \varepsilon.$$

If $Y^{(n)} \Rightarrow Y$ in $\prod_k D_{E_k}[0, \infty)$, and (2.1) has a weakly unique nonanticipating solution Z, then $Z^{(n)} \Rightarrow Z$ in $\prod_k D_{E_k}[0, \infty)$.

3.5 Remark **(a)** Note that (3.12) implies that the finite-dimensional distributions of $Y^{(n)}$ converge and that the Y_k are conditionally independent given $Y(0)$. See Remark 8.3(a) of Chapter 4 for conditions implying (3.12).

(b) If the $Y_k^{(n)}$ are Markov processes satisfying

$$(3.15)\quad E\left[\prod_{k\in I} f_k(Y_k^{(n)}(u_k+v_k))\,\middle|\,\mathscr{G}_u^{(n)}\right]=\prod_{k\in I} T_k^{(n)}(v_k)f_k(Y_k^{(n)}(u_k)),$$

then (3.12) is implied by

$$(3.16)\quad \lim_{n\to\infty} E[\,|T_k^{(n)}(t)f_k(Y_k^{(n)}(u)) - T_k(t)f_k(Y_k^{(n)}(u))|\,] = 0$$

for all $t, u \geq 0$ and $k = 1, 2, 3, \ldots$. □

Proof. **(a)** If $(Y^{(n)}, Z^{(n)}) \Rightarrow (Y, Z)$, then (3.13) and the continuity of the β_k imply $(Y^{(n)}, Z^{(n)}, \tau^{(n)}) \Rightarrow (Y, Z, \tau)$ in $\prod_k D_{E_k}[0, \infty) \times \prod_k D_{E_k}[0, \infty) \times [D_{[0,\infty)}[0, \infty)]^\infty$, where τ_k is as usual

$$(3.17)\quad \tau_k(t) = \int_0^t \beta_k(Z(s))\,ds.$$

It follows that $Z_k(t) = Y_k(\tau_k(t))$ or $Y_k^-(\tau_k(t))$. We need $Z_k(t) = Y_k(\tau_k(t))$. If Y_k is continuous, then (2.1) is satisfied; or if $\beta_k > 0$, then the fact that τ_k is (strictly) increasing and $Z_k(t)$ and $Y_k(\tau_k(t))$ are right continuous implies (2.1) is satisfied.

To see that Z is nonanticipating, note that with the parameters as in (2.19)

$$
\begin{aligned}
(3.18)\qquad & E\left[\prod_{k\in I} f_k(Y_k(u_k+v_k)) \prod_i g_{ik}(Y_k(s_{ik})) \cdot \prod_j h_j\left(\min_{l\in I}(u_l-\tau_l(t_j))\vee 0\right)\right] \\
&= \lim_{n\to\infty} E\left[\prod_{k\in I} f_k(Y_k^{(n)}(u_k+v_k)) \prod_i g_{ik}(Y_k^{(n)}(s_{ik})) \cdot \prod_j h_j\left(\min_{l\in I}(u_l-\tau_l^{(n)}(t_j))\vee 0\right)\right] \\
&= \lim_{n\to\infty} E\left[E\left[\prod_{k\in I} f_k(Y_k^{(n)}(u_k+v_k))\,\middle|\,\mathscr{G}_u^{(n)}\right] \prod_i g_{ik}(Y_k^{(n)}(s_{ik})) \cdot \prod_j h_j\left(\min_{l\in I}(u_l-\tau_l^{(n)}(t_j))\vee 0\right)\right] \\
&= \lim_{n\to\infty} E\left[\prod_{k\in I} T_k(v_k) f_k(Y_k^{(n)}(u_k)) \prod_i g_{ik}(Y_k^{(n)}(s_{ik})) \cdot \prod_j h_j\left(\min_{l\in I}(u_l-\tau_l^{(n)}(t_j))\vee 0\right)\right] \\
&= E\left[\prod_{k\in I} T_k(v_k) f_k(Y_k(u_k)) \prod_i g_{ik}(Y_k(s_{ik})) \cdot \prod_j h_j\left(\min_{l\in I}(u_l-\tau_l(t_j))\vee 0\right)\right].
\end{aligned}
$$

Observe that the τ_l are continuous and that $P\{Y_k(t) = Y_k(t-)\} = 1$ (cf. Theorem 3.12 of Chapter 4) for all t. Consequently all the finite-dimensional distributions of $(Y^{(n)}, \tau^{(n)})$ converge to those of (Y, τ). By Lemma 2.7, Z is a nonanticipating solution of (2.1).

(b) By part (a), it is enough to show that $\{(Y^{(n)}, Z^{(n)})\}$ is relatively compact, since any convergent subsequence must converge to the unique nonanticipating solution of (2.1). By Proposition 2.4 of Chapter 3, it is enough to verify the relative compactness of $\{Z_k^{(n)}\}$. Let

$$
(3.19)\qquad \gamma_k^{(n)}(t) = \int_0^t \beta_k(Z^{(n)}(s))\,ds.
$$

The monotonicity of $\gamma_k^{(n)}$ and $\tau_k^{(n)}$ and (3.14) imply the convergence in (3.13) is uniform in t on bounded intervals. For δ, $T > 0$, let

$$(3.20)\qquad \eta_k^{(n)}(\delta, T) = \sup_{t \le T} (\tau_k^{(n)}(t + \delta) - \tau_k^{(n)}(t)) + \sup_{t \le T} (\tau_k^{(n)}(t) - \tau_k^{(n)}(t-)).$$

Note that by the uniformity in t in (3.13) and (3.14), as $n \to \infty$ and $\delta \to 0$, $\eta_k^{(n)}(\delta, T) \xrightarrow{P} 0$. Finally

$$(3.21)\qquad w'(Z_k^{(n)}, \delta, T) \le w'(Y_k^{(n)}, \eta_k^{(n)}(\delta, T), \tau_k^{(n)}(T))$$

(see Problem 7), and hence for $\varepsilon > 0$ the relative compactness of $\{Y_k^{(n)}\}$ implies

$$(3.22)\qquad \begin{aligned} &\lim_{\delta \to 0} \overline{\lim_{n \to \infty}} P\{w'(Z_k^{(n)}, \delta, T) > \varepsilon\} \\ &\le \lim_{\delta \to 0} \overline{\lim_{n \to \infty}} P\{w'(Y_k^{(n)}, \eta_k^{(n)}(\delta, T), \tau_k^{(n)}(T)) > \varepsilon\} \\ &= 0 \end{aligned}$$

and the relative compactness of $\{Z_k^{(n)}\}$ follows. □

3.6 Corollary Let $Y_1, Y_2, \ldots$ be independent Markov processes (as above), let $\beta_k: E \to [0, \infty)$ be continuous and bounded, and assume either $\beta_k > 0$ or Y_k is continuous. Then (2.1) has a weak, nonanticipating solution.

Proof. Let $Y^{(n)} = Y$ and $W^{(n)}$ satisfy (3.7) with $h_n = 1/n$. Then $\{W_k^{(n)}\}$ is relatively compact by essentially the same estimates as in the proof of Proposition 3.1. Any limit point of $\{W^{(n)}\}$ is a nonanticipating solution of (2.1). □

4. MARKOV PROCESSES IN $\mathbb{Z}^d$

Let E be the one-point compactification of the d-dimensional integer lattice $\mathbb{Z}^d$, that is, $E = \mathbb{Z}^d \cup \{\Delta\}$. Let $\beta_l: \mathbb{Z}^d \to [0, \infty)$, $l \in \mathbb{Z}^d$, $\sum_l \beta_l(k) < \infty$ for each $k \in \mathbb{Z}^d$, and for f vanishing off a finite subset of $\mathbb{Z}^d$, set

$$(4.1)\qquad Af(x) = \begin{cases} \sum_l \beta_l(x)(f(x + l) - f(x)), & x \in \mathbb{Z}^d, \\ 0, & x = \Delta. \end{cases}$$

Let Y_l, $l \in \mathbb{Z}^d$, be independent Poisson processes, let $X(0)$ be nonrandom, and suppose X satisfies

$$(4.2)\qquad X(t) = X(0) + \sum_l l Y_l\left(\int_0^t \beta_l(X(s))\, ds\right), \qquad t < \tau_\infty,$$

and

$$X(t) = \Delta, \qquad t \geq \tau_\infty, \tag{4.3}$$

where $\tau_\infty = \inf\{t: X(t-) = \Delta\}$.

4.1 Theorem **(a)** Given $X(0)$, the solution of (4.2) and (4.3) is unique.

(b) X is a solution of the local-martingale problem for A. (Cf. Chapter 4, Section 7. Note, we have not assumed Af is bounded for each $f \in \mathscr{D}(A)$. If this is true, then X is a solution of the martingale problem for A.)

(c) If $\tilde{X}$ is a solution of the local-martingale problem for A with sample paths in $D_E[0, \infty)$ satisfying $\tilde{X}(t) = \Delta$ for $t \geq \tau_\infty$ (τ_∞ as above), then there is a version X of $\tilde{X}$ satisfying (4.2) and (4.3).

Proof. **(a)** Let $X_0(t) \equiv X(0)$ and set

$$X_k(t) = X(0) + \sum_l l Y_l\left(\int_0^t \beta_l(X_{k-1}(s))\, ds\right). \tag{4.4}$$

Then if τ_k is the kth jump time of X_k, $X_k(t) = X_{k-1}(t)$ for $t < \tau_k$. Therefore

$$X(t) = \lim_{k\to\infty} X_k(t), \qquad t < \lim_{k\to\infty} \tau_k, \tag{4.5}$$

exists and X satisfies (4.2). We leave the proof of uniqueness and the fact that $\lim_{k\to\infty} \tau_k = \tau_\infty$ to the reader.

(b) Let $\alpha(x) = 1 + \sum_l \beta_l(x)$ and

$$\int_0^{\eta(t)} \alpha(X(s))\, ds = t \tag{4.6}$$

(cf. Proposition 2.10). Then $X^0(t) \equiv X(\eta(t))$ is a solution of

$$X^0(t) = X(0) + \sum_l l Y_l\left(\int_0^t \beta_l^0(X^0(s))\, ds\right), \tag{4.7}$$

where $\beta_l^0 \equiv \beta_l/\alpha$. Note $\sum_l \beta_l^0 < 1$. If X^0 is a solution of the martingale problem for A^0 (defined as in (4.1) using the β_l^0), then by inverting the time change, we have that X is a solution of the local-martingale problem for A. For $z \in (\mathbb{Z}_+)^{\mathbb{Z}^d}$, let

$$\beta_k^1(z) = \begin{cases} \beta_k^0\left(X(0) + \sum_l l z_l\right), & \sum |l| z_l < \infty, \\ 0, & \sum |l| z_l = \infty, \end{cases} \tag{4.8}$$

and set

$$Z_l(t) = Y_l\left(\int_0^t \beta_l^0(X^0(s))\, ds\right) = Y_l\left(\int_0^t \beta_l^1(Z(s))\, ds\right). \tag{4.9}$$

Since Z is the unique solution of (4.9), it is nonanticipating by Theorem 2.2. Consequently, by Theorem 2.8(a), Z is a solution of the martingale problem for

$$B = \left\{\left(\prod_{l \in I} f_l, \sum_{k} \beta_k^1 (f_k(\cdot + e_k) - f_k) \prod_{l \neq k} f_l\right): I \subset \mathbb{Z}^d, I \text{ finite}, f_l \in B(\mathbb{Z}_+)\right\}.$$

The bp-closure of B contains $(f, \sum_{l \in I} \beta_l^1(f(\cdot + e_l) - f))$ where f is any bounded function depending only on the coordinates with indices in I (I finite). Consequently,

$$(4.10)\quad f\left(X(0) + \sum_{l \in I} lZ_l(t)\right) - \int_0^t \sum_{k \in I} \beta_k^0(X(s))\left(f\left(X(0) + \sum_{l \in I} lZ_l(s) + k\right)\right.$$
$$\left. - f\left(X(0) + \sum_{l \in I} lZ_l(s)\right)\right) ds$$

is a martingale for any finite I and any $f \in \mathscr{D}(A^0)$. Letting I increase to all of $\mathbb{Z}^d$, we see that X^0 is a solution of the martingale problem for A^0.

(c) As before let

$$(4.11)\quad \int_0^{\tilde{\eta}(t)} \alpha(\tilde{X}(s))\, ds = t.$$

Then $\tilde{X}^0(t) \equiv \tilde{X}(\tilde{\eta}(t))$ is a solution of the martingale problem for A^0. But A^0 is bounded so the solution is unique for each $\tilde{X}(0)$. Consequently, if $X(0) = \tilde{X}(0)$, then by part (b), X^0 must be a version of $\tilde{X}^0$ and X must be a version of $\tilde{X}$. □

5. DIFFUSION PROCESSES

Let $E = \mathbb{R}^d \cup \{\Delta\}$ be the one-point compactification of $\mathbb{R}^d$. For $k = 1, 2, \ldots$, let $\beta_k: \mathbb{R}^d \to [0, \infty)$ be measurable, $\alpha_k \in \mathbb{R}^d$, and suppose that for each compact $K \subset \mathbb{R}^d$, $\sup_{x \in K} \sum_{k=1}^{\infty} |\alpha_k|^2 \beta_k(x) < \infty$. Thinking of the α_k as column vectors, define

$$(5.1)\quad G(x) = ((G_{ij}(x))) = \sum_{k=1}^{\infty} \alpha_k \alpha_k^T \beta_k(x).$$

Let $F: \mathbb{R}^d \to \mathbb{R}^d$ be measurable and bounded on compact sets. For $f \in C_c^\infty(\mathbb{R}^d)$, extend f to E by setting $f(\Delta) = 0$, and define

$$(5.2)\quad Af(x) = \begin{cases} \frac{1}{2} \sum_{i,j} G_{ij}(x)\, \partial_i\, \partial_j f(x) + \sum_i F_i(x)\, \partial_i f(x), & x \neq \Delta, \\ 0, & x = \Delta. \end{cases}$$

Let W_i, $i = 1, 2, \ldots$, be independent standard Brownian motions, let $X(0)$ be nonrandom, and suppose X satisfies

$$(5.3)\quad X(t) = X(0) + \sum_{i=1}^{\infty} \alpha_i W_i\left(\int_0^t \beta_i(X(s))\, ds\right) + \int_0^t F(X(s))\, ds, \qquad t < \tau_\infty,$$

and

$$(5.4)\qquad X(t) = \Delta, \qquad t \geq \tau_\infty,$$

where $\tau_\infty = \inf\{t: X(t-) = \Delta\}$. The solution of (5.3) and (5.4) is not in general unique, so we again employ the notion of a nonanticipating solution. In this context X is nonanticipating if for each $t \geq 0$, $W_i^t \equiv W_i(\tau_i(t) + \cdot) - W_i(\tau_i(t))$, $i = 1, 2, \ldots$, are independent standard Brownian motions that are independent of $\mathscr{F}_t^X$.

5.1 Theorem If X is a nonanticipating solution of (5.3) and (5.4), then X is a solution of the martingale problem for A.

5.2 Remark **(a)** Note that uniqueness for the martingale problem for A implies uniqueness of nonanticipating solutions of (5.3) and (5.4).

(b) A converse for Theorem 5.1 can be obtained from Theorem 5.3 below and Theorem 3.3 of Chapter 5. □

Proof. The proof is essentially the same as for Theorem 4.1(b). □

To simplify the statement of the next result, we assume $\tau_\infty = \infty$ in (5.3) and (5.4).

5.3 Theorem **(a)** If X is a nonanticipating solution of (5.3) for all $t < \infty$ (i.e., $\tau_\infty = \infty$), then there is a version of X satisfying the stochastic integral equation

$$(5.5)\qquad Y(t) = Y(0) + \sum_{i=1}^{\infty} \alpha_i \int_0^t \sqrt{\beta_i(Y(s))}\, dB_i(s) + \int_0^t F(Y(s))\, ds.$$

(b) If Y is a solution of (5.5) for all $t < \infty$, then there is a version of Y that is a nonanticipating solution of (5.3).

Proof. **(a)** Since X is a solution of the martingale problem for A, (a) follows from Theorem 3.3 of Chapter 5.

(b) Let $\tilde{W}_i$, $i = 1, 2, \ldots$, be independent standard Brownian motions, independent of the B_i and Y. (It may be necessary to enlarge the sample space to obtain the $\tilde{W}_i$. See the proof of Theorem 3.3 in Chapter 5.)

Let

$$\tau_i(t) = \int_0^t \beta_i(Y(s))\, ds, \tag{5.6}$$

and let

$$\gamma_i(u) = \inf\left\{t: \int_0^t \beta_i(Y(s))\, ds > u\right\}, \qquad u \le \tau_i(\infty). \tag{5.7}$$

Define

$$W_i(u) = \begin{cases} \displaystyle\int_0^{\gamma_i(u)} \sqrt{\beta_i(Y(s))}\, dB_i(s), & u \le \tau_i(\infty), \\ \tilde{W}_i(u - \tau_i(\infty)) + W_i(\tau_i(\infty)), & \tau_i(\infty) < u < \infty. \end{cases} \tag{5.8}$$

Since $\gamma_i(u)$ is a stopping time, W_i is a martingale by the optional sampling theorem, as is $W_i^2(u) - u$. Consequently, W_i is a standard Brownian motion (Theorem 2.11 of Chapter 5). The independence of the W_i and the stopping properties of the τ_i follow by much the same argument as in the proof of the independence of the Y_k in Theorem 2.8(b). Finally, since $\int_0^t \sqrt{\beta_i Y(s)}\, dB_i(s)$ is constant on any interval on which $\beta_i(Y(s))$ is zero, it follows that Y is a solution of (5.3). □

The representations in Section 4 and in the present section combine to give a natural approach to diffusion approximations.

5.4 Theorem. Let $\beta_i^{(n)}: \mathbb{R}^d \to [0, \infty)$, $\alpha_i \in \mathbb{R}^d$, $i = 1, 2, \ldots$, satisfy

$$\sup_n \sup_{x \in K} \sum_i (1 \vee |\alpha_i|^2)\beta_i^{(n)}(x) < \infty \tag{5.9}$$

for each compact $K \subset \mathbb{R}^d$, and let $\lambda_n > 0$ satisfy $\lim_{n\to\infty} \lambda_n = \infty$. Let Y_i, $i = 1, 2, \ldots$, be independent unit Poisson processes and suppose X_n satisfies

$$X_n(t) = X_n(0) + \sum_i \lambda_n^{-1/2}\alpha_i\, Y_i\left(\lambda_n \int_0^t \beta_i^{(n)}(X_n(s))\, ds\right). \tag{5.10}$$

Define $W_i^{(n)}(u) = \lambda_n^{-1/2}(Y_i(\lambda_n u) - \lambda_n u)$ and

$$F_n(x) = \lambda_n^{1/2} \sum_i \alpha_i \beta_i^{(n)}(x). \tag{5.11}$$

Let $\beta_i: \mathbb{R}^d \to [0, \infty)$, $i = 1, 2, \ldots$, let $F: \mathbb{R}^d \to \mathbb{R}^d$ be continuous, and suppose for each compact $K \subset \mathbb{R}^d$ that

$$\lim_{n\to\infty} \sup_{x \in K} |\beta_i^{(n)}(x) - \beta_i(x)| = 0, \qquad i = 1, 2, \ldots, \tag{5.12}$$

$$\lim_{n\to\infty} \sup_{x \in K} |F_n(x) - F(x)| = 0, \tag{5.13}$$

and

$$\lim_{m\to\infty} \overline{\lim_{n\to\infty}} \sup_{x\in K} \sum_{i\geq m} |\alpha_i|^2 \beta_i^{(n)}(x) = 0. \tag{5.14}$$

Suppose that (5.3) and (5.4) have a unique nonanticipating solution and that $X_n(0) \to X(0)$. Let $\tau_a^n = \inf\{t: |X_n(t)| \geq a \text{ or } |X_n(t-)| \geq a\}$ and $\tau_a = \inf\{t: |X(t)| \geq a\}$. Then for all but countably many $a \geq 0$,

$$X_n(\cdot \wedge \tau_a^n) \Rightarrow X(\cdot \wedge \tau_a). \tag{5.15}$$

If $\lim_{a\to\infty} \tau_a = \infty$, then $X_n \Rightarrow X$.

5.5 Remark More-general results of this type can be obtained as corollaries to Theorem 3.4. □

Proof. Note that

$$X_n(t) = X_n(0) + \sum_i \alpha_i W_i^{(n)}\left(\int_0^t \beta_i^{(n)}(X_n(s))\, ds\right) + \int_0^t F_n(X_n(s))\, ds. \tag{5.16}$$

It follows from (5.12), (5.13), (5.14), the relative compactness of $\{W_i^{(n)}\}$, and (5.16), that $\{X_n(\cdot \wedge \tau_a^n)\}$ is relatively compact (cf. Proposition 3.1). Furthermore, if for $a_0 > 0$ and some subsequence $\{n_i\}$, $X_{n_i}(\cdot \wedge \tau_{a_0}^{n_i}) \Rightarrow Y_{a_0}$, then setting $\eta_a = \inf\{t: |Y_{a_0}(t)| \geq a \text{ or } |Y_{a_0}(t-)| \geq a\}$, $(X_{n_i}(\cdot \wedge \tau_a^{n_i}), \tau_a^{n_i}) \Rightarrow (Y_{a_0}(\cdot \wedge \eta_a), \eta_a)$ in $D_{\mathbb{R}^d}[0, \infty) \times [0, \infty]$ for all $a < a_0$ such that

$$P\left\{\lim_{b\to a} \eta_b = \eta_a\right\} = 1. \tag{5.17}$$

Note that the monotonicity of η_a implies (5.17) holds for all but countably many a.

Since a_0 is arbitrary, we can select the subsequence so that $\{(X_{n_i}(\cdot \wedge \tau_a^{n_i}), \tau_a^{n_i})\}$ converges in distribution for all but countably many a, and the limit has the form $(Y(\cdot \wedge \eta_a), \eta_a)$ for a fixed process Y with sample paths in $D_E[0, \infty)$ (η_a as before). (We may assume that $Y(t) = \Delta$ implies $Y(s) = \Delta$ for all $s > t$.) By the continuous mapping theorem (Corollary 1.9 of Chapter 3), Y satisfies

$$Y(t \wedge \eta_a) = Y(0) + \sum \alpha_i W_i\left(\int_0^{t\wedge\eta_a} \beta_i(Y(s))\, ds\right) + \int_0^{t\wedge\eta_a} F(Y(s))\, ds. \tag{5.18}$$

Here (5.14) allows the interchange of summation and limits. It follows as in the proof of Theorem 3.4 that Y is a nonanticipating solution of (5.3) and (5.4) and hence Y has the same distribution as X. The uniqueness of the possible limit point gives (5.15) for all a such that $\lim_{b\to a} \eta_b = \eta_a$ a.s. The final statement of the theorem is left to the reader. □

Equations of the form of (5.10) ordinarily arise after renormalization of space and time. For example, suppose

$$U_n(t) = U_n(0) + \sum_l lY_l\left(\int_0^t \beta_l^{(n)}(U_n(s))\, ds\right), \tag{5.19}$$

and set $X_n(t) = n^{-1/2}U_n(nt)$. Then X_n satisfies

$$X_n(t) = X_n(0) + \sum_l lW_l^{(n)}\left(\int_0^t \beta_l^{(n)}(n^{1/2}X_n(s))\, ds\right) + \int_0^t F_n(X_n(s))\, ds, \tag{5.20}$$

where $W_l^{(n)}(u) = n^{-1/2}(Y_l(nu) - nu)$ and

$$F_n(x) = n^{1/2} \sum_l l\beta_l^{(n)}(n^{1/2}x). \tag{5.21}$$

6. PROBLEMS

1. Let W be standard Brownian motion.
   (a) Show that for $0 < \alpha < 1$,

   $$\int_0^t \frac{1}{|W(s)|^\alpha}\, ds < \infty \quad \text{a.s.,} \qquad t \geq 0, \tag{6.1}$$

   and for $\alpha \geq 1$,

   $$\int_0^t \frac{1}{|W(s)|^\alpha}\, ds = \infty \quad \text{a.s.,} \qquad t > 0. \tag{6.2}$$

   (b) Show that for $\alpha \geq 1$ the solution of

   $$X(t) = X(0) + W\left(\int_0^t |X(s)|^\alpha\, ds\right) \tag{6.3}$$

   is unique, but it is not unique if $0 < \alpha < 1$.
   (c) Let $0 < \alpha < 1$ and $\gamma_0 = \sup\{t < 100\colon W(t) = 0\}$. Let $\tau(t)$ satisfy

   $$\begin{aligned} \int_0^{\tau(t)} \frac{1}{|W(s)|^\alpha}\, ds &= t, \qquad & t &< \int_0^{\gamma_0} \frac{1}{|W(s)|^\alpha}\, ds, \\ \tau(t) &= \gamma_0, \qquad & t &\geq \int_0^{\gamma_0} \frac{1}{|W(s)|^\alpha}\, ds. \end{aligned} \tag{6.4}$$

   Show that $X(t) = W(\tau(t))$ satisfies (6.3), but that it is not a solution of the martingale problem for $A = \{(f, \frac{1}{2}|x|^\alpha f'')\colon f \in C_c^\infty(\mathbb{R})\}$.

2. Let Y_1 and Y_2 be independent standard Brownian motions. Let β_1 and β_2 be nonnegative, measurable functions on $\mathbb{R}^2$ satisfying $\beta_i(x, y) \leq$

$K(1 + x^2 + y^2)$. Show that the random time change problem

$$Z_i(t) = Y_i\left(\int_0^t \beta_i(Z(s))\, ds\right) \tag{6.5}$$

is equivalent to the martingale problem for A given by

$$A = \{(f, \beta_1 f_{xx} + \beta_2 f_{yy}) : f \in C_c^2(\mathbb{R}^2)\}, \tag{6.6}$$

that is, any nonanticipating solution of (6.5) is a solution of the martingale problem for A, and any solution of the martingale problem for A has a version that is a weak, nonanticipating solution of (6.5).

3. State and prove a result analogous to that in Problem 2 in which Y_1 and Y_2 are Poisson processes.

4. Let Y be Brownian motion,

$$\beta_1(x) = \begin{cases} 1, & |x| \le 1, \\ 0, & |x| > 1, \end{cases} \tag{6.7}$$

and

$$\beta_2(x) = \begin{cases} 1, & |x| < 1, \\ 0, & |x| \ge 1. \end{cases} \tag{6.8}$$

Show that

$$Z(t) = Y\left(\int_0^t \beta_1(Z(s))\, ds\right) \tag{6.9}$$

has no solution but that

$$Z(t) = Y\left(\int_0^t \beta_2(Z(s))\, ds\right) \tag{6.10}$$

does. In the second case, what is the (strong) generator corresponding to Z?

5. Prove Proposition 3.2.

6. For $Y_1(t) = [t]$ and $Y_2(t) = t$, let $(Z_1^{(n)}, Z_2^{(n)})$ satisfy

$$\begin{aligned} Z_1^{(n)}(t) &= Y_1\left(\int_0^t (1 - n^{-1})\sqrt{(1 - Z_2^{(n)}(s)) \vee 0}\, ds\right), \\ Z_2^{(n)}(t) &= Y_2\left(\int_0^t (Z_1^{(n)}(s) + \sqrt{(1 - Z_2^{(n)}(s)) \vee 0})\, ds\right), \end{aligned} \tag{6.11}$$

and let (Z_1, Z_2) satisfy

$$Z_1(t) = Y_1\left(\int_0^t \sqrt{(1 - Z_2(s)) \vee 0}\, ds\right),$$

(6.12)

$$Z_2(t) = Y_2\left(\int_0^t (Z_1(s) + \sqrt{(1 - Z_2(s)) \vee 0})\, ds\right).$$

hat $\lim_{n\to\infty} (Z_1^{(n)}, Z_2^{(n)}) \neq (Z_1, Z_2)$.

7. Let $\tau(t)$ be nonnegative, nondecreasing, and right continuous. Let $y \in D_E[0, \infty)$ and $z = y(\tau(\cdot))$. Define

$$\eta(\delta, T) = \sup_{t \le T} (\tau(t + \delta) - \tau(t)) + \sup_{t \le T} (\tau(t) - \tau(t-)), \tag{6.13}$$

and $\gamma(t) = \inf \{u: \tau(u) \ge t\}$. Show that if $0 \le t_1 < t_2$ and $t_2 - t_1 > \eta(\delta, T)$, then $\gamma(t_2) - \gamma(t_1) > \delta$, and that

$$w'(z, \delta, T) \le w'(y, \eta(\delta, T), \tau(T)). \tag{6.14}$$

8. Suppose in (4.2) that $\sum |l| \beta_l(x) \le A + B|x|$. Show that $\tau_\infty = \infty$.

9. Let W and Y be independent, W a standard Brownian motion and Y a unit Poisson process. Show that

$$Z_n(t) = W\left(\int_0^t (2 + (-1)^{Y(ns)})\, ds\right) \tag{6.15}$$

and

$$\tilde{Z}_n(t) = \int_0^t \sqrt{2 + (-1)^{Y(ns)}}\, dW(s) \tag{6.16}$$

have the same distribution, that $\{Z_n\}$ converges a.s., but $\{\tilde{Z}_n\}$ does not converge a.s.

10. Let $E = \{(x, y):\ x,\ y \ge 0,\ x + y \le 1\}$. For $f \in C^\infty(E)$, define $Af = x(1 - x)f_{xx} - 2xyf_{xy} + y(1 - y)f_{yy}$. Show that if X is a solution of the martingale problem for A, then X satisfies (5.3) with $\alpha_i = 0, i \ge 4$.

11. Let f and g be locally absolutely continuous on $\mathbb{R}$.
   (a) Show that if h is bounded and Borel measurable, then

$$\int_a^b h(g(z))g'(z)\, dz = \int_{g(a)}^{g(b)} h(u)\, du, \qquad a, b \in \mathbb{R}, \tag{6.17}$$

   with the usual convention that $\int_a^b f(z)\, dz = -\int_b^a f(z)\, dz$ if $b < a$.
   Hint: Check (6.9) first for continuous h by showing both sides are locally absolutely continuous as functions of b and differentiating. Then apply a monotone class argument (see Appendix 4).

(b) Show that if $A \in \mathscr{B}(\mathbb{R})$ has Lebesgue measure zero, then

$$(6.18) \qquad \int_a^b \chi_{\{g(z) \in A\}} g'(z)\, dz = 0.$$

In particular, for each α, $m(\{g'(z) \neq 0\} \cap \{g(z) = \alpha\}) = 0$.

(c) Show that if g is nondecreasing, then $f \circ g$ is locally absolutely continuous.

(d) Define

$$(6.19) \qquad h(z) = \begin{cases} f'(g(z))g'(z) & \text{on } \{z: f'(g(z)) \text{ and } g'(z) \text{ exist}\}, \\ 0 & \text{otherwise.} \end{cases}$$

(Note that $m(\mathbb{R} - \{z: f'(g(z)) \text{ and } g'(z) \text{ exist}\} \cup \{z: g'(z) = 0\}) = 0$.) Show that $f \circ g$ is locally absolutely continuous if and only if h is locally L^1, and that under those conditions

$$(6.20) \qquad \frac{d}{dz} f(g(z)) = h(z) \quad \text{a.e.}$$

(e) Let $f(t) = \sqrt{|t|}$ and $g(t) = t^2 \cos^2 (1/t)$. Show that f and g are locally absolutely continuous, but $f \circ g$ is not.

Hint: Show that $f \circ g$ does not have bounded variation.

12. Let β be a nonnegative Borel measurable function on $[0, \infty)$ that is locally L^1. Define $\gamma(t) = \inf \{u: \int_0^u \beta(s)\, ds > t\}$.

(a) Show that γ is right continuous.

(b) Show that

$$(6.21) \qquad \int_a^b \beta(s)\, ds = \int_0^\infty \chi_{\{a \leq \gamma(t) < b\}}\, dt$$

for all $0 \leq a < b$.

(c) Show that if g is Borel measurable and βg is locally L^1, then

$$(6.22) \qquad \int_0^{\gamma(t)} \beta(s) g(s)\, ds = \int_0^t g(\gamma(u))\, du.$$

7. NOTES

Volkonski (1958) introduced the one-parameter random time change for Markov processes. See also Lamperti (1967b). Helland (1978) gave results similar to Theorem 1.5 with applications to branching processes (see Chapter 9, Section 1).

The multiparameter time changes were introduced by Helms (1974) and developed in Kurtz (1980a). Holley and Stroock (1976) use a slightly different approach.

Applications of multiparameter time changes to convergence theorems are given in Kurtz (1978a, 1981c, 1982). See Chapters 9 and 11.

Any diffusion with a uniformly elliptic generator with bounded coefficients can be obtained as a nonanticipating solution of an equation of the form of (5.3) with only finitely many nonzero α_i. See Kurtz (1980a).

7 INVARIANCE PRINCIPLES AND DIFFUSION APPROXIMATIONS

Let $\xi_1, \xi_2, \ldots$ be independent, identically distributed random variables with mean zero and variance one. Define

$$X_n(t) = n^{-1/2} \sum_{k=1}^{[nt]} \xi_k. \tag{0.1}$$

A simple application of Theorem 2.6 of Chapter 4 and Theorem 6.5 of Chapter 1 gives Donsker's (1951) invariance principle, Theorem 1.2(c) of Chapter 5, that is, that $X_n \Rightarrow W$ where W is standard Brownian motion. One noteworthy property of X_n is that it is a martingale. In Section 1 we show that the invariance principle can be extended to very general sequences of martingales.

Another direction in which the invariance principle has been extended is to processes satisfying mixing conditions, that is, some form of asymptotic independence. A large number of such conditions have been introduced. We consider some of these in Section 2 and give examples of related invariance principles in Section 3.

Section 4 is devoted to an extension of the results of Section 1 allowing the limiting process to be an arbitrary diffusion process.

Section 5 contains recent refinements of the invariance principle due to Komlós, Major, and Tusnády (1975, 1976) who showed how to construct X_n and W on the same sample space in such a way that

$$\sup_{t \le T} |X_n(t) - W(t)| = O\left(\frac{\log n}{\sqrt{n}}\right). \tag{0.2}$$

1. THE MARTINGALE CENTRAL LIMIT THEOREM

In this section we give the extension of Donsker's invariance principle to sequences of martingales in $\mathbb{R}^d$. The convergence results are based on the following martingale characterization of processes with independent increments.

1.1 Theorem Let $C = ((c_{ij}))$ be a continuous, symmetric, $d \times d$ matrix-valued function, defined on $[0, \infty)$, satisfying $C(0) = 0$ and

(1.1) $$\sum (c_{ij}(t) - c_{ij}(s))\xi_i \xi_j \geq 0, \qquad \xi \in \mathbb{R}^d, \qquad t > s \geq 0.$$

Then there exists a unique (in distribution) process X with sample paths in $C_{\mathbb{R}^d}[0, \infty)$ such that X_i, $i = 1, 2, \ldots, d$, and $X_i X_j - c_{ij}$, $i, j = 1, 2, \ldots, d$, are (local) martingales with respect to $\{\mathscr{F}_t^X\}$. The process X has independent Gaussian increments.

Proof. As in the proof of Theorem 2.12 of Chapter 5, if X is such a process, then for $\theta \in \mathbb{R}^d$

(1.2) $$f(t, X) = \exp \{i\theta \cdot X(t) + \tfrac{1}{2}\theta \cdot C(t)\theta\}$$

is a martingale, and hence

(1.3) $$E[\exp \{i\theta \cdot (X(t) - X(s))\} \mid \mathscr{F}_s] = \exp \{-\tfrac{1}{2}\theta \cdot (C(t) - C(s))\},$$

which implies X has independent Gaussian increments and determines the finite-dimensional distributions of X.

To obtain such an X set

(1.4) $$\gamma(t) = \sum_{i=1}^{d} c_{ii}(t).$$

Note that (1.1) implies

(1.5) $$\begin{aligned} |c_{ij}(t) - c_{ij}(s)| &\leq c_{ii}(t) - c_{ii}(s) + c_{jj}(t) - c_{jj}(s) \\ &\leq \gamma(t) - \gamma(s) \end{aligned}$$

(take $\xi_i = 1$, $\xi_j = \pm 1$, and $\xi_l = 0$ otherwise), and hence c_{ij} is of bounded variation and c_{ij} can be written as

(1.6) $$c_{ij}(t) = \int_0^t d_{ij}(s)\, d\gamma(s),$$

where $D(s) = ((d_{ij}(s)))$ is nonnegative definite. Let $D^{1/2}(s)$ denote the symmetric nonnegative-definite square root of $D(s)$, let W be d-dimensional standard Brownian motion, and set

$$M(t) = W(\gamma(t)). \tag{1.7}$$

Then

$$X(t) = \int_0^t D\, dM \tag{1.8}$$

is the desired process. □

1.2 Theorem Let C be as in Theorem 1.1. Suppose that X is a measurable process and that, for each $\theta \in \mathbb{R}^d$ and $f \in C_c^\infty(\mathbb{R})$,

$$f(\theta \cdot X(t)) - \int_0^t \tfrac{1}{2} f''(\theta \cdot X(s))\, dc_\theta(s) \tag{1.9}$$

is an $\{\mathscr{F}_t^X\}$-martingale, where

$$c_\theta(t) = \theta \cdot C(t)\theta. \tag{1.10}$$

Then X has independent Gaussian increments with mean zero and

$$E[X(t)X(t)^T] = C(t). \tag{1.11}$$

1.3 Remark Note that it is crucial that (1.9) be a martingale with respect to $\{\mathscr{F}_t^X\}$ and not just with respect to $\{\mathscr{F}_t^{\theta \cdot X}\} \equiv \sigma\{\theta \cdot X(s): s \le t\}$. See Problem 2. □

Proof. The collection of f for which (1.9) is an $\{\mathscr{F}_t^X\}$-martingale is closed under bp-convergence. Consequently,

$$\exp\{i\theta \cdot X(t)\} + \int_0^t \tfrac{1}{2} \exp\{i\theta \cdot X(s)\}\, dc_\theta(s) \tag{1.12}$$

is an $\{\mathscr{F}_t^X\}$-martingale, and hence, by Itô's formula, Theorem 2.9 of Chapter 5,

$$\exp\{i\theta \cdot X(t) + \tfrac{1}{2} c_\theta(t)\} \tag{1.13}$$

is an $\{\mathscr{F}_t^X\}$-martingale. The theorem follows as in the proof of Theorem 1.1. □

1.4 Theorem For $n = 1, 2, \ldots$, let $\{\mathscr{F}_t^n\}$ be a filtration and let M_n be an $\{\mathscr{F}_t^n\}$-local martingale with sample paths in $D_{\mathbb{R}^d}[0, \infty)$ and $M_n(0) = 0$. Let $A_n = ((A_n^{ij}))$ be symmetric $d \times d$ matrix-valued processes such that A_n^{ij} has sample paths in $D_{\mathbb{R}}[0, \infty)$ and $A_n(t) - A_n(s)$ is nonnegative definite for $t > s \ge 0$. Assume one of the following conditions holds:

(a) For each $T > 0$,

$$\lim_{n\to\infty} E\left[\sup_{t\le T} |M_n(t) - M_n(t-)|\right] = 0 \tag{1.14}$$

and

$$A_n^{ij} = [M_n^i, M_n^j]. \tag{1.15}$$

(b) For each $T > 0$ and $i, j = 1, 2, \ldots, d$,

$$\lim_{n\to\infty} E\left[\sup_{t\le T} |A_n^{ij}(t) - A_n^{ij}(t-)|\right] = 0, \tag{1.16}$$

$$\lim_{n\to\infty} E\left[\sup_{t\le T} |M_n(t) - M_n(t-)|^2\right] = 0, \tag{1.17}$$

and for $i, j = 1, 2, \ldots, d$,

$$M_n^i(t)M_n^j(t) - A_n^{ij}(t) \tag{1.18}$$

is an $\{\mathscr{F}_t^n\}$-local martingale.

Suppose that C satisfies the conditions of Theorem 1.1 and that, for each $t \ge 0$ and $i, j = 1, 2, \ldots, d$,

$$A_n^{ij}(t) \to c_{ij}(t) \tag{1.19}$$

in probability. Then $M_n \Rightarrow X$, where X is the process with independent Gaussian increments given by Theorem 1.1.

1.5 Remark In the discrete-time case, let $\{\xi_k^n\colon k = 1, 2, \ldots\}$ be a collection of $\mathbb{R}^d$-valued random variables and define

$$M_n(t) = \sum_{k=1}^{[\alpha_n t]} \xi_k^n \tag{1.20}$$

for some $\alpha_n \to \infty$. In condition (a)

$$A_n(t) = \sum_{k=1}^{[\alpha_n t]} \xi_k^n \xi_k^{nT} \tag{1.21}$$

(considering the ξ_k^n as column vectors), and for condition (b) one can take

$$A_n(t) = \sum_{k=1}^{[\alpha_n t]} E[\xi_k^n \xi_k^{nT} \mid \mathscr{F}_{k-1}^n] \tag{1.22}$$

where $\mathscr{F}_k^n = \sigma(\xi_l^n\colon l \le k)$. Of course M_n is a martingale if $E[\xi_k^n \mid \mathscr{F}_{k-1}^n] = 0$. □

Proof. Without loss of generality, we may assume the M_n are martingales. If not, there exist stopping times τ_n with $P\{\tau_n < n\} \le n^{-1}$ such that $M_n(\cdot \wedge \tau_n)$ is a martingale and $A_n(\cdot \wedge \tau_n)$ satisfies the conditions of the theorem with M_n

replaced by $M_n(\cdot \wedge \tau_n)$. Similarly, under condition (b) we may assume the processes in (1.18) are martingales.

(a) Assume condition (a). Let

$$\eta_n = \inf \{t: A_n^{ii}(t) > c_{ii}(t) + 1 \text{ for some } i \in \{1, 2, \ldots, d\}\}. \tag{1.23}$$

Since (1.19) implies $\eta_n \to \infty$ in probability, the convergence of M_n is equivalent to the convergence of $\tilde{M}_n = M_n(\cdot \wedge \eta_n)$.

Fix $\theta \in \mathbb{R}^d$ and define

$$Y_n(t) = \theta \cdot \tilde{M}_n(t), \tag{1.24}$$

$$A_n^\theta(t) = \sum_{i,j} A_n^{ij}(t)\theta_i \theta_j, \tag{1.25}$$

and

$$c_\theta(t) = \sum_{i,j} c_{ij}(t)\theta_i \theta_j. \tag{1.26}$$

Let $f \in C_c^\infty(\mathbb{R})$, $0 \le t = t_0 < t_1 < \cdots < t_m = t + s$, and $\xi_k = Y_n(t_{k+1}) - Y_n(t_k)$. Then

$$\begin{aligned} &E[f(Y_n(t + s)) - f(Y_n(t)) \mid \mathscr{F}_t^n] \\ &\quad = E\left[\sum_{k=0}^{m-1} (f(Y_n(t_{k+1})) - f(Y_n(t_k)) - f'(Y_n(t_k))\xi_k) \,\middle|\, \mathscr{F}_t^n\right]. \end{aligned} \tag{1.27}$$

Let

$$\gamma = \max \{k: t_k < \eta_n \wedge (t + s)\} \tag{1.28}$$

and

$$\zeta = \max \left\{k: t_k < \eta_n \wedge (t + s), \sum_{l-0}^{k} \xi_l^2 \le d \sum c_{ii}(t + s)\theta_i^2 + 2d|\theta|^2\right\}. \tag{1.29}$$

Note that by the definition of η_n, $\gamma = \zeta$ for max $(t_{k+1} - t_k)$ sufficiently small.

Then by (1.27),

$$\begin{aligned} &E[f(Y_n(t + s)) - f(Y_n(t)) \mid \mathscr{F}_t^n] \\ &\quad = E\left[\sum_{k=\zeta}^{\gamma} (f(Y_n(t_{k+1})) - f(Y_n(t_k)) - f'(Y_n(t_k))\xi_k) \,\middle|\, \mathscr{F}_t^n\right] \\ &\qquad + E\left[\sum_{k=0}^{\zeta-1} (f(Y_n(t_{k+1})) - f(Y_n(t_k)) \right. \\ &\qquad\qquad \left. - f'(Y_n(t_k))\xi_k - \tfrac{1}{2}f''(Y_n(t_k))\xi_k^2) \,\middle|\, \mathscr{F}_t^n\right] \\ &\qquad + E\left[\sum_{k=0}^{\zeta-1} \tfrac{1}{2}f''(Y_n(t_k))\xi_k^2 \,\middle|\, \mathscr{F}_t^n\right]. \end{aligned} \tag{1.30}$$

Setting $\Delta Y_n(u) = Y_n(u) - Y_n(u-)$, and letting $\max (t_{k+1} - t_k) \to 0$,

$$(1.31) \quad E[f(Y_n(t+s)) - f(Y_n(t)) \mid \mathscr{F}_t^n]$$

$$= E\Big[f(Y_n((t+s) \wedge \eta_n)) - f(Y_n((t+s) \wedge \eta_n -))$$

$$- f'(Y_n((t+s) \wedge \eta_n -)) \Delta Y_n((t+s) \wedge \eta_n) \Big| \mathscr{F}_t^n \Big]$$

$$+ E\Big[\sum_{t < u < (t+s) \wedge \eta_n} (f(Y_n(u)) - f(Y_n(u-))$$

$$- f'(Y_n(u-)) \Delta Y_n(u) - \tfrac{1}{2} f''(Y_n(u-))(\Delta Y_n(u))^2) \Big| \mathscr{F}_t^n \Big]$$

$$+ E\Big[\int_{(t, (t+s) \wedge \eta_n)} \tfrac{1}{2} f''(Y_n(u-)) \, dA_n^\theta(u) \Big| \mathscr{F}_t^n \Big].$$

Note that the second term on the right is bounded by

$$(1.32) \quad 6^{-1} \| f''' \| E\Big[\sup_{t < u < (t+s)} |\Delta Y_n(u)| \, d \sum (c_{ii}(t+s) + 1)\theta_i^2 \Big| \mathscr{F}_t^n \Big]$$

(by the definition of η_n), and hence

$$(1.33) \quad |E[f(Y_n(t+s)) - f(Y_n(t)) \mid \mathscr{F}_t^n]|$$

$$\leq C_f E\Big[\sup_{t < u \leq t+s} |\Delta Y_n(u)| \Big(1 + d \sum_i (c_{ii}(t+s) + 1)\theta_i^2 \Big)$$

$$+ A_n^\theta((t+s) \wedge \eta_n -) - A_n^\theta(t \wedge \eta_n -) \Big| \mathscr{F}_t^n \Big],$$

where C_f depends only on $\| f' \|$, $\| f'' \|$, and $\| f''' \|$. In particular (1.33) can be extended to all f (including unbounded f) whose first three derivatives are bounded.

Let φ be convex with $\varphi(0) = 0$, $\lim_{x \to \infty} \varphi(x) = \infty$, and φ', φ'', and φ''' bounded. Then

$$(1.34) \qquad P\left\{\sup_{t \le T} |Y_n(t)| > K\right\} \le \frac{E[\varphi(Y_n(t))]}{\varphi(K)}$$
$$\le C_\varphi E\left[\sup_{u \le T} |\Delta Y_n(u)| \left(1 + d \sum_i (c_{ii}(T) + 1)\theta_i^2\right) + d \sum_i (c_{ii}(T) + 1)\theta_i^2\right] \Big/ \varphi(K).$$

Furthermore, for $\delta > 0$, $0 \le t \le T$, and $0 \le u \le \delta$,

$$(1.35) \qquad E[(f(Y_n(t+u)) - f(Y_n(t)))^2 \,|\, \mathscr{F}_t^n]$$
$$= E[f^2(Y_n(t+u)) - f^2(Y_n(t)) \,|\, \mathscr{F}_t^n]$$
$$- 2f(Y_n(t))E[f(Y_n(t+u)) - f(Y_n(t)) \,|\, \mathscr{F}_t^n]$$
$$\le E[\gamma_n(\delta) \,|\, \mathscr{F}_t^n],$$

where

$$(1.36) \qquad \gamma_n(\delta) = (C_{f^2} + 2\|f\|C_f)\left[\sup_{s \le T+\delta} |\Delta Y_n(s)| \left(1 + d \sum_i (c_{ii}(T+\delta) + 1)\theta_i^2\right) + \sup_{t \le T} (A_n^\theta((t+\delta) \wedge \eta_n -) - A_n^\theta(t \wedge \eta_n -))\right].$$

By (1.14) and (1.19),

$$(1.37) \qquad \lim_{\delta \to 0} \lim_{\eta \to \infty} E[\gamma_n(\delta)] = \lim_{\delta \to 0} (C_{f^2} + 2\|f\|C_f) \sup_{t \le T} (c_\theta(t+\delta) - c_\theta(t))$$
$$= 0,$$

so for each $f \in C_c^\infty(\mathbb{R}^d)$, $\{f(Y_n)\}$ is relatively compact by Theorem 8.6 of Chapter 3. Consequently, since we have (1.34), the relative compactness of $\{Y_n\} = \{\theta \cdot \tilde{M}_n\}$ follows by Theorem 9.1 of Chapter 3. Since θ is arbitrary, $\{\tilde{M}_n\}$ must be relatively compact (cf. Problem 22 of Chapter 3).

The continuity of $c_\theta(t)$ and (1.19) imply

$$(1.38) \qquad \int_{(t,\,(t+s)\wedge \eta_n)} \tfrac{1}{2} f''(y(u-))\, dA_n^\theta(u) - \int_t^{t+s} \tfrac{1}{2} f''(y(u))\, dc_\theta(u) \to 0$$

in probability uniformly for y in compact subsets of $D_{\mathbb{R}}[0, \infty)$. Consequently the relative compactness of $\{Y_n\}$ implies

$$(1.39) \qquad E\left[\left|\int_{(t,\,(t+s)\wedge \eta_n)} \tfrac{1}{2} f''(Y_n(u-))\, dA_n^\theta(u) - \int_t^{t+s} \tfrac{1}{2} f''(Y_n(u))\, dc_\theta(u)\right|\right] \to 0.$$

The first two terms on the right of (1.32) go to zero in L^1 and it follows easily (cf. the proof of Theorem 8.10 in Chapter 4 and Problem 7 in this chapter) that if X is a limit point of $\{\tilde{M}_n\}$, then (1.9) is an $\{\mathscr{F}_t^X\}$-martingale. Since this uniquely characterizes X, the convergence of $\{\tilde{M}_n\}$ follows.

(b) The proof is similar to that of part (a). With η_n and $\tilde{M}_n$ defined as in (1.23), we have

$$A_n^{ii}(t \wedge \eta_n) \le c_{ii}(t) + 1 + \sup_{s \le t} (A_n^{ii}(s) - A_n^{ii}(s-)), \tag{1.40}$$

and the third term on the right goes to zero in L^1 by (1.16).

Setting $\tilde{A}_n^{ij}(t) = A_n^{ij}(t \wedge \eta_n)$, note that

$$E[|\tilde{M}_n(t+h) - \tilde{M}_n(t)|^2 | \mathscr{F}_t^n] = E\left[\sum_{i=1}^d (\tilde{A}_n^{ii}(t+h) - \tilde{A}_n^{ii}(t)) \middle| \mathscr{F}_t^n\right] \tag{1.41}$$

and to apply Theorem 8.6 of Chapter 3, fix $T > 0$ and define

$$\gamma_n(\delta) = \sup_{t \le T} \sum_{i=1}^d (\tilde{A}_n^{ii}(t+\delta) - \tilde{A}_n^{ii}(t)). \tag{1.42}$$

Since

$$\gamma_n(\delta) \le \sum_{i=1}^d \left(c_{ii}(T+\delta) + 1 + \sup_{s \le T+\delta} (A_n^{ii}(s) - A_n^{ii}(s-)) \right), \tag{1.43}$$

and since (1.16) implies the right side of (1.43) is convergent in L^1, we conclude that

$$\lim_{\delta \to 0} \lim_{n \to \infty} E[\gamma_n(\delta)] = \lim_{\delta \to 0} \sup_{t \le T} \sum_{i=1}^d (c_{ii}(t+\delta) - c_{ii}(t)) = 0. \tag{1.44}$$

Let X be any limit point of $\{\tilde{M}_n\}$. By (1.17), X is continuous. Since for each $T > 0$, $\sup_n E[|\tilde{M}_n(T)|^2] < \infty$, $\{\tilde{M}_n(T)\}$ is uniformly integrable, and hence X must be a martingale (see Problem 7). Since $X_i X_j - c_{ij}$ is the limit in distribution of a subsequence $\{\tilde{M}_{n_k}^i \tilde{M}_{n_k}^j - \tilde{A}_{n_k}^{ij}\}$, we can conclude that it is also a martingale if we show that $\{\tilde{M}_{n_k}^i(T)\tilde{M}_{n_k}^j(T) - \tilde{A}_{n_k}^{ij}(T)\}$ is uniformly integrable for each T. Since (1.40) and (1.16) imply that $\{\tilde{A}_{n_k}^{ij}(T)\}$ is uniformly integrable (recall $|\tilde{A}_n^{ij}(T)| \le \frac{1}{2}(\tilde{A}^{ii}(T) + \tilde{A}_n^{jj}(T))$), it is enough to consider $\{\tilde{M}_{n_k}^i(T)\tilde{M}_{n_k}^j(T)\}$, and since $|\tilde{M}_{n_k}^i(T)\tilde{M}_{n_k}^j(T)| \le \frac{1}{2}(\tilde{M}_{n_k}^i(T)^2 + \tilde{M}_{n_k}^j(T)^2)$, it is enough to consider $\{\tilde{M}_{n_k}^i(T)^2\}$. Since $\tilde{M}_{n_k}^i(T)^2 \Rightarrow X_i(T)^2$, $\{\tilde{M}_{n_k}^i(T)^2\}$ is uniformly integrable if (and only if) $E[\tilde{M}_{n_k}^i(T)^2] \to E[X_i(T)^2]$, that is, if

$$E[X_i(T)^2] = c_{ii}(T). \tag{1.45}$$

Let

$$\tau_n^\alpha = \inf \{t : \tilde{M}_n^i(t)^2 > \alpha\} \tag{1.46}$$

and

(1.47)
$$\tau^\alpha = \inf \{t\colon X_i(t)^2 > \alpha\}.$$

Since

(1.48)
$$\tilde{M}_n^i(T \wedge \tau_n^\alpha)^2 \le 2\left(\alpha + \sup_{s \le T} |\tilde{M}_n^i(s) - \tilde{M}_n^i(s-)|^2\right),$$

$\{\tilde{M}_n^i(T \wedge \tau_n^\alpha)\}$ is uniformly integrable by (1.17). For all but countably many α and T, $(\tau_{n_k}^\alpha, \tilde{M}_{n_k}^i(T \wedge \tau_{n_k}^\alpha)) \Rightarrow (\tau^\alpha, X^i(T \wedge \tau^\alpha))$ and, excluding the countably many α and T,

(1.49)
$$\begin{aligned} E[X_i(T \wedge \tau^\alpha)^2] &= \lim_{k \to \infty} E[\tilde{M}_{n_k}^i(T \wedge \tau_{n_k}^\alpha)^2] \\ &= \lim_{k \to \infty} E[\tilde{A}_{n_k}^{ii}(T \wedge \tau_{n_k}^\alpha)] \\ &= E[c_{ii}(T \wedge \tau^\alpha)]. \end{aligned}$$

Letting $\alpha \to \infty$ we have (1.45), and it follows that $X_i X_j - c_{ij}$ are martingales for $i, j = 1, \ldots, d$, and that X is the unique process characterized in Theorem 1.1. Therefore $M_n \Rightarrow X$. □

2. MEASURES OF MIXING

Measures of mixing are measures of the degree of independence of two σ-algebras. Let $(\Omega, \mathscr{F}, P)$ be a probability space, and let $\mathscr{G}$ and $\mathscr{H}$ be sub-σ-algebras of $\mathscr{F}$. Two kinds of measures of mixing are commonly used. The measure of *uniform mixing* is given by

(2.1)
$$\begin{aligned} \varphi(\mathscr{G} \mid \mathscr{H}) &\equiv \sup_{A \in \mathscr{G}} \sup_{\substack{B \in \mathscr{H} \\ P(B) > 0}} |P(A \mid B) - P(A)| \\ &= \sup_{A \in \mathscr{G}} \|P(A \mid \mathscr{H}) - P(A)\|_\infty, \end{aligned}$$

where $\|\cdot\|_p$ denotes the norm for $L^p(\Omega, \mathscr{F}, P)$. The proof of equality of the two expressions is left as a problem. The measure of *strong mixing* is given by

(2.2)
$$\begin{aligned} \alpha(\mathscr{G}, \mathscr{H}) &= \sup_{A \in \mathscr{G}} \sup_{B \in \mathscr{H}} |P(AB) - P(A)P(B)| \\ &= \tfrac{1}{2} \sup_{A \in \mathscr{G}} E[|P(A \mid \mathscr{H}) - P(A)|] \\ &= \tfrac{1}{2} \sup_{B \in \mathscr{H}} E[|P(B \mid \mathscr{G}) - P(B)|] \\ &= \tfrac{1}{2} \sup_{A \in \mathscr{G}} \|P(A \mid \mathscr{H}) - P(A)\|_1 . \end{aligned}$$

Again the equality of the four expressions is left as a problem.

A comparison of the right sides of (2.1) and (2.2) suggests the following general definition. For $1 \le p \le \infty$ set

$$\varphi_p(\mathscr{G} \mid \mathscr{H}) = \sup_{A \in \mathscr{G}} \| P(A \mid \mathscr{H}) - P(A) \|_p. \tag{2.3}$$

Note that $\varphi = \varphi_\infty$ and $\alpha = \frac{1}{2}\varphi_1$.

Let E_1 and E_2 be separable metric spaces. Let X be E_1-valued and $\mathscr{G}$-measurable, and let Y be E_2-valued and $\mathscr{H}$-measurable. The primary application of measures of mixing is to estimate differences such as

$$\left| E[\psi(X, Y)] - \int \psi(x, y) \mu_X(dx) \mu_Y(dy) \right| \tag{2.4}$$

where μ_X and μ_Y are the distributions of X and Y. Of course if X and Y are independent then (2.4) is zero. We need the following lemma.

2.1 Lemma Let μ_1 and μ_2 be measures on $\mathscr{G}$, and let $\|\mu_1 - \mu_2\|$ denote the total variation of $\mu_1 - \mu_2$. Let $r, s \in [1, \infty]$, $r^{-1} + s^{-1} = 1$. Then for g in $L^s(\Omega, \mathscr{G}, \mu_1 + \mu_2)$,

$$\left| \int g \, d\mu_1 - \int g \, d\mu_2 \right| \le \|\mu_1 - \mu_2\|^{1/r} \left[\int |g|^s \, d\mu_1 + \int |g|^s \, d\mu_2 \right]^{1/s}. \tag{2.5}$$

Proof. Let f_i be the Radon–Nikodym derivative of μ_i with respect to $\mu_1 + \mu_2$. Then

$$\begin{aligned} \|\mu_1 - \mu_2\| &\equiv \sup_{A \in \mathscr{G}} (\mu_1(A) - \mu_2(A)) + \sup_{A \in \mathscr{G}} (\mu_2(A) - \mu_1(A)) \\ &= \int |f_1 - f_2| \, d(\mu_1 + \mu_2), \end{aligned} \tag{2.6}$$

and

$$\begin{aligned} \left| \int g \, d\mu_1 - \int g \, d\mu_2 \right| &= \left| \int g(f_1 - f_2) \, d(\mu_1 + \mu_2) \right| \\ &\le \left[\int |g|^s |f_1 - f_2| \, d(\mu_1 + \mu_2) \right]^{1/s} \left[\int |f_1 - f_2| \, d(\mu_1 + \mu_2) \right]^{1/r} \\ &\le \left[\int |g|^s \, d\mu_1 + \int |g|^s \, d\mu_2 \right]^{1/s} \|\mu_1 - \mu_2\|^{1/r}. \end{aligned} \tag{2.7}$$

□

2.2 Proposition Let $1 \le r, s, p, q \le \infty$, $r^{-1} + s^{-1} = 1$, $p^{-1} + q^{-1} = 1$. Then for real-valued Y, Z with Y in $L^q(\Omega, \mathscr{H}, P)$ and Z in $L^{sp}(\Omega, \mathscr{G}, P)$,

$$|E[ZY] - E[Z]E[Y]| \le 2^{q \wedge 2} \varphi_p^{1/r}(\mathscr{G} \mid \mathscr{H}) \|Z\|_{sp} \|Y\|_q. \tag{2.8}$$

2.3 Remark Note that for $q > 2$ we may select $s = q/p$ so that (2.8) becomes

$$(2.9) \qquad |E[ZY] - E[Z]E[Y]| \le 4\varphi_p^{(q-p)/q}(\mathscr{G} \mid \mathscr{H})\|Z\|_q\|Y\|_q. \qquad \square$$

Proof. First assume $Y \ge 0$. For $A \in \mathscr{G}$, define $\mu_1(A) = E[\chi_A Y]$ and $\mu_2(A) = P(A)E[Y]$. Then

$$(2.10) \qquad \begin{aligned} \|\mu_1 - \mu_2\| &\le 2 \sup_{A \in \mathscr{G}} |\mu_1(A) - \mu_2(A)| \\ &= 2 \sup_{A \in \mathscr{G}} |E[(E[\chi_A \mid \mathscr{H}] - P(A))Y]| \\ &\le 2 \sup_{A \in \mathscr{G}} \|P(A \mid \mathscr{H}) - P(A)\|_p\|Y\|_q \\ &= 2\varphi_p(\mathscr{G} \mid \mathscr{H})\|Y\|_q. \end{aligned}$$

By Lemma 2.1,

$$(2.11) \qquad \begin{aligned} |E[ZY] - E[Z]E[Y]| &= \left| \int Z\, d\mu_1 - \int Z\, d\mu_2 \right| \\ &\le 2^{1/r}\varphi_p^{1/r}(\mathscr{G} \mid \mathscr{H})\|Y\|_q^{1/r}(E[|Z|^s Y] + E[|Z|^s]E[Y])^{1/s} \\ &\le 2\varphi_p^{1/r}(\mathscr{G} \mid \mathscr{H})\|Z\|_{sp}\|Y\|_q, \end{aligned}$$

since both $E[|Z|^s Y]$ and $E[|Z|^s]E[Y]$ are bounded by $E[|Z|^{sp}]^{1/p}\|Y\|_q$.

For general Y, apply (2.11) to Y^+ and Y^- and add to obtain (2.8). Note that $\|Y^+\|_q + \|Y^-\|_q \le 2\|Y\|_q$ for all q, and for $q < 2$ this can be improved to $\|Y^+\|_q + \|Y^-\|_q \le 2^{q-1}\|Y\|_q$. $\square$

2.4 Corollary Let $1 \le r,\ s \le \infty$, $r^{-1} + s^{-1} = 1$. Then for real-valued Z in $L^s(\Omega, \mathscr{G}, P)$,

$$(2.12) \qquad E[|E[Z \mid \mathscr{H}] - E[Z]|] \le 8\varphi_1^{1/r}(\mathscr{G} \mid \mathscr{H})\|Z\|_s.$$

Proof. Let Y_1 be the indicator of the event $\{E[Z \mid \mathscr{H}] - E[Z] \ge 0\}$ and $Y_2 = 1 - Y_1$. Then

$$(2.13) \qquad \begin{aligned} &E[|E[Z \mid \mathscr{H}] - E[Z]|] \\ &\quad = E[E[Z \mid \mathscr{H}]Y_1 - E[Z]Y_1] - E[E[Z \mid \mathscr{H}]Y_2 - E[Z]Y_2] \\ &\quad = E[ZY_1] - E[Z]E[Y_1] + |E[ZY_2] - E[Z]E[Y_2]| \end{aligned}$$

and (2.12) follows from (2.8). $\square$

2.5 Corollary Let $1 \le u,\ v,\ w \le \infty$, $u^{-1} + v^{-1} + w^{-1} = 1$. Then for real-valued Y, Z with Y in $L^w(\Omega, \mathscr{H}, P)$ and Z in $L^v(\Omega, \mathscr{G}, P)$,

$$(2.14) \qquad |E[ZY] - E[Z]E[Y]| \le 2^{v \wedge w \wedge 2 + 1}\alpha^{1/u}(\mathscr{G}, \mathscr{H})\|Z\|_v\|Y\|_w.$$

Proof. By the symmetry of α in $\mathscr{G}$ and $\mathscr{H}$, it is enough to consider the case $w \le v$. Let $q = w$, $p = q/(q-1)$, $s = v/p$, and $r = s/(s-1)$. Note that since $v^{-1} + w^{-1} = v^{-1} + q^{-1} \le 1$ we must have $v \ge p$ and hence $s \ge 1$, and that $u = pr$.

By (2.8),

$$|E[ZY] - E[Z]E[Y]| \le 2^{w \wedge 2}\varphi_p^{p/u}(\mathscr{G} \mid \mathscr{H})\|Z\|_v\|Y\|_w. \tag{2.15}$$

Finally note that

$$\varphi_p^{p/u}(\mathscr{G} \mid \mathscr{H}) = \sup_{A \in \mathscr{G}} E[|P(A \mid \mathscr{H}) - P(A)|^p]^{1/u} \tag{2.16}$$

is a decreasing function of p. Replacing p by 1 in (2.15) and φ_1 by 2α gives (2.14). □

In the uniform mixing case ($p = \infty$) much stronger results are possible. Note that for each $A \in \mathscr{G}$

$$|P(A \mid \mathscr{H}) - P(A)| \le \varphi_\infty(\mathscr{G} \mid \mathscr{H}) \qquad \text{a.s.} \tag{2.17}$$

where $\varphi_\infty(\mathscr{G} \mid \mathscr{H})$ is, of course, a constant. We relax this requirement by assuming the existence of an $\mathscr{H}$-measurable random variable Φ such that

$$|P(A \mid \mathscr{H}) - P(A)| \le \Phi \qquad \text{a.s.} \tag{2.18}$$

for each $A \in \mathscr{G}$. (See Problem 9.) To see why this generalization is potentially useful, consider a Markov process X with values in a complete, separable metric space, transition function $P(t, x, \Gamma)$, and initial distribution ν. Let $\mathscr{G} = \mathscr{F}^{t+s} = \sigma(X(u): u \ge t + s)$ and $\mathscr{H} = \mathscr{F}_t = \sigma(X(u): u \le t)$. By the Markov property, for $A \in \mathscr{F}^{t+s}$ there is a function h_A such that $E[\chi_A \mid \mathscr{F}_{t+s}] = h_A(X_{t+s})$. Therefore for $A \in \mathscr{F}^{t+s}$,

$$\begin{aligned} &|P(A \mid \mathscr{F}_t) - P(A)| \\ &\quad = |E[h_A(X_{t+s}) \mid \mathscr{F}_t] - E[h_A(X_{t+s})]| \\ &\quad = \left| \int h_A(y)P(s, X_t, dy) - \iint h_A(y)P(t+s, x, dy)\nu(dx) \right| \\ &\quad \le \psi_{t,s}(X_t) \end{aligned} \tag{2.19}$$

where

$$\psi_{t,s}(z) = \sup_\Gamma \left| P(s, z, \Gamma) - \int P(t+s, x, \Gamma)\nu(dx) \right|. \tag{2.20}$$

For examples, see Problem 10.

2.6 Proposition Let $1 < s \leq \infty$ and $r^{-1} + s^{-1} = 1$. Suppose that Φ is $\mathscr{H}$-measurable and satisfies (2.18). Then for real-valued Z in $L^s(\Omega, \mathscr{G}, P)$,

$$(2.21) \qquad |E[Z \mid \mathscr{H}] - E[Z]| \leq 2^{1/r}\Phi^{1/r}(E[|Z|^s \mid \mathscr{H}] + E[|Z|^s])^{1/s},$$

$$(2.22) \qquad \|E[Z \mid \mathscr{H}] - E[Z]\|_s \leq 2 \max (\|\Phi^{1/r}Z\|_s, \|\Phi^{1/r}\|_s\|Z\|_s),$$

and for $1 \leq p \leq \infty$,

$$(2.23) \qquad \|E[Z \mid \mathscr{H}] - E[Z]\|_p \leq 2\|\Phi\|_p^{1/r}\|E[|Z|^s \mid \mathscr{H}]\|_p^{1/s}.$$

Proof. Fix $B \in \mathscr{H}$ with $P(B) > 0$, and take $\mu_1(A) = P(A \mid B)$ and $\mu_2(A) = P(A)$, $A \in \mathscr{G}$. Then noting that $\|\mu_1 - \mu_2\| \leq 2P(B)^{-1} \int_B \Phi \, dP$, Lemma 2.1 gives

$$(2.24) \qquad \left| P(B)^{-1} \int_B E[Z \mid \mathscr{H}] \, dP - E[Z] \right|$$
$$\leq 2^{1/r}P(B)^{-1/r}\left[\int_B \Phi \, dP\right]^{1/r}\left(P(B)^{-1} \int_B E[|Z|^2 \mid \mathscr{H}] \, dP + E[|Z|^s]\right)^{1/s}.$$

For $\alpha, \beta > 0$, let $B = \{E[Z \mid \mathscr{H}] - E[Z] > 2^{1/r}\alpha\beta, \quad \Phi \leq \alpha^r, E[|Z|^s \mid \mathscr{H}] + E[|Z|^s] \leq \beta^s\}$. If $P(B) > 0$, then (2.24) is violated. Consequently, $P(B) = 0$ for all choices of α and β, which implies

$$(2.25) \qquad E[Z \mid \mathscr{H}] - E[Z] \leq 2^{1/r}\Phi^{1/r}(E[|Z|^s \mid \mathscr{H}] + E[|Z|^s])^{1/s}.$$

A similar argument gives the estimate for $E[Z] - E[Z \mid \mathscr{H}]$. Finally (2.21) and the Hölder inequality give (2.22) and (2.23). □

2.7 Corollary Let $1 < s \leq \infty$ and $r^{-1} + s^{-1} = 1$. Suppose that Φ is $\mathscr{H}$-measurable and satisfies (2.18). Then for real-valued Y, Z, with Y in $L^r(\Omega, \mathscr{H}, P)$ and Z in $L^s(\Omega, \mathscr{G}, P)$,

$$(2.26) \qquad |E[ZY] - E[Z]E[Y]| \leq 2 \max (\|\Phi^{1/r}Z\|_s, \|\Phi^{1/r}\|_s\|Z\|_s)\|Y\|_r$$

and

$$(2.27) \qquad |E[ZY] - E[Z]E[Y]| \leq 2\|Y\Phi^{1/r}\|_r\|Z\|_s.$$

In particular

$$(2.28) \qquad |E[ZY] - E[Z]E[Y]| \leq 2\varphi_\infty^{1/r}(\mathscr{G} \mid \mathscr{H})\|Z\|_s\|Y\|_r.$$

Proof. Use (2.21) to estimate $E[(E[Z \mid \mathscr{H}] - E[Z])Y]$ and apply the Hölder inequality. □

2.8 Corollary Let $1 < s \leq \infty$ and $r^{-1} + s^{-1} = 1$. Suppose that Φ is $\mathscr{H}$-measurable and satisfies (2.18). Let E_1 and E_2 be separable. Let X be $\mathscr{G}$-measurable and E_1-valued, let Y be $\mathscr{H}$-measurable and E_2-valued, and let μ_X

and μ_Y denote the distributions of X and Y. Then for ψ in $L^s(E_1 \times E_2, \mathscr{B}(E_1 \times E_2), \mu_X \times \mu_Y)$ such that $\psi(X, Y)$ is in $L^s(\Omega, \mathscr{F}, P)$,

$$(2.29) \quad \left| E[\psi(X, Y)] - \iint \psi(x, y)\mu_X(dx)\mu_Y(dy) \right| \leq 2(E[\Phi])^{1/r} \max\left(\|\psi(X, Y)\|_s, \left[\iint |\psi(x, y)|^s \mu_X(dx)\mu_Y(dy)\right]^{1/s} \right).$$

Proof. Since we can always approximate ψ by $\psi_n = n \wedge (\psi \vee (-n))$, we may as well assume that ψ is bounded, and since the collection of bounded ψ satisfying (2.28) is bp-closed, we may as well assume that ψ is continuous (see Appendix 4). Finally, if ψ is bounded and continuous, we can obtain (2.28) for arbitrary Y by approximating by discrete Y (recall E_2 is separable), so we may as well assume that Y is discrete.

There exists a $\mathscr{B}(E_2) \times \mathscr{H}$-measurable function $\varphi(y, \omega)$ such that $E[\psi(X, y) \mid \mathscr{H}] = \varphi(y, \cdot)$ for each y and $E[\psi(X, Y) \mid \mathscr{H}] = \varphi(Y(\cdot), \cdot)$. (See Appendix 4.) By (2.21)

$$(2.30) \quad \left| \varphi(y, \cdot) - \int \psi(x, y)\mu_X(dx) \right| \leq 2^{1/r}\Phi^{1/r}\left(E[|\psi(X, y)|^s \mid \mathscr{H}] + \int |\psi(x, y)|^s \mu_X(dx) \right)^{1/s},$$

and hence, since Y is discrete,

$$(2.31) \quad \left| E[\psi(X, Y) \mid \mathscr{H}] - \int \psi(x, Y)\mu_X(dx) \right| \leq 2^{1/r}\Phi^{1/r}\left(E[|\psi(X, Y)|^s \mid \mathscr{H}] + \int |\psi(x, Y)|^s \mu_X(dx) \right)^{1/s}.$$

Taking expectations in (2.31) and applying the Hölder inequality gives (2.29). □

3. CENTRAL LIMIT THEOREMS FOR STATIONARY SEQUENCES

In this section we apply the martingale central limit theorem of Section 1 to extend the invariance principle to stationary sequences of random variables. Let $\{Y_k, k \in \mathbb{Z}\}$ be $\mathbb{R}$-valued and stationary, and define $\mathscr{F}_n = \sigma(Y_k: k \leq n)$ and $\mathscr{F}^n = \sigma(Y_k: k \geq n)$. For $m \geq 0$, let

$$(3.1) \qquad \varphi_p(m) = \varphi_p(\mathscr{F}^{n+m} \mid \mathscr{F}_n).$$

The stationarity of $\{Y_k\}$ implies that the right side of (3.1) is independent of n. (See Problem 12.)

We are interested in

$$X_n(t) = \frac{1}{\sqrt{n}} \sum_{k=1}^{[nt]} Y_k. \tag{3.2}$$

3.1 Theorem Let $\{Y_k,\ k \in \mathbb{Z}\}$ be stationary with $E[Y_k] = 0$, and for some $\delta > 0$ suppose $E[|Y_k|^{2+\delta}] < \infty$. Let $p = (2 + \delta)/(1 + \delta)$ and suppose

$$\sum_m [\varphi_p(m)]^{\delta/(1+\delta)} < \infty. \tag{3.3}$$

Then the series

$$\sigma^2 = E[Y_1^2] + 2 \sum_{k=2}^{\infty} E[Y_1 Y_k] \tag{3.4}$$

is convergent and $X_n \Rightarrow X$, where X is Brownian motion with mean zero and variance parameter σ^2.

3.2 Remark **(a)** The assumption that $\{Y_k\}$ is indexed by all $k \in \mathbb{Z}$ is a convenience. Any stationary sequence indexed by $k = 1, 2, \ldots$ has a version that is part of a stationary sequence indexed by $\mathbb{Z}$. Specifically, given $\{X_k, k \geq 1\}$, if $\{X_k\}$ is stationary, then

$$\begin{aligned} &P\{Y_{l+1} \in \Gamma_1, Y_{l+2} \in \Gamma_2, \ldots, Y_{l+m} \in \Gamma_m\} \\ &= P\{X_1 \in \Gamma_1, X_2 \in \Gamma_2, \ldots, X_m \in \Gamma_m\}, \qquad l \in \mathbb{Z},\ m = 1, 2, \ldots,\ \Gamma_k \in \mathscr{B}(\mathbb{R}), \end{aligned} \tag{3.5}$$

determines a consistent family of finite-dimensional distributions.

(b) By (2.16), for $p = (2 + \delta)/(1 + \delta)$,

$$\sum_m [\varphi_p(m)]^{\delta/(1+\delta)} \leq \sum_m [\varphi_1(m)]^{\delta/(2+\delta)}. \tag{3.6}$$

Proof. By Corollary 2.4,

$$\begin{aligned} E[|E[Y_{n+m} | \mathscr{F}_n]|] &\leq 8\varphi_1^{(1+\delta)/(2+\delta)}(m) \| Y_{n+m} \|_{2+\delta} \\ &\leq 8\varphi_p^{(1+\delta)/(2+\delta)}(m) \| Y_1 \|_{2+\delta} \\ &\leq 8\varphi_p^{\delta/(1+\delta)}(m) \| Y_1 \|_{2+\delta}. \end{aligned} \tag{3.7}$$

Consequently the sum on the right of

$$M(l) = \sum_{k=1}^{l} Y_k + \sum_{m=1}^{\infty} E[Y_{l+m} | \mathscr{F}_l] \tag{3.8}$$

is convergent, and M is a martingale.

The convergence of the series in (3.4) follows from (2.9), which gives

$$E[Y_1 Y_k] \leq 4\varphi_p^{\delta/(1+\delta)}(k-1) \| Y_1 \|_{2+\delta} \| Y_k \|_{2+\delta}. \tag{3.9}$$

Note that

$$(3.10)\qquad M(l) - M(l-1) = Y_l + \sum_{m=1}^{\infty} E[Y_{l+m} | \mathscr{F}_l] - \sum_{m=1}^{\infty} E[Y_{l-1+m} | \mathscr{F}_{l-1}]$$

$$= \sum_{m=0}^{\infty} E[Y_{l+m} | \mathscr{F}_l] - \sum_{m=0}^{\infty} E[Y_{l+m} | \mathscr{F}_{l-1}]$$

is a stationary sequence. The sequence

$$(3.11)\qquad G_N = \sum_{m=0}^{N} E[Y_{l+m} | \mathscr{F}_l] - \sum_{m=0}^{N} E[Y_{l+m} | \mathscr{F}_{l-1}]$$

converges in L^2 (check that it is Cauchy), so

$$(3.12)\qquad E[(M(l) - M(l-1))^2] = \lim_{N\to\infty} E[G_N^2]$$

$$= \lim_{N\to\infty} \left(E\left[\left(\sum_{m=0}^{N} E[Y_{l+m} | \mathscr{F}_l] \right)^2 \right] - E\left[\left(\sum_{m=0}^{N} E[Y_{l+m} | \mathscr{F}_{l-1}] \right)^2 \right] \right)$$

$$= \lim_{N\to\infty} \left(E\left[\left(\sum_{m=0}^{N} E[Y_{l+m} | \mathscr{F}_l] \right)^2 \right] - E\left[\left(\sum_{m=1}^{N+1} E[Y_{l+m} | \mathscr{F}_l] \right)^2 \right] \right)$$

$$= \lim_{N\to\infty} \Bigg(E[Y_l^2] + 2 \sum_{m=1}^{N} E[Y_l Y_{l+m}] - E[E[Y_{l+N+1} | \mathscr{F}_l]^2]$$

$$- 2 \sum_{m=1}^{N} E[E[Y_{l+N+1} | \mathscr{F}_l] E[Y_{l+m} | \mathscr{F}_l]] \Bigg)$$

$$= \sigma^2.$$

We have used the stationarity here and the fact that

$$(3.13)\qquad | E[E[Y_{l+N+1} | \mathscr{F}_l] E[Y_{l+m} | \mathscr{F}_l]] |$$

$$= | E[Y_{l+N+1} E[Y_{l+m} | \mathscr{F}_l]] |$$

$$\leq 4\varphi_p^{\delta/(1+\delta)}(N+1) \| Y_{l+N+1} \|_{2+\delta} \| Y_{l+m} \|_{2+\delta}.$$

The fact that $(N+1)\varphi_p^{\delta/(1+\delta)}(N+1) \to 0$ as $N \to \infty$ follows from (3.3) and the monotonicity of $\varphi_p(m)$. Since $\{Y_k\}$ is mixing, it is ergodic (see, e.g., Lamperti (1977), page 96), and the ergodic theorem gives

$$(3.14)\qquad \lim_{n\to\infty} \frac{1}{n} \sum_{l=1}^{[nt]} (M(l) - M(l-1))^2 = \sigma^2 t \qquad \text{a.s.}$$

Define $M_n(t) = n^{-1/2} M([nt])$. Then (3.14) gives (1.19).

To obtain (1.14), the stationarity of $\{M(l) - M(l-1)\}$ implies that for each $\varepsilon > 0$,

$$E\Big[\sup_{t \le T} |M_n(t) - M_n(t-)|\Big] \tag{3.15}$$
$$= \int_0^\infty P\Big\{\sup_{t \le T} |M_n(t) - M_n(t-)| > x\Big\}\, dx$$
$$\le \varepsilon + \int_\varepsilon^\infty [nT] P\{|M(1) - M(0)| > \sqrt{n}x\}\, dx,$$
$$\le \varepsilon + T\varepsilon^{-1} E[|M(1) - M(0)|^2 \chi_{\{|M(1)-M(0)| > \sqrt{n}\varepsilon\}}].$$

By Theorem 1.4(a), $M_n \Rightarrow X$.

Finally note that $\sup_{t \le T} |X_n(t) - M_n(t)| \to 0$ in probability by the same type of estimate used in (3.15), so $X_n \Rightarrow X$. □

Now let $\Phi_n(m)$ be a random variable satisfying

$$|P(A \mid \mathscr{F}_n) - P(A)| \le \Phi_n(m) \quad \text{a.s.} \tag{3.16}$$

for each $A \in \mathscr{F}^{n+m}$. Without loss of generality we can assume that for each m, $\{\Phi_n(m)\}$ is stationary and $\Phi_n(m) \le 1$ a.s.

3.3 Theorem Let $\{Y_k, k \in \mathbb{Z}\}$ be stationary with $E[Y_k] = 0$. Let $1 < s \le \infty$ and $s^{-1} + r^{-1} = 1$. Suppose $E[|Y_k|^{r \vee s}] < \infty$,

$$\sum_{m=1}^\infty \|\Phi_0^{1/r}(m) Y_m\|_s < \infty, \tag{3.17}$$

and

$$\sum_{m=1}^\infty \|\Phi_0^{1/r}(m)\|_s < \infty. \tag{3.18}$$

Then the series in (3.4) converges, and $X_n \Rightarrow X$, where X is Brownian motion with mean zero and variance parameter σ^2.

3.4 Remark If

$$\sum_{m=1}^\infty \varphi_\infty^{1/r}(m) < \infty, \tag{3.19}$$

then (3.17) and (3.18) hold. □

Proof. The proof is essentially the same as for Theorem 3.1 using (2.22) to estimate the left side of (3.7) and (2.26) to estimate the left sides of (3.9) and (3.13). □

4. DIFFUSION APPROXIMATIONS

We now give conditions analogous to those of Theorem 1.4 for the convergence to general diffusion processes.

4.1 Theorem Let $a = ((a_{ij}))$ be a continuous, symmetric, nonnegative definite, $d \times d$ matrix-valued function on $\mathbb{R}^d$ and let $b: \mathbb{R}^d \to \mathbb{R}^d$ be continuous. Let

$$A = \{(f, Gf \equiv \tfrac{1}{2} \sum a_{ij}\, \partial_i\, \partial_j f + \sum b_i\, \partial_i f): f \in C_c^\infty(\mathbb{R}^d)\},$$

and suppose that the $C_{\mathbb{R}^d}[0, \infty)$ martingale problem for A is well-posed.

For $n = 1, 2, \ldots$, let X_n and B_n be processes with sample paths in $D_{\mathbb{R}^d}[0, \infty)$, and let $A_n = ((A_n^{ij}))$ be a symmetric $d \times d$ matrix-valued process such that A_n^{ij} has sample paths in $D_{\mathbb{R}}[0, \infty)$ and $A_n(t) - A_n(s)$ is nonnegative definite for $t > s \geq 0$. Set $\mathscr{F}_t^n = \sigma(X_n(s), B_n(s), A_n(s): s \leq t)$.

Let $\tau_n^r = \inf\{t: |X_n(t)| \geq r \text{ or } |X_n(t-)| \geq r\}$, and suppose that

$$M_n \equiv X_n - B_n \tag{4.1}$$

and

$$M_n^i M_n^j - A_n^{ij}, \qquad i, j = 1, 2, \ldots, d, \tag{4.2}$$

are $\{\mathscr{F}_t^n\}$-local martingales, and that for each $r > 0$, $T > 0$, and $i, j = 1, \ldots, d$,

$$\lim_{n\to\infty} E\left[\sup_{t \leq T \wedge \tau_n^r} |X_n(t) - X_n(t-)|^2\right] = 0, \tag{4.3}$$

$$\lim_{n\to\infty} E\left[\sup_{t \leq T \wedge \tau_n^r} |B_n(t) - B_n(t-)|^2\right] = 0, \tag{4.4}$$

$$\lim_{n\to\infty} E\left[\sup_{t \leq T \wedge \tau_n^r} |A_n^{ij}(t) - A_n^{ij}(t-)|\right] = 0, \tag{4.5}$$

$$\sup_{t \leq T \wedge \tau_n^r} \left| B_n^i(t) - \int_0^t b_i(X_n(s))\, ds \right| \xrightarrow{p} 0, \tag{4.6}$$

and

$$\sup_{t \leq T \wedge \tau_n^r} \left| A_n^{ij}(t) - \int_0^t a_{ij}(X_n(s))\, ds \right| \xrightarrow{p} 0. \tag{4.7}$$

Suppose that $PX_n(0)^{-1} \Rightarrow \nu \in \mathscr{P}(\mathbb{R}^d)$. Then $\{X_n\}$ converges in distribution to the solution of the martingale problem for (A, ν).

Proof. Let

$$\eta_n^r = \inf\left\{t: A_n^{ii}(t) > t \sup_{|x| \leq r} a_{ii}(x) + 1 \text{ for some } i\right\}, \tag{4.8}$$

and set $\tilde{M}_n^r = M_n(\cdot \wedge \eta_n^r \wedge \tau_n^r)$. Relative compactness of $\{\tilde{M}_n^r\}$ follows as in the proof of Theorem 1.4(b), which in turn implies relative compactness for $\{X_n(\cdot \wedge \tau_n^r)\}$ since $\{B_n(\cdot \wedge \tau_n^r)\}$ is relatively compact by (4.6). Fix $r_0 > 0$ and let $\{X_{n_k}(\cdot \wedge \tau_n^{r_0})\}$ be a convergent subsequence with limit X^{r_0}. For all but countably many $r < r_0$,

$$(X_{n_k}(\cdot \wedge \tau_n^r), \tau_n^r) \Rightarrow (X^{r_0}(\cdot \wedge \tau^r), \tau^r),$$

where $\tau^r = \inf\{t: |X^{r_0}(t)| \geq r\}$ (i.e., for all $r < r_0$ such that $P\{\lim_{s \to r} \tau^s = \tau^r\} = 1$).

Again as in the proof of Theorem 1.4(b),

$$M^{r_0}(t \wedge \tau_r) \equiv X^{r_0}(t \wedge \tau^r) - \int_0^{t \wedge \tau^r} b(X^{r_0}(s))\, ds \tag{4.9}$$

and

$$M_i^{r_0}(t \wedge \tau^r) M_j^{r_0}(t \wedge \tau^r) - \int_0^{t \wedge \tau^r} a_{ij}(X^{r_0}(s))\, ds \tag{4.10}$$

are martingales, and by Itô's formula, Theorem 2.9 of Chapter 5,

$$f(X^{r_0}(t \wedge \tau^r)) - \int_0^{t \wedge \tau^r} Gf(X^{r_0}(s))\, ds \tag{4.11}$$

is a martingale for each $f \in C_c^\infty(\mathbb{R}^d)$. Uniqueness for the martingale problem for (A, ν) implies uniqueness for the stopped martingale problem for $(A, \nu, \{x: |x| < r\})$. Consequently, if X is a solution of the martingale problem for (A, ν), then $X_n(\cdot \wedge \tau_n^r) \Rightarrow X(\cdot \wedge \tau^r)$ for all r such that $P\{\lim_{s \to r} \tau^s = \tau^r\} = 1$ (here $\tau^r = \inf\{t: |X(t)| \geq r\}$). But $\tau_r \to \infty$ as $r \to \infty$ (since X has sample paths in $C_{\mathbb{R}^d}[0, \infty)$), so $X_n \Rightarrow X$. □

4.2 Corollary Let a, b, and A be as in Theorem 4.1, and suppose the martingale problem for (A, ν) has a unique solution for each $\nu \in \mathcal{P}(\mathbb{R}^d)$. Let $\mu_n(x, \Gamma)$, $n = 1, 2, \ldots$, be a transition function on $\mathbb{R}^d$, and set

$$b_n(x) = n \int_{|y - x| \leq 1} (y - x) \mu_n(x, dy) \tag{4.12}$$

and

$$a_n(x) = n \int_{|y - x| \leq 1} (y - x)(y - x)^T \mu_n(x, dy). \tag{4.13}$$

Suppose for each $r > 0$ and $\varepsilon > 0$,

$$\sup_{|x| \leq r} |a_n(x) - a(x)| \to 0, \tag{4.14}$$

$$\sup_{|x| \leq r} |b_n(x) - b(x)| \to 0, \tag{4.15}$$

and

(4.16)
$$\sup_{|x| \le r} n\mu_n(x, \{y: |y - x| \ge \varepsilon\}) \to 0.$$

Let Y_n be a Markov chain with transition function $\mu_n(x, \Gamma)$ and define $X_n(t) = Y_n([nt])$. If $PY_n(0)^{-1} \Rightarrow \nu$, then $\{X_n\}$ converges in distribution to the solution of the martingale problem for (A, ν).

Proof. Let τ_n^r be as in the proof of Theorem 4.1, and let $\gamma_n = \inf\{t: |X_n(t) - X_n(t-)| > 1\}$. Then (4.16) implies $P\{\gamma_n < \tau_n^r \wedge T\} \to 0$ for each $r > 0$ and $T > 0$. Therefore (see Problem 13), we may as well assume $\mu_n(x, \{y: |y - x| > 1\}) = 0$. Let

(4.17)
$$B_n(t) = \int_0^{[nt]/n} b_n(X_n(s))\, ds$$

and

(4.18)
$$A_n(t) = \int_0^{[nt]/n} (a_n(X_n(s)) - n^{-1} b_n(X_n(s)) b_n^T(X_n(s)))\, ds,$$

and (4.3)–(4.7) follow easily. □

5. STRONG APPROXIMATION THEOREMS

In this section we present, without proof, strong approximation theorems of Komlós, Major, and Tusnády for sums of independent identically distributed random variables. We obtain as a corollary a result on the approximation of the Poisson process by Brownian motion. To understand the significance of a strong approximation theorem, it may be useful to recall Theorem 1.2 of Chapter 3. This theorem can be restated to say that if $\mu, \nu \in \mathscr{P}(S)$ and $\rho(\mu, \nu) < \varepsilon$, then there exist a probability space $(\Omega, \mathscr{F}, P)$ and random variables X and Y defined on $(\Omega, \mathscr{F}, P)$, X with distribution μ, Y with distribution ν, such that $P\{d(X, Y) \ge \varepsilon\} \le \varepsilon$.

5.1 Theorem Let $\mu \in \mathscr{P}(\mathbb{R})$ satisfy $\int e^{\alpha x}\mu(dx) < \infty$ for $|\alpha| \le \alpha_0$, some $\alpha_0 > 0$. Then there exist a sequence of independent identically distributed random variables $\{\xi_k\}$ with distribution μ, a Brownian motion W with $m \equiv E[W(1)] = E[\xi_1]$ and $\sigma^2 \equiv \operatorname{var}(W(1)) = \operatorname{var}(\xi_1)$ (defined on the same sample space), and positive constants C, K, and λ depending only on μ, such that

(5.1)
$$P\left\{\max_{1 \le k \le n} |S_k - W(k)| > C \log n + x\right\} < Ke^{-\lambda x}$$

for each $n \ge 1$ and $x > 0$, where $S_k = \sum_{i=1}^k \xi_i$.

Proof. See Komlós, Major, and Tusnády (1975, 1976). □

5.2 Corollary Let $\{\xi_k\}$ and W be as in Theorem 5.1. Set $X_n(t) = n^{-1/2} \sum_{k=1}^{[nt]} (\xi_k - E[\xi_k])$ and $W_n(t) = n^{-1/2}(W(nt) - E[W(nt)])$. (Note that $W_n(t)$ is a Brownian motion with mean zero and var $(W_n(t)) = t$ var (ξ_1).) Then there exist positive constants C, K, γ, and λ, depending only on μ, such that for $T \geq 1$, $n \geq 1$, and $x > 0$,

$$P\left\{\sup_{t \leq T} |X_n(t) - W_n(t)| > Cn^{-1/2} \log n + x\right\} \leq KT^\gamma e^{-\lambda\sqrt{nx}}. \tag{5.2}$$

It follows that there exists a $\beta > 0$ such that for $n \geq 2$, $\rho(PX_n^{-1}, PW_n^{-1}) \leq \beta n^{-1/2} \log n$, where ρ is the Prohorov metric on $\mathscr{P}(D_{\mathbb{R}}[0, \infty))$.

Proof. Let C_1, K_1, and λ_1 be the C, K, and λ guaranteed by Theorem 5.1 and set $C = 2C_1$. Then defining $\tilde{W}(t) = W(t) - t$, the left side of (5.2) is bounded by

$$\begin{aligned} &P\left\{\sup_{k \leq nT} |S_k - W(k)| > C_1 \log [nT] - C_1 \log T + \frac{\sqrt{nx}}{2}\right\} \\ &\quad + P\left\{\sup_{k \leq nT} \sup_{0 \leq s \leq 1} |\tilde{W}(k+s) - \tilde{W}(k)| > C_1 \log n + \frac{\sqrt{nx}}{2}\right\}. \end{aligned} \tag{5.3}$$

The second term in (5.3) is bounded by

$$nTP\left\{\sup_{0 \leq s \leq 1} |\tilde{W}(s)| > C_1 \log n + \frac{\sqrt{nx}}{2}\right\}, \tag{5.4}$$

and for any $\alpha > 0$,

$$P\left\{\sup_{0 \leq s \leq 1} |\tilde{W}(s)| > z\right\} \leq 2e^{(1/2)\alpha^2\sigma^2} e^{-\alpha z} \tag{5.5}$$

(see Problem 17). Selecting $\alpha > \lambda_1$ so that $\alpha C_1 > 1$, (5.2) is bounded by

$$\begin{aligned} &K_1 \exp\left\{-\lambda_1\left(-C_1 \log T + \frac{\sqrt{nx}}{2}\right)\right\} + K_2 T \exp\left\{\frac{-\lambda_1\sqrt{nx}}{2}\right\} \\ &\qquad \leq KT^\gamma \exp\{-\lambda\sqrt{nx}\}, \end{aligned} \tag{5.6}$$

for $\gamma = (\lambda_1 C_1) \vee 1$, $\lambda = \lambda_1/2$, and $K = K_1 + K_2$.

For $a_k > 0$, $k = 1, 2, \ldots$, with $\sum_k a_k = 1$, $A > 0$ to be determined, and $n \geq 2$,

$$(5.7)\qquad \begin{aligned} P\{d(X_n, W_n) > An^{-1/2} \log n\} &\leq P\left\{\int_0^\infty e^{-t} \sup_{s \leq t} |X_n(s) - W_n(s)|\, dt > An^{-1/2} \log n\right\} \\ &\leq \sum_k P\left\{\int_{k-1}^k e^{-t} \sup_{s \leq t} |X_n(s) - W_n(s)|\, dt > a_k An^{-1/2} \log n\right\} \\ &\leq \sum_k P\left\{\sup_{s \leq k} |X_n(s) - W_n(s)| > e^{k-1} a_k An^{-1/2} \log n\right\} \\ &\leq \sum_k Kk^\gamma \exp\{-\lambda(e^{k-1} a_k A - C) \log n\} \\ &\leq n^{-1} \sum_k Kk^\gamma \exp\{-\lambda(e^{k-1} a_k A - C - \lambda^{-1}) \log 2\}, \end{aligned}$$

provided $e^{k-1} a_k A - C - \lambda^{-1} > 0$ and the sum is finite. Note the a_k and A can be selected so these conditions are satisfied. Finally, select $\beta \geq A$ so that $\beta n^{-1/2} \log n$ bounds the right side for all $n \geq 2$. □

5.3 Corollary Let $\mu \in \mathscr{P}(\mathbb{R})$ be infinitely divisible and $\int e^{\alpha x} \mu(dx) < \infty$ for $|\alpha| \leq \alpha_0$, some $\alpha_0 > 0$. Then there exists a process X with stationary independent increments, $X(1)$ with distribution μ, a Brownian motion W with the same mean and variance as X, and positive constants C, K, and λ depending only on μ such that for $T \geq 1$ and $x > 0$,

$$(5.8)\qquad P\left\{\sup_{t \leq T} |X(t) - W(t)| > C \log T + x\right\} \leq Ke^{-\lambda x}.$$

5.4 Remark Note that if we replace x by $x + \gamma \log T$, then (5.8) becomes

$$(5.9)\qquad P\left\{\sup_{t \leq T} |X(t) - W(t)| > (C + \gamma) \log T + x\right\} \leq KT^{-\gamma\lambda} e^{-\lambda x}.$$ □

Proof. Let $\{\xi_k\}$ and W be as in Theorem 5.1. (Note that the C, K, and λ of the corollary differ from those of the theorem.) Let $\{X_k\}$ be independent processes with stationary independent increments with the distribution of $X_k(1)$ being μ. Since the distribution on $\mathbb{R}^\infty$ of $\{X_k(1)\}$ is the same as the distribution of $\{\xi_k\}$, by Lemma 5.15 of Chapter 4 we may assume $\{X_k\}$, $\{\xi_k\}$, and W are defined on the same sample space and that $X_k(1) = \xi_k$. Finally define

$$(5.10)\qquad X(t) = \sum_{i=1}^{k-1} \xi_i + X_k(t - k + 1), \qquad k - 1 \leq t < k,$$

and note that the left side of (5.8) is bounded by

$$P\left\{\max_{k\le T}|S_k - W(k)| > 3^{-1}C\log[T] + 3^{-1}x\right\} \tag{5.11}$$
$$+ P\left\{\max_{k\le T}\sup_{s\le 1}|\tilde{X}_k(s)| > 3^{-1}C\log T + 3^{-1}x\right\}$$
$$+ P\left\{\max_{k\le T}\sup_{s\le 1}|\tilde{W}(s+k) - \tilde{W}(k)| > 3^{-1}C\log T + 3^{-1}x\right\}$$

where $\tilde{W}(t) = W(t) - E[W(t)]$ and $\tilde{X}(t) = X(t) - E[X(t)]$. The result follows from (5.1) and Problem 17. □

5.5 Corollary Let X and W be as in Corollary 5.3. Then

$$\sup_t \frac{|X(t) - W(t)|}{\log(2 \vee t)} < \infty \quad \text{a.s.} \tag{5.12}$$

Proof. Take $\gamma = 1$ in (5.9). Then

$$P\left\{\sup_t \frac{|X(t) - W(t)|}{\log(2\vee t)} > (C+1)2\log 2 + \frac{x}{\log 2}\right\} \tag{5.13}$$
$$\le \sum_{n=2}^{\infty} P\left\{\sup_{t\le 2^n}\frac{|X(t) - W(t)|}{\log(2^{n-1}\vee t)} > \frac{(C+1)\log 2^n}{\log 2^{n-1}} + \frac{x}{\log 2}\right\}$$
$$\le \sum_{n=2}^{\infty} P\left\{\sup_{t\le 2^n}|X(t) - W(t)| > (C+1)\log 2^n + x\right\}$$
$$\le K(1 - 2^{-\lambda})^{-1}e^{-\lambda x}.$$
□

The construction in Corollary 5.5 is best possible in the following sense.

5.6 Theorem Suppose X is a process with stationary independent increments, sample paths in $D_{\mathbb{R}}[0, \infty)$, and $X(0) = 0$ a.s., W is a Brownian motion, and

$$\lim_{t\to\infty} \frac{|X(t) - W(t)|}{\log(2\vee t)} = 0 \quad \text{a.s.} \tag{5.14}$$

Then X is a Brownian motion with the same mean and variance as W.

Proof. See Bártfai (1966). □

6. PROBLEMS

1. (a) Let N be a counting process (i.e., N is right continuous and constant except for jumps of $+1$) with $N(0) = 0$. Suppose that C is a continuous nondecreasing function with $C(0) = 0$ and that $N(t) - C(t)$ is a martingale. Show that N has independent Poisson distributed increments, and that the distribution of N is uniquely determined.

(b) Let $\{N_n\}$ be a sequence of counting processes, with $N_n(0) = 0$, and let A_n, $n = 1, 2, \ldots$, be a process with nondecreasing sample paths such that $\sup_t (A_n(t) - A_n(t-)) \le 1$ and $N_n - A_n$ is a martingale. Let C be a continuous nondecreasing function with $C(0) = 0$, and suppose for each $t \ge 0$ that $A_n(t) \xrightarrow{p} C(t)$. Show that $N_n \Rightarrow N$ where N is the process characterized in part (a).

Remark. In regard to the condition $\sup_t (A_n(t) - A_n(t-)) \le 1$, consider the following example. Let Y_1, Y_2, and A be independent processes, Y_1 and Y_2 Poisson with parameter one and A nondecreasing. Let $N_n(t) = Y_1(A(t))$ and $A_n(t) = nY_2(A(t)/n)$. Then $N_n - A_n$ is a martingale and $A_n(t) \to 0$ in probability.

2. Let W_1 and W_2 be independent standard Brownian motions and define

$$X(t) = \begin{pmatrix} W_1\left(\frac{t}{3}\right) - W_2\left(\frac{2t}{3}\right) \\ W_1\left(\frac{2t}{3}\right) + W_2\left(\frac{t}{3}\right) \end{pmatrix}. \tag{6.1}$$

Show that for each $\theta \in \mathbb{R}^2$, $\theta \cdot X(t)$ is Brownian motion with variance $|\theta|^2 t$ and hence

$$f(\theta \cdot X(t)) - \int_0^t \tfrac{1}{2}|\theta|^2 f''(\theta \cdot X(s))\, ds \tag{6.2}$$

is a martingale with respect to $\mathscr{F}_t^{\theta \cdot X} = \sigma(\theta \cdot X(s)\colon s \le t)$ (cf. Theorem 1.2), but that (6.2) is not (in general) an $\{\mathscr{F}_t^X\}$-martingale.

3. Let N be a Poisson process with parameter 1, and define $V(t) = (-1)^{N(t)}$. For $n = 1, 2, \ldots$, let

$$X_n(t) = n^{-1} \int_0^{n^2 t} V(s)\, ds. \tag{6.3}$$

Show that

$$M(t) = V(t) + 2\int_0^t V(s)\, ds \tag{6.4}$$

is a martingale and use this fact to show $X_n \Rightarrow W$ where W is standard Brownian motion.

4. Develop and prove the analogue of the result in Problem 3 in which V is an Ornstein–Uhlenbeck process, that is, V is a diffusion in $\mathbb{R}$ with generator $Af = \frac{1}{2}af'' - bxf'$, $a, b > 0$, $f \in C_c^\infty(\mathbb{R})$.

5. Let $\xi_1, \xi_2, \ldots$ be independent and identically distributed with $\xi_k \geq 0$ a.s., $E[\xi_k] = \mu > 0$, and $\text{var}\,(\xi_k) = \sigma^2 < \infty$. Let $N(t) = \max\,\{k\colon \sum_{i=1}^k \xi_i \leq t\}$, and define

$$X_n(t) = n^{-1/2}\left(N(nt) - \frac{nt}{\mu}\right). \tag{6.5}$$

(a) Show that

$$M(t) = \sum_{k=1}^{N(t)+1} \xi_k - (N(t) + 1)\mu \tag{6.6}$$

is a martingale.

(b) Apply Theorem 1.4 to show that $X_n \Rightarrow X$, where X is a Brownian motion with mean zero and variance parameter σ^2/μ^3.

6. Let E be the unit sphere in $\mathbb{R}^3$. Let $\mu(x, \Gamma)$ be a transition function on $E \times \mathscr{B}(E)$ satisfying

$$\int y\mu(x, dy) = \rho x, \qquad x \in E, \tag{6.7}$$

for some $\rho \in (-1, 1)$.

Define $Tf(x) = \int f(y)\mu(x, dy)$. Suppose there exists $\nu \in \mathscr{P}(E)$ such that

$$\lim_{n\to\infty} n^{-1} \sum_{k=1}^{n} T^n f(x) = \int f\, d\nu \tag{6.8}$$

for each $x \in E$ and $f \in C(E)$. Let $\{Y(k),\ k = 0, 1, 2, \ldots\}$ be a Markov chain with transition function $\mu(x, \Gamma)$ and define

$$X_n(t) = \frac{1}{\sqrt{n}} \sum_{k=1}^{[nt]} Y_k. \tag{6.9}$$

Show that

$$M_n(t) = X_n(t) + \rho(1 - \rho)^{-1} Y_{[nt]}/\sqrt{n} \tag{6.10}$$

is a martingale, and use Theorem 1.4 to prove $X_n \Rightarrow X$ where X is a three-dimensional Brownian motion with mean zero and covariance

$$C(t) = t\,\frac{1+\rho}{1-\rho}\left(\left(\int y_i y_j\, \nu(dy)\right)\right). \tag{6.11}$$

7. For $n = 1, 2, \ldots$, let X_n be a process with sample paths in $D_E[0, \infty)$, and let M_n be a process with sample paths in $D_{\mathbb{R}}[0, \infty)$. Suppose that M_n is an $\{\mathscr{F}_t^{X_n}\}$-martingale, and that $(X_n, M_n) \Rightarrow (X, M)$. Show that if M is $\{\mathscr{F}_t^X\}$-adapted and for each $t \geq 0$ $\{M_n(t)\}$ is uniformly integrable, then M is an $\{\mathscr{F}_t^X\}$-martingale.

8. Verify the identities in (2.1) and (2.2).

9. Let Γ be a collection of nonnegative random variables, and suppose there is a random variable η such that for each $\xi \in \Gamma$, $\xi \leq \eta$ a.s. Show that there is a minimal such η, that is, show that there is a random variable η_0 such that for each $\xi \in \Gamma$, $\xi \leq \eta_0$ a.s. and there exist $\xi_i \in \Gamma$, $i = 1, 2, \ldots$, such that $\eta_0 = \sup_i \xi_i$.

   Hint: First assume $E[\eta] < \infty$ and consider

$$\sup_{\{\xi_i\} \subset \Gamma} E\left[\sup_i \xi_i\right]. \tag{6.12}$$

10. (a) Let $E = \{1, 2, \ldots, d\}$. Let $\{Y(k), k = 0, 1, 2, \ldots\}$ be a Markov chain in E with transition matrix $P = ((p_{ij}))$, that is, $P\{Y(k+1) = j \mid Y(k) = i\} = p_{ij}$. Suppose P is irreducible and aperiodic, that is, P^k has all elements positive for k sufficiently large. Let $\mathscr{F}_n = \sigma\{Y(k)\colon k \leq n\}$ and $\mathscr{F}^n = \sigma\{Y(k)\colon k \geq n\}$, and define $\varphi(m) = \sup_n \varphi_\infty(\mathscr{F}^{n+m} \mid \mathscr{F}_n)$. Show that $\lim_{m\to\infty} m^{-1} \log \varphi(m) = \alpha < 0$ exists, and characterize α in terms of P.

    (b) Let X be an Ornstein–Uhlenbeck process with generator $Af = \frac{1}{2}af'' - bxf'$, $a, b > 0$. Suppose $PX(0)^{-1} = \nu$ is the stationary distribution for X. Compute $\psi_{t,s} = \psi_{0,s}$ given by (2.20).

    (c) Let X be a reflecting Brownian motion on $[0, 1]$ with generator $Af = \frac{1}{2}\sigma^2 f''$. Suppose $PX(0)^{-1} = m$ (Lebesgue measure) and compute $\psi_{t,s} = \psi_{0,s}$ given by (2.20).

11. For $n = 1, 2, \ldots$, let $\{\xi_k^n, k = 1, 2, \ldots\}$ be a stationary sequence of random variables with $P\{\xi_k^n = 1\} = p_n$ and $P\{\xi_k^n = 0\} = 1 - p_n$. Let $\mathscr{F}_k^n = \sigma(\xi_i^n\colon i \leq k)$, $\mathscr{G}_k^n = \sigma(\xi_i^n\colon i \geq k)$, and define $\varphi_p^n(m) = \sup_k \varphi_p(\mathscr{G}_{k+m}^n \mid \mathscr{F}_k^n)$.

    Suppose $np_n \to \lambda > 0$ and $\max_{k \leq n} P\{\xi_{k+1}^n = 1 \mid \mathscr{F}_k^n\} \overset{p}{\to} 0$ as $n \to \infty$, and define

$$N_n(t) = \sum_{k=1}^{[nt]} \xi_k^n. \tag{6.13}$$

    Give conditions on $\varphi_p^n(m)$ that imply $N_n \Rightarrow N$, where N is a Poisson process with parameter λ.

12. Let $\{Y_k, k \in \mathbb{Z}\}$ be stationary and define $\mathscr{F}_n^0 = \sigma(Y_k\colon 1 \leq k \leq n)$ and $\mathscr{F}^n = \sigma(Y_k\colon k \geq n)$. Show that for each m, $\varphi_p(\mathscr{F}^{n+m} \mid \mathscr{F}_n^0)$ is a nondecreasing function of n and $\varphi_p(m) = \lim_{n\to\infty} \varphi_p(\mathscr{F}^{n+m} \mid \mathscr{F}_n^0)$, where $\varphi_p(m)$ is given by (3.1).

13. For $n = 1, 2, \ldots$, let $\mu_n(x, \Gamma)$ and $\nu_n(x, \Gamma)$ be transition functions on $\mathbb{R}^d \times \mathscr{B}(\mathbb{R}^d)$. Suppose $\mu_n(x, \Gamma \cap B(x, 1)) = \nu_n(x, \Gamma \cap B(x, 1))$, $x \in \mathbb{R}^d$, $\Gamma \in \mathscr{B}(\mathbb{R}^d)$, and $\lim_{n\to\infty} \sup_x n\mu_n(x, B(x, 1)^c) = 0$. Show that for each $\nu \in \mathscr{P}(\mathbb{R}^d)$ there exist Markov chains $\{Y_n(k), k = 0, 1, 2, \ldots\}$ and $\{Z_n(k), k = 0, 1, 2, \ldots\}$ with $PY_n(0)^{-1} = PX_n(0)^{-1} = \nu$, such that Y_n corresponds to $\mu_n(x, \Gamma)$ and Z_n corresponds to $\nu_n(x, \Gamma)$, and for each $K > 0$, $\lim_{n\to\infty} P\{Y_n(k) \neq Z_n(k) \text{ for some } k \leq nK\} = 0$.

14. For $n = 1, 2, \ldots$ let Y_n be a Markov chain in $E_n = \{k/n: k = 0, 1, \ldots, n\}$ with transition function given by

$$P\left\{Y_n(k+1) = \frac{j}{n} \,\middle|\, Y_n(k) = x\right\} = \binom{n}{j} x^j(1-x)^{n-j}. \tag{6.14}$$

Apply Corollary 4.2 to obtain a diffusion approximation for $Y_n([nt])$ (cf. Chapter 10).

15. Let $E = \{0, 1, 2, \ldots\}$, and let Z_n be a continuous-time branching process with generator

$$Af(k) = \lambda k \sum_{l=1}^{\infty} p_l(f(k+l-1) - f(k)) \tag{6.15}$$

for $f \in C_c(E)$, where $p_l \geq 0$, $l = 0, 1, 2, \ldots$, $\lambda > 0$, and $\sum_{l=1}^{\infty} p_l = 1$. Define $X_n(t) = Z_n(nt)/n$, and assume $PX_n(0)^{-1} \Rightarrow \nu \in \mathscr{P}([0, \infty))$. Suppose $\sum_{l=0}^{\infty} lp_l = 1$ and $\sum_{l=0}^{\infty} l^2 p_l < \infty$. Apply Theorem 4.1 to obtain a diffusion approximation for X_n (cf. Chapter 9).

16. Let N_1 and N_2 be independent Poisson processes and let $F, G \in C^1(\mathbb{R})$. Apply Theorem 4.1 to obtain a diffusion approximation for X_n satisfying

$$\dot{X}_n(t) = (-1)^{N_1(n^2t)} nF(X_n(t)) + (-1)^{N_2(n^2t)} nG(X_n(t)) \tag{6.16}$$

(cf. Chapter 12).

Hint: Find the analogue of the martingale defined in (6.4).

17. Let X be a process with stationary independent increments satisfying $E[X(t)] = 0$ for $t \geq 0$, and suppose there exists $\alpha_0 > 0$ such that $e^{\psi(\alpha)} \equiv E[e^{\alpha X(1)}] < \infty$ for $|\alpha| \leq \alpha_0$. Show that $\exp\{\alpha X(t) - t\psi(\alpha)\}$ is a martingale for each α, $|\alpha| \leq \alpha_0$, and that for $0 < \alpha \leq \alpha_0$

$$P\left\{\sup_{s \leq t} |X(s)| \geq x\right\} \leq [\exp\{t\psi(\alpha)\} + \exp\{t\psi(-\alpha)\}] \exp\{-\alpha x\}. \tag{6.17}$$

7. NOTES

The invariance principle for independent random variables was given by Donsker (1951). For discrete time, the fact that the conditions of Theorem 1.4(b) with A_n given by (1.22) imply asymptotic normality was observed by Lévy (see Doob (1953), page 383) and developed by Dvoretzky (1972). Brown (1971) gave the corresponding invariance principle. McLeish (1974) gave the discrete-time version of Theorem 1.4(a). Various authors have extended and refined these results, for example Rootzen (1977, 1980) and Gänsler and Häusler (1979). Rebolledo (1980) extended the results to continuous time. The version we have presented is not quite the most general. See also Hall and Heyde (1980) and the survey article by Helland (1982).

Uniform mixing was introduced by Ibragimov (1959) and strong mixing by Rosenblatt (1956). For $p = r = 1$, (2.8) is due to Volkonskii and Rozanov (1959), and (2.14) is due to Davydov (1968). For $\Phi = \varphi_\infty(\mathscr{G} \mid \mathscr{H})$, (2.26) appears in Ibragimov (1962). A variety of other mixing conditions are discussed in Withers (1981) and Peligrad (1982).

A vast literature exists on central limit theorems and invariance principles under mixing conditions. See Hall and Heyde (1980), Chapter 5, for a recent survey and Ibragimov and Linnik (1971). Central limit theorems under the hypotheses of Theorems 3.1 and 3.3 (assuming (3.19)) were given by Ibragimov (1962). Weak convergence assuming (3.19) was established by Billingsley (1968). The proof given here is due to Heyde (1974).

Theorem 4.1, in the form given here, is due to Rebolledo (1979). Corollary 4.2 is due to Stroock and Varadhan (1969). See Stroock and Varadhan (1979), page 266. Skorohod (1965), Borovkov (1970), and Kushner (1974) give other approaches to diffusion approximations.

Theorem 5.1 is due to Komlós, Major, and Tusnády (1975, 1976). See also Major (1976) and Csörgö and Révész (1981). Theorem 5.6 is due to Bártfai (1966).

The characterization of the Poisson process given in Problem 1(a) is due to Watanabe (1964). Various authors have given results along the lines of Problem 1(b), Brown (1978), Kabanov, Lipster, and Shiryaev (1980), Grigelionis and Mikulevičios (1981), and Kurtz (1982).

The example in Problem 2 is due to Hardin (1985).

There is also a vast literature on central limit theorems and related invariance principles for Markov processes (Problems 3, 4, and 6). The martingale approach to these results has been taken by Maigret (1978), Bhattacharya (1982), and Kurtz (1981b).

8 EXAMPLES OF GENERATORS

The purpose of this chapter is to list conditions under which certain linear operators, corresponding (at least intuitively) to specific Markov processes, generate Feller semigroups or have the property that the associated martingale problem is well-posed. In contrast to other chapters, here we reference other sources wherever possible.

Section 1 contains results for nondegenerate diffusions. These include classical, one-dimensional diffusions with local boundary conditions, diffusions in bounded regions with absorption or oblique reflection at the boundary, diffusions in $\mathbb{R}^d$ with Hölder continuous coefficients, and diffusions in $\mathbb{R}^d$ with continuous, time-dependent coefficients.

Section 2 concerns degenerate diffusions. Results are given for one-dimensional diffusions with smooth coefficients and with Lipschitz continuous, time-dependent coefficients, diffusions in $\mathbb{R}^d$ with smooth diffusion matrix and with diffusion matrix having a Lipschitz continuous, time-dependent square root, and a class of diffusions in a subset of $\mathbb{R}^d$ occurring in population genetics.

In Section 3, other processes are considered. Results are included for jump processes with unbounded generators, processes with Lévy generators including independent-increment processes, and two classes of infinite particle systems, namely, spin-flip systems and exclusion processes.

1. NONDEGENERATE DIFFUSIONS

We begin with the one-dimensional case, where one can explicitly characterize the generator of the semigroup. Given $-\infty \le r_0 < r_1 \le \infty$, let I be the closed interval in $\mathbb{R}$ with endpoints r_0 and r_1, I° its interior, and $\bar{I}$ its closure in $[-\infty, \infty]$. In other words,

$$I = [r_0, r_1] \cap \mathbb{R}, \qquad I^\circ = (r_0, r_1), \qquad \bar{I} = [r_0, r_1]. \tag{1.1}$$

We identify $C(\bar{I})$ with the space of $f \in C(I^\circ)$ for which $\lim_{x \to r_i} f(x)$ exists and is finite for $i = 0, 1$. Suppose $a, b \in C(I^\circ)$ and $a > 0$ on I°. Then there is at least one restriction of

$$G = \tfrac{1}{2}a(x)\frac{d^2}{dx^2} + b(x)\frac{d}{dx} \tag{1.2}$$

acting on $\{f \in C(\bar{I}) \cap C^2(I^\circ): Gf \in C(\bar{I})\}$ that generates a Feller semigroup on $C(\bar{I})$.

To specify the appropriate restrictions, we need to introduce Feller's boundary classification. Fix $r \in (r_0, r_1)$ and define $B, m, p \in C(I^\circ)$ by

$$B(x) = \int_r^x 2a(y)^{-1}b(y)\,dy, \tag{1.3}$$

$$p(x) = \int_r^x e^{-B(y)}\,dy, \qquad m(x) = \int_r^x 2a(y)^{-1}e^{B(y)}\,dy, \tag{1.4}$$

so that

$$G = \tfrac{1}{2}a(x)e^{-B(x)}\frac{d}{dx}\left(e^{B(x)}\frac{d}{dx}\right) = \frac{d}{dm(x)}\left(\frac{d}{dp(x)}\right). \tag{1.5}$$

Define u and v on $\bar{I}$ by

$$u(x) = \int_r^x m\,dp, \qquad v(x) = \int_r^x p\,dm. \tag{1.6}$$

Then, for $i = 0, 1$, the boundary r_i is said to be

$$\begin{array}{llll} \textit{regular} & \text{if } u(r_i) < \infty & \text{and} & v(r_i) < \infty, \\ \textit{exit} & \text{if } u(r_i) < \infty & \text{and} & v(r_i) = \infty, \\ \textit{entrance} & \text{if } u(r_i) = \infty & \text{and} & v(r_i) < \infty, \\ \textit{natural} & \text{if } u(r_i) = \infty & \text{and} & v(r_i) = \infty. \end{array} \tag{1.7}$$

Regular and exit boundaries are said to be *accessible*; entrance and natural boundaries are termed *inaccessible*. See Problem 1 for some motivation for this terminology.

Let

$$\mathscr{D} = \{f \in C(\bar{I}) \cap C^2(I^\circ): Gf \in C(\bar{I})\}, \tag{1.8}$$

and for $i = 0, 1$, define

$$\mathscr{D}_i = \mathscr{D}, \qquad r_i \text{ inaccessible}, \tag{1.9}$$

$$\mathscr{D}_i = \left\{f \in \mathscr{D}: \lim_{x \to r_i} Gf(x) = 0\right\}, \qquad r_i \text{ exit}, \tag{1.10}$$

and

$$\mathscr{D}_i = \left\{f \in \mathscr{D}: q_i \lim_{x \to r_i} Gf(x) = (-1)^i(1 - q_i) \lim_{x \to r_i} e^{B(x)} f'(x)\right\}, \qquad r_i \text{ regular}, \tag{1.11}$$

where $q_i \in [0, 1]$ and B is given by (1.3).

1.1 Theorem Given $-\infty \le r_0 < r_1 \le \infty$, define I, I°, and $\bar{I}$ by (1.1). Suppose $a, b \in C(I^\circ)$ with $a > 0$ on I°, and define G, $\mathscr{D}_0$, and $\mathscr{D}_1$ by (1.2) and (1.8)–(1.11), where $q_i \in [0, 1]$ is fixed if r_i is regular, $i = 0, 1$. Then $\{(f, Gf): f \in \mathscr{D}_0 \cap \mathscr{D}_1\}$ generates a Feller semigroup on $C(\bar{I})$.

Proof. See Mandl (1968) (except for the exit case). □

1.2 Corollary Suppose, in addition to the hypotheses of Theorem 1.1, that infinite boundaries of I are natural. Then $\{(f, Gf): f \in \hat{C}(I) \cap \mathscr{D}_0 \cap \mathscr{D}_1, Gf \in \hat{C}(I)\}$ generates a Feller semigroup on $\hat{C}(I)$.

Proof. See Problem 1. □

1.3 Remark We note a simple, sufficient (but not necessary) condition for the extra hypothesis in Corollary 1.2. If there exists a constant K such that

$$a(x) \le K(1 + x^2), \; |b(x)| \le K(1 + |x|), \qquad x \in I^\circ, \tag{1.12}$$

then infinite boundaries of I are natural. The proof is left to the reader (Problem 2). □

For some applications, it is useful to find a core (such as $C_c^\infty(I)$) for the generator in Corollary 1.2. In Section 2 we do this under certain assumptions on the coefficients a and b.

We turn next to the case of a bounded, connected, open set $\Omega \subset \mathbb{R}^d$, where $d \ge 2$. Before stating any results, we need to introduce a certain amount of notation and terminology.

Let $0 < \mu \le 1$. A function $f \in \bar{C}(\Omega)$ is said to satisfy a *Hölder condition* with exponent μ and Hölder constant M if

$$\sup \rho^{-\mu}\left\{\sup_{y \in V} f(y) - \inf_{y \in V} f(y)\right\} = M, \tag{1.13}$$

where the supremum is over $0 < \rho \le \rho_0$ (ρ_0 fixed), $x \in \mathbb{R}^d$, and components V of $\Omega \cap B(x, \rho)$. We denote M by $|f|_\mu$.

For $m = 0, 1, \ldots$, we define

$$C^{m,\mu}(\bar{\Omega}) = \{f \in \bar{C}^m(\Omega): |D^\alpha f|_\mu < \infty \quad \text{whenever} \quad |\alpha| = m\}, \tag{1.14}$$

where $\bar{C}^0(\Omega) = \bar{C}(\Omega)$, $\alpha \in (\mathbb{Z}_+)^d$, $D^\alpha = \partial_1^{\alpha_1} \cdots \partial_d^{\alpha_d}$, and $|\alpha| = \alpha_1 + \cdots + \alpha_d$. Observe that functions in $C^{0,\mu}(\bar{\Omega})$ need not have continuous extensions to $\bar{\Omega}$, as such extensions might have to be multivalued on $\partial\Omega$.

Regarding elements of $\mathbb{R}^d$ as column vectors, a mapping of $\mathbb{R}^d$ onto $\mathbb{R}^d$ of the form $y = U(x - x_0)$, where $x_0 \in \partial\Omega$ and U is an orthogonal matrix ($UU^T = I_d$), is said to be a *local Cartesian coordinate system with origin at* x_0 if the outer normal to $\partial\Omega$ at x_0 is mapped onto the nonnegative y_d axis. For $m = 1, 2, \ldots$, we say that $\partial\Omega$ is of class $C^{m,\mu}$ if there exists $\rho > 0$ such that for every $x_0 \in \partial\Omega$, $\partial\Omega \cap B(x_0, \rho)$ is a connected surface that, in terms of the local Cartesian coordinate system $(y_1, \ldots, y_d)$ with origin at x_0, is of the form $y_d = v(y_1, \ldots, y_{d-1})$, where $v \in C^{m,\mu}(\bar{D})$, D being the projection of $\partial\Omega \cap B(x_0, \rho)$ (in the local Cartesian coordinate system) onto $y_d = 0$. Assuming $\partial\Omega$ is of class $C^{m,\mu}$, a function $\varphi: \partial\Omega \to \mathbb{R}$ is said to belong to $C^{m,\mu}(\partial\Omega)$ if, for every $x_0 \in \partial\Omega$, φ as a function of $(y_1, \ldots, y_{d-1})$ belongs to $C^{m,\mu}(\bar{D})$, where the notation is as above. Note that if $\partial\Omega$ is of class $C^{m,\mu}$ and if $0 \le k \le m$, then each function in $C^{k,\mu}(\bar{\Omega})$ has a (unique) continuous extension to $\bar{\Omega}$, and its restriction to $\partial\Omega$ belongs to $C^{k,\mu}(\partial\Omega)$.

We consider

$$G = \frac{1}{2}\sum_{i,j=1}^{d} a_{ij}(x)\,\partial_i\,\partial_j + \sum_{i=1}^{d} b_i(x)\,\partial_i, \tag{1.15}$$

treating separately the cases of absorption and oblique reflection at the boundary. S_d denotes the space of $d \times d$ nonnegative-definite matrices.

1.4 Theorem Let $d \ge 2$ and $0 < \mu \le 1$, and let $\Omega \subset \mathbb{R}^d$ be bounded, connected, and open with $\partial\Omega$ of class $C^{2,\mu}$. Suppose $a: \Omega \to S_d$, $b: \Omega \to \mathbb{R}^d$, a_{ij}, $b_i \in C^{0,\mu}(\bar{\Omega})$ for $i, j = 1, \ldots, d$, and

$$\inf_{x \in \Omega} \inf_{|\theta| = 1} \theta \cdot a(x)\theta > 0. \tag{1.16}$$

Then, with G defined by (1.15), the closure of

$$A \equiv \{(f, Gf): f \in C^{2,\mu}(\bar{\Omega}),\ Gf = 0 \text{ on } \partial\Omega\} \tag{1.17}$$

is single-valued and generates a Feller semigroup on $C(\bar{\Omega})$.

Proof. We apply Theorem 2.2 of Chapter 4. Clearly, A satisfies the positive maximum principle. If $\lambda > 0$ and $g \in C^{2,\mu}(\bar{\Omega})$, then, by Theorem 3.13 of Ladyzhenskaya and Ural'tseva (1968), the equation $\lambda f - Gf = g$ has a solution $f \in C^{2,\mu}(\bar{\Omega})$ with $f = \lambda^{-1}g$ on $\partial\Omega$. It follows that $Gf = 0$ on $\partial\Omega$, so $f \in \mathscr{D}(A)$, proving that $\mathscr{R}(\lambda - A) \supset C^{2,\mu}(\bar{\Omega})$ for every $\lambda > 0$.

To show that $\mathscr{D}(A)$ is dense in $C(\bar{\Omega})$, let $f \in C^{2,\mu}(\bar{\Omega})$. For each $\lambda > 0$, choose $h_\lambda \in \mathscr{D}(A)$ such that $(\lambda - A)h_\lambda = \lambda f - Gf$. Then, as $\lambda \to \infty$,

(1.18) $$\|f - h_\lambda\| = \sup_{x \in \partial\Omega} |f(x) - h_\lambda(x)| = \sup_{x \in \partial\Omega} \lambda^{-1} |Gf(x)| \to 0,$$

where the first equality is due to the weak maximum principle for elliptic operators (Friedman (1975), Theorem 6.2.2). □

1.5 Theorem Suppose, in addition to the hypotheses of Theorem 1.4, that $c: \partial\Omega \to \mathbb{R}^d$, $c_i \in C^{1,\mu}(\partial\Omega)$ for $i = 1, \ldots, d$, and

(1.19) $$\inf_{x \in \partial\Omega} c(x) \cdot n(x) > 0,$$

where $n(x)$ denotes the outward unit normal to $\partial\Omega$ at x. Then, with G defined by (1.15), the closure of

(1.20) $$A \equiv \{(f, Gf): f \in C^{2,\mu}(\bar{\Omega}),\ c \cdot \nabla f = 0 \text{ on } \partial\Omega\}$$

is single-valued and generates a Feller semigroup on $C(\bar{\Omega})$.

Proof. Again, we apply Theorem 2.2 of Chapter 4. Because $c(x) \cdot n(x) \neq 0$ for all $x \in \partial\Omega$, A satisfies the positive maximum principle. (If $f \in \mathscr{D}(A)$ has a positive maximum at $x \in \partial\Omega$, then $\nabla f(x) = 0$.) By Theorem 3.3.2 of Ladyzhenskaya and Ural'tseva (1968), there exists $\lambda > 0$ such that for every $g \in C^{2,\mu}(\bar{\Omega})$ the equation $\lambda f - Gf = g$ has a solution $f \in C^{2,\mu}(\bar{\Omega})$ with $c \cdot \nabla f = 0$ on $\partial\Omega$. Thus $\mathscr{R}(\lambda - A) \supset C^{2,\mu}(\bar{\Omega})$.

It remains to show that $\mathscr{D}(A)$ is dense in $C(\bar{\Omega})$, or equivalently (by Remark 3.2 of Chapter 1), that the bp-closure of $\mathscr{D}(A)$ contains $C(\bar{\Omega})$. By Stroock and Varadhan (1971) there exists for each $x \in \bar{\Omega}$ a solution X_x of the $C_{\bar{\Omega}}[0, \infty)$ martingale problem for (A, δ_x). Consequently, for each $f \in C(\bar{\Omega})$,

(1.21) $$f(x) = \operatorname*{bp-lim}_{n \to \infty} n \int_0^\infty e^{-nt} E[f(X_x(t))]\, dt.$$

Since the right side of (1.21) belongs to the bp-closure of $\mathscr{D}(\bar{A})$, the proof is complete. □

Let us now consider the case in which $\Omega = \mathbb{R}^d$.

1.6 Theorem Let $a\colon \mathbb{R}^d \to S_d$ and $b\colon \mathbb{R}^d \to \mathbb{R}^d$ be bounded, $0 < \mu \le 1$, and $K > 0$, and suppose that

$$(1.22) \qquad |a(x) - a(y)| + |b(x) - b(y)| \le K|x - y|^\mu, \qquad x, y \in \mathbb{R}^d,$$

and

$$(1.23) \qquad \inf_{x \in \mathbb{R}^d} \inf_{|\theta| = 1} \theta \cdot a(x)\theta > 0.$$

Then, with G defined by (1.15), the closure of $\{(f, Gf)\colon f \in C_c^\infty(\mathbb{R}^d)\}$ is single-valued and generates a Feller semigroup on $\hat{C}(\mathbb{R}^d)$.

Proof. According to Theorem 5.11 of Dynkin (1965), there exists a strongly continuous, positive, contraction semigroup $\{T(t)\}$ on $\hat{C}(\mathbb{R}^d)$ whose generator A extends $G|_{C_c^2(\mathbb{R}^d)}$ (and therefore $G|_{\hat{C}^2(\mathbb{R}^d)}$). It suffices to show that A is conservative and that $C_c^\infty(\mathbb{R}^d)$ is a core for A.

Given $f \in \hat{C}(\mathbb{R}^d)$ and $t > 0$, the estimate in part 2° of the proof of Theorem 5.11 of Dynkin (1965), with (0.41) and (0.42) in place of (0.40), implies that $\partial_i T(t)f$ and $\partial_i \partial_j T(t)f$ exist and belong to $\hat{C}(\mathbb{R}^d)$. Thus, $T(t)\colon \hat{C}^2(\mathbb{R}^d) \to \hat{C}^2(\mathbb{R}^d)$ for all $t \ge 0$, so $\hat{C}^2(\mathbb{R}^d)$ is a core for A by Proposition 3.3 of Chapter 1. Let $h \in C_c^2(\mathbb{R}^d)$ satisfy $\chi_{B(0,1)} \le h \le \chi_{B(0,2)}$ and approximate $f \in C^2(\mathbb{R}^d)$ by $\{fh_n\} \subset C_c^2(\mathbb{R}^d)$, where $h_n(x) = h(x/n)$, to show that $C_c^2(\mathbb{R}^d)$ is a core for A. Similarly, using the sequence $\{(h_n, Gh_n)\}$, we find that A is conservative. Finally, choose $\varphi \in C_c^\infty(\mathbb{R}^d)$ with $\varphi \ge 0$ and $\int \varphi(x)\,dx = 1$, and approximate $f \in C_c^2(\mathbb{R}^d)$ by $\{f * \varphi_n\} \subset C_c^\infty(\mathbb{R}^d)$, where $\varphi_n(x) = n^d \varphi(nx)$, to show that $C_c^\infty(\mathbb{R}^d)$ is a core for A. (Note that $f * \varphi_n$ has compact support because both f and φ_n do.) □

If one is satisfied with uniqueness of solutions of the martingale problem, the assumptions of Theorem 1.6 can be weakened considerably. Moreover, time-dependent coefficients are permitted. Consider therefore

$$(1.24) \qquad G = \frac{1}{2} \sum_{i,j=1}^{d} a_{ij}(t, x)\,\partial_i \partial_j + \sum_{i=1}^{d} b_i(t, x)\,\partial_i.$$

1.7 Theorem Let $a\colon [0, \infty) \times \mathbb{R}^d \to S_d$ and $b\colon [0, \infty) \times \mathbb{R}^d \to \mathbb{R}^d$ be locally bounded and Borel measurable. Suppose that, for every $x \in \mathbb{R}^d$ and $t_0 > 0$,

$$(1.25) \qquad \inf_{0 \le t \le t_0} \inf_{|\theta| = 1} \theta \cdot a(t, x)\theta > 0$$

and

$$(1.26) \qquad \lim_{y \to x} \sup_{0 \le t \le t_0} |a(t, y) - a(t, x)| = 0.$$

Suppose further that there exists a constant K such that

$$(1.27) \qquad |a(t, x)| \le K(1 + |x|^2), \qquad t \ge 0,\ x \in \mathbb{R}^d,$$

and

$$x \cdot b(t, x) \leq K(1 + |x|^2), \qquad t \geq 0,\ x \in \mathbb{R}^d. \tag{1.28}$$

Then, with G defined by (1.24), the martingale problem for $\{(f, Gf): f \in C_c^\infty(\mathbb{R}^d)\}$ is well-posed.

Proof. Recalling Proposition 3.5 of Chapter 5, the result follows from Theorem 10.2.2 of Stroock and Varadhan (1979) and the discussion following their Corollary 10.1.2. □

2. DEGENERATE DIFFUSIONS

Again we treat the one-dimensional case separately. We begin by obtaining sufficient conditions for $C_c^\infty(I)$ to be a core for the generator in Corollary 1.2.

2.1 Theorem Given $-\infty \leq r_0 < r_1 \leq \infty$, define I and I° by (1.1). Suppose $a \in C^2(I)$, $a \geq 0$, a'' is bounded, $b: I \to \mathbb{R}$ is Lipschitz continuous (that is, $\sup_{x, y \in I, x \neq y} |b(y) - b(x)|/|x - y| < \infty$), and

$$a(r_i) = 0 \leq (-1)^i b(r_i) \text{ if } |r_i| < \infty, \qquad i = 0, 1. \tag{2.1}$$

Then with G defined by (1.2), the closure of $\{(f, Gf): f \in C_c^\infty(I)\}$ is single-valued and generates a Feller semigroup $\{T(t)\}$ on $\hat{C}(I)$. If $a > 0$ on I°, then $\{T(t)\}$ coincides with the semigroup defined in terms of a and b by Corollary 1.2 (with $q_i = 0$ if r_i is regular, $i = 0, 1$).

The proof depends on a lemma, which involves the function space $\hat{C}^m_{-\gamma}(I)$, defined for $m = 0, 1, \ldots$ and $\gamma \geq 0$ by

$$\hat{C}^m_{-\gamma}(I) = \{f \in C^m(I): \varphi_\gamma f^{(k)} \in \hat{C}(I),\ k = 0, \ldots, m\}, \tag{2.2}$$

where $\varphi_\gamma(x) = (1 + x^2)^{\gamma/2}$. Note that $\hat{C}^m_0(I) = \hat{C}^m(I)$.

2.2 Lemma Assume, in addition to the hypotheses of Theorem 2.1, that $b \in C^2(I)$ and b'' is bounded. Then there exists a (unique) strongly continuous, positive, contraction semigroup $\{T(t)\}$ on $\hat{C}(I)$ whose generator A is an extension of $\{(f, Gf): f \in \hat{C}(I) \cap C^2(I), Gf \in \hat{C}(I)\}$; moreover:

- **(a)** $T(t): \hat{C}^m_{-\gamma}(I) \to \hat{C}^m_{-\gamma}(I)$ for all $t \geq 0$, $m = 1, 2$, and $\gamma \geq 0$.
- **(b)** $\|(T(t)f)'\| \leq \exp(\|b'\|t)\|f'\|$ for all $f \in \hat{C}^1(I)$ and $t \geq 0$.

Proof. This is a special case of a result of Ethier (1978). □

Proof of Theorem 2.1. $\hat{C}^2_{-2}(I) \subset \hat{C}(I) \cap C^2(I) \cap A^{-1}\hat{C}(I)$, so under the additional assumptions of Lemma 2.2, $\hat{C}^2_{-2}(I)$ is a core for A by Proposition 3.3 of

Chapter 1. To obtain this conclusion without the additional assumptions, choose $\{b_n\} \subset C^2(I)$ such that each b_n satisfies the conditions of Lemma 2.2, $\lim_{n\to\infty} \|b_n - b\| = 0$, and $\sup_n \|b_n'\| \equiv M < \infty$. This can be done via convolutions. For each n, let $\{T_n(t)\}$ be associated with a and b_n as in Lemma 2.2. We apply Proposition 3.7 of Chapter 1 with $D_0 = \mathscr{D}(A) = \hat{C}^2_{-2}(I)$, $D_1 = \hat{C}^1(I)$, $|||f||| = \|f\| + \|f'\|$, $\omega = M$, and $\varepsilon_n = \|b_n - b\|$, concluding that $\hat{C}^2_{-2}(I)$ is a core for A under the assumptions of the theorem. The remainder of the proof that $C_c^\infty(I)$ is a core for A (and the proof that A is conservative) is analogous to that of Theorem 1.6. For the proof of the second conclusion of the theorem, see Ethier (1978). □

Of course, one can get uniqueness of solutions of the martingale problem under conditions that are much weaker than those of Theorem 2.1. One of the assumptions in Theorem 2.1 that is often too restrictive in applications is the requirement (when $b \in C^1(I)$) that b' be bounded, because, in the context of Theorem 1.1, infinite boundaries are often entrance. We permit time-dependent coefficients and so we consider

$$G = \tfrac{1}{2}a(t, x)\frac{d^2}{dx^2} + b(t, x)\frac{d}{dx}. \tag{2.3}$$

2.3 Theorem Given $-\infty \le r_0 < r_1 \le \infty$, define I by (1.1), and let a and b be real-valued, locally bounded, and Borel measurable on $[0, \infty) \times I$ with $a \ge 0$. Suppose that, for each $n \ge 1$ and $t_0 > 0$, a and b are Lipschitz continuous in $|x| \le n$, uniformly in $0 \le t \le t_0$. Suppose further that there exists a constant K such that (1.27) and (1.28) hold with $\mathbb{R}^d$ replaced by I, and

$$a(t, r_i) = 0 \le (-1)^i b(t, r_i) \quad \text{if} \quad |r_i| < \infty \qquad t \ge 0,\ i = 0, 1. \tag{2.4}$$

Then, with G defined by (2.3), the martingale problem for $\{(f, Gf): f \in C_c^\infty(I)\}$ is well-posed.

Proof. Existence of solutions follows from Theorem 3.10 of Chapter 5, together with (2.4). (Extend a to be zero outside of I and b by setting $b(t, x) = b(t, r_0)$, $x < r_0$, and $b(t, x) = b(t, r_1)$, $x > r_1$.) Uniqueness is a consequence of Theorem 3.8, Proposition 3.5, and Corollary 3.4, all from Chapter 5. □

Unfortunately, the extent to which Theorem 2.1 can be generalized to d dimensions is unknown. However, results can be obtained in a few special cases.

2.4 Proposition Let $\sigma: \mathbb{R}^d \to \mathbb{R}^d \otimes \mathbb{R}^d$ and $b: \mathbb{R}^d \to \mathbb{R}^d$ satisfy $\sigma_{ij}, b_i \in \bar{C}^2(\mathbb{R}^d)$ for $i, j = 1, \ldots, d$, and put $a = \sigma\sigma^T$. Then, with G defined by (1.15), the closure of $\{(f, Gf): f \in C_c^\infty(\mathbb{R}^d)\}$ is single-valued and generates a Feller semigroup on $\hat{C}(\mathbb{R}^d)$.

Proof. A proof has been outlined by Roth (1977). The details are left to the reader (Problem 4). □

The following result generalizes Proposition 2.4.

2.5 Theorem Let $a: \mathbb{R}^d \to S_d$ satisfy $a_{ij} \in C^2(\mathbb{R}^d)$ with $\partial_k \partial_l a_{ij}$ bounded for $i, j, k, l = 1, \ldots, d$, and let $b: \mathbb{R}^d \to \mathbb{R}^d$ be Lipschitz continuous (i.e., $\sup_{x, y \in \mathbb{R}^d, x \neq y} |b(x) - b(y)|/|x - y| < \infty$). Then, with G defined by (1.15), the closure of $\{(f, Gf): f \in C_c^\infty(\mathbb{R}^d)\}$ is single-valued and generates a Feller semigroup on $\hat{C}(\mathbb{R}^d)$.

Proof. We simply outline the proof, leaving it to the reader to fill in a number of details. First, some additional notation is needed. For $\gamma \geq 0$ define $\varphi_\gamma: \mathbb{R}^d \to (0, \infty)$ by $\varphi_\gamma(x) = (1 + |x|^2)^{\gamma/2}$ and

$$\hat{C}_{-\gamma}(\mathbb{R}^d) = \{f \in C(\mathbb{R}^d): \varphi_\gamma f \in \hat{C}(\mathbb{R}^d)\}. \tag{2.5}$$

For $m = 1, 2, \ldots$ and $\gamma \geq 0$ define

$$\hat{C}^m_{-\gamma}(\mathbb{R}^d) = \{f \in C^m(\mathbb{R}^d): \varphi_\gamma D^\alpha f \in \hat{C}(\mathbb{R}^d) \quad \text{if} \quad |\alpha| \leq m\}. \tag{2.6}$$

A useful norm on $\hat{C}^m_{-\gamma}(\mathbb{R}^d)$ is given by

$$\|f\|_{\hat{C}^m_{-\gamma}(\mathbb{R}^d)} = \left\| \varphi_\gamma \left\{ \sum_{1 \leq |\alpha| \leq m} (D^\alpha f)^2 \right\}^{1/2} \right\|. \tag{2.7}$$

Finally, we define

$$\hat{C}^{0,m}_{-\gamma}([0, T] \times \mathbb{R}^d) = \{f \in C^{0,m}([0, T] \times \mathbb{R}^d): \varphi_\gamma D^\alpha f \in \hat{C}([0, T] \times \mathbb{R}^d) \quad \text{if} \quad |\alpha| \leq m\}. \tag{2.8}$$

Suppose, in addition to the hypotheses of the theorem, that $b_i \in C^2(\mathbb{R}^d)$ and $\partial_k b_i$, $\partial_k \partial_l b_i$ are bounded for $i, k, l = 1, \ldots, d$. Then there exist sequences $\sigma^{(n)}: \mathbb{R}^d \to \mathbb{R}^d \otimes \mathbb{R}^d$ and $b^{(n)}: \mathbb{R}^d \to \mathbb{R}^d$ with the following properties, where $a^{(n)} = \sigma^{(n)}(\sigma^{(n)})^T$: $\sigma_{ij}^{(n)}, b_i^{(n)} \in \bar{C}^\infty(\mathbb{R}^d)$, $a_{ij}^{(n)} \to a_{ij}$ and $b_i^{(n)} \to b_i$ uniformly on compact sets, and $a_{ij}^{(n)}/\varphi_2$, $b_i^{(n)}/\varphi_1$, $\partial_k \partial_l a_{ij}^{(n)}$, $\partial_k b_i^{(n)}$, $\partial_k \partial_l b_i^{(n)}$ are uniformly bounded, $i, j, k, l = 1, \ldots, d$.

Fix n for now. Letting G_n be defined in terms of $a^{(n)}$ and $b^{(n)}$ as in (1.15), one can show as in Problem 5(b) that the closure of $\{(f, G_n f): f \in C_c^\infty(\mathbb{R}^d)\}$ is single-valued and generates a Feller semigroup $\{T_n(t)\}$ on $\hat{C}(\mathbb{R}^d)$. (This also follows from Proposition 2.4.) Moreover, a slight extension of this argument shows that for each $m \geq 1$, $T_n(t): \hat{C}^m(\mathbb{R}^d) \to \hat{C}^m(\mathbb{R}^d)$ for all $t \geq 0$ and $\{T_n(t)\}$ restricted to $\hat{C}^m(\mathbb{R}^d)$ is strongly continuous in the norm $\|\cdot\|_{\hat{C}_0^m(\mathbb{R}^d)}$ (recall (2.7)). A simple argument involving Gronwall's inequality implies that for each $\gamma > 0$, there exists $\lambda_{n,\gamma} > 0$, such that $T_n(t)\varphi_{-\gamma} \leq \exp(\lambda_{n,\gamma} t)\varphi_{-\gamma}$ for all $t \geq 0$. Using the fact that $\bar{C}^3(\mathbb{R}^d) \cap C_{-\gamma}(\mathbb{R}^d) \subset \hat{C}^2_{-3}(\mathbb{R}^d)$ for γ sufficiently large, we conclude that if $f \in C_c^\infty(\mathbb{R}^d)$, $t > 0$, and $u_n(s, x) \equiv T_n(t - s)f(x)$, then $u_n \in \hat{C}^{0,2}_{-3}([0, t] \times \mathbb{R}^d) \cap \hat{C}^{0,4}([0, t] \times \mathbb{R}^d)$ and $\partial u_n/\partial s + G_n u_n = 0$.

We can therefore apply Oleinik's (1966) a priori estimate (or actually a simple extension of it to $C^m_{-\gamma}(\mathbb{R}^d)$) to conclude that there exists $\omega \geq 0$, depending (continuously) only on the uniform bounds on $a_{ij}^{(n)}/\varphi_2$, $b_i^{(n)}/\varphi_1$, $\partial_k \partial_l a_{ij}^{(n)}$, $\partial_k b_i^{(n)}$, and $\partial_k \partial_l b_i^{(n)}$, such that

$$\|T_n(t)f\|_{\hat{C}^2_{-3}(\mathbb{R}^d)} \leq e^{\omega t}\|f\|_{\hat{C}^2_{-3}(\mathbb{R}^d)} \tag{2.9}$$

for all $f \in C_c^\infty(\mathbb{R}^d)$, all n, and all $t \geq 0$. (The proof is essentially as in Stroock and Varadhan (1979), Theorem 3.2.4.) It follows that for each n and $t \geq 0$, $T_n(t)\colon \hat{C}^2_{-3}(\mathbb{R}^d) \to \hat{C}^2_{-3}(\mathbb{R}^d)$ and (2.9) holds for all $f \in \hat{C}^2_{-3}(\mathbb{R}^d)$. Since

$$\begin{aligned} \|(G_n - G)f\| &\leq \frac{1}{2}\sum_{i,j=1}^d \left\|\frac{a_{ij}^{(n)} - a_{ij}}{\varphi_3}\right\| \|\varphi_3 \partial_i \partial_j f\| \\ &\quad + \sum_{i=1}^d \left\|\frac{b_i^{(n)} - b_i}{\varphi_3}\right\| \|\varphi_3 \partial_i f\| \\ &\leq \varepsilon_n \|f\|_{\hat{C}^2_{-3}(\mathbb{R}^d)} \end{aligned} \tag{2.10}$$

for all $f \in \hat{C}^2_{-3}(\mathbb{R}^d)$ and all n, where $\lim_{n\to\infty} \varepsilon_n = 0$, the stated conclusion follows from Proposition 3.7 of Chapter 1 with $D_0 = C_c^\infty(\mathbb{R}^d)$ and $\mathscr{D}(A) = D_1 = \hat{C}^2_{-3}(\mathbb{R}^d)$, at least under the additional hypotheses noted in the second paragraph of the proof. But these are removed just as in the proof of Theorem 2.1, the analogue of Lemma 2.2 following as in (2.9) from Oleinik's result. □

To get uniqueness of solutions of the martingale problem in general, we need to assume that a has a Lipschitz continuous square root.

2.6 Theorem Let $\sigma\colon [0, \infty) \times \mathbb{R}^d \to \mathbb{R}^d \otimes \mathbb{R}^d$ and $b\colon [0, \infty) \times \mathbb{R}^d \to \mathbb{R}^d$ satisfy the conditions of Theorem 3.10 of Chapter 5, and put $a = \sigma\sigma^T$. Then, with G defined by (1.24), the martingale problem for $\{(f, Gf)\colon f \in C_c^\infty(\mathbb{R}^d)\}$ is well-posed.

Proof. The result is a consequence of Theorems 3.10 and 3.6, Proposition 3.5, and Corollary 3.4, all from Chapter 5. □

2.7 Remark Typically, it is a, rather than σ, that is given, and with this in mind we give sufficient conditions for $a^{1/2}$, the S_d-valued square root of a, to be Lipschitz continuous. Let $a\colon [0, \infty) \times \mathbb{R}^d \to S_d$ be Borel measurable, and assume either of the following conditions:

(a) There exist $C > 0$ and $\delta > 0$ such that

$$|a(t, y) - a(t, x)| \leq C|y - x|, \qquad t \geq 0,\ x, y \in \mathbb{R}^d, \tag{2.11}$$

and

$$\inf_{(t,x)\in[0,\infty)\times\mathbb{R}^d} \ \inf_{|\theta|=1} \theta \cdot a(t, x)\theta > \delta. \tag{2.12}$$

(b) $a_{ij}(t, \cdot) \in C^2(\mathbb{R}^d)$ for $i, j = 1, \ldots, d$ and all $t \geq 0$, and there exists $\lambda > 0$ such that

$$\sup_{(t, x) \in [0, \infty) \times \mathbb{R}^d} \max_{1 \leq i \leq d} \sup_{|\theta| = 1} |\theta \cdot (\partial_i^2 a)(t, x)\theta| \leq \lambda. \tag{2.13}$$

Then $a^{1/2}$ is Borel measurable and

$$|a^{1/2}(t, y) - a^{1/2}(t, x)| \leq K|y - x|, \qquad t \geq 0,\ x, y \in \mathbb{R}^d, \tag{2.14}$$

where $K = C/(2\delta^{1/2})$ if (a) holds and $K = d(2\lambda)^{1/2}$ if (b) holds. See Section 5.2 of Stroock and Varadhan (1979). □

We conclude this section by considering a special class of generators, which arise in population genetics (see Chapter 10).

2.8 Theorem Let

$$K_d = \left\{x \in [0, 1]^d : \sum_{i=1}^d x_i \leq 1\right\}, \tag{2.15}$$

define $a: K_d \to S_d$ by $a_{ij}(x) = x_i(\delta_{ij} - x_j)$, and let $b: K_d \to \mathbb{R}^d$ be Lipschitz continuous and satisfy

$$\begin{aligned} b_i(x) &\geq 0 \quad \text{if} \quad x \in K_d \quad \text{and} \quad x_i = 0, \qquad i = 1, \ldots, d, \\ \sum_{i=1}^d b_i(x) &\leq 0 \quad \text{if} \quad x \in K_d \quad \text{and} \quad \sum_{i=1}^d x_i = 1. \end{aligned} \tag{2.16}$$

Then, with G defined by (1.15), the closure of $\{(f, Gf): f \in C^2(K_d)\}$ is single-valued and generates a Feller semigroup on $C(K_d)$. Moreover, the space of polynomials on K_d is a core for the generator.

The proof is quite similar to that of Theorem 2.1. It depends on the following lemma from Ethier (1976).

2.9 Lemma Assume, in addition to the hypotheses of Theorem 2.8, that $b_1, \ldots, b_d \in C^4(K_d)$. Then the first conclusion of the theorem holds, as do the following two assertions:

(a) $T(t): C^m(K_d) \to C^m(K_d)$ for all $t \geq 0$ and $m = 1, 2$.

(b) $\sum_{i=1}^d \|\partial_i T(t)f\| \leq e^{\lambda t} \sum_{i=1}^d \|\partial_i f\|$ for all $f \in C^1(K_d)$ and $t \geq 0$, where

$$\lambda = \max_{1 \leq i \leq d} \sum_{j=1}^d \|\partial_j b_i\|. \tag{2.17}$$

Proof of Theorem 2.8. Choose $b^{(n)}: K_d \to \mathbb{R}^d$ for $n = 1, 2, \ldots$ satisfying the conditions of Lemma 2.9 such that $\lim_{n\to\infty} \|b^{(n)} - b\| = 0$ and

(2.18) $$\sup_{n \geq 1} \max_{1 \leq i \leq d} \sum_{j=1}^{d} \|\partial_j b_i^{(n)}\| < \infty.$$

The latter two conditions follow using convolutions. To get (2.16), it may be necessary to add $\varepsilon_n(1 - (d + 1)x_i)$ to the $b_i^{(n)}(x)$ thus obtained, where $\varepsilon_n \to 0+$. The first conclusion now follows from Proposition 3.7 of Chapter 1 in the same way that Theorem 2.1 did. The second conclusion is a consequence of the fact that the space of polynomials on K_d is dense in $C^2(K_d)$ with respect to the norm

(2.19) $$|||f|||_{C^2(K_d)} = \sum_{|\alpha| \leq 2} \|D^\alpha f\|$$

(see Appendix 7). □

3. OTHER PROCESSES

We begin by considering jump Markov processes in a locally compact, separable metric space E, the generators for which have the form

(3.1) $$Af(x) = \lambda(x) \int (f(y) - f(x))\mu(x, dy),$$

where $\lambda \in B_{\mathrm{loc}}(E)$ is nonnegative and $\mu(x, \Gamma)$ is a transition function on $E \times \mathscr{B}(E)$. We assume, among other things, that λ and the mapping $x \to \mu(x, \cdot)$ are continuous on E. Thus, if E is compact, then A is a bounded linear operator on $C(E)$ and generates a Feller semigroup on $C(E)$. We can therefore assume without further loss of generality that E is noncompact. The case in which $E = \{0, 1, \ldots\}$ is treated separately as a corollary.

3.1 Theorem Let E be a locally compact, noncompact, separable metric space and let $E^\Delta = E \cup \{\Delta\}$ be its one-point compactification. Let $\lambda \in C(E)$ be nonnegative and let $\mu(x, \Gamma)$ be a transition function on $E \times \mathscr{B}(E)$ such that the mapping $x \to \mu(x, \cdot)$ of E into $\mathscr{P}(E)$ is continuous. Let γ and η be positive functions in $C(E)$ such that $1/\gamma$ and $1/\eta$ belong to $\hat{C}(E)$ and

(3.2) $$\sup_{x \in E} \frac{\lambda(x)}{\gamma(x)} \equiv C_1 < \infty,$$

(3.3) $$\lim_{x \to \Delta} \lambda(x)\mu(x, K) = 0 \quad \text{for every compact} \quad K \subset E,$$

(3.4) $$\sup_{x \in E} \lambda(x) \int \frac{\gamma(x) - \gamma(y)}{\gamma(y)} \mu(x, dy) \equiv C_2 < \infty,$$

(3.5) $$\sup_{x \in E} \lambda(x) \int \frac{\eta(y) - \eta(x)}{\eta(x)} \mu(x, dy) \equiv C_3 < \infty.$$

Then, with A defined by (3.1), the closure of $\{(f, Af): f \in \hat{C}(E),\ \gamma f \in \bar{C}(E),\ Af \in \hat{C}(E)\}$ is single-valued and generates a Feller semigroup on $\hat{C}(E)$. Moreover, $C_c(E)$ is a core for this generator.

Proof. Consider A as a linear operator on $\bar{C}(E)$ with domain $\mathscr{D}(A) = \{f \in \bar{C}(E): \gamma f \in \bar{C}(E)\} \subset \hat{C}(E)$. To see that $A: \mathscr{D}(A) \to \bar{C}(E)$, let $f \in \mathscr{D}(A)$ and observe that $Af \in C(E)$ and

(3.6) $$\begin{aligned} |Af(x)| &\leq \lambda(x)\left\{\int \frac{1}{\gamma(y)} \mu(x, dy) + \frac{1}{\gamma(x)}\right\} \|\gamma f\| \\ &\leq \left\{\frac{C_2 + \lambda(x)}{\gamma(x)} + C_1\right\} \|\gamma f\| \\ &\leq (C_2\|1/\gamma\| + 2C_1)\|\gamma f\|, \qquad x \in E, \end{aligned}$$

by (3.2) and (3.4). Using the idea of Lemma 2.1 of Chapter 4, we also find that A is dissipative.

We claim that $\overline{\mathscr{R}(\alpha - A)} \supset \mathscr{D}(A)$ for all α sufficiently large. Given $n \geq 1$, define A_n on $\hat{C}(E)$ as in (3.1) but with $\lambda(x)$ replaced by $\lambda(x) \wedge n$. By (3.3), $A_n: C_c(E) \to \hat{C}(E)$, and hence A_n is a bounded linear operator on $\hat{C}(E)$ satisfying the positive maximum principle. It therefore generates a strongly continuous, positive, contraction semigroup $\{T_n(t)\}$ on $\hat{C}(E)$. By (3.4), there exists $\omega \geq 0$ not depending on n such that $\gamma A_n(1/\gamma) \leq \omega$, so

(3.7) $$e^{-\omega t} T_n(t) \frac{1}{\gamma} - \frac{1}{\gamma} = \int_0^t e^{-\omega s} T_n(s) \left(A_n \frac{1}{\gamma} - \frac{\omega}{\gamma}\right) ds \leq 0$$

for all $t \geq 0$. Let $f \in \mathscr{D}(A)$. By (3.7), $T_n(t)f \in \mathscr{D}(A)$ and

(3.8) $$\|\gamma T_n(t) f\| \leq \left\|\gamma T_n(t) \frac{1}{\gamma}\right\| \|\gamma f\| \leq e^{\omega t} \|\gamma f\|$$

for all $t \geq 0$. Hence, if $\alpha > \omega$, then $(\alpha - A_n)^{-1}f \in \mathscr{D}(A)$ and

$$\|\gamma(\alpha - A_n)^{-1}f\| \leq (\alpha - \omega)^{-1}\|\gamma f\|, \tag{3.9}$$

so

$$\begin{aligned} \|(A_n - A)(\alpha - A_n)^{-1}f\| &\leq \left\|\frac{\lambda \wedge n - \lambda}{(\lambda \wedge n)\gamma}\right\| \|\gamma A_n(\alpha - A_n)^{-1}f\| \\ &\leq \frac{C_1}{n}\left(\frac{\alpha}{\alpha - \omega} + 1\right)\|\gamma f\| \end{aligned} \tag{3.10}$$

by (3.2). Since f and n were arbitrary, we conclude from Lemma 3.6 of Chapter 1 that $\mathscr{R}(\lambda - A) \supset \mathscr{D}(A)$.

Thus, by Theorem 4.3 of Chapter 1,

$$A_0 \equiv \{(f, g) \in \bar{A} : g \in \hat{C}(E)\} \tag{3.11}$$

is single-valued and generates a strongly continuous, positive, contraction semigroup $\{T(t)\}$ on $\hat{C}(E)$. Clearly, if $f \in \mathscr{D}(A_0)$, then $A_0 f$ is given by the right side of (3.1) and $A_n f \to A_0 f$ as $n \to \infty$. It follows from Theorem 6.1 of Chapter 1 that $T_n(t)f \to T(t)f$ for all $f \in \hat{C}(E)$ and $t \geq 0$. In particular, by (3.7), $T(t)(1/\gamma) \leq e^{\omega t}(1/\gamma)$ for all $t \geq 0$, so $T(t)\colon \mathscr{D}(A) \to \mathscr{D}(A)$ for every $t \geq 0$. We conclude from Proposition 3.3 of Chapter 1 that $\mathscr{D}(A_0) \cap \mathscr{D}(A)$ is a core for A_0, that is, the closure of $\{(f, Af) : f \in \hat{C}(E), \gamma f \in \bar{C}(E), Af \in \hat{C}(E)\}$ generates $\{T(t)\}$.

Let $f \in \mathscr{D}(A_0) \cap \mathscr{D}(A)$ and choose $\{h_n\} \subset C_c(E)$ such that $\chi_{\{x \in E : |f(x)| > 1/n\}} \leq h_n \leq 1$ for each n, and observe that $\{f h_n\} \subset C_c(E)$, $f h_n \to f$ uniformly, and $A(f h_n) \to Af$ boundedly (by (3.7)) and pointwise. Recalling Remark 3.2 of Chapter 1, this implies that $C_c(E)$ is a core for A_0.

It remains to show that A_0 is conservative. Fix $x \in E$, and let X be a Markov process corresponding to $\{T^{\Delta}(t)\}$ (see Lemma 2.3 of Chapter 4) with sample paths in $D_{E^{\Delta}}[0, \infty)$ and initial distribution δ_x. Extend η from E to E^{Δ} by setting $\eta(\Delta) = \infty$. Let $n \geq 1$ and define

$$\tau_n = \inf\{t \geq 0 : \eta(X(t)) > n\}. \tag{3.12}$$

Then, approximating η monotonically from below by functions in $C_c(E)$, we find that

$$\begin{aligned} E[\eta(X(t \wedge \tau_n))] &= \eta(x) + E\left[\int_0^{t \wedge \tau_n} A\eta(X(s))\, ds\right] \\ &\leq \eta(x) + C_3 E\left[\int_0^{t \wedge \tau_n} \eta(X(s))\, ds\right] \\ &\leq \eta(x) + C_3 tn, \end{aligned} \tag{3.13}$$

for all $t \geq 0$ by (3.5), so $E[\eta(X(t \wedge \tau_n))]$ is bounded in t on bounded intervals. By the first inequality in (3.13),

$$E[\eta(X(t \wedge \tau_n))] \leq \eta(x) + C_3 \int_0^t E[\eta(X(s \wedge \tau_n))] \, ds, \tag{3.14}$$

and thus

$$nP\{\tau_n \leq t\} \leq E[\eta(X(t \wedge \tau_n))] \leq \eta(x)e^{C_3 t} \tag{3.15}$$

for all $t \geq 0$ by Gronwall's inequality. It follows that $\lim_{n \to \infty} \tau_n = \infty$ a.s. and hence X has almost all sample paths in $D_E[0, \infty)$. By Corollary 2.8 of Chapter 4, we conclude that A_0 is conservative. □

3.2 Corollary Let $E = \{0, 1, \ldots\}$ and

$$Af(i) = \sum_{j \geq 0} q_{ij} f(j), \tag{3.16}$$

where the matrix $(q_{ij})_{i, j \geq 0}$ has nonnegative off-diagonal entries and row sums equal to zero. Assume also

$$\sup_{i \geq 0} \frac{|q_{ii}|}{i + 1} < \infty, \tag{3.17}$$

$$\lim_{i \to \infty} q_{ij} = 0, \qquad j \geq 0, \tag{3.18}$$

$$\sup_{i \geq 0} \sum_{j \geq 0} \frac{i + 1}{j + 1} q_{ij} < \infty, \tag{3.19}$$

$$\sup_{i \geq 0} \frac{1}{i + 1} \sum_{j \geq 0} (j - i) q_{ij} < \infty. \tag{3.20}$$

Then the closure of $\{(f, Af): f \in C_c(E)\}$ is single-valued and generates a Feller semigroup on $\hat{C}(E)$.

Proof. Apply Theorem 3.1 with $\lambda(i)\mu(i, \{j\}) = q_{ij}$ for $i \neq j$, $\mu(i, \{i\}) = 0$, and $\gamma(i) = \eta(i) = i + 1$. □

We next state a uniqueness result for processes in $\mathbb{R}^d$ with Lévy generators, that is, generators of the form

$$Gf(x) = \frac{1}{2} \sum_{i, j=1}^{d} a_{ij}(t, x) \, \partial_i \, \partial_j f(x) + \sum_{i=1}^{d} b_i(t, x) \, \partial_i f(x) + \int_{\mathbb{R}^d} \left(f(x + y) - f(x) - \frac{y \cdot \nabla f(x)}{1 + |y|^2} \right) \mu(t, x; dy). \tag{3.21}$$

3.3 Theorem Let a: $[0, \infty) \times \mathbb{R}^d \to S_d$ be bounded, continuous, and positive-definite-valued, b: $[0, \infty) \times \mathbb{R}^d \to \mathbb{R}^d$ bounded and Borel measurable, and

$\mu\colon [0, \infty) \times \mathbb{R}^d \to \mathcal{M}(\mathbb{R}^d)$ such that $\int_\Gamma |y|^2(1 + |y|^2)^{-1}\mu(t, x; dy)$ is bounded and continuous in (t, x) for every $\Gamma \in \mathscr{B}(\mathbb{R}^d)$. Then, with G defined by (3.21), the martingale problem for $\{(f, Gf)\colon f \in C_c^\infty(\mathbb{R}^d)\}$ is well-posed.

Proof. By Corollary 3.7 and Theorem 3.8, both from Chapter 4, every solution of the martingale problem has a modification with sample paths in $D_{\mathbb{R}^d}[0, \infty)$. The result therefore follows from Stroock (1975). □

When a, b, and μ in (3.22) are independent of (t, x), A becomes

$$Gf(x) = \frac{1}{2}\sum_{i,j=1}^{d} a_{ij}\,\partial_i\,\partial_j f(x) + \sum_{i=1}^{d} b_i\,\partial_i f(x) + \int_{\mathbb{R}^d}\left(f(x+y) - f(x) - \frac{y \cdot \nabla f(x)}{1 + |y|^2}\right)\mu(dy). \tag{3.22}$$

Every stochastically continuous process in $\mathbb{R}^d$ with homogeneous, independent increments has a generator of this form, where $a \in S_d$, $b \in \mathbb{R}^d$, $\mu \in \mathcal{M}(\mathbb{R}^d)$, and $\int_{\mathbb{R}^d} |y|^2(1 + |y|^2)^{-1}\mu(dy) < \infty$ (see Gihman and Skorohod (1969), Theorem VI.3.3). In this case we can strengthen the conclusion of Theorem 3.3.

3.4 Theorem Let $a \in S_d$, $b \in \mathbb{R}^d$, and $\mu \in \mathcal{M}(\mathbb{R}^d)$, and assume that $\int_{\mathbb{R}^d} |y|^2(1 + |y|^2)^{-1}\mu(dy) < \infty$. Then, with G defined by (3.22), the closure of $\{(f, Gf)\colon f \in \hat{C}^2(\mathbb{R}^d)\}$ is single-valued and generates a Feller semigroup on $\hat{C}(\mathbb{R}^d)$. Moreover, $C_c^\infty(\mathbb{R}^d)$ is a core for this generator.

Proof. If a is positive definite, then by Theorem 3.3, the martingale problem for $\{(f, Gf)\colon f \in C_c^\infty(\mathbb{R}^d)\}$ is well-posed. For each $x \in \mathbb{R}^d$, denote by X_x a solution with initial distribution δ_x, and note that since $(Gf)^x = G(f^x)$ for all $f \in C_c^\infty(\mathbb{R}^d)$, where $f^x(y) \equiv f(x + y)$, we have

$$E[f(X_x(t))] = E[f(x + X_0(t))] \tag{3.23}$$

for all $f \in B(E)$ and $t \geq 0$. It follows that we can define a strongly continuous, positive, contraction semigroup $\{T(t)\}$ on $\hat{C}(\mathbb{R}^d)$ by letting $T(t)f(x)$ be given by (3.23). Denoting the generator of $\{T(t)\}$ by A, we have $\{(f, Gf)\colon f \in C_c^\infty(\mathbb{R}^d)\} \subset A$, hence $\{(f, Gf)\colon f \in \hat{C}^2(\mathbb{R}^d)\} \subset A$. Moreover, by (3.23), $T(t)\colon \hat{C}^\infty(\mathbb{R}^d) \to \hat{C}^\infty(\mathbb{R}^d)$ for all $t \geq 0$, so $\hat{C}^\infty(\mathbb{R}^d)$ is a core for A by Proposition 3.3 of Chapter 1.

Let $h \in C_c^\infty(\mathbb{R}^d)$ satisfy $\chi_{B(0,1)} \leq h \leq \chi_{B(0,2)}$, and approximate $f \in \hat{C}^\infty(\mathbb{R}^d)$ by $\{fh_n\} \subset C_c^\infty(\mathbb{R}^d)$, where $h_n(x) = h(x/n)$, to show that $C_c^\infty(\mathbb{R}^d)$ is a core for A. (To check that $\text{bp-lim}_{n\to\infty} A(fh_n) = Af$, it suffices to split the integral in (3.22) into two parts, $|y| \leq 1$ and $|y| > 1$.) Similarly, using $\{(h_n, Ah_n)\}$, we find that A is conservative. The case in which a is only nonnegative definite can be handled by approximating a by $a + \varepsilon I$, $\varepsilon > 0$. □

We conclude this section with two results from the area of infinite particle systems. The first concerns spin-flip systems and the second exclusion processes. For further background on these processes, see Liggett (1977, 1985).

3.5 Theorem Let S be a countable set, and give $\{-1, 1\}$ the discrete topology and $E \equiv \{-1, 1\}^S$ the product topology. For each $i \in S$, define the difference operator Δ_i on $C(E)$ by $\Delta_i f(\eta) = f({}_i\eta) - f(\eta)$, where $({}_i\eta)_j = (1 - 2\delta_{ij})\eta_j$ for all $j \in S$. For each $i \in S$, let $c_i \in C(E)$ be nonnegative, and assume that

(3.24) $$\sup_{i \in S} \|c_i\| < \infty, \qquad \sup_{i \in S} \sum_{j \in S} \|\Delta_j c_i\| < \infty.$$

Then, with

(3.25) $$A = \sum_{i \in S} c_i(\eta)\Delta_i,$$

the closure of $\{(f, Af): f \in C(E), \sum_{i \in S} \|\Delta_i f\| < \infty\}$ is single-valued and generates a Feller semigroup on $C(E)$. Moreover, the space of (continuous) functions on E depending on only finitely many coordinates is a core for this generator.

Proof. The first assertion is essentially a special case of a more general result of Liggett (1972). The second is left to the reader (Problem 8). □

3.6 Theorem Let S be a countable set, and give $\{0, 1\}$ the discrete topology and $E \equiv \{0, 1\}^S$ the product topology. For each $i, j \in S$, define the difference operator Δ_{ij} on $C(E)$ by $\Delta_{ij} f(\eta) = f({}_{ij}\eta) - f(\eta)$, where

(3.26) $$({}_{ij}\eta)_k = \begin{cases} \eta_k, & k \neq i, j \\ \eta_j, & k = i \\ \eta_i, & k = j. \end{cases}$$

For each $i, j \in S$, let $c_{ij} \in C(E)$ be nonnegative and γ_{ij} be a nonnegative number, and assume that $c_{ii} \equiv 0$, $c_{ij} \equiv c_{ji}$, $c_{ij} \le \gamma_{ij}$, and $\gamma_{ij} = \gamma_{ji}$ for all $i, j \in S$,

(3.27) $$\sup_{i \in S} \sum_{j \in S} \gamma_{ij} < \infty,$$

and

(3.28) $$\sum_{k \in S} \sup_{\eta \in E} |c_{ij}({}_k\eta) - c_{ij}(\eta)| \le K\gamma_{ij}, \qquad i, j \in S,$$

where $({}_k\eta)_l = \delta_{kl} + (1 - 2\delta_{kl})\eta_l$ for all $l \in S$ and K is a constant. Then, with

(3.29) $$A = \sum_{i, j \in S} c_{ij}(\eta)\, \Delta_{ij},$$

the closure of $\{(f, Af): f \in C(E), \sum_{i, j \in S} \gamma_{ij}\|\Delta_{ij} f\| < \infty\}$ is single-valued and generates a Feller semigroup on $C(E)$. Moreover, the space of (continuous) functions on E depending on only finitely many coordinates is a core for this generator.

Proof. The references given for the preceding theorem apply here. □

4. PROBLEMS

1. For each $x \in \bar{I} = [r_0, r_1]$, let $P_x \in \mathscr{P}(C_{\bar{I}}[0, \infty))$ be the distribution of the diffusion process in $\bar{I}$ with initial distribution δ_x corresponding to the semigroup of Theorem 1.1. Let X be the coordinate process on $C_{\bar{I}}[0, \infty)$, and define $\tau_y = \inf\{t \geq 0: X(t) = y\}$ for $y \in \bar{I}$.

   (a) Show that r_1 is accessible if and only if there exist $x \in I^\circ$ and $t > 0$ such that

   $$\inf_{y \in (x, r_1)} P_x\{\tau_y \leq t\} > 0. \tag{4.1}$$

   (b) Suppose r_1 is inaccessible. Show that r_1 is entrance if and only if there exist $y \in I^\circ$ and $t > 0$ such that

   $$\inf_{x \in (y, r_1)} P_x\{\tau_y \leq t\} > 0. \tag{4.2}$$

   (c) Prove Corollary 1.2.

2. Suppose, in addition to the hypotheses of Theorem 1.1, that there exists a constant K such that (1.12) holds. Show that infinite boundaries of I are natural.

3. Use Proposition 3.4 of Chapter 1 to establish Theorem 2.1 in the special case in which $I = [0, \infty)$ and

   $$a(x) = ax, \qquad b(x) = bx, \qquad x \in I, \tag{4.3}$$

   where $0 < a < \infty$ and $-\infty < b < \infty$. (The resulting diffusion occurs in Chapter 9.)

   Hint: Look for solutions of the form $u(t, x) = e^{-\lambda(t)x}$.

4. Assume the hypotheses of Proposition 2.4, and for each $t \geq 0$ define the linear contraction $S(t)$ on $\hat{C}(\mathbb{R}^d)$ by

   $$S(t)f(x) = E[f(x + \sqrt{t}\sigma(x)Z + tb(x))], \tag{4.4}$$

   where Z is $N(0, I_d)$. ($\{S(t)\}$ is not necessarily a semigroup.) Given $t \geq 0$ and a partition $\pi = (0 = t_0 \leq t_1 \leq \cdots \leq t_n = t)$ of $[0, t]$, define $\mu(\pi) = \max_{1 \leq i \leq n} (t_i - t_{i-1})$ and

   $$S_\pi = S(t_n - t_{n-1}) \cdots S(t_1 - t_0), \tag{4.5}$$

and note that S_π: $\hat{C}^2(\mathbb{R}^d) \to \hat{C}^2(\mathbb{R}^d)$. Define the norm $|||\cdot|||$ on $\hat{C}^2(\mathbb{R}^d)$ by

$$(4.6)\qquad |||f||| = \left\| \left\{ \sum_{i=1}^{d} (\partial_i f)^2 + \sum_{1 \le i \le j \le d} (\partial_i \partial_j f)^2 \right\}^{1/2} \right\|.$$

Prove Proposition 2.4 by verifying each of the following assertions:

(a) There exists $K > 0$ such that

$$(4.7)\qquad \|S(t)S(s)f - S(t+s)f\| \le Ks\sqrt{t}\,|||f|||$$

for all $f \in \hat{C}^2(\mathbb{R}^d)$ and $s, t \in [0, 1]$.

(b) There exists $K > 0$ such that

$$(4.8)\qquad |||S(t)f||| \le (1 + Kt)\,|||f|||$$

for all $f \in \hat{C}^2(\mathbb{R}^d)$ and $0 \le t \le 1$.

(c) By parts (a) and (b), there exists $K > 0$ such that

$$(4.9)\qquad \|S_{\pi_1} f - S_{\pi_2} f\| \le Kt\sqrt{\mu(\pi_1) \vee \mu(\pi_2)}\,|||f|||$$

for all $f \in \hat{C}^2(\mathbb{R}^d)$, $0 \le t \le 1$, and partitions π_1, π_2 of $[0, t]$.

(d) Choose $\varphi \in C_c^\infty(\mathbb{R}^d)$ with $\varphi \ge 0$ and $\int \varphi(x)\,dx = 1$, and define $\{\varphi_n\} \subset C_c^\infty(\mathbb{R}^d)$ by $\varphi_n(x) = n^d \varphi(nx)$. Then there exists $K > 0$ such that

$$(4.10)\qquad \|S(t)(f * \varphi_n) - (S(t)f) * \varphi_n\| \le \frac{Kt\,|||f|||}{n}$$

for all $f \in \hat{C}^2(\mathbb{R}^d)$, $0 \le t \le 1$, and n.

(e) By parts (b)–(d), for each $f \in \hat{C}(\mathbb{R}^d)$ and $t \ge 0$,

$$(4.11)\qquad T(t)f = \lim_{n \to \infty} S\left(\frac{t}{n}\right)^n f$$

exists and defines a Feller semigroup $\{T(t)\}$ on $\hat{C}(\mathbb{R}^d)$ whose generator is the closure of $\{(f, Gf): f \in C_c^\infty(\mathbb{R}^d)\}$, where G is given by (1.15).

5. (a) Use Corollary 3.8 of Chapter 1 to prove the following result.

Let E be a closed convex set in $\mathbb{R}^d$ with nonempty interior, let a: $E \to S_d$ and b: $E \to \mathbb{R}^d$ be bounded and continuous, and for every $x \in E$ let $\xi(x)$ be an $\mathbb{R}^d$-valued random variable with mean vector 0 and covariance matrix $a(x)$. Suppose that $E[|\xi(x)|^3]$ is bounded in x and that, for some $t_0 > 0$,

$$(4.12)\qquad x + \sqrt{t}\,\xi(x) + tb(x) \in E \qquad \text{a.s.}$$

whenever $x \in E$ and $0 \le t \le t_0$. Suppose further that, for $0 \le t \le t_0$, the equation

(4.13) $$S(t)f(x) = E[f(x + \sqrt{t}\xi(x) + tb(x))]$$

defines a linear contraction $S(t)$ on $\hat{C}(E)$ that maps $\hat{C}^3(E)$ into $\hat{C}^3(E)$, and that there exists $K > 0$ and a norm $|||\cdot|||$ on $\hat{C}^3(E)$ with respect to which it is a Banach space such that

(4.14) $$|||S(t)f||| \le (1 + Kt)\,|||f|||$$

for all $f \in \hat{C}^3(E)$ and $0 \le t \le t_0$. Then, with G defined by (1.15), the closure of $\{(f, Gf): f \in C_c^\infty(E)\}$ is single-valued and generates a Feller semigroup on $\hat{C}(E)$.

(b) Use part (a) to prove Proposition 2.4 under the additional assumption that $\sigma_{ij}, b_i \in \bar{C}^3(\mathbb{R}^d)$ for $i, j = 1, \ldots, d$.

(c) Use part (a) to prove Theorem 2.8 under the additional assumption that $b_1, \ldots, b_d \in C^3(K_d)$.

(d) Use part (a) to prove Theorem 2.1 under the additional assumptions that $-\infty < r_0 < r_1 < \infty$, $a, b \in C^3(I)$, and $a = \sigma_0\sigma_1$, where $\sigma_i \in C^3(I)$, $\sigma_i(r_i) = 0$, and $\sigma_i > 0$ on I° for $i = 0, 1$, and $\sigma_0/(\sigma_0 + \sigma_1)$ is nondecreasing on I° and extends to an element of $C^3(I)$.

6. Fix integers $r, s \ge 2$ and index the coordinates of elements of $\mathbb{R}^{rs-1}$ by

(4.15) $$J = \{(i, j): i = 1, \ldots, r, j = 1, \ldots, s, (i, j) \ne (r, s)\}.$$

Fix $\gamma \ge 0$ and, using the notation (2.15), define $G: C^2(K_{rs-1}) \to C(K_{rs-1})$ by

(4.16) $$G = \frac{1}{2} \sum_{(i,j),(k,l) \in J} x_{ij}(\delta_{ik}\delta_{jl} - x_{kl})\,\partial_{ij}\,\partial_{kl} - \gamma \sum_{(i,j) \in J} (x_{ij} - x_{i\cdot}x_{\cdot j})\,\partial_{ij},$$

where $x_{i\cdot} = \sum_{j=1}^{s} x_{ij}$, $x_{\cdot j} = \sum_{i=1}^{r} x_{ij}$, and $x_{rs} = 1 - \sum_{(i,j) \in J} x_{ij}$. It follows from Theorem 2.8 that the closure of $\{(f, Gf): f \in C^2(K_{rs-1})\}$ is single-valued and generates a Feller semigroup on $C(K_{rs-1})$. Use Proposition 3.5 of Chapter 1 to give a direct proof of this result.

Hint: Make the change of variables

(4.17) $$p_i = x_{i\cdot}, \qquad q_j = x_{\cdot j}, \qquad D_{ij} = x_{ij} - x_{i\cdot}x_{\cdot j},$$

where $i = 1, \ldots, r - 1$ and $j = 1, \ldots, s - 1$, and define

(4.18) $$\text{degree}\left(\prod_{i=1}^{r-1} p_i^{k_i} \prod_{j=1}^{s-1} q_j^{l_j} \prod_{i=1}^{r-1}\prod_{j=1}^{s-1} D_{ij}^{m_{ij}}\right) = \sum_{i=1}^{r-1} k_i + \sum_{j=1}^{s-1} l_j + 2\sum_{i=1}^{r-1}\sum_{j=1}^{s-1} m_{ij}.$$

Let L_n be the space of polynomials of "degree" less than or equal to n.

7. Let S be a countable set, give $[0, 1]^S$ the product topology, and define

$$K = \left\{x \in [0, 1]^S : \sum_{i \in S} x_i \le 1\right\}. \tag{4.19}$$

Suppose the matrix $(q_{ij})_{i,\, j \in S}$ has nonnegative off-diagonal entries and row sums equal to zero, and

$$\sup_{j \in S} \sum_{i \in S} |q_{ij}| < \infty. \tag{4.20}$$

Show that, with

$$G = \frac{1}{2} \sum_{i,\, j \in S} x_i(\delta_{ij} - x_j)\, \partial_i\, \partial_j + \sum_{i \in S} \left(\sum_{j \in S} q_{ji} x_j\right) \partial_i, \tag{4.21}$$

the closure of $\{(f, Gf) : f \in C(K), f$ depends on only finitely many coordinates and is twice continuously differentiable$\}$ is single-valued and generates a Feller semigroup on $C(K)$.

8. Use Problem 8 of Chapter 1 to prove Theorems 3.5 and 3.6.

5. NOTES

Theorem 1.1 is a very special case of Feller's (1952) theory of one-dimensional diffusions. (Our treatment follows Mandl (1968).) Theorems 1.4, 1.5, and 1.6 are based, respectively, on partial differential equation results of Schauder (1934), Fiorenza (1959), and Il'in, Kalashnikov, and Oleinik (1962). The first two of these results are presented in Ladyzhenskaya and Ural'tseva (1968) and the latter in Dynkin (1965). Theorem 1.7 is due to Stroock and Varadhan (1979).

Essentially Theorem 2.1 appears in Ethier (1978). Theorem 2.3 is due primarily to Yamada and Watanabe (1971). Roth (1977) is responsible for Proposition 2.4, while Theorem 2.5 is based on Oleinik (1966). Remark 2.7 is due to Freidlin (1968) and Phillips and Sarason (1968). Theorem 2.8 is a slight improvement of a result of Ethier (1976).

Theorem 3.3 was obtained by Stroock (1975), and Theorems 3.5 and 3.6 by Liggett (1972).

Problem 4 is Roth's (1977) proof. Problem 5(c) generalizes Norman (1971) and Problem 5(d) is due to Norman (1972). Problem 6 can be traced to Littler (1972) and Serant and Villard (1972). Problem 7 is due to Ethier (1981).

9 BRANCHING PROCESSES

Because of their independence properties, branching processes provide a rich source of weak convergence results. Here we give four examples. Section 1 considers the classical Galton–Watson process and the Feller diffusion approximation. Section 2 gives an analogous result for two-type branching models, and Section 3 does the same for a branching process in random environments. In Section 4 conditions are given for a sequence of branching Markov processes to converge to a measure-valued process.

1. GALTON–WATSON PROCESSES

In this section we consider approximations for the Galton–Watson branching process, which can be described as follows: Let independent, nonnegative integer-valued random variables Z_0, ξ_k^n, k, $n = 1, 2, \ldots$, be given, and assume the ξ_k^n are identically distributed. Define $Z_1, Z_2, \ldots$ recursively by

(1.1)
$$Z_n = \sum_{k=1}^{Z_{n-1}} \xi_k^n.$$

Then Z_n gives the number of particles in the nth generation of a population in which individual particles reproduce independently and in the same manner. The distribution of ξ is called the *offspring distribution*, and that of Z_0 the *initial distribution.* We are interested in approximations of this process when Z_0 is large. The first such approximations are given by the law of large numbers and the central limit theorem.

1.1 Theorem Let Z_n, ξ_k^n be as above and assume $E[\xi_k^n] \equiv m < \infty$. Then

$$\lim_{Z_0 \to \infty} \frac{Z_n}{Z_0} = m^n \qquad \text{a.s.} \tag{1.2}$$

In addition let $\operatorname{var}(\xi_k^n) = E[(\xi_k^n - m)^2] \equiv \sigma^2 < \infty$. Then as $Z_0 \to \infty$ the joint distributions of

$$W_n = Z_0^{-1/2}(Z_n - m^n Z_0) \tag{1.3}$$

converge to those of

$$W_n^\infty = \sum_{l=1}^{n} m^{n-(l+1)/2} V_l \tag{1.4}$$

where the V_l are independent normal random variables with $E[V_l] = 0$ and $\operatorname{var}(V_l) = \sigma^2$.

1.2 Remark Note that

$$W_n^\infty = m W_{n-1}^\infty + m^{(n-1)/2} V_n. \tag{1.5}$$

□

Proof. The limit in (1.2) is obtained by induction. The law of large numbers gives

$$\lim_{Z_0 \to \infty} \frac{Z_1}{Z_0} = \lim_{Z_0 \to \infty} Z_0^{-1} \sum_{k=1}^{Z_0} \xi_k^1 = m \qquad \text{a.s.,} \tag{1.6}$$

and assuming $\lim_{Z_0 \to \infty} Z_{n-1}/Z_0 = m^{n-1}$ a.s.,

$$\lim_{Z_0 \to \infty} \frac{Z_n}{Z_0} = \left(\frac{Z_{n-1}}{Z_0}\right) Z_{n-1}^{-1} \sum_{k=1}^{Z_{n-1}} \xi_k^n = m^n \qquad \text{a.s.} \tag{1.7}$$

Rewriting W_n as

$$\begin{aligned} W_n &= Z_0^{-1/2}(Z_n - m^n Z_0) \\ &= Z_0^{-1/2} \sum_{l=1}^{n} (m^{n-l} Z_l - m^{n-l+1} Z_{l-1}) \\ &= \sum_{l=1}^{n} m^{n-l} \left(\frac{Z_{l-1}}{Z_0}\right)^{1/2} Z_{l-1}^{-1/2} \sum_{k=1}^{Z_{l-1}} (\xi_k^l - m), \end{aligned} \tag{1.8}$$

we see from (1.8) that it is enough to show the random variables

$$U_l = Z_{l-1}^{-1/2} \sum_{k=1}^{Z_{l-1}} (\xi_k^l - m) \tag{1.9}$$

converge in distribution to independent $N(0, \sigma^2)$ random variables. Let $\mathscr{F}_n = \sigma(Z_0, \xi_k^l: 1 \le l \le n, 1 \le k < \infty)$. Then as in the usual proof of the central limit theorem

$$\text{(1.10)} \qquad \lim_{Z_0 \to \infty} E[\exp\{i\theta U_l\} \mid \mathscr{F}_{l-1}]$$

$$= \lim_{Z_{l-1} \to \infty} E_\xi[\exp\{i\theta Z_{l-1}^{-1/2}(\xi - m)\}]^{Z_{l-1}} = \exp\{-\tfrac{1}{2}\sigma^2\theta^2\} \qquad \text{a.s.,}$$

where the expectation on the right is with respect to ξ. Therefore

$$\text{(1.11)} \qquad \lim_{Z_0 \to \infty} E\left[\prod_{l=1}^{n} \exp\{i\theta_l U_l\}\right] = \prod_{l=1}^{n} \exp\{-\tfrac{1}{2}\sigma^2\theta_l^2\}$$

and the convergence in distribution follows from the convergence of the characteristic functions. □

Of more interest is the Feller diffusion approximation.

1.3 Theorem Let $E[\xi_k^n] = 1$ (the critical case) and let $\text{var}(\xi_k^n) = \sigma^2 < \infty$. Let $Z_k^{(n)}$ be a sequence of Galton–Watson processes defined as in (1.1), and suppose $Z_0^{(m)}/m$ converges in distribution. Then W_m defined by

$$\text{(1.12)} \qquad W_m(t) \equiv \frac{Z_{[mt]}^{(m)}}{m}$$

converges in distribution in $D_{[0,\infty)}[0, \infty)$ to a diffusion with generator

$$\text{(1.13)} \qquad Af(x) = \tfrac{1}{2}\sigma^2 x f''(x), \qquad f \in C_c^\infty([0, \infty)).$$

Proof. By Theorem 2.1 of Chapter 8, $C_c^\infty([0, \infty))$ is a core for A. Note that $Z_n^{(m)}/m$ is a Markov chain with values in $E_m = \{l/m: l = 0, 1, 2, \ldots\}$, and define

$$\text{(1.14)} \qquad T_m f(x) = E\left[f\left(m^{-1} \sum_{k=1}^{mx} \xi_k\right)\right],$$

where the ξ_k are independent and distributed as the ξ_k^n. By Theorem 6.5 of Chapter 1 and Corollary 8.9 of Chapter 4, it is sufficient to show

$$\text{(1.15)} \quad \lim_{m \to \infty} \sup_{x \in E_m} |m(T_m f(x) - f(x)) - \tfrac{1}{2}\sigma^2 x f''(x)| = 0, \qquad f \in C_c^\infty([0, \infty)).$$

For $x \in E_m$, define

$$\text{(1.16)} \qquad \varepsilon_m(x) = m(T_m f(x) - f(x)) - \tfrac{1}{2}\sigma^2 x f''(x)$$

$$= mE\left[f\left(m^{-1}\sum_{k=1}^{mx} \xi_k\right) - f(x)\right] - \tfrac{1}{2}E\left[\left(\frac{1}{\sqrt{m}}\sum_{k=1}^{mx}(\xi_k - 1)\right)^2\right] f''(x)$$

$$= E\left[\int_0^1 S_{mx}^2 x(1 - v)\left(f''\left(x + v\sqrt{\frac{x}{m}}\, S_{mx}\right) - f''(x)\right) dv\right],$$

where $S_n = n^{-1/2} \sum_{k=1}^{n} (\xi_k - 1)$. Suppose the support of f is contained in $[0, c]$. Since

$$x + v\sqrt{\frac{x}{m}}\, S_{mx} = x + v\frac{1}{m}\sum_{k=1}^{mx} (\xi_k - 1) \geq x(1 - v), \tag{1.17}$$

the integrand on the right of (1.16) is zero if $v \leq 1 - c/x$. Consequently,

$$\begin{aligned} &\left|\int_0^1 S_{mx}^2\, x(1-v)\left(f''\left(x + v\sqrt{\frac{x}{m}}\, S_{mx}\right) - f''(x)\right) dv\right| \\ &\qquad \leq \int_{0 \vee (1-c/x)}^{1} S_{mx}^2\, x(1-v) 2\|f''\|\, dv \\ &\qquad = x\|f''\|((c/x) \wedge 1)^2 S_{mx}^2 . \end{aligned} \tag{1.18}$$

To show that $\lim_{m\to\infty} \sup_x \varepsilon_m(x) = 0$, it is enough to show $\lim_{m\to\infty} \varepsilon_m(x_m) = 0$ for every convergent sequence x_m, where we allow $\lim_{m\to\infty} x_m = \infty$. Since $E[S_{mx}^2] = \sigma^2$ for all m, x, inequality (1.18) implies $\lim_{m\to\infty} \varepsilon_m(x_m) = 0$ if either $\lim_{m\to\infty} x_m = 0$ or $\lim_{m\to\infty} x_m = \infty$. Therefore we need only consider the case $x_m \to x$, for $0 < x < \infty$. Replacing x by x_m in (1.18), $S_{mx_m} \Rightarrow V$, where V is $N(0, \sigma^2)$, and hence the left side of (1.18) converges in probability to zero and the right side converges in distribution to $x\|f''\|((c/x) \wedge 1)^2 V^2$. Since

$$\lim_{m\to\infty} E\left[x_m \|f''\|\left(\left(\frac{c}{x_m}\right) \wedge 1\right)^2 S_{mx_m}^2\right] = E\left[x\|f''\|\left(\left(\frac{c}{x}\right) \wedge 1\right)^2 V^2\right], \tag{1.19}$$

the dominated convergence theorem implies $\lim_{m\to\infty} \varepsilon_m(x_m) = 0$. □

Let ξ_k be a sequence of nonnegative integer-valued, independent, identically distributed random variables. Given Z_0, independent of the ξ_k, define

$$S(l) = Z_0 + \sum_{k=1}^{l} (\xi_k - 1) \tag{1.20}$$

and for $n > 0$,

$$Z_n = S\left(\sum_{i=0}^{n-1} Z_i\right). \tag{1.21}$$

Let

$$Y_n = \sum_{i=0}^{n-1} Z_i . \tag{1.22}$$

Since

$$Z_n - Z_{n-1} = \sum_{k=Y_{n-1}+1}^{Y_n} (\xi_k - 1) = \sum_{k=1}^{Z_{n-1}} \xi_{Y_{n-1}+k} - Z_{n-1}, \tag{1.23}$$

we see that Z_n is a Galton–Watson branching process (i.e., the joint distribution of the Z_n defined by (1.21) is the same as in (1.1) for the same offspring and initial distributions). We use the representation in (1.21) and Theorem 1.5 of Chapter 6 to give a generalization of Theorem 1.3.

Let $Z_n^{(m)}$ be a sequence of branching processes represented as in (1.21) with offspring distributions that may depend on m. Let $c_m > 0$ be a sequence of constants with $\lim_{m\to\infty} c_m = \infty$ and define

$$Y_m(u) = c_m^{-1}\left(\sum_{k=1}^{[mc_m u]} (\xi_k^m - 1) + Z_0^{(m)}\right) \tag{1.24}$$

and

$$W_m(t) = c_m^{-1} Z_{[mt]}^{(m)}. \tag{1.25}$$

Then

$$W_m(t) = Y_m\left(\int_0^{[mt]/m} W_m(s)\, ds\right) = Y_m\left(\int_0^{[mt]/m} \beta(W_m(s))\, ds\right), \tag{1.26}$$

where $\beta(x) = x \vee 0$.

1.4 Theorem Suppose (for simplicity) that $Y_m(0)$ is a constant, that $\lim_{m\to\infty} Y_m(0) = Y(0) > 0$, and that $\{Y_m(1)\}$ converges in distribution. Then $Y_m \Rightarrow Y$ where Y is a process with independent increments, and $W_m \Rightarrow W$ in $D_{[0,\infty]}[0, \infty)$ where W satisfies

$$W(t) = Y\left(\int_0^t \beta(W(s))\, ds\right) \tag{1.27}$$

for $t < \tau_\infty \equiv \lim_{n\to\infty} \inf\{s\colon W(s) > n\}$ and $W(t) = \infty$ for $t \geq \tau_\infty$.

1.5 Remark If $W(t) < \infty$ for all $t \geq 0$ a.s., then $W_m \Rightarrow W$ in $D_{[0,\infty)}[0, \infty)$. □

Proof. For simplicity we treat only the case where $\alpha = \sup_m E[|Y_m(1)|] < \infty$, which in particular implies $E[W_m(t)] \leq Y_m(0)e^{\alpha t}$ and hence $\tau_\infty = \infty$.

Let $X_m(t) = Y_m(t) - Y_m(0)$. Since $\{Y_m(1)\}$ converges in distribution and $Y_m(0) \to Y(0)$, $\{X_m(1)\}$ converges in distribution, and we have

$$\lim_{m\to\infty} E[\exp\{i\theta X_m(m^{-1}c_m^{-1})\}]^{mc_m} = \lim_{m\to\infty} E[\exp\{i\theta X_m(1)\}] \equiv \psi(\theta). \tag{1.28}$$

It follows that

$$\lim_{m\to\infty} E[\exp\{i\theta X_m(t)\}] = \lim_{m\to\infty} E[\exp\{i\theta X_m(m^{-1}c_m^{-1})\}]^{[mc_m t]} = \psi(\theta)^t. \tag{1.29}$$

The independence of the ξ_k implies the finite-dimensional distributions of

X_m converge to those of the process X having independent increments and satisfying $E[e^{i\theta X(t)}] = \psi(\theta)^t$. To see that the sequence $\{X_m\}$ is relatively compact, define $\mathscr{F}_t^{(m)} = \sigma(X_m(s): s \le t)$ and note that the independence implies

$$(1.30) \qquad E[\,|X_m(t+u) - X_m(t)| \wedge 1 \,|\, \mathscr{F}_t^{(m)}] \le E[\,|X_m(u)| \wedge 1] \vee E\left[\left|X_m\left(u + \frac{1}{m}\right)\right| \wedge 1\right],$$

and the relative compactness follows from Theorem 8.6 of Chapter 3.

Under the assumption that $\sup_m E[\,|Y_m(1)|\,] < \infty$, the theorem follows from Theorem 1.5 of Chapter 6 if we verify that $\tau_0 = \tau_1$ a.s. (defined as in (1.6) and (1.9), both of Chapter 6).

Note that $X_m(t) - X_m(t-) \ge -c_m^{-1}$ and it follows from Theorem 10.2 of Chapter 3 that $M(t) \equiv \inf_{s \le t} X(s)$ is continuous (see Problem 26 of Chapter 3). Consequently $X(\tau_0) = 0$ if $\tau_0 < \infty$, and by the strong Markov property the following lemma implies $\tau_0 = \tau_1$ a.s. and completes the proof. □

1.6 Lemma Let X be a right continuous process with stationary, independent increments (in particular, the distribution of $X(t+u) - X(t)$ does not depend on u) and let $X(0) = 0$. Suppose $\inf_{s \le t} X(s)$ is continuous. Then

$$(1.31) \qquad P\left\{\int_0^\varepsilon (0 \vee X(t))^{-1}\,dt = \infty\right\} = 1$$

for all $\varepsilon > 0$.

Proof. The process X may be written as a sum $X = X_1 + X_2$ where X_1 and X_2 are independent, X_2 is a compound Poisson process, and X_1 has finite expectation. Then (1.31) is equivalent to

$$(1.32) \qquad P\left\{\int_0^\varepsilon (0 \vee X_1(t))^{-1}\,dt = \infty\right\} = 1$$

for all $\varepsilon > 0$, which in turn is equivalent to showing

$$(1.33) \qquad \tau(t) \equiv \inf\left\{s: \int_0^s (0 \vee X_1(u))^{-1}\,du \ge t\right\} = 0 \quad \text{a.s.}$$

for all $t \ge 0$. Let $Z(t) = X_1(\tau(t))$. Then (cf. (1.4) of Chapter 6)

$$(1.34) \qquad \tau(t) = \int_0^t 0 \vee Z(s)\,ds = \int_0^t Z(s)\,ds$$

since Z is nonnegative by the continuity of $\inf_{s \le t} X(s)$. Let $c = E[X_1(1)]$. Since $\tau(t)$ is a stopping time and $X_1(t) - ct$ is a martingale, we have

$$(1.35) \qquad E[\tau(t)] = \int_0^t E[Z(s)]\, ds = \int_0^t E[X_1(\tau(s))]\, ds$$

$$= \int_0^t cE[\tau(s)]\, ds,$$

and it follows that $E[\tau(t)] = 0$. Hence we have (1.33). □

2. TWO-TYPE MARKOV BRANCHING PROCESSES

We alter the model considered in Section 1 in two ways. First, we assume that there are two types of particles (designated type 1 and type 2) with each particle capable, at death, of producing particles not only of its type but of the other type as well. Second, we assume each particle lives a random length of time that has an exponential distribution with parameter λ_i, $i = 1, 2$, depending on the type of the particle. Let p_{kl}^i be the probability that at death a particle of type i produces k offspring of type 1 and l offspring of type 2.

The generator for the two-type process then has the form

$$(2.1) \qquad Bf(z_1, z_2) = \lambda_1 z_1 \sum_{k,l} p_{kl}^1 (f(z_1 - 1 + k, z_2 + l) - f(z_1, z_2))$$
$$+ \lambda_2 z_2 \sum_{k,l} p_{kl}^2 (f(z_1 + k, z_2 - 1 + l) - f(z_1, z_2)).$$

Let (γ_1, γ_2) have joint distribution p_{kl}^1 and let (ψ_1, ψ_2) have joint distribution p_{kl}^2. Assume that $E[\gamma_i^3] < \infty$ and $E[\psi_i^3] < \infty$, $i = 1, 2$. Let $m_{1j} = E[\gamma_j]$ and $m_{2j} = E[\psi_j]$. We assume $m_{ij} > 0$ for all i, j; that is, there is positive probability of offspring of each type regardless of the type of the parent. We also assume that the process is critical; in other words the matrix

$$(2.2) \qquad M = \begin{pmatrix} m_{11} & m_{12} \\ m_{21} & m_{22} \end{pmatrix}$$

has largest eigenvalue 1. This implies that the matrix

$$(2.3) \qquad \Lambda = \begin{pmatrix} \lambda_1(m_{11} - 1) & \lambda_1 m_{12} \\ \lambda_2 m_{21} & \lambda_2(m_{22} - 1) \end{pmatrix}$$

has eigenvalues 0 and $-\eta$ for some $\eta > 0$. Let $(v_1, v_2)^T$ denote the eigenvector corresponding to 0 and $(\mu_1, \mu_2)^T$ the eigenvector corresponding to $-\eta$. We can take $v_1, v_2 > 0$ and μ_1 and μ_2 will have opposite signs.

Let (z_1, z_2) be fixed. We consider a sequence of processes $\{(Z_1^{(n)}, Z_2^{(n)})\}$ with generator B and $(Z_1^{(n)}(0), Z_2^{(n)}(0)) = ([nz_1], [nz_2])$. Define

$$X_n(t) = \frac{v_1 Z_1^{(n)}(nt) + v_2 Z_2^{(n)}(nt)}{n} \tag{2.4}$$

and

$$Y_n(t) = \frac{\mu_1 Z_1^{(n)}(nt) + \mu_2 Z_2^{(n)}(nt)}{n}. \tag{2.5}$$

Then X_n and $Y_n e^{n\eta t}$ are martingales (Problem 8). Since for $t > 0$, $e^{n\eta t} \to \infty$ as $n \to \infty$, the fact that $Y_n e^{n\eta t}$ is a martingale suggests that $Y_n(t) \to 0$ and, consequently, that $Z_2^{(n)}(nt)/n \approx -\mu_1 \mu_2^{-1} Z_1^{(n)}(nt)/n$ so that $Z_1^{(n)}(nt)/n \approx \mu_2 X_n(t)/(v_1 \mu_2 - v_2 \mu_1)$ and $Z_2^{(n)}(nt)/n \approx \mu_1 X_n(t)/(v_2 \mu_1 - v_1 \mu_2)$. This is indeed the case, so the limiting behavior of $\{X_n\}$ gives the limiting behavior of $(Z_1^{(n)}(nt)/n, Z_2^{(n)}(nt)/n)$ for $t > 0$.

We describe the limiting behavior of $\{Y_n\}$ in terms of

$$W_n(t) = Y_n(t) + \int_0^t n\eta Y_n(s)\, ds, \tag{2.6}$$

which is also a martingale.

Define

$$\begin{aligned} \xi_1 &= v_1(\gamma_1 - 1) + v_2 \gamma_2, & \xi_2 &= \mu_1(\gamma_1 - 1) + \mu_2 \gamma_2, \\ \varphi_1 &= v_1 \psi_1 + v_2(\psi_2 - 1), & \varphi_2 &= \mu_2 \psi_1 + \mu_2(\psi_2 - 1). \end{aligned} \tag{2.7}$$

Let $\alpha_{ij}^1 = E[\xi_i \xi_j]$ and $\alpha_{ij}^2 = E[\varphi_i \varphi_j]$, and note that $E[\xi_1] = E[\varphi_1] = 0$, $E[\xi_2] = -\eta\mu_1 \lambda_1^{-1}$, and $E[\varphi_2] = -\eta\mu_2 \lambda_2^{-1}$.

2.1 Theorem **(a)** The sequence $\{(X_n, W_n)\}$ converges in distribution to a diffusion process (X, W) with generator

$$Af(x, w) = \frac{x}{2}(a_{11} f_{xx}(x, w) + 2a_{12} f_{xw}(x, w) + a_{22} f_{ww}(x, w)) \tag{2.8}$$

where $a_{ij} = (\lambda_1 \mu_2 \alpha_{ij}^1 - \lambda_2 \mu_1 \alpha_{ij}^2)/(v_1 \mu_2 - \mu_1 v_2)$.

(b) For $T > 0$, $\sup_{t \le T} | Y_n(t) - Y_n(0)e^{-n\eta t} |$ converges to zero in probability, and hence for $0 < t_1 < t_2$, $\int_{t_1}^{t_2} n\eta Y_n(s)\, ds$ converges in distribution to $W(t_2) - W(t_1)$.

(c) For $T > 0$,

$$\sup_{t \le T} | Z_1^{(n)}(nt)/n - (\mu_2 X_n(t) - v_2 Y_n(0)e^{-n\eta t})/(v_1 \mu_2 - v_2\mu_1) |$$

converges to zero in probability.

Proof. It is not difficult to show that $C_c^\infty([0, \infty) \times (-\infty, \infty))$ is a core for A (see Problem 3). With reference to Corollary 8.6 of Chapter 4, let $f \in C_c^\infty([0, \infty) \times (-\infty, \infty))$ and set

$$(2.9) \qquad \tilde{f}_n(t) = f(X_n(t), W_n(t)).$$

To find $\tilde{g}_n$ so that $(\tilde{f}_n, \tilde{g}_n) \in \hat{\mathcal{A}}_n$ we calculate $\lim_{\varepsilon \to 0} \varepsilon^{-1} E[\tilde{f}_n(t + \varepsilon) - \tilde{f}_n(t) | \mathscr{F}_t^n]$ and obtain

(2.10)

$$\tilde{g}_n(t) = n\lambda_1 Z_1^{(n)}(nt) E_\xi \left[f\left(X_n(t) + \frac{1}{n} \xi_1, W_n(t) + \frac{1}{n} \xi_2 \right) - f(X_n(t), W_n(t)) \right]$$

$$+ n\lambda_2 Z_2^{(n)}(nt) E_\varphi \left[f\left(X_n(t) + \frac{1}{n} \varphi_1, W_n(t) + \frac{1}{n} \varphi_2 \right) - f(X_n(t), W_n(t)) \right]$$

$$+ n\eta Y_n(t) f_w(X_n(t), W_n(t)),$$

where E_ξ and E_φ only involve the ξ_i and φ_i. Recalling that $E[\xi_1] = E[\varphi_1] = 0$, $E[\xi_2] = -\eta\mu_1 \lambda_1^{-1}$, $E[\varphi_2] = -\eta\mu_2 \lambda_2^{-1}$, and $Y_n(t) = (\mu_1 Z_1^{(n)}(nt) + \mu_2 Z_2^{(n)}(nt))/n$, we see that if we expand f in a Taylor series about $(X_n(t), W_n(t))$, then the terms involving f_x have zero coefficients and the terms involving f_w cancel. Hence we have

$$(2.11) \qquad \tilde{g}_n(t) = \lambda_1 2^{-1} n^{-1} Z_1^{(n)}(nt)[\alpha_{11}^1 f_{xx}(X_n(t), W_n(t))$$

$$+ 2\alpha_{12}^1 f_{xw}(X_n(t), W_n(t)) + \alpha_{22}^1 f_{ww}(X_n(t), W_n(t))]$$

$$+ \lambda_2 2^{-1} n^{-1} Z_2^{(n)}(nt)[\alpha_{11}^2 f_{xx}(X_n(t), W_n(t))$$

$$+ 2\alpha_{12}^2 f_{xw}(X_n(t), W_n(t)) + \alpha_{22}^2 f_{ww}(X_n(t), W_n(t))]$$

$$+ O(n^{-1} X_n(t))$$

$$= Af(X_n(t), W_n(t)) + Y_n(t)h(X_n(t), W_n(t)) + O(n^{-1} X_n(t)),$$

where h is smooth and does not depend on n. The last step follows by replacing $n^{-1} Z_1^{(n)}(nt)$ by $(\mu_2 X_n(t) - \nu_1 Y_n(t))/(\nu_1 \mu_2 - \nu_2 \mu_1)$ and $n^{-1} Z_2^{(n)}(nt)$ by $(\mu_1 X_n(t) - \nu_1 Y_n(t))/(\nu_2 \mu_1 - \nu_1 \mu_2)$ and collecting terms. The error is $O(n^{-1} X_n(t))$ since $n^{-1} Z_1^{(n)}(nt)$ and $n^{-1} Z_2^{(n)}(nt)$ are bounded by a constant times $X_n(t)$.

Finally, setting

$$(2.12) \qquad f_n(t) = f(X_n(t), W_n(t)) + n^{-1} \eta^{-1} Y_n(t)h(X_n(t), W_n(t)),$$

and calculating g_n so that $(f_n, g_n) \in \hat{\mathscr{A}}_n$, we obtain

$$
\begin{aligned}
(2.13) \qquad g_n(t) &= \tilde{g}_n(t) + \eta^{-1}\lambda_1 Z^{(n)}(nt) \\
&\quad \times E_\xi\left[\left(Y_n(t) + \frac{1}{n}\xi_2\right)h\left(X_n(t) + \frac{1}{n}\xi_1, W_n(t) + \frac{1}{n}\xi_2\right) \right. \\
&\qquad\qquad \left. - Y_n(t)h(X_n(t), W_n(t))\right] \\
&\quad + \eta^{-1}\lambda_2 Z_2^{(n)}(nt) \\
&\quad \times E_\varphi\left[\left(Y_n(t) + \frac{1}{n}\varphi_2\right)h\left(X_n(t) + \frac{1}{n}\varphi_1, W_n(t) + \frac{1}{n}\varphi_2\right) \right. \\
&\qquad\qquad \left. - Y_n(t)h(X_n(t), W_n(t))\right] \\
&\quad + Y_n^2(t)h_w(X_n(t), W_n(t)) \\
&= \tilde{g}_n(t) - n^{-1}Z_1^{(n)}(nt)\mu_1 h(X_n(t), W_n(t)) \\
&\quad - n^{-1}Z_2^{(n)}(nt)\mu_2 h(X_n(t), W_n(t)) + O(n^{-1}X_n^2(t)) \\
&= Af(X_n(t), W_n(t)) + O(n^{-1}X_n(t), n^{-1}X_n^2(t)).
\end{aligned}
$$

Since X_n is a martingale and

$$
\begin{aligned}
(2.14) \qquad E[X_n^2(t)] &= X_n^2(0) + E\left[\int_0^t (\lambda_1 n^{-1}Z_1^{(n)}(ns)E_\xi[\xi_1^2] \right. \\
&\qquad \left. + \lambda_2 n^{-1}Z_2^{(n)}(ns)E_\varphi[\varphi_1^2])\, ds\right] \\
&\le X_n^2(0) + (\lambda_1 \nu_1^{-1} + \lambda_2 \nu_2^{-1})\int_0^t E[X_n(s)]\, ds \\
&= X_n^2(0) + t(\lambda_1 \nu_1^{-1} + \lambda_2 \nu_2^{-1})X_n(0),
\end{aligned}
$$

we have

$$
(2.15) \qquad E\left[\sup_{t \le T} X_n^2(t)\right] \le 4E[X_n^2(T)] < \infty.
$$

Consequently,

$$
(2.16) \qquad \lim_{n\to\infty} E\left[\sup_{t \le T} |f(X_n(t), W_n(t)) - f_n(t)|\right] = 0
$$

and

$$(2.17) \qquad \lim_{n\to\infty} E\left[\sup_{t\le T} |Af(X_n(t), W_n(t)) - g_n(t)|\right] = 0,$$

so part (a) follows by Corollary 8.6 of Chapter 4. Observe that (2.6) can be inverted to give

$$(2.18) \qquad Y_n(t) - e^{-n\eta t} Y_n(0) = \int_0^t n\eta e^{-n\eta(t-s)}(W_n(t) - W_n(s))\, ds + e^{-n\eta t}(W_n(t) - W_n(0)).$$

Let $U_n(s) = \sup_{t\le T} |W_n(t+s) - W_n(t)|$. Then for $t \le T$,

$$(2.19) \qquad |Y_n(t) - e^{-n\eta t} Y_n(0)| \le \int_0^t n\eta e^{-n\eta s} U_n(s)\, ds + e^{-n\eta t} U_n(t).$$

Part (b) follows from the fact that $U_n \Rightarrow U$ ($U(s) = \sup_{t\le T} |W(t+s) - W(t)|$) and $\lim_{s\to 0} U(s) = 0$. Part (c) follows from part (b) and the definitions of X_n and Y_n. □

3. BRANCHING PROCESSES IN RANDOM ENVIRONMENTS

We now consider continuous-time processes in which the splitting intensities are themselves stochastic processes. Let X be the population size and Λ_k, $k \ge 0$, be nonnegative stochastic processes. We want

$$(3.1) \qquad P\{X(t+\Delta t) = X(t) - 1 + k \mid \mathscr{F}_t\} = E\left[\int_t^{t+\Delta t} \Lambda_k(s) X(s)\, ds \,\middle|\, \mathscr{F}_t\right] + o(\Delta t),$$

that is (essentially), we want $\Lambda_k(t)\,\Delta t$ to be the probability that a given particle dies and is replaced by k particles. The simplest way to construct such a process is to take independent standard Poisson processes Y_k, independent of the Λ_k, and solve

$$(3.2) \qquad X(t) = X(0) + \sum_{k=0}^{\infty} (k-1) Y_k\left(\int_0^t \Lambda_k(s) X(s)\, ds\right).$$

We assume that $\sum k \int_0^t \Lambda_k(s)\, ds < \infty$ a.s. for all $t > 0$ to assure that a solution of (3.2) exists for all time. In fact, we take (3.2) to define X rather than (3.1). We leave the verification that (3.1) holds for X satisfying (3.2) as a problem (Problem 4).

By analogy with the results of Sections 1 and 2, we consider a sequence of processes X_n with corresponding intensity processes $\Lambda_k^{(n)}$ and define $Z_n(t) = X_n(nt)/n$. Assuming $X_n(0) = n$ and defining $A_n(t) = \sum_{k=0}^{\infty} (k-1)\Lambda_k^{(n)}(t)$, we get

$$(3.3) \qquad \begin{aligned} Z_n(t) &= 1 + \sum_{k=0}^{\infty} (k-1)n^{-1} Y_k\left(n^2 \int_0^t \Lambda_k^{(n)}(ns) Z_n(s)\, ds\right) \\ &= 1 + \sum_{k=0}^{\infty} (k-1)n^{-1} \tilde{Y}_k\left(n^2 \int_0^t \Lambda_k^{(n)}(ns) Z_n(s)\, ds\right) \\ &\quad + \int_0^t nA_n(ns) Z_n(s)\, ds \\ &\equiv U_n(t) + \int_0^t nA_n(ns) Z_n(s)\, ds. \end{aligned}$$

Set

$$(3.4) \qquad B_n(t) = \int_0^t nA_n(ns)\, ds.$$

Then

$$(3.5) \qquad Z_n(t) e^{-B_n(t)} = 1 + \int_0^t e^{-B_n(s)}\, dU_n(s).$$

Note that U_n is (at least) a local martingale. However, since B_n is continuous and U_n has bounded variation, no special definition of the stochastic integral is needed.

3.1 Theorem Let $D_n(t) = \int_0^t \sum_{k=0}^{\infty} (k-1)^2 \Lambda_k^{(n)}(ns)\, ds$. Suppose that $(B_n, D_n) \Rightarrow (B, D)$ and that there exist α_n satisfying $\alpha_n/n \to 0$ and

$$(3.6) \qquad \lim_{n\to\infty} \int_0^t \sum_{k>\alpha_n} (k-1)^2 \Lambda_k^{(n)}(ns)\, ds = 0 \text{ a.s.}$$

for all $t > 0$. Then Z_n converges in distribution to the unique solution of

$$(3.7) \qquad Z(t) = e^{B(t)}\left[1 + W\left(\int_0^t e^{-2B(s)} Z(s)\, dD(s)\right)\right],$$

where W is standard Brownian motion independent of B and D.

Proof. We begin by verifying the uniqueness of the solution of (3.7). Let $\gamma(t)$ satisfy

$$(3.8) \qquad \int_0^{\gamma(t)} e^{-B(s)}\, dD(s) = t$$

for $t < \Gamma = \int_0^\infty e^{-B(s)}\, dD(s)$. Then

$$(3.9)\qquad Z(\gamma(t)) = e^{B(\gamma(t))}\left[1 + W\left(\int_0^{\gamma(t)} e^{-2B(s)}Z(s)\, dD(s)\right)\right]$$
$$= e^{B(\gamma(t))}\left[1 + W\left(\int_0^t e^{-B(\gamma(s))}Z(\gamma(s))\, ds\right)\right].$$

It follows that $Z(t) = e^{B(t)}\tilde{Z}(\int_0^t e^{-B(s)}\, dD(s))$, where $\tilde{Z}$ is the unique solution of

$$(3.10)\qquad \tilde{Z}(u) = 1 + W\left(\int_0^u \tilde{Z}(s)\, ds\right).$$

Note that $\tilde{Z}$ is the diffusion arising in Theorem 1.3, with $\sigma^2 = 1$. See Theorem 1.1 of Chapter 6. By Corollary 1.9 of Chapter 3, we may assume for all $t > 0$,

$$(3.11)\qquad \lim_{n\to\infty} \sup_{s\le t} |B_n(s) - B(s)| = 0 \quad \text{a.s.},$$
$$\lim_{n\to\infty} \sup_{s\le t} |D_n(s) - D(s)| = 0 \quad \text{a.s.},$$
$$\lim_{n\to\infty} \int_0^t \sum_{k>\alpha_n} (k-1)^2\Lambda_k^n(ns)\, ds = 0 \quad \text{a.s.}$$

Since the Λ_k^n are independent of the Y_k, it is enough to prove the theorem under the assumption that the Λ_k^n are deterministic and satisfy (3.11). This amounts to conditioning on the Λ_k^n. With this assumption we have that

$$(3.12)\qquad E[Z_n(t)] = e^{B_n(t)}$$

and $V_n \equiv \int_0^t e^{-B_n(s)}\, dU_n(s)$ is a square integrable martingale with

$$(3.13)\qquad \langle V_n, V_n\rangle_t = \int_0^t e^{-2B_n(s)}Z_n(s)\, dD_n(s).$$

Fix $T > 0$. Let $\tau_n(t)$ satisfy

$$(3.14)\qquad \int_0^{\tau_n(t)} e^{-2B_n(s)}Z_n(s)\, dD_n(s) = t$$

for $t < \Gamma_n = \int_0^T e^{-2B_n(s)}Z_n(s)\, dD_n(s)$, let W_0 be a Brownian motion independent of all the other processes (we can always enlarge the sample space to obtain W_0), and define

$$(3.15)\qquad W_n(t) = \begin{cases} V_n(\tau_n(t)), & t < \Gamma_n, \\ W_0(t - \Gamma_n) + V_n(T), & \Gamma_n \le t < \infty. \end{cases}$$

Then W_n is a square integrable martingale with $\langle W_n, W_n\rangle_t = t$, and

$$(3.16)\qquad Z_n(t)e^{-B_n(t)} = 1 + W_n\left(\int_0^t e^{-2B_n(s)}Z_n(s)\, dD_n(s)\right), \qquad t \le T.$$

Since $\langle W_n, W_n \rangle_t = t$ for all n, to show that $W_n \Rightarrow W$ using Theorem 1.4(b) of Chapter 7 we need only verify

$$\lim_{n\to\infty} E\left[\sup_{s\le t} |W_n(s) - W_n(s-)|^2\right] = 0 \tag{3.17}$$

for all $t > 0$. Setting $b_n = \sup_{0\le t\le T} |B_n(t)|$, we have

$$\begin{aligned} E\left[\sup_s |W_n(s) - W_n(s-)|^2\right] &= E\left[\sup_{0\le s\le T} |V_n(s) - V_n(s-)|^2\right] \\ &\le e^{2b_n}\alpha_n^2 n^{-2} + e^{2b_n} E\left[\sum_{k>\alpha_n} \frac{(k-1)^2}{n^2} Y_k\left(n^2 \int_0^T \Lambda_k^{(n)}(ns) Z_n(s)\, ds\right)\right] \\ &= e^{2b_n}\alpha_n^2 n^{-2} + e^{2b_n} \int_0^T \sum_{k>\alpha_n} (k-1)^2 \Lambda_k^{(n)}(ns) E[Z_n(s)]\, ds \\ &\le e^{2b_n}\alpha_n^2 n^{-2} + e^{3b_n} \int_0^T \sum_{k>\alpha_n} (k-1)^2 \Lambda_k^{(n)}(ns)\, ds. \end{aligned} \tag{3.18}$$

The right side goes to zero by (3.11) and the hypotheses on α_n. Since $Z_n(t)e^{-B_n(t)}$ is a martingale,

$$P\left\{\sup_{s\le T} Z_n(t)e^{-B_n(t)} > z\right\} \le z^{-1}, \tag{3.19}$$

and relative compactness for $\{Z_n\}$ (restricted to $[0, T]$) follows easily from (3.16) and the relative compactness for $\{W_n\}$. If a subsequence $\{Z_{n_k}\}$ converges in distribution to Z, then a subsequence of

$$\{(W_{n_k}, \textstyle\int_0^t \exp\{-2B_{n_k}(s)\} Z_{n_k}(s)\, dD_{n_k}(s))\}$$

converges in distribution to $(W, \int_0^t \exp\{-2B(s)\} Z(s)\, dD(s))$, and (3.16) and the continuous mapping theorem (Corollary 1.9 of Chapter 3) imply

$$Z(t) = e^{B(t)}\left[1 + W\left(\int_0^t e^{-2B(s)} Z(s)\, dD(s)\right)\right] \tag{3.20}$$

for $t \le T$. The theorem now follows from the uniqueness of the solution of (3.20) and the fact that T is arbitrary. □

3.2 Example Let $\xi(t)$ be a standard Poisson process. Let $\Lambda_0^n(t) \equiv 1$, $\Lambda_2^n(t) \equiv 1 + n^{-1/2}(-1)^{\xi(t)}$, and $\Lambda_k^n = 0$ for $k \ne 0, 2$. This gives

$$B_n(t) = \int_0^t n^{1/2}(-1)^{\xi(ns)}\, ds \tag{3.21}$$

and

$$D_n(t) = \int_0^t (2 + n^{-1/2}(-1)^{\xi(ns)})\, ds. \tag{3.22}$$

Then $(B_n, D_n) \Rightarrow (B, D)$ where B is a standard Brownian motion and $D(t) = 2t$. (See Problem 3 of Chapter 7). The limit Z then satisfies

$$Z(t) = e^{B(t)}\left[1 + W\left(\int_0^t 2e^{-2B(s)}Z(s)\, ds\right)\right]. \tag{3.23}$$

Note that B and $M(\cdot) = W(\int_0^{\cdot} 2e^{-2B(s)}Z(s)\, ds)$ are martingales with $\langle B \rangle_t = t$, $\langle B, M \rangle_t = 0$, and $\langle M \rangle_t = \int_0^t 2e^{-2B(s)}Z(s)\, ds$. For $f \in C_c^2(\mathbb{R})$ define $g(x, y) = f(e^x(1 + y))$. Then by Itô's formula,

$$\begin{aligned} f(Z(t)) &= g(B(t), M(t)) \\ &= f(1) + \int_0^t g_x(B(s), M(s))\, dB(s) \\ &\quad + \int_0^t g_y(B(s), M(s))\, dM(s) \\ &\quad + \int_0^t (\tfrac{1}{2}g_{xx}(B(s), M(s)) + e^{-2B(s)}Z(s)g_{yy}(B(s), M(s))\, ds \\ &= f(1) + \tilde{M}(t) \\ &\quad + \int_0^t (\tfrac{1}{2}Z(s)f'(Z(s)) + (Z(s) + \tfrac{1}{2}Z(s)^2)f''(Z(s)))\, ds, \end{aligned} \tag{3.24}$$

and we see that Z is a solution of the martingale problem for A with $Af(z) = \frac{1}{2}zf'(z) + (z + \frac{1}{2}z^2)f''(z)$. □

4. BRANCHING MARKOV PROCESSES

We begin with an example of the type of process we are considering. Take the number of particles $\{N(t), t \geq 0\}$ in a collection to be a continuous-time Markov branching process; that is, each particle lives an exponentially distributed lifetime with some parameter α and at death is replaced by a random number of offspring, where the lifetimes and numbers of offspring are independent random variables. Note that N has generator (on an appropriate domain)

$$Af(k) = \sum_l \alpha k p_l(f(k - 1 + l) - f(k)) \tag{4.1}$$

where p_l is the probability that a particle has l offspring.

In addition, we assume each particle has a location in $\mathbb{R}^d$ and moves as a Brownian motion with generator $\frac{1}{2}\Delta$, and the motions are taken to be independent and independent of the lifetimes and numbers of offspring. We also assume that the offspring are initially given their parent's location at the time of birth. The state space for the process can be described as

$$E = \{(k, x_1, x_2, \ldots, x_k): k = 0, 1, 2, \ldots, x_i \in \mathbb{R}^d\}; \tag{4.2}$$

that is, k is the number of particles and the x_i are the locations. However, we modify this description later.

Of course it may not be immediately clear that such a process exists or that the above conditions uniquely determine the behavior of the process. Consequently, in order to make the above description precise, we specify the generator for the process on functions of the form $f(k, x_1, x_2, \ldots, x_k) = \prod_{i=1}^k g(x_i)$ where $g \in \mathscr{D}(\Delta)$ and $\|g\| < 1$. If the particles were moving without branching, then the generator would be

$$A_1 = \left\{\left(\prod_{i=1}^k g(x_i), \frac{1}{2} \sum_{j=1}^k \Delta g(x_j) \prod_{i \neq j} g(x_i)\right): g \in \mathscr{D}(\Delta), \|g\| < 1\right\}. \tag{4.3}$$

If there were branching but not motion, then the generator would be

$$A_2 = \left\{\left(\prod_{i=1}^k g(x_i), \sum_{j=1}^k \alpha(\varphi(g(x_j)) - g(x_j)) \prod_{i \neq j} g(x_i)\right): \|g\| < 1\right\} \tag{4.4}$$

where $\varphi(z) = \sum_{l=0}^\infty p_l z^l$, that is, φ is the generating function of the offspring distribution. The assumption that the motion and branching are independent suggests that the desired generator is $A_1 + A_2$.

More generally we consider processes in which the particles are located in a separable, locally compact, metric space E_0, move as Feller processes with generator B, die with a location dependent intensity $\alpha \in \bar{C}(E_0)$ (that is, a particle located at x at time t dies before time $t + \Delta t$ with probability $\alpha(x)\,\Delta t + o(\Delta t)$), and in which the offspring distribution is location dependent (that is, a particle that dies at x produces l offspring with probability $p_l(x)$). We assume that $p_l \in \bar{C}(E_0)$ and define

$$\varphi(z) = \sum_l p_l z^l, \qquad |z| \leq 1. \tag{4.5}$$

Note that for fixed z, $\varphi(z) \in \bar{C}(E_0)$. We also assume $\sum_l lp_l \in \bar{C}(E_0)$, that is, the mean number of offspring is finite and depends continuously on the location of the parent. We denote $(\partial/\partial z)\varphi(z)$ by $\varphi'(z)$. In particular $\varphi'(1) = \sum_l lp_l$.

The order of the x_i in the state $(k, x_1, x_2, \ldots, x_k)$ is not really important and causes some notational difficulty. Consequently, we take for the state space, not (4.2), but the space of measures

$$E = \left\{\sum_{i=1}^k \delta_{x_i}: k = 0, 1, 2, \ldots, x_i \in E_0\right\} \tag{4.6}$$

where δ_x denotes the measure with mass one at x. Of course E is a subset of the space $\mathscr{M}(E_0)$ of finite, positive, Borel measures on E_0. We topologize

$\mathcal{M}(E_0)$ (and hence E) with the weak topology. In other words, $\lim_{n\to\infty} \mu_n = \mu$ if and only if

$$\lim_{n\to\infty} \int f\, d\mu_n = \int f\, d\mu \tag{4.7}$$

for every $f \in \bar{C}(E_0)$. The weak topology is metrizable by a natural extension of the Prohorov metric (Problem 6). Note that in E, convergence in the weak topology just means the number and locations of the particles converge.

Let $\bar{C}^+(E_0) = \{f \in \bar{C}(E_0): \inf f > 0\}$. Define

$$\langle g, \mu \rangle = \int g\, d\mu, \qquad g \in \bar{C}(E_0), \qquad \mu \in E, \tag{4.8}$$

and note that for $\mu = \sum_{i=1}^k \delta_{x_i}$ and $g \in \bar{C}^+(E_0)$,

$$\prod_{i=1}^k g(x_i) = e^{\langle \log g, \mu \rangle}. \tag{4.9}$$

Extend B to the span of the constants and the original domain, by defining $B1 = 0$, so that the martingale problem for $\{(f, Bf): f \in \mathscr{D}(B) \cap \bar{C}^+(E_0)\}$ is still well-posed. With reference to (4.3) and (4.4), the generator for the process will be

$$A = \left\{\left(e^{\langle \log g, \mu \rangle}, e^{\langle \log g, \mu \rangle} \left\langle \frac{Bg + \alpha(\varphi(g) - g)}{g}, \mu \right\rangle\right): g \in \mathscr{D}(B) \cap \bar{C}^+(E_0), \|g\| < 1\right\}. \tag{4.10}$$

Let $\{S(t)\}$ denote the semigroup generated by B. By Lemma 3.4 of Chapter 4, if X is a solution of the martingale problem for A, then for g satisfying the conditions in (4.10)

$$\exp\{\langle \log S(T-t)g, X(t) \rangle\} - \int_0^t \exp\{\langle \log S(T-s)g, X(s) \rangle\} \times \left\langle \frac{\alpha(\varphi(S(T-s)g) - S(t-s)g)}{S(T-s)g}, X(s) \right\rangle ds \tag{4.11}$$

is a martingale for $0 \le t \le T$. Note that

$$\frac{d}{dt} \exp\{\langle \log S(T-t)g, \mu \rangle\} = \exp\{\langle \log S(T-t)g, \mu \rangle\} \times \left\langle -\frac{BS(T-t)g}{S(T-t)g}, \mu \right\rangle. \tag{4.12}$$

4.1 Lemma Let X be a solution of the martingale problem for (A, δ_μ). Then setting $|X(t)| = \langle 1, X(t)\rangle$ (i.e., $|X|$ is the total population size),

$$E[|X(t)|] \leq |\mu| \exp\{t\|\alpha(\varphi'(1) - 1)\|\}, \tag{4.13}$$

and

$$P\left\{\sup_t |X(t)| \exp\{-t\|\alpha(\varphi'(1) - 1)\|\} > x\right\} \leq \frac{|\mu|}{x}. \tag{4.14}$$

Proof. Let $\lambda > 0$ be a constant. Take $g = e^{-\lambda}$. Then

$$M_\lambda(t) \equiv e^{-\lambda|X(t)|} - \int_0^t e^{-\lambda|X(s)|}\langle \alpha(\varphi(e^{-\lambda}) - e^{-\lambda})e^{\lambda}, X(s)\rangle\, ds \tag{4.15}$$

is a martingale, and hence

$$E[e^{-\lambda|X(t)|}] = e^{-\lambda|\mu|} + \int_0^t E[e^{-\lambda|X(s)|}\langle \alpha(\varphi(e^{-\lambda}) - e^{-\lambda})e^{\lambda}, X(s)\rangle]\, ds, \tag{4.16}$$

so

$$\begin{aligned} E[e^{\;\lambda|X(t)|}\lambda|X(t)|] &\leq E[1 - e^{-\lambda|X(t)|}] \\ &= 1 - e^{-\lambda|\mu|} + \int_0^t E[e^{-\lambda|X(s)|}\langle \alpha(1 - e^{\lambda}\varphi(e^{-\lambda})), X(s)\rangle]\, ds \\ &\leq 1 - e^{-\lambda|\mu|} + \int_0^t E[e^{-\lambda|X(s)|}\langle \alpha(\varphi'(1) - \varphi(e^{-\lambda}))\lambda, X(s)\rangle]\, ds \\ &\leq 1 - e^{-\lambda|\mu|} + \int_0^t \|\alpha(\varphi'(1) - \varphi(e^{-\lambda}))\|\, E[e^{-\lambda|X(s)|}\lambda|X(s)|]\, ds. \end{aligned} \tag{4.17}$$

By Gronwall's inequality

$$\begin{aligned} &E[\exp\{-\lambda|X(t)|\}|X(t)|] \\ &\qquad \leq \lambda^{-1}(1 - \exp\{-\lambda|\mu|\}) \exp\{t\|\alpha(\varphi'(1) - \varphi(e^{-\lambda}))\|\}. \end{aligned} \tag{4.18}$$

Letting $\lambda \to 0$ gives (4.13).

Let

$$M(t) = \lim_{\lambda \to 0} \lambda^{-1}(1 - M_\lambda(t)) = |X(t)| - \int_0^t \langle \alpha(\varphi'(1) - 1), X(s)\rangle\, ds. \tag{4.19}$$

From (4.13) it follows that the convergence in (4.19) is in L^1 and hence M is a martingale and

$$|X(t)| \exp\{-t\|\alpha(\varphi'(1) - 1)\|\} \tag{4.20}$$

is a supermartingale. Consequently (4.14) follows from Proposition 2.16 of Chapter 2. □

4.2 Theorem Let B, α, and φ be as above, and let A be given by (4.10). Then for $\nu \in \mathscr{P}(E)$ the martingale problem for (A, ν) has a unique solution.

Proof. Existence is considered in Problem 7. To obtain uniqueness, we apply Theorem 4.2 of Chapter 4. Let X be a solution of the martingale problem for (A, ν) and define

$$u(t, g) = E[\exp\{\langle \log g, X(t)\rangle\}]. \tag{4.21}$$

Note that $u(t, \cdot)$ is a bounded continuous function on $E' = \{g \in \bar{C}^+(E_0): \|g\| < 1\}$. For $H \in \bar{C}(E')$ define

$$\Gamma H(g) = \lim_{\varepsilon \to 0+} \varepsilon^{-1}(H(e^{-\alpha\varepsilon}g + (1 - e^{-\alpha\varepsilon})\varphi(g)) - H(g)) \tag{4.22}$$

if the limit exists uniformly in g. Observe that Γ is dissipative, since $\Gamma H = \lim_{\varepsilon \to 0+} \varepsilon^{-1}(Q_\varepsilon - I)H$ where Q_ε is a contraction. We claim that $u(t, \cdot) \in \mathscr{D}(\Gamma)$ and

$$\Gamma u(t, g) = E\left[\exp\{\langle \log g, X(t)\rangle\}\left\langle \frac{\alpha(\varphi(g) - g)}{g}, X(t)\right\rangle\right]. \tag{4.23}$$

To see this write

$$\begin{aligned} &\varepsilon^{-1}(u(t, (e^{-\alpha\varepsilon}g + (1 - e^{-\alpha\varepsilon})\varphi(g))) - u(t, g)) \\ &= \varepsilon^{-1}\int_0^\varepsilon E\Bigg[\exp\{\langle \log(e^{-\alpha s}g + (1 - e^{-\alpha s})\varphi(g)), X(t)\rangle\} \\ &\qquad \times \left\langle \frac{\alpha(\varphi(g) - g)}{g + (e^{\alpha s} - 1)\varphi(g)}, X(t)\right\rangle\Bigg]\, ds. \end{aligned} \tag{4.24}$$

The expression inside the expectation on the right of (4.24) is dominated by

$$\|\alpha(\varphi(g) - g)\|\, |X(t)| \le 2\|\alpha\|\, |X(t)|, \tag{4.25}$$

so by (4.13) and the dominated convergence theorem it is enough to show

$$\begin{aligned} &\Bigg|\exp\{\langle \log(e^{-\alpha s}g + (1 - e^{-\alpha s})\varphi(g)), \mu\rangle\}\left\langle \frac{\alpha(\varphi(g) - g)}{g + (e^{\alpha s} - 1)\varphi(g)}, \mu\right\rangle \\ &\qquad - \exp\{\langle \log g, \mu\rangle\}\left\langle \frac{\alpha(\varphi(g) - g)}{g}, \mu\right\rangle\Bigg| \end{aligned} \tag{4.26}$$

converges to zero as $s \to 0$ uniformly in g for each $\mu \in E$. To check that this convergence holds, calculate the derivative of (4.26) and show that it is bounded. Finally, define

$$\mathscr{S}(t)H(g) = H(S(t)g) \tag{4.27}$$

and note that $\{\mathscr{S}(t)\}$ is a contraction semigroup on $\bar{C}(E')$. The fact that (4.11) is a martingale gives (for $T = t$)

$$u(t, g) = E[\exp\{\langle \log S(t)g, X(0)\rangle\}] + \int_0^t \Gamma u(s, S(t-s)g)\, ds \tag{4.28}$$

$$= \mathscr{S}(t)u_0(g) + \int_0^t \mathscr{S}(t-s)\Gamma u(s, g)\, ds.$$

By Proposition 5.4 of Chapter 1 there is at most one solution of this equation, so $E[\exp\{\langle \log g, X(t)\rangle\}]$ is uniquely determined. Since the linear space generated by functions of the form $\exp\{\langle \log g, \mu\rangle\}$ for $g \in \bar{C}^+(E_0)$ is an algebra that separates points in $\mathscr{M}(E_0)$, it follows that the distribution of $X(t)$ is determined, and since ν was arbitrary, the solution of the martingale problem for (A, ν) is unique by Theorem 4.2 of Chapter 4. □

We now consider a sequence of branching Markov processes X_n, $n = 1, 2, 3, \ldots$, with death intensities α_n, and offspring generating functions φ_n, in which the particles move as Feller processes in E_0 with generators B_n, extended as before. We define

$$Z_n = n^{-1}X_n. \tag{4.29}$$

Note that the state space for Z_n is

$$E_n = \left\{\mu \in \mathscr{M}(E_0) \colon \mu = n^{-1}\sum_{i=1}^{k} \delta_{x_i},\ x_i \in E_0\right\}, \tag{4.30}$$

and that Z_n is a solution of the martingale problem for

(4.31)

$$A_n = \left\{\left(\exp\{\langle n \log g, \mu\rangle\}, \exp\{\langle n \log g, \mu\rangle\}\left\langle \frac{nB_n g + n\alpha_n(\varphi_n(g) - g)}{g}, \mu\right\rangle\right) \colon g \in \mathscr{D}(B_n), \inf g > 0, \|g\| < 1\right\}.$$

Define

$$F_n(h) = n\alpha_n(1 - \varphi_n(1 - n^{-1}h) - n^{-1}h) \tag{4.32}$$

for $h \in \bar{C}^+(E_0)$, $\|h\| < n$. If $h \in \mathscr{D}(B_n) \cap \bar{C}^+(E_0)$ and $\|h\| < n$, then setting $g = 1 - n^{-1}h$ we have

$$\left(\exp\{\langle n \log(1 - n^{-1}h), \mu\rangle\}, \exp\{\langle n \log(1 - n^{-1}h), \mu\rangle\} \times \left\langle \frac{-B_n h - F_n(h)}{1 - n^{-1}h}, \mu\right\rangle\right) \in A_n. \tag{4.33}$$

For simplicity, we assume E_0 is compact (otherwise replace E_0 by its one-point compactification).

4.3 Theorem Let E_0 be compact. Let B be the generator for a Feller semigroup extended as before, and let $F(\cdot)\colon \bar{C}^+(E_0) \to \bar{C}(E_0)$. Suppose

$$\text{ex-}\lim_{n\to\infty} B_n = B, \tag{4.34}$$

$$\sup_n \|\alpha_n(\varphi_n'(1) - 1)\| < \infty, \tag{4.35}$$

and for each $k > 0$,

$$\lim_{n\to\infty} \sup_{\substack{h \in \bar{C}^+(E_0)\\ \|h\| \le k}} \|F_n(h) - F(h)\| = 0. \tag{4.36}$$

If $\{Z_n(0)\}$ has limiting distribution $\nu \in \mathscr{P}(\mathscr{M}(E_0))$, then $Z_n \Rightarrow Z$ where Z is the unique solution of the martingale problem for (A, ν) with

(4.37)

$$A = \{\exp\{-\langle h, \mu\rangle\}, \exp\{-\langle h, \mu\rangle\}\langle -Bh - F(h), \mu\rangle : h \in \mathscr{D}(B) \cap \bar{C}^+(E_0)\}.$$

4.4 Remark From Taylor's formula it follows that

$$F_n(h) = \alpha_n(\varphi_n'(1) - 1)h - n^{-1}\alpha_n \int_0^1 (1 - v)\varphi_n''(1 - n^{-1}hv)\, dv\, h^2, \tag{4.38}$$

so typically $F(h) = ah - bh^2$, where

$$a = \lim_{n\to\infty} \alpha_n(\varphi_n'(1) - 1) \tag{4.39}$$

and

$$b = \lim_{n\to\infty} \frac{\alpha_n}{2n} \varphi_n''(1). \tag{4.40}$$

In particular, if $\alpha_n \equiv n$ and $\varphi_n(z) \equiv \frac{1}{2} + \frac{1}{2}z^2$, then $F(h) = -\frac{1}{2}h^2$. Since the integral expression multiplying h^2 in (4.38) is decreasing in h, (4.35) and the existence of the limit in (4.36) imply there exist positive constants C_k, $k = 1, 2, 3$, such that

$$-(C_1 h + C_2 h^2) \le F(h) \le C_3 h. \tag{4.41}$$

□

Proof. We apply Corollary 8.16 of Chapter 4. For $h \in \mathscr{D}(B) \cap \bar{C}^+(E_0)$, there exist $h_n \in \mathscr{D}(B_n) \cap \bar{C}^+(E_0)$ such that $\lim_{n\to\infty} h_n = h$ and $\lim_{n\to\infty} B_n h_n = Bh$.

For n sufficiently large, $\|h_n\| < n$ and $h_n \geq \frac{1}{2} \inf h \equiv \varepsilon > 0$. Consequently, taking $g = (1 - n^{-1}h_n)$ in A_n,

$$(4.42)\qquad \sup_{\mu \in E_n} |\exp\{\langle n \log(1 - n^{-1}h_n), \mu\rangle\} - \exp\{-\langle h, \mu\rangle\}|$$
$$\leq \sup_{\mu \in E_n} \exp\{-\varepsilon\langle 1, \mu\rangle\}\langle 1, \mu\rangle \|n \log(1 - n^{-1}h_n) - h\|$$
$$\leq \varepsilon^{-1}\|n \log(1 - n^{-1}h_n) - h\|$$

and

$$(4.43)\qquad \sup_{\mu \in E_n} \left| \exp\{\langle n \log(1 - n^{-1}h_n), \mu\rangle\} \left\langle \frac{-B_n h_n - F_n(h_n)}{1 - n^{-1}h_n}, \mu \right\rangle \right.$$
$$\left. - e^{-\langle h, \mu\rangle}\langle -Bh - F(h), \mu\rangle \right|$$
$$\leq \sup_{\mu \in E_n} \exp\{-\varepsilon\langle 1, \mu\rangle\}\langle 1, \mu\rangle \left(\left\| \frac{-B_n h_n - F_n(h_n)}{1 - n^{-1}h_n} + Bh + F(h) \right\| \right.$$
$$\left. + \|-Bh - F(h)\| \|n \log(1 - n^{-1}h_n) - h\| \right).$$

Therefore, condition (f) of Corollary 8.7 in Chapter 4 is satisfied with $G_n = E_n$.

The compact containment condition follows from (4.14), and it remains only to verify uniqueness for the martingale problem. Uniqueness can be obtained by the same argument used in the proof of Theorem 4.2, in this case defining

$$(4.44)\qquad \Gamma H(h) = \lim_{\varepsilon \to 0+} \varepsilon^{-1}(H((h + \varepsilon F(h)) \vee 0) - H(h)).$$

The estimates in (4.41) ensure that the limit

$$(4.45)\qquad \Gamma E[\exp\{-\langle h, X(t)\rangle\}]$$
$$= \lim_{\varepsilon \to 0+} \varepsilon^{-1} E[\exp\{-\langle (h + \varepsilon F(h)) \vee 0, X(t)\rangle\} - \exp\{-\langle h, X(t)\rangle\}]$$

exists uniformly in h. □

5. PROBLEMS

1. State and prove an analogue of Theorem 1.3 for a Galton–Watson process in independent random environments. That is, let $\eta_1, \eta_2, \ldots$ be independent and uniform on [0, 1]. Suppose the ξ_k^n are conditionally independent given $\mathscr{E} = \sigma(\eta_i, i = 1, 2, \ldots)$ and $P\{\xi_k^n = l \mid \mathscr{E}\} = P_l(\eta_n)$. Define Z_n as in (1.1).

Consider a sequence of such processes $\{Z^{(m)}\}$ determined by $\{P_l^{(m)}\}$ and $Z^{(m)}(0)$, and give conditions under which $Z^{(m)}([m\cdot])/m$ converges in distribution.

2. Let $\{X_n\}$ be as in Section 2. Represent $(Z_1^{(n)}, Z_2^{(n)})$ using multiple random time changes (see Chapter 6), and use the representation to prove the convergence of $\{X_n\}$.

3. Show that $D = C_c^\infty([0, \infty) \times (-\infty, \infty))$ is a core for A given by (2.8).
Hint: Begin by looking for solutions of $u_t = Au$ of the form $e^{-a(t)x} \sin(b(t)x + cy)$ and $e^{-a(t)x} \cos(b(t)x + cy)$. Show that the bounded pointwise closure of $A|_D$ contains (f, Af) for $f = e^{-ax} \sin(bx + cy)$ and $f = e^{-ax} \cos(bx + cy)$, and the bp-closure of $\mathscr{R}(\lambda - A|_D)$ contains $e^{-ax} \sin(bx + cy)$ and $e^{-ax} \cos(bx + cy)$, and hence all of $\bar{C}([0, \infty) \times (-\infty, \infty))$. See Chapter 1, Section 3.

4. In (3.2), assume $\sum k \int_0^t \Lambda_k(s)\, ds < \infty$ a.s. for all $t > 0$.
   (a) Show that the solution of (3.2) exists for all time.
   (b) Show that the solution of (3.2) satisfies (3.1).

5. (a) In (3.23) suppose B is a Brownian motion with generator $\frac{1}{2}af'' + bf'$. Show that Z is a Markov process and find its generator.
   (b) In (3.23) suppose B is a diffusion process with generator $\frac{1}{2}\sigma^2(x)f'' + m(x)f'$. Show that (Z, B) is a Markov process and find its generator.

6. Let $\mathscr{M}(E_0)$ be the space of finite, positive Borel measures on a metric space E_0. Let $|\mu| = \mu(E_0)$ and define $\tilde{\rho}(\mu, \nu) = \rho(\mu/|\mu|, \nu/|\nu|) + ||\mu| - |\nu||$ where ρ is the Prohorov metric. Show that $\tilde{\rho}$ is a metric on $\mathscr{M}(E_0)$ giving the weak topology and that $(\mathscr{M}(E_0), \tilde{\rho})$ is complete and separable if (E_0, r) is.

7. Let B, α, and φ be as in (4.10), and let $\varepsilon > 0$. Let $B_\varepsilon \equiv B(I - \varepsilon B)^{-1}$ be the Yosida approximation of B and let $|\mu| = \mu(E_0)$, that is, the total number of particles. Set

$$A_\varepsilon = \left\{\left(e^{\langle \log g, \mu \rangle}, e^{\langle \log g, \mu \rangle}\left\langle \frac{B_\varepsilon g + \alpha e^{-\varepsilon|\mu|}(\varphi(g) - g)}{g}, \mu\right\rangle\right): g \in \bar{C}(E_0), \inf g > 0, \|g\| < 1\right\}.$$

   (a) Show that A_ε extends to an operator of the form of (2.1) in Chapter 4 and hence for each $\mu \in E$, the martingale problem for $(A_\varepsilon, \delta_\mu)$ has a unique solution. Describe the behavior of this process.
   (b) Let $\mu \in E$ and let X_ε be a solution of the martingale problem for $(A_\varepsilon, \delta_\mu)$ with sample paths in $D_E[0, \infty)$. Show that $\{X_\varepsilon, 0 < \varepsilon < 1\}$ is relatively compact and that any limit in distribution of a sequence

$\{X_{\varepsilon_n}\}$, $\varepsilon_n \to 0$, is a solution of the martingale problem for (A, δ_μ), A given by (4.10).

8. For X_n and Y_n defined by (2.4) and (2.5) show that X_n and $Y_n e^{n\eta t}$ are martingales.

6. NOTES

For a general introduction to branching processes see Athreya and Ney (1972). The diffusion approximation for the Galton–Watson process was formulated by Feller (1951) and proved by Jiřina (1969) and Lindvall (1972). These results have been extended to the age-dependent case by Jagers (1971). Theorem 1.4 is due to Grimvall (1974). The approach taken here is from Helland (1978). Work of Lamperti (1967a) is closely related.

Theorem 2.1 is from Kurtz (1978b) and has been extended by Joffe and Metivier (1984). Keiding (1975) formulated a diffusion approximation for a Galton–Watson process in a random environment that was made rigorous by Helland (1981). The Galton–Watson analogue of Theorem 3.1 is in Kurtz (1978b). See also Barbour (1976).

Branching Markov processes were extensively studied by Ikeda, Nagasawa, and Watanabe (1968, 1969). The measure diffusion approximation was given by Watanabe (1968) and Dawson (1975). Also see Wang (1982b). The limiting measure-valued process has been studied by Dawson (1975, 1977, 1979), Dawson and Hochberg (1979), and Wang (1982b).

10 GENETIC MODELS

Diffusion processes have been used to approximate discrete stochastic models in population genetics for over fifty years. In this chapter we describe several such models and show how the results of earlier chapters can be used to justify these approximations mathematically.

In Section 1 we give a fairly careful formulation of the so-called Wright–Fisher model, defining the necessary terminology from genetics; we then obtain a diffusion process as a limit in distribution. Specializing to the case of two alleles in Section 2, we give three applications of this diffusion approximation, involving stationary distributions, mean absorption times, and absorption probabilities. Section 3 is concerned with more complicated genetic models, in which the gene-frequency process may be non-Markovian. Nevertheless limiting diffusions are obtained as an application of Theorem 7.6 of Chapter 1. Finally, in Section 4, we consider the infinitely-many-neutral-alleles model with uniform mutation, and we characterize the stationary distribution of the limiting (measure-valued) diffusion process. We conclude with a derivation of Ewens' sampling formula.

1. THE WRIGHT–FISHER MODEL

We begin by introducing a certain amount of terminology from population genetics.

Every organism is initially, at the time of conception, just a single cell. It is this cell, called a *zygote* (and others formed subsequently that have the same genetic makeup), that contains all relevant genetic information about an individual and influences that of its offspring. Thus, when discussing the genetic composition of a population, it is understood that by the genetic properties of an individual member of the population one simply means the genetic properties of the zygote from which the individual developed.

Within each cell are a certain fixed number of *chromosomes*, threadlike objects that govern the inheritable characteristics of an organism. Arranged in linear order at certain positions, or *loci*, on the chromosomes, are *genes*, the fundamental units of heredity. At each locus there are several alternative types of genes that can occur; the various alternatives are called *alleles*.

We restrict our attention to *diploid* organisms, those for which the chromosomes occur in *homologous* pairs, two chromosomes being homologous if they have the same locus structure. An individual's genetic makeup with respect to a particular locus, as indicated by the unordered pair of alleles situated there (one on each chromosome), is referred to as its *genotype*. Thus, if there are r alleles, $A_1, \ldots, A_r$, at a given locus, then there are $r(r+1)/2$ possible genotypes, $A_i A_j$, $1 \le i \le j \le r$.

We also limit our discussion to *monoecious* populations, those in which each individual can act as either a male or a female parent. While many populations (e.g., plants) are in fact monoecious, this is mainly a simplifying assumption. Several of the problems at the end of the chapter deal with models for *dioecious* populations, those in which individuals can act only as male or as female parents.

To describe the Wright–Fisher model, we first propose a related model. Let $A_1, \ldots, A_r$ be the various alleles at a particular locus in a population of N adults. We assume, in effect, that generations are nonoverlapping. Let P_{ij} be the (relative) frequency of $A_i A_j$ genotypes just prior to reproduction, $1 \le i \le j \le r$. Then

$$p_i = P_{ii} + \frac{1}{2} \sum_{j:\, j>i} P_{ij} + \frac{1}{2} \sum_{j:\, j<i} P_{ji} \tag{1.1}$$

is the frequency of the allele A_i, $1 \le i \le r$.

For our purposes, the reproductive process can be roughly described as follows. Each individual has a large number of *germ* cells, cells of the same genotype (neglecting mutation) as that of the zygote. These germ cells split into *gametes*, cells containing one chromosome from each homologous pair in the original cell, thus half the usual number. We assume that the gametes are produced without *fertility* differences, that is, that all genotypes have equal probabilities of transmitting gametes in this way. The gametes then fuse at random, forming the zygotes of the next generation. We suppose that the number of such zygotes is (effectively) infinite, and so the genotypic frequencies among zygotes are $(2 - \delta_{ij})p_i p_j$, where δ_{ij} is the Kronecker delta. These are the so-called *Hardy–Weinberg* proportions.

Typically, certain individuals have a better chance than others of survival to reproductive age. Letting w_{ij} denote the *viability* of $A_i A_j$ individuals, that is, the relative likelihood that an $A_i A_j$ zygote will survive to maturity, we find that, after taking into account this viability *selection*, the genotypic frequencies become

$$P_{ij}^* = \frac{(2 - \delta_{ij})w_{ij} p_i p_j}{\sum_{k \le l} (2 - \delta_{kl}) w_{kl} p_k p_l}, \tag{1.2}$$

and the allelic frequencies have the form

$$p_i^* = \frac{\sum_j w_{ij} p_i p_j}{\sum_{k,l} w_{kl} p_k p_l}, \tag{1.3}$$

where $w_{ji} \equiv w_{ij}$ for $1 \le i < j \le r$. The population size remains infinite.

We next consider the possibility of mutations. Letting u_{ij} denote the probability that an A_i gene mutates to an A_j gene ($u_{ii} \equiv 0$), and assuming that the two genes carried by an individual mutate independently, we find that the genotypic frequencies after mutation are given by

$$P_{ij}^{**} = (1 - \tfrac{1}{2}\delta_{ij}) \sum_{k \le l} (u_{ki}^* u_{lj}^* + u_{kj}^* u_{li}^*) P_{kl}^*, \tag{1.4}$$

where

$$u_{ij}^* = \left(1 - \sum_k u_{ik}\right)\delta_{ij} + u_{ij}, \tag{1.5}$$

the latter denoting the probability that an A_i gene in a zygote appears as A_j in a gamete. The corresponding allelic frequencies have the form

$$p_i^{**} = \sum_k u_{ki}^* p_k^*, \tag{1.6}$$

as shown by the calculation

$$\begin{aligned} p_i^{**} &= \sum_j \tfrac{1}{2}(1 + \delta_{ij}) P_{i \wedge j, i \vee j}^{**} \\ &= \frac{1}{2} \sum_j \sum_{k \le l} (u_{ki}^* u_{lj}^* + u_{kj}^* u_{li}^*) P_{kl}^* \\ &= \frac{1}{2} \sum_{k \le l} (u_{ki}^* + u_{li}^*) P_{kl}^* \\ &= \frac{1}{2} \sum_{k,l} (u_{ki}^* + u_{li}^*) \tfrac{1}{2}(1 + \delta_{kl}) P_{k \wedge l, k \vee l}^* \end{aligned} \tag{1.7}$$

$$= \sum_{k,l} u_{ki}^* \tfrac{1}{2}(1 + \delta_{kl}) P_{k \wedge l, k \vee l}^*$$

$$= \sum_k u_{ki}^* p_k^*.$$

Again, the population size remains infinite.

Finally, we provide for chance fluctuations in genotypic frequencies, known as *random genetic drift*, by reducing the population to its original size N through random sampling. The genotypic frequencies P_{ij}' in the next generation just prior to reproduction have the joint distribution specified by

$$(P_{ij}')_{i \le j} \sim N^{-1} \text{ multinomial } (N, (P_{ij}^{**})_{i \le j}). \tag{1.8}$$

This is simply a concise notation for the statement that $(NP_{ij}')_{i \le j}$ has a multinomial distribution with sample size N and mean vector $(NP_{ij}^{**})_{i \le j}$. In terms of probability generating functions,

$$E\left[\prod_{i \le j} \zeta_{ij}^{NP_{ij}'}\right] = \left(\sum_{i \le j} P_{ij}^{**} \zeta_{ij}\right)^N. \tag{1.9}$$

We summarize our description of the model in the following diagram:

adult	$\xrightarrow{\text{reproduction}}$	zygote	$\xrightarrow{\text{selection}}$	adult	$\xrightarrow{\text{mutation}}$	adult	$\xrightarrow{\text{regulation}}$	adult
N, P_{ij}, p_i		$\infty, (2 - \delta_{ij}) p_i p_j, p_i$		∞, P_{ij}^*, p_i^*		$\infty, P_{ij}^{**}, p_i^{**}$		N, P_{ij}', p_i'

Observe that (1.8), (1.4), (1.5), (1.2), and (1.1) define the transition function of a homogeneous Markov chain, in that the distribution of $(P_{ij}')_{i \le j}$ is completely specified in terms of $(P_{ij})_{i \le j}$. We have more to say about this chain in Section 3. For now we simply note that if the frequencies P_{ij}^{**} are in Hardy–Weinberg form, that is, if

$$P_{ij}^{**} = (2 - \delta_{ij}) p_i^{**} p_j^{**} \tag{1.10}$$

for all $i \le j$, then

$$E\left[\prod_i \tau_i^{2Np_i'}\right] = E\left[\prod_{i \le j} (\tau_i \tau_j)^{NP_{ij}'}\right] \tag{1.11}$$

$$= \left(\sum_{i \le j} (2 - \delta_{ij}) p_i^{**} p_j^{**} \tau_i \tau_j\right)^N$$

$$= \left(\sum_i p_i^{**} \tau_i\right)^{2N}$$

by (1.9), implying that

$$(p_1', \ldots, p_r') \sim (2N)^{-1} \text{ multinomial } (2N, (p_1^{**}, \ldots, p_r^{**})). \tag{1.12}$$

One can check that (1.10) holds (for all $(P_{ij})_{i \leq j}$) in the absence of selection (i.e., $w_{ij} = 1$ for all $i \leq j$) and, more generally, when viabilities are multiplicative (i.e., there exist $v_1, \ldots, v_r$ such that $w_{ij} = v_i v_j$ for all $i \leq j$), but not in general.

Nevertheless, whether or not (1.10) necessarily holds, (1.12), (1.6), (1.5), and (1.3) define the transition function of a homogeneous Markov chain, in that the distribution of $(p'_1, \ldots, p'_{r-1})$ is completely specified in terms of $(p_1, \ldots, p_{r-1})$. (Note that $p_r = 1 - \sum_{i=1}^{r-1} p_i$.) This chain, which may or may not be related to the previously described chain by (1.1), is known as the Wright–Fisher model. Although its underlying biological assumptions are somewhat obscure, the Wright–Fisher model is probably the best-known discrete stochastic model in population genetics. Nevertheless, because of the complicated nature of its transition function, it is impractical to study this Markov chain directly. Instead, it is typically approximated by a diffusion process. Before indicating in the next section the usefulness of such an approach, we formulate the diffusion approximation precisely.

Put $\mathbb{Z}_+ = \{0, 1, \ldots\}$ and

$$(1.13) \qquad K_N = \left\{ (2N)^{-1}\alpha: \alpha \in (\mathbb{Z}_+)^{r-1}, \sum_{i=1}^{r-1} \alpha_i \leq 2N \right\}.$$

Given constants $\mu_{ij} \geq 0$ (with $\mu_{ii} = 0$) and $\sigma_{ij} (= \sigma_{ji})$ real for $i, j = 1, \ldots, r$, let $\{Z^N(k), k = 0, 1, \ldots\}$ be a homogeneous Markov chain in K_N whose transition function, starting at $(p_1, \ldots, p_{r-1}) \in K_N$, is specified by (1.12), (1.6), (1.5), (1.3), and

$$(1.14) \qquad u_{ij} = [(2N)^{-1}\mu_{ij}] \wedge r^{-1}, \qquad w_{ij} = [1 + (2N)^{-1}\sigma_{ij}] \vee \tfrac{1}{2},$$

[margin note: $\sigma_{ij} = 2N(w_{ij} - 1)$]

[margin note: $= 2Ns$]

and let T_N be the associated transition operator on $C(K_N)$, that is,

[margin note: if $w_{ij} = 1+s$]

$$(1.15) \qquad T_N f(p_1, \ldots, p_{r-1}) = E[f(p'_1, \ldots, p'_{r-1})].$$

Let

[margin note: should be negative $\sigma_{ii} = 0$]

[margin note: if $w_{ii} = 1$]

$$(1.16) \qquad K = \left\{ p = (p_1, \ldots, p_{r-1}) \in [0, 1]^{r-1}: \sum_{i=1}^{r-1} p_i \leq 1 \right\},$$

and form the differential operator

$$(1.17) \qquad G = \frac{1}{2} \sum_{i,j=1}^{r-1} a_{ij}(p) \frac{\partial^2}{\partial p_i \, \partial p_j} + \sum_{i=1}^{r-1} b_i(p) \frac{\partial}{\partial p_i},$$

where

$$(1.18) \qquad a_{ij}(p) = p_i(\delta_{ij} - p_j)$$

and

$$(1.19) \qquad b_i(p) = - \sum_{j=1}^{r} \mu_{ij} p_i + \sum_{j=1}^{r} \mu_{ji} p_j + p_i \left(\sum_{j=1}^{r} \sigma_{ij} p_j - \sum_{k,l=1}^{r} \sigma_{kl} p_k p_l \right).$$

Let $\{T(t)\}$ be the Feller semigroup on $C(K)$ generated by the closure of $A \equiv \{(f, Gf): f \in C^2(K)\}$ (see Theorem 2.8 of Chapter 8), and let X be a diffusion process in K with generator A (i.e., a Markov process with sample paths in $C_K[0, \infty)$ corresponding to $\{T(t)\}$).

Finally, let X^N be the process with sample paths in $D_K[0, \infty)$ defined by

$$X^N(t) = Z^N([2Nt]). \tag{1.20}$$

1.1 Theorem. Under the above conditions,

$$\lim_{N\to\infty} \sup_{0\le t\le t_0} \sup_{p\in K_N} |T_N^{[2Nt]}f(p) - T(t)f(p)| = 0 \tag{1.21}$$

for every $f \in C(K)$ and $t_0 \ge 0$. Consequently, if $X^N(0) \Rightarrow X(0)$ in K, then $X^N \Rightarrow X$ in $D_K[0, \infty)$.

Proof. To prove (1.21), it suffices by Theorem 6.5 of Chapter 1 to show that

$$\lim_{N\to\infty} \sup_{p\in K_N} |2N(T_N - I)f(p) - Gf(p)| = 0 \tag{1.22}$$

for all $f \in C^2(K)$. By direct calculation,

$$2NE[p_i' - p_i] = b_i(p) + O(N^{-1}), \tag{1.23}$$

$$2N \operatorname{cov}(p_i', p_j') = a_{ij}(p) + O(N^{-1}), \tag{1.24}$$

$$2NE[(p_i' - p_i)^4] = O(N^{-1}), \tag{1.25}$$

and hence

$$2NE[(p_i' - p_i)(p_j' - p_j)] = a_{ij}(p) + O(N^{-1}), \tag{1.26}$$

$$2NP\{|p_i' - p_i| > \varepsilon\} = O(N^{-1}), \tag{1.27}$$

as $N \to \infty$, uniformly in $p \in K_N$, for $i, j = 1, \ldots, r - 1$ and $\varepsilon > 0$. We leave to the reader the proof that (1.23), (1.26), and (1.27) imply (1.22) (Problem 1).

The second assertion of the theorem is a consequence of (1.21) and Corollary 8.9 of Chapter 4. □

2. APPLICATIONS OF THE DIFFUSION APPROXIMATION

In this section we describe three applications of Theorem 1.1. We obtain diffusion approximations of stationary distributions, mean absorption times, and absorption probabilities of the one-locus, two-allele Wright–Fisher model. Moreover, we justify these approximations mathematically by proving appropriate limit theorems.

Let $\{Z^N(k), k = 0, 1, \ldots\}$ be a homogeneous Markov chain in

(2.1) $$K_N = \left\{\frac{i}{2N} : i = 0, 1, \ldots, 2N\right\}$$

whose transition function, starting at $p \in K_N$, is specified by (1.12), (1.6), (1.5), (1.3), and (1.14) in the special case $r = 2$. Concerning the parameters μ_{12}, μ_{21}, σ_{11}, σ_{12}, and σ_{22} in (1.14), we assume that $\sigma_{12} = 0$ and relabel the remaining parameters as μ_1, μ_2, σ_1, and σ_2 to reduce the number of subscripts. (Since all viabilities can be multiplied by a constant without affecting (1.3), it involves no real loss of generality to take $w_{12} = 1$, i.e., $\sigma_{12} = 0$.) Then Z^N satisfies

(2.2) $$P\left\{Z^N(k+1) = \frac{j}{2N} \,\middle|\, Z^N(k) = p\right\} = \binom{2N}{j}(p^{**})^j(1 - p^{**})^{2N-j},$$

where

(2.3) $$p^{**} = (1 - u_1)p^* + u_2(1 - p^*),$$

(2.4) $$p^* = \frac{w_1 p^2 + p(1-p)}{w_1 p^2 + 2p(1-p) + w_2(1-p)^2},$$

and

(2.5) $$u_i = [(2N)^{-1}\mu_i] \wedge \tfrac{1}{2}, \qquad w_i = [1 + (2N)^{-1}\sigma_i] \vee \tfrac{1}{2}, \qquad i = 1, 2.$$

Recalling the other notation that is needed, T_N is the transition operator on $C(K_N)$ defined by (1.15),

(2.6) $$K = [0, 1],$$

and X^N is the process with sample paths in $D_K[0, \infty)$ defined by (1.20). Finally, $\{T(t)\}$ is the strongly continuous semigroup on $C(K)$ generated by the closure of $A \equiv \{(f, Gf): f \in C^2(K)\}$, where

(2.7) $$G = \tfrac{1}{2}a(p)\frac{\partial^2}{\partial p^2} + b(p)\frac{\partial}{\partial p},$$

(2.8) $$a(p) = p(1-p),$$

and

(2.9) $$b(p) = -\mu_1 p + \mu_2(1-p) + p(1-p)[\sigma_1 p - \sigma_2(1-p)],$$

and X is a diffusion process in K with generator A. Clearly, the conclusions of Theorem 1.1 are valid in this special case.

As a first application, we consider the problem of approximating stationary distributions. Note that, if $\mu_1, \mu_2 > 0$, then $0 < p^{**} < 1$ for all $p \in K_N$, so Z^N is an irreducible, finite Markov chain. Hence it has a unique stationary distribution $\nu_N \in \mathscr{P}(K_N)$. (Of course, we may also regard ν_N as an element of $\mathscr{P}(K)$.) Because ν_N cannot be effectively evaluated, we approximate it by the stationary distribution of X.

2.1 Lemma Let $\mu_1, \mu_2 > 0$. Then X has one and only one stationary distribution $\nu \in \mathscr{P}(K)$. Moreover, ν is absolutely continuous with respect to Lebesgue measure on K, and its density h_0 is the unique $C^2(0, 1)$ solution of the equation

$$\tag{2.10} \tfrac{1}{2}(ah_0)'' - (bh_0)' = 0$$

with $\int_0^1 h_0(p)\, dp = 1$. Consequently, there is a constant $\beta > 0$ such that

$$\tag{2.11} h_0(p) = \beta p^{2\mu_2 - 1}(1 - p)^{2\mu_1 - 1} \exp\{\sigma_1 p^2 + \sigma_2(1 - p)^2\}$$

for $0 < p < 1$.

Proof. We first prove existence. Define h_0 by (2.11), where β is such that $\int_0^1 h_0(p)\, dp = 1$, and define $\nu \in \mathscr{P}(K)$ by $\nu(dp) = h_0(p)\, dp$. Since $(ah_0)(0+) = (ah_0)(1-) = 0$ and (2.10) holds, integration by parts yields

$$\tag{2.12} \int_K Gf\, d\nu = 0, \qquad f \in C^2(K).$$

It follows that $\int_K \bar{A}f\, d\nu = 0$ for all $f \in \mathscr{D}(\bar{A})$, and hence

$$\tag{2.13} \int_K T(t)f\, d\nu = \int_K f\, d\nu, \qquad f \in C(K),\ t \geq 0.$$

Thus, ν is a stationary distribution for X. (See Chapter 4, Section 9.)

Turning to uniqueness, let $\nu \in \mathscr{P}(K)$ be a stationary distribution for X, and define

$$\tag{2.14} c(p) = \int_{[0,\, p]} b(q)\nu(dq).$$

Since (2.13) holds, so does (2.12). In particular, $c(1) = 0$ (take $f(p) = p$), so

$$\tag{2.15} \begin{aligned} 0 &= \int_K \tfrac{1}{2}af''\, d\nu - \int_K b(q)\left[\int_q^1 f''(p)\, dp\right]\nu(dq) \\ &= \int_K \tfrac{1}{2}af''\, d\nu - \int_0^1 f''(p)\left[\int_{[0,\, p]} b(q)\nu(dq)\right] dp \\ &= \int_K f''(p)[\tfrac{1}{2}a(p)\nu(dp) - c(p)\, dp] \end{aligned}$$

for every $f \in C^2(K)$. Therefore,

$$\tag{2.16} \tfrac{1}{2}a(p)\nu(dp) = c(p)\, dp$$

as Borel measures on K. Since $a > 0$ on $(0, 1)$, we have $\nu(dp) \ll dp$ on $(0, 1)$, so by (2.14), c is continuous on $[0, 1)$. By (2.16), $\nu(dp)/dp$ has a continuous version h_0 on $(0, 1)$, and $\frac{1}{2}ah_0 = c$ there. Thus, by (2.14),

$$\frac{1}{2}a(p)h_0(p) = c(0) + \int_0^p b(q)h_0(q)\,dq \tag{2.17}$$

for $0 < p < 1$. It follows that $h_0 \in C^2(0, 1)$ and (2.10) holds, so since h_0 is Lebesgue integrable, it is easily verified that h_0 has the form (2.11) for some constant β. To verify that β is such that $\int_0^1 h_0(p)\,dp = 1$, and to complete the proof that ν is uniquely determined, it suffices to show that $\nu(\{0\}) = \nu(\{1\}) = 0$. By (2.11), $(ah_0)(0+) = 0$, so by (2.17), $c(0) = 0$. Since $b(0) = \mu_2 > 0$, we have $\nu(\{0\}) = 0$ by (2.14). Similarly, $c(1-) = 0$, so $\nu(\{1\}) = 0$, completing the proof. □

We now show that ν_N, the stationary distribution of Z^N, can be approximated by ν, the stationary distribution of X.

2.2 Theorem. Let $\mu_1, \mu_2 > 0$. Then $\nu_N \Rightarrow \nu$ on K.

Proof We could essentially quote Theorem 9.10 of Chapter 4, but instead we give a self-contained proof. By Prohorov's theorem, $\{\nu_N\}$ is relatively compact in $\mathscr{P}(K)$, so every subsequence of $\{\nu_N\}$ has a further subsequence $\{\nu_{N'}\}$ that converges weakly to some limit $\tilde{\nu} \in \mathscr{P}(K)$. Consequently, for all $f \in C(K)$ and $t \geq 0$,

$$\begin{aligned} \int_K T(t)f\,d\tilde{\nu} &= \lim_{N' \to \infty} \int_{K_{N'}} T(t)f\,d\nu_{N'} \\ &= \lim_{N' \to \infty} \int_{K_{N'}} T_{N'}^{[2N't]}f\,d\nu_{N'} \\ &= \lim_{N' \to \infty} \int_{K_{N'}} f\,d\nu_{N'} \\ &= \int_K f\,d\tilde{\nu}, \end{aligned} \tag{2.18}$$

so $\tilde{\nu}$ is a stationary distribution for X. (This gives, incidentally, an alternative proof of the existence of a stationary distribution for X.) By Lemma 2.1, $\tilde{\nu} = \nu$. Hence the original sequence converges weakly to ν. □

2.3 Remark If $\sigma_1 = \sigma_2 = 0$, then the stationary distribution of Theorem 2.2 belongs to the beta family. In particular, its mean is $\mu_2/(\mu_1 + \mu_2)$, which also happens to be the stable equilibrium of the corresponding deterministic model $\dot{p} = -\mu_1 p + \mu_2(1 - p)$. □

If $\mu_1 = 0$ (respectively, if $\mu_2 = 0$), then 1 (respectively, 0) is an absorbing state for the Markov chain Z^N, so interest centers on the time until absorption and (if $\mu_1 = \mu_2 = 0$) the probability of ultimate absorption at 1. Of the three cases, $\mu_1 = \mu_2 = 0$, $\mu_1 = 0 < \mu_2$, and $\mu_1 > 0 = \mu_2$, it will suffice (by symmetry) to treat the first two. Let

$$F = \{0, 1\} \quad \text{if} \quad \mu_1 = \mu_2 = 0, \qquad F = \{1\} \quad \text{if} \quad \mu_1 = 0,\ \mu_2 > 0. \tag{2.19}$$

Then F is the set of absorbing states of Z^N (and hence X^N), and it is easily seen (by uniqueness, e.g.) that F is also the set of absorbing states of the diffusion X.

Define $\zeta: D_K[0, \infty) \to [0, \infty]$ by

$$\zeta(x) = \inf \{t \geq 0: x(t) \in F \text{ or } x(t-) \in F\} \tag{2.20}$$

where $\inf \varnothing = \infty$. Then ζ is Borel measurable, so we can define

$$\tau_N = \zeta(X^N), \qquad \tau = \zeta(X). \tag{2.21}$$

In order to study the mean absorption time $E[\tau_N]$ and (if $\mu_1 = \mu_2 = 0$) the absorption probability $P\{X^N(\tau_N) = 1\}$, we regard $E[\tau]$ and $P\{X(\tau) = 1\}$ as approximations, the latter two quantities being quite easy to evaluate. The following theorem is used to provide a justification for these approximations.

2.4 Theorem Let $\mu_1 = 0$, $\mu_2 \geq 0$. If $X^N(0) \Rightarrow X(0)$ in K, then $(X^N, \tau_N) \Rightarrow (X, \tau)$ in $D_K[0, \infty) \times [0, \infty]$, and the sequence $\{\tau_N\}$ is uniformly integrable.

We postpone the proof to the end of this section.

2.5 Remark It follows from Theorem 2.4 that, for each $N \geq 1$, τ_N and τ have finite expectations, hence they are a.s. finite, and therefore $X^N(\tau_N)$ and $X(\tau)$ are defined a.s. and equal to 0 or 1 a.s. These facts are needed in the corollaries that follow. It is also worth noting that the first assertion of Theorem 2.4 is *not* a consequence of Corollary 1.9 of Chapter 3 because the function $x \mapsto (x, \zeta(x))$ on $D_K[0, \infty)$ is discontinuous at every $x \in D_K[0, \infty)$ for which $\zeta(x) < \infty$, hence discontinuous a.s. with respect to the distribution of X. □

2.6 Corollary Let $\mu_1 = 0$, $\mu_2 \geq 0$. If $X^N(0) \Rightarrow X(0)$, then

$$\lim_{N \to \infty} E[\tau_N] = E[\tau]. \tag{2.22}$$

Proof. By Theorem 2.4, $\tau_N \Rightarrow \tau$, so (2.22) follows from the uniform integrability of $\{\tau_N\}$. □

2.7 Corollary Let $\mu_1 = \mu_2 = 0$. If $X^N(0) \Rightarrow X(0)$, then

$$\lim_{N \to \infty} P\{X^N(\tau_N) = 1\} = P\{X(\tau) = 1\}. \tag{2.23}$$

Proof. Define $\xi: D_K[0, \infty) \times [0, \infty] \to K$ by $\xi(x, t) = x(t)$ for $0 \le t < \infty$ and $\xi(x, \infty) = \frac{1}{2}$, say. Then ξ is continuous at each point of $C_K[0, \infty) \times [0, \infty)$, hence continuous a.s. with respect to the distribution of (X, τ). By Theorem 2.4 of this chapter and Corollary 1.9 of Chapter 3, $X^N(\tau_N) \Rightarrow X(\tau)$, so (2.23) follows from Remark 2.5. □

To evaluate the right sides of equations (2.22) and (2.23), we introduce the notation $P_p\{\cdot\}$ and $E_p[\cdot]$, where the subscript p denotes the starting point of the process involved in the probability or expectation.

2.8 Proposition Suppose first that $\mu_1 = \mu_2 = 0$. Let f_0 be the unique $C^2(K)$ solution of the differential equation $Gf_0 = 0$ with boundary conditions $f_0(0) = 0, f_0(1) = 1$. Then $P_p\{X(\tau) = 1\} = f_0(p)$ for all $p \in K$. Consequently,

$$P_p\{X(\tau) = 1\} = \frac{\int_0^p e^{-\lambda(q)}\,dq}{\int_0^1 e^{-\lambda(q)}\,dq}, \tag{2.24}$$

where $\lambda(q) = \sigma_1 q^2 + \sigma_2(1 - q)^2$.

Now suppose that $\mu_1 = 0$, $\mu_2 \ge 0$. Let g_0 be the unique $C(K) \cap C^2(K - F)$ solution of the differential equation $Gg_0 = -1$ with boundary conditions

$$\begin{aligned} &g_0(0) = g_0(1) = 0, && \text{if} \quad \mu_2 = 0,\\ &g_0'(0+) \text{ finite}, \quad g_0(1) = 0, && \text{if} \quad \mu_2 > 0. \end{aligned} \tag{2.25}$$

Then $E_p[\tau] = g_0(p)$ for all $p \in K$. Consequently, if $\mu_2 = 0$, then

$$\begin{aligned} E_p[\tau] = &\int_0^p e^{-\lambda(q)} \int_q^{1/2} \frac{2e^{\lambda(r)}}{r(1-r)}\,dr\,dq \\ &- f_0(p) \int_0^1 e^{-\lambda(q)} \int_q^{1/2} \frac{2e^{\lambda(r)}}{r(1-r)}\,dr\,dq, \end{aligned} \tag{2.26}$$

and, if $\mu_2 > 0$, then

$$E_p[\tau] = \int_p^1 q^{-2\mu_2} e^{-\lambda(q)} \int_0^q 2r^{2\mu_2 - 1}(1 - r)^{-1} e^{\lambda(r)}\,dr\,dq. \tag{2.27}$$

Proof. For each $f \in C^2(K)$, $p \in K$, and $t \ge 0$, the optional sampling theorem implies that

$$E_p[f(X(\tau \wedge t))] = f(p) + E_p\left[\int_0^{\tau \wedge t} Gf(X(s))\,ds\right]. \tag{2.28}$$

Replacing f by f_0 and letting $t \to \infty$, we get $P_p\{X(\tau) = 1\} = f_0(p)$ for all $p \in K$. Here we are using Remark 2.5.

Given $h \in C(K)$, let g be the unique $C(K) \cap C^2(K - F)$ solution of the differential equation $Gg = -h$ with boundary conditions analogous to (2.25). Then $g = Bh$, where

$$(2.29) \qquad Bh(p) = \int_0^p e^{-\lambda(q)} \int_q^{1/2} \frac{2e^{\lambda(r)}}{r(1-r)} h(r)\, dr\, dq - f_0(p) \int_0^1 e^{-\lambda(q)} \int_q^{1/2} \frac{2e^{\lambda(r)}}{r(1-r)} h(r)\, dr\, dq$$

if $\mu_2 = 0$, and

$$(2.30) \qquad Bh(p) = \int_p^1 q^{-2\mu_2} e^{-\lambda(q)} \int_0^q 2r^{2\mu_2 - 1}(1-r)^{-1} e^{\lambda(r)} h(r)\, dr\, dq$$

if $\mu_2 > 0$. Consequently, $Bh \in C^2(K)$ if $h \in C^1(K)$ and $h = 0$ on F. Thus, we choose $\{h_n\} \subset C^1(K)$ with $h_n \geq 0$ and $h_n \nearrow 1 - \chi_F$. Replacing f by $g_n = Bh_n$ in (2.28), and noting that bp-lim $g_n = g_0$, we obtain

$$(2.31) \qquad E_p[g_0(X(\tau \wedge t))] = g_0(p) - E_p[\tau \wedge t]$$

for all $p \in K$ and $t \geq 0$. Letting $t \to \infty$ gives $E_x[\tau] = g_0(p)$ by (2.25).

We leave it to the reader to verify (2.24), (2.26), and (2.27). □

2.9 Remark As simple special cases of Proposition 2.8, one can check that

$$(2.32) \qquad P_p\{X(\tau) = 1\} = \begin{cases} p, & \sigma = 0, \\ \dfrac{1 - e^{-2\sigma p}}{1 - e^{-2\sigma}}, & \sigma \neq 0, \end{cases}$$

if $\mu_1 = \mu_2 = 0$ and $\sigma_1 = -\sigma_2 = \sigma$, and

$$(2.33) \qquad E_p[\tau] = -2[p \log p + (1-p) \log (1-p)]$$

if $\mu_1 = \mu_2 = \sigma_1 = \sigma_2 = 0$. To get some idea of the effect that selection can have, observe that, when $p = \frac{1}{2}$, the right side of (2.32) becomes $1/(1 + e^{-\sigma})$, and in view of (2.5), $|\sigma|$ may differ significantly from zero. When interpreting (2.33) (or, more generally, (2.26) or (2.27)), one must keep in mind that, because of (1.20), time is measured in units of $2N$ generations. □

In order to prove Theorem 2.4, we need the following lemma.

2.10 Lemma Let $\mu_1 = 0$, $\mu_2 \geq 0$, and define the function g_0 as in Proposition 2.8. Then there exist positive integers, κ and N_0, depending only on μ_2, σ_1, and σ_2, such that

$$(2.34) \qquad E_p[\tau_N] \leq \kappa g_0(p), \qquad p \in K_N,\ N \geq N_0.$$

Proof. Define the operator G_N on $C(K_N)$ by

$$G_N f(p) = 2N\{E_p[f(Z^N(1))] - f(p)\}. \tag{2.35}$$

For $0 \le \varepsilon < \frac{1}{2}$, let

$$V(\varepsilon) = \begin{cases} (\varepsilon, 1-\varepsilon) & \text{if } \mu_2 = 0 \\ [0, 1-\varepsilon) & \text{if } \mu_2 > 0, \end{cases} \tag{2.36}$$

and put $V_N(\varepsilon) = K_N \cap V(\varepsilon)$. (Note that $V(0) = K - F$.) The first step in the proof is to show that

$$\lim_{m\to\infty} \overline{\lim_{N\to\infty}} \sup_{p \in V_N(m/2N)} G_N g_0(p) \le -1. \tag{2.37}$$

A fourth-order Taylor expansion yields

$$G_N g_0(p) = 2N\left\{\sum_{k=1}^{3} \frac{1}{k!} E_p[(Z^N(1) - p)^k] g_0^{(k)}(p) + \frac{1}{3!} E_p\left[(Z^N(1) - p)^4 \int_0^1 (1-t)^3 g_0^{(4)}(p + t(Z^N(1) - p))\, dt\right]\right\} \tag{2.38}$$

for all $p \in V_N(0)$ and $N \ge 1$. (We note that the integral under the fourth expectation exists, as does the expectation itself.) Expanding each of the moments about p^{**}, which we temporarily denote by γ, we obtain

$$\begin{aligned} G_N g_0(p) = {} & 2N(\gamma - p)g_0'(p) + \frac{2N}{2}\left[\frac{\gamma(1-\gamma)}{2N} + (\gamma - p)^2\right] g_0''(p) \\ & + \frac{2N}{6}\left[\frac{\gamma(1-\gamma)(1-2\gamma)}{(2N)^2} + \frac{3\gamma(1-\gamma)(\gamma - p)}{2N} + (\gamma - p)^2\right] g_0^{(3)}(p) \\ & + \theta_{N,p}\frac{2N}{6}\left[\frac{3\gamma^2(1-\gamma)^2}{(2N)^2} + \frac{\gamma(1-\gamma)(1 - 6\gamma + 6\gamma^2)}{(2N)^3}\right. \\ & \left. + \frac{4\gamma(1-\gamma)(1-2\gamma)(\gamma - p)}{(2N)^2} + \frac{6\gamma(1-\gamma)(\gamma-p)^2}{2N} + (\gamma - p)^2\right] \\ & \times \int_0^1 (1-t)^3 \sup_{0 \le q \le 1} |g_0^{(4)}(p + t(q - p))|\, dt \end{aligned} \tag{2.39}$$

for all $p \in V_N(0)$ and $N \ge 1$, where $|\theta_{N,p}| \le 1$. Now one can easily check that

$$2N(p^{**} - p) = b(p)(1 + O(N^{-1})) + O(\mu_2(1-p)^2 N^{-1}) \tag{2.40}$$

and

$$p^{**}(1 - p^{**}) = a(p)(1 + O(N^{-1})) + O(\mu_2(1-p)^2 N^{-1}) \tag{2.41}$$

as $N \to \infty$, uniformly in $p \in K_N$. Also, by direct calculation, there exist constants $M_1, \ldots, M_4$, depending only on μ_2, σ_1, and σ_2, such that

$$(2.42)\qquad \begin{aligned} |g_0'(p)| &\le M_1 \log\left[\frac{1+\mu_2}{(p+\mu_2)(1-p)}\right], \\ |g_0^{(k)}(p)| &\le M_k\left[\frac{1+\mu_2}{(p+\mu_2)(1-p)}\right]^{k-1}, \end{aligned}$$

for all $p \in V(0)$ and $k = 2, 3, 4$. Finally, we note that

$$(2.43)\qquad \min_{0\le q\le 1} \frac{[p+t(q-p)+\mu_2][1-p-t(q-p)]}{1+\mu_2} \ge (1-t)\frac{(p+\mu_2)(1-p)}{1+\mu_2}$$

since the minimum occurs at $q = 0$ or $q = 1$, and therefore

$$(2.44)\qquad \int_0^1 (1-t)^3 \sup_{0\le q\le 1} |g_0^{(4)}(p+t(q-p))|\, dt \le M_4\left[\frac{1+\mu_2}{(p+\mu_2)(1-p)}\right]^3$$

for all $p \in V(0)$. By (2.39)–(2.42) and (2.44), we have

$$(2.45)\qquad \begin{aligned} G_N g_0(p) \le{}& \tfrac{1}{2}a(p)g_0''(p) + b(p)g_0'(p) \\ &+ \left(\tfrac{1}{12}M_3 + \tfrac{1}{4}M_4\right)\left[\frac{N(p+\mu_2)(1-p)}{1+\mu_2}\right]^{-1} \\ &+ \tfrac{1}{24}M_4\left[\frac{N(p+\mu_2)(1-p)}{1+\mu_2}\right]^{-2} + O(N^{-1}) \end{aligned}$$

as $N \to \infty$, uniformly in $p \in V_N(0)$, which implies (2.37) since $Gg_0 = -1$.

Next, we show that

$$(2.46)\qquad \overline{\lim_{N\to\infty}}\, G_N g_0\left(1 - \frac{m}{2N}\right) < 0, \qquad m = 1, 2, \ldots.$$

Fix $m \ge 1$, and let $p_N = (1 - m/2N) \vee 0$. Since g_0'' is bounded on $(0, \tfrac{1}{2})$ if $\mu_2 > 0$, there exists a constant M_0, depending only on μ_2, σ_1, and σ_2, such that

$$(2.47)\qquad \begin{aligned} g_0''(p) &= \frac{-2}{p(1-p)} - \left[\frac{2\mu_2}{p} + 2(\sigma_1 p - \sigma_2(1-p))\right]g_0'(p) \\ &\le \frac{-2}{1-p} + M_0 \log\left[\frac{1+\mu_2}{(p+\mu_2)(1-p)}\right] \end{aligned}$$

for all $p \in V(0)$. Consequently, a second-order Taylor expansion yields

$$(2.48)\qquad \begin{aligned} G_N g_0(p_N) &= 2NE_{p_N}[Z^N(1) - p_N]g_0'(p_N) \\ &\quad + 2NE_{p_N}\left[(Z^N(1) - p_N)^2 \int_0^1 (1-t)g_0''(Y)\,dt\right] \\ &\le 2NE_{p_N}[Z^N(1) - p_N]g_0'(p_N) \\ &\quad + 2NM_0 E_{p_N}\left[(Z^N(1) - p_N)^2 \int_0^1 (1-t) \right. \\ &\qquad\qquad \left. \times \log\left(\frac{1+\mu_2}{(Y+\mu_2)(1-Y)}\right) dt\right] \\ &\quad - 4NE_{p_N}\left[(Z^N(1) - p_N)^2 \int_0^1 (1-t)(1-Y)^{-1}\,dt\right] \end{aligned}$$

for each $N \ge 1$, where $Y = p_N + t(Z^N(1) - p_N)$. Using (2.42) and (2.43), the first two terms on the right side of (2.48) are $O(N^{-1}\log N)$ as $N \to \infty$, so (2.46) is equivalent to

$$(2.49)\qquad \underline{\lim}_{N\to\infty} NE_{p_N}\left[(Z^N(1) - p_N)^2 \int_0^1 (1-t)(1-Y)^{-1}\,dt\right] > 0.$$

Denoting p^{**} by p_N^{**} when $p = p_N$, the expectation in (2.49) can be expressed as

$$(2.50)\qquad \sum_{l=0}^{2N}\left(\frac{l}{2N} - \frac{m}{2N}\right)^2 \int_0^1 (1-t)\left(\frac{m}{2N} + t\left(\frac{l}{2N} - \frac{m}{2N}\right)\right)^{-1} dt \times \binom{2N}{l}(1-p_N^{**})^l (p_N^{**})^{2N-l},$$

and since $1 - p_N^{**} = m/2N + O(N^{-2})$, an application of Fatou's lemma shows that the left side of (2.49) is at least as large as

$$(2.51)\qquad \sum_{l=0}^{\infty}(l-m)^2 \int_0^1 (1-t)[m + t(l-m)]^{-1}\,dt\,\frac{m^l e^{-m}}{l!},$$

which of course is positive; here we have used the familiar Poisson approximation of the binomial distribution. This proves (2.46), and, by symmetry,

$$(2.52)\qquad \overline{\lim}_{N\to\infty} G_N g_0\left(\frac{m}{2N}\right) < 0, \qquad m = 1, 2, \ldots,$$

if $\mu_2 = 0$. Combining (2.37), (2.46), and (if $\mu_2 = 0$) (2.52), we conclude that there exist κ and N_0 such that

$$(2.53)\qquad G_N g_0(p) \le -\frac{1}{\kappa}, \qquad p \in V_N(0),\ N \ge N_0.$$

Finally, to complete the proof of the lemma, we note that

(2.54) $$\left\{g_0(Z^N(n)) - \frac{1}{2N}\sum_{m=0}^{n-1} G_N g_0(Z^N(m)), \qquad n = 0, 1, \ldots\right\}$$

is a martingale, so by the optional sampling theorem and (2.53),

(2.55) $$\begin{aligned} E_p[g_0(X^N(\tau_N \wedge t))] &= g_0(p) + E_p\left[\frac{1}{2N}\sum_{m=0}^{2N(\tau_N \wedge t)-1} G_N g_0\left(X^N\left(\frac{m}{2N}\right)\right)\right] \\ &\le g_0(p) - \frac{1}{\kappa} E_p[\tau_N \wedge t] \end{aligned}$$

for all $p \in V_N(0)$, $t = 0, 1/2N, 2/2N, \ldots$, and $N \ge N_0$, and this implies (2.34). □

Proof of Theorem 2.4. For $0 \le \varepsilon < \frac{1}{2}$, define $\zeta^\varepsilon: D_K[0, \infty) \to [0, \infty]$ by

(2.56) $$\zeta^\varepsilon(x) = \inf\{t \ge 0: x(t) \notin V(\varepsilon) \text{ or } x(t-) \notin V(\varepsilon)\}$$

where $V(\varepsilon)$ is given by (2.36). (Note that $\zeta^0 = \zeta$; see (2.20).) Then $\zeta^\varepsilon(x) \to \zeta(x)$ and $\varepsilon \to 0$ for every $x \in C_K[0, \infty)$, hence a.s. with respect to the distribution of X. In addition, we leave it to the reader to show that ζ^ε is continuous a.s. with respect to the distribution of X for $0 < \varepsilon < \frac{1}{2}$ (Problem 3).

We apply the result of Problem 5 of Chapter 3 with $S = D_K[0, \infty)$, $S' = D_K[0, \infty) \times [0, \infty]$, $h(x) = (x, \zeta(x))$, $h_k(x) = (x, \zeta^{\varepsilon_k}(x))$, where $0 < \varepsilon_k < \frac{1}{2}$ and $\varepsilon_k \to 0$ as $k \to \infty$. To conclude that $h(X^N) \Rightarrow h(X)$, that is, $(X^N, \tau_N) \Rightarrow (X, \tau)$, we need only show that

(2.57) $$\lim_{\varepsilon \to 0} \overline{\lim_{N \to \infty}} P\{\rho(\zeta^\varepsilon(X^N), \zeta(X^N)) > \delta\} = 0$$

for every $\delta > 0$, where $\rho(t, t') = |\tan^{-1} t - \tan^{-1} t'|$. By the strong Markov property, the inequality $|\tan^{-1} t - \tan^{-1}(t + s)| \le s$ for $s, t \ge 0$, and Lemma 2.10, we have

(2.58) $$\begin{aligned} P\{\rho(\zeta^\varepsilon(X^N), \zeta(X^N)) > \delta\} &\le E[\chi_{\{\tau_N^\varepsilon < \infty\}} P_{X^N(\tau_N^\varepsilon)}\{\tau_N > \delta\}] \\ &\le \delta^{-1} \sup_{p \in K_N \cap V(\varepsilon)^c} E_p[\tau_N] \\ &\le \delta^{-1}\kappa \sup_{p \in K \cap V(\varepsilon)^c} g_0(p) \end{aligned}$$

for all $\delta > 0$, $N \ge N_1$, and $0 < \varepsilon < \frac{1}{2}$, where $\tau_N^\varepsilon = \zeta^\varepsilon(X^N)$. Since $g_0 = 0$ on F, (2.57) follows from (2.58).

Finally, we claim that the uniform integrability of $\{\tau_N\}$ is also a consequence of Lemma 2.10. Let g_0, κ, and N_0 be as in that lemma. Then

(2.59) $$\sup_{p \in K_N} P_p\{\tau_N > t\} \le t^{-1} \sup_{p \in K_N} E_p[\tau_N] \le t^{-1}\kappa \sup_{p \in K} g_0(p)$$

for all $N \geq N_0$ and $t > 0$, so there exist $t_0 > 0$ and $\eta < 1$ such that

$$\sup_{N \geq 1} \sup_{p \in K_N} P_p\{\tau_N > t_0\} < \eta. \tag{2.60}$$

Letting $E_m^N = \{\tau_N > m/2N\}$, we conclude from the strong Markov property that, if $n \geq [2Nt_0]$, then

$$P_p(E_{m+n}^N) = E_p[\chi_{E_m^N} P_{X^N(m/2N)}(E_n^N)] \leq \eta P_p(E_m^N) \tag{2.61}$$

for each $m \geq 0$, $p \in K_N$, and $N \geq 1$. Consequently, for arbitrary $l > 1$,

$$\begin{aligned} E_p[(\tau_N)^l] &= \sum_{k=0}^{\infty} \sum_{j=kn+1}^{(k+1)n} \left(\frac{j}{2N}\right)^l P_p\left\{\tau_N = \frac{j}{2N}\right\} \\ &\leq \sum_{k=0}^{\infty} (k+1)^l t_0^l P_p(E_{kn}^N) \\ &\leq t_0^l \sum_{k=0}^{\infty} (k+1)^l \eta^k < \infty \end{aligned} \tag{2.62}$$

for all $p \in K_N$ and $N \geq 1$, where $n = [2Nt_0]$. Since the bound in (2.62) is uniform in p and N, the uniform integrability of $\{\tau_N\}$ follows, and the proof is complete. □

3. GENOTYPIC-FREQUENCY MODELS

There are several one-locus genetic models in which the successive values (from generation to generation, typically) of the vector $(P_{ij})_{i \leq j}$ of genotypic frequencies form a Markov chain, but the successive values of the vector $(p_1, \ldots, p_{r-1})$ of allelic frequencies do not; nevertheless, the genotypic frequencies rapidly converge to Hardy–Weinberg proportions, while, at a slower rate, the allelic frequencies converge to a diffusion process. Thus, in this section, we formulate a limit theorem for diffusion approximations of Markov chains with two "time scales," and we apply it to two models. Further applications are mentioned in the problems.

Let K and H be compact, convex subsets of $\mathbb{R}^m$ and $\mathbb{R}^n$, respectively, having nonempty interiors, and assume that $0 \in H$. We begin with two lemmas involving first-order differential and difference equations, in which the zero solution is globally asymptotically stable.

3.1 Lemma Let $c: K \times \mathbb{R}^n \to \mathbb{R}^n$ be of class C^2 and such that the solution $Y(t, x, y)$ of the differential equation

$$\frac{d}{dt} Y(t, x, y) = c(x, Y(t, x, y)), \qquad Y(0, x, y) = y, \tag{3.1}$$

exists for all $(t, x, y) \in [0, \infty) \times K \times H$ and satisfies

$$\lim_{t \to \infty} \sup_{(x, y) \in K \times H} |Y(t, x, y)| = 0. \tag{3.2}$$

Then there exists a compact set E, with $K \times H \subset E \subset K \times \mathbb{R}^n$, such that $(x, y) \in E$ implies $(x, Y(t, x, y)) \in E$ for all $t \geq 0$, and the formula

$$S(t)h(x, y) = h(x, Y(t, x, y)) \tag{3.3}$$

defines a strongly continuous semigroup $\{S(t)\}$ on $C(E)$ (with sup norm). The generator B of $\{S(t)\}$ has $C^2(E) \equiv \{f|_E : f \in C^2(\mathbb{R}^m \times \mathbb{R}^n)\}$ as a core, and

$$Bh(x, y) = \sum_{l=1}^{n} c_l(x, y) \frac{\partial}{\partial y_l} h(x, y) \text{ on } K \times H, \qquad h \in C^2(E). \tag{3.4}$$

Finally,

$$\lim_{t \to \infty} \sup_{(x, y) \in E} |S(t)h(x, y) - h(x, 0)| = 0, \qquad h \in C(E). \tag{3.5}$$

Proof. Let $E = \{(x, Y(t, x, y)): (t, x, y) \in [0, \infty) \times K \times H\}$. By (3.2), E is bounded, and E is easily seen to be closed. If $(x, y) \in E$, then $y = Y(s, x, y_0)$ for some $s \geq 0$ and $y_0 \in H$. Hence $(x, Y(t, x, y)) = (x, Y(t + s, x, y_0)) \in E$ for all $t \geq 0$, and

$$\lim_{t \to \infty} \sup_{(x, y) \in E} |Y(t, x, y)| \leq \lim_{t \to \infty} \sup_{s \geq 0} \sup_{(x, y_0) \in K \times H} |Y(t + s, x, y_0)| = 0 \tag{3.6}$$

by (3.2). It is straightforward to check that $\{S(t)\}$ is a strongly continuous semigroup on $C(E)$. By the mean value theorem, $C^2(E) \subset \mathscr{D}(B)$ and (3.4) holds. Since $S(t): C^2(E) \to C^2(E)$ for all $t \geq 0$, $C^2(E)$ is a core for B. Finally, (3.5) is a consequence of (3.6). □

3.2 Remark If $c(x, y) = \varphi(x)y$ for all $(x, y) \in K \times \mathbb{R}^n$, where $\varphi: K \to \mathbb{R}^n \otimes \mathbb{R}^n$ is of class C^2, and if for each $x \in K$ all eigenvalues of $\varphi(x)$ have negative real parts, then c satisfies the hypotheses of Lemma 3.1. In this case, $Y(t, x, y) \equiv e^{t\varphi(x)}y$. □

3.3 Lemma Given $\delta_\infty > 0$, let $c: K \times \mathbb{R}^n \to \mathbb{R}^n$ be continuous, such that the solution $Y(k, x, y)$ of the difference equation

$$\delta_\infty^{-1}\{Y(k + 1, x, y) - Y(k, x, y)\} = c(x, Y(k, x, y)), \qquad Y(0, x, y) = y, \tag{3.7}$$

which exists for all $(k, x, y) \in \mathbb{Z}_+ \times K \times H$, satisfies

$$\lim_{k \to \infty} \sup_{(x, y) \in K \times H} |Y(k, x, y)| = 0. \tag{3.8}$$

Then there exists a compact set E, with $K \times H \subset E \subset K \times \mathbb{R}^n$, such that $(x, y) \in E$ implies $(x, Y(k, x, y)) \in E$ for $k = 0, 1, \ldots,$ and the formula

$$S(t)h(x, y) = E[h(x, Y(V(t), x, y))], \tag{3.9}$$

where V is a Poisson process with parameter δ_∞^{-1}, defines a strongly continuous semigroup $\{S(t)\}$ on $C(E)$. The generator B of $\{S(t)\}$ is the bounded linear operator

$$B = \delta_\infty^{-1}(Q - I), \tag{3.10}$$

where Q is defined on $C(E)$ by $Qh(x, y) = h(x, y + \delta_\infty c(x, y))$. Finally,

$$\lim_{k\to\infty} \sup_{(x,\, y)\in E} |Q^k h(x, y) - h(x, 0)| = 0, \qquad h \in C(E). \tag{3.11}$$

Proof. Let $E = \{(x, Y(k, x, y)): (k, x, y) \in \mathbb{Z}_+ \times K \times H\}$. The details of the proof are left to the reader. □

3.4 Remark If $c(x, y) = \varphi(x)y$ for all $(x, y) \in K \times \mathbb{R}^n$, where $\varphi: K \to \mathbb{R}^n \otimes \mathbb{R}^n$ is continuous, and if for each $x \in K$ all eigenvalues of $\varphi(x)$ belong to $\{\zeta \in \mathbb{C}: |\zeta + \delta_\infty^{-1}| < \delta_\infty^{-1}\}$, then c satisfies the hypotheses of Lemma 3.3. In this case, $Y(k, x, y) \equiv (I + \delta_\infty \varphi(x))^k y$. □

The preceding lemmas allow us to state the following theorem. Recall the assumptions on K and H in the second paragraph of this section.

3.5 Theorem For $N = 1, 2, \ldots$, let $\{Z^N(k), k = 0, 1, \ldots\}$ be a Markov chain in a metric space E_N with a transition function $\mu_N(z, \Gamma)$, and denote $\int f(z')\mu_N(z, dz')$ by $E_z[f(Z^N(1))]$. Suppose both $\Phi_N: E_N \to K$ and $\Psi_N: E_N \to H$ are Borel measurable, define $X^N(k) = \Phi_N(Z^N(k))$ and $Y^N(k) = \Psi_N(Z^N(k))$ for each $k \geq 0$, and let $\varepsilon_N > 0$ and $\delta_N > 0$. Assume that $\lim_{N\to\infty} \delta_N = \delta_\infty \in [0, \infty)$ and $\lim_{N\to\infty} \varepsilon_N/\delta_N = 0$. Let each of the functions $a: K \times \mathbb{R}^n \to \mathbb{R}^m \otimes \mathbb{R}^m$, $b: K \times \mathbb{R}^n \to \mathbb{R}^m$, and $c: K \times \mathbb{R}^n \to \mathbb{R}^n$ be continuous, and suppose that, for i, $j = 1, \ldots, m$ and $l = 1, \ldots, n$,

$$\varepsilon_N^{-1} E_z[X_i^N(1) - x_i] = b_i(x, y) + o(1), \tag{3.12}$$

$$\varepsilon_N^{-1} E_z[(X_i^N(1) - x_i)(X_j^N(1) - x_j)] = a_{ij}(x, y) + o(1), \tag{3.13}$$

$$\varepsilon_N^{-1} E_z[(X_i^N(1) - x_i)^4] = o(1), \tag{3.14}$$

$$\delta_N^{-1} E_z[Y_l^N(1) - y_l] = c_l(x, y) + o(1), \tag{3.15}$$

$$\delta_N^{-1} E_z[(Y_l^N(1) - E_z[Y_l(1)])^2] = o(1), \tag{3.16}$$

as $N \to \infty$, uniformly in $z \in E_N$, where $x = \Phi_N(z)$ and $y = \Psi_N(z)$. Let

$$G = \frac{1}{2} \sum_{i,\, j=1}^{m} a_{ij}(x, 0) \frac{\partial^2}{\partial x_i\, \partial x_j} + \sum_{i=1}^{m} b_i(x, 0) \frac{\partial}{\partial x_i}, \tag{3.17}$$

and assume that the closure of $\{(f, Gf): f \in C^2(K)\}$ is single-valued and generates a Feller semigroup $\{U(t)\}$ on $C(K)$ corresponding to a diffusion process X in K. Suppose further that c satisfies the hypotheses of Lemma 3.1 if $\delta_\infty = 0$ and of Lemma 3.3 if $\delta_\infty > 0$. Then the following conclusions hold:

(a) If $X^N(0) \Rightarrow X(0)$ in K, then $X^N([\cdot/\varepsilon_N]) \Rightarrow X$ in $D_K[0, \infty)$.

(b) If $\{t_N\} \subset [0, \infty)$ satisfies $\lim_{N\to\infty} t_N = \infty$, then $Y^N([t_N/\delta_N]) \Rightarrow 0$ in H.

3.6 Remark **(a)** Observe that (3.12)–(3.14) are analogous to (1.23), (1.26), and (1.25), except that the right sides of (3.12) and (3.13) depend on y. But because of (3.15), (3.16), and the conditions on c, it is clear (at least intuitively) that conclusion (b) holds, and hence that $Y^N([t/\varepsilon_N]) \Rightarrow 0$ for each $t > 0$. Thus, in the "slow" time scale (i.e., t/ε_N), the Y^N process is "approximately" zero, and therefore the limiting generator has the form (3.17).

(b) We note that (3.14) implies

$$\varepsilon_N^{-1} P_z\{|X_i^N(1) - x_i| > \gamma\} = o(1), \qquad \gamma > 0, \tag{3.18}$$

for $i = 1, \ldots, m$. (Here and below, we omit the phrase, "as $N \to \infty$, uniformly in $z \in E_N$, where $x = \Phi_N(z)$ and $y = \Psi_N(z)$.") In fact the latter condition suffices in the proof. □

Proof. Let E be as in Lemma 3.1 if $\delta_\infty = 0$ and as in Lemma 3.3 if $\delta_\infty > 0$, and apply Theorem 7.6(b) of Chapter 1 with $L_N = B(E_N)$ (with sup norm), $L = C(E)$, and $\pi_N: L \to L_N$ defined by $\pi_N f(z) = f(x, y)$, where $x = \Phi_N(z)$ and $y = \Psi_N(z)$. Define the linear operator A on L by

$$A = \frac{1}{2} \sum_{i,j=1}^{m} a_{ij}(x, y) \frac{\partial^2}{\partial x_i \, \partial x_j} + \sum_{i=1}^{m} b_i(x, y) \frac{\partial}{\partial x_i}, \qquad \mathscr{D}(A) = C^2(E), \tag{3.19}$$

and let B be the generator of the semigroup $\{S(t)\}$ on L defined in Lemma 3.1 if $\delta_\infty = 0$ and in Lemma 3.3 if $\delta_\infty > 0$. Define P on L by $Ph(x, y) = h(x, 0)$, and let $D = \mathscr{D}(A) \cap \mathscr{R}(P)$ and $D' = C^2(E)$. By the lemmas, D' is a core for B, and (7.15) of Chapter 1 holds if $\delta_\infty = 0$, while (7.16) of that chapter holds (where Q is as in Lemma 3.3) if $\delta_\infty > 0$. Let $A_N = \varepsilon_N^{-1}(T_N - I)$, where T_N is defined on L_N by $T_N f(z) = E_z[f(Z^N(1))]$, and let $\alpha_N = \delta_N/\varepsilon_N$.

Given $f \in D$,

$$\begin{aligned} A_N \pi_N f(z) &= \varepsilon_N^{-1} E_z[f(X^N(1), y) - f(x, y)] \\ &= \sum_{i=1}^{m} \varepsilon_N^{-1} E_z[X_i^N(1) - x_i] f_{x_i}(x, y) \\ &\quad + \frac{1}{2} \sum_{i,j=1}^{m} \varepsilon_N^{-1} E_z[(X_i^N(1) - x_i)(X_j^N(1) - x_j)] f_{x_i x_j}(x, y) \\ &\quad + \sum_{i,j=1}^{m} \varepsilon_N^{-1} E_z\Big[(X_i^N(1) - x_i)(X_j^N(1) - x_j) \\ &\quad \cdot \int_0^1 (1 - u)\{f_{x_i x_j}(x + u(X^N(1) - x), y) - f_{x_i x_j}(x, y)\}\, du\Big] \\ &= Af(x, y) + o(1), \end{aligned} \tag{3.20}$$

where the first equality uses the fact that $f \in \mathscr{R}(P)$, the second uses the convexity of K, and the third depends on (3.12), (3.13), and (3.18). (To show that the remainder term in the Taylor expansion is $o(1)$, integrate separately over $|X^N(1) - x| \leq \gamma$ and $|X^N(1) - x| > \gamma$, using the Schwarz inequality, (3.13), and (3.18).) This implies (7.17) of Chapter 1.

Given $h \in D'$,

$$(3.21)\quad \delta_N^{-1} E_z[h(X^N(1), Y^N(1)) - h(x, E_z[Y^N(1)])]$$
$$= \sum_{i=1}^{m} \delta_N^{-1} E_z[X_i^N(1) - x_i] h_{x_i}(x, y)$$
$$+ \frac{1}{2} \sum_{i,j=1}^{m} \delta_N^{-1} E_z[(X_i^N(1) - x_i)(X_j^N(1) - x_j)\eta^N(h_{x_i x_j}, z, Z^N(1))]$$
$$+ \sum_{i=1}^{m} \sum_{j=1}^{n} \delta_N^{-1} E_z[(X_i^N(1) - x_i)(Y_j^N(1) - E_z[Y_j^N(1)])\eta^N(h_{x_i y_j}, z, Z^N(1))]$$
$$+ \frac{1}{2} \sum_{i,j=1}^{n} \delta_N^{-1} E_z[(Y_i^N(1) - E_z[Y_i^N(1)])$$
$$\times (Y_j^N(1) - E_z[Y_j^N(1)])\eta^N(h_{y_i y_j}, z, Z^N(1))]$$

where

$$(3.22)\quad \eta^N(g, z, Z^N(1))$$
$$\equiv \int_0^1 (1 - u) g(x + u(X^N(1) - x), E_z[Y^N(1)] + u(Y^N(1) - E_z[Y^N(1)]))\, du.$$

(Here the convexity of H and of $K \times H$ is used.) But the right side of (3.21) is $o(1)$ by the Schwarz inequality, (3.12), (3.13), and (3.16). Consequently,

$$(3.23)\quad \alpha_N^{-1} A_N \pi_N h(z) = \delta_N^{-1}\{h(x, E_z[Y^N(1)]) - h(x, y)\} + o(1)$$
$$= Bh(x, y) + o(1)$$

by (3.15) and either (3.4) or (3.10). This implies (7.18) of Chapter 1.

Finally, define $\rho: K \to E$ by $\rho(x) = (x, 0)$, and observe that $G(f \circ \rho) = (PAf) \circ \rho$ for all $f \in D$. Since the closure of $\{(f, Gf): f \in C^2(K)\}$ is single-valued and generates a Feller semigroup $\{U(t)\}$ on $C(K)$, the Feller semigroup $\{T(t)\}$ on $\bar{D} = \mathscr{R}(P)$ satisfying

$$(3.24)\quad U(t)(f \circ \rho) = [T(t)f] \circ \rho, \qquad f \in \bar{D},\ t \geq 0,$$

is generated by the closure of $PA|_D$. Theorem 7.6(b) of Chapter 1, together with Corollary 8.9 of Chapter 4, yields conclusion (a) of the theorem, and Corollary 7.7 of Chapter 1 (with $h(x, y) \equiv |y|$) yields conclusion (b). □

3.7 Remark **(a)** Since $\lim_{N\to\infty} \varepsilon_N = 0$, (3.12) implies that (3.13) is equivalent to

(3.25) $$\varepsilon_N^{-1} E_z[(X_i^N(1) - E_z[X_i^N(1)])(X_j^N(1) - E_z[X_j^N(1)])] = a_{ij}(x, y) + o(1)$$

for $i, j = 1, \ldots, m$ and that (3.14) is equivalent to

(3.26) $$\varepsilon_N^{-1} E_z[(X_i^N(1) - E_z[X_i^N(1)])^4] = o(1)$$

for $i = 1, \ldots, m$. It is often more convenient to verify (3.25) and (3.26). We note also that, if $\lim_{N\to\infty} \delta_N = 0$, then (3.15) implies that (3.16) is equivalent to

(3.27) $$\delta_N^{-1} E_z[(Y_l^N(1) - y_l)^2] = o(1)$$

for $l = 1, \ldots, n$.

(b) It is sometimes possible to avoid explicit calculation of (3.16) by using the following inequalities. Let ξ and η be real random variables with means $\bar{\xi}$ and $\bar{\eta}$ such that $|\xi| \leq M$ a.s. and $|\eta| \leq M$ a.s. Then

(3.28) $$\text{var}(\xi + \eta) \leq 2(\text{var}\,\xi + \text{var}\,\eta)$$

and

(3.29) $$\begin{aligned} \text{var}(\xi\eta) &\leq E[(\xi\eta - \bar{\xi}\bar{\eta})^2] \\ &\leq 2E[(\xi - \bar{\xi})^2\eta^2] + 2\bar{\xi}^2 E[(\eta - \bar{\eta})^2] \\ &\leq 2M^2(\text{var}\,\xi + \text{var}\,\eta). \end{aligned}$$ □

In the remainder of this section we consider two genetic models in detail, showing that Theorem 3.5 is applicable to both of them. Although the models differ substantially, they have several features in common, and it may be worthwhile pointing these out explicitly beforehand.

Adopting the convention that coordinates of elements of $\mathbb{R}^{r(r+1)/2}$ are to be indexed by $\{(i, j): 1 \leq i \leq j \leq r\}$, the state space E_N of the underlying Markov chain Z^N in both cases is the space of genotypic frequencies

(3.30) $$E_N = \left\{N^{-1}\nu: \nu \in (\mathbb{Z}_+)^{r(r+1)/2}, \sum_{i \leq j} \nu_{ij} = N\right\}.$$

In applying Theorem 3.5, the transformations $\Phi_N: E \to \mathbb{R}^{r-1}$ and $\Psi_N: E_N \to \mathbb{R}^{r(r+1)/2}$ are given by

(3.31) $$\Phi_N((P_{ij})_{i\leq j}) = (p_1, \ldots, p_{r-1})$$

and

(3.32) $$\Psi_N((P_{ij})_{i\leq j}) = (Q_{ij})_{i\leq j},$$

where p_i is the allelic frequency (1.1) and Q_{ij} is the Hardy–Weinberg deviation

(3.33) $$Q_{ij} = P_{ij} - (2 - \delta_{ij})p_i p_j.$$

Observe that $\Phi_N(E_N) \subset K$, where K is defined by (1.16). As we see, in both of our examples, the functions $a: K \times \mathbb{R}^{r(r+1)/2} \to \mathbb{R}^{r-1} \otimes \mathbb{R}^{r-1}$ and $b: K \times \mathbb{R}^{r(r+1)/2} \to \mathbb{R}^{r-1}$ are such that $a_{ij}(p, 0)$ and $b_i(p, 0)$ are given by the right sides of (1.18) and (1.19). Consequently, the condition on G in Theorem 3.5 is satisfied. In addition, the function $c: K \times \mathbb{R}^{r(r+1)/2} \to \mathbb{R}^{r(r+1)/2}$ is seen to trivially satisfy the conditions of either Remark 3.2 or Remark 3.4. (Hence H can be taken to be an arbitrary compact, convex set containing $\bigcup_{N \geq 1} \Psi_N(E_N)$.)

Thus, to apply Theorem 3.5, it suffices in each case to specify the transition function, starting at $(P_{ij})_{i \leq j} \in E_N$, of the Markov chain Z^N, and to verify the five moment conditions (3.12)–(3.16) for appropriately chosen sequences $\{\varepsilon_N\}$ and $\{\delta_N\}$.

Before proceeding, we introduce a useful computational device, which already appeared in (1.7) without explicit mention. Given $(d_{ij})_{i \leq j} \in \mathbb{R}^{r(r+1)/2}$, we define

$$\tilde{d}_{ij} = \tfrac{1}{2}(1 + \delta_{ij})\, d_{i \wedge j,\, i \vee j}, \qquad i, j = 1, \ldots, r. \tag{3.34}$$

We apply this symmetrization to P_{ij}, P_{ij}^*, P_{ij}^{**}, P'_{ij}, Q_{ij}, Q'_{ij}, and so on. The point is that (1.1) can be expressed more concisely as $p_i = \sum_j \tilde{P}_{ij}$. For later reference, we isolate the following simple identity. With $(d_{ij})_{i \leq j}$ as above,

$$\begin{aligned} &\sum_{k,l} (\delta_{ij}\delta_{kl} \vee \delta_{il}\delta_{jk})(1 + \delta_{jl})\, \tilde{d}_{ik} \\ &\quad = \sum_{k,l} [\delta_{ij}\delta_{kl} + \delta_{il}\delta_{jk}(1 - \delta_{ij}\delta_{kl})](1 + \delta_{jl})\, \tilde{d}_{ik} \\ &\quad = \delta_{ij} \sum_k \tilde{d}_{ik} + \tilde{d}_{ij}, \qquad i, j = 1, \ldots, r. \end{aligned} \tag{3.35}$$

3.8 Example We consider first the multinomial-sampling model described in Section 1. The transition function of Z^N, starting at $(P_{ij})_{i \leq j} \in E_N$, is specified by (1.8), (1.4), (1.5), (1.2), (1.1), and (1.14).

Since $E[P'_{ij}] = P_{ij}^{**}$, we have

$$2NE[p'_i - p_i] = 2N(p_i^{**} - p_i) = b_i(p) + O(N^{-1}), \tag{3.36}$$

where $b: K \to \mathbb{R}^{r-1}$ is given by (1.19). (Throughout, all O and o terms are uniform in the genotypic frequencies.) The relation $\operatorname{cov}(P'_{ik}, P'_{jl}) = N^{-1}P_{ik}^{**}(\delta_{ij}\delta_{kl} - P_{jl}^{**})$ implies

$$\operatorname{cov}(\tilde{P}'_{ik}, \tilde{P}'_{jl}) = N^{-1}[\tfrac{1}{2}(\delta_{ij}\delta_{kl} \vee \delta_{il}\delta_{jk})(1 + \delta_{jl})\tilde{P}_{ik}^{**} - \tilde{P}_{ik}^{**}\tilde{P}_{jl}^{**}], \tag{3.37}$$

and therefore, by (3.35),

$$\begin{aligned} 2N \operatorname{cov}(p'_i, p'_j) &= \sum_{k,l} 2N \operatorname{cov}(\tilde{P}'_{ik}, \tilde{P}'_{jl}) \\ &= \delta_{ij} p_i^{**} + \tilde{P}_{ij}^{**} - 2p_i^{**}p_j^{**} \\ &= p_i^{**}(\delta_{ij} - p_j^{**}) + \tilde{P}_{ij}^{**} - p_i^{**}p_j^{**}. \end{aligned} \tag{3.38}$$

This shows, incidentally, that (1.10) is not only sufficient for (1.12) but necessary as well. Now observe that $p_i^{**} = p_i + O(N^{-1})$ by (3.36) and

$$\tilde{P}_{ij}^{**} = \frac{1}{2} \sum_{k \le l} (\delta_{ki}\delta_{lj} + \delta_{kj}\delta_{li})(2 - \delta_{kl})p_k p_l + O(N^{-1}) \tag{3.39}$$

$$= p_i p_j + O(N^{-1}),$$

so

$$2N \operatorname{cov}(p_i', p_j') = p_i(\delta_{ij} - p_j) + O(N^{-1}). \tag{3.40}$$

Next, we note that

$$E[Q_{ij}' - Q_{ij}] = P_{ij}^{**} - (2 - \delta_{ij})[\operatorname{cov}(p_i', p_j') + p_i^{**}p_j^{**}] - Q_{ij} \tag{3.41}$$

$$= -Q_{ij} + O(N^{-1})$$

by (3.39) and (3.40).

Finally,

$$2NE[(p_i' - E[p_i'])^4] \le 2Nr^3 \sum_{j=1}^{r} E[(\tilde{P}_{ij}' - E[\tilde{P}_{ij}'])^4] \tag{3.42}$$

$$= O(N^{-1})$$

since $P_{ij}' \sim N^{-1}$ binomial $(N, E[P_{ij}'])$ for each $i \le j$, and the fourth central moment of N^{-1} binomial (N, p) is $O(N^{-2})$, uniformly in p. Also,

$$\operatorname{var}(Q_{ij}') \le 2 \operatorname{var}(P_{ij}') + 2(2 - \delta_{ij})^2 \operatorname{var}(p_i' p_j') \tag{3.43}$$

$$\le O(N^{-1}) + 4(2 - \delta_{ij})^2(\operatorname{var}(p_i') + \operatorname{var}(p_j'))$$

$$= O(N^{-1})$$

by (3.28) and (3.29). This completes the verification of conditions (3.12)–(3.16) of Theorem 3.5 (see Remark 3.7(a)) with $\varepsilon_N = (2N)^{-1}$ and $\delta_N = 1$.

We note that the limits as $N \to \infty$ of the right sides of (3.36) and (3.40) depend only on $p_1, \ldots, p_{r-1}$. (For this reason, Theorem 3.5 could easily be avoided hcre.) However, this is not typical, as other examples suggest. □

3.9 Example The next genetic model we consider is a generalization of a model due to Moran. Its key feature is that, in contrast to the multinomial-sampling model of Example 3.8, generations are overlapping. A single step of the Markov chain here corresponds to the death of an individual and its replacement by the birth of another.

Suppose the genotype $A_i A_j$ produces gametes with fertility $w_{ij}^{(1)}$ and has mortality rate $w_{ij}^{(2)}$. If $(P_{ij})_{i \le j} \in E_N$ is the initial vector of genotypic frequencies, then the probability that an $A_i A_j$ individual dies is

$$\gamma_{ij} = \frac{w_{ij}^{(2)} P_{ij}}{\sum_{k \le l} w_{kl}^{(2)} P_{kl}}. \tag{3.44}$$

The frequency of A_i in gametes before mutation reads

$$p_i^* = \frac{\sum_j w_{ij}^{(1)} \tilde{P}_{ij}}{\sum_{k,l} w_{kl}^{(1)} \tilde{P}_{kl}}, \tag{3.45}$$

where $w_{ji}^{(1)} \equiv w_{ij}^{(1)}$ for $1 \le i < j \le r$. With mutation rates u_{ij} (where $u_{ii} = 0$), this becomes

$$p_i^{**} = \left(1 - \sum_j u_{ij}\right) p_i^* + \sum_j u_{ji} p_j^* \tag{3.46}$$

after mutation, so the probability that an $A_i A_j$ individual is born has the form

$$\beta_{ij} = (2 - \delta_{ij}) p_i^{**} p_j^{**}. \tag{3.47}$$

Consequently, the joint distribution of genotypic frequencies P'_{ij} after a birth-death event is specified as follows. For each $(k, l) \neq (m, n)$ (with $1 \le k \le l \le r$, $1 \le m \le n \le r$),

$$P'_{ij} = \begin{cases} P_{ij} + N^{-1} & \text{if } (i, j) = (m, n) \\ P_{ij} - N^{-1} & \text{if } (i, j) = (k, l) \\ P_{ij} & \text{otherwise} \end{cases} \tag{3.48}$$

with probability $\gamma_{kl} \beta_{mn}$, and $P'_{ij} = P_{ij}$ for all $i \le j$ with probability $\sum_{k \le l} \gamma_{kl} \beta_{kl}$. If we further require that

$$u_{ij} = (N^{-1} \mu_{ij}) \wedge r^{-1}, \qquad w_{ij}^{(k)} = (1 + N^{-1} \sigma_{ij}^{(k)}) \vee \tfrac{1}{2}, \qquad k = 1, 2, \tag{3.49}$$

where $\mu_{ij} \ge 0$ (with $\mu_{ii} = 0$) and $\sigma_{ij}^{(k)} (= \sigma_{ji}^{(k)})$ is real for $i, j = 1, \ldots, r$, then the transition function of Z^N, starting at $(P_{ij}) \in E_N$, is specified by (3.44)–(3.49).

To evaluate the appropriate moments, observe that

$$E[P'_{ij} - P_{ij}] = N^{-1}[-\gamma_{ij}(1 - \beta_{ij}) + (1 - \gamma_{ij})\beta_{ij}] = N^{-1}(\beta_{ij} - \gamma_{ij}) \tag{3.50}$$

and

$$E[(P'_{ik} - P_{ik})(P'_{jl} - P_{jl})] = N^{-2}[-(\gamma_{ik} \beta_{jl} + \gamma_{jl} \beta_{ik}) + \delta_{ij} \delta_{kl}(\gamma_{ik} + \beta_{ik})]. \tag{3.51}$$

Noting that $\tilde{\gamma}_{ij} = w_{ij}^{(2)} \tilde{P}_{ij} / \sum_{k,l} w_{kl}^{(2)} \tilde{P}_{kl}$ and $\tilde{\beta}_{ij} = p_i^{**} p_j^{**}$, where $w_{ji}^{(2)} \equiv w_{ij}^{(2)}$ for $1 \le i < j \le r$, and letting $\gamma_i = \sum_j \tilde{\gamma}_{ij}$, we have

$$\begin{aligned} N^2 E[p'_i - p_i] &= N^2 \sum_j E[\tilde{P}'_{ij} - \tilde{P}_{ij}] \\ &= N \sum_j (\tilde{\beta}_{ij} - \tilde{\gamma}_{ij}) \\ &= N(p_i^{**} - \gamma_i) \\ &= -\sum_j \mu_{ij} p_i + \sum_j \mu_{ji} p_j + \sum_j \sigma_{ij} \tilde{P}_{ij} \\ &\quad - p_i \sum_{k,l} \sigma_{kl} \tilde{P}_{kl} + O(N^{-1}), \end{aligned} \tag{3.52}$$

where $\sigma_{ij} = \sigma_{ij}^{(1)} - \sigma_{ij}^{(2)}$ is the difference between the scaled fertility and mortality, and

$$(3.53)\quad \begin{aligned} N^2E[(p_i' - p_i)(p_j' - p_j)] &= N^2 \sum_{k,l} E[(\tilde{P}_{ik}' - \tilde{P}_{ik})(\tilde{P}_{jl}' - \tilde{P}_{jl})] \\ &= \sum_{k,l} [-(\tilde{\gamma}_{ik}\tilde{\beta}_{jl} + \tilde{\gamma}_{jl}\tilde{\beta}_{ik}) \\ &\quad + \tfrac{1}{2}(\delta_{ij}\delta_{kl} \vee \delta_{il}\delta_{jk})(1 + \delta_{jl})(\tilde{\gamma}_{ik} + \tilde{\beta}_{ik})] \\ &= -(\gamma_i p_j^{**} + \gamma_j p_i^{**}) + \tfrac{1}{2}[\delta_{ij}(\gamma_i + p_i^{**}) + \tilde{\gamma}_{ij} + \tilde{\beta}_{ij}] \\ &= -2p_i p_j + \tfrac{1}{2}(2\delta_{ij} p_i + \tilde{P}_{ij} + p_i p_j) + O(N^{-1}) \\ &= p_i(\delta_{ij} - p_j) + \tfrac{1}{2}\tilde{Q}_{ij} + O(N^{-1}), \end{aligned}$$

where the third equality uses (3.35).

We also have

$$(3.54)\quad \begin{aligned} NE[Q_{ij}' - Q_{ij}] &= NE[P_{ij}' - P_{ij}] + O(N^{-1}) \\ &= \beta_{ij} - \gamma_{ij} + O(N^{-1}) \\ &= -Q_{ij} + O(N^{-1}), \end{aligned}$$

so since $|P_{ij}' - P_{ij}| \leq N^{-1}$ with probability one, the conditions (3.12)–(3.16) of Theorem 3.5 are satisfied with $\varepsilon_N = N^{-2}$ and $\delta_N = N^{-1}$ (recall (3.27)). Thus, the theorem is applicable to this model as well. □

4. INFINITELY-MANY-ALLELE MODELS

In the absence of selection, the Wright–Fisher model (defined by (1.12), (1.6) with $p_k^* = p_k$, and (1.5)) can be described as follows. Each of the $2N$ genes in generation $k + 1$ selects a "parent" gene at random (with replacement) from generation k. If the "parent" gene is of allelic type A_i, then its "offspring" gene is of allelic type A_j with probability u_{ij}^*.

In this section we consider a generalization of this model, as well as its diffusion limit. Let E be a compact metric space. E is the space of "types." For each positive integer M, let $P_M(x, \Gamma)$ be a transition function on $E \times \mathscr{B}(E)$, and define a Markov chain $\{Y^M(k), k = 0, 1, \ldots\}$ in $E^M = E \times \cdots \times E$ (M factors) as follows, where $Y_i^M(k)$ represents the type of the ith individual in generation k. Each of the M individuals in generation $k + 1$ selects a parent at random (with replacement) from generation k. If the parent is of type x, then its offspring's type belongs to Γ with probability $P_M(x, \Gamma)$. In particular,

$$(4.1)\quad E\left[\prod_{i=1}^{l} f_i(Y_{j_i}^M(k+1)) \,\middle|\, Y^M(k) = (x_1, \ldots, x_M)\right] = \prod_{i=1}^{l} \left\{\frac{1}{M}\sum_{j=1}^{M} \int f_i(y) P_M(x_j, dy)\right\}$$

if $j_1, \ldots, j_l \in \{1, \ldots, M\}$ are distinct and $f_1, \ldots, f_l \in B(E)$, since the components of $Y^M(k+1)$ are conditionally independent given $Y^M(k)$.

Observe that the process

(4.2) $$X^M(t) = \sum_{i=1}^{M} \frac{1}{M} \delta_{Y_i^M([Mt])}, \qquad t \geq 0,$$

has sample paths in $D_{\mathscr{P}(E)}[0, \infty)$. Our first result gives conditions under which there exists a Markov process X with sample paths in $D_{\mathscr{P}(E)}[0, \infty)$ such that $X^M \Rightarrow X$.

Suppose that B is a linear operator on $C(E)$, and let

(4.3) $$\mathscr{D} = \left\{\varphi = \prod_{i=1}^{l} \langle f_i, \cdot \rangle : l \geq 1, f_1, \ldots, f_l \in \mathscr{D}(B)\right\} \subset C(\mathscr{P}(E)),$$

where $\langle f, \mu \rangle$ denotes $\int f \, d\mu$ for $f \in B(E)$ and $\mu \in \mathscr{P}(E)$. Given $\varphi = \prod_{i=1}^{l} \langle f_i, \cdot \rangle \in \mathscr{D}$, define

(4.4) $$G\varphi(\mu) = \frac{1}{2} \sum_{i \neq j} (\langle f_i f_j, \mu \rangle - \langle f_i, \mu \rangle \langle f_j, \mu \rangle) \prod_{m:\, m \neq i,\, j} \langle f_m, \mu \rangle + \sum_{i=1}^{l} \langle Bf_i, \mu \rangle \prod_{j:\, j \neq i} \langle f_j, \mu \rangle,$$

and let

(4.5) $$A = \{(\varphi, G\varphi) : \varphi \in \mathscr{D}\}.$$

4.1 Theorem Suppose that the linear operator B on $C(E)$ is dissipative, $\mathscr{D}(B)$ is dense in $C(E)$, and the closure of B (which is single-valued) contains $(1, 0)$. Then, for each $\tilde{\nu} \in \mathscr{P}(\mathscr{P}(E))$, there exists a solution of the $D_{\mathscr{P}(E)}[0, \infty)$ martingale problem for $(A, \tilde{\nu})$. If the closure of B generates a Feller semigroup on $C(E)$, then the martingale problem for A is well-posed.

For $M = 1, 2, \ldots$, define X^M in terms of $P_M(x, \Gamma)$ as in (4.2), and define Q_M and B_M on $B(E)$ by

(4.6) $$Q_M f(x) = \int f(y) P_M(x, dy), \qquad B_M = M(Q_M - I).$$

If the closure of B generates a Feller semigroup on $C(E)$, if

(4.7) $$B \subset \underset{M \to \infty}{\text{ex-lim}} \, B_M,$$

and if X is a solution of the $D_{\mathscr{P}(E)}[0, \infty)$ martingale problem for A, then $X^M(0) \Rightarrow X(0)$ in $\mathscr{P}(E)$ implies $X^M \Rightarrow X$ in $D_{\mathscr{P}(E)}[0, \infty)$.

Proof. Under the first set of hypotheses on B, Lemma 5.3 of Chapter 4 implies the existence of a sequence of transition functions $P_M(x, \Gamma)$ on $E \times \mathscr{B}(E)$ such that the operators B_M, defined by (4.6), satisfy $\lim_{M\to\infty} B_M f = Bf$ for all $f \in \mathscr{D}(B)$. In particular, (4.7) holds.

Hence it suffices to prove existence of solutions of the martingale problem for A assuming that $\mathscr{D}(B)$ is dense in $C(E)$ and (4.7) holds. Let $\varphi = \prod_{i=1}^k \langle f_i, \cdot\rangle \in \mathscr{D}$, choose $f_1^M, \dots, f_k^M \in B(E)$ such that $f_i^M \to f_i$ and $B_M f_i^M \to Bf_i$ for $i = 1, \dots, k$, and put $\varphi_M = \prod_{i=1}^k \langle f_i^M, \cdot\rangle$. Given $\mu = M^{-1} \sum_{j=1}^M \delta_{x_j}$, where $(x_1, \dots, x_M) \in E^M$, we have

$$(4.8) \quad E\left[\varphi_M\left(X^M\left(\frac{1}{M}\right)\right)\middle| X^M(0) = \mu\right]$$

$$= M^{-k} E\left[\prod_{i=1}^k \sum_{j=1}^M f_i^M(Y_j^M(1)) \middle| Y^M(0) = (x_1, \dots, x_M)\right]$$

$$= M^{-k} E\left[\sum_{j_1=1}^M \cdots \sum_{j_k=1}^M \prod_{i=1}^k f_i^M(Y_{j_i}^M(1)) \middle| Y^M(0) = (x_1, \dots, x_M)\right]$$

$$= M^{-k} \frac{M!}{(M-k)!} \prod_{i=1}^k \langle Q_M f_i^M, \mu\rangle$$

$$+ M^{-k} \frac{M!}{(M-k+1)!} \sum_{l<m} \langle Q_M f_l^M f_m^M, \mu\rangle \prod_{n:\, n \neq l, m} \langle Q_M f_n^M, \mu\rangle$$

$$+ O(M^{-2}),$$

where the factor $M!/(M-k)!$ is the number of ways of selecting $j_1, \dots, j_k$ so that they are all distinct, and $M!/(M-k+1)!$ is the number of ways of selecting $j_1, \dots, j_k$ so that $j_l = j_m$ (l, m fixed) but they are otherwise distinct. Hence

$$(4.9) \quad ME\left[\varphi_M\left(X^M\left(\frac{1}{M}\right)\right) - \varphi_M(\mu) \middle| X^M(0) = \mu\right]$$

$$= M^{-k} \frac{M!}{(M-k)!} \sum_{i=1}^k \left(\prod_{j:\, j<i} \langle f_j^M, \mu\rangle\right) \langle B_M f_i^M, \mu\rangle \left(\prod_{j:\, j>i} \langle Q_M f_j^M, \mu\rangle\right)$$

$$+ M^{-k+1} \frac{M!}{(M-k+1)!}$$

$$\times \frac{1}{2} \sum_{l \neq m} \left\{\langle Q_M f_l^M f_m^M, \mu\rangle \prod_{n:\, n \neq l, m} \langle Q_M f_n^M, \mu\rangle - \prod_{i=1}^k \langle f_i^M, \mu\rangle\right\}$$

$$+ O(M^{-1})$$

$$= G\varphi(\mu) + o(1),$$

and the convergence is uniform in μ of the form $\mu = M^{-1} \sum_{j=1}^{M} \delta_{x_j}$, where $(x_1, \ldots, x_M) \in E^M$. Here we are using the fact that, since $\mathscr{D}(B)$ is dense in $C(E)$, $Q_M f \to f$ for every $f \in C(E)$. As in Remark 5.2 of Chapter 4, it follows from Theorems 9.1 and 9.4, both of Chapter 3, that $\{X^M\}$ is relatively compact. As in Lemma 5.1 of Chapter 4 we conclude that for each $\tilde{\nu} \in \mathscr{P}(\mathscr{P}(E))$, there exists a solution of the martingale problem for $(A, \tilde{\nu})$.

Fix $\tilde{\nu} \in \mathscr{P}(\mathscr{P}(E))$. To complete the proof, it will suffice by Corollary 8.17 of Chapter 4 to show that the martingale problem for $(A, \tilde{\nu})$ has a unique solution, assuming that $\bar{B}$ generates a Feller semigroup $\{S(t)\}$ on $C(E)$. Let X be a solution of the martingale problem for $(A, \tilde{\nu})$. By Corollary 3.7 of Chapter 4, X has a modification X^* with sample paths in $D_{\mathscr{P}(E)}[0, \infty)$. Let $f_1, \ldots, f_k \in \mathscr{D}(\bar{B})$. Then

$$(4.10) \quad \frac{\partial}{\partial t} \prod_{i=1}^{k} \langle S(u - t)f_i, \mu \rangle = - \sum_{i=1}^{k} \langle \bar{B}S(u - t)f_i, \mu \rangle \prod_{j: j \neq i} \langle S(u - t)f_j, \mu \rangle$$

for $0 \leq t \leq u$ and all $\mu \in \mathscr{P}(E)$, so by Lemma 3.4 of Chapter 4,

$$(4.11) \quad \prod_{i=1}^{k} \langle S(u - t \wedge u)f_i, X^*(t \wedge u) \rangle$$

$$- \int_0^{t \wedge u} \frac{1}{2} \sum_{l \neq m} \{\langle S(u - s)f_l S(u - s)f_m, X^*(s) \rangle - \langle S(u - s)f_l, X^*(s) \rangle$$

$$\cdot \langle S(u - s)f_m, X^*(s) \rangle\} \prod_{n: n \neq l, m} \langle S(u - s)f_n, X^*(s) \rangle \, ds$$

is a martingale in t for each $u \geq 0$. Hence

$$(4.12) \quad E\left[\prod_{i=1}^{k} \langle f_i, X(u) \rangle \right] = \int \prod_{i=1}^{k} \langle S(u)f_i, \mu \rangle \tilde{\nu}(d\mu)$$

$$+ \frac{1}{2} \int_0^u \sum_{l \neq m} E\left[\langle S(u - s)f_l S(u - s)f_m, X(s) \rangle \right.$$

$$\times \left. \prod_{n: n \neq l, m} \langle S(u - s)f_n, X(s) \rangle \right] ds$$

$$- \binom{k}{2} \int_0^u E\left[\prod_{i=1}^{k} \langle S(u - s)f_i, X(s) \rangle \right] ds.$$

Moreover, (4.12) holds for all $f_1, \ldots, f_k \in C(E)$ since $\mathscr{D}(B)$ is dense in $C(E)$.

Let Y be another solution of the martingale problem for $(A, \tilde{\nu})$, and put

$$(4.13) \quad \beta_k(u) = \sup_{f_1, \ldots, f_k \in C(E), \, \|f_i\| \leq 1} \left| E\left[\prod_{i=1}^{k} \langle f_i, X(u) \rangle - \prod_{i=1}^{k} \langle f_i, Y(u) \rangle \right] \right|.$$

Then, since (4.12) holds with X replaced by Y,

(4.14) $$\beta_k(u) \le k(k-1) \int_0^u \beta_k(s)\, ds, \qquad u \ge 0.$$

We conclude that $\beta_k(u) = 0$ for all $k \ge 1$ and $u \ge 0$, and hence that X and Y have the same one-dimensional distributions. Uniqueness then follows from Theorem 4.2 of Chapter 4. □

The process X of Theorem 4.1 is therefore characterized by its type space E and the linear operator B on $C(E)$. Let $E_1, E_2, \ldots$ and E be compact metric spaces. For $n = 1, 2, \ldots$, let $\eta_n: E_n \to E$ be continuous, and define π_n: $C(E) \to C(E_n)$ by $\pi_n f = f \circ \eta_n$, $\hat{\eta}_n: \mathscr{P}(E_n) \to \mathscr{P}(E)$ by $\hat{\eta}_n \mu = \mu\eta_n^{-1}$, and $\hat{\pi}_n$: $C(\mathscr{P}(E)) \to C(\mathscr{P}(E_n))$ by $\hat{\pi}_n f = f \circ \hat{\eta}_n$.

4.2 Proposition Let $B_1, B_2, \ldots$ and B be linear operators on $C(E_1), C(E_2), \ldots$ and $C(E)$, respectively, satisfying the conditions in the first sentence of Theorem 4.1. Define $A_1, A_2, \ldots$ and A in terms of $E_1, E_2, \ldots$ and E and $B_1, B_2, \ldots$ and B as in (4.3)–(4.5). For $n = 1, 2, \ldots$, let X_n be a solution of the $D_{\mathscr{P}(E_n)}[0, \infty)$ martingale problem for A_n. If the closure of B generates a Feller semigroup on $C(E)$, if

(4.15) $$B \subset \underset{n\to\infty}{\text{ex-lim}}\, B_n \text{ (with respect to } \{\pi_n\}),$$

and if X is a solution of the $D_{\mathscr{P}(E)}[0, \infty)$ martingale problem for A, then $\hat{\eta}_n(X_n(0)) \Rightarrow X(0)$ in $\mathscr{P}(E)$ implies $\hat{\eta}_n \circ X_n \Rightarrow X$ in $D_{\mathscr{P}(E)}[0, \infty)$.

Proof. By (4.15),

(4.16) $$A \subset \underset{n\to\infty}{\text{ex-lim}}\, A_n \text{ (with respect to } \{\hat{\pi}_n\}),$$

so the result follows from Corollary 8.16 of Chapter 4. □

We give two examples of Proposition 4.2. In both, E_n is a subset of E, η_n is an inclusion map, and hence $\hat{\eta}_n$ can be regarded as an inclusion map (that is, elements of $\mathscr{P}(E_n)$ can be regarded as belonging to $\mathscr{P}(E)$). With this understanding, we can suppress the notation η_n and $\hat{\eta}_n$.

4.3 Example For $n = 1, 2, \ldots$ let $E_n = \{k/\sqrt{n}: k \in \mathbb{Z}\} \cup \{\Delta\}$ (the one-point compactification), define B_n to be the bounded linear operator on $C(E_n)$ given by

$$B_n f\left(\frac{k}{\sqrt{n}}\right) = \sigma^2 n \left\{\frac{1}{2} f\left(\frac{k-1}{\sqrt{n}}\right) + \frac{1}{2} f\left(\frac{k+1}{\sqrt{n}}\right) - f\left(\frac{k}{\sqrt{n}}\right)\right\} \tag{4.17}$$

and $B_n f(\Delta) = 0$, where $\sigma^2 \geq 0$, and let X_n be a solution of the $D_{\mathscr{P}(E_n)}[0, \infty)$ martingale problem for A_n (defined as in (4.3)–(4.5)). X_1 is known as the Ohta–Kimura model.

Let $E = \mathbb{R} \cup \{\Delta\}$ (the one-point compactification), define B to be the linear operator on $C(E)$ given by

$$Bf(x) = \tfrac{1}{2}\sigma^2 f''(x) \tag{4.18}$$

and $Bf(\Delta) = 0$, where $\mathscr{D}(B) = \{f \in C(E): (f - f(\Delta))|_{\mathbb{R}} \in C_c^2(\mathbb{R})\}$, and let X be a solution of the $D_{\mathscr{P}(E)}[0, \infty)$ martingale problem for A (defined by (4.3)–(4.5)). X is known as the Fleming–Viot model.

By Proposition 1.1 of Chapter 5 and Proposition 4.2 of this chapter, $X_n(0) \Rightarrow X(0)$ in $\mathscr{P}(E)$ implies $X_n \Rightarrow X$ in $D_{\mathscr{P}(E)}[0, \infty)$. (Recall that we are regarding $\mathscr{P}(E_n)$ as a subset of $\mathscr{P}(E)$.) The use of one-point compactifications here is only so that Theorem 4.1 and Proposition 4.2 will apply. It is easy to see that, for example, $P\{X(0)(\mathbb{R}) = 1\} = 1$ implies $P\{X(t)(\mathbb{R}) = 1$ for all $t > 0\} = 1$. □

4.4 Example For $n = 2, 3, \ldots$, let $E_n = \{1/n, 2/n, \ldots, 1\}$, define B_n to be the bounded linear operator on $C(E_n)$ given by

$$B_n f\left(\frac{i}{n}\right) = \frac{1}{2}\frac{\theta}{n-1} \sum_{j=1}^{n} \left\{f\left(\frac{j}{n}\right) - f\left(\frac{i}{n}\right)\right\}, \tag{4.19}$$

where $\theta > 0$, and let X_n be a solution of the martingale problem for A_n (defined as in (4.3)–(4.5)). Observe that $X_r(t) = \sum_{i=1}^{r} p_i(t)\delta_{i/r}$, where $(p_1(t), \ldots, p_{r-1}(t))$ is the diffusion process of Section 1 with $\sum_{j:\, j \neq i} \mu_{ij} = \theta/2$ for $i = 1, \ldots, r$, μ_{ij} $(i \neq j)$ independent of i and j, and $\sigma_{ij} = 0$ for $i, j = 1, \ldots, r$. Thus, X_r could be called the neutral r-allele model with uniform mutation. (The term "neutral" refers to the lack of selection.)

Let $E = [0, 1]$, define B to be the bounded linear operator on $C[0, 1]$ given by

$$Bf(x) = \tfrac{1}{2}\theta \int_0^1 (f(y) - f(x))\, dy = \tfrac{1}{2}\theta(\langle f, \lambda \rangle - f(x)), \tag{4.20}$$

where λ denotes Lebesgue measure on $[0, 1]$, and let X be a solution of the $D_{\mathscr{P}[0,1]}[0, \infty)$ martingale problem for A (defined by (4.3)–(4.5)). We call X the infinitely-many-neutral-alleles model with uniform mutation.

By Proposition 4.2, $X_n(0) \Rightarrow X(0)$ in $\mathscr{P}[0, 1]$ implies $X_n \Rightarrow X$ in $D_{\mathscr{P}[0, 1]}[0, \infty)$. (Again, we are regarding $\mathscr{P}(E_n)$ as a subset of $\mathscr{P}[0, 1]$.) Of course, with $E = [0, 1]$ and

M individuals.

(4.21) $\qquad P_M(x, dy) = \left(1 - \dfrac{\theta}{2M}\right)\delta_x(dy) + \dfrac{\theta}{2M}\lambda(dy), \qquad M \geq \theta,$

This is trans. funct. for X^m giving

in Theorem 4.1, we have $X^M(0) \Rightarrow X(0)$ in $\mathscr{P}[0, 1]$ implies $X^M \Rightarrow X$ in $X^m \to X$. $D_{\mathscr{P}[0, 1]}[0, \infty)$. Thus, X can be thought of as either a limit in distribution of certain $(n - 1)$-dimensional diffusions as $n \to \infty$, or as a limit in distribution of a certain sequence of infinite-dimensional Markov chains. $\qquad\square$

The remainder of this section is devoted to a more detailed examination of the infinitely-many-neutral-alleles model with uniform mutation.

4.5 Theorem Given $\tilde{v} \in \mathscr{P}(\mathscr{P}[0, 1])$, let X be as in Example 4.4 with initial distribution $\tilde{v}$. (In other words, X is the process of Theorem 4.1 with $E = [0, 1]$, with B defined on $C[0, 1]$ by $Bf = \frac{1}{2}\theta(\langle f, \lambda \rangle - f)$, where $\theta > 0$ and λ is Lebesgue measure on $[0, 1]$, and with initial distribution $\tilde{v}$.) Then almost all sample paths of X belong to $C_{\mathscr{P}[0, 1]}[0, \infty)$, and

in the limiting process, all values are in the atomic

(4.22) $\qquad P\{X(t) \in \mathscr{P}_a[0, 1] \text{ for all } t > 0\} = 1,$

where $\mathscr{P}_a[0, 1]$ denotes the set of purely atomic Borel probability measures on *pr. meas.* $[0, 1]$.

in 6,1

Proof. Using the notation of Example 4.4, let X_n have initial distribution $\tilde{v}_n \in \mathscr{P}(\mathscr{P}(E_n))$, where the sequence $\{\tilde{v}_n\}$ is chosen so that $\tilde{v}_n \Rightarrow \tilde{v}$ on $\mathscr{P}[0, 1]$. Then $X_n \Rightarrow X$ in $D_{\mathscr{P}[0, 1]}[0, \infty)$, and since $C_{\mathscr{P}[0, 1]}[0, \infty)$ is a closed subset of $D_{\mathscr{P}[0, 1]}[0, \infty)$, we have

(4.23) $\qquad 1 = \overline{\lim_{n \to \infty}} P\{X_n \in C_{\mathscr{P}[0, 1]}[0, \infty)\} \leq P\{X \in C_{\mathscr{P}[0, 1]}[0, \infty)\}$

by Theorem 3.1 of Chapter 3.

The proof of the second assertion is more complicated. Observe first that for $f \in C[0, 1]$,

(4.24) $\qquad M_f(t) \equiv \langle f, X(t) \rangle - \langle f, X(0) \rangle - \int_0^t \frac{1}{2}\theta(\langle f, \lambda \rangle - \langle f, X(s) \rangle)\, ds$

is a continuous, square-integrable martingale, and (see Problem 29 of Chapter 2) its increasing process has the form

(4.25) $\qquad \langle M_f \rangle_t = \int_0^t (\langle f^2, X(s) \rangle - \langle f, X(s) \rangle^2)\, ds.$

Consequently, if $\gamma \geq 2$ and $f, g \in C[0, 1]$ with $f, g \geq 0$, Itô's formula (Theorem 2.9 of Chapter 5) implies that

$$M_f^\gamma(t) \equiv \langle f, X(t)\rangle^\gamma - \langle f, X(0)\rangle^\gamma - \int_0^t \left\{ \binom{\gamma}{2} (\langle f^2, X(s)\rangle - \langle f, X(s)\rangle^2) \langle f, X(s)\rangle^{\gamma-2} + \frac{\gamma}{2}\theta(\langle f, \lambda\rangle - \langle f, X(s)\rangle)\langle f, X(s)\rangle^{\gamma-1} \right\} ds \tag{4.26}$$

is a continuous, square-integrable martingale, and

$$\langle M_f^\gamma, M_g^\gamma\rangle_t = \gamma^2 \int_0^t \langle f, X(s)\rangle^{\gamma-1}(\langle fg, X(s)\rangle - \langle f, X(s)\rangle\langle g, X(s)\rangle)\langle g, X(s)\rangle^{\gamma-1}\, ds. \tag{4.27}$$

(Note that $\langle \cdot, \cdot \rangle$ is used in two different ways here.) Let us define $\varphi_{n,\gamma}$: $\mathscr{P}[0, 1] \to [0, \infty)$ for each $\gamma > 0$ and $n = 1, 2, \ldots$ by

$$\varphi_{n,\gamma}(\mu) = \mu\left(\left[0, \frac{1}{2^n}\right]\right)^\gamma + \mu\left(\left(\frac{1}{2^n}, \frac{2}{2^n}\right]\right)^\gamma + \cdots + \mu\left(\left(1 - \frac{1}{2^n}, 1\right]\right)^\gamma. \tag{4.28}$$

It follows that, for each $\gamma \geq 2$ and $n = 1, 2, \ldots,$

$$Z_{n,\gamma}(t) \equiv \varphi_{n,\gamma}(X(t)) - \varphi_{n,\gamma}(X(0)) + \int_0^t \left\{ \binom{\gamma}{2} (\varphi_{n,\gamma-1} - \varphi_{n,\gamma})(X(s)) + \frac{\gamma}{2}\theta(2^{-n}\varphi_{n,\gamma-1} - \varphi_{n,\gamma})(X(s)) \right\} ds \tag{4.29}$$

is a continuous, square-integrable martingale with increasing process

$$I_{n,\gamma}(t) = \gamma^2 \int_0^t (\varphi_{n,2\gamma-1} - \varphi_{n,\gamma}^2)(X(s))\, ds; \tag{4.30}$$

in fact, this holds for each $\gamma > 1$ as can be seen by approximating the function x^γ by the $C^2[0, \infty)$ function $(x + \varepsilon)^\gamma$. Defining φ_γ: $\mathscr{P}[0, 1] \to [0, \infty]$ for each $\gamma > 0$ by

$$\varphi_\gamma(\mu) = \begin{cases} \sum_{0 \leq x \leq 1} \mu(\{x\})^\gamma, & \text{if } \gamma \neq 1, \\ 1, & \text{if } \gamma = 1, \end{cases} \tag{4.31}$$

we have $\text{bp-lim}_{n\to\infty}\varphi_{n,\gamma} = \varphi_\gamma$ for each $\gamma \geq 1$, while $\varphi_{n,\gamma} \nearrow \varphi_\gamma$ as $n \nearrow \infty$ for $0 < \gamma < 1$. We conclude that, for each $\gamma > 1$,

$$(4.32) \qquad Z_\gamma(t) \equiv \varphi_\gamma(X(t)) - \varphi_\gamma(X(0)) - \int_0^t \left\{\binom{\gamma}{2}\varphi_{\gamma-1} - \left[\binom{\gamma}{2} + \frac{\gamma}{2}\theta\right]\varphi_\gamma\right\}(X(s))\, ds$$

is a continuous, square-integrable martingale with increasing process

$$(4.33) \qquad I_\gamma(t) = \gamma^2 \int_0^t (\varphi_{2\gamma-1} - \varphi_\gamma^2)(X(s))\, ds;$$

here we are using $\lim_{n\to\infty} E[Z_{n,\gamma}(t)^2] = \lim_{n\to\infty} E[I_{n,\gamma}(t)] = E[I_\gamma(t)]$ and the monotone convergence theorem to show that, when $1 < \gamma < 2$,

$$(4.34) \qquad E\left[\left(\int_0^t \varphi_{\gamma-1}(X(s))\, ds\right)^2\right] < \infty, \qquad t \geq 0.$$

Letting $\varphi_{1+}(\mu) = \text{bp-lim}_{\gamma\to 1+}\varphi_\gamma(\mu) = \sum_{0\leq x\leq 1}\mu(\{x\})$, we have

$$(4.35) \qquad 0 = \lim_{\gamma\to 2+} E[Z_\gamma(t) - Z_2(t)] = E\left[\int_0^t (1 - \varphi_{1+})(X(s))\, ds\right]$$

for all $t \geq 0$, so $P\{X(t) \in \mathscr{P}_a[0, 1]$ for almost every $t > 0\} = 1$. To remove the word "almost," observe that

$$(4.36) \qquad \lim_{\gamma\to 1+} E[Z_\gamma(t)^2] = \lim_{\gamma\to 1+} E[I_\gamma(t)] = E\left[\int_0^t \varphi_{1+}(1 - \varphi_{1+})(X(s))\, ds\right] = 0$$

for each $t \geq 0$ by (4.33) and (4.35). Fix $t_0 > 0$. By Doob's martingale inequality, $\sup_{0\leq t\leq t_0}|Z_\gamma(t)| \to 0$ in probability as $\gamma \to 1+$, so there exists a sequence $\gamma_n \to 1+$ such that $\sup_{0\leq t\leq t_0}|Z_{\gamma_n}(t)| \to 0$ a.s. Letting

$$(4.37) \qquad \eta(t) = \overline{\lim_{n\to\infty}} \binom{\gamma_n}{2} \int_0^t \varphi_{\gamma_n - 1}(X(s))\, ds,$$

we obtain from (4.32) and (4.35) that, almost surely,

$$(4.38) \qquad \varphi_{1+}(X(t)) - \varphi_{1+}(X(0)) - \eta(t) + \tfrac{1}{2}\theta t = 0, \qquad 0 \leq t \leq t_0.$$

Since $\eta(t)$ is nondecreasing in t, we conclude that $P\{X(t) \in \mathscr{P}_a[0, 1]$ for all $t > 0\} = 1$, as required. □

4.6 Theorem The measure-valued diffusions X_n, $n = 2, 3, \ldots$, and X defined as in Example 4.4, have unique stationary distributions $\tilde{\mu}_n$, $n = 2, 3, \ldots$, and $\tilde{\mu}$, respectively. In fact, $\tilde{\mu}_n$ is the distribution of $\sum_{i=1}^n \xi_i^n \delta_{i/n}$, where $(\xi_1^n, \ldots, \xi_n^n)$ has a symmetric Dirichlet distribution with parameter $\theta/(n-1)$ (defined below).

Moreover, there exist random variables $\xi_1 \geq \xi_2 \geq \cdots \geq 0$ with $\sum_{i=1}^{\infty} \xi_i = 1$ such that

prob. 13 p. 450 gives r.ved.
$\stackrel{d}{=}$ to $(\xi_{1}, \dots)$

(4.39)
$$(\xi_{(1)}^n, \dots, \xi_{(k)}^n) \Rightarrow (\xi_1, \dots, \xi_k)$$

as $n \to \infty$ for each $k \geq 1$, where $\xi_{(1)}^n, \dots, \xi_{(n)}^n$ denote the descending order statistics of $\xi_1^n, \dots, \xi_n^n$. Finally, $\tilde{\mu}$ is the distribution of $\sum_{i=1}^{\infty} \xi_i \delta_{u_i}$, where u_1, $u_2, \dots$ is a sequence of independent, uniformly distributed random variables on $[0, 1]$, independent of $\xi_1, \xi_2, \dots$.

Proof. Fix $n \geq 2$. Let $E_n = \{1/n, 2/n, \dots, 1\}$ and define $f_1, \dots, f_n \in C(E_n)$ by $f_i(j/n) = \delta_{ij}$. Let $(\xi_1^n, \dots, \xi_n^n)$ have a symmetric Dirichlet distribution with parameter $\varepsilon_n \equiv \theta/(n-1)$, that is, $(\xi_1^n, \dots, \xi_{n-1}^n)$ is a K-valued random variable (recall (1.16) with $r = n$) with Lebesgue density $\Gamma(n\varepsilon_n)\Gamma(\varepsilon_n)^{-n}(p_1 \cdots p_n)^{\varepsilon_n - 1}$, and $\xi_n^n = 1 - \sum_{i=1}^{n-1} \xi_i^n$. Let $\tilde{\nu}_n \in P(\mathcal{P}(E_n))$ be the distribution of $\sum_{i=1}^{n} \xi_i^n \delta_{i/n}$. To show that $\tilde{\nu}_n$ is a stationary distribution for X_n, it suffices by Theorem 9.17 of Chapter 4 to show that

(4.40)
$$\int_{\mathcal{P}(E_n)} G_n\left(\prod_{i=1}^{n} \langle f_i, \mu \rangle^{m_i}\right) \tilde{\nu}_n(d\mu) = 0$$

for all integers $m_1, \dots, m_n \geq 0$, where G_n is as in (4.4) with B_n given by (4.19). (Actually, this can be proved without the aid of Theorem 9.17 of Chapter 4 by checking that the span of the functions within parentheses in (4.40) forms a core.)

But with $|m| = m_1 + \cdots + m_n$, the left side of (4.40) becomes

(4.41)

$$\int_{\mathcal{P}(E_n)} \left\{ \frac{1}{2} \sum_{i,j=1}^{n} (\langle f_i f_j, \mu \rangle - \langle f_i, \mu \rangle \langle f_j, \mu \rangle) m_i(m_j - \delta_{ij}) \prod_{l=1}^{n} \langle f_l, \mu \rangle^{m_l - \delta_{il} - \delta_{jl}} \right.$$
$$\left. + \frac{1}{2} \theta \sum_{i=1}^{n} \left(\frac{1}{n-1} - \frac{n}{n-1} \langle f_i, \mu \rangle \right) m_i \prod_{l=1}^{n} \langle f_l, \mu \rangle^{m_l - \delta_{il}} \right\} \tilde{\nu}_n(d\mu)$$

$$= E\left[\frac{1}{2} \sum_{i,j=1}^{n} \xi_i^n(\delta_{ij} - \xi_j^n) m_i(m_j - \delta_{ij}) \prod_{l=1}^{n} (\xi_l^n)^{m_l - \delta_{il} - \delta_{jl}} \right.$$
$$\left. + \frac{1}{2} \theta \sum_{i=1}^{n} \left(\frac{1}{n-1} - \frac{n}{n-1} \xi_i^n \right) m_i \prod_{l=1}^{n} (\xi_l^n)^{m_l - \delta_{il}} \right]$$

$$= E\left[\frac{1}{2} \sum_{i=1}^{n} m_i(m_i - 1 + \varepsilon_n) \prod_{l=1}^{n} (\xi_l^n)^{m_l - \delta_{il}} \right.$$
$$\left. - \frac{1}{2} |m|(|m| - 1 + n\varepsilon_n) \prod_{l=1}^{n} (\xi_l^n)^{m_l} \right],$$

and this is zero because

$$(4.42) \qquad E[(\xi_1^n)^{m_1} \cdots (\xi_n^n)^{m_n}] = \frac{\Gamma(m_1 + \varepsilon_n) \cdots \Gamma(m_n + \varepsilon_n)}{\Gamma(|m| + n\varepsilon_n)}$$

and $\Gamma(u) = (u - 1)\Gamma(u - 1)$ if $u > 1$. As for uniqueness, suppose $\tilde{\mu}_n$ is a stationary distribution for X_n, and note that the left side of (4.41) with $\tilde{\nu}_n$ replaced by $\tilde{\mu}_n$ is zero. Inducting on the degree of $\prod_{i=1}^{n} \langle f_i, \mu \rangle^{m_i}$ (namely, $|m|$), we find that $\int \prod_{i=1}^{n} \langle f_i, \mu \rangle^{m_i} \tilde{\mu}_n(d\mu)$ is uniquely determined for all $m_1, \ldots, m_n \geq 0$. Hence $\tilde{\mu}_n$ is uniquely determined.

By Theorem 9.3 of Chapter 4, X has a stationary distribution $\tilde{\mu}$. Noting that

$$(4.43) \qquad \int G\left(\prod_{i=1}^{k} \langle f_i, \mu \rangle \right) \tilde{\mu}(d\mu) = 0$$

for all $k \geq 1$ and $f_1, \ldots, f_k \in C[0, 1]$, we obtain the uniqueness of $\tilde{\mu}$ as above, except here we induct on k. It follows from Theorem 9.12 of Chapter 4 that $\tilde{\mu}_n \Rightarrow \tilde{\mu}$ on $\mathscr{P}[0, 1]$.

Theorem 4.5 immediately implies that $\tilde{\mu}(\mathscr{P}_a[0, 1]) = 1$. We leave it to the reader to check that therefore there exist random variables $\xi_1 \geq \xi_2 \geq \cdots \geq 0$ with $\sum_{i=1}^{\infty} \xi_i = 1$ and $u_1, u_2, \ldots$ with distinct values in $[0, 1]$ such that $\tilde{\mu}$ is the distribution of $\sum_{i=1}^{\infty} \xi_i \delta_{u_i}$. The assertion that (4.39) holds says simply that the joint $\tilde{\mu}_n$-distribution of the sizes of the k largest atoms converges to the joint $\tilde{\mu}$-distribution. Unfortunately, this cannot be deduced merely from the fact that $\tilde{\mu}_n \Rightarrow \tilde{\mu}$ on $\mathscr{P}[0, 1]$. However, by giving a stronger topology to $\mathscr{P}[0, 1]$, we can obtain the desired conclusion.

We define the metric ρ^* on $\mathscr{P}[0, 1]$ as follows: given $\mu, \nu \in \mathscr{P}[0, 1]$, let F_μ, F_ν be the corresponding (right-continuous) cumulative distribution functions, and put

$$(4.44) \qquad \rho^*(\mu, \nu) = d_{D[0, 1]}(F_\mu, F_\nu),$$

where $d_{D[0, 1]}$ denotes a metric that induces the Skorohod topology on $D[0, 1]$ (see Billingsley (1968)). The separability of $(\mathscr{P}[0, 1], \rho^*)$ follows as does the separability of $D[0, 1]$. We note that $\sum_{i=1}^{n} \xi_i^n \delta_{i/n}$, $n = 2, 3, \ldots$, and $\sum_{i=1}^{\infty} \xi_i \delta_{u_i}$ are $(\mathscr{P}[0, 1], \rho^*)$-valued random variables, so we regard their distributions $\tilde{\mu}_n$, $n = 2, 3, \ldots$, and $\tilde{\mu}$ as belonging to $\mathscr{P}(\mathscr{P}[0, 1], \rho^*)$. We claim that

$$(4.45) \qquad \tilde{\mu}_n \Rightarrow \tilde{\mu} \text{ on } (\mathscr{P}[0, 1], \rho^*).$$

Letting

$$(4.46) \qquad F_n(t) = \sum_{i/n \leq t} \xi_i^n, \qquad F(t) = \sum_{u_i \leq t} \xi_i, \qquad 0 \leq t \leq 1,$$

we see from the definition of ρ^* that it suffices to show that $F_n \Rightarrow F$ in $D[0, 1]$. We verify this using Theorem 15.6 of Billingsley (1968), which is the analogue for $D[0, 1]$ of Theorem 8.8 of Chapter 3.

Let $C = \bigcup_{i=1}^{\infty}\{t \in [0, 1]: P\{u_i = t\} > 0\}$. Then C is at most countable, and for $t_1, \ldots, t_k \in [0, 1] - C$, the function $\mu \to (\mu([0, t_1]), \ldots, \mu([0, t_k]))$ is $\tilde{\mu}$-a.s. continuous on $(\mathscr{P}[0, 1], \rho)$ (ρ being the Prohorov metric), so

$$(F_n(t_1), \ldots, F_n(t_k)) \Rightarrow (F(t_1), \ldots, F(t_k)) \tag{4.47}$$

by Corollary 1.9 of Chapter 3. In particular, if $t \in [0, 1] - C$, then

$$\begin{aligned} E[F(t)] &= \lim_{n\to\infty} E[F_n(t)] \\ &= \lim_{n\to\infty} E[\xi_1^n + \cdots + \xi_{[nt]}^n] \\ &= \lim_{n\to\infty} \frac{[nt]}{n} = t, \end{aligned} \tag{4.48}$$

so $E[F(t) - F(t-)] = 0$, and hence $C = \varnothing$. It follows that (4.47) holds for all $t_1, \ldots, t_k \in [0, 1]$. Finally, let $0 \le t_1 \le t \le t_2 \le 1$. Then for $n = 1, 2, \ldots,$

$$\begin{aligned} &E[(F_n(t) - F_n(t_1))^2(F_n(t_2) - F_n(t))^2] \\ &\quad = E[(\xi_{[nt_1]+1}^n + \cdots + \xi_{[nt]}^n)^2(\xi_{[nt]+1}^n + \cdots + \xi_{[nt_2]}^n)^2] \\ &\quad = E[(Z_1^n)^2(Z_2^n)^2], \end{aligned} \tag{4.49}$$

where (Z_1^n, Z_2^n) is a K-valued random variable (recall (1.16) with $r = 3$) with Lebesgue density $\Gamma(\alpha_n + \beta_n + \gamma_n)\{\Gamma(\alpha_n)\Gamma(\beta_n)\Gamma(\gamma_n)\}^{-1}p_1^{\alpha_n - 1}p_2^{\beta_n - 1}p_3^{\gamma_n - 1}$ and $(\alpha_n, \beta_n, \gamma_n) = (([nt] - [nt_1])\varepsilon_n, ([nt_2] - [nt])\varepsilon_n, (n - [nt_2] + [nt_1])\varepsilon_n)$. Hence (4.49) becomes

$$\begin{aligned} \frac{\alpha_n(\alpha_n + 1)\beta_n(\beta_n + 1)}{(\alpha_n + \beta_n + \gamma_n) \cdots (\alpha_n + \beta_n + \gamma_n + 3)} &\le \left(\frac{[nt] - [nt_1]}{n}\right)\left(\frac{[nt_2] - [nt]}{n}\right) \\ &\le (t_2 - t_1)^2. \end{aligned} \tag{4.50}$$

We conclude that $F_n \Rightarrow F$ in $D[0, 1]$, and hence (4.45) holds.

Let us say that $x \in D[0, 1]$ has a jump of size $\delta > 0$ at a location $t \in [0, 1]$ if $|x(t) - x(t-)| = \delta$. For each $x \in D[0, 1]$ and $i \ge 1$, we define $s_i(x)$ and $l_i(x)$ to be the size and location of the ith largest jump of x. If $s_i(x) = s_{i+1}(x)$, we adopt the convention that $l_i(x) < l_{i+1}(x)$ (that is, ties are labeled from left to right). If x has only k jumps, we define $s_i(x) = l_i(x) = 0$ for each $i > k$. We leave it to the reader to check that $s_1, s_2, \ldots$ and $l_1, l_2, \ldots$ are Borel measurable on $D[0, 1]$. Suppose $\{x_n\} \subset D[0, 1]$, $x \in D[0, 1]$, and $d_{D[0, 1]}(x_n, x) \to 0$. Then

$$(s_1(x_n), \ldots, s_k(x_n)) \to (s_1(x), \ldots, s_k(x)) \tag{4.51}$$

and, if $s_1(x) > s_2(x) > \cdots$, then

$$(s_1(x_n), \ldots, s_k(x_n), l_1(x_n), \ldots, l_k(x_n)) \to (s_1(x), \ldots, s_k(x), l_1(x), \ldots, l_k(x)). \tag{4.52}$$

It follows from the definition of ρ^* that

$$(s_1(F_\mu), \ldots, s_k(F_\mu)) \tag{4.53}$$

is a ρ^*-continuous function of $\mu \in \mathscr{P}[0, 1]$, where F_μ is as in (4.44). Now since the $\tilde{\mu}_n$-distribution of (4.53) is the distribution of $(\xi^n_{(1)}, \ldots, \xi^n_{(k)})$ for each $n \geq k$, and since the $\tilde{\mu}$-distribution of (4.53) is the distribution of $(\xi_1, \ldots, \xi_k)$, we obtain (4.39) from (4.45).

We leave it as a problem to show that

$$P\{\xi_1 > \xi_2 > \cdots\} = 1 \tag{4.54}$$

(Problem 12). It follows that

$$(s_1(F_\mu), \ldots, s_k(F_\mu), l_1(F_\mu), \ldots, l_k(F_\mu)) \tag{4.55}$$

is a $\tilde{\mu}$-a.s. ρ^*-continuous function of $\mu \in \mathscr{P}[0, 1]$. Now the $\tilde{\mu}_n$-distribution of (4.55) is the distribution of

$$(\xi^n_{(1)}, \ldots, \xi^n_{(k)}, u^n_1, \ldots, u^n_k) \tag{4.56}$$

for each $n \geq k$, where $(nu^n_1, \ldots, nu^n_n)$ is independent of $(\xi^n_1, \ldots, \xi^n_n)$ and takes on each of the permutations of $(1, 2, \ldots, n)$ with probability $1/n!$. By Corollary 1.9 of Chapter 3, (4.56) converges in distribution as $n \to \infty$ to the $\tilde{\mu}$-distribution of (4.55), that is, to the distribution of $(\xi_1, \ldots, \xi_k, u_1, \ldots, u_k)$. This allows us to conclude that $u_1, u_2, \ldots$ is a sequence of independent, uniformly distributed random variables on $[0, 1]$, independent of $\xi_1, \xi_2, \ldots$. □

We close this section with a derivation of Ewens' sampling formula. Given a positive integer r, a vector $\beta = (\beta_1, \ldots, \beta_r)$ belonging to the finite set

$$\Gamma_r = \left\{\alpha \in (\mathbb{Z}_+)^r : \sum_{j=1}^{r} j\alpha_j = r\right\}, \tag{4.57}$$

and $\tilde{\nu} \in \mathscr{P}(\mathscr{P}_a[0, 1])$, let $P(\beta, \tilde{\nu})$ denote the probability that in a random sample of size r from a population whose "type" frequencies are random and distributed according to $\tilde{\nu}$, β_j "types" are represented j times $(j = 1, \ldots, r)$.

4.7 Theorem Let $\tilde{\mu}$ be as in Theorem 4.6, let $r \geq 1$, and let $\beta \in \Gamma_r$. Then

$$P(\beta, \tilde{\mu}) = \prod_{j=1}^{r} \frac{j}{j + \theta - 1} \frac{\theta^{\beta_j}}{j^{\beta_j}\beta_j!}. \tag{4.58}$$

Proof. Observe that for each $\tilde{\nu} \in \mathscr{P}(\mathscr{P}_a[0, 1])$,

$$P(\beta, \tilde{\nu}) = \int_{\mathscr{P}_a[0,1]} P(\beta, \delta_\mu)\tilde{\nu}(d\mu) \tag{4.59}$$

and

$$P(\beta, \delta_\mu) = \sum \frac{r!}{m_1!\, m_2! \cdots} s_1(F_\mu)^{m_1} s_2(F_\mu)^{m_2} \cdots, \tag{4.60}$$

where the sum ranges over all sequences $(m_1, m_2, \ldots)$ of nonnegative integers for which $\sum_{i=1}^{\infty} m_i = r$ and β_j is the cardinality of $\{i \geq 1: m_i = j\}$ $(j = 1, \ldots, r)$. Denote (4.60) by $\varphi_\beta(\mu)$. Then φ_β is lower semicontinuous with respect to ρ^* and

$$\sum_{\alpha \in \Gamma_r} \varphi_\alpha(\mu) = (s_1(F_\mu) + s_2(F_\mu) + \cdots)^r = 1, \tag{4.61}$$

implying that $\varphi_\beta = 1 - \sum_{\alpha \in \Gamma_r - \{\beta\}} \varphi_\alpha$ is also upper semicontinuous with respect to ρ^*, hence ρ^*-continuous. We conclude that

$$\lim_{n \to \infty} \int \varphi_\beta \, d\tilde{\mu}_n = \int \varphi_\beta \, d\tilde{\mu}. \tag{4.62}$$

The proof is completed by showing that the left side of (4.62) equals the right side of (4.58). That is,

$$\begin{aligned}
&\int \sum \frac{r!}{m_1! \, m_2! \cdots} s_1(F_\mu)^{m_1} s_2(F_\mu)^{m_2} \cdots d\tilde{\mu}_n \qquad (4.63)\\
&= \frac{r!}{(1!)^{\beta_1}(2!)^{\beta_2} \cdots (r!)^{\beta_r}} \sum E[(\xi_{(1)}^n)^{m_1} \cdots (\xi_{(n)}^n)^{m_n}]\\
&= \prod_{j=1}^{r} \frac{j}{(j!)^{\beta_j}} \sum E[(\xi_1^n)^{m_1} \cdots (\xi_n^n)^{m_n}]\\
&= \prod_{j=1}^{r} \frac{j}{(j!)^{\beta_j}} \sum \frac{\Gamma(m_1 + \varepsilon_n) \cdots \Gamma(m_n + \varepsilon_n)}{\Gamma(r + n\varepsilon_n)} \frac{\Gamma(n\varepsilon_n)}{\Gamma(\varepsilon_n)^n}\\
&= \left(\prod_{j=1}^{r} \frac{j}{(j!)^{\beta_j}} \right) \frac{n! \, \Gamma(\varepsilon_n)^{n - \Sigma\beta_j} \Gamma(1 + \varepsilon_n)^{\beta_1} \cdots \Gamma(r + \varepsilon_n)^{\beta_r}}{\beta_1! \cdots \beta_r! (n - \Sigma\beta_j)! \, \Gamma(r + n\varepsilon_n)} \frac{\Gamma(n\varepsilon_n)}{\Gamma(\varepsilon_n)^n}\\
&\to \prod_{j=1}^{r} \frac{j}{j + \theta - 1} \frac{\theta^{\beta_j}}{j^{\beta_j} \beta_j!},
\end{aligned}$$

where the sums are as in (4.60); here we are using $\Gamma(\varepsilon_n) = \Gamma(1 + \varepsilon_n)/\varepsilon_n$. □

5. PROBLEMS

1. Show that (1.23), (1.26), and (1.27) imply (1.22).

2. Let X be the diffusion process in K of Section 1 in the special case in which $\mu_{ij} = \gamma_j > 0$ for $i, j = 1, \ldots, r$. Show that the measure $\mu \in \mathscr{P}(K)$, defined by

$$\mu(dp) = \beta p_1^{2\gamma_1 - 1} \cdots p_r^{2\gamma_r - 1} \exp\left(\sum_{i,j=1}^{r} \sigma_{ij} p_i p_j \right) dp_1 \cdots dp_{r-1} \tag{5.1}$$

looks like Wright's formula for K-allele stationary dist.

for some constant $\beta > 0$, is a stationary distribution for X. (This generalizes parts of Theorems 2.1 and 4.6.)

3. Let X be as in Theorem 2.4. Show that ζ^ε, defined by (2.56), is continuous a.s. with respect to the distribution of X for $0 < \varepsilon < \frac{1}{2}$.

4. Let X be the diffusion process in K of Section 1 in the special case in which $\mu_{ir} = \gamma > 0$ for $i = 1, \ldots, r - 1$, $\mu_{ij} = 0$ otherwise, and $\sigma_{ij} = 0$ for $i, j = 1, \ldots, r$. Let $\tau = \inf\{t \geq 0: \min_{1 \leq i \leq r-1} X_i(t) = 0\}$. If $p \in K$ and $P\{X(0) = p\} = 1$, show that, for $i = 1, \ldots, r - 1$,

$$(5.2) \quad P\{X_i(\tau) = 0\} = 1 - p_i\Bigg\{\sum_{\substack{1 \leq j \leq r-1 \\ j \neq i}} (p_i + p_j)^{-1} - \sum_{\substack{1 \leq j < k \leq r-1 \\ j, k \neq i}} (p_i + p_j + p_k)^{-1} + \cdots + (-1)^{r-1}(1 - p_r)^{-1}\Bigg\}.$$

5. Put $I = \{1, \ldots, r\}$, and let X be the diffusion process in $E = [0, 1]^I$ with generator $A \equiv \{(f, Gf): f \in C^2(E)\}$, where

$$(5.3) \quad G = \frac{1}{2}\sum_{i \in I} \kappa_i^{-1} p_i(1 - p_i)\, \partial_i^2 + \sum_{i \in I}\Bigg\{\sigma p_i(1 - p_i) + \sum_{j \in I} \mu_{ij} p_j\Bigg\}\, \partial_i,$$

$(\mu_{ij})_{i, j \in I}$ is the infinitesimal matrix of an irreducible jump Markov process in I with stationary distribution $(\kappa_i)_{i \in I}$, and σ is real. (Specifically, $\mu_{ij} \geq 0$ for $i \neq j$, there does not exist a nonempty proper subset J of I with $\mu_{ij} = 0$ for all $i \in J$ and $j \notin J$, $\kappa_i > 0$ for each $i \in I$, $\sum_{j \in I} \kappa_j = 1$, and $\sum_{i \in I} \kappa_i \mu_{ij} = 0$ for each $j \in I$.)

(a) Formulate a geographically structured Wright–Fisher model of which X is the diffusion limit, proving an appropriate limit theorem (cf. Problem 7).

(b) Let $\tau = \inf\{t \geq 0: X(t) = (0, \ldots, 0) \text{ or } X(t) = (1, \ldots, 1)\}$. Show that $E[\tau] < \infty$, regardless of what the initial distribution of X may be.

(c) If $p \in E$ and $P\{X(0) = p\} = 1$, show that

$$(5.4) \quad P\{X(\tau) = (1, \ldots, 1)\} = \frac{1 - \exp\{-2\sigma \sum_{i \in I} \kappa_i p_i\}}{1 - \exp\{-2\sigma\}}.$$

6. Construct a sequence of diffusion processes X^N in $[0, 1]$, and a diffusion X in $[0, 1]$ with the following properties: 0 and 1 are absorbing boundaries for X^N and X, and $X^N \Rightarrow X$ in $D_{[0,1]}[0, \infty)$; but, defining ζ by (2.20) with $F = \{0, 1\}$ and τ_N and τ by (2.21), τ_N fails to converge in distribution to τ.

7. Apply Theorem 3.5 to Nagylaki's (1980) geographically structured Wright–Fisher model. In fact, this is already done in the given reference, so the problem consists merely of verifying the analysis appearing there and checking the technical conditions.

8. Apply Theorem 3.5 to the dioecious multinomial-sampling model described by Watterson (1964).

   Hint: Rather than numbering the genotypes arbitrarily, it simplifies matters to let $P_{ij}^{(s)}$ be the frequency of $A_i A_j$ $(i \leq j)$ in sex $s(s = 1, 2)$. This suggests, incidentally, that the general case of r alleles is no more difficult than the special case $r = 2$. Finally, we remark that the assumption that mutation rates and selection intensities are equal in the two sexes is unnecessarily restrictive.

9. Apply Theorem 3.5 to the dioecious overlapping-generation model described by Watterson (1964).

10. Karlin and Levikson (1974) state some results concerning diffusion limits of genetic models with random selection intensities. Prove these results.

    Hint: As a first step, one must specify the discrete models precisely.

11. Let X_1 (respectively, X) be the Ohta–Kimura model (respectively, the Fleming–Viot model) of Example 4.3, regarded as taking values in $\mathscr{P}(\mathbb{Z})$ (respectively, $\mathscr{P}(\mathbb{R})$). Show that X_1 (respectively, X) has no stationary distribution.

12. Let $\xi_1 \geq \xi_2 \geq \cdots$ be as in Theorem 4.6. Show that $P\{\xi_1 > \xi_2 > \cdots\} = 1$.

13. Let $\xi_1 \geq \xi_2 \geq \cdots$ be as in Theorem 4.6. Consider an inhomogeneous Poisson process on $(0, \infty)$ with rate function $\rho(x) = \theta x^{-1} e^{-x}$. In particular, the number of points in the interval (a, b) is Poisson distributed with parameter $\int_a^b \rho(x)\, dx$. Because $\int_a^\infty \rho(x)\, dx < \infty$ $(= \infty)$ if $a > 0$ $(= 0)$, the points of the Poisson process can be labeled as $\eta_1 > \eta_2 > \cdots$. Moreover, $\sum_{i=1}^\infty \eta_i$ has expectation $\int_0^\infty x\rho(x)\, dx = \theta$, and is therefore finite a.s. Show that $(\xi_1, \xi_2, \ldots)$ has the same distribution as

$$\left(\frac{\eta_1}{\sum_{i=1}^\infty \eta_i}, \frac{\eta_2}{\sum_{i=1}^\infty \eta_i}, \ldots\right). \tag{5.5}$$

14. Let X be the stationary infinitely-many-neutral-alleles model with uniform mutation (see Theorem 4.6), with its time-parameter set extended to $(-\infty, \infty)$.

    (a) Show that $\{X(t), -\infty < t < \infty\}$ and $\{X(-t), -\infty < t < \infty\}$ induce the same distribution on $C_{\mathscr{P}[0,1]}(-\infty, \infty)$. (Because of this, X is said to be *reversible*.)

    (b) Using (a), show that the probability that the most frequent allele (or "type") at time 0, say, is oldest equals the probability that the most frequent allele at time 0 will survive the longest.

(c) Show that the second probability in (b) is just $E[\xi_1]$, where ξ_1 is as in Theorem 4.6. (See Watterson and Guess (1977) for an evaluation of this expectation.)

6. NOTES

The best general reference on mathematical population genetics is Ewens (1979). Also useful is Kingman (1980).

The genetic model described in Section 1 is a variation of a model of Moran (1958c) due to Ethier and Nagylaki (1980). The Wright–Fisher model was formulated implicitly by Fisher (1922) and explicitly by Wright (1931). Various versions of Theorem 1.1 have been obtained by various authors. Trotter (1958) treated the neutral diallelic case, Norman (1972) the general diallelic case, Littler (1972) the neutral multi-allelic case, and Sato (1976) the general multi-allelic case.

The proof of Lemma 2.1 follows Norman (1975b). Theorem 2.4 is essentially from Ethier (1979), but a special case had earlier been obtained by Guess (1973). Corollary 2.7 is due to Norman (1972).

Section 3 comes from Ethier and Nagylaki (1980). Earlier work on diffusion approximations of non-Markovian models includes that of Watterson (1962) and Norman (1975a). Example 3.8, as noted above, is similar to a model of Moran (1958c). Example 3.9 is essentially due to Moran (1958a, b).

Theorem 4.1 is due to Kurtz (1981a). The characterization of X had earlier been obtained in certain cases by Fleming and Viot (1979). The processes of Example 4.3 are those of Ohta and Kimura (1973) and Fleming and Viot (1979). Example 4.4 was motivated by Watterson (1976), but the model goes back to Kimura and Crow (1964). Theorem 4.5 is analogous to a result of Ethier and Kurtz (1981). The main conclusion of Theorem 4.6, namely (4.39), is due to Kingman (1975). Finally, Theorem 4.7 is Ewens' (1972) sampling formula; our proof is based on Watterson (1976) and Kingman (1977).

Problem 2 is essentially Wright's (1949) formula. The reader is referred to Shiga (1981) for uniqueness. Problem 4 comes from Littler and Good (1978). See Nagylaki (1982) for Problem 5(a)(c). Problem 11 (for X_1) is due to Shiga (1982), who obtains much more general results. Problem 13 is a theorem of Kingman (1975), while Problem 14 is adapted from Watterson and Guess (1977).

11 DENSITY DEPENDENT POPULATION PROCESSES

By a population process we mean a stochastic model for a system involving a number of similar particles. We use the term "particles" broadly to include molecules in a chemical reaction model and infected individuals in an epidemic model. The branching and genetic models of the previous two chapters are examples of what we have in mind.

In this chapter we consider certain one-parameter families of processes that arise in a variety of applications. Section 1 gives examples that motivate the general formulation. Section 2 gives the basic law of large numbers and central limit theorem and Section 3 examines the corresponding diffusion approximation. Asymptotics for hitting distributions are considered in Section 4.

1. EXAMPLES

We are interested in certain families of jump Markov processes depending on a parameter that has different interpretations in different contexts, for example, total population size, area, or volume. We always denote this parameter by n. To motivate and identify the structure of these particular families, we give some examples:

A. Logistic Growth

In this context we interpret n as the area of a region occupied by a certain population. If the population size is k, then the population density is k/n. For simplicity we assume births and deaths occur singly. The intensities for births and deaths should be approximately proportional to the population size. We assume, however, that crowding affects the birth and death rates, which therefore depend on the population density. Hence the intensities can be written

$$\begin{aligned} q^{(n)}_{k,\,k+1} &= \lambda\left(\frac{k}{n}\right)k = n\lambda\left(\frac{k}{n}\right)\frac{k}{n} \\ q^{(n)}_{k,\,k-1} &= \mu\left(\frac{k}{n}\right)k = n\mu\left(\frac{k}{n}\right)\frac{k}{n}. \end{aligned} \tag{1.1}$$

If we take $\lambda(x) \equiv a$ and $\mu(x) = b + cx$, we have a stochastic model analogous to the deterministic logistic model given by

$$\dot{X} = (a - b)X - cX^2. \tag{1.2}$$ □

B. Epidemics

Here we interpret n as the total population size, which remains constant. In the population at any given time there are a number of individuals i that are susceptible to a particular disease and a number of individuals j who have the disease and can pass it on. A susceptible individual encounters diseased individuals at a rate proportional to the fraction of the total population that is diseased. Consequently, the intensity for a new infection is

$$q_{(i,\,j)(i-1,\,j+1)} = \lambda i \frac{j}{n} = n\lambda \frac{i}{n}\frac{j}{n}. \tag{1.3}$$

We assume diseased individuals recover and become immune independently of each other, which leads to the assumption

$$q_{(i,\,j)(i,\,j-1)} = \mu j = n\mu \frac{j}{n}. \tag{1.4}$$

The analogous deterministic model in this case is

$$\begin{aligned} \dot{X}_1 &= -\lambda X_1 X_2, \\ \dot{X}_2 &= \lambda X_1 X_2 - \mu X_2. \end{aligned} \tag{1.5}$$ □

C. Chemical Reactions

We now interpret n as the volume of a chemical system containing d chemical reactants $R_1, R_2, \ldots, R_d$ undergoing r chemical reactions

$$(1.6)\qquad b_{1j}R_1 + b_{2j}R_2 + \cdots + b_{dj}R_d \rightleftharpoons c_{1j}R_1 + c_{2j}R_2 + \cdots + c_{dj}R_d, \qquad j = 1, 2, \ldots, r,$$

that is, for example, when the jth reaction occurs in the forward direction, b_{1j} molecules of reactant R_1, b_{2j} molecules of reactant R_2, and so on, react to form c_{1j} molecules of R_1, c_{2j} molecules of R_2, and so on. Let $b_j = (b_{1j}, b_{2j}, \ldots, b_{dj})$, $c_j = (c_{1j}, c_{2j}, \ldots, c_{dj})$, and define

$$(1.7)\qquad |b_j| = b_{1j} + b_{2j} + b_{3j} + \cdots + b_{dj}, \qquad x^{b_j} = \prod_{i=1}^{d} x_i^{b_{ij}}.$$

The stochastic analogue of the "law of mass action" suggests that the intensity for the occurrence of the forward reaction should be

$$(1.8)\qquad n^{-|b_j|+1}\lambda_j \prod_{i=1}^{d} \binom{k_i}{b_{ij}} = n\left[\frac{\lambda_j}{\prod_{i=1}^{d} b_{ij}!}\left(\frac{k}{n}\right)^{b_j} + O\left(\frac{1}{n}\right)\right],$$

where $k = (k_1, k_2, \ldots, k_d)$ are the numbers of molecules of the reactants. The intensity for the reverse reaction is

$$(1.9)\qquad n^{-|c_j|+1}\mu_j \prod_{i=1}^{d} \binom{k_i}{c_{ij}} = n\left[\frac{\mu_j}{\prod_{i=1}^{d} c_{ij}!}\left(\frac{k}{n}\right)^{c_j} + O\left(\frac{1}{n}\right)\right].$$

If we take as the state the numbers of molecules of the reactants, then the transition intensities become

$$(1.10)\qquad q_{k,k+l}^{(n)} = \sum_{c_j - b_j = l} n^{-|b_j|+1}\lambda_j \prod_{i=1}^{d}\binom{k_i}{b_{ij}} + \sum_{b_j - c_j = l} n^{-|c_j|+1}\mu_j \prod_{i=1}^{d}\binom{k_i}{c_{ij}}$$
$$= n\left[\sum_{c_j - b_j = l} \frac{\lambda_j}{\prod_{i=1}^{d} b_{ij}!}\left(\frac{k}{n}\right)^{b_j} + \sum_{b_j - c_j = l} \frac{\mu_j}{\prod_{i=1}^{d} c_{ij}!}\left(\frac{k}{n}\right)^{c_j} + O\left(\frac{1}{n}\right)\right]$$

and the analogous deterministic model is

$$(1.11)\qquad \dot{X}_i = \sum_{j=1}^{r} \left((c_{ij} - b_{ij})\tilde{\lambda}_j X^{b_j} + (b_{ij} - c_{ij})\tilde{\mu}_j X^{c_j}\right)$$

where

$$\tilde{\lambda}_j = \frac{\lambda_j}{\prod_{i=1}^{d} b_{ij}!} \quad \text{and} \quad \tilde{\mu}_j = \frac{\mu_j}{\prod_{i=1}^{d} c_{ij}!}. \qquad \square$$

In the first two examples the transition intensities are of the form

(1.12) $$q_{k,k+l}^{(n)} = n\beta_l\left(\frac{k}{n}\right).$$

In the last example, (1.12) is the correct form if c_{ij} and b_{ij} only assume the values 0 and 1, while in general we have

(1.13) $$q_{k,k+l}^{(n)} = n\left[\beta_l\left(\frac{k}{n}\right) + O\left(\frac{1}{n}\right)\right].$$

We consider families with transition intensities of the form (1.12) and observe that the results usually carry over to the more general form with little additional effort.

To be precise, we assume we are given a collection of nonnegative functions β_l, $l \in \mathbb{Z}^d$, defined on a subset $E \subset \mathbb{R}^d$. Setting

(1.14) $$E_n = E \cap \{n^{-1}k: k \in \mathbb{Z}^d\},$$

we require that $x \in E_n$ and $\beta_l(x) > 0$ imply $x + n^{-1}l \in E_n$. By a *density dependent family* corresponding to the β_l we mean a sequence $\{X_n\}$ of jump Markov processes such that X_n has state space E_n and transition intensities

(1.15) $$q_{x,y}^{(n)} = n\beta_{n(y-x)}(x), \qquad x, y \in E_n.$$

2. LAW OF LARGE NUMBERS AND CENTRAL LIMIT THEOREM

By Theorem 4.1 of Chapter 6 we see that the Markov process $\hat{X}_n$ with intensities $q_{k,k+l}^{(n)} = n\beta_l(k/n)$, satisfies, for t less than the first infinity of jumps,

(2.1) $$\hat{X}_n(t) = \hat{X}_n(0) + \sum_l lY_l\left(n\int_0^t \beta_l\left(\frac{\hat{X}_n(s)}{n}\right) ds\right),$$

where the Y_l are independent standard Poisson processes.

Setting

(2.2) $$F(x) = \sum_l l\beta_l(x)$$

and $X_n = n^{-1}\hat{X}_n$, we have

(2.3) $$X_n(t) = X_n(0) + \sum_l ln^{-1}\tilde{Y}_l\left(n\int_0^t \beta_l(X_n(s))\, ds\right) + \int_0^t F(X_n(s))\, ds$$

where $\tilde{Y}_l(u) = Y_l(u) - u$ is the Poisson process centered at its expectation. The state space for X_n is E_n given by (1.14), and the form of the generator for X_n is

$$A_n f(x) = \sum_l n\beta_l(x)(f(x + n^{-1}l) - f(x)) \tag{2.4}$$

$$= \sum_l n\beta_l(x)(f(x + n^{-1}l) - f(x) - n^{-1}l \cdot \nabla f(x))$$

$$+ F(x) \cdot \nabla f(x), \qquad x \in E_n.$$

Observing that

$$\lim_{n\to\infty} \sup_{u \le v} |n^{-1}\tilde{Y}_l(nu)| = 0 \qquad \text{a.s.}, \qquad v \ge 0, \tag{2.5}$$

we have the following theorem.

2.1 Theorem Suppose that for each compact $K \subset E$,

$$\sum_l |l| \sup_{x \in K} \beta_l(x) < \infty \tag{2.6}$$

and there exists $M_K > 0$ such that

$$|F(x) - F(y)| \le M_K |x - y|, \qquad x, y \in K. \tag{2.7}$$

Suppose X_n satisfies (2.3), $\lim_{n\to\infty} X_n(0) = x_0$, and X satisfies

$$X(t) = x_0 + \int_0^t F(X(s))\, ds, \qquad t \ge 0. \tag{2.8}$$

Then for every $t \ge 0$,

$$\lim_{n\to\infty} \sup_{s \le t} |X_n(s) - X(s)| = 0 \qquad \text{a.s.} \tag{2.9}$$

2.2 Remark Implicitly we are assuming global existence for $\dot{X} = F(X)$, $X(0) = x_0$. Of course (2.7) guarantees uniqueness. □

Proof. Since for fixed $t \ge 0$ the validity of (2.9) depends only on the values of the β_l in some small neighborhood of $\{X(s): s \le t\}$, we may as well assume that $\bar{\beta}_l \equiv \sup_{x \in E} \beta_l(x)$ satisfies

$$\sum |l| \bar{\beta}_l < \infty \tag{2.10}$$

and that there exists a fixed $M > 0$ such that

$$|F(x) - F(y)| \le M |x - y|, \qquad x, y \in E. \tag{2.11}$$

Then

$$\begin{aligned}(2.12)\qquad \varepsilon_n(t) &\equiv \sup_{u\le t}\left|X_n(u) - X_n(0) - \int_0^u F(X_n(s))\,ds\right| \\ &\le \sum_l |l|\,n^{-1}\sup_{u\le t}|\tilde{Y}_l(n\bar{\beta}_l u)| \\ &\le \sum_l |l|\,n^{-1}(Y_l(n\bar{\beta}_l t) + n\bar{\beta}_l t),\end{aligned}$$

where the last inequality is term by term. Note that the process on the right is a process with independent increments, and the law of large numbers implies

$$\begin{aligned}(2.13)\qquad \lim_{n\to\infty}\sum_l |l|\,n^{-1}(Y_l(n\bar{\beta}_l t) + n\bar{\beta}_l t) &= \sum_l 2|l|\,\bar{\beta}_l t \\ &= \sum_l \lim_{n\to\infty} |l|\,n^{-1}(Y_l(n\bar{\beta}_l t) + n\bar{\beta}_l t),\end{aligned}$$

that is, we can interchange the limit and the summation. But the term by term inequality in (2.12) implies we can interchange the limit and summation for the middle expression as well, and we conclude from (2.5) that

$$(2.14)\qquad \lim_{n\to\infty}\varepsilon_n(t) = 0 \qquad \text{a.s.}$$

Now (2.11) implies

$$(2.15)\qquad |X_n(t) - X(t)| \le |X_n(0) - x_0| + \varepsilon_n(t) + \int_0^t M\,|X_n(s) - X(s)|\,ds,$$

and hence by Gronwall's inequality (Appendix 5),

$$(2.16)\qquad |X_n(t) - X(t)| \le (|X_n(0) - x_0| + \varepsilon_n(t))e^{Mt}$$

and (2.9) follows. □

Set $W_l^{(n)}(u) = n^{-1/2}\tilde{Y}_l(nu)$. The fact that $W_l^{(n)} \Rightarrow W_l$, standard Brownian motion, immediately suggests a central limit theorem for the deviation of X_n from X. Let $V_n(t) = \sqrt{n}(X_n(t) - X(t))$. Then

$$(2.17)\qquad V_n(t) = V_n(0) + \sum_l l W_l^{(n)}\left(\int_0^t \beta_l(X_n(s))\,ds\right) + \int_0^t \sqrt{n}\,(F(X_n(s)) - F(X(s)))\,ds.$$

Observing that $X_n = X + n^{-1/2}V_n$, (2.17) suggests the following limiting equation:

(2.18) $$V(t) = V(0) + \sum_l l W_l\left(\int_0^t \beta_l(X(s))\,ds\right) + \int_0^t \partial F(X(s))V(s)\,ds$$

$$\equiv V(0) + U(t) + \int_0^t \partial F(X(s))V(s)\,ds,$$

where $\partial F(x) = ((\partial_j F_i(x)))$.

Let Φ be the solution of the matrix equation

(2.19) $$\frac{\partial}{\partial t}\Phi(t, s) = \partial F(X(t))\Phi(t, s), \qquad \Phi(s, s) = I.$$

Then

(2.20) $$V(t) = \Phi(t, 0)V(0) + \int_0^t \Phi(t, s)\,dU(s)$$

$$= \Phi(t, 0)V(0) + U(t) + \int_0^t \Phi(t, s)\,\partial F(X(s))U(s)\,ds$$

$$= \Phi(t, 0)(V(0) + U(t)) + \int_0^t \Phi(t, s)\,\partial F(X(s))(U(s) - U(t))\,ds.$$

Since U is Gaussian (in fact, a time-inhomogeneous Brownian motion), V is Gaussian with mean $\Phi(t, 0)V(0)$ (we assume $V(0)$ is nonrandom) and covariance matrix

(2.21) $$\operatorname{cov}(V(t), V(r)) = \int_0^{t \wedge r} \Phi(t, s)G(X(s))[\Phi(r, s)]^T\,ds,$$

where

(2.22) $$G(x) = \sum_l l l^T \beta_l(x).$$

2.3 Theorem Suppose for each compact $K \subset E$,

(2.23) $$\sum_l |l|^2 \sup_{x \in K} \beta_l(x) < \infty,$$

and that the β_l and ∂F are continuous. Suppose X_n satisfies (2.3), X satisfies (2.8), $V_n = \sqrt{n}(X_n - X)$, and $\lim_{n\to\infty} V_n(0) = V(0)$ ($V(0)$ constant). Then $V_n \Rightarrow V$ where V is the solution of (2.18).

Proof. Comparing (2.17) to (2.18), set

(2.24) $$U_n(t) = \sum_l l W_l^{(n)}\left(\int_0^t \beta_l(X_n(s))\,ds\right)$$

and

$$\varepsilon_n(t) = \int_0^t \sqrt{n}(F(X_n(s)) - F(X(s)) - n^{-1/2}\,\partial F(X(s))V_n(s))\,ds. \tag{2.25}$$

Theorem 2.1 implies $\sup_{s \le t} |\varepsilon_n(s)| \to 0$ a.s. and $U_n \Rightarrow U$ in $D_{\mathbb{R}^d}[0, \infty)$. But, as in (2.20),

$$V_n(t) = \Phi(t, 0)V_n(0) + U_n(t) + \varepsilon_n(t) + \int_0^t \Phi(t, s)\,\partial F(X(s))(U_n(s) + \varepsilon_n(s))\,ds, \tag{2.26}$$

and $V_n \Rightarrow V$ by the continuous mapping theorem, Corollary 1.9 of Chapter 3. □

3. DIFFUSION APPROXIMATIONS

The basic implication of Theorem 2.3 is that X_n can be approximated (in distribution) by $\tilde{Z}_n = X + n^{-1/2}V$. An alternative approximation is the "diffusion approximation" Z_n, whose generator is obtained heuristically by expanding f in (2.4) in a Taylor series and dropping terms beyond the second order. This gives

$$B_n f(x) = \frac{1}{2n}\sum_{i,j} G_{ij}(x)\,\partial_i\,\partial_j f(x) + \sum_i F_i(x)\,\partial_i f(x). \tag{3.1}$$

The statement that $\tilde{Z}_n$ approximates X_n is, of course, justified by the central limit theorem. No similar limit theorem can justify the statement that Z_n approximates X_n, since the Z_n are not expressible in terms of any sort of limiting process. To overcome this problem we use the coupling theorem of Komlos, Major and Tusnady, Corollary 5.5 of Chapter 7, to obtain a direct comparison between X_n and Z_n.

Suppose the β_l are continuous and the solution of the martingale problem for $(B_n, \delta_{X_n(0)})$ is unique. Then it follows from Theorem 5.1 of Chapter 6 that the solution can be obtained as a solution of

$$Z_n(t) = X_n(0) + \sum_l l n^{-1} W_l\left(n \int_0^t \beta_l(Z_n(s))\,ds\right) + \int_0^t F(Z_n(s))\,ds \tag{3.2}$$

where the W_l are independent standard Brownian motions. By Corollary 5.5 and Remark 5.4 both of Chapter 7, we can assume the existence of centered Poisson processes $\tilde{Y}_l$ such that

$$\sup_t \frac{|\tilde{Y}_l(t) - W_l(t)|}{\log(2 \vee t)} < \infty \tag{3.3}$$

and for $\beta, \gamma > 0$ there exist $\lambda, K, C > 0$ such that

$$P\left\{\sup_{t \le \beta n} |\tilde{Y}_l(t) - W_l(t)| > C \log n + x\right\} \le K n^{-\gamma} e^{-\lambda x}. \tag{3.4}$$

Since the W_l are independent, we can take the $\tilde{Y}_l$ to be independent as well. Let X_n satisfy (2.3) using the $\tilde{Y}_l$ constructed from the W_l. Then X_n and Z_n are defined on the same sample space and we can consider the difference $|X_n(t) - Z_n(t)|$. (Note that the pair (X_n, Z_n) is *not* a Markov process even though each component is.)

3.1 Theorem Let X, X_n, and Z_n be as above, and assume $\lim_{n\to\infty} X_n(0) = X(0)$. Fix $\varepsilon, T > 0$, and set $N_\varepsilon = \{y \in E: \inf_{t \le T} |X(t) - y| \le \varepsilon\}$. Let $\bar{\beta}_l \equiv \sup_{x \in N_\varepsilon} \beta_l(x) < \infty$ and suppose $\bar{\beta}_l = 0$ except for finitely many l. Suppose $M > 0$ satisfies

$$|\beta_l(x) - \beta_l(y)| \le M|x - y|, \qquad x, y \in N_\varepsilon, \tag{3.5}$$

and

$$|F(x) - F(y)| \le M|x - y|, \qquad x, y \in N_\varepsilon. \tag{3.6}$$

Let $\tau_n = \inf\{t: X_n(t) \notin N_\varepsilon \text{ or } Z_n(t) \notin N_\varepsilon\}$. (Note $P\{\tau_n > T\} \to 1$.) Then for $n \ge 2$ there is a random variable Γ_n^T and positive constants λ_T, C_T, and K_T depending on T, on M, and on the $\bar{\beta}_l$, but *not* on n, such that

$$\sup_{t \le T \wedge \tau_n} |X_n(t) - Z_n(t)| \le \Gamma_n^T \frac{\log n}{n}, \tag{3.7}$$

and

$$P\{\Gamma_n^T > C_T + x\} \le K_T n^{-2} \exp\left\{-\lambda_T x^{1/2} - \frac{\lambda_T x}{\log n}\right\}. \tag{3.8}$$

Proof. Again we can assume $\bar{\beta}_l = \sup_{x \in E} \beta_l(x) < \infty$ and (3.5) and (3.6) are satisfied for all $x, y \in E$. Under these hypotheses we can drop the τ_n in (3.7).

By (3.4) there exist constants C_l^1, K_l^1 independent of n and nonnegative random variables $L_{l,n}^1$ such that

$$\sup_{t \le n\bar{\beta}_l T} |\tilde{Y}_l(t) - W_l(t)| \le C_l^1 \log n + L_{l,n}^1 \tag{3.9}$$

and

$$P\{L_{l,n}^1 > x\} \le K_l^1 n^{-2} e^{-\lambda x}. \tag{3.10}$$

Define

$$\Lambda_{l,n}(z) = \sup_{u \le n\bar{\beta}_l T} \sup_{v \le z \log n} |W_l(u+v) - W_l(u)|. \tag{3.11}$$

Let $k_n = [n\bar{\beta}_l T/z \log n] + 1$. Then

$$\Lambda_{l,n}(z) \le 3 \sup_{k \le k_n} \sup_{v \le z \log n} |W_l(kz \log n + v) - W_l(kz \log n)|, \tag{3.12}$$

and hence for $z, c, x \ge 0$

$$\begin{aligned} P\{\Lambda_{l,n}(z) &\ge z^{1/2}(c \log n + x)\} \\ &\le k_n P\left\{3 \sup_{v \le z \log n} |W_l(v)| \ge z^{1/2}(c \log n + x)\right\} \\ &\le 2k_n P\{|W_l(1)| \ge \tfrac{1}{3}(c \log n + x)(\log n)^{-1/2}\} \\ &\le ak_n \exp\left\{-\frac{(c^2 \log n + 2cx + x^2/\log n)}{18}\right\} \end{aligned} \tag{3.13}$$

where $a = \sup_{x \ge 0} e^{x^2/2} P\{|W_l(1)| \ge x\}$. Since $\Lambda_{l,n}(z)$ is monotone in z,

$$\begin{aligned} \bar{\Lambda}_{l,n} &\equiv \sup_{0 \le z \le n\bar{\beta}_l T} (z+1)^{-1/2} \Lambda_{l,n}(z) \\ &\le \sup_{1 \le m \le n\bar{\beta}_l T + 1} m^{-1/2} \Lambda_{l,n}(m) \end{aligned} \tag{3.14}$$

for integer m. Therefore

$$\begin{aligned} P\{\bar{\Lambda}_{l,n} &> c \log n + x\} \\ &\le (1 + n\bar{\beta}_l T) ak_n \exp\left\{-\frac{(c^2 \log n + 2cx + x^2/\log n)}{18}\right\}. \end{aligned} \tag{3.15}$$

Since $k_n = O(n)$ there exist constants C_l^2 and K_l^2, independent of n, such that

$$P\{\bar{\Lambda}_{l,n} > C_l^2 \log n + x\} \le K_l^2 n^{-2} \exp\left\{-\lambda x - \frac{x^2}{18 \log n}\right\}. \tag{3.16}$$

Setting $L_{l,n}^2 = (\bar{\Lambda}_{l,n} - C_l^2 \log n) \vee 0$, we have

$$P\{L_{l,n}^2 > x\} \le K_l^2 n^{-2} \exp\left\{-\lambda x - \frac{x^2}{18 \log n}\right\}. \tag{3.17}$$

Taking the difference between (2.3) and (3.2) (recall only finitely many β_l are nonzero),

$$
\begin{aligned}
(3.18)\qquad & |X_n(t) - Z_n(t)| \\
&\le n^{-1} \sum_l |l| \left| \tilde{Y}_l\left(n \int_0^t \beta_l(X_n(s))\,ds\right) - W_l\left(n \int_0^t \beta_l(X_n(s))\,ds\right)\right| \\
&\quad + n^{-1} \sum_l |l| \left| W_l\left(n \int_0^t \beta_l(X_n(s))\,ds\right) - W_l\left(n \int_0^t \beta_l(Z_n(s))\,ds\right)\right| \\
&\quad + \int_0^t |F(X_n(s)) - F(Z_n(s))|\,ds \\
&\le n^{-1} \sum_l |l| (C_l^1 \log n + L_{l,n}^1) \\
&\quad + n^{-1} \sum_l |l| \Lambda_{l,n}\left(\frac{n}{\log n}\left|\int_0^t (\beta_l(X_n(s)) - \beta_l(Z_n(s)))\,ds\right|\right) \\
&\quad + \int_0^t M|X_n(s) - Z_n(s)|\,ds.
\end{aligned}
$$

Let $\gamma_n(t) = n|X_n(t) - Z_n(t)|/\log n$. Then

$$
\begin{aligned}
(3.19)\qquad \gamma_n(t) \le e^{Mt}\Bigg[&\sum_l |l| \left(C_l^1 + \frac{L_{l,n}^1}{\log n}\right) \\
&+ \left(1 + M \int_0^t \gamma_n(s)\,ds\right)^{1/2} \sum_l |l| \left(C_l^2 + \frac{L_{l,n}^2}{\log n}\right)\Bigg],
\end{aligned}
$$

and setting $\bar{\gamma}_n = \sup_{t \le T} \gamma_n(t)$ we have

$$
\begin{aligned}
(3.20)\qquad \bar{\gamma}_n \le e^{MT} &\sum_l |l| \left(C_l^1 + \frac{L_{l,n}^1}{\log n}\right) \\
&+ \left(\frac{1}{MT} + \bar{\gamma}_n\right)^{1/2} e^{MT}(MT)^{1/2} \sum_l |l| \left(C_l^2 + \frac{L_{l,n}^2}{\log n}\right).
\end{aligned}
$$

The inequality $\gamma \le a + \gamma^{1/2} b$ implies $\gamma \le 2a + b^2$. Hence there is a constant C_T such that

$$
\begin{aligned}
(3.21)\qquad \bar{\gamma}_n &\le \frac{1}{MT} + 2e^{MT} \sum_l |l| \left(C_l^1 + \frac{L_{l,n}^1}{\log n}\right) + MTe^{2MT}\left(\sum_l |l| \left(C_l^2 + \frac{L_{l,n}^2}{\log n}\right)\right)^2 \\
&\le C_T + \frac{2e^{MT}}{\log n} \sum_l |l| L_{l,n}^1 + \frac{MTe^{2MT}}{(\log n)^2}\left(\sum_l |l| L_{l,n}^2\right)^2 \\
&\equiv C_T + L_n \equiv \Gamma_n^T.
\end{aligned}
$$

Since the sums in (3.21) are finite, (3.10) and (3.17) imply there exist constants $K_T, \lambda_T > 0$ such that

$$
(3.22)\qquad P\{L_n > x\} \le K_T n^{-2} \exp\left\{-\lambda_T x^{1/2} - \frac{\lambda_T x}{\log n}\right\},
$$

and (3.8) follows. □

We now have two approximations for X_n, namely Z_n and $\tilde{Z}_n$. The question arises as to which is "best." The answer is that, at least asymptotically, for bounded time intervals they are essentially equivalent. In order to make this precise, we consider Z_n and $\tilde{Z}_n$ as solutions of stochastic integral equations:

$$Z_n(t) = Z_n(0) + n^{-1/2} \sum_l \int_0^t l\beta_l^{1/2}(Z_n(s))\, dW_l(s) + \int_0^t F(Z_n(s))\, ds \tag{3.23}$$

and

$$\tilde{Z}_n(t) = X(0) + n^{-1/2} \sum_l \int_0^t l\beta_l^{1/2}(X(s))\, dW_l(s) + \int_0^t (F(X(s)) + \partial F(X(s))(\tilde{Z}_n(s) - X(s)))\, ds. \tag{3.24}$$

3.2 Theorem In addition to the assumptions of Theorem 3.1, suppose that the $\beta_l^{1/2}$ are continuously differentiable and that F is twice continuously differentiable in N_ε. Let Z_n and $\tilde{Z}_n$ satisfy (3.23) and (3.24). Let $n(X_n(0) - X(0)) \to 0$. Then

$$\sup_{t \le T} |n(Z_n(t) - \tilde{Z}_n(t)) - \hat{V}(t)| \xrightarrow{p} 0, \tag{3.25}$$

where $\hat{V}$ satisfies

$$\hat{V}(t) = \sum_l l \int_0^t \nabla\beta_l^{1/2}(X(s)) \cdot V(s)\, dW_l(s) + \int_0^t \partial F(X(s))\hat{V}(s)\, ds + \int_0^t \sum_{i,j} \partial_i\, \partial_j F(X(s)) V_i(s) V_j(s)\, ds \tag{3.26}$$

and V satisfies

$$V(t) = \sum_l l \int_0^t \beta_l^{1/2}(X(s))\, dW_l(s) + \int_0^t \partial F(X(s)) V(s)\, ds. \tag{3.27}$$

3.3 Remark The assumption that $\bar{\beta}_l > 0$ for only finitely many l can be replaced by

$$\sum |l|^2 \bar{\beta}_l < \infty \quad \text{and} \quad \sum |l|^2 \sup_{x \in N_\varepsilon} |\nabla\beta_l^{1/2}(x)|^2 < \infty. \qquad \square$$

Proof. Let $U_n = \sqrt{n}(Z_n - X)$. A simple argument gives

$$E\left[\sup_{t \le T} |U_n(t) - V(t)|^2\right] \to 0. \tag{3.28}$$

We have

$$(3.29)\qquad \begin{aligned} \hat{V}_n(t) &\equiv n(Z_n(t) - \tilde{Z}_n(t)) \\ &= n(X_n(0) - X(0)) + \sum_l l \int_0^t \sqrt{n}(\beta_l^{1/2}(Z_n(s)) - \beta_l^{1/2}(X(s)))\, dW_l(s) \\ &\quad + \int_0^t \partial F(X(s))\hat{V}_n(s)\, ds \\ &\quad + \int_0^t \sqrt{n}(\sqrt{n}(F(Z_n(s)) - F(X(s))) - n^{-1/2}\, \partial F(X(s))U_n(s))\, ds. \end{aligned}$$

By hypothesis $n(X_n(0) - X(0)) \to 0$. The second term on the right converges to the first term on the right of (3.26) by (3.28) of this chapter and (2.19) of Chapter 5. The limit in (3.28) also implies the last term in (3.29) converges to the last term in (3.26), and the theorem follows. □

By the construction in Theorem 3.1, for each $T > 0$, there is a constant $C_T > 0$ such that

$$(3.30)\qquad \lim_{n\to\infty} P\left\{\sup_{t\le T} |X_n(t) - Z_n(t)| > \frac{C_T \log n}{n}\right\} = 0,$$

whereas by Theorem 3.2, for any sequence $a_n \to \infty$,

$$(3.31)\qquad \lim_{n\to\infty} P\left\{\sup_{t\le T} |Z_n(t) - \tilde{Z}_n(t)| > \frac{a_n}{n}\right\} = 0.$$

Since Bartfai's theorem, Theorem 5.6 of Chapter 7, implies that (3.30) is best possible, at least in some cases, we see that asymptotically the two approximations are essentially the same.

4. HITTING DISTRIBUTIONS

The time and location of the first exit of a process from a region are frequently of interest. The results of Section 2 give the asymptotic behavior for these quantities as well. We characterize the region of interest as the set on which a given function φ is positive.

4.1 Theorem Let φ be continuously differentiable on $\mathbb{R}^d$. Let X_n and X satisfy (2.3) and (2.8), respectively, with $\varphi(X(0)) > 0$, and suppose the conditions of Theorem 2.3 hold. Let

$$(4.1)\qquad \tau_n = \inf\{t: \varphi(X_n(t)) \le 0\}$$

and

(4.2) $$\tau = \inf \{t: \varphi(X(t)) \le 0\}.$$

Suppose $\tau < \infty$ and

(4.3) $$\nabla\varphi(X(\tau)) \cdot F(X(\tau)) < 0.$$

Then

(4.4) $$\sqrt{n}(\tau_n - \tau) \Rightarrow -\frac{\nabla\varphi(X(\tau)) \cdot V(\tau)}{\nabla\varphi(X(\tau)) \cdot F(X(\tau))}$$

and

(4.5) $$\sqrt{n}(X_n(\tau_n) - X(\tau)) \Rightarrow V(\tau) - \frac{\nabla\varphi(X(\tau)) \cdot V(\tau)}{\nabla\varphi(X(\tau)) \cdot F(X(\tau))} F(X(\tau)).$$

4.2 Remark One example of interest is the number of susceptibles remaining in the population when the last infective is removed in the epidemic model described in Section 1. This situation is not covered directly by the theorem. (In particular, $\tau = \infty$). However see Problem 5. □

Proof. Note that

(4.6) $$\frac{\partial}{\partial t} \varphi(X(t)) = \nabla\varphi(X(t)) \cdot F(X(t))$$

so that (4.3) implies $\varphi(X(\tau - \varepsilon)) > 0$ and $\varphi(X(\tau + \varepsilon)) < 0$ for $0 < \varepsilon < \tau$.

Since $X_n \to X$ a.s. uniformly on bounded time intervals, it follows that $\tau_n \to \tau$ a.s. Since $\varphi(X_n(\tau_n)) \le 0$ and $\varphi(X_n(\tau_n -)) \ge 0$,

(4.7) $$\begin{aligned} |\sqrt{n}\varphi(X_n(\tau_n))| &\le |\sqrt{n}(\varphi(X_n(\tau_n)) - \varphi(X_n(\tau_n -)))| \\ &= |\nabla\varphi(\theta_n) \cdot (V_n(\tau_n) - V_n(\tau_n -))| \end{aligned}$$

for some θ_n on the line between $X_n(\tau_n)$ and $X_n(\tau_n -)$, and since $V_n \Rightarrow V$ and V is continuous, the right side of (4.7) converges in distribution to zero. By the continuity of X, $\varphi(X(\tau)) = 0$ and

(4.8) $$\begin{aligned} &\sqrt{n}(\varphi(X(\tau)) - \varphi(X(\tau_n))) \\ &= \sqrt{n}(\varphi(X_n(\tau_n)) - \varphi(X(\tau_n))) - \sqrt{n}\varphi(X_n(\tau_n)) \\ &= \sqrt{n}(\varphi(X(\tau_n) + n^{-1/2}V_n(\tau_n)) - \varphi(X(\tau_n))) - \sqrt{n}\varphi(X_n(\tau_n)) \\ &\Rightarrow \nabla\varphi(X(\tau)) \cdot V(\tau). \end{aligned}$$

But the left side of (4.8) is asymptotic to

(4.9) $$-\nabla\varphi(X(\tau)) \cdot F(X(\tau))\sqrt{n}(\tau_n - \tau),$$

and (4.4) follows.

Finally

$$\begin{aligned}\sqrt{n}(X_n(\tau_n) - X(\tau)) &= V_n(\tau_n) + \sqrt{n}(X(\tau_n) - X(\tau)) \\ &\Rightarrow V(\tau) - \frac{\nabla\varphi(X(\tau)) \cdot V(\tau)}{\nabla\varphi(X(\tau)) \cdot F(X(\tau))} F(X(\tau)). \end{aligned} \tag{4.10}$$

□

5. PROBLEMS

1. Let X_n be the logistic growth model described in Section 1.
   (a) Compute the parameters of the limiting Gaussian process V given by Theorem 2.3.
   (b) Let Z_n and $\tilde{Z}_n$ be the approximations of X_n discussed in Section 3. Assuming $Z_n(0) = \tilde{Z}_n(0) \neq 0$, show that Z_n eventually absorbs at zero, but that $\tilde{Z}_n$ is asymptotically stationary (and nondegenerate).

2. Consider the chemical reaction model for $R_1 + R_2 \rightleftharpoons R_3$ with parameters given by (1.10).
   (a) Compute the parameters of the limiting Gaussian process V given by Theorem 2.3.
   (b) Let $X(0)$ be the fixed point of the limiting deterministic model (so $X(t) = X(0)$ for all $t \geq 0$). Then V_n is a Markov process with stationary transition probabilities. Apply Theorem 9.14 of Chapter 4 to show that the stationary distribution for V_n converges to the stationary distribution for V.

3. Use the fact that, under the assumptions of Theorem 2.1,

$$X_n(t) - X_n(0) - \int_0^t F(X_n(s))\, ds \tag{5.1}$$

is a local martingale and Gronwall's inequality to estimate $P\{\sup_{s \leq t} |X_n(s) - X(s)| \geq \varepsilon\}$.

4. Under the hypotheses of Theorems 3.1 and 3.2, show that for any bounded $U \subset \mathbb{R}^d$ with smooth boundary,

$$|P\{V_n(t) \in U\} - P\{V(t) \in U\}| = O\left(\frac{\log n}{\sqrt{n}}\right). \tag{5.2}$$

5. Let $X_n = (S_n, I_n)$ be the epidemic model described in Section 1 and let $X = (S, I)$ denote the limiting deterministic model (S for susceptible, I for infectious). Let $\tau_n = \inf\{t: I_n(t) = 0\}$.

(a) Show that if $I(0) > 0$, then $I(t) > 0$ for all $t \geq 0$, but that $\lim_{t \to \infty} I(t) = 0$ and $S(\infty) \equiv \lim_{t \to \infty} S(t)$ exists.

(b) Show that if $\sqrt{n}(X_n(0) - X(0))$ converges, then $\sqrt{n}(S_n(\tau_n) - S(\infty))$ converges in distribution.

Hint: Let γ_n satisfy

$$\int_0^{\gamma_n(t)} I_n(s)\, ds = t, \qquad t < \int_0^\infty I_n(s)\, ds, \tag{5.3}$$

and show that $X_n(\gamma_n(\cdot))$ extends to a process satisfying the conditions of Theorem 4.1 with $\varphi(x_1, x_2) = x_2$.

6. NOTES

Most of the material in this chapter is from Kurtz (1970b, 1971, 1978a). Norman (1974) gives closely related results including conditions under which the convergence of $V_n(t)$ to $V(t)$ is uniform for all t (see Problem 2). Barbour (1974, 1980) studies the same class of processes giving rates of convergence for the distributions of certain functionals of V_n in the first paper and for the stationary distributions in the second. Berry–Esseen type results have been given by Allain (1976) and Alm (1978). Analogous results for models with age dependence have been given by Wang (1977).

Darden and Kurtz (1985) study the situation in which the limiting deterministic model has a stable fixed point, extending the uniformity results of Norman and obtaining asymptotic exponentiality for the distribution of the exit time from a neighborhood of the stable fixed point.

12 RANDOM EVOLUTIONS

The aim in this chapter is to study the asymptotic behavior of certain random evolutions as a small parameter tends to zero. We do not attempt to achieve the greatest generality, but only enough to be able to treat a variety of examples. Section 1 introduces the basic ideas and terminology in terms of perhaps the simplest example. Sections 2 and 3 consider the case in which the underlying process (or driving process) is Markovian and ergodic, while Section 4 requires it to be stationary and uniform mixing.

1. INTRODUCTION

One of the simplest examples of a random evolution can be described as follows. Let N be a Poisson process with parameter λ, and fix $\alpha > 0$. Given $(x, y) \in \mathbb{R} \times \{-1, 1\}$, define the pair of processes (X, Y) by

$$X(t) = x + \alpha \int_0^t Y(s)\, ds, \qquad Y(t) = (-1)^{N(t)} y. \tag{1.1}$$

$X(t)$ and $\alpha Y(t)$ represent the position and velocity at time t of a particle moving in one dimension at constant speed α, but subject to reversals of direction at the jump times of N, given initial position x and velocity αy.

Let us first observe that (X, Y) is a Markov process in $\mathbb{R} \times \{-1, 1\}$. For if we define for each $t \geq 0$ the linear operator $T(t)$ on $\hat{C}(\mathbb{R} \times \{-1, 1\})$ by

$$(1.2) \qquad T(t)f(x, y) = E[f(X(t), Y(t))],$$

where (X, Y) is given by (1.1), then the Markov property of Y implies that

$$(1.3) \qquad E[f(X(t), Y(t)) \mid \mathscr{F}_s^Y] = T(t-s)f(X(s), Y(s))$$

for all $f \in \hat{C}(\mathbb{R} \times \{-1, 1\})$, $(x, y) \in \mathbb{R} \times \{-1, 1\}$, and $t > s \geq 0$. It follows easily that $\{T(t)\}$ is a Feller semigroup on $\hat{C}(\mathbb{R} \times \{-1, 1\})$ and (X, Y) is a Markov process in $\mathbb{R} \times \{-1, 1\}$ corresponding to $\{T(t)\}$.

Clearly, however, X itself is non-Markovian. Nevertheless, while Y visits $y \in \{-1, 1\}$, X evolves according to the Feller semigroup $\{T_y(t)\}$ on $\hat{C}(\mathbb{R})$ defined by

$$(1.4) \qquad T_y(t)f(x) = f(x + \alpha t y).$$

Consequently, letting $\tau_1, \tau_2, \ldots$ denote the jump times of Y, the evolution of X over the time interval $[s, t]$ is described by the operator-valued random variable

$$(1.5) \qquad \mathscr{T}(s, t) = T_{Y(0)}((\tau_1 \vee s) \wedge t - s) T_{Y(\tau_1)}((\tau_2 \vee s) \wedge t - (\tau_1 \vee s) \wedge t) \cdots$$

in the sense that

$$(1.6) \qquad E[f(X(t), Y(t)) \mid \mathscr{F}_s^X \vee \mathscr{F}_t^Y] = \mathscr{T}(s, t)\{f(\cdot, Y(t))\}(X(s))$$

for all $f \in \hat{C}(\mathbb{R} \times \{-1, 1\})$. The family $\{\mathscr{T}(s, t), t \geq s \geq 0\}$ satisfies

$$(1.7) \qquad \mathscr{T}(s, t)\mathscr{T}(t, u) = \mathscr{T}(s, u), \qquad s \leq t \leq u,$$

and is therefore called a *random evolution*. Because Y "controls" the development of X, we occasionally refer to Y as the *driving process* and to X as the *driven process*. Observe that (1.3) and (1.6) specify the relationship between $\{T(t)\}$ and $\{\mathscr{T}(t)\}$. (Of course, in the special case of (1.1), the left side of (1.6) can be replaced by $f(X(t), Y(t))$ because $\mathscr{F}_t^X \subset \mathscr{F}_t^Y$. In general, however, X need not evolve deterministically while Y visits y.)

To determine the generator of the semigroup $\{T(t)\}$, let $f \in C^{1,0}(\mathbb{R} \times \{-1, 1\})$ and $t_2 > t_1 \geq 0$. Then

$$(1.8) \qquad f(X(t_2), Y(t_2)) - f(X(t_1), Y(t_2)) = \int_{t_1}^{t_2} \alpha Y(s) f_x(X(s), Y(t_2))\, ds$$

and

$$(1.9) \qquad E[f(X(t_1), Y(t_2)) - f(X(t_1), Y(t_1)) \mid \mathscr{F}_{t_1}^Y]$$

$$= E\left[\int_{t_1}^{t_2} \lambda\{f(X(t_1), -Y(s)) - f(X(t_1), Y(s))\}\, ds \,\middle|\, \mathscr{F}_{t_1}^Y\right],$$

so by Lemma 3.4 of Chapter 4,

$$f(X(t), Y(t)) - \int_0^t Af(X(s), Y(s))\, ds \tag{1.10}$$

is an $\{\mathscr{F}_t^Y\}$-martingale, where

$$Af(x, y) = \alpha y f_x(x, y) + \lambda\{f(x, -y) - f(x, y)\}. \tag{1.11}$$

Identifying $\hat{C}(\mathbb{R} \times \{-1, 1\})$ with $\hat{C}(\mathbb{R}) \times \hat{C}(\mathbb{R})$, we can rewrite A as

$$A = \alpha \begin{pmatrix} \dfrac{d}{dx} & 0 \\ 0 & -\dfrac{d}{dx} \end{pmatrix} + \lambda \begin{pmatrix} -1 & 1 \\ 1 & -1 \end{pmatrix} \tag{1.12}$$

with $\mathscr{D}(A) = \hat{C}^1(\mathbb{R}) \times \hat{C}^1(\mathbb{R})$. Since $\mathscr{R}(A) \subset \hat{C}(\mathbb{R}) \times \hat{C}(\mathbb{R})$, it follows from the martingale property and the strong continuity of $\{T(t)\}$ that the generator of $\{T(t)\}$ extends A. But by Problem 1 and Corollary 7.2, both of Chapter 1, A generates a strongly continuous semigroup on $\hat{C}(\mathbb{R}) \times \hat{C}(\mathbb{R})$. We conclude from Proposition 4.1 of Chapter 1 that A is precisely the generator of $\{T(t)\}$.

This has an interesting consequence. Let $f \in \hat{C}^2(\mathbb{R})$ and define

$$\begin{pmatrix} g(t, x) \\ h(t, x) \end{pmatrix} = T(t) \begin{pmatrix} f \\ f \end{pmatrix} (x) \tag{1.13}$$

for all $(t, x) \in [0, \infty) \times \mathbb{R}$. Then g and h belong to $\hat{C}^2([0, \infty) \times \mathbb{R})$ and satisfy the system of partial differential equations

$$\begin{aligned} g_t &= \alpha g_x - \lambda(g - h), \\ h_t &= -\alpha h_x + \lambda(g - h). \end{aligned} \tag{1.14}$$

Letting $u = \frac{1}{2}(g + h)$ and $v = \frac{1}{2}(g - h)$, we have

$$\begin{aligned} u_t &= \alpha v_x, \\ v_t &= \alpha u_x - 2\lambda v. \end{aligned} \tag{1.15}$$

Hence $u_{tt} = \alpha v_{tx} = \alpha u_{xx} - 2\lambda v_x = \alpha u_{xx} - (2\lambda/\alpha)u_t$, or

$$u_t = \frac{\alpha^2}{2\lambda} u_{xx} - \frac{\alpha}{2\lambda} u_{tt}. \tag{1.16}$$

This is a hyperbolic equation known as the telegrapher's equation. Random evolutions can be used to represent the solutions of certain of these equations probabilistically, though that is not our concern here.

In the context of the present example, we are interested instead in the asymptotic distribution of X as $\alpha \to \infty$ and $\lambda \to \infty$ with $\alpha^2 = \lambda$. Let $0 < \varepsilon < 1$ and observe that with $\alpha = 1/\varepsilon$ and $\lambda = 1/\varepsilon^2$, (1.16) becomes

$$u_t = \tfrac{1}{2}u_{xx} - \frac{\varepsilon}{2} u_{tt}. \tag{1.17}$$

This suggests that as $\varepsilon \to 0$, we should have $X \Rightarrow x + W$ in $C_{\mathbb{R}}[0, \infty)$, where W is a standard one-dimensional Brownian motion. To make this precise, let N be a Poisson process with parameter 1. Given $(x, y) \in \mathbb{R} \times \{-1, 1\}$, define $(X^\varepsilon, Y^\varepsilon)$ for $0 < \varepsilon < 1$ by

$$X^\varepsilon(t) = x + \frac{1}{\varepsilon}\int_0^t Y^\varepsilon(s)\, ds, \qquad Y^\varepsilon(t) = (-1)^{N(t/\varepsilon^2)}y. \tag{1.18}$$

By the Markov property of Y^ε,

$$\begin{aligned} M^\varepsilon(t) &\equiv \frac{\varepsilon}{2} Y^\varepsilon(t) - \frac{\varepsilon}{2} y + \frac{1}{\varepsilon}\int_0^t Y^\varepsilon(s)\, ds \\ &= X^\varepsilon(t) - x + \frac{\varepsilon}{2} Y^\varepsilon(t) - \frac{\varepsilon}{2} y \end{aligned} \tag{1.19}$$

is a zero-mean $\{\mathscr{F}_t^{Y^\varepsilon}\}$-martingale, and $M^\varepsilon(t)^2 - t$ is also an $\{\mathscr{F}_t^{Y^\varepsilon}\}$-martingale. It follows immediately from the martingale central limit theorem (Theorem 1.4 of Chapter 7) that $M^\varepsilon \Rightarrow W$ in $D_{\mathbb{R}}[0, \infty)$, hence by (1.19) and Problem 25 of Chapter 3, that $X^\varepsilon \Rightarrow x + W$ in $C_{\mathbb{R}}[0, \infty)$.

There is an alternative proof of this result that generalizes much more readily. Let $f \in \hat{C}^2(\mathbb{R})$, and define $f_\varepsilon \in \hat{C}^{1,0}(\mathbb{R} \times \{-1, 1\})$ for $0 < \varepsilon < 1$ by

$$f_\varepsilon(x, y) = f(x) + \frac{\varepsilon}{2} yf'(x). \tag{1.20}$$

Then, defining A_ε by (1.11) with $\alpha = 1/\varepsilon$ and $\lambda = 1/\varepsilon^2$, we have

$$A_\varepsilon f_\varepsilon(x, y) = \frac{1}{\varepsilon} yf'(x) + \tfrac{1}{2}y^2 f''(x) - \frac{1}{\varepsilon} yf'(x) = \tfrac{1}{2}f''(x) \tag{1.21}$$

for all $(x, y) \in \mathbb{R} \times \{-1, 1\}$. The desired conclusion now follows from Proposition 1.1 of Chapter 5 and Corollary 8.7 of Chapter 4.

It is the purpose of this chapter to obtain limit theorems of the above type under a variety of assumptions. More specifically, given $F, G: \mathbb{R}^d \times E \to \mathbb{R}^d$, a process Y with sample paths in $D_E[0, \infty)$, and $x \in \mathbb{R}^d$, we consider the solution X^ε, where $0 < \varepsilon < 1$, of the differential equation

$$\frac{d}{dt} X^\varepsilon(t) = F\left(X^\varepsilon(t), Y\left(\frac{t}{\varepsilon^2}\right)\right) + \frac{1}{\varepsilon} G\left(X^\varepsilon(t), Y\left(\frac{t}{\varepsilon^2}\right)\right) \tag{1.22}$$

with initial condition $X^\varepsilon(0) = x$. Of course, F and G must satisfy certain smoothness and growth assumptions in order for X^ε to be well defined. In

Section 2, we consider the case in which E is a compact metric space and the driving process Y is Markovian and ergodic. (This clearly generalizes (1.18).) In Section 3, we allow E to be a locally compact, separable metric space. In Section 4, we again require that E be compact but allow Y to be stationary and uniform mixing instead of Markovian.

2. DRIVING PROCESS IN A COMPACT STATE SPACE

The argument following (1.20) provided the motivation for Corollary 7.8 of Chapter 1. We include essentially a restatement of that result in a form that is suitable for application to random evolutions with driving process in a compact state space.

First, however, we need to generalize the Riemann integral of Chapter 1, Section 1.

2.1 Lemma Let E be a metric space, let L be a separable Banach space, and let $\mu \in \mathscr{P}(E)$. If $f: E \to L$ is Borel measurable and

$$\int \|f(y)\| \mu(dy) < \infty, \tag{2.1}$$

then there exists a sequence $\{f_n\}$ of Borel measurable simple functions from E into L such that

$$\lim_{n \to \infty} \int \|f_n(y) - f(y)\| \mu(dy) = 0. \tag{2.2}$$

The separability assumption on L is unnecessary if E is σ-compact and f is continuous.

Proof. If L is separable, let $\{g_n\}$ be dense in L; if E is σ-compact and f is continuous, then $f(E)$ is σ-compact, hence separable, so let $\{g_n\}$ be dense in $f(E)$. For $m, n = 1, 2, \ldots$ define $A_{n,m} = \{g \in L: \|g - g_n\| < 1/m\} - \bigcup_{k=1}^{n-1} A_{k,m}$ and

$$h_{n,m}(y) = \sum_{k=1}^{n} \chi_{f^{-1}(A_{k,m})}(y) g_k. \tag{2.3}$$

Then, letting $B_{n,m} = \bigcup_{k=1}^{n} f^{-1}(A_{k,m})$, we have

$$\int \|h_{n,m}(y) - f(y)\| \mu(dy) \le \frac{1}{m} \mu(B_{n,m}) + \int_{B_{n,m}^c} \|f(y)\| \mu(dy) \tag{2.4}$$

for $m, n = 1, 2, \ldots$, and hence

$$\lim_{m\to\infty} \overline{\lim_{n\to\infty}} \int \|h_{n,m}(y) - f(y)\| \mu(dy) = 0. \tag{2.5}$$

We conclude that there exists $\{m_n\}$ such that (2.2) holds with $f_n = h_{n, m_n}$. □

Let E, L, and μ be as in Lemma 2.1, and let $f: E \to L$ be a Borel measurable simple function, that is,

$$f(y) = \sum_{k=1}^{n} \chi_{B_k}(y) g_k, \tag{2.6}$$

where $B_1, \ldots, B_n \in \mathscr{B}(E)$ are disjoint, $g_1, \ldots, g_n \in L$, and $n \geq 1$. Then we define

$$\int f\, d\mu = \sum_{k=1}^{n} \mu(B_k) g_k. \tag{2.7}$$

More generally, suppose $f: E \to L$ is Borel measurable and (2.1) holds. Let $\{f_n\}$ be as in Lemma 2.1. Then we define the *(Bochner) integral of f with respect to μ* by

$$\int f\, d\mu = \lim_{n\to\infty} \int f_n\, d\mu. \tag{2.8}$$

It is easily checked that this limit exists and is independent of the choice of the approximating sequence $\{f_n\}$.

In particular, if E is compact, L is arbitrary, and $\mu \in \mathscr{P}(E)$, then $\int f\, d\mu$ exists for all f belonging to $C_L(E)$, the space of continuous functions from E into L. We note that $C_L(E)$ is a Banach space with norm $|||f||| = \sup_{y \in E} \|f(y)\|$.

2.2 Proposition Let E be a compact metric space, L a Banach space, $P(t, y, \Gamma)$ a transition function on $[0, \infty) \times E \times \mathscr{B}(E)$, and $\mu \in \mathscr{P}(E)$. Assume that the formula

$$S(t)g(y) = \int g(z) P(t, y, dz) \tag{2.9}$$

defines a Feller semigroup $\{S(t)\}$ on $C(E)$ satisfying

$$\lim_{\lambda\to 0+} \lambda \int_0^\infty e^{-\lambda t} S(t) g\, dt = \int g\, d\mu \tag{2.10}$$

for all $g \in C(E)$, and let B_0 denote its generator. Observe that (2.9) also defines a strongly continuous contraction semigroup $\{S(t)\}$ on $C_L(E)$, and let B denote its generator.

Let D be a dense subspace of L, and for each $y \in E$, let Π_y and A_y be linear operators on L with domains containing D such that the functions $y \to \Pi_y f$

and $y \to A_y f$ belong to $C_L(E)$ for each $f \in D$. Define linear operators Π and A on $C_L(E)$ with

$$\mathscr{D}(\Pi) = \{f \in C_L(E): f(y) \in \mathscr{D}(\Pi_y) \text{ for every } y \in E, \quad \text{and } y \to \Pi_y(f(y)) \text{ belongs to } C_L(E)\} \tag{2.11}$$

and

$$\mathscr{D}(A) = \{f \in C_L(E): f(y) \in \mathscr{D}(A_y) \text{ for every } y \in E \quad \text{and } y \to A_y(f(y)) \text{ belongs to } C_L(E)\} \tag{2.12}$$

by $(\Pi f)(y) = \Pi_y(f(y))$ and $(Af)(y) = A_y(f(y))$. Let $\mathscr{D}$ be a subspace of $C_L(E)$ such that

$$\mathscr{D}_0 \equiv \left\{\sum_{i=1}^{n} g_i(\cdot)f_i: g_1, \ldots, g_n \in \mathscr{D}(B_0), f_1, \ldots, f_n \in D, n \geq 1\right\} \subset \mathscr{D} \subset \mathscr{D}(\Pi) \cap \mathscr{D}(A) \cap \mathscr{D}(B), \tag{2.13}$$

and assume that, for $0 < \varepsilon < 1$, an extension of $\{(f, \Pi f + \varepsilon^{-1}Af + \varepsilon^{-2}Bf): f \in \mathscr{D}\}$ generates a strongly continuous contraction semigroup $\{T_\varepsilon(t)\}$ on $C_L(E)$.

Suppose there is a linear operator V on $C_L(E)$ such that $Af \in \mathscr{D}(V)$ and $VAf \in \mathscr{D}$ for all $f \in D$ and $BVg = -g$ for all $g \in \mathscr{D}(V)$. (Here and below, we identify elements of L with constant functions in $C_L(E)$.) Put

$$C = \left\{\left(f, \int (\Pi f + AVAf)(y)\mu(dy)\right): f \in D\right\}. \tag{2.14}$$

Then C is dissipative, and if $\bar{C}$, which is single-valued, generates a strongly continuous contraction semigroup $\{T(t)\}$ on L, then, for each $f \in L$, $\lim_{\varepsilon \to 0} T_\varepsilon(t)f = T(t)f$ for all $t \geq 0$, uniformly on bounded intervals.

2.3 Remark Suppose that (2.10) can be strengthened to

$$\int_0^\infty \left|\left|\left| S(t)g - \int g\, d\mu \right|\right|\right| dt < \infty, \qquad g \in C_L(E). \tag{2.15}$$

By the uniform boundedness principle, there exists a constant M such that

$$\left|\left|\left| \int_0^\infty \left(S(t)g - \int g\, d\mu\right) dt \right|\right|\right| \leq M \,|||g|||, \qquad g \in C_L(E), \tag{2.16}$$

and hence for each $y \in E$ there exists a finite signed Borel measure $\nu(y, \cdot)$ such that

$$\int_0^\infty \left(S(t)g(y) - \int g\, d\mu\right) dt = \int g(z)\nu(y, dz), \qquad g \in C_L(E), \tag{2.17}$$

where the right side is defined using (2.8). If $\int (Af)(z)\nu(\cdot, dz) \in \mathscr{D}$ for all $f \in D$, then, by Remark 7.9(b) of Chapter 1, $V: \{Af: f \in D\} \to \mathscr{D}$ is given by

$$(Vg)(y) = \int g(z)\nu(y, dz). \tag{2.18}$$

□

Proof. We claim first that

$$\left\{\sum_{i=1}^{n} g_i(\cdot)f_n: g_1, \ldots, g_n \in C(E), f_1, \ldots, f_n \in L, n \geq 1\right\} \tag{2.19}$$

is dense in $C_L(E)$. To see this, let $\varepsilon > 0$ and choose $y_1, \ldots, y_n \in E$ such that $E = \bigcup_{i=1}^n B(y_i, \varepsilon)$. Let $g_1, \ldots, g_n \in C(E)$ be a partition of unity, that is, for $i = 1, \ldots, n$, $g_i \geq 0$, $\operatorname{supp} g_i \subset B(y_i, \varepsilon)$, and $\sum_{i=1}^n g_i \equiv 1$ (Rudin (1974), Theorem 2.13). Given $f \in C_L(E)$, let $f_\varepsilon = \sum_{i=1}^n g_i(\cdot)f(y_i)$. Then f_ε belongs to (2.19) and

$$|||f_\varepsilon - f||| \leq \sup \{\|f(x) - f(y)\|: r(x, y) < \varepsilon\}, \tag{2.20}$$

where r is the metric for E. But the right side of (2.20) tends to zero as $\varepsilon \to 0$, so the claim is proved. Since $\mathscr{D}(B_0)$ is dense in $C(E)$ and D is dense in L, we also have $\mathscr{D}_0$ dense in $C_L(E)$.

It follows that $\{S(t)\}$ is strongly continuous on $C_L(E)$, (2.10) holds for all $f \in C_L(E)$, and (2.16) implies (2.17). Note also that $S(t): \mathscr{D}_0 \to \mathscr{D}_0$, so $\mathscr{D}_0$ is a core for B by Proposition 3.3 of Chapter 1.

We apply Corollary 7.8 of Chapter 1 (see Remark 7.9(c) in that chapter) with the roles of Π and A played by the restrictions of Π and A to $\mathscr{D}$. □

2.4 Theorem Let E be a compact metric space, let $F, G \in C_{\mathbb{R}^d}(\mathbb{R}^d \times E)$, and suppose that for each $n \geq 1$ there exists a constant M_n for which

$$|F(x, y) - F(x', y)| \leq M_n|x - x'|, \qquad |x| \vee |x'| \leq n,\ y \in E, \tag{2.21}$$

that $G_1, \ldots, G_d \in C^{1,0}(\mathbb{R}^d \times E)$, and that

$$\sup_{(x, y) \in \mathbb{R}^d \times E} \frac{|F(x, y)| \vee |G(x, y)|}{1 + |x|} < \infty. \tag{2.22}$$

Let $\{S(t)\}$ be a Feller semigroup on $C(E)$, let $\mu \in \mathscr{P}(E)$, and assume that

$$\lim_{\lambda \to 0+} \lambda \int_0^\infty e^{-\lambda t}S(t)g\, dt = \int g\, d\mu, \qquad g \in C(E). \tag{2.23}$$

Let B_0 denote the generator of $\{S(t)\}$.

Suppose that $\int G(x, y)\mu(dy) = 0$ for all $x \in \mathbb{R}^d$ and that there exists for each $y \in E$ a finite, signed, Borel measure $\nu(y, \cdot)$ on E such that the function $H: R^d \times E \to \mathbb{R}^d$, defined by

$$H(x, y) = \int G(x, z)\nu(y, dz), \tag{2.24}$$

satisfies, for $i = 1, \ldots, d$, $H_i \in C^{1,0}(\mathbb{R}^d \times E)$, $H_i(x, \cdot) \in \mathscr{D}(B_0)$ for each $x \in \mathbb{R}^d$, and $B_0[H_i(x, \cdot)](y) = -G_i(x, y)$ for all $(x, y) \in \mathbb{R}^d \times E$. Fix $\mu_0 \in \mathscr{P}(E)$, and let Y be a Markov process corresponding to $\{S(t)\}$ with sample paths in $D_E[0, \infty)$ and initial distribution μ_0. Fix $x_0 \in \mathbb{R}^d$, and define X^ε for $0 < \varepsilon < 1$ to be the solution of the differential equation

$$\text{(2.25)} \qquad \frac{d}{dt} X^\varepsilon(t) = F\left(X^\varepsilon(t), Y\left(\frac{t}{\varepsilon^2}\right)\right) + \frac{1}{\varepsilon} G\left(X^\varepsilon(t), Y\left(\frac{t}{\varepsilon^2}\right)\right)$$

with initial condition $X^\varepsilon(0) = x_0$. Put

$$\text{(2.26)} \qquad C = \left\{\left(f, \frac{1}{2} \sum_{i,j=1}^{d} a_{ij}\, \partial_i\, \partial_j f + \sum_{i=1}^{d} b_i\, \partial_i f\right): f \in C_c^2(\mathbb{R}^d)\right\},$$

where

$$\text{(2.27)} \qquad a_{ij}(x) = \int G_i(x, y) H_j(x, y) \mu(dy) + \int G_j(x, y) H_i(x, y) \mu(dy)$$

and

$$\text{(2.28)} \qquad b_i(x) = \int F_i(x, y) \mu(dy) + \int G(x, y) \cdot \nabla_x H_i(x, y) \mu(dy).$$

Then C is dissipative. Assume that $\bar{C}$ generates a Feller semigroup $\{T(t)\}$ on $\hat{C}(\mathbb{R}^d)$, and let X be a Markov process corresponding to $\{T(t)\}$ with sample paths in $C_{\mathbb{R}^d}[0, \infty)$ and initial distribution δ_{x_0}. Then $X^\varepsilon \Rightarrow X$ in $C_{\mathbb{R}^d}[0, \infty)$ as $\varepsilon \to 0$.

2.5 Remark Suppose that (2.23) can be strengthened to

$$\text{(2.29)} \qquad \int_0^\infty \sup_{(x, y) \in \mathbb{R}^d \times E} \left| S(t)[g(x, \cdot)](y) - \int g(x, z) \mu(dz) \right| dt < \infty, \qquad g \in \hat{C}(\mathbb{R}^d \times E).$$

Then $v(y, dz)$ is as in (2.17) with $L = \hat{C}(\mathbb{R}^d)$. □

Proof. We identify $C_{\hat{C}(\mathbb{R}^d)}(E)$ with $\hat{C}(\mathbb{R}^d \times E)$ and apply Proposition 2.2 with $L = \hat{C}(\mathbb{R}^d)$, $D = C_c^2(\mathbb{R}^d)$, $\Pi_y = F(\cdot, y) \cdot \nabla$, $A_y = G(\cdot, y) \cdot \nabla$, $\mathscr{D}(\Pi_y) = \mathscr{D}(A_y) = C_c^1(\mathbb{R}^d)$, and

$$\text{(2.30)} \qquad \mathscr{D} = \{f \in C_c^{1,0}(\mathbb{R}^d \times E): f(x, \cdot) \in \mathscr{D}(B_0) \text{ for all } x \in \mathbb{R}^d,$$
$$\text{and } (x, y) \to B_0[f(x, \cdot)](y) \text{ belongs to } \hat{C}(\mathbb{R}^d \times E)\}.$$

Clearly, $\mathscr{D} \subset D(\Pi) \cap \mathscr{D}(A)$. We claim that $\mathscr{D} \subset \mathscr{D}(B)$. To see this, let

$$\text{(2.31)} \qquad \hat{B} = \{(f, g) \in \hat{C}(\mathbb{R}^d \times E) \times \hat{C}(\mathbb{R}^d \times E):$$
$$f(x, \cdot) \in \mathscr{D}(B_0) \text{ for all } x \in \mathbb{R}^d,$$
$$\text{and } g(x, y) = B_0[f(x, \cdot)](y) \text{ for all } (x, y) \in \mathbb{R}^d \times E\},$$

and observe that $\hat{B}$ is a dissipative linear extension of the generator B of $\{S(t)\}$ on $\hat{C}(\mathbb{R}^d \times E)$, and hence $\hat{B} = B$ by Proposition 4.1 of Chapter 1.

Next, fix $\varepsilon \in (0, 1)$, and define the contraction semigroup $\{T_\varepsilon(t)\}$ on $B(\mathbb{R}^d \times E)$ by

$$T_\varepsilon(t)f(x, y) = E\left[f\left(X^\varepsilon(t), Y\left(\frac{t}{\varepsilon^2}\right)\right)\right], \tag{2.32}$$

where Y is a Markov process corresponding to $\{S(t)\}$ with sample paths in $D_E[0, \infty)$ and initial distribution δ_y, and X^ε satisfies the differential equation (2.25) with initial condition $X^\varepsilon(0) = x$. The semigroup property follows from the identity

$$\begin{aligned} E\left[f\left(X^\varepsilon(t), Y\left(\frac{t}{\varepsilon^2}\right)\right)\middle| \mathscr{F}^Y_{s/\varepsilon^2}\right] &= E\left[f\left(X^\varepsilon(s) + \int_s^t \left(F + \frac{1}{\varepsilon} G\right)\left(X^\varepsilon(u), Y\left(\frac{u}{\varepsilon^2}\right)\right) du, Y\left(\frac{t}{\varepsilon^2}\right)\right)\middle| \mathscr{F}^Y_{s/\varepsilon^2}\right] \\ &= T_\varepsilon(t - s)f\left(X^\varepsilon(s), Y\left(\frac{s}{\varepsilon^2}\right)\right), \end{aligned} \tag{2.33}$$

valid for all $t > s \geq 0$ and $f \in B(\mathbb{R}^d \times E)$ by the Markov property of Y and the fact that $X^\varepsilon(s + \cdot)$ solves the differential equation (2.25) with $(X^\varepsilon(0), Y(\cdot/\varepsilon^2))$ replaced by $(X^\varepsilon(s), Y((s + \cdot)/\varepsilon^2))$. We leave it to the reader to check that $T_\varepsilon(t)$: $\bar{C}(\mathbb{R}^d \times E) \to \bar{C}(\mathbb{R}^d \times E)$ for all $t \geq 0$. Using (2.22) we conclude that $T_\varepsilon(t)$: $\hat{C}(\mathbb{R}^d \times E) \to \hat{C}(\mathbb{R}^d \times E)$ for every $t \geq 0$. Let $f \in \mathscr{D}$ and $t_2 > t_1 \geq 0$. Then

$$\begin{aligned} f\left(X^\varepsilon(t_2), Y\left(\frac{t_2}{\varepsilon^2}\right)\right) - f\left(X^\varepsilon(t_1), Y\left(\frac{t_2}{\varepsilon^2}\right)\right) &= \int_{t_1}^{t_2} \left(F + \frac{1}{\varepsilon} G\right)\left(X^\varepsilon(s), Y\left(\frac{t_2}{\varepsilon^2}\right)\right) \cdot \nabla_x f\left(X^\varepsilon(s), Y\left(\frac{t_2}{\varepsilon^2}\right)\right) ds \end{aligned} \tag{2.34}$$

and

$$\begin{aligned} E\left[f\left(X^\varepsilon(t_1), Y\left(\frac{t_2}{\varepsilon^2}\right)\right) - f\left(X^\varepsilon(t_1), Y\left(\frac{t_1}{\varepsilon^2}\right)\right)\middle| \mathscr{F}^Y_{t_1/\varepsilon^2}\right] &= E\left[\int_{t_1}^{t_2} \frac{1}{\varepsilon^2} B_0[f(X^\varepsilon(t_1), \cdot)]\left(Y\left(\frac{s}{\varepsilon^2}\right)\right) ds \middle| \mathscr{F}^Y_{t_1/\varepsilon^2}\right], \end{aligned} \tag{2.35}$$

so by Lemma 3.4 of Chapter 4,

$$f\left(X^\varepsilon(t), Y\left(\frac{t}{\varepsilon^2}\right)\right) - \int_0^t \left(\Pi f + \frac{1}{\varepsilon} Af + \frac{1}{\varepsilon^2} Bf\right)\left(X^\varepsilon(s), Y\left(\frac{s}{\varepsilon^2}\right)\right) ds \tag{2.36}$$

is a martingale. It follows that $\{T(t)\}$ is strongly continuous on $\mathscr{D}$, hence on $\hat{C}(\mathbb{R}^d \times E)$. We conclude therefore that the generator of $\{T_\varepsilon(t)\}$ extends $\{(f, \Pi f + \varepsilon^{-1}Af + \varepsilon^{-2}Bf): f \in \mathscr{D}\}$.

We define V on $\mathscr{D}(V) \equiv \{Af: f \in C_c^2(\mathbb{R}^d)\}$ by $(Vg)(x, y) = \int g(x, z)\nu(y, dz)$ and note that $V: \mathscr{D}(V) \to \mathscr{D}$ and

$$(2.37) \qquad \begin{aligned}(BVAf)(x, y) &= B_0[H(x, \cdot) \cdot \nabla f(x)](y) \\ &= -G(x, y) \cdot \nabla f(x) = -Af(x, y)\end{aligned}$$

for all $f \in C_c^2(\mathbb{R}^d)$ and $(x, y) \in \mathbb{R}^d \times E$. It is immediate that C, defined by (2.14), has the form (2.26)–(2.28). Under the assumptions of the theorem, we infer from Proposition 2.2 that, for each $f \in \hat{C}(\mathbb{R}^d)$, $T_\varepsilon(t)f \to T(t)f$ as $\varepsilon \to 0$ for all $t \geq 0$, uniformly on bounded intervals. By (2.33), $(X^\varepsilon(\cdot), Y(\cdot/\varepsilon^2))$ is a Markov process corresponding to $\{T_\varepsilon(t)\}$ with sample paths in $D_{\mathbb{R}^d \times E}[0, \infty)$ and initial distribution $\delta_{x_0} \times \mu_0$ and therefore, by Corollary 8.7 of Chapter 4 and Problem 25 of Chapter 3, $X^\varepsilon \Rightarrow X$ in $C_{\mathbb{R}^d}[0, \infty)$ as $\varepsilon \to 0$. □

2.6 Example Let E be finite, and define

$$(2.38) \qquad B_0\, g(i) = \sum_{j \in E} q_{ij}\, g(j),$$

where $Q = (q_{ij})_{i,\, j \in E}$ is an irreducible, infinitesimal matrix (i.e., $q_{ij} \geq 0$ for all $i \neq j$, $\sum_{j \in E} q_{ij} = 0$ for all $i \in E$, and there does not exist a nonempty, proper subset J of E such that $q_{ij} = 0$ for all $i \in J$ and $j \notin J$). Let $\mu = (\mu_i)_{i \in E}$ denote the unique stationary distribution. It is well known that

$$(2.39) \qquad \lim_{t \to \infty} P(t, i, \{j\}) = \mu_j, \qquad i, j \in E,$$

and (2.23) follows from this. By the existence of generalized inverses (Rao (1973), p. 25) and Lemma 7.3(d) of Chapter 1, there exists a real matrix $\nu = (\nu_{ij})_{i,\, j \in E}$ such that $Q\nu\lambda = -\lambda$ for all real column vectors $\lambda = (\lambda_i)_{i \in E}$ for which $\lambda \cdot \mu = 0$. It follows that the function H of Theorem 2.4 is given by

$$(2.40) \qquad H(x, i) = \sum_{j \in E} \nu_{ij}\, G(x, j).$$

Alternatively, using the fact that the convergence in (2.39) is exponentially fast (Doob (1953), Theorem VI.1.1), Remark 2.5 gives (2.40) with

$$(2.41) \qquad \nu_{ij} = \int_0^\infty (P(t, i, \{j\}) - \mu_j)\, dt.$$

This generalizes the example of Section 1. □

2.7 Example Let $E = [0, 1]$, and define

$$(2.42) \qquad B_0 = \{(g, \tfrac{1}{2}g''): g \in C^2[0, 1],\, g'(0) = g'(1) = 0\}.$$

We claim that the Feller semigroup $\{S(t)\}$ on $C[0, 1]$ generated by B_0 (see Problem 6(a) in Chapter 1) satisfies

$$\lim_{t\to\infty} \sup_{0\le y\le 1} \left| S(t)g(y) - \int_0^1 g(z)\, dz \right| = 0, \qquad g \in C[0, 1]. \tag{2.43}$$

This follows from the fact that $\{S(t)\}$ has the form

$$S(t)g(y) = \int_0^1 g(z)\tilde{p}(t, y, z)\, dz, \tag{2.44}$$

where

$$\tilde{p}(t, y, z) = \sum_{n=-\infty}^{\infty} p(t, y, 2n + z) + \sum_{n=-\infty}^{\infty} p(t, y, 2n - z) \tag{2.45}$$

and $p(t, y, z) = (2\pi t)^{-1/2} \exp\{-(z - y)^2/2t\}$, together with the crude inequality

$$\sup_{0\le z\le 1} \tilde{p}(t, y, z) - \inf_{0\le z\le 1} \tilde{p}(t, y, z) \le \frac{2}{\sqrt{2\pi t}}, \tag{2.46}$$

valid for $0 \le y \le 1$ and $t > 0$. In particular, μ is Lebesgue measure on $[0, 1]$. The function H of Theorem 2.4 can be defined by

$$H(x, y) = -2 \int_0^y \int_0^z G(x, w)\, dw\, dz. \tag{2.47}$$

Note that $H_i(x, \cdot) \in \mathscr{D}(B_0)$ for each $x \in \mathbb{R}^d$ and $i = 1, \ldots, d$ since $\int_0^1 G(x, w)\, dw = 0$ for all $x \in \mathbb{R}^d$ by assumption. □

3. DRIVING PROCESS IN A NONCOMPACT STATE SPACE

Let Y be an Ornstein–Uhlenbeck process, that is, a diffusion process in $\mathbb{R}$ with generator

$$B_0 = \{(g, \mathscr{G}g): g \in \hat{C}(\mathbb{R}) \cap C^2(\mathbb{R}), \mathscr{G}g \in \hat{C}(\mathbb{R})\}, \tag{3.1}$$

where $\mathscr{G}g(y) = g''(y) - yg'(y)$. The analogue of (1.18) is

$$X^\varepsilon(t) = x + \frac{1}{\varepsilon} \int_0^t Y^\varepsilon(s)\, ds, \qquad Y^\varepsilon(t) = Y\left(\frac{t}{\varepsilon^2}\right), \tag{3.2}$$

and one might ask whether the analogous conclusion holds. Even if Theorem 2.4 could be extended to the case of E locally compact (with $C(E)$ replaced by $\hat{C}(E)$), it would still be inadequate for at least three reasons. First, (2.22) is not satisfied. Second, convergence in (2.23) cannot be uniform if the right side is nonzero. Third, with $G(x, y) = y$, we have $H(x, y) = y$, which is not even bounded in $y \in \mathbb{R}$, much less an element of $\mathscr{D}(B_0)$. This last problem causes the most difficulty. We may be able to find an operator V that formally satisfies $B_0 Vg = -g$ (e.g., if B_0 is given by a differential operator $\mathscr{G}$, we may be able to solve $\mathscr{G}h = -g$ for a large class of g), but $Vg \notin \mathscr{D}(B_0)$ for the functions g in which we are interested.

There are several ways to proceed. One is to prove an analogue of Proposition 2.2 with the role of $C_L(E)$ played by the space of (equivalence classes of) Borel measurable functions $f: E \to L$ with $\|f(\cdot)\| \in L^1(\mu)$, where μ is the stationary distribution of Y. However, this approach seems to require that Y have initial distribution μ.

Instead, we apply Corollary 8.7 (or 8.16) of Chapter 4, which was formulated with problems such as this in mind. The basic idea in the theorem is to "cut off" unbounded Vg by multiplying by a function $\phi_\varepsilon \in C_c(E)$, with $\phi_\varepsilon = 1$ on a large compact set, selected so that $\phi_\varepsilon Vg \in \mathscr{D}(B_0)$ and $B_0(\phi_\varepsilon Vg)$ is approximately $-g$. We show, in the case of (3.2), that $X^\varepsilon \Rightarrow x + \sqrt{2}W$ in $C_{\mathbb{R}}[0, \infty)$ as $\varepsilon \to 0+$.

3.1 Theorem Let E be a locally compact, separable metric space, let $F, G \in C_{\mathbb{R}^d}(\mathbb{R}^d \times E)$, and suppose that $F_1, \ldots, F_d \in C^{1,0}(\mathbb{R}^d \times E)$, that $G_1, \ldots, G_d \in C^{2,0}(\mathbb{R}^d \times E)$, and that

$$\sup_{(x, y) \in \mathbb{R}^d \times K} \frac{|F(x, y)| \vee |G(x, y)|}{1 + |x|} < \infty \tag{3.3}$$

for every compact set $K \subset E$. Let $\{S(t)\}$ be a Feller semigroup on $\hat{C}(E)$ with generator B_0 and let $\mu \in \mathscr{P}(E)$. Let $\rho: E \to (0, \infty)$ satisfy $1/\rho \in \hat{C}(E)$, let $\phi \in C_c^2[0, \infty)$ satisfy $\chi_{[0,1]} \le \phi \le \chi_{[0,2]}$, fix $0 < \theta < 1$, and define $\phi_\varepsilon \in C_c(E)$ and $K_\varepsilon \subset E$ by

$$\phi_\varepsilon(y) = \phi(\varepsilon\rho(y)) \quad \text{and} \quad K_\varepsilon = \{y \in E: \varepsilon^\theta \rho(y) \le 1\}. \tag{3.4}$$

Assume that $\phi_\varepsilon \in \mathscr{D}(B_0)$ for each $\varepsilon \in (0, 1)$ and

$$\sup_{y \in K_\varepsilon} |B_0 \phi_\varepsilon(y)| = o(\varepsilon^2) \quad \text{as} \quad \varepsilon \to 0. \tag{3.5}$$

Define

$$\mathscr{M} = \left\{ g \in C(\mathbb{R}^d \times E): \int \sup_{|x| \le l} |g(x, y)| \mu(dy) < \infty \quad \text{for} \quad l = 1, 2, \ldots \right\}, \tag{3.6}$$

and let V be a linear operator on $\mathscr{M}$ with $\mathscr{D}(V) \subset \{g \in \mathscr{M}: \int g(x, y)\mu(dy) = 0$ for all $x \in \mathbb{R}^d\}$ such that if $g \in \mathscr{D}(V)$, then $(Vg)(x, \cdot)\phi_\varepsilon(\cdot) \in \mathscr{D}(B_0)$ for every $x \in \mathbb{R}^d$ and $0 < \varepsilon < 1$ and

$$\sup_{|x| \le l,\, y \in K_\varepsilon} |B_0[(Vg)(x, \cdot)\phi_\varepsilon(\cdot)](y) + g(x, y)| = o(1) \quad \text{as} \quad \varepsilon \to 0 \tag{3.7}$$

for $l = 1, 2, \ldots$. Assume that $f \in C(\mathbb{R}^d)$ and $g \in \mathscr{D}(V)$ imply $fg \in \mathscr{D}(V)$ and $V(fg) = f\,Vg$.

The following assumptions and definitions are made for $i, j, k = 1, \ldots, d$. Suppose $G_i \in \mathscr{D}(V)$ and the left side of (3.7) with $g = G_i$ is $o(\varepsilon)$ as $\varepsilon \to 0$ for $l = 1, 2, \ldots$. Assume $H_i \equiv VG_i \in C^{2,0}(\mathbb{R}^d \times E)$ and $F_i, G_i H_j, G_i\, \partial H_j/\partial x_i \in \mathscr{M}$. Suppose a_{ij} and b_i, defined by (2.27) and (2.28), belong to $C^1(\mathbb{R}^d)$. Suppose $G_i H_j + G_j H_i - a_{ij} \in \mathscr{D}(V)$, $F_i + G \cdot \nabla_x H_i - b_i \in \mathscr{D}(V)$, $\hat{a}_{ij} \equiv V(G_i H_j + G_j H_i - a_{ij}) \in C^{1,0}(\mathbb{R}^d \times E)$, and $\hat{b}_i \equiv V(F_i + G \cdot \nabla_x H_i - b_i) \in C^{1,0}(\mathbb{R}^d \times E)$. Assume that H_i, $F_i H_j$, $F_i\, \partial H_j/\partial x_i$, $G_i \hat{a}_{jk}$, $G_i\, \partial \hat{a}_{jk}/\partial x_i$, $G_i \hat{b}_j$, $G_i\, \partial \hat{b}_j/\partial x_i$, when multiplied by the function $(x, y) \to \chi_{[0, l]}(|x|)/\rho(y)$, are bounded on $\mathbb{R}^d \times E$ for $l = 1, 2, \ldots$. Assume further that $\hat{a}_{ij}$, $\hat{b}_i$, $F_i \hat{a}_{jk}$, $F_i\, \partial \hat{a}_{jk}/\partial x_i$, $F_i \hat{b}_j$, $F_i\, \partial \hat{b}_j/\partial x_i$, when multiplied by the function $(x, y) \to \chi_{[0, l]}(|x|)/\rho(y)^2$, are bounded on $\mathbb{R}^d \times E$ for $l = 1, 2, \ldots$.

Fix $\mu_0 \in \mathscr{P}(E)$, and let Y be a Markov process corresponding to $\{S(t)\}$ with sample paths in $D_E[0, \infty)$ and initial distribution μ_0. Assume that

$$\lim_{\varepsilon \to 0} P\left\{Y\left(\frac{t}{\varepsilon^2}\right) \in K_\varepsilon \quad \text{for} \quad 0 \le t \le T\right\} = 1 \tag{3.8}$$

for each $T > 0$. Fix $x_0 \in \mathbb{R}^d$ and define X^ε for $0 < \varepsilon < 1$ to be the solution of the differential equation (2.25) with initial condition $X^\varepsilon(0) = x_0$. Put

$$C = \left\{\left(f, \frac{1}{2} \sum_{i,j=1}^{d} a_{ij}\, \partial_i\, \partial_j f + \sum_{i=1}^{d} b_i\, \partial_i f\right): f \in C_c^3(\mathbb{R}^d)\right\}. \tag{3.9}$$

Then C is dissipative. Assume that $\bar{C}$, which is single-valued, generates a Feller semigroup $\{T(t)\}$ on $\hat{C}(\mathbb{R}^d)$, and let X be a Markov process corresponding to $\{T(t)\}$ with sample paths in $C_{\mathbb{R}^d}[0, \infty)$ and initial distribution δ_{x_0}. Then $X^\varepsilon \Rightarrow X$ in $C_{\mathbb{R}^d}[0, \infty)$ as $\varepsilon \to 0$.

3.2 Remark Instead of assuming an ergodicity condition such as (2.29), which would be rather difficult to exploit here (and may be rather difficult to verify), we assume the existence of a linear operator V such that (essentially) $B_0 V = -I$. □

Proof. For each $\varepsilon \in (0, 1)$, exactly the same argument as used in the proof of Theorem 2.4 shows that $\{(X^\varepsilon(t), Y(t/\varepsilon^2)), t \ge 0\}$ is a progressive Markov process in $\mathbb{R}^d \times E$ corresponding to a measurable contraction semigroup with full generator that extends $\{(f, A_\varepsilon f): f \in \mathscr{D}\}$, where

$$\mathscr{D} = \{f \in C_c^{1,0}(\mathbb{R}^d \times E): f(x, \cdot) \in \mathscr{D}(B_0) \quad \text{for all} \quad x \in \mathbb{R}^d\} \tag{3.10}$$

and

$$A_\varepsilon f(x, y) = \{F(x, y) + \varepsilon^{-1} G(x, y)\} \cdot \nabla_x f(x, y) + \varepsilon^{-2} B_0[f(x, \cdot)](y). \tag{3.11}$$

(Note that if $f \in \mathscr{D}$, the function $(x, y) \to B_0[f(x, \cdot)](y)$ is automatically jointly measurable.) Let $(f, g) \in C$ and define

$$h_1 = V(G \cdot \nabla f) = H \cdot \nabla f \tag{3.12}$$

and

(3.13)
$$h_2 = V(F \cdot \nabla f + G \cdot \nabla_x h_1 - g)$$
$$= \frac{1}{2} \sum_{i,j=1}^{d} \hat{a}_{ij}\, \partial_i\, \partial_j f + \sum_{i=1}^{d} \hat{b}_i\, \partial_i f.$$

For each $\varepsilon \in (0, 1)$, define $f_\varepsilon \in \mathscr{D}$ by

(3.14)
$$f_\varepsilon(x, y) = (f(x) + \varepsilon h_1(x, y) + \varepsilon^2 h_2(x, y))\phi_\varepsilon(y),$$

and observe that

(3.15)
$$\begin{aligned} A_\varepsilon f_\varepsilon(x, y) = {} & \varepsilon^{-2} f(x) B_0\, \phi_\varepsilon(y) \\ & + \varepsilon^{-1}\{G(x, y) \cdot \nabla f(x)\phi_\varepsilon(y) + B_0[h_1(x, \cdot)\phi_\varepsilon(\cdot)](y)\} \\ & + \{F(x, y) \cdot \nabla f(x) + G(x, y) \cdot \nabla_x h_1(x, y)\}\phi_\varepsilon(y) \\ & + B_0[h_2(x, \cdot)\phi_\varepsilon(\cdot)](y) \\ & + \varepsilon\{F(x, y) \cdot \nabla_x h_1(x, y) + G(x, y) \cdot \nabla_x h_2(x, y)\}\phi_\varepsilon(y) \\ & + \varepsilon^2 F(x, y) \cdot \nabla_x h_2(x, y)\phi_\varepsilon(y) \end{aligned}$$

for all $(x, y) \in \mathbb{R}^d \times E$. By (3.5) and the other assumptions,

(3.16)
$$\sup_{0<\varepsilon<1}\ \sup_{(x,\, y) \in \mathbb{R}^d \times E} |f_\varepsilon(x, y)| < \infty,$$

(3.17)
$$\lim_{\varepsilon \to 0}\ \sup_{(x,\, y) \in \mathbb{R}^d \times K_\varepsilon} |f_\varepsilon(x, y) - f(x)| = 0,$$

(3.18)
$$\lim_{\varepsilon \to 0}\ \sup_{(x,\, y) \in \mathbb{R}^d \times K_\varepsilon} |A_\varepsilon f_\varepsilon(x, y) - g(x)| = 0.$$

In view of (3.8), the result follows from Corollary 8.7 of Chapter 4. □

3.3 Example Let $E = \mathbb{R}$ and define B_0 by (3.1), where $\mathscr{G}g(y) = g''(y) - yg'(y)$. It is quite easy to show that the Feller semigroup $\{S(t)\}$ on $\hat{C}(\mathbb{R})$ generated by B_0 has a unique stationary distribution μ, that μ is $N(0, 1)$, the standard normal distribution, and that

(3.19)
$$\operatorname*{bp-lim}_{t \to \infty} \left| \frac{1}{t} \int_0^t S(s)g\, ds - \int g(z)\mu(dz) \right| = 0, \qquad g \in \hat{C}(\mathbb{R}).$$

However, these results are not explicitly needed.

For each $n \ge 1$, define $\phi_n: \mathbb{R} \to (0, \infty)$ by $\phi_n(y) = (1 + y^2)^{n/2}$,

(3.20)
$$\mathscr{M}_n = \left\{ g \in C(\mathbb{R}^d \times \mathbb{R}):\ \sup_{|x| \le l,\, y \in \mathbb{R}} \frac{|g(x, y)|}{\phi_n(y)} < \infty \quad \text{for} \quad l = 1, 2, \ldots \right\},$$

and $\mathscr{M}_\infty = \bigcup_{n=1}^\infty \mathscr{M}_n$. Define V on

(3.21)
$$\mathscr{D}(V) = \left\{ g \in \mathscr{M}_\infty:\ \int_{-\infty}^{\infty} g(x, y) e^{-y^2/2}\, dy = 0 \quad \text{for all} \quad x \in \mathbb{R}^d \right\}$$

by

$$Vg(x, y) = \int_0^y e^{z^2/2} \int_z^\infty g(x, w)e^{-w^2/2}\, dw\, dz, \tag{3.22}$$

and note that $V: \mathscr{D}(V) \cap \mathscr{M}_n \to \mathscr{M}_n$ for each $n \geq 1$. Also, if $g \in \mathscr{D}(V) \cap C^{1,0}(\mathbb{R}^d \times \mathbb{R})$ and $|\nabla_x g| \in \mathscr{M}_\infty$, then $Vg \in C^{1,0}(\mathbb{R}^d \times \mathbb{R})$ and, for $i = 1, \dots, d$, $\partial g/\partial x_i \in \mathscr{D}(V)$ and $\partial(Vg)/\partial x_i = V(\partial g/\partial x_i)$.

Fix $m \geq 1$ and let $\rho = \phi_{3m}$ and $\theta = \frac{1}{2}$. Observe that V satisfies the required conditions (in fact (3.7) is zero). Assume, in addition to the assumptions on F and G in the first sentence of the theorem, that $G_i \in \mathscr{D}(V) \cap \mathscr{M}_m$ and F_i, $\partial F_i/\partial x_j$, G_i, $\partial G_i/\partial x_j$, $\partial^2 G_i/\partial x_j\, \partial x_k \in \mathscr{M}_m$ for $i, j, k = 1, \dots, d$. If $\bar{C}$ satisfies the condition of the theorem, the only condition that remains to be verified is (3.8). For this it suffices to show that

$$\lim_{\varepsilon \to 0} \sup_{0 \leq s \leq t} \varepsilon^{1/2} \left(1 + Y\left(\frac{s}{\varepsilon^2}\right)^2\right)^{3m/2} = 0 \quad \text{a.s.,} \qquad t \geq 0. \tag{3.23}$$

For the latter it is enough to show that for each $\lambda > 0$ there exists a random variable η such that

$$P\{|Y(t)| \leq \eta + t^\lambda \quad \text{for all} \quad t \geq 0\} = 1. \tag{3.24}$$

To verify (3.24), we need only show that $\lim_{t \to \infty} |Y(t)|/t^\lambda = 0$ a.s. for every $\lambda > 0$, which follows from the representation

$$Y(t) = e^{-t}Y(0) + e^{-t}W(e^{2t} - 1), \tag{3.25}$$

where W is a standard one-dimensional Brownian motion, and the law of the iterated logarithm for W. □

4. NON-MARKOVIAN DRIVING PROCESS

We again consider the limit in distribution as $\varepsilon \to 0+$ of the solution X^ε of the differential equation

$$\frac{d}{dt} X^\varepsilon(t) = F\left(X^\varepsilon(t), Y\left(\frac{t}{\varepsilon^2}\right)\right) + \frac{1}{\varepsilon} G\left(X^\varepsilon(t), Y\left(\frac{t}{\varepsilon^2}\right)\right) \tag{4.1}$$

driven by $Y(\cdot/\varepsilon^2)$, where Y is a process in a compact state space. However, instead of assuming that Y is Markovian and ergodic as in Section 2, we require that Y be stationary and uniform mixing.

4.1 Theorem Let E be a compact metric space, let $F, G: \mathbb{R}^d \times E \to \mathbb{R}^d$, and suppose that $F_1, \ldots, F_d \in C^{1,0}(\mathbb{R}^d \times E)$, $G_1, \ldots, G_d \in C^{2,0}(\mathbb{R}^d \times E)$, and

$$(4.2) \qquad \sup_{(x, y) \in \mathbb{R}^d \times E} \frac{|F(x, y)| \vee |G(x, y)|}{1 + |x|} < \infty.$$

Let Y be a stationary process with sample paths in $D_E[0, \infty)$, and for each $t \geq 0$, let $\mathscr{F}_t$ and $\mathscr{F}^t$ denote the completions of the σ-algebras $\mathscr{F}_t^Y$ and $\sigma\{Y(s): s \geq t\}$, respectively. Assume that the filtration $\{\mathscr{F}_t\}$ is right continuous, and that

$$(4.3) \qquad \varphi(u) \equiv \sup_{t \geq 0} \sup_{A \in \mathscr{F}_t, B \in \mathscr{F}^{t+u}} |P(B | A) - P(B)|$$

satisfies

$$(4.4) \qquad \int_0^\infty u\varphi(u)\, du < \infty.$$

Suppose that

$$(4.5) \qquad E[G(x, Y(0))] = 0, \qquad x \in \mathbb{R}^d.$$

Fix $x_0 \in \mathbb{R}^d$, and define X^ε for $0 < \varepsilon < 1$ to be the solution of the differential equation (4.1) with initial condition $X^\varepsilon(0) = x_0$. Put

$$(4.6) \qquad C = \left\{\left(f, \frac{1}{2} \sum_{i, j=1}^{d} a_{ij}\, \partial_i\, \partial_j f + \sum_{i=1}^{d} b_i\, \partial_i f\right): f \in C_c^3(\mathbb{R}^d)\right\},$$

where

$$(4.7) \quad a_{ij}(x) = \int_0^\infty E[G_i(x, Y(0))G_j(x, Y(t))]\, dt + \int_0^\infty E[G_j(x, Y(0))G_i(x, Y(t))]\, dt$$

and

$$(4.8) \qquad b_i(x) = E[F_i(x, Y(0))] + \int_0^\infty E[G(x, Y(0)) \cdot \nabla_x G_i(x, Y(t))]\, dt.$$

Then C is dissipative. Assume that $\bar{C}$, which is single-valued, generates a Feller semigroup $\{T(t)\}$ on $\hat{C}(\mathbb{R}^d)$, and let X be a Markov process corresponding to $\{T(t)\}$ with sample paths in $C_{\mathbb{R}^d}[0, \infty)$ and initial distribution δ_{x_0}. Then $X^\varepsilon \Rightarrow X$ in $C_{\mathbb{R}^d}[0, \infty)$ as $\varepsilon \to 0$.

Proof. Let $t, u \geq 0$ and let X be essentially bounded and $\mathscr{F}^{t+u}$-measurable. Then by Proposition 2.6 of Chapter 7 ($r = 1$, $p = \infty$),

(4.9)
$$\|E[X \mid \mathscr{F}_t] - E[X]\|_\infty \le 2\varphi(u)\|X\|_\infty,$$

where $\varphi(u)$ is defined by (4.3). For example, conditioning on $\mathscr{F}_0$ and using (4.5), we find that

(4.10)
$$|E[G_i(x, Y(0))G_j(x, Y(t))]| \le \sup_y |G_i(x, y)| 2\varphi(t) \sup_y |G_j(x, y)|$$

for all $x \in \mathbb{R}^d$, $t \ge 0$, and $i, j = 1, \dots, d$. The same inequality holds when G_i and/or G_j are replaced by any of their first- or second-order partial x-derivatives, and therefore the coefficients (4.7) and (4.8) are continuously differentiable on $\mathbb{R}^d$.

We also observe that the diffusion matrix $(a_{ij}(x))$ is nonnegative definite for each $x \in \mathbb{R}^d$. For if $x, \xi \in \mathbb{R}^d$ and $T > 0$,

(4.11)
$$\begin{aligned}
&\frac{1}{T} E\left[\left(\xi \cdot \int_0^T G(x, Y(t))\, dt\right)^2\right] \\
&= \sum_{i,j=1}^d \xi_i \xi_j \frac{1}{T} E \int_0^T \int_0^t [G_i(x, Y(s))G_j(x, Y(t)) \\
&\qquad + G_j(x, Y(s))G_i(x, Y(t))]\, ds\, dt \\
&= \sum_{i,j=1}^d \xi_i \xi_j \frac{1}{T} \int_0^T \int_0^t E[G_i(x, Y(0))G_j(x, Y(t-s)) \\
&\qquad + G_j(x, Y(0))G_i(x, Y(t-s))]\, ds\, dt \\
&= \sum_{i,j=1}^d \xi_i \xi_j \frac{1}{T} \int_0^T \left\{E\left[G_i(x, Y(0)) \int_0^t G_j(x, Y(s))\, ds\right]\right. \\
&\qquad \left. + E\left[G_j(x, Y(0)) \int_0^t G_i(x, Y(s))\, ds\right]\right\} dt.
\end{aligned}$$

As $T \to \infty$, (4.11), which is nonnegative, converges to

(4.12)
$$\sum_{i,j=1}^d \xi_i \xi_j a_{ij}(x).$$

Thus, C satisfies the positive maximum principle, hence C is dissipative (Lemma 2.1 of Chapter 4) and $\bar{C}$ is single-valued (Lemma 4.2 of Chapter 1).

The growth condition (4.2) guarantees the global existence of the solution X^ε of (4.1). Denote by $\{\mathscr{F}_t^\varepsilon\}$ the filtration given by $\mathscr{F}_t^\varepsilon = \mathscr{F}_{t/\varepsilon^2}$, and let $\hat{\mathscr{A}}^\varepsilon$ be the full generator of the associated semigroup of conditioned shifts (Chapter 2, Section 7). By Theorem 8.2 of Chapter 4, the finite-dimensional distributions of X^ε will converge weakly to those of X if for each $(f, g) \in C$, we can find $(f^\varepsilon, g^\varepsilon) \in \hat{\mathscr{A}}^\varepsilon$ for every $\varepsilon \in (0, 1)$ such that

$$\sup_{\varepsilon} \sup_{t \le T} E[\,|f^\varepsilon(t)|\,] < \infty, \qquad T > 0, \tag{4.13}$$

$$\sup_{\varepsilon} \sup_{t \le T} E[\,|g^\varepsilon(t)|\,] < \infty, \qquad T > 0, \tag{4.14}$$

$$\lim_{\varepsilon \to 0} E[\,|f^\varepsilon(t) - f(X^\varepsilon(t))|\,] = 0, \qquad t \ge 0, \tag{4.15}$$

and

$$\lim_{\varepsilon \to 0} E[\,|g^\varepsilon(t) - g(X^\varepsilon(t))|\,] = 0, \qquad t \ge 0. \tag{4.16}$$

By Corollary 8.6 of Chapter 4 and Problem 25 of Chapter 3, we have $X^\varepsilon \Rightarrow X$ in $C_{\mathbb{R}^d}[0, \infty)$ as $\varepsilon \to 0$ if (4.14) and (4.15) can be replaced by the stronger conditions

$$\sup_{\varepsilon} E\left[\operatorname*{ess\,sup}_{t \le T} |g^\varepsilon(t)|\right] < \infty, \qquad T > 0, \tag{4.17}$$

and

$$\lim_{\varepsilon \to 0} E\left[\sup_{t \in \mathbb{Q} \cap [0, T]} |f^\varepsilon(t) - f(X^\varepsilon(t))|\right] = 0, \qquad T > 0. \tag{4.18}$$

Fix $(f, g) \in C$, and let $\varepsilon \in (0, 1)$ be arbitrary. We let

$$f^\varepsilon(t) = f(X^\varepsilon(t)) + \varepsilon \hat{h}_1^\varepsilon(t) + \varepsilon^2 \hat{h}_2^\varepsilon(t), \tag{4.19}$$

where the correction terms $\hat{h}_1^\varepsilon$, $\hat{h}_2^\varepsilon \in \mathscr{D}(\hat{\mathscr{A}}^\varepsilon)$ are chosen by analogy with (3.12) and (3.13). Let us first consider $\hat{h}_1^\varepsilon$. We define $f_1^\varepsilon: \mathbb{R}^d \times [0, \infty) \times \Omega \to \mathbb{R}$ by

$$f_1^\varepsilon(x, t, \omega) = G\left(x, Y\left(\frac{t}{\varepsilon^2}, \omega\right)\right) \cdot \nabla f(x). \tag{4.20}$$

Clearly, f_1^ε is $\mathscr{B}(\mathbb{R}^d) \times \mathscr{B}[0, \infty) \times \mathscr{F}$-measurable and is C_c^2 in x for fixed (t, ω). In fact, there is a constant k_1 such that $f_1^\varepsilon(x, t, \omega) = 0$ for all $|x| \ge k_1$, $t \ge 0$, and $\omega \in \Omega$, and

$$\|f_1^\varepsilon(\cdot, t, \omega)\|_{C^2} \le \sup_{y} \|G(\cdot, y) \cdot \nabla f(\cdot)\|_{C^2} \equiv \gamma < \infty \tag{4.21}$$

for all $t \geq 0$ and $\omega \in \Omega$, where $\|f\|_{C^m} \equiv \sum_{|\alpha| \leq m} \|D^\alpha f\|$. By Corollary 4.5 of Chapter 2 there exists $g_1^\varepsilon \colon \mathbb{R}^d \times [0, \infty) \times [0, \infty) \times \Omega \to \mathbb{R}$, $\mathscr{B}(\mathbb{R}^d) \times \mathscr{B}[0, \infty) \times \mathscr{O}$-measurable, C_c^2 in x for fixed (s, t, ω), such that

$$g_1^\varepsilon(x, s, t, \omega) = E_t^\varepsilon[f_1^\varepsilon(x, t + s, \cdot)](\omega) \tag{4.22}$$

for all $x \in \mathbb{R}^d$ and $s, t \geq 0$, where E_t^ε denotes conditional expectation given $\mathscr{F}_t^\varepsilon$ here and below. Moreover, g_1^ε may be chosen so that $g_1(x, s, t, \omega) = 0$ for all $|x| \geq k_1$, $s, t \geq 0$, and $\omega \in \Omega$, and

$$\|g_1^\varepsilon(\cdot, s, t, \omega)\|_{C^2} \leq 2\gamma\phi\left(\frac{s}{\varepsilon^2}\right) \tag{4.23}$$

for all $s, t \geq 0$ and $\omega \in \Omega$. The latter can be deduced from (4.5), (4.9), and (4.21). We now define $h_1^\varepsilon \colon \mathbb{R}^d \times [0, \infty) \times \Omega \to \mathbb{R}$ by

$$h_1^\varepsilon(x, t, \omega) = \varepsilon^{-2} \int_0^\infty g_1^\varepsilon(x, s, t, \omega)\, ds. \tag{4.24}$$

Clearly, h_1^ε is $\mathscr{B}(\mathbb{R}^d) \times \mathscr{O}$-measurable and is C_c^2 in x for fixed (t, ω). In fact, $h_1^\varepsilon(x, t, \omega) = 0$ for all $|x| \geq k_1$, $t \geq 0$, and $\omega \in \Omega$, and

$$\|h_1^\varepsilon(\cdot, t, \omega)\|_{C^2} \leq 2\gamma \int_0^\infty \varphi(s)\, ds \tag{4.25}$$

for all $t \geq 0$ and $\omega \in \Omega$. Finally, we define $\hat{h}_1^\varepsilon \colon [0, \infty) \times \Omega \to \mathbb{R}$ by

$$\hat{h}_1^\varepsilon(t, \omega) = h_1^\varepsilon(X^\varepsilon(t, \omega), t, \omega). \tag{4.26}$$

It follows that $\hat{h}_1^\varepsilon$ is optional (hence progressive).

To show that $\hat{h}_1^\varepsilon \in \mathscr{D}(\hat{\mathscr{A}}^\varepsilon)$, we apply Lemma 3.4 of Chapter 4. Fix $t_2 > t_1 \geq 0$. Clearly,

$$\begin{aligned} &h_1^\varepsilon(X^\varepsilon(t_2), t_2) - h_1^\varepsilon(X^\varepsilon(t_1), t_2) \\ &\quad = \int_{t_1}^{t_2} \nabla_x h_1^\varepsilon(X^\varepsilon(s), t_2) \cdot \frac{d}{ds} X^\varepsilon(s)\, ds \\ &\quad = \int_{t_1}^{t_2} \left\{F\left(X^\varepsilon(s), Y\left(\frac{s}{\varepsilon^2}\right)\right) + \frac{1}{\varepsilon} G\left(X^\varepsilon(s), Y\left(\frac{s}{\varepsilon^2}\right)\right)\right\} \\ &\qquad \cdot \nabla_x h_1^\varepsilon(X^\varepsilon(s), t_2)\, ds. \end{aligned} \tag{4.27}$$

For each $x \in \mathbb{R}^d$ and $s \geq 0$, we have

$$\begin{aligned} E_{t_1}^\varepsilon[g_1^\varepsilon(x, s, t_2, \cdot)](\omega) &= E_{t_1}^\varepsilon[E_{t_2}^\varepsilon[f_1^\varepsilon(x, t_2 + s, \cdot)]](\omega) \\ &= E_{t_1}^\varepsilon[f_1^\varepsilon(x, t_2 + s, \cdot)](\omega) \\ &= g_1^\varepsilon(x, s + t_2 - t_1, t_1, \omega), \end{aligned} \tag{4.28}$$

and therefore

$$(4.29)\quad \begin{aligned} &E_{t_1}^\varepsilon[h_1^\varepsilon(X^\varepsilon(t_1), t_2) - h_1^\varepsilon(X^\varepsilon(t_1), t_1)] \\ &= \varepsilon^{-2}\left\{E_{t_1}^\varepsilon\left[\int_0^\infty g_1^\varepsilon(X^\varepsilon(t_1), s, t_2)\, ds\right] - \int_0^\infty g_1^\varepsilon(X^\varepsilon(t_1), s, t_1)\, ds\right\} \\ &= \varepsilon^{-2}\left\{\int_0^\infty g_1^\varepsilon(X^\varepsilon(t_1), s + t_2 - t_1, t_1)\, ds - \int_0^\infty g_1^\varepsilon(X^\varepsilon(t_1), s, t_1)\, ds\right\} \\ &= -\varepsilon^{-2}\int_0^{t_2 - t_1} g_1^\varepsilon(X^\varepsilon(t_1), s, t_1)\, ds \\ &= -\varepsilon^{-2}E_{t_1}^\varepsilon\left[\int_{t_1}^{t_2} f_1^\varepsilon(X^\varepsilon(t_1), s)\, ds\right]. \end{aligned}$$

Finally, we must verify condition (3.15) of Chapter 5, which amounts to showing that, for each $t \geq 0$,

$$(4.30)\qquad \lim_{\delta \to 0+} E[|\nabla_x h_1^\varepsilon(X^\varepsilon(t), t + \delta) - \nabla_x h_1^\varepsilon(X^\varepsilon(t), t)|] = 0.$$

(We can ignore the factor $F + \varepsilon^{-1}G$ because $\nabla_x h_1^\varepsilon(x, t, \omega)$ has compact support in x, uniformly in (t, ω).) Using the bound (4.23), the dominated convergence theorem reduces the problem to one of showing that, for each s, $t \geq 0$,

$$(4.31)\qquad \lim_{\delta \to 0+} E[|\nabla_x g_1^\varepsilon(X^\varepsilon(t), s, t + \delta, \cdot) - \nabla_x g_1^\varepsilon(X^\varepsilon(t), s, t, \cdot)|] = 0$$

or

$$(4.32)\qquad \lim_{\delta \to 0+} E[|E_{t+\delta}^\varepsilon[\nabla_x f_1^\varepsilon(X^\varepsilon(t), t + \delta + s)] - E_t^\varepsilon[\nabla_x^\varepsilon f_1^\varepsilon(X^\varepsilon(t), t + s)]|] = 0.$$

But (4.32) follows easily from the right continuity of Y and the right continuity of the filtration $\{\mathscr{F}_t^\varepsilon\}$. We conclude from Lemma 3.4 of Chapter 4 that

$$(4.33)\quad (\hat{h}_1^\varepsilon(t), \{F(x, y) + \varepsilon^{-1}G(x, y)\} \cdot \nabla_x h_1^\varepsilon(x, t) - \varepsilon^{-2}G(x, y) \cdot \nabla f(x)) \in \hat{\mathscr{A}}^\varepsilon,$$

where $x = X^\varepsilon(t)$ and $y = Y(t/\varepsilon^2)$.

We turn now to the definition of $\hat{h}_2^\varepsilon(t)$. We define $f_2^\varepsilon \colon \mathbb{R}^d \times [0, \infty) \times \Omega \to \mathbb{R}$ by

$$(4.34)\quad f_2^\varepsilon(x, t, \omega) = F\left(x, Y\left(\frac{t}{\varepsilon^2}, \omega\right)\right) \cdot \nabla f(x) + G\left(x, Y\left(\frac{t}{\varepsilon^2}, \omega\right)\right) \cdot \nabla_x h_1^\varepsilon(x, t, \omega) - g(x).$$

Observe that f_2^ε is $\mathscr{B}(\mathbb{R}^d) \times \mathscr{O}$-measurable and is C_c^1 in x for fixed (t, ω). In fact, there is a constant k_2 such that $f_2(x, t, \omega) = 0$ for all $|x| \geq k_2$, $t \geq 0$, and $\omega \in \Omega$, and

$$\|f_2^\varepsilon(\cdot, t, \omega)\|_{C^1} \le \sup_y \|F(\cdot, y) \cdot \nabla f(x)\|_{C^1} + \sup_{y, t, \omega} \|G(\cdot, y) \cdot \nabla_x h_1^\varepsilon(\cdot, t, \omega)\|_{C^1} + \|g\|_{C^1} \equiv \eta < \infty \tag{4.35}$$

by (4.25). We now define g_2^ε, h_2^ε, and $\hat{h}_2^\varepsilon$ by analogy with g_1^ε, h_1^ε, and $\hat{h}_1^\varepsilon$. The only thing that needs to be checked is the analogue of (4.23), which is that, for appropriate constants $c_1, c_2 > 0$,

$$\|g_2^\varepsilon(\cdot, s, t, \omega)\|_{C^1} \le c_1 \varphi\left(\frac{s}{\varepsilon^2}\right) + c_2 \int_0^\infty \varphi\left(\left(\frac{s}{\varepsilon^2}\right) \vee s'\right) ds' \tag{4.36}$$

for all $s, t \ge 0$ and $\omega \in \Omega$. Observe that the right side of (4.36) is Lebesgue integrable on $[0, \infty)$ by (4.4).

To justify (4.36), fix $x \in \mathbb{R}^d$ and $s, t \ge 0$. Then

$$\begin{aligned} g_2^\varepsilon(x, s, t) &= E_t^\varepsilon[f_2^\varepsilon(x, t + s)] \\ &= E_t^\varepsilon\left[F\left(x, Y\left(\frac{t+s}{\varepsilon^2}\right)\right) \cdot \nabla f(x)\right] \\ &\quad - E\left[F\left(x, Y\left(\frac{t+s}{\varepsilon^2}\right)\right) \cdot \nabla f(x)\right] \\ &\quad + E_t^\varepsilon\left[G\left(x, Y\left(\frac{t+s}{\varepsilon^2}\right)\right) \cdot \nabla_x h_1^\varepsilon(x, t + s)\right] \\ &\quad - E\left[G\left(x, Y\left(\frac{t+s}{\varepsilon^2}\right)\right) \cdot \nabla_x h_1^\varepsilon(x, t + s)\right] \end{aligned} \tag{4.37}$$

by the definition of C. Consequently, a similar equation holds for $\nabla_x g_2^\varepsilon(x, s, t, \cdot)$ with each integrand replaced by its x-gradient. By (4.9),

$$\left\| E_t^\varepsilon\left[F\left(x, Y\left(\frac{t+s}{\varepsilon^2}\right)\right) \cdot \nabla f(x)\right] - E\left[F\left(x, Y\left(\frac{t+s}{\varepsilon^2}\right)\right) \cdot \nabla f(x)\right]\right\|_\infty \le 2\varphi\left(\frac{s}{\varepsilon^2}\right) \sup_y |F(x, y) \cdot \nabla f(x)|. \tag{4.38}$$

Since $\nabla_x h_1^\varepsilon(x, t + s) = \varepsilon^{-2} \int_0^\infty \nabla_x g_1^\varepsilon(x, s', t + s)\, ds'$, we need to consider

$$\left\| E_t^\varepsilon\left[G\left(x, Y\left(\frac{t+s}{\varepsilon^2}\right)\right) \cdot \nabla_x \left\{G\left(x, Y\left(\frac{t+s+s'}{\varepsilon^2}\right)\right) \cdot \nabla f(x)\right\}\right] - E\left[G\left(x, Y\left(\frac{t+s}{\varepsilon^2}\right)\right) \cdot \nabla_x \left\{G\left(x, Y\left(\frac{t+s+s'}{\varepsilon^2}\right)\right) \cdot \nabla f(x)\right\}\right]\right\|_\infty \tag{4.39}$$

for fixed $s' \geq 0$. By (4.9), this is bounded by

$$2\varphi\left(\frac{s}{\varepsilon^2}\right) \sup_{y, z} | G(x, y) \cdot \nabla_x\{G(x, z) \cdot \nabla f(x)\} |. \tag{4.40}$$

Moreover, conditioning on $\mathscr{F}^\varepsilon_{t+s}$ and applying (4.9) and (4.5), each of the two expectations in (4.39) is bounded a.s. by

$$\sup_y | G(x, y) | 2\varphi\left(\frac{s'}{\varepsilon^2}\right) \sup_y | \nabla_x\{G(x, y) \cdot \nabla f(x)\} |. \tag{4.41}$$

Thus, (4.39) is bounded by $c_3 \varphi((s \vee s')/\varepsilon^2)$ for an appropriate constant c_3. Similar bounds hold when all integrands are replaced by their x-gradients, and thus (4.36) can be assumed to hold. It follows from (4.36) that

$$\|h_2^\varepsilon(\cdot, t, \omega)\|_{C^1} \leq c_1 \int_0^\infty \varphi(s)\, ds + 2c_2 \int_0^\infty s\varphi(s)\, ds \tag{4.42}$$

for all $t \geq 0$ and $\omega \in \Omega$.

The argument used to show that $\hat{h}_1^\varepsilon \in \mathscr{D}(\hat{\mathscr{A}}^\varepsilon)$ now applies, almost word-for-word, to show that

$$\begin{aligned} (\hat{h}_2^\varepsilon(t), \{F(x, y) + \varepsilon^{-1}G(x, y)\} \cdot \nabla_x h_2^\varepsilon(x, t) \\ - \varepsilon^{-2}\{F(x, y) \cdot \nabla f(x) + G(x, y) \cdot \nabla_x h_1^\varepsilon(x, t) - g(x)\}) \in \hat{\mathscr{A}}^\varepsilon, \end{aligned} \tag{4.43}$$

where $x = X^\varepsilon(t)$ and $y = Y(t/\varepsilon^2)$. The only point that should be made is that, in proving the analogue of (4.32), $\nabla_x f_2^\varepsilon(X^\varepsilon(t), t + \delta + s)$ no longer converges pointwise in ω, but only in L^1. However, this suffices.

Clearly,

$$(f(x), \{F(x, y) + \varepsilon^{-1}G(x, y)\} \cdot \nabla f(x)) \in \hat{\mathscr{A}}^\varepsilon \tag{4.44}$$

where $x = X^\varepsilon(t)$ and $y = Y(t/\varepsilon^2)$. Recalling (4.19), we obtain from (4.44), (4.33), and (4.43) that

$$\begin{aligned} (f^\varepsilon(t), g(x) + \varepsilon F(x, y) \cdot \nabla_x h_1^\varepsilon(x, t) + \varepsilon G(x, y) \cdot \nabla_x h_2^\varepsilon(x, t) \\ + \varepsilon^2 F(x, y) \cdot \nabla_x h_2^\varepsilon(x, t)) \in \hat{\mathscr{A}}^\varepsilon, \end{aligned} \tag{4.45}$$

where $x = X^\varepsilon(t)$ and $y = Y(t/\varepsilon^2)$. By (4.25) and (4.42), together with the fact that $\nabla_x h_1^\varepsilon(x, t, \omega)$ and $\nabla_x h_2^\varepsilon(x, t, \omega)$ have compact support in x, uniformly in (t, ω), we see that (4.13)–(4.18) are satisfied, and hence the proof is complete. □

5. PROBLEMS

1. Formulate and prove a discrete-parameter analogue of Theorem 2.4. (Both X and Y are discrete-parameter processes, and the differential equation (2.25) is a difference equation.)

2. Give a simpler proof that $X^\varepsilon \Rightarrow x + \sqrt{2}W$ in (3.2) by using the representation

$$X^\varepsilon(t) = x + \varepsilon Y(0) + \sqrt{2}\varepsilon W\left(\frac{t}{\varepsilon^2}\right) - \varepsilon Y\left(\frac{t}{\varepsilon^2}\right), \qquad t \geq 0, \tag{5.1}$$

   where W is a one-dimensional Brownian motion.

3. Generalize Example 3.3 to the case in which $Y(t) \equiv UZ(t)$, where

$$dZ(t) = S\, dW(t) + NZ(t)\, dt \tag{5.2}$$

   and U, S, and N are (constant) $d \times d$ matrices with the eigenvalues of N having negative real parts, and W is a d-dimensional Brownian motion.

4. Extend Theorem 4.1 to noncompact E. The extension should include the case in which Y is a stationary Gaussian process.

6. NOTES

Random evolutions, introduced by Griego and Hersh (1969), are surveyed by Hersh (1974) and Pinsky (1974).

The derivation of the telegrapher's equation in Section 1 is due to Goldstein (1951) and Kac (1956).

The results of Section 2 were motivated by work of Pinsky (1968), Griego and Hersh (1971), Hersh and Papanicolaou (1972), and Kurtz (1973).

Theorem 4.1 is due essentially to Kushner (1979) (see also Kushner (1984)), though the problem had earlier been treated by Stratonovich (1963, 1967), Khas'minskii (1966), Papanicolaou and Varadhan (1973), and Papanicolaou and Kohler (1974).

APPENDIXES

1. CONVERGENCE OF EXPECTATIONS

Recall that $X_n \xrightarrow{\text{a.s.}} X$ implies $X_n \xrightarrow{p} X$ implies $X_n \Rightarrow X$, so the following results, which are stated in terms of convergence in distribution, apply to the other types of convergence as well.

1.1 Proposition (Fatou's Lemma) Let $X_n \geq 0$, $n = 1, 2, \ldots$, and $X_n \Rightarrow X$. Then

(1.1) $$\varliminf_{n \to \infty} E[X_n] \geq E[X].$$

Proof. For $M > 0$

(1.2) $$\varliminf_{n \to \infty} E[X_n] \geq \lim_{n \to \infty} E[X_n \wedge M] = E[X \wedge M],$$

where the equality holds by definition. Letting $M \to \infty$ we have (1.1). □

1.2 Theorem (Dominated Convergence Theorem) Suppose

$$|X_n| \leq Y_n, \qquad n = 1, 2, \ldots, \qquad X_n \Rightarrow X, \qquad Y_n \Rightarrow Y$$

and $\lim_{n \to \infty} E[Y_n] = E[Y]$. Then

(1.3) $$\lim_{n \to \infty} E[X_n] = E[X].$$

Proof. It is not necessarily the case that $(X_n, Y_n) \Rightarrow (X, Y)$. However, by Proposition 2.4 of Chapter 3, every subsequence of $\{(X_n, Y_n)\}$ has a further subsequence such that $(X_{n_k}, Y_{n_k}) \Rightarrow (\tilde{X}, \tilde{Y})$, where $\tilde{X}$ and X have the same distribution, as do $\tilde{Y}$ and Y. Consequently, $Y_{n_k} + X_{n_k} \Rightarrow \tilde{Y} + \tilde{X}$ and $Y_{n_k} - X_{n_k} \Rightarrow \tilde{Y} - \tilde{X}$, so by Fatou's lemma,

$$\underline{\lim}_{k \to \infty} (E[Y_{n_k}] + E[X_{n_k}]) \geq E[Y] + E[X] \tag{1.4}$$

and

$$\underline{\lim}_{k \to \infty} (E[Y_{n_k}] - E[X_{n_k}]) \geq E[Y] - E[X]. \tag{1.5}$$

Therefore $\lim_{k \to \infty} E[X_{n_k}] = E[X]$, and (1.3) follows. □

2. UNIFORM INTEGRABILITY

A collection of real-valued random variables $\{X_\alpha\}$ is *uniformly integrable* if $\sup_\alpha E[|X_\alpha|] < \infty$, and for every $\varepsilon > 0$ there exists a $\delta > 0$ such that for every α, $P(A_\alpha) < \delta$ implies $|E[X_\alpha \chi_{A_\alpha}]| < \varepsilon$.

2.1 Proposition The following are equivalent:

- **(a)** $\{X_\alpha\}$ is uniformly integrable.
- **(b)** $\lim_{N \to \infty} \sup_\alpha E[\chi_{\{|X_\alpha| > N\}} |X_\alpha|] = 0$.
- **(c)** $\lim_{N \to \infty} \sup_\alpha E[|X_\alpha| - N \wedge |X_\alpha|] = 0$.

Proof. Since $P\{|X_\alpha| > N\} \leq N^{-1} E[|X_\alpha|]$, it is immediate that (a) implies (b). More precisely,

$$NP\{|X_\alpha| > N\} \leq E[\chi_{\{|X_\alpha| > N\}} |X_\alpha|],$$

and since

$$\begin{aligned} E[|X_\alpha| - N \wedge |X_\alpha|] &= E[\chi_{\{|X_\alpha| > N\}}(|X_\alpha| - N)] \\ &= E[\chi_{\{|X_\alpha| > N\}} |X_\alpha|] - NP\{|X_\alpha| > N\}, \end{aligned} \tag{2.1}$$

(b) implies (c). Finally note that if $P(A_\alpha) \leq N^{-2}$, then

$$\begin{aligned} E[\chi_{A_\alpha} |X_\alpha|] &\leq E[|X_\alpha| - N \wedge |X_\alpha|] + NP(A_\alpha) \\ &\leq E[|X_\alpha| - N \wedge |X_\alpha|] + N^{-1}, \end{aligned} \tag{2.2}$$

and (c) implies (a). □

2.2 Proposition A collection of real-valued random variables $\{X_\alpha\}$ is uniformly integrable if and only if there exists an increasing convex function φ on $[0, \infty)$ such that $\lim_{x\to\infty} \varphi(x)/x = \infty$ and $\sup_\alpha E[\varphi(|X_\alpha|)] < \infty$.

Proof. We can assume $\varphi(0) = 0$. Then $\varphi(x)/x$ is increasing and

$$E[\chi_{\{|X_\alpha| > N\}} |X_\alpha|] \leq \frac{NE[\varphi(|X_\alpha|)]}{\varphi(N)}. \tag{2.3}$$

Therefore sufficiency follows from Proposition 2.1(b).

By (b) there exists an increasing sequence $\{N_k\}$ such that

$$\sup_\alpha \sum_{k=1}^{\infty} kE[\chi_{\{|X_\alpha| > N_k\}} |X_\alpha|] < \infty. \tag{2.4}$$

Assume $N_0 = 0$ and define $\varphi(0) = 0$ and

$$\varphi'(x) = \left(k - \frac{N_{k+1} - x}{N_{k+1} - N_k}\right), \qquad N_k \leq x < N_{k+1}. \tag{2.5}$$

□

2.3 Proposition If $X_n \Rightarrow X$ and $\{X_n\}$ is uniformly integrable, then $\lim_{n\to\infty} E[X_n] = E[X]$. Conversely, if the X_n are integrable, $X_n \Rightarrow X$, and $\lim_{n\to\infty} E[|X_n|] = E[|X|]$, then $\{X_n\}$ is uniformly integrable.

Proof. If $\{X_n\}$ is uniformly integrable, the first term on the right of

$$E[|X_n|] = E[|X_n| - N \wedge |X_n|] + E[N \wedge |X_n|] \tag{2.6}$$

can be made small uniformly in n. Consequently $\lim_{n\to\infty} E[|X_n|] = E[|X|]$, and hence $\lim_{n\to\infty} E[X_n] = E[X]$ by Theorem 1.2.

Conversely, since

$$\lim_{n\to\infty} E[|X_n| - N \wedge |X_n|] = E[|X|] - E[N \wedge |X|] \tag{2.7}$$

and the right side of (2.7) can be made arbitrarily small by taking N large, (b) of Proposition 2.1 follows. □

2.4 Proposition Let $\{X_n\}$ be a uniformly integrable sequence of random variables defined on $(\Omega, \mathcal{F}, P)$. Then there exists a subsequence $\{X_{n_k}\}$ and a random variable X such that

$$\lim_{k\to\infty} E[X_{n_k} Z] = E[XZ] \tag{2.8}$$

for every bounded random variable Z defined on $(\Omega, \mathcal{F}, P)$.

2.5 Remark The converse also holds. See Dunford and Schwartz (1957), page 294. □

Proof. Since we can consider $\{X_n \vee 0\}$ and $\{X_n \wedge 0\}$ separately, we may as well assume $X_n \geq 0$. Let $\mathscr{A}$ be the algebra generated by $\{\{X_n < a\}: n = 1, 2, \ldots, a \in \mathbb{Q}\}$. Note that $\mathscr{A}$ is countable so there exists a subsequence $\{X_{n_k}\}$ such that

$$\mu(A) \equiv \lim_{k \to \infty} E[X_{n_k} \chi_A] \tag{2.9}$$

exists for every $A \in \mathscr{A}$. Let $\mathscr{G}$ be the collection of sets $A \in \mathscr{F}$ for which the limit in (2.9) exists. Then $\mathscr{A} \subset \mathscr{G}$. If A, $B \in \mathscr{G}$ and $A \subset B$, then $B - A \in \mathscr{G}$. The uniform integrability implies that, if $\{A_k\} \subset \mathscr{G}$ and $A_1 \subset A_2 \subset \cdots$, then $\bigcup_k A_k \in \mathscr{G}$. ($P(\bigcup_k A_k - A_m)$ can be made arbitrarily small by choosing m large.) Therefore the Dynkin class theorem (Appendix 4) implies $\mathscr{G} \supset \sigma(\mathscr{A})$.

Clearly μ is finitely additive on $\sigma(\mathscr{A})$, and the uniform integrability implies μ is countably additive.

Clearly $\mu \ll P$ on $\sigma(\mathscr{A})$, so there exists a $\sigma(\mathscr{A})$-measurable random variable X such that $\mu(A) = E[X\chi_A]$, $A \in \sigma(\mathscr{A})$.

By (2.9),

$$\lim_{k \to \infty} E[X_{n_k} Z] = E[XZ] \tag{2.10}$$

for all simple $\sigma(\mathscr{A})$-measurable random variables and the uniform integrability allows the extension of that conclusion to all bounded, $\sigma(\mathscr{A})$-measurable random variables. Finally, for any bounded random variable Z,

$$\begin{aligned} \lim_{k \to \infty} E[X_{n_k} Z] &= \lim_{k \to \infty} E[X_{n_k} E[Z \mid \sigma(\mathscr{A})]] \\ &= E[XE[Z \mid \sigma(\mathscr{A})]] = E[XZ]. \end{aligned} \tag{2.11}$$

□

3. BOUNDED POINTWISE CONVERGENCE

Let E be a metric space and let $V(E)$ denote the space of finite signed Borel measures on E with total variation norm

$$\|v\| = \sup_{A \in \mathscr{B}(E)} (|v(A)| + |v(E - A)|). \tag{3.1}$$

A sequence $\{f_n\} \subset B(E)$ converges in the weak* topology to f (denoted by w*-$\lim_{n \to \infty} f_n = f$) if $\lim_{n \to \infty} \int f_n \, dv = \int f \, dv$ for each $v \in V(E)$. A sequence $\{f_n\}$ converges boundedly and pointwise to f (denoted by bp-$\lim_{n \to \infty} f_n = f$) if $\sup_n \|f_n\| < \infty$ and $\lim_{n \to \infty} f_n(x) = f(x)$ for each $x \in E$.

3.1 Proposition Let f_n, $n = 1, 2, \ldots$, and f belong to $B(E)$. w*-$\lim_{n \to \infty} f_n = f$ if and only if bp-$\lim_{n \to \infty} f_n = f$.

3.2 Remark This result holds only for sequences, not for nets. □

Proof. If w*-$\lim_{n\to\infty} f_n = f$, then $\sup_n |\int f_n\, d\nu| < \infty$ for each $\nu \in V(E)$ and the uniform boundedness theorem (see e.g., Rudin (1974), page 104) implies $\sup_n \|f_n\| < \infty$. Of course, taking $\nu = \delta_x$ implies $\lim_{n\to\infty} f_n(x) = f(x)$.

The converse follows by the dominated convergence theorem. □

Let $H \subset B(E)$. The *bp-closure* of H is the smallest subset $\tilde{H}$ of $B(E)$ containing H such that $\{f_n\} \subset \tilde{H}$ and bp-$\lim_{n\to\infty} f_n = f$ imply $f \in \tilde{H}$. Note the bp-closure of H is not necessarily the same as the weak* closure. For example, let $E = [0, 1]$ and $H = \{n\chi_{[k/n^2, (k+1)/n^2)}: 0 \le k < n^2, n = 1, 2, \ldots\}$. Then H is bp-closed, but it is not closed in the weak* topology.

4. MONOTONE CLASS THEOREMS

Let Ω be a set. A collection $\mathscr{M}$ of subsets of Ω is a *monotone class* if

(M1) $\{A_n\} \subset \mathscr{M}$ and $A_1 \subset A_2 \subset \cdots$ imply $\cup A_n \in \mathscr{M}$

and

(M2) $\{A_n\} \subset \mathscr{M}$ and $A_1 \supset A_2 \supset \cdots$ imply $\cap A_n \in \mathscr{M}$.

A collection $\mathscr{D}$ of subsets of Ω is a *Dynkin class* if

(D1) $\Omega \in \mathscr{D}$,

(D2) $A, B \in \mathscr{D}$ and $A \subset B$ imply $B - A \in \mathscr{D}$,

(D3) $\{A_n\} \subset \mathscr{D}$ and $A_1 \subset A_2 \subset \cdots$ imply $\cup A_n \in \mathscr{D}$.

4.1 Theorem (Monotone Class Theorem) If $\mathscr{A}$ is an algebra and $\mathscr{M}$ is a monotone class with $\mathscr{A} \subset \mathscr{M}$, then $\sigma(\mathscr{A}) \subset \mathscr{M}$.

Proof. Let $M(\mathscr{A})$ be the smallest monotone class containing $\mathscr{A}$. We want to show $M(\mathscr{A}) = \sigma(\mathscr{A})$. Clearly it will be sufficient to show that $M(\mathscr{A})$ is a σ-algebra.

First note that $\{A \in M(\mathscr{A}): A^c \in M(\mathscr{A})\}$ is a monotone class that contains $\mathscr{A}$ and hence $M(\mathscr{A})$, that is, $A \in M(\mathscr{A})$ implies $A^c \in M(\mathscr{A})$. Next note that for $A \in \mathscr{A}$, $\{B: A \cup B \in M(A)\}$ is a monotone class containing $\mathscr{A}$ and hence $M(\mathscr{A})$, that is, $A \in \mathscr{A}$ and $B \in M(\mathscr{A})$ imply $A \cup B \in M(\mathscr{A})$. Finally, by this last observation, if $A \in M(\mathscr{A})$ then $\{B: A \cup B \in M(\mathscr{A})\}$ is a monotone class containing $\mathscr{A}$ and hence $M(\mathscr{A})$, that is, $M(\mathscr{A})$ is closed under finite unions. Since $M(\mathscr{A})$ is closed under finite unions, by (M1) it is closed under countable unions, and hence is a σ-algebra. □

4.2 Theorem (Dynkin Class Theorem) Let $\mathscr{S}$ be a collection of subsets of Ω such that $A, B \in \mathscr{S}$ implies $A \cap B \in \mathscr{S}$. If $\mathscr{D}$ is a Dynkin class with $\mathscr{S} \subset \mathscr{D}$, then $\sigma(\mathscr{S}) \subset \mathscr{D}$.

Proof. Let $D(\mathscr{S})$ be the smallest Dynkin class that contains $\mathscr{S}$. It is sufficient to show that $D(\mathscr{S})$ is a σ-algebra. This will follow from (D3) if we show $D(\mathscr{S})$ is closed under finite unions.

If $A, B \in \mathscr{S}$, then A^c, B^c, and $A^c \cup B^c = \Omega - A \cap B$ are in $D(\mathscr{S})$. Consequently $A^c \cup B^c - A^c = A \cap B^c$, $A^c \cup B = \Omega - A \cap B^c$, $A^c \cap B^c = A^c \cup B - B$, and $A \cup B = \Omega - A^c \cap B^c$ are in $D(\mathscr{S})$.

For $A \in \mathscr{S}$, $\{B: A \cup B \in D(\mathscr{S})\}$ is a Dynkin class containing $\mathscr{S}$, and hence $A \in \mathscr{S}$ and $B \in D(\mathscr{S})$ imply $A \cup B \in D(\mathscr{S})$. Consequently, for $A \in D(\mathscr{S})$, $\{B: A \cup B \in D(\mathscr{S})\}$ is a Dynkin class containing $\mathscr{S}$, and hence $A, B \in D(\mathscr{S})$ implies $A \cup B \in D(\mathscr{S})$. □

4.3 Theorem Let H be a linear space of bounded functions on Ω that contains constants, and let $\mathscr{S}$ be a collection of subsets of Ω such that $A, B \in \mathscr{S}$ implies $A \cap B \in \mathscr{S}$. Suppose $\chi_A \in H$ for all $A \in \mathscr{S}$, and $\{f_n\} \subset H, f_1 \le f_2 \le \cdots$, and $\sup_n f_n \le c$ for some constant c imply $f \equiv \text{bp-lim}_{n\to\infty} f_n \in H$. Then H contains all bounded $\sigma(\mathscr{S})$-measurable functions.

Proof. Note that $\{A: \chi_A \in H\}$ is a Dynkin class containing $\mathscr{S}$ and hence $\sigma(\mathscr{S})$. Since H is linear, H contains all simple $\sigma(\mathscr{S})$-measurable functions. Since any bounded $\sigma(\mathscr{S})$-measurable function is the pointwise limit of an increasing sequence of simple functions, the theorem follows. □

4.4 Corollary Let H be a linear space of bounded functions on Ω containing constants that is closed under uniform convergence and under bounded pointwise convergence of nondecreasing sequences (as in Theorem 4.3). Suppose $H_0 \subset H$ is closed under multiplication ($f, g \in H_0$ implies $fg \in H_0$). Then H contains all bounded $\sigma(H_0)$-measurable functions.

Proof. Let $F \in C(\mathbb{R})$. Then on any bounded interval, F is the uniform limit of polynomials, and hence $f \in H_0$ implies $F(f) \in H$. In particular, $f_n = [1 \wedge (f - a) \vee 0]^{1/n}$ is in H. Note that $f_1 \le f_2 \le \cdots \le 1$, and hence

$$\chi_{\{f > a\}} = \lim_{n\to\infty} f_n \in H. \tag{4.1}$$

Similarly, for $f_1, \ldots, f_m \in H_0$,

$$\chi_{\{f_1 > a_1, \ldots, f_m > a_m\}} \in H \tag{4.2}$$

and, since $\sigma(H_0) = \sigma(\{\{f_1 > a_1, \ldots, f_m > a_m\}: f_i \in H_0,\ a_i \in \mathbb{R}\})$, the corollary follows. □

We give an additional application of Theorem 4.3.

4.5 Proposition Let E_1 and E_2 be separable metric spaces. Let X be an E_1-valued random variable defined on $(\Omega, \mathscr{F}, P)$, and let $\mathscr{H}$ be a sub-σ-algebra of $\mathscr{F}$. Then for each $\psi \in B(E_1 \times E_2)$ there is a bounded $\mathscr{B}(E_2) \times \mathscr{H}$-measurable function φ such that

$$E[\psi(X, Y) \mid \mathscr{H}](\omega) = \varphi(Y(\omega), \omega), \tag{4.3}$$

for every $\mathscr{H}$-measurable, E_2-valued random variable Y. If X is independent of $\mathscr{H}$, then φ does not depend on ω. Specifically,

$$\varphi(y, \omega) = \varphi(y) = E[\psi(X, y)]. \tag{4.4}$$

Proof. If $\psi(x, y) = g(x)h(y)$, then $\varphi(y, \cdot) = h(y)E[g(X) \mid \mathscr{H}]$. Let H be the collection of $\psi \in B(E_1 \times E_2)$ for which the conclusion of the proposition is valid, and let $\mathscr{S} = \{A \times B: A \in \mathscr{B}(E_1), B \in \mathscr{B}(E_2)\}$. Since $\chi_{A \times B}(x, y) = \chi_A(x)\chi_B(y)$, the proposition follows by Theorem 4.3. □

5. GRONWALL'S INEQUALITY

5.1 Theorem Let μ be a Borel measure on $[0, \infty)$, let $\varepsilon \geq 0$, and let f be a Borel measurable function that is bounded on bounded intervals and satisfies

$$0 \leq f(t) \leq \varepsilon + \int_{[0, t)} f(s)\mu(ds), \qquad t \geq 0. \tag{5.1}$$

Then

$$f(t) \leq \varepsilon e^{\mu[0, t)}, \qquad t \geq 0. \tag{5.2}$$

In particular, if $M > 0$ and

$$0 \leq f(t) \leq \varepsilon + M \int_0^t f(s)\, ds, \qquad t \geq 0, \tag{5.3}$$

then

$$f(t) \leq \varepsilon e^{Mt}, \qquad t \geq 0. \tag{5.4}$$

Proof. Iterating (5.1) gives

$$\begin{aligned} f(t) &\leq \varepsilon + \varepsilon \sum_{k=1}^{\infty} \int_{[0, t)} \int_{[0, s_1)} \cdots \int_{[0, s_{k-1})} \mu(ds_k) \cdots \mu(ds_1) \\ &\leq \varepsilon + \varepsilon \sum_{k=1}^{\infty} \frac{1}{k!} (\mu[0, t))^k \\ &= \varepsilon e^{\mu[0, t)}. \end{aligned} \tag{5.5}$$

□

6. THE WHITNEY EXTENSION THEOREM

For $x \in \mathbb{R}^d$ and $\alpha \in \mathbb{Z}_+^d$, let $x^\alpha = \prod_{k=1}^d x_k^{\alpha_k}$, $|\alpha| = \sum_{k=1}^d \alpha_k$, and $\alpha! = \prod_{k=1}^d \alpha_k!$. Similarly, if D_k denotes differentiation in the kth variable,

$$D^\alpha f = \prod_{k=1}^{d} D_k^{\alpha_k} f, \tag{6.1}$$

and if f is r times continuously differentiable on a convex, open set in $\mathbb{R}^d$, then by Taylor's theorem

$$\begin{aligned} D^\alpha f(y) = &\sum_{|\beta| \le r-|\alpha|} \frac{1}{\beta!} D^{\alpha+\beta} f(x)(y-x)^\beta \\ &+ \sum_{|\beta| = r-|\alpha|} \frac{r-|\alpha|}{\beta!} \int_0^1 (1-u)^{r-|\alpha|-1} [D^{\alpha+\beta} f(x+u(y-x)) \\ &\qquad - D^{\alpha+\beta} f(x)]\, du\, (y-x)^\beta. \end{aligned} \tag{6.2}$$

6.1 Theorem Let $E \subset \mathbb{R}^d$ be closed. Suppose a collection of functions $\{f_\alpha \colon \alpha \in \mathbb{Z}_+^d, |\alpha| \le r\}$ satisfies $f_\alpha \colon E \to \mathbb{R}$ for each α,

$$f_\alpha(y) = \sum_{|\beta| \le r-|\alpha|} \frac{1}{\beta!} f_{\alpha+\beta}(x)(y-x)^\beta + R_\alpha(x, y) \tag{6.3}$$

for all $x, y \in E$, and for each compact set $K \subset \mathbb{R}^d$,

$$\lim_{\delta \to 0} \sup \left\{ \frac{|R_\alpha(x, y)|}{|x-y|^{r-|\alpha|}} \colon x, y \in E \cap K, |x-y| < \delta \right\} = 0. \tag{6.4}$$

Then there exists $f \in C^r(\mathbb{R}^d)$ such that $f|_E = f_0$.

6.2 Remark Essentially the theorem states that a function f_0 that is r times continuously differentiable on E can be extended to a function that is r times continuously differentiable on $\mathbb{R}^d$. □

Proof. The theorem is due to Whitney (1934). For a more recent exposition see Abraham and Robbin (1967), Appendix A. □

6.3 Corollary Let E be convex, and suppose E is the closure of its interior E°. Suppose f_0 is r times continuously differentiable on E° and that the derivatives $D^\alpha f_0$ are uniformly continuous on E°. Then there exists $f \in C^r(\mathbb{R}^d)$ such that $f|_{E^\circ} = f_0$.

Proof. Let $R_\alpha(x, y)$ be the remainder (second) term on the right of (6.2). There exists a constant C such that for $x, y \in E^\circ$,

$$|R_\alpha(x, y)| \le C|x-y|^{r-|\alpha|} w(|x-y|), \tag{6.5}$$

where

$$w(\delta) = \max_{|\alpha|=r} \sup_{\substack{x, y \in E^\circ \\ |x-y|<\delta}} |D^\alpha f(y) - D^\alpha f(x)|. \tag{6.6}$$

By continuity (6.5) extends to all $x, y \in E$. Since $\lim_{\delta \to 0} w(\delta) = 0$, the corollary follows. □

7. APPROXIMATION BY POLYNOMIALS

In a variety of contexts it is useful to approximate a function $f \in C^r(\mathbb{R}^d)$ by polynomials in such a way that not only f but all of its derivatives of order less than or equal to r are approximated uniformly on compact sets. To obtain such approximations, one need only construct a sequence of polynomials $\{\rho_n\}$ that are approximate delta functions in the sense that for every $f \in C_c(\mathbb{R}^d)$,

$$\lim_{n\to\infty} \int f(y)\rho_n(x-y)\, dy = f(x), \tag{7.1}$$

and the convergence is uniform for x in compact sets. Such a sequence can be constructed in a variety of ways. A simple example is

$$\rho_n(z) = n^d \left(1 - \frac{|z|^2}{n^2}\right)^{n^4} \pi^{-d/2}. \tag{7.2}$$

To see that this sequence has the desired property, first note that

$$\sup_{|u| \le n^2} e^{|u|^2} \left(1 - \frac{|u|^2}{n^4}\right)^{n^4} = 1. \tag{7.3}$$

For x in a fixed compact set and n sufficiently large

$$\begin{aligned} \int_{\mathbb{R}^d} f(y)\rho_n(x-y)\, dy &= \int_{\mathbb{R}^d} f\left(x - \frac{1}{n}u\right)\left(1 - \frac{|u|^2}{n^4}\right)^{n^4} \pi^{-d/2}\, du \\ &= \int_{|u| \le n^2} f\left(x - \frac{1}{n}u\right) e^{-|u|^2} e^{|u|^2} \left(1 - \frac{|u|^2}{n^4}\right)^{n^4} \pi^{-d/2}\, du. \end{aligned} \tag{7.4}$$

The second equality follows from the fact that f has compact support, and (7.1) follows by the dominated convergence theorem.

7.1 Proposition Let $f \in C^r(\mathbb{R}^d)$. Then for each compact set K and $\varepsilon > 0$, there exists a polynomial p such that for every $|\alpha| \le r$,

$$\sup_{x \in K} |D^\alpha f(x) - D^\alpha p(x)| \le \varepsilon. \tag{7.5}$$

Proof. Without loss of generality we can assume f has compact support. (Replace f by $\zeta \cdot f$, where $\zeta \in C_c^\infty(\mathbb{R}^d)$ and $\zeta = 1$ on K.) Take

$$p_n(x) = \int_{\mathbb{R}^d} f(y)\rho_n(x - y)\, dy = \int_{\mathbb{R}^d} f(x - y)\rho_n(y)\, dy \tag{7.6}$$

and note

$$D^\alpha p_n(x) = \int_{\mathbb{R}^d} D^\alpha f(x - y)\rho_n(y)\, dy = \int_{\mathbb{R}^d} D^\alpha f(y)\rho_n(x - y)\, dy. \tag{7.7}$$

For n sufficiently large, p_n will have the desired properties. □

As an application of the previous result we have the following.

7.2 Proposition Let φ be convex on $\mathbb{R}^d$. Then for each compact, convex set K and $\varepsilon > 0$, there exists a polynomial p such that p is convex on K and

$$\sup_{x \in K} |\varphi(x) - p(x)| \le \varepsilon. \tag{7.8}$$

Proof. Let $\rho \in C_c^\infty(\mathbb{R}^d)$ be nonnegative and

$$\int_{\mathbb{R}^d} \rho(y)\, dy = 1. \tag{7.9}$$

Then for n sufficiently large,

$$\varphi_1(x) \equiv \int_{\mathbb{R}^d} \varphi(y) n^d \rho(n(x - y))\, dy \tag{7.10}$$

is infinitely differentiable, convex, and satisfies

$$\sup_{x \in K} |\varphi(x) - \varphi_1(x)| \le \frac{\varepsilon}{3}. \tag{7.11}$$

For δ sufficiently small, $\varphi_2(x) \equiv \varphi_1(x) + \delta |x|^2$ satisfies

$$\sup_{x \in K} |\varphi(x) - \varphi_2(x)| \le \frac{2\varepsilon}{3}. \tag{7.12}$$

Recall that a function $\psi \in C^2(\mathbb{R}^d)$ is convex on K if and only if the Hessian matrix $((D_i D_j \psi))$ is nonnegative definite. Note that $((D_i D_j \varphi_2))$ is positive definite. In particular

$$\sum z_i z_j D_i D_j \varphi_2(x) \geq \delta |z|^2. \tag{7.13}$$

By Proposition 7.1 there exists a polynomial p such that

$$\sup_{x \in K} |\varphi_2(x) - p(x)| \leq \frac{\varepsilon}{3}, \tag{7.14}$$

and $D_i D_j p$ approximates $D_i D_j \varphi_2$ closely enough so that

$$\sum z_i z_j D_i D_j p(x) \geq \frac{\delta |z|^2}{2}, \qquad x \in K. \tag{7.15}$$

Consequently, p is convex on K, and (7.12) and (7.14) imply p satisfies (7.8). □

8. BIMEASURES AND TRANSITION FUNCTIONS

Let $(M, \mathcal{M})$ be a measurable space and (E, r) a complete, separable metric space. A function $\nu_0(A, B)$ defined for $A \in \mathcal{M}$ and $B \in \mathcal{B}(E)$ is a *bimeasure* if for each $A \in \mathcal{M}$, $\nu_0(A, \cdot)$ is a measure on $\mathcal{B}(E)$ and for each $B \in \mathcal{B}(E)$, $\nu_0(\cdot, B)$ is a measure on $\mathcal{M}$.

8.1 Theorem Let ν_0 be a bimeasure on $\mathcal{M} \times \mathcal{B}(E)$ such that $0 < \nu_0(M, E) < \infty$, and define $\mu = \nu_0(\cdot, E)$. Then there exists $\eta: M \times \mathcal{B}(E) \to [0, \infty)$ such that for each $x \in M$, $\eta(x, \cdot)$ is a measure on $\mathcal{B}(E)$, for each $B \in \mathcal{B}(E)$, $\eta(\cdot, B)$ is $\mathcal{M}$-measurable, and

$$\nu_0(A, B) = \int_A \eta(x, B)\mu(dx), \qquad A \in \mathcal{M},\ B \in \mathcal{B}(E). \tag{8.1}$$

Furthermore,

$$\nu(C) \equiv \iint \chi_C(x, y)\eta(x, dy)\mu(dx) \tag{8.2}$$

defines a measure on the product σ-algebra $\mathcal{M} \times \mathcal{B}(E)$ satisfying $\nu(A \times B) = \nu_0(A, B)$ for all $A \in \mathcal{M}$, $B \in \mathcal{B}(E)$.

8.2 Remark The first part of the theorem is essentially just the existence of a regular conditional distribution. The observation that a bimeasure (as defined by Kingman (1967)) determines a measure on the product σ-algebra is due to Morando (1969), page 224.

Proof. Without loss of generality, we can assume $\nu_0(M, E) = 1$ (otherwise replace $\nu_0(A, B)$ by $\nu_0(A, B)/\nu_0(M, E)$). Let $\{x_i\}$ be a countable dense subset of E, and let $B_1, B_2, \ldots$ be an ordering of $\{B(x_i, k^{-1})\colon i = 1, 2, \ldots, k = 1, 2, \ldots\}$.

For each $B \in \mathscr{B}(E)$, $\nu_0(\cdot, B) \ll \mu$, so there exists $\eta_0(\cdot, B)$, $\mathscr{M}$-measurable, such that

$$(8.3) \qquad \nu_0(A, B) = \int_A \eta_0(x, B)\mu(dx), \qquad A \in \mathscr{M}.$$

We can always assume $\eta_0(x, B) \le 1$, and for fixed B, C, with $B \subset C$, we can assume $\eta_0(x, B) \le \eta_0(x, C)$ for all x. Therefore we may define $\eta_0(x, E) = 1$, select $\eta_0(x, B_1)$ satisfying (8.3) (with $B = B_1$), and define $\eta_0(x, B_1^c) = 1 - \eta_0(x, B_1)$, which satisfies (8.3) with $B = B_1^c$. For any sequence $C_1, C_2, \ldots$ where C_i is B_i or B_i^c, working recursively we can select $\eta_0(x, C_1 \cap C_2 \cap \cdots \cap C_k \cap B_{k+1})$ satisfying (8.3) with $B = C_1 \cap C_2 \cap \cdots \cap C_k \cap B_{k+1}$ and $\eta_0(x, C_1 \cap C_2 \cap \cdots \cap C_k \cap B_{k+1}) \le \eta_0(x, C_1 \cap C_2 \cap \cdots \cap C_k)$, and define

$$\eta_0(x, C_1 \cap C_2 \cap \cdots \cap C_k \cap B_{k+1}^c) = \eta_0(x, C_1 \cap C_2 \cap \cdots \cap C_k) - \eta_0(x, C_1 \cap C_2 \cap \cdots \cap C_k \cap B_{k+1}),$$

which satisfies (8.3) with $B = C_1 \cap C_2 \cap \cdots \cap C_k \cap B_{k+1}^c$. For $B \in \mathscr{F}_n = \sigma(B_1, \ldots, B_n)$, define $\eta_0(x, B) = \sum \eta_0(x, C_1 \cap C_2 \cap \cdots \cap C_n)$ where the sum is over $\{C_1 \cap C_2 \cap \cdots \cap C_n\colon C_i \text{ is } B_i \text{ or } B_i^c,\ C_1 \cap C_2 \cap \cdots \cap C_n \subset B\}$. Then $\eta_0(x, B)$ satisfies (8.3) and $\eta_0(x, \cdot)$ is finitely additive on $\bigcup_n \mathscr{F}_n$.

Let $\Gamma_n = \{C_1 \cap C_2 \cap \cdots \cap C_n\colon C_i \text{ is } B_i \text{ or } B_i^c\}$, and for $C \in \Gamma_n$ such that $C \ne \emptyset$, let $z_C \in C$. Define $\eta_n(x, \cdot) \in \mathscr{P}(E)$ by

$$(8.4) \qquad \eta_n(x, B) = \sum_{C \in \Gamma_n} \delta_{z_C}(B)\eta_0(x, C).$$

Note that for $B \in \mathscr{F}_n$, $\eta_n(x, B) = \eta_0(x, B)$. For $m = 1, 2, \ldots$ let K_m be compact and satisfy $\nu_0(M, K_m) \ge 1 - 2^{-m}$. For each m, there exists N_m such that for $n \ge N_m$ there is a $B \in \mathscr{F}_n$ satisfying $K_m \subset B \subset K_m^{1/m}$. Hence

$$(8.5) \qquad \int \inf_{n \ge N_m} \eta_n(x, K_m^{1/m})\mu(dx) \ge \int \eta_0(x, B)\mu(dx) \ge \nu_0(M, K_m) \ge 1 - 2^{-m}.$$

Therefore

$$(8.6) \qquad \mu\left\{x\colon \inf_{n \ge N_m} \eta_n(x, K_m^{1/m}) < 1 - m^{-1}\right\} \le m2^{-m},$$

and hence by Borel–Cantelli

$$(8.7) \qquad G = \left\{x\colon \varliminf_{n \to \infty} \eta_n(x, K_m^{1/m}) \ge 1 - m^{-1} \text{ for all but finitely many } m\right\}$$

satisfies $\mu(G) = 1$. It follows that for each $x \in G$, $\{\eta_n(x, \cdot)\}$ is relatively compact. Since $\lim_{n \to \infty} \eta_n(x, B) = \eta_0(x, B)$ for every $B \in \bigcup_n \mathscr{F}_n$, for $x \in G$ there exists

$\eta(x, \cdot)$ such that $\eta_n(x, \cdot) \Rightarrow \eta(x, \cdot)$. (See Problem 27 in Chapter 3.) By Theorem 3.1 of Chapter 3

$$(8.8) \qquad \lim_{n \to \infty} \int_A \eta_n(x, B)\mu(dx) = \int_A \eta(x, B)\mu(dx)$$

for all $B \in \mathcal{B}(E)$ such that

$$(8.9) \qquad \int_A \eta(x, \partial B)\mu(dx) = 0.$$

Since for $B \in \bigcup_n \mathcal{F}_n$

$$(8.10) \qquad \lim_{n \to \infty} \int_A \eta_n(x, B)\mu(dx) = \nu_0(A, B),$$

it follows from Problem 27 of Chapter 3 that (8.1) holds. $\qquad\qquad\square$

9. TULCEA'S THEOREM

9.1 Theorem   Let $(\Omega_k, \mathcal{F}_k)$, $k = 1, 2, \ldots$, be measurable spaces, $\Omega = \Omega_1 \times \Omega_2 \times \cdots$ and $\mathcal{F} = \mathcal{F}_1 \times \mathcal{F}_2 \times \cdots$. Let P_1 be a probability measure on $\mathcal{F}_1$, and for $k = 2, 3, \ldots$ let $P_k: \Omega_1 \times \cdots \times \Omega_{k-1} \times \mathcal{F}_k \to [0, 1]$ be such that for each $(\omega_1, \ldots, \omega_{k-1}) \in \Omega_1 \times \cdots \times \Omega_{k-1}$, $P_k(\omega_1, \ldots, \omega_{k-1}, \cdot)$ is a probability measure on $\mathcal{F}_k$, and for each $A \in \mathcal{F}_k$, $P_k(\cdot, A)$ is $\mathcal{F}_1 \times \cdots \times \mathcal{F}_{k-1}$-measurable. Then there is a probability measure P on $\mathcal{F}$ such that for $A \in \mathcal{F}_1 \times \cdots \times \mathcal{F}_k$,

$$(9.1) \quad P(A \times \Omega_{k+1} \times \Omega_{k+2} \times \cdots)$$

$$= \int_{\Omega_1} \cdots \int_{\Omega_k} \chi_A(\omega_1, \ldots, \omega_k) P_k(\omega_1, \ldots, \omega_{k-1}, d\omega_k) \cdots P_1(d\omega_1).$$

Proof.   The collection of sets

$$\mathcal{A} = \{A \times \Omega_{k+1} \times \Omega_{k+2} \times \cdots : A \in \mathcal{F}_1 \times \cdots \times \mathcal{F}_k, \quad k = 1, 2, \ldots\}$$

is an algebra. Clearly P defined by (9.1) is finitely additive on $\mathcal{A}$. To apply the Caratheodory extension theorem to extend P to a measure on $\sigma(\mathcal{A}) = \mathcal{F}$, we must show that P is countably additive on $\mathcal{A}$. (See Billingsley (1979), Theorem 3.1.)

   To verify countable additivity it is enough to show that $\{B_n\} \subset \mathcal{A}$, $B_1 \supset B_2 \supset \cdots$ and $\lim_{n \to \infty} P(B_n) > 0$ imply $\bigcap_n B_n \neq \emptyset$. Let $B_n = A_n \times \Omega_{k_n + 1}$

$\times\, \Omega_{k_n+2} \times \cdots$ for $A_n \in \mathscr{F}_1 \times \cdots \times \mathscr{F}_{k_n}$, $B_1 \supset B_2 \supset \cdots$, and $\lim_{n\to\infty} P(B_n) > 0$. (We can assume $k_n \to \infty$.) For $n = 1, 2, \ldots$ and $k < k_n$, define

$$(9.2) \qquad f_{k,n}(\omega_1, \ldots, \omega_k) = \int_{\Omega_{k+1}} \cdots \int_{\Omega_{k_n}} \chi_{A_n}(\omega_1, \ldots, \omega_{k_n}) \times P_{k_n}(\omega_1, \ldots, \omega_{k_n-1}, d\omega_{k_n}) \cdots P_{k+1}(\omega_1, \ldots, \omega_k, d\omega_{k+1}),$$

and for $k \geq k_n$, $f_{k,n}(\omega_1, \ldots, \omega_k) = \chi_{A_n}(\omega_1, \ldots, \omega_{k_n})$. Note that

$$(9.3) \quad f_{k,n}(\omega_1, \ldots, \omega_k) = \int f_{k+1,n}(\omega_1, \ldots, \omega_{k+1}) P_{k+1}(\omega_1, \ldots, \omega_k, d\omega_{k+1})$$

and

$$(9.4) \qquad P(B_n) = \int f_{1,n}(\omega_1) P_1(d\omega_1).$$

Furthermore note that $f_{k,n} \geq f_{k,n+1}$ so $g_k \equiv \text{bp-}\lim_{n\to\infty} f_{k,n}$ exists, and by the monotone convergence theorem,

$$(9.5) \qquad \lim_{n\to\infty} P(B_n) = \int g_1(\omega_1) P_1(d\omega_1)$$

and

$$(9.6) \quad g_k(\omega_1, \ldots, \omega_k) = \int g_{k+1}(\omega_1, \ldots, \omega_{k+1}) P_{k+1}(\omega_1, \ldots, \omega_k, d\omega_{k+1}).$$

If $\lim_{n\to\infty} P(B_n) > 0$, there must be $\tilde{\omega}_1 \in \Omega_1$ such that $g_1(\tilde{\omega}_1) > 0$ and by induction a sequence $\tilde{\omega}_1, \tilde{\omega}_2, \ldots$ such that $g_k(\tilde{\omega}_1, \ldots, \tilde{\omega}_k) > 0$, $k = 1, 2, \ldots$. Finally, since

$$(9.7) \qquad \begin{aligned} g_{k_n}(\tilde{\omega}_1, \ldots, \tilde{\omega}_{k_n}) &\leq f_{k_n,n}(\tilde{\omega}_1, \ldots, \tilde{\omega}_{k_n}) \\ &= \chi_{A_n}(\tilde{\omega}_1, \ldots, \tilde{\omega}_{k_n}), \end{aligned}$$

$(\tilde{\omega}_1, \tilde{\omega}_2, \ldots) \in B_n$ for every n, and hence $(\tilde{\omega}_1, \tilde{\omega}_2, \ldots) \in \bigcap_n B_n$. □

10. MEASURABLE SELECTIONS AND MEASURABILITY OF INVERSES

Let $(M, \mathscr{M})$ be a measurable space and (S, ρ) a complete, separable metric space. Suppose for each $x \in M$, $\Gamma_x \subset S$. A measurable selection of $\{\Gamma_x\}$ is an $\mathscr{M}$-measurable function $f\colon M \to S$ such that $f(x) \in \Gamma_x$ for every $x \in M$.

10.1 Theorem Suppose for each $x \in M$, Γ_x is a closed subset of S and that for every open set $U \subset S$, $\{x \in M\colon \Gamma_x \cap U \neq \varnothing\} \in \mathscr{M}$. Then there exist $f_n\colon M \to S$, $n = 1, 2, \ldots$, such that f_n is $\mathscr{M}$-measurable, $f_n(x) \in \Gamma_x$ for every $x \in M$, and Γ_x is the closure of $\{f_1(x), f_2(x), \ldots\}$.

10.2 Remark Regarding $x \to \Gamma_x$ as a set-valued function, if $\{x \in M: \Gamma_x \cap U \neq \varnothing\} \in \mathscr{M}$ for every open U, the function is said to be *weakly measurable.* The function is *measurable* if "open" can be replaced by "closed." The theorem not only gives the existence of a measurable selection, but also shows that any closed-set-valued, weakly measurable function has the representation (known as the Castaing representation) $\Gamma_x = \text{closure}\{f_1(x), f_2(x), \ldots\}$ for some countable collection of $\mathscr{M}$-measurable functions. □

Proof. See Himmelberg (1975), Theorem 5.6. Earlier versions of the result are in Castaing (1967) and Kuratowski and Ryll-Nardzewski (1965). □

10.3 Corollary Suppose $(M, \mathscr{M}) = (E, \mathscr{B}(E))$ for a metric space E. If $y_n \in \Gamma_{x_n}$, $n = 1, 2, \ldots$, and $\lim_{n\to\infty} x_n = x$ imply that $\{y_n\}$ has a limit point in Γ_x, then there is a measurable selection of $\{\Gamma_x\}$.

10.4 Remark The assumptions of the corollary imply that for $K \subset E$ compact, $\bigcup_{x \in K} \Gamma_x$ is compact. □

Proof. Note that for a closed set F, $\{x: \Gamma_x \cap F \neq \varnothing\}$ is closed, hence measurable. If U is open, then $U = \bigcup_n F_n$ for some sequence of closed sets $\{F_n\}$, and hence $\{x: \Gamma_x \cap U = \varnothing\} = \bigcup_n \{x: \Gamma_x \cap F_n = \varnothing\}$ is measurable. □

For a review of results on measurable selections, see Wagner (1977).

One source of set-valued functions is the inverse mapping of a given function $\varphi: E_1 \to E_2$, that is, for $x \in E_2$ take $\Gamma_x = \varphi^{-1}(x) = \{y \in E_1: \varphi(y) = x\}$. If φ is one-to-one, then the existence of a measurable selection is precisely the measurability of the inverse function. The following theorem of Kuratowski gives conditions for this measurability.

10.5 Theorem Let (S_1, ρ_1) and (S_2, ρ_2) be complete, separable metric spaces. Let $E_1 \in \mathscr{B}(S_1)$, and let $\varphi: E_1 \to S_2$ be Borel measurable and one-to-one. Then $E_2 \equiv \varphi(E_1) \equiv \{\varphi(x): x \in E_1\}$ is a Borel subset of S_2 and φ^{-1} is a Borel measurable function from E_2 onto E_1.

Proof. See Theorem 3.9 and Corollary 3.3 of Chapter I of Parthasarathy (1967). □

11. ANALYTIC SETS

Let $\mathbb{N}$ denote the set of positive integers and $\mathscr{N} = \mathbb{N}^\infty$. We give $\mathbb{N}$ the discrete topology and $\mathscr{N}$ the corresponding product topology. Let (S, ρ) be a complete, separable metric space. A subset $A \subset S$ is *analytic* if there exists a continuous function φ mapping $\mathscr{N}$ onto A.

11.1 Proposition Every Borel subset of a complete, separable metric space is analytic.

Proof. See Theorem 2.5 of Parthasarathy (1967). □

Analytic sets arise most naturally as images of Borel sets.

11.2 Proposition Let (S_1, ρ_1) and (S_2, ρ_2) be complete, separable metric spaces and let $\varphi: S_1 \to S_2$ be Borel measurable. If $A \in \mathscr{B}(S_1)$, then $\varphi(A) = \{\varphi(x): x \in A\}$ is an analytic subset of S_2.

Proof. See Theorem 3.4 of Parthasarathy (1967). □

11.3 Theorem Let (S, ρ) be a complete, separable metric space and let $(\Omega, \mathscr{F}, P)$ be a complete probability space. If Y is an S-valued random variable defined on $(\Omega, \mathscr{F}, P)$ and A is an analytic subset of S, then $\{Y \in A\} \in \mathscr{F}$.

Proof. See Dellacherie and Meyer (1978), page 58. The definition of analytic set used there is more general than that given above. The role of the paved set $(F, \mathscr{F})$ in the definition in Dellacherie and Meyer (page 41) is taken by $(S, \mathscr{B}(S))$, and the auxiliary compact space E is $(\mathbb{N}^\Delta)^\infty$, where $\mathbb{N}^\Delta$ is the one-point compactification of $\mathbb{N}$. Let $B \subset E \times S$ be given by $B = \{(x, \varphi(x)): x \in \mathbb{N}^\infty\}$, where φ is continuous on $\mathbb{N}^\infty$. Then for $\{z_i\}$ dense in S, $B = \bigcap_n \bigcup_i \bigcup_m cl\{x \in \mathbb{N}^\infty: |x_j| \le m, j = 1, \ldots, n, \varphi(x) \in B(z_i, n^{-1})\} \times B(z_i, n^{-1})$, where cl denotes the closure in $(\mathbb{N}^\Delta)^\infty$. Consequently $B \in (\mathscr{K}(E) \times \mathscr{B}(S))_{\sigma\delta}$ ($\mathscr{K}(E)$ is the class of compact subsets of E) and A is the projection onto S of B, so A is $\mathscr{B}(S)$-analytic in the terminology of Dellacherie and Meyer. □

REFERENCES

Abraham, Ralph and Robbin, Joel (1967). *Transversal Mappings and Flows.* Benjamin, New York.

Aldous, David (1978). Stopping times and tightness. *Ann. Probab.* **6**, 335–340.

Alexandroff, A. D. (1940–1943). Additive set functions in abstract spaces. *Mat. Sb.* **8**, 307–348; **9**, 563–628; **13**, 169–238.

Allain, Marie-France (1976). Étude de la vitesse de convergence d'une suite de processus de Markov de saut pur. *C. R. Acad. Sci. Paris* **282**, 1015–1018.

Alm, Sven Erick (1978). On the rate of convergence in diffusion approximation of jump Markov processes. Technical Report, Department of Mathematics, Uppsala University, Sweden.

Anderson, Robert F. (1976). Diffusion with second order boundary conditions, I, II. *Indiana Univ. Math. J.* **25**, 367–395, 403–441.

Artstein, Zvi (1983). Distributions of random sets and random selections. *Israel J. Math.* **46**, 313–324.

Athreya, Krishna B. and Ney, Peter E. (1972). *Branching Processes.* Springer-Verlag, Berlin.

Barbour, Andrew D. (1974). On a functional central limit theorem for Markov population processes. *Adv. Appl. Probab.* **6**, 21–39.

Barbour, Andrew D. (1976). Second order limit theorems for the Markov branching process in random environments. *Stochastic Process. Appl.* **4**, 33–40.

Barbour, Andrew D. (1980). Equilibrium distributions for Markov population processes. *Adv. Appl. Probab.* **12**, 591–614.

Bártfai, Pál (1966). Die Bestimmung des zu einem wiederkehrenden Prozess gehorenden Verteilungsfunktion aus den mit Fehlern behaften Daten eiher Einziger Relation. *Stud. Sci. Math. Hung.* **1**, 161–168.

Beneš, Václav Edvard (1968). Finite regular invariant measures for Feller processes. *J. Appl. Probab.* **5**, 203–209.

Bhattacharya, Rabi N. (1982). On the functional central limit theorem and the law of the iterated logarithm for Markov processes. *Z. Wahrsch. verw. Gebiete* **60**, 185–201.

Billingsley, Patrick (1968). *Convergence of Probability Measures.* Wiley, New York.

Billingsley, Patrick (1979). *Probability and Measure.* Wiley, New York.

Blackwell, David and Dubins, Lester E. (1983). An extension of Skorohod's almost sure representation theorem. *Proc. Amer. Math. Soc.* **89**, 691–692.

Blankenship, Gilmer L. and Papanicolaou, George C. (1978). Stability and control of stochastic systems with wide band noise disturbances I. *SIAM J. Appl. Math.* **34**, 437–476.

Borovkov, A. A. (1970). Theorems on the convergence to Markov diffusion processes. *Z. Wahrsch. verw. Gebiete* **16**, 47–76.

Breiman, Leo (1968). *Probability.* Addison-Wesley, Reading, Mass.

Brown, Bruce M. (1971). Martingale central limit theorems. *Ann. Math. Statist.* **42**, 59–66.

Brown, Timothy C. (1978). A martingale approach to the Poisson convergence of simple point processes. *Ann. Probab.* **6**, 615–628.

Castaing, Charles (1967). Sur les multi-applications mesurables. *Rev. Francaise Inf. Rech. Opéra.* **1**, 91–126.

Chenčov, N. N. (1956). Weak convergence of stochastic processes whose trajectories have no discontinuities of the second kind and the heuristic approach to the Kolmogorov–Smirnov tests. *Theory Probab. Appl.* **1**, 140–149.

Chernoff, Herman (1956). Large sample theory: parametric case. *Ann. Math. Statist.* **27**, 1–22.

Chernoff, Paul R. (1968). Note on product formulas for operator semigroups. *J. Funct. Anal.* **2**, 238–242.

Chow, Yuah Shih (1960). Martingales in a σ-finite measure space indexed by directed sets. *Trans. Amer. Math. Soc.* **97**, 254–285.

Chung, Kai Lai and Williams, Ruth J. (1983). *Introduction to Stochastic Integration.* Birkhauser, Boston.

Cohn, Donald L. (1980). *Measure Theory.* Birkhauser, Boston.

Costantini, Cristina, Gerardi, Anna and Nappo, Giovanna (1982). On the convergence of sequences of stationary jump Markov processes. *Statist. Probab. Lett.* **1**, 155–160.

Costantini, Cristina and Nappo, Giovanna (1982). Some results on weak convergence of jump Markov processes and their stability properties. *Systems Control Lett.* **2**, 175–183.

Courrège, Philippe (1963). Intégrales stochastiques et martingales de carré integrable. Séminaire Brelot-Choquet-Deny, 7th year.

Cox, E. Theodore, and Rösler, Uwe (1982). A duality relation for entrance and exit laws for Markov processes. *Stochastic Process. Appl.* **16**, 141–151.

Crandall, Michael G. and Liggett, Thomas M. (1971). Generation of semi-groups of nonlinear transformations on general Banach spaces. *Amer. J. Math.* **93**, 265–298.

Csörgő, Miklós and Révész, Pál (1981). *Strong Approximations in Probability and Statistics.* Academic, New York.

Darden, Thomas and Kurtz, Thomas G. (1986). Nearly deterministic Markov processes near a stable point. To appear.

Davies, Edward Brian (1980). *One-Parameter Semigroups.* Academic, London.

Davydov, Yu. A. (1968). Convergence of distributions generated by stationary stochastic processes. *Theory Probab. Appl.* **13**, 691–696.

Dawson, Donald A. (1975). Stochastic evolution equations and related measure processes. *J. Multivar. Anal.* **5**, 1–52.

Dawson, Donald A. (1977). The critical measure diffusion process. *Z. Wahrsch. verw. Gebiete* **40**, 125–145.

Dawson, Donald A. (1979). Stochastic measure diffusion processes. *Canad. Math. Bull.* **22**, 129–138.

Dawson, Donald A. and Hochberg, Kenneth J. (1979). The carrying dimension of a measure diffusion process. *Ann. Probab.* **7**, 693–703.

Dawson, Donald A. and Kurtz, Thomas G. (1982). Applications of duality to measure-valued processes. *Advances in Filtering and Optimal Stochastic Control. Lect. Notes Cont. Inf. Sci.* **42**, Springer-Verlag, Berlin, pp. 177–191.

Dellacherie, Claude and Meyer, Paul-André (1978). *Probabilities and Potential*. North-Holland, Amsterdam.

Dellacherie, Claude and Meyer, Paul-André (1982). *Probabilities and Potential B*. North-Holland, Amsterdam.

Doleans-Dade, Catherine (1969). Variation quadratique des martingales continues à droite. *Ann. Math. Statist.* **40**, 284–289.

Doleans-Dade, Catherine and Meyer, Paul-André (1970). Intégrales stochastiques par rapport aux martingales locales. *Séminaire de Probabilités IV, Lect. Notes Math.*, **124**, Springer-Verlag, Berlin.

Donsker, Monroe D. (1951). An invariance principle for certain probability limit theorems. *Mem. Amer. Math. Soc.* **6**.

Doob, Joseph L. (1953). *Stochastic Processes*. Wiley, New York.

Dudley, Richard M. (1968). Distances of probability measures and random variables. *Ann. Math. Statist.* **39**, 1563–1572.

Dunford, Nelson and Schwartz, Jacob T. (1957). *Linear Operators Part I: General Theory*. Wiley-Interscience, New York.

Dvoretzky, Aryeh (1972). Asymptotic normality for sums of dependent random variables. *Proc. Sixth Berkeley Symp. Math. Statist. Prob.* **2**, University of California Press, pp. 513–535.

Dynkin, E. B. (1961). *Theory of Markov Processes*. Prentice-Hall, Englewood Cliffs, N.J.

Dynkin, E. B. (1965). *Markov Processes I, II*. Springer-Verlag, Berlin.

Echeverria, Pedro E. (1982). A criterion for invariant measures of Markov processes. *Z. Wahrsch. verw. Gebiete* **61**, 1–16.

Elliot, Robert J. (1982). *Stochastic Calculus and Applications*. Springer-Verlag, New York.

Ethier, Stewart N. (1976). A class of degenerate diffusion processes occurring in population genetics. *Comm. Pure Appl. Math.* **29**, 483–493.

Ethier, Stewart N. (1978). Differentiability preserving properties of Markov semigroups associated with one-dimensional diffusions. *Z. Wahrsch. verw. Gebiete* **45**, 225–238.

Ethier, Stewart N. (1979). Limit theorems for absorption times of genetic models. *Ann. Probab.* **7**, 622–738.

Ethier, Stewart N. (1981). A class of infinite-dimensional diffusions occurring in population genetics. *Indiana Univ. Math. J.* **30**, 925–935.

Ethier, Stewart N. and Kurtz, Thomas G. (1981). The infinitely-many-neutral-alleles diffusion model. *Adv. Appl. Probab.* **13**, 429–452.

Ethier, Stewart N. and Kurtz, Thomas G. (1986). The infinitely-many-alleles model with selection as a measure-valued diffusion. To appear.

Ethier, Stewart N. and Nagylaki, Thomas (1980). Diffusion approximation of Markov chains with two time scales and applications to population genetics. *Adv. Appl. Probab.* **12**, 14–49.

Ewens, Warren J. (1972). The sampling theory of selectively neutral alleles. *Theor. Pop. Biol.* **3**, 87–112.

Ewens, Warren J. (1979). *Mathematical Population Genetics*. Springer-Verlag, Berlin.

Feller, William (1951). Diffusion processes in genetics. *Proc. Second Berkeley Symp. Math. Statist. Prob.*, University of California Press, Berkeley, pp. 227–246.

Feller, William (1952). The parabolic differential equations and the associated semi-groups of transformations. *Ann. Math.* **55**, 468–519.

Feller, William (1953). On the generation of unbounded semi-groups of bounded linear operators. *Ann. Math.* **58**, 166–174.

Feller, William (1971). *An Introduction to Probability Theory and Its Applications II*, 2nd ed., Wiley, New York.

Fiorenza, Renato (1959). Sui problemi di derivata obliqua per le equazioni ellittiche. *Ric. Mat.* **8**, 83–110.

Fisher, R. A. (1922). On the dominance ratio. *Proc. Roy. Soc. Edin.* **42**, 321–431.

Fleming, Wendall H. and Viot, Michel (1979). Some measure-valued Markov processes in population genetics theory. *Indiana Univ. Math. J.* **28**, 817–843.

Freidlin, M. I. (1968). On the factorization of non-negative definite matrices. *Theory Probab. Appl.* **13**, 354–356.

Friedman, Avner (1975). *Stochastic Differential Equations I*. Academic, New York.

Gänsler, Peter and Häusler, Erich (1979). Remarks on the functional central limit theorem for martingales. *Z. Wahrsch. verw. Gebiete* **50**, 237–243.

Gihman, I. I. and Skorohod, A. V. (1969). *Introduction to the Theory of Random Processes*. W. B. Saunders Co., Philadelphia.

Gihman, I. I. and Skorohod, A. V. (1972). *Stochastic Differential Equations*. Springer-Verlag, Berlin.

Gihman, I. I. and Skorohod, A. V. (1974). *The Theory of Stochastic Processes I*. Springer-Verlag, Berlin.

Goldstein, Jerome A. (1976). Semigroup-theoretic proofs of the central limit theorem and other theorems of analysis. *Semigroup Forum* **12**, 189–206.

Goldstein, Sydney (1951). On diffusion by discontinuous movements, and on the telegraph equation. *Quart. J. Mech. Appl. Math.* **4**, 129–156.

Gray, Lawrence and Griffeath, David (1977a). On the uniqueness and nonuniqueness of proximity processes. *Ann. Probab.* **5**, 678–692.

Gray, Lawrence and Griffeath, David (1977b). Unpublished manuscript.

Griego, Richard J. and Hersh, Reuben (1969). Random evolutions, Markov chains, and systems of partial differential equations. *Proc. Nat. Acad. Sci. USA* **62**, 305–308.

Griego, Richard J. and Hersh, Reuben (1971). Theory of random evolutions with applications to partial differential equations. *Trans. Amer. Math. Soc.* **156**, 405–418.

Grigelionis, Bronius and Mikulevičius, R. (1981). On the weak convergence of random point processes. *Lithuanian Math. Trans.* **21**, 49–55.

Grimvall, Anders (1974). On the convergence of sequences of branching processes. *Ann. Probab.* **2**, 1027–1045.

Guess, Harry A. (1973). On the weak convergence of Wright–Fisher models. *Stochastic Process. Appl.* **1**, 287–306.

Gustafson, Karl (1966). A perturbation lemma. *Bull. Amer. Math. Soc.* **72**, 334–338.

Hall, P. (1935). On representatives of subsets. *J. London Math. Soc.* **10**, 26–30.

Hall, Peter and Heyde, C. C. (1980). *Martingale Limit Theory and Its Applications*. Academic, New York.

Hardin, Clyde (1985). A spurious Brownian motion. *Proc. Amer. Math. Soc.* **93**, 350.

Harris, Theodore E. (1976). On a class of set-valued Markov processes. *Ann. Probab.* **4**, 175–194.

Hasegawa, Minoru (1964). A note on the convergence of semi-groups of operators. *Proc. Japan Acad.* **40**, 262–266.

Helland, Inge S. (1978). Continuity of a class of random time transformations. *Stochastic Process. Appl.* **7**, 79–99.

Helland, Inge S. (1981). Minimal conditions for weak convergence to a diffusion process on the line. *Ann. Probab.* **9**, 429–452.

Helland, Inge S. (1982). Central limit theorems for martingales with discrete or continuous time. *Scand. J. Statist.* **9**, 79–94.

Helms, Lester L. (1974). Ergodic properties of several interacting Poisson particles. *Adv. Math.* **12**, 32–57.

Hersh, Reuben (1974). Random evolutions: a survey of results and problems. *Rocky Mt. J. Math.* **4**, 443–477.

Hersh, Reuben and Papanicolaou, George C. (1972). Non-commuting random evolutions, and an operator-valued Feynman–Kac formula. *Comm. Pure Appl. Math.* **25**, 337–367.

Heyde, C. C. (1974). On the central limit theorem for stationary processes. *Z. Wahrsch. verw. Gebiete* **30**, 315–320.

Hille, Einar (1948). *Functional Analysis and Semi-groups*, Am. Math. Soc. Colloq. Publ. **31**, New York.

Hille, Einar and Phillips, Ralph (1957). *Functional Analysis and Semi-groups*, rev. ed., Amer. Math. Soc. Colloq. Publ. **31**, Providence, R.I.

Himmelberg, C. J. (1975). Measurable relations. *Fund. Math.* **87**, 53–72.

Holley, Richard A. and Liggett, Thomas M. (1975). Ergodic theorems for weakly interacting systems and the voter model. *Ann. Probab.* **3**, 643–663.

Holley, Richard A. and Stroock, Daniel W. (1976). A martingale approach to infinite systems of interacting processes. *Ann. Probab.* **4**, 195–228.

Holley, Richard A. and Stroock, Daniel W. (1978). Generalized Ornstein–Uhlenbeck processes and infinite particle branching Brownian motions. *Publ. RIMS, Kyoto Univ.* **14**, 741–788.

Holley, Richard A. and Stroock, Daniel W. (1979). Dual processes and their applications to infinite interacting systems. *Adv. Math.* **32**, 149–174.

Holley, Richard A., Stroock, Daniel W., and Williams, David (1977). Applications of dual processes to diffusion theory. *Proc. Symp. Pure Math.* **31**, AMS, Providence, R.I., pp. 23–36.

Ibragimov, I. A. (1959). Some limit theorems for strict-sense stationary stochastic processes. *Dokl. Akad. Nauk SSSR*, **125**, 711–714.

Ibragimov, I. A. (1962). Some limit theorems for stationary processes. *Theory Probab. Appl.* **7**, 349–382.

Ibragimov, I. A. and Linnik, Yu. V. (1971). *Independent and Stationary Sequences of Random Variables.* Wolters-Noordhoff, Groningen.

Ikeda, Nobuyuki, Nagasawa, Masao, and Watanabe, Shinzo (1968, 1969). Branching Markov processes I, II, III. *J. Math. Kyoto* **8**, 233–278, 365–410; **9**, 95–160.

Ikeda, Nobuyuki and Watanabe, Shinzo (1981). *Stochastic Differential Equations and Diffusion Processes.* North Holland, Amsterdam.

Il'in, A. M., Kalashnikov, A. S., and Oleinik, O. A. (1962). Linear equations of the second order of parabolic type. *Russ. Math. Surveys* **17**, 1–143.

Itô, Kiyosi (1951). On stochastic differential equations. *Mem. Amer. Math. Soc.* **4**.

Itô, Kiyosi and Watanabe, Shinzo (1965). Transformations of Markov processes by multiplicative functionals. *Ann. Inst. Fourier* **15**, 15–30.

Jacod, Jean, Mémin, Jean, and Métivier, Michel (1983). On tightness and stopping times. *Stochastic Process. Appl.* **14**, 109–146.

Jagers, Peter (1971). Diffusion approximations of branching processes. *Ann. Math. Statist.* **42**, 2074–2078.

Jiřina, Miloslav (1969). On Feller's branching diffusion processes. *Časopis Pěst. Mat.* **94**, 84–90.

Joffe, Anatole and Metivier, Michel (1984). Weak convergence of sequences of semimartingales with applications to multitype branching processes. Tech. Rep., Université de Montréal.

Kabanov, Yu. M., Lipster, R. Sh., and Shiryaev, A. N. (1980). Some limit theorems for simple point processes (a martingale approach). *Stochastics* **3**, 203–216.

Kac, Mark (1956). Some stochastic problems in physics and mathematics. Magnolia Petroleum Co. Colloq. Lect. **2**.

Kac, Mark (1974). A stochastic model related to the telegrapher's equation. *Rocky Mt. J. Math.*, **4**, 497–509.

Kallman, Robert R. and Rota, Gian-Carlo (1970). On the inequality $\|f'\|^2 \le 4\|f\| \cdot \|f''\|$. *Inequalities, Vol. II*, Oved Shisha, Ed. Academic, New York, pp. 187–192.

Karlin, Samuel and Levikson, Benny (1974). Temporal fluctuations in selection intensities: Case of small population size. *Theor. Pop. Biol.* **6**, 383–412.

Kato, Tosio (1966). *Perturbation Theory for Linear Operators.* Springer-Verlag, New York.

Keiding, Niels (1975). Extinction and exponential growth in random environments. *Theor. Pop. Biol.* **8**, 49–63.

Kertz, Robert P. (1974). Perturbed semigroup limit theorems with applications to discontinuous random evolutions. *Trans. Amer. Math. Soc.* **199**, 29–53.

Kertz, Robert P. (1978). Limit theorems for semigroups with perturbed generators, with applications to multiscaled random evolutions. *J. Funct. Anal.* **27**, 215–233.

Khas'minskii, R. Z. (1960). Ergodic properties of recurrent diffusion processes and stabilization of the solution of the Cauchy problem for parabolic equations. *Theory Probab. Appl.* **5**, 179–196.

Khas'minskii, R. Z. (1966). A limit theorem for the solutions of differential equations with random right-hand sides. *Theory Probab. Appl.* **11**, 390–406.

Khas'minskii, R. Z. (1980). *Stochastic Stability of Differential Equations.* Sijthoff and Nordhoff.

Khintchine, A. (1933). *Asymptotische Gesetze der Wahrscheinlichkeitsrechnung.* Springer-Verlag, Berlin.

Kimura, Motoo and Crow, James F. (1964). The number of alleles that can be maintained in a finite population. *Genetics* **49**, 725–738.

Kingman, J. F. C. (1967). Completely random measures. *Pacific J. Math.* **21**, 59–78.

Kingman, J. F. C. (1975). Random discrete distributions. *J. R. Statist. Soc. B* **37**, 1–22.

Kingman, J. F. C. (1977). The population structure associated with the Ewens sampling formula. *Theor. Pop. Biol.* **11**, 274–283.

Kingman, J. F. C. (1980). *Mathematics of Genetic Diversity. CBMS-NSF Regional Conf. Series in Appl. Math.*, **34**. SIAM, Philadelphia.

Kolmogorov, A. N. (1956). On Skorohod convergence. *Theory Probab. Appl.* **1**, 213–222.

Komlós, János, Major, Peter, and Tusnády, Gábor (1975, 1976). An approximation of partial sums of independent RV's and the sample DF I, II. *Z. Wahrsch. verw. Gebiete* **32**, 111–131; **34**, 33–58.

Krylov, N. V. (1973). The selection of a Markov process from a system of processes and the construction of quasi-diffusion processes. *Math. USSR Izvestia* **7**, 691–709.

Kunita, Hiroshi and Watanabe, Shinzo (1967). On square integrable martingales. *Nagoya Math. J.* **30**, 209–245.

Kuratowski, K. and Ryll-Nardzewski, C. (1965). A general theorem on selectors. *Bull. Acad. Polon. Sci. Ser. Sci. Math. Astronom. Phys.* **13**, 397–403.

Kurtz, Thomas G. (1969). Extensions of Trotter's operator semigroup approximation theorems. *J. Funct. Anal.* **3**, 354–375.

Kurtz, Thomas G. (1970a). A general theorem on the convergence of operator semigroups. *Trans. Amer. Math. Soc.* **148**, 23–32.

Kurtz, Thomas G. (1970b). Solutions of ordinary differential equations as limits of pure jump Markov processes. *J. Appl. Probab.* **7**, 49–58.

Kurtz, Thomas G. (1971). Limit theorems for sequences of jump Markov processes approximating ordinary differential processes. *J. Appl. Probab.* **8**, 344–356.

Kurtz, Thomas G. (1973). A limit theorem for perturbed operator semigroups with applications to random evolutions. *J. Funct. Anal.* **12**, 55–67.

Kurtz, Thomas G. (1975). Semigroups of conditioned shifts and approximation of Markov processes. *Ann. Probab.* **3**, 618–642.

Kurtz, Thomas G. (1977). Applications of an abstract perturbation theorem to ordinary differential equations. *Houston J. Math.* **3**, 67–82.

Kurtz, Thomas G. (1978a). Strong approximation theorems for density dependent Markov chains. *Stochastic Process. Appl.* **6**, 223–240.

Kurtz, Thomas G. (1978b). Diffusion approximations for branching processes. *Branching Processes, Adv. Prob.* **5**, Anatole Joffe and Peter Ney, Eds., Marcel Dekker, New York, pp. 262–292.

Kurtz, Thomas G. (1980a). Representation of Markov processes as multiparameter time changes. *Ann. Probab.* **8**, 682–715.

Kurtz, Thomas G. (1980b). The optional sampling theorem for martingales indexed by directed sets. *Ann. Probab.* **8**, 675–681.

Kurtz, Thomas G. (1981a). *Approximation of Population Processes. CBMS-NSF Regional Conf. Series in Appl. Math.* **36**, SIAM, Philadelphia.

Kurtz, Thomas G. (1981b). The central limit theorem for Markov chains. *Ann. Probab.* **9**, 557–560.

Kurtz, Thomas G. (1981c). Approximation of discontinuous processes by continuous processes. *Stochastic Nonlinear Systems*, L. Arnold and R. Lefever, Eds. Springer-Verlag, Berlin, pp. 22–35.

Kurtz, Thomas G. (1982). Representation and approximation of counting processes. *Advances in Filtering and Optimal Stochastic Control, Lect. Notes Cont. Inf. Sci.* **42**, W. H. Fleming and L. G. Gorostiza, Eds., Springer-Verlag, Berlin.

Kushner, Harold J. (1974). On the weak convergence of interpolated Markov chains to a diffusion. *Ann. Probab.* **2**, 40–50.

Kushner, Harold J. (1979). Jump-diffusion approximation for ordinary differential equations with wide-band random right hand sides. *SIAM J. Control Optim.* **17**, 729–744.

Kushner, Harold J. (1980). A martingale method for the convergence of a sequence of processes to a jump-diffusion process on the line. *Z. Wahrsch. verw. Gebiete* **53**, 207–219.

Kushner, Harold J. (1982). Asymptotic distributions of solutions of ordinary differential equations with wide band noise inputs; approximate invariant measures. *Stochastics* **6**, 259–277.

Kushner, Harold J. (1984). *Approximation and Weak Convergence Methods for Random Processes.* MIT Press, Cambridge, Massachusetts.

Ladyzhenskaya, O. A. and Ural'tseva, N. N. (1968). *Linear and Quasilinear Elliptic Partial Differential Equations.* Academic, New York.

Lamperti, John (1967a). The limit of a sequence of branching processes. *Z. Wahrsch. verw. Gebiete* **7**, 271–288.

Lamperti, John (1967b). On random time substitutions and the Feller property. *Markov Processes and Potential Theory*, Joshua Chover, Ed., Wiley, New York, pp. 87–101.

Lamperti, John (1977). *Stochastic Processes.* Springer-Verlag, New York.

Lévy, Paul (1948). *Processus Stochastique et Mouvement Brownien.* Gauthier-Villars, Paris.

Liggett, Thomas M. (1972). Existence theorems for infinite particle systems. *Trans. Amer. Math. Soc.* **165**, 471–481.

Liggett, Thomas M. (1977). The stochastic evolution of infinite systems of interacting particles. *Lect. Notes Math.* **598**, 187–248. Springer-Verlag, New York.

Liggett, Thomas M. (1985). *Interacting Particle Systems.* Springer-Verlag, New York.

Lindvall, Torgny (1972). Convergence of critical Galton-Watson branching processes. *J. Appl. Probab.* **9**, 445–450.

Lindvall, Torgny (1974). Limit theorems for some functionals of certain Galton–Watson branching processes. *Adv. Appl. Probab.* **6**, 309–321.

Littler, Raymond A. (1972). Multidimensional stochastic models in genetics. Ph.D thesis, Monash Univ.

Littler, Raymond A. and Good, A. J. (1978). Ages, extinction times, and first passage probabilities for a multiallele diffusion model with irreversible mutation. *Theor. Pop. Biol.* **13**, 214–225.

Mackevičius, V. (1974). On the question of the weak convergence of random processes in the space $D[0, \infty)$. *Lithuanian Math. Trans.* **14**, 620–623.

Maigret, Nelly (1978). Théorème de limite centrale functionnel pour une chaine de Markov recurrente au sens de Harris et positive. *Ann. Inst. Henri Poincaré* **14**, 425–440.

Major, Peter (1976). The approximation of partial sums of independent RV's. *Z. Wahrsch. verw. Gebiete* **35**, 213–220.

Malek-Mansour, M., Van Den Broeck, C., Nicolis, G., and Turner, J. W. (1981). Asymptotic properties of Markovian master equations. *Ann. Phys.* **131**, 283–313.

Mann, Henry B. and Wald, Abraham (1943). On stochastic limit and order relations. *Ann. Math. Statist.* **14**, 217–226.

Mandl, Petr (1968). *Analytical Treatment of One-Dimensional Markov Processes.* Springer-Verlag, Berlin.

McKean, Henry P., Jr. (1969). *Stochastic Integrals.* Academic, New York.

McLeish, D. L. (1974). Dependent central limit theorems and invariance principles. *Ann. Probab.* **2**, 608–619.

Métivier, Michel (1982). *Semimartingales: A Course on Stochastic Processes.* Walter de Gruyter, Berlin.

Meyer, Paul-André (1966). *Probability and Potentials.* Blaisdell, Waltham, Mass.

Meyer, Paul-André (1967). Intégrales stochastiques I, II, III, IV. *Séminaire de Probabilités I. Lect. Notes Math.* **39**, 72–162.

Meyer, Paul-André (1968). Guide detaille de la theorie generale des processus. *Séminaire de Probabilités* II. *Lect. Notes Math.* **51**, 140–165.

Mirando, Carlo (1970). *Partial Differential Equations of Elliptic Type.* Springer-Verlag, Berlin.

Moran, P. A. P. (1958a). A general theory of the distribution of gene frequencies I. Overlapping generations. *Proc. Roy. Soc. London B* **149**, 102–111.

Moran, P. A. P. (1958b). A general theory of the distribution of gene frequencies. II. Non-overlapping generations. *Proc. Roy. Soc. London B* **149**, 113–116.

Moran, P. A. P. (1958c). Random processes in genetics. *Proc. Camb. Phil. Soc.* **54**, 60–71.

Morando, Philippe (1969). Mesures aleatoires. *Séminaire de Probabilités III, Lect. Notes Math.* **88**, 190–229. Springer-Verlag, Berlin.

Morkvenas, R. (1974), Convergence of Markov chains to solution of martingale problem. *Lithuanian Math. Trans.* **14**, 460–466.

Nagylaki, Thomas (1980). The strong-migration limit in geographically structured populations. *J. Math. Biol.* **9**, 101–114.

Nagylaki, Thomas (1982). Geographical invariance in population genetics. *J. Theor. Biol.* **99**, 159–172.

Neveu, J. (1958). Théorie des semi-groups de Markov. *Univ. California Publ. Statist.* **2**, 319–394.

Norman, M. Frank (1971). Slow learning with small drift in two-absorbing-barrier models. *J. Math. Psych.* **8**, 1–21.

Norman, M. Frank (1972). *Markov Processes and Learning Models.* Academic, New York.

Norman, M. Frank (1974). A central limit theorem for Markov processes that move by small steps. *Ann. Probab.* **2**, 1065–1074.

Norman, M. Frank (1975a). Diffusion approximation of non-Markovian processes. *Ann. Probab.* **3**, 358–364.

Norman, M. Frank (1975b). Limit theorems for stationary distributions. *Adv. Appl. Probab.* **7**, 561–575.

Norman, M. Frank (1977). Ergodicity of diffusion and temporal uniformity of diffusion approximation. *J. Appl. Probab.* **14**, 399–404.

Ohta, Tomoko and Kimura, Motoo (1969). Linkage disequilibrium due to random genetic drift. *Genet. Res. Camb.* **13**, 47–55.

Ohta, Tomoko and Kimura, Motoo (1973). A model of mutation appropriate to estimate the number of electrophoretically detectable alleles in a finite population. *Genet. Res. Camb.* **22**, 201–204.

Oleinik, Olga A. (1966). Alcuni risultati sulle equazioni lineari e quasi lineari ellitico-paraboliche a derivate parziali del second ordine. *Atti Accad. Naz. Lincei Rend. Cl. Sci. Fis. Mat. Natur. (8)* **40**, 775–784.

Papanicolaou, George C. and Kohler, W. (1974). Asymptotic theory of mixing stochastic ordinary differential equations. *Comm. Pure Appl. Math.* **27**, 641–668.

Papanicolaou, George C., Stroock, Daniel W., and Varadhan, S. R. S. (1977). Martingale approach to some limit theorems. *Conference on Statistical Mechanics, Dynamical Systems, and Turbulence, Duke University*, M. Reed, Ed., Duke Univ. Math. Series **3**.

Papanicolaou, George C. and Varadhan, S. R. S. (1973). A limit theorem with strong mixing in Banach space and two applications to stochastic differential equations. *Comm. Pure Appl. Math.* **26**, 497–524.

Parthasarathy, K. R. (1967). *Probability Measures on Metric Spaces.* Academic, New York.

Pazy, Amnon (1983). *Semigroups of Linear Operators and Applications to Partial Differential Equations.* Springer-Verlag, New York.

Peligrad, Magda (1982). Invariance principles for mixing sequences of random variables. *Ann. Probab.* **10**, 968–981.

Phillips, Ralph S. and Sarason, Leonard (1968). Elliptic-parabolic equations of the second order. *J. Math. Mech.* **17**, 891–917.

Pinsky, Mark A. (1968). Differential equations with a small parameter and the central limit theorem for functions defined on a finite Markov chain. *Z. Wahrsch. verw. Gebiete* **9**, 101–111.

Pinsky, Mark A. (1974). Multiplicative operator functionals and their asymptotic properties. *Advances in Probability* **3**, Dekker, New York.

Priouret, P. (1974). Processus de diffusion et equations différentielles stochastiques. *Lect. Notes Math.* **390**. Springer-Verlag, Berlin.

Prohorov, Yu. V. (1956). Convergence of random processes and limit theorems in probability theory. *Theory Probab. Appl.* **1**, 157–214.

Rao, C. Radhakrishna (1973). *Linear Statistical Inference and Its Applications*, 2nd ed. Wiley, New York.

Rebolledo, Rolando (1979). La méthod des martingales appliquée à l'étude de la convergence en loi de processus. *Bull. Soc. Math. France Mem.*, **62**.

Rebolledo, Rolando (1980). Central limit theorems for local martingales. *Z. Wahrsch. verw. Gebiete* **51**, 269–286.

Rishel, Raymond (1970). Necessary and sufficient dynamic programming conditions for continuous time stochastic control. *SIAM J. Control* **8**, 559–571.

Rootzén, Holger (1977). On the functional central limit theorem for martingales. *Z. Wahrsch. verw. Gebiete* **38**, 199–210.

Rootzén, Holger (1980). On the functional central limit theorem for martingales II. *Z. Wahrsch. verw. Gebiete* **51**, 79–93.

Rosenblatt, Murray (1956). A central limit theorem and a strong mixing condition. *Proc. Nat. Acad. Sci. U.S.A.* **42**, 43–47.

Rosenkrantz, Walter A. (1974). Convergent family of diffusion processes whose diffusion coefficients diverge. *Bull. Amer. Math. Soc.* **80**, 973–976.

Rosenkrantz, Walter A. (1975). Limit theorems for solutions to a class of stochastic differential equations. *Indiana Univ. Math. J.* **24**, 613–625.

Rosenkrantz, Walter A. and Dorea, C. C. Y. (1980). Limit theorems for Markov processes via a variant of the Trotter–Kato theorem. *J. Appl. Probab.* **17**, 704–715.

Roth, Jean-Pierre (1976). Opérateurs dissipatifs et semigroups dans les espaces de fonctions continues. *Ann. Inst. Fourier, Grenoble* **26**, 1–97.

Roth, Jean-Pierre (1977). Opérateurs elliptiques comme générateurs infinitésimaux de semigroupes de Feller. *C. R. Acad. Sci. Paris* **284**, 755–757.

Rudin, Walter (1973). *Functional Analysis.* McGraw-Hill, New York.

Rudin, Walter (1974). *Real and Complex Analysis*, 2nd Ed. McGraw-Hill, New York.

Sato, Ken-iti (1976). Diffusion processes and a class of Markov chains related to population genetics. *Osaka J. Math.* **13**, 631–659.

Sato, Ken-iti (1977). A note on convergence of probability measures on C and D. *Ann. Sci. Kanazawa Univ.* **14**, 1–5.

Schauder, J. (1934). Uber lineare elliptische Differentialgleichungen zweiter Ordnung. *Math. Z.* **38**, 257–282.

Schlögl, F. (1972). Chemical reaction models for non-equilibrium phase transitions. *Z. Physik* **253**, 147–161.

Serant, Daniel and Villard, Michel (1972). Linearization of crossing-over and mutation in a finite random-mating population. *Theor. Pop. Biol.* **3**, 249–257.

Shiga, Tokuzo (1980). An interacting system in population genetics. *J. Math. Kyoto Univ.* **20**, 213–242.

Shiga, Tokuzo (1981). Diffusion processes in population genetics. *J. Math. Kyoto Univ.* **21**, 133–151.

Shiga, Tokuzo (1982). Wandering phenomena in infinite allelic diffusion models. *Adv. Appl. Probab.* **14**, 457–483.

Siegmund, David (1976). The equivalence of absorbing and reflecting barrier problems for stochastically monotone Markov processes. *Ann. Probab.* **4**, 914–924.

Skorohod, A. V. (1956). Limit theorems for stochastic processes. *Theory Probab. Appl.* **1**, 261–290.

Skorohod, A. V. (1958). Limit theorems for Markov processes. *Theory Probab. Appl.* **3**, 202–246.

Skorohod, A. V. (1965). *Studies in the Theory of Random Processes.* Addison-Wesley, Reading, Mass.

Sova, Miroslav (1967). Convergence d'opérations linéaires non bornées. *Rev. Roumaine Math. Pures Appliq.* **12**, 373–389.

Spitzer, Frank (1970). Interaction of Markov processes. *Adv. Math.* **5**, 246–290.

Stone, Charles (1963). Weak convergence of stochastic processes defined on a semi-infinite time interval. *Proc. Amer. Math. Soc.* **14**, 694–696.

Stratonovich, R. L. (1963, 1967). *Topics in the Theory of Random Noise I, II.* Gordon and Breach, New York.

Strassen, Volker (1965). The existence of probability measures with given marginals. *Ann. Math. Statist.* **36**, 423–439.

Stroock, Daniel W. (1975). Diffusion processes associated with Levy generators. *Z. Wahrsch. verw. Gebiete* **32**, 209–244.

Stroock, Daniel W. and Varadhan, S. R. S. (1969). Diffusion processes with continuous coefficients I, II. *Comm. Pure Appl. Math.* **22**, 345–400, 479–530.

Stroock, Daniel W. and Varadhan, S. R. S. (1971). Diffusion processes with boundary conditions. *Comm. Pure Appl. Math.* **24**, 147–225.

Stroock, Daniel W. and Varadhan, S. R. S. (1972). On the support of diffusion processes with applications to the strong maximum principle. *Proc. Sixth Berkeley Symp. Math. Statist. Prob.* **3**, 333–359.

Stroock, Daniel W. and Varadhan, S. R. S. (1979). *Multidimensional Diffusion Processes.* Springer-Verlag, Berlin.

Trotter, Hale F. (1958). Approximation of semi-groups of operators. *Pacific J. Math.* **8**, 887–919.

Trotter, Hale F. (1959). On the product of semi-groups of operators. *Proc. Amer. Math. Soc.* **10**, 545–551.

Vasershtein, L. N. (1969). Markov processes over denumerable products of spaces describing large systems of automata. *Probl. Peredachi Inform.* **5**(3), 64–73.

Vasershtein, L. N. and Leontovich, A. M. (1970). Invariant measures of certain Markov operators describing a homogeneous random medium. *Probl. Peredachi Inform.* **6**(1), 71–80.

Volkonskii, V. A. (1958). Random substitution of time in strong Markov processes. *Theory Probab. Appl.* **3**, 310–326.

Volkonskii, V. A. and Rozanov, Yu. A. (1959). Some limit theorems for random functions I. *Theory Probab. Appl.* **4**, 178–197.

Wagner, Daniel H. (1977). Survey of measurable selection theorems. *SIAM J. Control Optim.* **15**, 859–903.

Wang, Frank J. S. (1977). A central limit theorem for age and density dependent population processes. *Stochastic Process. Appl.* **5**, 173–193.

Wang, Frank J. S. (1982a). Probabilities of extinction of multiplicative measure diffusion processes with absorbing boundary. *Indiana Univ. Math. J.* **31**, 97–107.

Wang, Frank J. S. (1982b). Diffusion approximations of age-and-position dependent branching processes. *Stochastic Process. Appl.* **13**, 59–74.

Watanabe, Shinzo (1964). Additive functionals of Markov processes and Lévy systems. *Japanese J. Math.* **34**, 53–70.

Watanabe, Shinzo (1968). A limit theorem of branching processes and continuous state branching processes. *J. Math. Kyoto Univ.* **8**, 141–167.

Watanabe, Shinzo (1971). On stochastic differential equations for multi-dimensional diffusion processes with boundary conditions. *J. Math. Kyoto Univ.* **11**, 169–180.

Watterson, G. A. (1962). Some theoretical aspects of diffusion theory in population genetics. *Ann. Math. Statist.* **33**, 939–957.

Watterson, G. A. (1964). The application of diffusion theory to two population genetic models of Moran. *J. Appl. Probab.* **1**, 233–246.

Watterson, G. A. (1970). On the equivalence of random mating and random union of gametes in finite, monoecious populations. *Theor. Pop. Biol.* **1**, 233–250.

Watterson, G. A. (1976). The stationary distribution of the infinitely-many neutral alleles diffusion model. *J. Appl. Probab.* **13**, 639–651.

Watterson, G. A. and Guess, Harry A. (1977). Is the most frequent allele the oldest? *Theor. Pop. Biol.* **11**, 141–160.

Weiss, Alan (1981). Invariant measures of diffusions in bounded domains. Ph.D. dissertation, New York University.

Whitney, Hassler (1934). Analytic extensions of differentiable functions defined on closed sets. *Trans. Amer. Math. Soc.* **36**, 369–387.

Williams, David (1979). *Diffusions, Markov Processes, and Martingales.* Wiley, New York.

Withers, C. S. (1981). Central limit theorems for dependent variables. I. *Z. Wahrsch. verw. Gebiete* **57**, 509–534.

Wonham, W. M. (1966). Lyapunov criteria for weak stochastic stability. *J. Differential Equations* **2**, 195–207.

Wright, Sewall (1931). Evolution in Mendelian populations. *Genetics* **16**, 97–159.

Wright, Sewall (1949). Adaptation and selection. *Genetics, Paleontology, and Evolution*, G. L. Jepson, E. Mayr, and G. G. Simpson, Eds. Princeton University Press, Princeton, pp. 365–389.

Yamada, Toshio and Watanabe, Shinzo (1971). On the uniqueness of solutions of stochastic differential equations. *J. Math. Kyoto Univ.* **11**, 155–167.

Yosida, Kosaku (1948). On the differentiability and the representation of one-parameter semi-groups of linear operators. *J. Math. Soc. Japan* **1**, 15–21.

Yosida, Kosaku (1980). *Functional Analysis*, 6th ed. Springer-Verlag, Berlin.

Zakai, Moshe (1969). A Lyapunov criterion for the existence of stationary probability distributions for systems perturbed by noise. *SIAM J. Control* **7**, 390–397.

INDEX

Note: * indicates definition

FLOWCHART

This table indicates the relationships between theorems, corollaries, and so on. For example, the entry **C2.8** P2.1 P2.7 *T6.9 T4.2.2* under Chapter 1 means that Corollary 2.8 of Chapter 1 requires Propositions 2.1 and 2.7 of that chapter for its proof and is used in the proofs of Theorem 6.9 of Chapter 1 and Theorem 2.2 of Chapter 4.

Chapter 1

P1.1 *C1.2 P1.5b C1.6* **C1.2** P1.1 **R1.3** **L1.4a** Prob.3 **L1.4b** Prob.3 *P2.1 P3.4* **L1.4c** Prob.3 *P1.5c L2.5 P5.4 L6.2* **P1.5a** *C1.6* **P1.5b** P1.1 *P2.1 P3.3 P3.7 T10.4.1* **P1.5c** L1.4c *C1.6 T2.6 P2.7 P3.4 L6.2 T6.11 R4.2.10 P4.9.2 T8.3.1* **C1.6** P1.1 P1.5ac *P2.1 T2.6 P2.7 P4.9.2 T10.4.1* **P2.1** L1.4b L1.5b C1.6 *T2.6 P2.7 C2.8 P3.3 P3.7 P4.1 T6.1 L6.3 T6.5 T6.9 T7.1 R7.9b T2.7.1 C4.2.8* **L2.2** *L2.3 T2.6 L2.11* **L2.3** L2.2 *T2.6 T4.3 T6.9 T6.11 T4.5.19a* **L2.4a** *T2.6 P2.7 T6.1* **L2.4b** *T2.6 P2.7* **L2.4c** *T2.6 P2.7 T6.1 C6.8 T7.1* **L2.5** L1.4c *T2.6 P2.7* **T2.6** P1.5c C1.6 P2.1 L2.2 L2.3 L2.4abc L2.5 *T2.12 P3.4 T4.3 T7.1 T4.4.1* **P2.7** P1.5c C1.6 P2.1 L2.4abc L2.5 P2.9 *C2.8 T6.1 T4.2.7* **C2.8** P2.1 P2.7 *T6.9 T4.2.2* **P2.9** P2.10 *P2.7* **P2.10** *P2.9 P3.4* **L2.11** L2.2 *T2.12 P3.1 P3.4* **T2.12** T2.6 L2.11 *P3.1 P3.5 P3.7 T4.2.2* **P3.1** L2.11 T2.12 *P3.3 T6.1 L6.3 T6.5* **R3.2** *T8.1.5 T8.3.1* **P3.3** P1.5b P2.1 P3.1 *P5.1.1 T8.1.6 T8.2.1 T8.3.1 T8.3.4 L10.3.1 P12.2.2* **P3.4** L1.4b P1.5c T2.6 P2.10 L2.11 **P3.5** T2.12 **L3.6** *P3.7 T8.3.1* **P3.7** P1.5b P2.1 T2.12 L3.6 *C3.8 T8.2.1 T8.2.5 T8.2.8* **C3.8** P3.7 **P4.1** P2.1 *T7.1 R7.9c T12.2.4* **L4.2** *T4.3 T4.4.1 T12.4.1* **T4.3** L2.3 T2.6 L4.2 *T8.3.1* **P5.1** *C4.8.7 C4.8.16* **P5.2** *P2.7.5* **P5.3** **P5.4** L1.4c *T9.4.3* **R5.5** **T6.1** P2.1 L2.4ac P2.7 P3.1 L6.2 L6.3 *T6.5 T7.6a C7.7a T4.2.5 T4.2.11 R4.8.8a T8.3.1* **L6.2** P1.4c P1.5c *T6.1* **L6.3** P2.1 P3.1 *T6.1* **L6.4** *T6.5 T6.11* **T6.5** P2.1 P3.1 L6.4 T6.1 *C6.6 C6.7 C6.8 T7.6b C7.7b T4.2.6 T4.2.12 T5.1.2c T9.1.3 T10.1.1* **C6.6** T6.5 *C6.7 C6.8* **C6.7** T6.5 C6.6 **C6.8** L2.4c T6.5 C6.6 *T2.7.1 T4.4.1 P4.9.2 T4.9.3* **T6.9** P2.1 L2.3 C2.8 *T6.11 T4.8.2* **R6.10** Prob.16 **T6.11** P1.5c L2.3 L6.4 T6.9 **T7.1** P2.1 L2.4c T2.6 P4.1 *C7.2* **C7.2** T7.1 **L7.3a** Prob.18 *C7.7ab* **L7.3b** Prob.18 **L7.3c** Prob.18 **L7.3d** Prob.18 *T7.6ab R7.9a E12.2.6* **R7.4** *T10.3.5ab* **R7.5** *T10.3.5ab* **T7.6a** T6.1 L7.3d *C7.7a C7.8* **T7.6b** T6.5 L7.3d *C7.7b T10.3.5ab* **C7.7a** T6.1 L7.3a T7.6a **C7.7b** T6.5 L7.3a T7.6b *T10.3.5b* **C7.8** T7.6a *P12.2.2* **R7.9a** L7.3d **R7.9b** P2.1 *R12.2.3* **R7.9c** P4.1 *P12.2.2* **R7.9d**

Chapter 2

L1.1 *P1.2d P1.5b L4.1* **P1.2a** **P1.2b** *P1.4g* **P1.2c** **P1.2d** L1.1 **P1.3** *T2.13 R3.8.5a R4.1.4 T4.2.7 T5.1.2a* **P1.4a** **P1.4b** *P1.4def P2.15 T4.2 C4.4 C4.5* **P1.4c** *P1.4ef L2.2 T2.13 P3.2 L4.1* **P1.4d** P1.4b *L2.2 T2.13 P3.2 L4.1 R4.3 P4.1.5 T4.3.12 T4.4.2bc L5.2.4* **P1.4e** P1.4bc *P1.4f* **P1.4f** P1.4bce **P1.4g** P1.2b *L3.8.4* **P1.5a** *P3.2 P3.4 L3.5 P3.6 T5.1 T4.6.1 T4.6.2 T4.6.3 C4.6.4 L4.6.5 L4.10.6 T5.2.9 T5.3.7 T5.3.11* **P1.5b** L1.1 *P2.15 P2.16b C5.3 C5.4 T4.3.8 C4.3.13 T7.1.4ab T7.4.1 T8.3.1* **T1.6** **P2.1a** *C2.17 P3.4 L3.5 P3.6 P6.2 T7.1.4a T9.2.1a* **P2.1b** *P2.9 T2.13 L7.2* **L2.2** P1.4cd *L2.3 L2.5 T2.13* **L2.3** L2.2 *C2.4* **C2.4** L2.3 *P2.9 C2.11 R2.12 P2.16a* **L2.5** L2.2 *C2.6* **C2.6** L2.5 *P2.9 C2.11 R2.12* **L2.7** *P2.9* **L2.8** *P2.9 T4.3.6* **P2.9** P2.1b C2.4 C2.6 L2.7 L2.8 Prob.8 Prob.9 Prob.10a *C2.10 L4.1 R7.3 T4.3.6* **C2.10** P2.9 Prob.10a *C2.11 R2.14 T12.4.1* **C2.11** C2.4 C2.6 C2.10 Prob.9 *T5.1 R5.2a* **R2.12** C2.4 C2.6 Prob.9 *P3.4 L3.5 P3.6 T5.1 L4.6.5* **T2.13** P1.3 P1.4cd P2.1b L2.2 Prob.10a *R2.14 P2.13 P2.16b P3.1 P3.2 P3.4 L3.5 P3.6 T4.2 C4.4 C4.5 T5.1 R5.2b C5.4 P4.2.9 T4.3.8 P4.3.9 P4.3.10 T4.3.12 C4.4.14 T4.5.11b L4.5.13 T4.6.1 T4.6.2 T4.6.3 C4.6.4 L4.6.5 T4.10.1 L4.10.6 L5.2.4 T5.3.7 T6.1.3 T6.1.4 T6.2.8b T6.5.3b T7.1.4ab T7.4.1 T8.3.1 L9.1.6 P10.2.8 L10.2.10* **R2.14** C2.10 T2.13 *P2.15 T4.3.8 P4.3.9 P4.3.10 C4.3.13 L4.5.13 T6.1.3 T6.1.4 T6.2.8b T6.5.3b T7.1.4ab T7.4.1 T8.3.1* **P2.15** P1.4b P1.5b T2.13 R2.14 *P4.2.4* **P2.16a** C2.4 *C2.17 P3.4 P3.6 T7.1.4a T9.3.1 L9.4.1* **P2.16b** P1.5b T2.13 *C2.17 T5.2.3 T5.3.11 T9.2.1a* **C2.17** P2.1a P2.16ab *P3.4 L3.5 P3.6 T10.4.5* **P3.1** T2.13 **P3.2** P1.4cd P1.5a T2.13 *C3.3 L4.3.2* **C3.3** P3.2 **P3.4** P1.5a P2.1a R2.12 T2.13 P2.16a C2.17 L3.5 Prob.8 Prob.10a *P6.1 T5.2.9 T7.1.4a* **L3.5** P1.5a P2.1a R2.12 T2.13 C2.17 Prob.10a *P3.4* **P3.6** P1.5a P2.1a R2.12 T2.13 P2.16a C2.17 Prob.10a **L4.1** L1.1 P1.4cd P2.9 *T4.2 C4.4 C4.5* **T4.2** P1.4b T2.13 L4.1 TA.4.2 *P7.5* **R4.3** P1.4d **C4.4** P1.4b T2.13 L4.1 TA.4.2 **C4.5** P1.4b T2.13 L4.1 TA.4.2 *T12.4.1* **P4.6** TA.4.3 *R4.7* **R4.7** P4.6 *T7.1 T4.8.2 R4.8.3b C4.8.4 C4.8.5 C4.8.12 C4.8.13* **T5.1** P1.5a C2.11 R2.12 T2.13 Prob.15 PA.2.1 PA.2.4 *C5.3 C5.4 P6.2 L7.2 T5.2.3 L5.2.4* **R5.2a** C2.11 **R5.2b** T2.13 **R5.2c** **C5.3** P1.5b T5.1 **C5.4** T2.13 T5.1 *P6.2 T5.2.9* **P6.1** P3.4 *P6.2 T5.2.9* **P6.2** P2.1a T5.1 C5.4 P6.1 Prob.10c *T5.2.3 L5.2.4* **T7.1** P1.2.1 C1.6.8 R4.7 *P7.6 T4.8.2 R4.8.3a T4.8.10* **L7.2** P2.1b T5.1 Prob.10b *C7.4* **R7.3** P2.9 **C7.4** L7.2 **P7.5** P1.5.2 T4.2 *P7.6 R4.8.3b* **P7.6** T7.1 P7.5 **P8.1a** **P8.1b** *P8.6* **P8.2** *P8.5c* **R8.3** **P8.4** *T8.7* **P8.5a** **P8.5b** **P8.5c** P8.2 **P8.6** P8.1b *P6.2.10* **T8.7** P8.4 *T6.2.8a*

Chapter 3

L1.1 **T1.2** L1.3 *C1.6* **L1.3** C1.5 *T1.2 T1.7 T1.8* **L1.4** *C1.5* **C1.5** L1.4 *L1.3* **C1.6** T1.2 *T1.7 C1.9* **T1.7** L1.3 C1.6 Prob.3 *T2.2 T4.4.6 T4.6.3* **T1.8** L1.3 *C1.9 T7.8a T6.1.5 T6.3.3* **C1.9** C1.6 T1.8 *T4.5b T6.3.4a T6.5.4 T7.4.1 T9.2.1b T9.3.1 C10.2.6 C10.2.7 T10.4.6 T11.2.3 TA.1.2* **L2.1** T2.2 *T4.5ab C7.4 C8.10 C9.2 T4.1.1 P4.4.7 TA.8.1* **T2.2** T1.7 L2.1 *C2.3 T4.5b P4.6b T7.2 R7.3 L7.5 L4.5.3 T4.5.11b L4.5.15 R4.9.4 T4.9.9 T4.9.10 T10.2.2 TA.1.2 TA.8.1* **C2.3** T2.2 *L4.9.13* **P2.4** *P4.6b T4.1.1 T3.3.4b TA.1.2* **T3.1** *C3.2 C3.3 P4.4 T4.5b C8.10 C9.2 T10.2b P10.4 T4.5.11c L4.5.17 T10.4.5 TA.8.1* **C3.2** T3.1 *C9.3 R9.1.5* **C3.3**

T3.1 *P10.4 T7.3.1 T7.3.3 T7.4.1 T11.2.3* **R3.4** *L8.1* **L4.1** *P4.2* **P4.2** L4.1 TA.4.3 *T4.5a T4.4.6 T4.5.19a C7.2.8* **L4.3** *T4.5b P4.6b L4.8.1* **P4.4** T3.1 **T4.5a** L2.1 P4.2 *T4.5b* **T4.5b** C1.9 L2.1 T2.2 T3.1 L4.3 T4.5a **P4.6a** *P4.1.6 T4.4.2a C4.4.3 L4.8.1* **P4.6b** T2.2 P2.4 L4.3 **L5.1** *P5.2 P6.5 P7.1 T6.1.1b* **P5.2** L5.1 *P6.5 P7.1 T7.8a T4.5.11c* **P5.3** *C5.5 T5.6 L6.2b P6.5 C9.2 L10.1 L4.5.10 T4.5.19a* **R5.4** *T5.6 P6.5* **C5.5** P5.3 **T5.6** P5.3 R5.4 Prob.14 *P7.1 T7.2 R7.3 T7.8a C9.2 T4.4.6 T4.6.3* **L6.1** *T6.3 T7.2 L7.5* **L6.2a** *L6.2b T6.3 P6.5 C7.4* **L6.2b** P5.3 L6.2a *L6.2c T6.3* **L6.2c** L6.2b **T6.3** L6.1 L6.2ab Prob.16 *T7.2 T6.3.3* **R6.4** Prob.16 *R7.3 C9.2* **P6.5** L5.1 P5.2 P5.3 R5.4 L6.2a *P6.3.2* **P7.1** L5.1 P5.2 T5.6 *T7.8b C4.4.3* **T7.2** T2.2 T5.6 L6.1 T6.3 *C7.4 T8.6 T9.1* **R7.3** T2.2 T5.6 R6.4 **C7.4** L2.1 L6.2a T7.2 *T6.1.5 P6.3.1 T6.3.4b C6.3.6 T6.5.4 T9.3.1* **L7.5** T2.2 L6.1 Prob.15 *T7.6* **T7.6** L7.5 **L7.7** *T7.8ab L4.5.1 T4.8.10* **T7.8a** T1.8 P5.2 T5.6 L7.7 *T7.8b C8.10 C9.2 L4.5.1 T4.8.10* **T7.8b** P7.1 L7.7 T7.8a TA.4.2 *C9.3 C4.8.6 C4.8.15* **L8.1** R3.4 *P8.3* **L8.2** *P8.3* **P8.3** L8.1 L8.2 *T8.6* **L8.4** P2.1.4g *R8.5b T8.6* **R8.5a** P2.1.3 **R8.5b** L8.4 *T8.6* **T8.6** Prob.2.25 T7.2 P8.3 L8.4 R8.5b *R8.7a C8.10 T9.1 T9.4 T7.1.4ab T7.4.1 T9.1.4* **R8.7a** T8.6 *T9.4* **R8.7b** **T8.8** *C8.10* **R8.9a** **R8.9b** **C8.10** L2.1 T3.1 T7.8a T8.6 T8.8 **T9.1** T7.2 T8.6 Prob.13 *C9.2 C9.3 R4.5.2 L4.5.17 C4.8.6 C4.8.15 T4.9.17 T7.1.4a T10.4.1* **C9.2** L2.1 T3.1 P5.3 T5.6 R6.4 T7.8a T9.1 *C4.8.6* **C9.3** C3.2 T7.8b T9.1 *T4.2.5 T4.2.6 T4.2.11 T4.2.12* **T9.4** T8.6 R8.7a Prob.23 *T4.2.5 T4.2.6 T4.2.11 T4.2.12 R4.5.2 L4.5.17 C4.8.6 C4.8.15 T4.9.17 T10.4.1* **R9.5a** **R9.5b** **L10.1** P5.3 *T10.2b* **T10.2a** *P10.4 T7.1.4b T9.1.4 T11.4.1* **T10.2b** T3.1 L10.1 **P10.3** **P10.4** T3.1 C3.3 T10.2a

Chapter 4

T1.1 Prob.2.27 L3.2.1 P3.2.4 TA.9.1 **P1.2** TA.4.2 *P1.5 T5.19c* **P1.3** *T2.7* **R1.4** P2.1.3 **P1.5** P2.1.4d P1.2 TA.4.2 **P1.6** P3.4.6a *T4.1* **P1.7** *P10.2.8* **L2.1** *T2.2 T12.4.1* **T2.2** C1.2.8 T1.2.12 L2.1 *T8.1.4 T8.1.5 T8.3.1* **L2.3** *P2.4 T2.5 T2.6 T2.11 T2.12 T8.3.1* **P2.4** P2.2.15 L2.3 *T2.5 T2.6 T2.11 T2.12* **T2.5** T1.6.1 C3.9.3 T3.9.4 L2.3 P2.4 *T2.7 T11.2.3* **T2.6** T1.6.5 C3.9.3 T3.9.4 L2.3 P2.4 *T5.1.2c* **T2.7** P1.2.7 P2.1.3 P1.3 T2.5 *C2.8 P2.9 T5.1.2 T8.3.1 T10.1.1 T12.2.4 T12.3.1 T12.4.1* **C2.8** P1.2.1 T2.7 *T8.3.1* **P2.9** T2.2.13 T2.7 *T5.1.2 T10.2.4 T12.2.4 T12.3.1 T12.4.1* **R2.10** P1.1.5c **T2.11** T1.6.1 C3.9.3 T3.9.4 L2.3 P2.4 **T2.12** T1.6.5 C3.9.3 T3.9.4 L2.3 P2.4 **P3.1** *T4.1* **L3.2** P2.3.2 Prob.2.22 Prob.14 *P3.3 P3.5 T4.1 C4.4 L5.6 L5.18 T5.19a P9.2 T9.3 T10.1 L6.2.6 T6.2.8ab L9.4.1* **P3.3** L3.2 **L3.4** *T10.3 L6.2.8b T9.4.3 T10.4.1 T12.2.4 T12.3.1 T12.4.1* **P3.5** L3.2 *T5.19a* **T3.6** L2.2.8 P2.2.9 Prob.3.7 *C3.7* **C3.7** T3.6 *P5.3.5 T8.3.3 T10.4.1* **T3.8** P2.1.5b T2.2.13 R2.2.14 *R5.5 T8.3.3* **P3.9** P2.1.5b T2.2.13 R2.2.14 *P5.3.5 P5.3.10* **P3.10** P2.1.5b T2.2.13 R2.2.14 **R3.11** **T3.12** P2.1.4d T2.2.13 Prob.3.7 *C3.13 T5.11c R6.1.6 T6.3.4a* **C3.13** P2.1.5b R2.2.14 T3.12 **T4.1** T1.2.6 L1.4.2 C1.6.8 P1.6 P3.1 L3.2 *T5.19c* **T4.2a** P3.4.6a *P4.7 T10.1 T10.3 T9.4.3 T10.4.1* **T4.2b** P2.1.4d Prob.2.11b **T4.2c** P2.1.4d Prob.2.11b *T5.11c P9.19 T9.14* **C4.3** P3.4.6a P3.7.1 *C4.4 T6.2 P9.19* **C4.4** L3.2 C4.3 **R4.5** Prob.22 **T4.6** T3.1.7 P3.4.2 T3.5.6 TA.10.5 **P4.7** L3.2.1 T4.2a **R4.8a** **R4.8b** **E4.9** Prob.23 **L4.10** *R4.12* **T4.11** *C4.13* **R4.12** L4.10 **C4.13** T4.11 **C4.14** T2.2.13 *C4.15* **C4.15** C4.14 **R4.16** **L5.1** L3.7.7 T3.7.8a *T5.4 L5.17 T9.17* **R5.2** T3.9.1 T3.9.4 *T5.4* **L5.3** T3.2.2 TA.10.1 *T5.4 T10.4.1* **T5.4** L5.1 R5.2

L5.3 *T5.3.10* **R5.5** T3.8 **L5.6** L3.2 **L5.8** **L5.9** *L5.10* **L5.10** P3.5.3 L5.9 *T5.11b* **T5.11a** L5.16 *T5.11b* **T5.11b** T2.2.13 T3.2.2 L5.10 T5.11a L5.17 *T5.11c* **T5.11c** T3.3.1 P3.5.2 T3.12 T4.2c T5.11b **R5.12** **L5.13** T2.2.13 R2.2.14 **E5.14** **L5.15** Prob.2.26 T3.2.2 *L5.16 T5.3.6 C7.5.3* **L5.16** L5.15 *T5.11a T6.1 L6.5 T6.1.4 T6.2.8b* **L5.17** T3.3.1 T3.9.1 T3.9.4 L5.1 *T5.11b* **L5.18** L3.2 *T5.19a* **T5.19a** L1.2.3 P3.4.2 P3.5.3 L3.2 P3.5 L5.18 L5.20 **T5.19b** **T5.19c** T4.1 P1.2 **T5.19d** L5.20 **L5.20** P3.5.3 L5.10 *T5.19ad* **T6.1** P2.1.5a T2.2.13 L5.16 *C6.4 T7.4.1* **T6.2** P2.1.5a T2.2.13 C4.3 Prob.27 **T6.3** P2.1.5a T2.2.13 T3.1.7 T3.5.6 *C6.4* **C6.4** P2.1.5a T2.2.13 T6.1 T6.3 **L6.5** P2.1.5a R2.2.12 T2.2.13 L5.16 *T6.6* **T6.6** L6.5 **T7.1** *P5.3.5 T5.3.10* **L7.2** **T7.3** Prob.29 **L8.1** L3.4.3 P3.4.6a *T8.2 R6.3.5a* **T8.2** T1.6.9 R2.4.7 T2.7.1 L8.1 *C8.4 C8.5 C8.6* **R8.3a** T2.7.1 **R8.3b** R2.4.7 P2.7.5 **R8.3c** *C8.7 C8.9* **R8.3d** **C8.4** R2.4.7 T8.2 **C8.5** R2.4.7 T8.2 **C8.6** T3.7.8b T3.9.1 C3.9.2 T3.9.4 T8.2 *C8.7 C8.9 T9.2.1a T12.4.1* **C8.7** P1.5.1 R8.3c C8.6 *T12.2.4 T12.3.1* **R8.8a** T1.6.1 **R8.8b** **C8.9** R8.3c C8.6 *T9.1.3 T10.1.1 T10.3.5a* **T8.10** T2.7.1 L3.7.7 T3.7.8a *C8.12 C8.13 C8.15* **R8.11** *C8.16 C8.17* **C8.12** R2.4.7 T8.10 **C8.13** R2.4.7 T8.10 **R8.14** **C8.15** T3.7.8b T3.9.1 T3.9.4 T8.10 *C8.16 C8.17* **C8.16** P1.5.1 R8.11 C8.15 *T9.4.3 P10.4.2* **C8.17** R8.11 C8.15 *T10.4.1* **P8.18** **L9.1** *P9.2* **P9.2** P1.1.5c C1.1.6 C1.6.8 L3.2 L9.1 *T9.3 L10.2.1* **T9.3** C1.6.8 L3.2 P9.2 *T10.4.6* **R9.4** T3.2.2 **L9.5** **R9.6** **L9.7** T3.2.2 **C9.8** **T9.9** T3.2.2 **T9.10** T3.2.2 **R9.11** **T9.12** *T10.4.6* **L9.13** C3.2.3 **T9.14** **R9.15** **L9.16** *T9.17* **T9.17** T3.9.1 T3.9.4 L5.1 L9.16 Prob.41 TA.8.1 *P9.19 T10.4.6* **P9.18** **P9.19** T4.2c C4.3 T9.17 T10.3 **T10.1** T2.2.13 Prob.2.23 L3.2 L4.2a *T10.3* **P10.2** PA.4.5 *T10.3* **T10.3** L3.4 T4.2a T10.1 P10.2 L10.5 L10.6 *P9.19* **R10.4** **L10.5** *T10.3* **L10.6** P2.1.5a T2.2.13 *T10.3*

Chapter 5

P1.1 P1.3.3 *T1.2 T1.2c T9.3.1 E10.4.3 T11.2.3* **T1.2** T4.2.7 P4.2.9 P1.1 *P3.1 T9.3.1* **T1.2a** P2.1.3 *T1.2bc T11.2.3* **T1.2b** T1.2a *C3.4 T3.7 T3.8 T3.11 E9.3.2* **T1.2c** T1.6.5 T4.2.6 P1.1 T1.2a **L2.1** *T2.3* **L2.2** *T2.3* **T2.3** P2.2.16b T2.5.1 P2.6.2 L2.1 L2.2. *L2.4 T2.6 T3.11* **L2.4** P2.1.4d T2.2.13 T2.5.1 P2.6.2 T2.3 *T2.6 L2.7* **L2.5** Prob.11 *L2.8* **T2.6** T2.3 L2.4 *L2.7 L2.8 T2.9 T3.1 T3.3 T3.7 T7.1.1* **L2.7** L2.4 T2.6 *T2.9* **L2.8** L2.5 T2.6 *P3.1 T3.3* **T2.9** P2.1.5a P2.3.4 C2.5.4 P2.6.1 T2.6 L2.7 L2.11 Prob.12 *T2.12 P3.1 C3.4 T3.8 T7.1.1 T7.1.2 T7.4.1 E9.3.2 T10.4.5* **R2.10** **L2.11** *T2.9* **T2.12** T2.9 *P3.1 T3.3 T6.5.3b* **P3.1** T1.2b T2.6 L2.8 T2.9 T2.12 **L3.2** TA.10.1 *T3.3* **T3.3** T2.6 L2.8 T2.12 L3.2 *C3.4 T6.5.3a* **C3.4** T1.2b T2.9 T3.3 *T3.10 T8.2.3 T8.2.6* **P3.5** C4.3.7 P4.3.9 T4.7.1 Prob.4.19 *T3.10 T8.1.7 T8.2.3 T8.2.6* **T3.6** L4.5.15 *T8.2.6* **T3.7** P2.1.5a P2.2.13 T1.2b T2.6 TA.5.1 *T3.11* **T3.8** T1.2b T2.9 TA.5.1 *T8.2.3* **R3.9** **T3.10** P4.3.9 T4.5.4 T4.7.1 Prob.4.19 C3.4 P3.5 *T8.2.3 T8.2.6* **T3.11** P2.1.5a P2.2.16b T1.2b T2.3 T3.7 *T11.3.2*

Chapter 6

T1.1a *T1.1b* **T1.1b** L3.5.1 T1.1a *T1.5* **R1.2** **T1.3** T2.2.13 R2.2.14 *T9.3.1* **T1.4** T2.2.13 R2.2.14 L4.5.16 Prob.4.45 Prob.12 **T1.5** T3.1.8 C3.7.4 T1.1.b *T9.1.4* **R1.6** T4.3.12 **L2.1** TA.11.3 **T2.2a** TA.11.3 *T2.2b* **T2.2b** T2.2a *T4.1b T5.1* **R2.3** Prob.1 **P2.4** TA.11.3 **R2.5** **L2.6** L4.3.2 *L2.7 T2.8ab P2.10* **L2.7** L2.6 *T3.4a T5.4* **T2.8a** T2.8.7 L4.3.2 L2.6 *T4.1b T5.1* **T2.8b** Prob.1.23 T2.2.13 R2.2.14 Prob.2.24 L4.3.2 L4.3.4 L4.5.16 Prob.4.45 L2.6 **R2.9a** **R2.9b** **P2.10**

P2.8.6 L2.6 *T4.1b T5.1* **P3.1** C3.7.4 **P3.2** P3.6.5 Prob.5 **T3.3** T3.1.8 T3.6.3 **T3.4a** C3.1.9 T4.3.12 L2.7 *T3.4b C3.6* **T3.4b** P3.2.4 C3.7.4 T3.4a Prob.7 **R3.5a** L4.8.1 **R3.5b** **C3.6** C3.7.4 T3.4a **T4.1a** **T4.1b** T2.2b T2.8a P2.10 *T4.1c* **T4.1c** T4.1b *T11.2.1* **T5.1** T2.2b T2.8a P2.10 *R5.2a T5.3a T11.3.1* **R5.2a** T5.1 **R5.2b** **T5.3a** T5.3.3 T5.1 **T5.3b** T2.2.13 R2.2.14 T5.2.12 **T5.4** C3.1.9 C3.7.4 L2.7 **R5.5**

Chapter 7

T1.1 T5.2.6 T5.2.9 *T1.4b* **T1.2** T5.2.9 *T1.4a* **R1.3** Prob.2 **T1.4a** P2.1.5b P2.2.1a T2.2.13 R2.2.14 P2.2.16a P2.3.4 T3.8.6 T3.9.1 Prob.3.22c T1.2 Prob.7 *T3.1 T3.3* **T1.4b** P2.1.5b T2.2.13 R2.2.14 T3.8.6 T3.10.2a T1.1 Prob.7 PA.2.2 PA.2.3 *T9.3.1* **R1.5** **L2.1** *P2.2 P2.6* **P2.2** L2.1 *R2.3 C2.4 C2.5* **R2.3** P2.2 *T3.1* **C2.4** P2.2 *T3.1* **C2.5** P2.2 **P2.6** L2.1 *C2.7 C2.8 T3.3 T12.4.1* **C2.7** P2.6 *T3.3* **C2.8** P3.4.2 P2.6 PA.4.5 **T3.1** C3.3.3 T1.4a R2.3 C2.4 **R3.2a** **R3.2b** **T3.3** C3.3.3 T1.4a P2.6 C2.7 **R3.4** **T4.1** P2.1.5b T2.2.13 R2.2.14 C3.1.9 T3.3.3 T3.8.6 T4.6.1 T5.2.9 *C4.2* **C4.2** T4.1 Prob.13 **T5.1** *C5.2 C5.3* **C5.2** T5.1 Prob.17 **C5.3** L4.5.15 T5.1 Prob.17 **R5.4** *C5.5 T11.3.1* **C5.5** R5.4 *T11.3.1* **T5.6**

Chapter 8

T1.1 *E12.3.3* **C1.2** Prob.1 **R1.3** Prob.2 **T1.4** T4.2.2 **T1.5** R1.3.2 T4.2.2 **T1.6** P1.3.3 **T1.7** P5.3.5 **T2.1** P1.3.3 P1.3.7 L2.2 *T9.1.3* **L2.2** *T2.1* **T2.3** C5.3.4 P5.3.5 T5.3.8 T5.3.10 **P2.4** Prob.4 *T2.5* **T2.5** P1.3.7 P2.4 TA.5.1 **T2.6** C5.3.4 P5.3.5 T5.3.6 T5.3.10 **R2.7** **T2.8** P1.3.7 L2.9 PA.7.1 *T10.1.1 L10.2.1 T10.3.5ab* **L2.9** *T2.8* **T3.1** P1.1.5c R1.3.2 P1.3.3 L1.3.6 T1.4.3 T1.6.1 P2.1.5b T2.2.13 R2.2.14 T4.2.2 L4.2.3 T4.2.7 C4.2.8 TA.5.1 *C3.2* **C3.2** T3.1 **T3.3** C4.3.7 C4.3.8 *T3.4* **T3.4** P1.3.3 T3.3 **T3.5** Prob.8 **T3.6** Prob.8

Chapter 9

T1.1 PA.4.5 **R1.2** **T1.3** T1.6.5 C4.8.9 T8.2.1 TA.1.2 **T1.4** T3.8.6 T3.10.2a Prob.3.26 T4.4.2c T6.1.5 L1.6 **R1.5** C3.3.2 **L1.6** T2.2.13 *T1.4* **T2.1a** P2.2.1a P2.2.16b C4.8.6 Prob.3 *T2.1b* **T2.1b** C3.1.9 T2.1a *T2.1c* **T2.1c** T2.1b **T3.1** T2.2.16a C3.1.9 C3.7.4 P5.1.1 T5.1.2 T6.1.3 T7.1.4b *E3.2* **E3.2** T5.1.2b T5.2.9 Prob.7.3 T3.1 **L4.1** P2.2.16a L4.3.2 TA.5.1 *T4.3* **T4.2** P1.5.4 L4.3.4 T4.4.2a L4.1 Prob.7 **T4.3** P1.5.4 L4.3.4 T4.4.2a C4.8.16 L4.1 **R4.4**

Chapter 10

T1.1 T1.6.5 T4.2.7 C4.8.9 T8.2.8 Prob.1 *T2.2* **L2.1** P4.9.2 T8.2.8 *T2.2 R2.3* **T2.2** T3.2.2 T1.1 L2.1 **R2.3** L2.1 **T2.4** Prob.3.5 P4.2.9 L2.10 Prob.3 PA.2.2 *C2.6 C2.7* **R2.5** PA.2.3 *C2.7 P2.8* **C2.6** C3.1.9 T2.4 PA.2.3 **C2.7** C3.1.9 T2.4 R2.5 **P2.8** T2.2.13 P4.1.7 R2.5 *R2.9* **R2.9** P2.8 **L2.10** T2.2.13 *T2.4* **L3.1** P1.3.3 *T3.5ab* **R3.2** **L3.3** *T3.5ab* **R3.4** **T3.5a** R1.7.4 R1.7.5 T1.7.6b C4.8.9 T8.2.8 L3.1 L3.3 R3.6a *E3.8 E3.9* **T3.5b** R1.7.4 R1.7.5 T1.7.6b C1.7.7b T8.2.8 L3.1 L3.3 R3.6a **R3.6a** *T3.5ab* **R3.6b** **R3.7a** *E3.8 E3.9* **R3.7b** **E3.8** T3.5a R3.7a **E3.9** T3.5a R3.7a **T4.1** P1.1.5b C1.1.6 T3.9.1 T3.9.4 L4.3.4 C4.3.7 T4.4.2a L4.5.3 C4.8.17 TA.5.1 *P4.2 E4.4* **P4.2** C4.8.16 T4.1 *E4.3 E4.4 T4.5* **E4.3** P5.1.1 P4.2 **E4.4** T4.1 P4.2 *T4.5 T4.6* **T4.5** C2.2.17 Prob.2.29 T3.3.1 T5.2.9 P4.2 E4.4

T4.6 **T4.6** C3.1.9 T4.9.3 T4.9.12 T4.9.17 E4.4 T4.5 Prob.12 *T4.7* **T4.7** T4.6

Chapter 11

T2.1 T6.4.1c TA.5.1. *T2.3* *T4.1* **R2.2** **T2.3** C3.1.9 C3.3.3 T4.2.5 P5.1.1 T5.1.2a T2.1 *T4.1* **T3.1** T6.5.1 R7.5.4 C7.5.5 TA.5.1 **T3.2** T5.3.11 **R3.3** **T4.1** T3.10.2a T2.1 T2.3 **R4.2** Prob.5

Chapter 12

L2.1 **P2.2** P1.3.3 C1.7.8 R1.7.9c R2.3 *T2.4* **R2.3** R1.7.9b *P2.2* *R2.5* **T2.4** P1.4.1 Prob.3.25 T4.2.7 P4.2.9 L4.3.4 C4.8.7 P2.2 *E2.6* *E2.7* **R2.5** R2.3 *E2.6* **E2.6** L1.7.3d T2.4 R2.5 **E2.7** Prob1.6a T2.4 **T3.1** Prob.3.25 T4.2.7 P4.2.9 L4.3.4 C4.8.7 *E3.3* **R3.2** **E3.3** T8.1.1 T3.1 **T4.1** L1.4.2 C2.2.10 C2.4.5 Prob.3.25 L4.2.1 T4.2.7 P4.2.9 L4.3.4 C4.8.6 P7.2.6

Appendixes

P1.1 *T1.2* **T1.2** C3.1.9 T3.2.2 P3.2.4 P1.1 *T9.1.3* *P2.3* **P2.1** *T2.5.1* *P2.2* *P2.3* **P2.2** P2.1 *T7.1.4b* *T10.2.4* **P2.3** T1.2 P2.1 *T7.1.4b* *R10.2.5* *C10.2.6* **P2.4** T4.2 *T2.5.1* **R2.5** **P3.1** **P3.2** **T4.1** **T4.2** *T2.4.2* *C2.4.4* *C2.4.5* *T3.7.8b* *P4.1.2* *P4.1.5* *P2.4* *T4.3* **T4.3** T4.2 *P2.4.6* *P3.4.2* *C4.4* *P4.5* **C4.4** T4.3 **P4.5** T4.3 *P4.10.2* *C7.2.8* *T9.1.1* **T5.1** *T5.3.7* *T5.3.8* *T8.2.5* *T8.3.1* *L9.4.1* *T10.4.1* *T11.2.1* *T11.3.1* **T6.1** *C6.3* **R6.2** **C6.3** T6.1 **P7.1** *T8.2.8* *P7.2* **P7.2** P7.1 **T8.1** L3.2.1 T3.2.2 T3.3.1 Prob.3.27 *T4.9.17* **R8.2** **T9.1** *T4.1.1* **T10.1** *L4.5.3* *L5.3.2* *C10.3* **R10.2** **C10.3** T10.1 **R10.4** **T10.5** *T4.4.6* **P11.1** **P11.2** **T11.3** *L6.2.1* *T6.2.2a* *P6.2.4*

Applied Probability and Statistics (Continued)

(continued from front)

HOEL and JESSEN • Basic Statistics for Business and Economics, *Third Edition*

HOGG and KLUGMAN • Loss Distributions

HOLLANDER and WOLFE • Nonparametric Statistical Methods

IMAN and CONOVER • Modern Business Statistics

JAGERS • Branching Processes with Biological Applications

JESSEN • Statistical Survey Techniques

JOHNSON and KOTZ • Distributions in Statistics
- Discrete Distributions
- Continuous Univariate Distributions—1
- Continuous Univariate Distributions—2
- Continuous Multivariate Distributions

JOHNSON and KOTZ • Urn Models and Their Application: An Approach to Modern Discrete Probability Theory

JOHNSON and LEONE • Statistics and Experimental Design in Engineering and the Physical Sciences, Volumes I and II, *Second Edition*

JUDGE, HILL, GRIFFITHS, LÜTKEPOHL and LEE • Introduction to the Theory and Practice of Econometrics

JUDGE, GRIFFITHS, HILL, LÜTKEPOHL and LEE • The Theory and Practice of Econometrics, *Second Edition*

KALBFLEISCH and PRENTICE • The Statistical Analysis of Failure Time Data

KISH • Survey Sampling

KUH, NEESE, and HOLLINGER • Structural Sensitivity in Econometric Models

KEENEY and RAIFFA • Decisions with Multiple Objectives

LAWLESS • Statistical Models and Methods for Lifetime Data

LEAMER • Specification Searches: Ad Hoc Inference with Nonexperimental Data

LEBART, MORINEAU, and WARWICK • Multivariate Descriptive Statistical Analysis: Correspondence Analysis and Related Techniques for Large Matrices

LINHART and ZUCCHINI • Model Selection

McNEIL • Interactive Data Analysis

MAINDONALD • Statistical Computation

MANN, SCHAFER and SINGPURWALLA • Methods for Statistical Analysis of Reliability and Life Data

MARTZ and WALLER • Bayesian Reliability Analysis

MIKÉ and STANLEY • Statistics in Medical Research: Methods and Issues with Applications in Cancer Research

MILLER • Beyond ANOVA, Basics of Applied Statistics

MILLER • Survival Analysis

MILLER, EFRON, BROWN, and MOSES • Biostatistics Casebook

MONTGOMERY and PECK • Introduction to Linear Regression Analysis

NELSON • Applied Life Data Analysis

OSBORNE • Finite Algorithms in Optimization and Data Analysis

OTNES and ENOCHSON • Applied Time Series Analysis: Volume I, Basic Techniques

OTNES and ENOCHSON • Digital Time Series Analysis

PANKRATZ • Forecasting with Univariate Box-Jenkins Models: Concepts and Cases

PIELOU • Interpretation of Ecological Data: A Primer on Classification and Ordination

POLLOCK • The Algebra of Econometrics

PRENTER • Splines and Variational Methods

RAO and MITRA • Generalized Inverse of Matrices and Its Applications

RIPLEY • Spatial Statistics

SCHUSS • Theory and Applications of Stochastic Differential Equations

SEAL • Survival Probabilities: The Goal of Risk Theory

SEARLE • Linear Models